Biographical Dictionary of American and Canadian Naturalists and Environmentalists

BIOGRAPHICAL DICTIONARY OF AMERICAN AND CANADIAN NATURALISTS AND ENVIRONMENTALISTS

EDITED BY

Keir B. Sterling, Richard P. Harmond, George A. Cevasco, and Lorne F. Hammond

GREENWOOD PRESS
Westport, Connecticut • London

Library of Congress Cataloging-in-Publication Data

Biographical dictionary of American and Canadian naturalists and environmentalists / edited by Keir B. Sterling . . . [et al.].
p. cm.
Includes bibliographical references and index.
ISBN 0-313-23047-1 (alk. paper)
1. Naturalists—United States—Biography. 2. Naturalists—Canada—Biography. 3. Environmentalists—United States—Biography. 4. Environmentalists—Canada—Biography. I. Sterling, Keir B. (Keir Brooks)
QH26.B535 1997
508'.092'273—dc20
[B] 96-156

British Library Cataloguing in Publication Data is available.

Library of Congress Catalog Card Number: 96-156
ISBN: 0-313-23047-1

First published in 1997

Greenwood Press, 88 Post Road West, Westport, CT 06881
An imprint of Greenwood Publishing Group, Inc.

Printed in the United States of America

The paper used in this book complies with the Permanent Paper Standard issued by the National Information Standards Organization (Z39.48-1984).

10 9 8 7 6 5 4 3 2 1

To Our Parents

Henry S. and Noel de W. Sterling
William and Violet Harmond
George and Anna Cevasco
Gary and Dorothea Hammond

Contents

Acknowledgments

This work has had a very long gestation period since Keir Sterling proposed it to James T. Sabin, executive vice president of Greenwood Press, in the early 1980s. We thank him and Cynthia Harris, Greenwood's executive editor for reference books, for their enormous patience, flexibility, support, and understanding as the book gradually took shape. We also extend our thanks to the many contributors to this volume for their expertise and for their efforts on behalf of this project, particularly those who originally drafted entries more than a dozen years ago. These colleagues have, for the most part, been active and retired university faculty, college and university librarians, graduate students, active and retired government officials, environmentalists, scientists, naturalists, authors of biographies and scientific works, and, in a few cases, family members of the biographical subjects. Every effort has been made to review the sketches and bibliographies for accuracy. The editors naturally assume responsibility for errors of omission and commission. We would be grateful for any corrections that are brought to our attention by discerning readers.

A number of colleagues recommended possible subjects, helped to identify possible contributors, recommended possible published sources, and assisted in a number of other ways. Some also served as contributors themselves. They include Kraig Adler, Walter Bock, Stephen Bocking, Charles Boewe, Michael Brodhead, William R. Burk, Martha Coolidge, Bill Deiss, the late Ralph Dexter, Janet Foster, John W. Frederick, Al Gardner, Susan Glenn, Janice Goldblum, the late Joseph Hickey, C. Stuart Houston, Richard Jarrell, Sally Gregory Kohlstedt, Ann Loxterman, Mark Lytle, the late Newell B. Mack, Rina Mathless, Martin McNicholl, Jeffrey S. Murray, the late Paul Oehser, John Opie, Michael Osborne, James S. Pringle, Nina Root, the late Ann H. W. Rudolph, Oscar Sanchez-Herrera, Anne D. Sterling, Duncan D. Sterling, Warner S. Sterling,

Theodore C. Sterling, Ronald L. Stuckey, Margaret Studier, John Wadland, William A. Waiser, and Ronald S. Wilkinson.

The staffs of the following libraries aided and assisted the editors at various times: Academy of Natural Sciences of Philadelphia; American Museum of Natural History, New York; American Philosophical Society, Philadelphia; Archives Canada; Bard College; Bay Shore-Brightwaters Public Library, Brightwaters, New York; Columbia University Libraries, New York City; Davenport, Iowa, Public Library; Duke University Library; Harford County (Maryland) Public Libraries; Harford Community College Library, Bel Air, Maryland; Harvard University Libraries; Henrico County (Virginia) Public Libraries; Library of Congress (including the Manuscripts Division); Milton S. Eisenhower Library, Johns Hopkins University; National Academy of Sciences Archives, Washington, D.C.; National Archives and Records Administration; New York Public Library; McKeldin and Science Libraries, University of Maryland; Blacker-Wood Library, McGill University; Richmond, Virginia, Public Library; University of California Library System; University of Montreal/McPherson Library; University of North Carolina Libraries; North Carolina State University Libraries; Pace University Libraries, New York and Westchester; University of Richmond Libraries; Rhinebeck (New York) Public Library; St. John's University Library; Taylor Library, Vassar College; University of Ottawa Library; University of Victoria (British Columbia) Library; Virginia Commonwealth University Libraries; University of Virginia Libraries; University of Wisconsin Libraries; and the Sterling Memorial Library, Yale University.

This project could not have succeeded without the support of five other individuals. Glennda C. Leslie and Jane E. Jenkins, who were both with the Saskatchewan Archives Board, joined us as research associates at the beginning of 1987 and spent considerable time, with the aid of a number of Canadian colleagues, in developing a list of Canadian entries. They subsequently identified and helped to enlist the services of a number of our Canadian contributors. Unfortunately, new commitments obliged them to leave the project at the end of 1989. Juliette Fernan, a graduate assistant in the Department of English at St. John's University, was of considerable aid in checking information, keeping track of contributors and their sketches, handling numerous revisions, preparing some of the frontmatter and the appendix for this volume and helping us in a number of other ways during the latter half of 1995. Mary Rahn and Karen Capece of St. John's capably provided additional secretarial services. We thank them all for their indispensable assistance. We take this opportunity to thank all those whose names may have been inadvertently omitted who helped to bring this project to completion.

Members of the editorial staff at Greenwood Publishing had the difficult task of translating the numerous corrections and changes incorporated in our typescripts and computer printouts into published text. With their aid, we were able to avoid many errors and oversights. They include: Jason Azze, Meg Fergusson, Cynthia Harris, Leanne Jisonna, Carol Lucas, and Gary Aleksiewicz. They and their other colleagues at Greenwood have our warm thanks for their patience and persistence, together with their dedicated professionalism.

Finally, we thank our families for their patience and understanding over the years as we strove to complete this project.

Introduction

This project had its inception in the early 1980s, when Keir Sterling approached Greenwood Press with a proposal for a biographical dictionary that would cover environmentalists and ecologists active in Canada, the United States, and Mexico from the beginning of the colonial period to the end of 1970. After some years of effort, it became apparent that, for practical reasons, the Mexican entries would have to be the subject of a separate volume to be published at a later date. It was also decided to move the terminal date for coverage up to the present, specifically the summer of 1997. Coverage was broadened to include individuals whose work contributed to an understanding of North American environment and natural history, though they themselves may not have made broader environmental concerns their first priority. Richard Harmond and George A. Cevasco, senior faculty members of the St. John's University History and English Departments, respectively, joined the project in the early 1990s, contributing valuable additional perspectives and resources. Lorne Hammond, then at the University of Ottawa and now an Historian at the Royal British Columbia Museum, Victoria, British Columbia, subsequently joined forces with the other three editors and contributed a great deal by helping to round out the focus on the Canadian side.

NATURE AND OBJECTIVES OF THIS VOLUME

A number of reference works have been published in recent years that include, but are not restricted to, biographical information about naturalists and environmentalists, especially those who have worked in the United States in the past several centuries. This volume, however, is designed somewhat differently and is intended to meet several specific needs. It provides personal and professional information concerning some 445 representative American and Canadian naturalists and environmentalists, including some whose accomplishments are undeservedly obscure and who lived between the late fifteenth and late twentieth centuries. We have intentionally cast a wide net in defining these two categories of subjects. We have included explorers who were responsible for published works dealing with the natural history or geography of North America, conservationists, ecologists, environmentalists, wildlife management specialists, park planners, administrators of national parks and forests, political figures, government officials, including museum administrators whose work dealt with environmental issues, field and laboratory biologists, zoologists, botanists, biogeographers, agriculturists, geologists, marine biologists, microbiologists, paleontologists, mapmakers, natural historians, artists and writers concerned with nature and the environment, and a few promoters from Europe who made extended visits to North America from the colonial period into the mid-nineteenth century. Academics and laypersons, professionals and amateurs, all figure in these pages.

Each entry contains a succinct but careful evaluation of a subject's career and contributions, as well as an up-to-date bibliography and list of pertinent manuscript sources, where known to be available. There is some variation in the length of each sketch, depending, in large part, on a subject's accomplishments, the complexity of his or her professional commitments, interest in the individual's life and accomplishments as reflected in what has been published about him or her, and the extent to which the available information about the individual has been accessible to both authors and editors. In some instances, it has not been possible to prepare as complete a biographical sketch as we would have liked.

The reader should clearly understand that the editors consider this biographical dictionary to be a contribution to the available information concerning the larger community of naturalists and environmentalists. This volume makes no pretense to being an all-inclusive or definitive work of reference; a single volume of 445 entries could hardly be that. The exponential growth of most of the occupational fields represented here in the twentieth century alone would necessitate the compilation of a number of volumes. Rather, we have made an effort to include a selection of representative figures who worked in a diversified range of activities and disciplines over a period of nearly 500 years. We have sought and received advice about possible subjects from a large number of colleagues in the United States and Canada and regret that space did not permit us to include every deserving person. The choices of biographical subjects in this book naturally reflect the professional interests and concerns of the editors, to some extent, and we naturally assume responsibility for the selections. We fully appreciate the likelihood that other sets of editors might well have made very different choices. In some instances it was impossible to find sufficient information concerning potential subjects, who had perforce to be eliminated from consideration. We hope that this book will prompt researchers in both the

United States and Canada to build on what we have done here, so that a comprehensive history of the accomplishments of North American naturalists and environmentalists may eventually be written.

For certain of our subjects, particularly in the colonial period, natural history in its various manifestations was secondary to their other objectives. Others were dedicated observers, artists, and writers who learned as they went along. Scientific professionalism in North America did not begin to become a factor until the latter stages of the eighteenth century. The role of government in matters of environmental concern is a much more recent development, dating back little more than 125 years in both the United States and Canada. Whatever their particular contributions, however, we consider all of the subjects profiled in this volume to be naturalists and/or environmentalists and have placed the broadest possible construction on both terms. These men and women have either advanced our knowledge of some aspect of nature or enhanced our appreciation of the value, beauty, and vulnerability of the natural world. Some individuals, of course, have managed to encompass both of these approaches in their work. There will be many well-known names in the pages that follow. On the other hand, the accomplishments of a substantial number of the individuals covered in this volume are little known. They are receiving a measure of their due here for the first time.

This biographical dictionary will make known to residents of the United States information about a considerable number of Canadian naturalists and environmentalists, while, at the same time, Canadian readers will have an opportunity to familiarize themselves with many environmental practitioners who lived in the States. All extended our knowledge or deepened our appreciation of nature. Not surprisingly, there are some subjects whose activities took them into both countries, underscoring the fact that the study of the geographical distribution and conservation of animals, plants, and other natural productions and of the environment transcends political boundaries.

Readers will normally use this volume for information concerning specific individuals. But surveying them as a group, as we have done in compiling the appendix, one cannot help being struck by the ethnic and socioeconomic backgrounds of the biographees. The large majority of them were of Northern and Western European ancestry and, so far as we have been able to determine from the available evidence, came from the middle (sometimes rural middle), upper-middle, and upper classes. Moreover, an impressive number received college and even postgraduate training.

As work on this volume progressed, we noted heightened interest in, and the development of new perspectives on, the role of women in natural history, environmentalism, and the biological sciences. This is reflected in the large number of recently published studies, including a number of reference works, focusing on women who have been active in these areas. We anticipate a similar impact with the emerging interest in the contribution of African Americans, the traditional ecological knowledge preserved by so many influential Native American elders in the United States and First Nations elders in Canada and the involvement of other important elements in North American society whose work has not heretofore received scholarly emphasis. It is the editors' hope that the contributions of these and other groups will soon receive the attention they deserve from younger scholars.

A few comments may be in order here concerning our rationale for attempting to cover five centuries of natural history, ecological thinking, and environmentalism in Canada and the United States. Certainly much of what was later done in these two countries reflected certain Old World antecedents. Interest in natural history, for example, and instances of plants and animals being placed in private or semi-public exhibition areas date back to ancient times. Modern European exemplars included the earliest museum of natural history, established near Florence, Italy, before the end of the sixteenth century. The Royal Society of London (1663) was an important pioneering organization made up of individuals committed to scientific interests. A good many of its members and agents concerned themselves with identifying, describing, collecting, and propagating North American plants and animals.

Many European travellers to the North American colonies, and residents of what are now the United States and Canada, were also actively involved in natural history. Colonists often worked in concert with European colleagues. Collecting activities have continued down to the present. The earliest New World museum containing natural history specimens, for example, was established in Charlestown, South Carolina, in 1773. Several early state natural history collections, begun in the 1830s, were turned over to colleges and universities such as Amherst and Yale; New York created the first state museum in 1843. Active government sponsorship of natural history research, specimen collections, and museums in Canada and the United States dates from the mid-nineteenth century. Some North American institutional collections developed during the past century and a half are among the world leaders in their various disciplines. University-level training of biologists, geographers, geologists, environmentalists, ecologists, and other natural scientists has been an important twentieth-century development in the United States and Canada.

Ecological thinking probably had its beginnings in the early nineteenth century. Conservation policies were being put in place from the mid- to late-nineteenth century, and environmentalism has been developing over the past half century and more. This is not the place for an extended discussion of the origins and development of these movements, but it is important to place them in a broader historical perspective.

Many of our modern ideas on these subjects have much earlier European roots. Restrictions on hunting and forest use in Great Britain, for example, date back at least to the time of the Normans. The first legislative enactments by Parliament date from 1390. Wild animals (*ferae naturae*) were considered "the property of no one" and hence common property. The Crown, propertied classes, and individuals favored by the Crown had certain hunting rights denied to the general population. In modified form, parliamentary enactments dating from 1671 extended many of these privileges into the nineteenth century. By the beginning of the 1800s, however, the importance of rank and wealth in the crafting of game legislation had declined, and principles of forest and game management more nearly in line with modern practices had their inception. English laws restricting individual rights to convert wild animals to private property through capture or hunting have long influenced legislation in North America.[1] The English, of course, were not the only Europeans with an interest in protecting their natural resources. We will cite just three other examples here. Phillip II of Spain expressed concern about "the conservation of forests and their increase" as early as 1582. In 1765, Charles III of Spain obliged any individual cutting trees on private property or public lands to secure licenses and to plant three trees for every one cut.[2]

Peter the Great of Russia, who reigned from 1696 to 1725, designed conservation measures intended to "promote the well-being of the whole Russian state and not simply that of the ruler's personal estate." He also initiated what today would be called a "sustained yield" basis for logging.[3]

During their colonial and early national periods, Americans and Canadians experienced what they saw as a limitless abundance of natural resources. Their numbers grew, but so, for many years, did the territory available for their use. Mercantilism and resource extraction were the norms. Careful farming, logging, hunting, fishing, and land use seemed unnecessary. There were a few farsighted colonial enactments designed to conserve animals, plants, and the environment, which began with efforts to protect cahows and green turtles in Bermuda (1616), but these generally had limited public support. The concepts of conservation and ecological thought, when they did finally manifest themselves in the mid-nineteenth century, were, in large part, transplants from abroad. When, therefore, men such as George Catlin and Henry David Thoreau first began writing on the values of wilderness preservation in the 1830s, they were in touch with, and borrowing from, much older European traditions. We have felt it important to examine and touch base with some of these lines of thinking as translated into the North American experience through the lives and careers of the subjects in this volume.[4]

It is quite possible that the first modern conservation legislation represented an example of early melding in European and North American thinking on the subject. Closed seasons for "shootable" game and protected status for certain animals, including endangered species, were established in the English Cape Colony of South Africa in 1822. Still other species were placed in a "vermin" class, which could be hunted at any time. Later modifications made it possible for animals to be added and removed from these categories as their demographics changed.[5]

We believe conservation should be thought of as reflecting deep cultural and intellectual traditions, as well as an ongoing process of constructing and reforming our visions of nature.

SCOPE

The *Biographical Dictionary of American and Candian Naturalists and Environmentalists* includes individuals whose birth dates span the period from the late 1400s to the mid-twentieth century and who died as late as July 1996. We have not included Mexico in this volume because it has not yet been possible to gather sufficient source material, but we plan to devote a subsequent volume to that nation. In collecting information about, and evaluating, our choices, we have benefited significantly from consulting a number of knowledgeable historians of science, biologists, and environmentalists in the United States, Canada, and Western Europe. In addition we have turned to a number of published sources for background information that are listed in the Selected Bibliography of this volume.

ENTRIES

Each of the entries in the *Biographical Dictionary of American and Canadian Naturalists and Environmentalists* is comprised of the following elements:

1. Full name of the subject; birth date and place of birth; death date and place of death; principal occupation or avocation for which known.
2. Family background; parentage and number of siblings, where known; education; names of spouse(s), if applicable, with dates of marriage(s); number of children; details concerning honorary degrees, where known.
3. Positions held or principal avocations pursued and their location, in chronological order.
4. Career, including a summary of major activities during subject's life and either mention or listing of any major publications. The standard biographical dictionaries and other sources on which we have relied do not always agree on the specifics of an individual's career. In some instances, necessary information is lacking. We have made every effort to resolve such difficulties and to make each biographical sketch as accurate as possible. Within the limitations of space, an individual's principal publications are listed. Where this is not done, reference is made to the research areas for which the subject is best known. Bibliographies in the publications listed for each subject should, of course, be mined for additional source materials.
5. Major contributions. Provides a clear and succinct summing up of what is most noteworthy about the subject's career as an environmentalist, together with the long-term significance of the subject's work.
6. Bibliography sources: these are selective, not exhaustive, and list important biographies and other useful works about the subject, to include repositories where unpublished manuscript materials may be found. In general, shorter biographical sketches appearing in compendiums such as the *Dictionary of American Biography* or obituaries in newspapers and journals are usually listed only when full-length biographies, autobiographies, or monographic studies of an

individual's career and contributions are not available. Where known, sources indicating where a bibliography of subject's publications may be found are listed. Readers are reminded that the *National Union Catalog of Manuscript Collections* (1959–) is an ongoing, multivolume publication wherein the location of collections of personal papers and manuscripts that have been recorded from U.S. repositories may be found. The comparable Canadian reference is Robert S. Gordon and others, *Union List of Manuscripts in Canadian Repositories/Catalogue collectif des manuscrits des archives canadiennes*, 2 volumes, 1968. A revised edition, edited by E. Grace Maurice, appeared in two volumes (1975), with three supplements, edited by Grace Maurice Hyam, Victorin Chabot, and Peter Yirkiw, having been published in 1976, 1979, and 1982. In addition, there are published catalogs, guides, or inventories to the National Archives of Canada, to collections in the individual provinces, to collections held by Canadian universities, and to the archives of various museums, religious organizations, and private agencies.

APPENDIX

The appendix includes information about the national origins of our 445 subjects, together with the occupations or avocations in which they were active and to which they made their principal contributions. We have listed as many as three fields for each person. As will be noted by referring to the individual entries, a modest number of individuals were active in four or more fields. We have also indicated whether or not subjects had the benefit of formal college or university training and also their inclusive dates of birth and death. This should enable the reader to gain some impression of the range of preparation and expertise represented by the subjects in this volume, together with the period of history during which they were active.

NOTES

1. James A. Tober, *Who Owns the Wildlife?: The Political Economy of Conservation in Nineteenth Century America* (1981), 146–47.
2. Otto T. Solbrig and Dorothy J. Solbrig, *So Shall You Reap: Farming and Crops in Human Affairs* (1994), xii; Lane Simonian, *Defending the Land of the Jaguar: A History of Conservation in Mexico* (1995), 226.
3. Douglas R. Weiner, *Models of Nature: Ecology, Conservation and Cultural Revolution in Soviet Russia* (1988), 7.
4. Craig W. Allin, *The Politics of Wilderness Preservation* (1982), 12–15, 17.
5. Paul Johnson, *The Birth of the Modern World Society, 1815–1830* (1991), 284.

A

ABBOT, JOHN. Born London, England, 31 May/1 June 1751; died Bulloch County, Georgia, December 1840 or January 1841. Entomologist, ornithologist, artist.

FAMILY AND EDUCATION

Eldest child of James, well-to-do attorney-at-law, and Ann Clousinger Abbot. Tutored at home. Studied art with Jacob Bonneau during the 1760s. Married Penelope Warren, sometime prior to 1779; one son.

POSITIONS

Law clerk to father in London, 1769–1773. In 1773, Royal Society of London, together with private group of naturalists headed by Dru Drury, naturalist, entomologist, silversmith, fellow of the Linnaean Society (1725–1803), commissioned Abbot to study and collect natural history specimens in Virginia. Lived there September 1773–December 1775, then moved to Georgia, where he lived until his death. Important gaps remain today in what is known of his life and work.

CAREER

Developed interest in insects early in life. Encouraged in his artwork and insect collecting by father. Developed considerable competence with his drawing in a comparatively short time. Some of his drawings done between 1767 and 1773 "can be classed as among the best eighteenth-century work of their kind." Abbot had no interest in pursuing his father's profession and chose to sell his

English insect collections and go to the American colonies. Lived in Jamestown area for several years but was not satisfied with collecting opportunities there and moved to Georgia with family named Goodall, with whom he had lodged in Virginia and continued to lodge for some years south of Atlanta. Revolutionary War disrupted his work for several years. He served for a time as a private in 3rd Georgia Battalion, Continental Army, under Lt. Col. James McIntosh. Was given bounty of 575 acres for his services. Following the war, did field study and explored region of the Savannah and Ogeechee Rivers as well as port of Savannah and adjacent seacoast. Until c.1795, lived comfortably as planter but may then have lost his wife and apparently encountered financial difficulties for some time. Taught school to supplement income at one stage. His skill in preserving specimens of insects and birds and the scientific precision of his watercolors made his work highly prized. Various metamorphic stages of insects and their food plants were often represented. He did not, however, begin preparing herbaria until the 1820s. Abbot attempted to depict birds in characteristic poses but made no real attempt to portray them life-size.

Probably began his watercolor drawings of birds in the early 1790s and may have completed more than 1,300 of these over a thirty-five year period. There was much duplication, but Abbot probably limned between 206 and 217 species. Some of these he copied from other contemporary artists, but the bulk of it was his own work. Between 1791 and 1810, Abbot completed at least four sets of watercolor drawings, three of which are now at Harvard, Knowsley Hall in Prescot, England, and at the British Library in London. A fourth set was sold at auction in 1980 and "subsequently dispersed." In preparing the nomenclature for some of this work, Abbot was influenced by the Scots-American ornithologist Alexander Wilson, whom he met in 1809. The two men are known to have exchanged a good deal of information. Most of Abbot's birds are painted perched on small trees or shrubs. The backgrounds in most of these watercolors are more detailed than in his later work. Abbot completed four more sets of bird paintings in the 1820s. Two, both entitled *Birds of Georgia*, one in 1823 and the other in 1827, totaled some 189 watercolors between them. One is now in the British Museum (Natural History), the other in a private collection in Georgia. Two other sets, totaling some 250 watercolors, are in the Smithsonian and University of Georgia collections. John Latham (1740–1837), a British physician-naturalist, used many of Abbot's descriptions and specimens in compiling his own multivolume works on birds. Small collections of bird specimens known to have been prepared by Abbot are at Harvard, in Liverpool, and in Berlin.

Much of this material reached various American and European collectors and institutions over the years, but several shipments were lost at sea, resulting in Abbot's temporary discouragement and the abandonment of his work. Other manuscripts and drawings were lost after his death. Abbot often retouched the work of his assistants. He completed a substantial, but unknown, number of insect and plant drawings during his career, continuing to work into his eighties. John Francillon, a London-based entomologist and silversmith, was, for many years,

Abbot's agent in dealings with the latter's English and European customers. One shipment sent to Francillon by Abbot in 1793 contained 1,021 sheets with 1,833 figures depicting 1,664 distinct species of insects. Shy and retiring, Abbot published nothing himself but supplied others with essential scientific information on life histories with his drawings and watercolors and left most taxonomic details to others. In 1797, James Smith published *The Natural History of the Rarer Lepidopterous Insects of Georgia Collected from Observations by John Abbot* in two volumes. Smith included Abbot's drawings and notes, sometimes editing the latter. On this basis, Smith identified species, a number of them new. Smith stressed, "The materials for [this] work have been collected on the spot by a faithful observer, Mr. John Abbot, many years resident in Georgia." Several reprints appeared over a period of at least thirty years. A third volume was projected by William Swainson in 1818 but was not published, owing to Swainson's unhappiness with the size of the drawings and their quality. J. A. Boisduval and J. E. LeConte's *Histoire Générale et Iconographie des Lépidoptères at les Chenilles de l'Amérique Septrionale* (1833) included Abbot's drawings but did not credit them to him.

Other works used Abbot's data but omitted his watercolors: Charles Athanase, Baron Waickenaer's *Histoire Naturelle des Insects; Aptères* (1837–1847); John Latham's *Supplement to the General Synopsis of Birds* (1787–1801); *Index Ornithologicus* (1790) and a *General History of Birds* (1821–1828); Alexander Wilson's *American Ornithology* (1808–1814); and Stephen Elliott's *A Sketch of the Botany of South-Carolina and Georgia* (1821–1824). Substantial numbers of Abbot's insect and plant drawings and specimens repose in European and American collections today.

MAJOR CONTRIBUTIONS

Abbot's detailed field observations and watercolors influenced the developing sciences of entomology, arachnology, ornithology, and botany by providing significant information concerning species native to Georgia. Others had depicted insects before him, but his was the first systematic effort by a competent field naturalist to record the insects and birds of the southeastern United States. Ronald S. Wilkinson, Wilson's most recent biographer, has stated that "his ornithological work reveals that he was the most significant field observer of North American birds before Alexander Wilson."

BIBLIOGRAPHY

Allen, Elsa G. *The History of American Ornithology Before Audubon* (1951).

Allen, Elsa G. "John Abbot, Pioneer Naturalist of Georgia." *Georgia Historical Quarterly* 41 (1957).

Mallis, Arnold. *American Entomologists*, 1971.

Rogers-Price, Vivian. *John Abbot's Birds of Georgia: Selected Drawings from Houghton Library, Harvard University* (1997).

Simpson, Marcus B. "The Artist-Naturalist John Abbot (1751–ca. 1840): Contributions to the Ornithology of the Southeastern United States." *North Carolina Historical Review* 84 (1984).

Wilkinson, Ronald S. "John Abbot's Birth Data." *Entomologist's Record* (*Ent. Rec.*) 87 (1975).
Wilkinson, Ronald S. "Smith and Abbot, *The Natural History of the Rarer Lepidopterous Insects of Georgia* (1797): Its Authorship and Later History." *Ent. Rec.* 93 (1981).
Wilkinson, Ronald S. "John Abbot's Drawings and Notes for a Proposed Supplement to Smith and Abbot . . ." *Ent. Rec.* 94 (1982).
Wilkinson, Ronald S. "Nineteenth Century Issues of Smith and Abbot . . ." *Ent. Rec.* 94 (1982).
Wilkinson, Ronald S. "John Abbot's London Years: Some Addenda." *Ent. Rec.* 97 (1985).

UNPUBLISHED SOURCES

See John Abbot, "Notes on My Life," Museum of Comparative Zoology, Harvard University; Elsa G. Allen, unpublished manuscript on Abbot, E. G. Allen Papers, University Archives, Olin Library, Cornell University; Watercolors and Notes of Abbot, Harvard University; British Museum (Natural History), London; British Library, London; Bibliothèque Centrale du Muséum National d'Histoire Naturelle, Paris; Carnegie Museum of Natural History, Pittsburgh; University of Georgia Libraries, Athens, Georgia; Turnbull Library, Wellington, New Zealand.

Vivian Rogers-Price

ABBOTT, CHARLES CONRAD. Born Trenton, New Jersey, 4 June 1843; died Bristol, Pennsylvania, 27 July 1919. Geologist.

FAMILY AND EDUCATION

Son of Timothy Abbott, farmer, banker, and Susan Conrad Abbott. Grandson of Solomon White Conrad, sometime lecturer in botany and mineralogy at University of Pennsylvania. Graduated Trenton Academy, 1858; M.D. degree from University of Pennsylvania, 1865. Married Julia Olden, 1867. Three children.

POSITIONS

Private, N.J. National Guard, 1863; field assistant, Peabody Museum of Archaeology and Ethnology, Harvard University, 1876–1889; curator of archeology, University Museum, University of Pennsylvania, 1889–1893. Did not practice medicine.

CAREER

Developed enthusiasm for nature studies as youth with special interest in birds and mammals. Most of work was as independent investigator and writer with focus on area of family homestead, Three Beeches, at juncture of Crosswicks Creek with Delaware River in New Jersey. Began to write quite early but first papers rejected for publication, a fact Abbott resented throughout his life. Nevertheless, first publications by 1859 and in 1868 produced ''Catalogue of the Vertebrate Animals of New Jersey,'' Appendix E of state-prepared *Geology of New Jersey*, republished by Julius Nelson, with additions and corrections, as ''Descriptive Catalog of the Vertebrates of New Jersey'' (1890). In 1870s, turned increasingly to archeology after discovery of stone tools on the Abbott farm that, Abbott was convinced, were of the Pleistocene glacial period. Supported by Frederic Ward Putnam of Harvard, made collection of 20,000 stone tools, which were deposited at Peabody Museum at Harvard, and later collection deposited at Princeton University. Published *Primitive Industry* (1881), in which he posited three human occupations of Delaware Valley, beginning with early man. For time, seemed destined to become America's Boucher de Perthes, but by late 1880s claims were strongly disputed by several leading geologists and archeologists. In 1884 began to produce popular books on personal experiences in study of nature. *Upland and Meadow* (1886) was described by one English critic as the ''most delightful book of its kind which America has given us.'' In 1895, published first of several unsuccessful novels.

Writings number over 150 titles, including *A Naturalist's Rambles about Home* (1884), *Cyclopedia of Natural History* (1887; reissued under different titles); *Waste-land Wanderings* (1887), *Recent Rambles* (1892), *Days Out of Doors* (1889), *Outings at Odd Times* (1890), *The Birds about Us* (1895), *Birdland Echoes* (1896), *The Freedom of the Fields* (1898), *Clear Skies and Cloudy* (1899), *In Nature's Realm* (1900), *Rambles of an Idler* (1906), *Archaeologica Nova Caesarea* (1907–1908), and *Ten Years' Digging in Lenape Land* (1912).

Apparently of a highly sensitive nature, reflected in his personal essays. But also combative, stubborn, clinging to error even when presented clearly contrary facts. Contemptuous of ''dry-as-dustical'' naturalists. Qualities tended to isolate him from the mainstream of natural history and prevented correction of erroneous ideas.

MAJOR CONTRIBUTIONS

Abbott's popular writing included accounts of personal experiences in nature studies. Much natural history work in Delaware Valley and New Jersey in general, but marred by errors. Claims to have found traces of Paleolithic man in America—his own main claim to distinction—are historically important because of the prolonged debate that followed, but generally they have been rejected by archeologists.

BIBLIOGRAPHY

National Cyclopaedia of American Biography.

Palmer, T. S., ed. *Biographies of Members of the American Ornithologists' Union* (1954).

Popular Science Monthly (1887).

Richards, Horace G. ''Reconsideration of the Dating of the Abbott Farm Site at Trenton, New Jersey.'' *American Journal of Science* (1939).

Stone, Witmore. *Dictionary of American Biography.*

Wissler, Clark. *American Anthropologist* (1920).
Wright, G. Frederick. *Science* (1919).

UNPUBLISHED SOURCES

See letters in Abbott family papers in the Princeton University Library and correspondence with Julia S. Roberts in Academy of Natural Sciences, Philadelphia. Many of Abbott's original materials were destroyed when the Abbott farmhouse burned around 1914.

James R. Glenn

ADAMS, ANSEL EASTON. Born San Francisco, 20 February 1902; died San Francisco, 22 April 1984. Teacher, artist, mountaineer, and environmentalist.

FAMILY AND EDUCATION

Son of Charles Hitchcock Adams, a prosperous businessman. Mother, Olive Bray, was from Carson City, Nevada, though the Brays originally came from Baltimore. An only child, was close to his father, who taught him to appreciate the beauty of central California coast and gave him a sense of duty with stress on the Puritan work ethic. Was hyperactive in grammar school. With help of private tutors, graduated from local grade school in 1917. Received an honorary doctorate of fine arts from University of California in 1961; in 1981, an honorary doctorate of fine arts from Harvard University. In 1928, married Virginia Best, the daughter of Harry Best of Best's Photographic Studio in Yosemite, California. They had two children.

POSITIONS

In 1919 was custodian at the headquarters of Sierra Club at Yosemite National Park. In 1920s led mountaineering trips for the Sierra Club, which began to publish—in *The Sierra Club Bulletin*—Adams's photographs, along with articles on his hikes. In 1930 opened his own photography establishment and sought commercial assignments. Clients included Pacific Gas and Electric (PG&E), American Telephone and Telegraph, American Trust, Eastman-Kodak, Hills Brothers Coffee, and Yosemite Park and Curry Company. In 1934, with other photographers, including Imogen Cunningham and Edward Weston, created Group *f*/64, an association to advance photography as an art form. Was elected the same year to Board of Directors of the Sierra Club. Lobbied for club in Washington, D.C. Resigned in 1971 in a dispute over his role with PG&E, though elected an honorary vice president in 1979. In 1940 began teaching in the Art Center School, Los Angeles, where he developed the zone system—from total black, zone zero, to pure white, zone ten. Taught practical photography at Fort Ord during World War II; also printed top-secret negatives of Japanese military installations. During World War II, became official photographer of the Mural Project, taking pictures of Indian reservations, national parks, and other facilities controlled by Department of the Interior.

CAREER

By 1931 wrote photography reviews for *The Fortnightly*. In 1933 had exhibition in New York at gallery called Delphic Studios. Met Alfred Stieglitz, who admired his work, and in 1936 had one-man show at Stieglitz's New York gallery. Returning home, he opened Ansel Adams Gallery in San Francisco. In 1935 published *Making a Photograph: An Introduction to Photography*. In 1937 he and family moved to Yosemite and took over his deceased father-in-law's shop while continuing San Francisco establishment. His photographs of Sierra Mountains and the Southwest were published in first historical survey of photography by the Museum of Modern Art (New York City). In 1938 published *Sierra Nevada: The John Muir Trail. Born Free and Equal*, published in 1944, reflected, in photo-essays, his view of imprisoned Japanese Americans interned at Manzanar Relocation Center. In 1946 (renewed in 1948) received Guggenheim Fellowship, which permitted him to go to Alaska. In 1949 became a consultant for the Polaroid Corporation. In 1950 received another Guggenheim Fellowship for photo study of Hawaii. In 1964 published *An Introduction to Hawaii*. Following year joined President Lyndon Johnson's environmental task force. In 1966 was elected a fellow of the American Academy of Arts and Sciences. In 1967 founded and was president of Friends of Photography, Carmel. In 1974 took first trip to Europe. Taught at the Arles, France, photography festival. Following year established the Center for Creative Photography at the University of Arizona, Tucson. In 1980 was recipient of Presidential Medal of Freedom from President Jimmy Carter. One year after his death, an 11,760–foot peak in Yosemite National Park was named Mt. Ansel Adams, in his honor. Made more than 40,000 negatives, signed 10,000 fine prints, held over 500 exhibitions of work. His books sold over 1 million copies.

MAJOR CONTRIBUTIONS

Photographic artist who created in photos (mainly in black and white) reflection of his love affair with nature as well as plea for preservation of wilderness. Called his technique "visualization," wherein the photographic artist intuitively searches visual and emotional qualities of completed picture. Wrote of himself as one who has "been able to have had some small effect on the increasing awareness of the world situation through both my photographs and my social assertions." Was among the first to regard photography as an expressive art form. Generally rejected documentation photography, finding it negative. For that reason, among others, did not particularly care for photographs of Edward Steichen.

BIBLIOGRAPHY

Alinder, James. *The Unknown Ansel Adams* (1982).

Alinder, Mary S. *Ansel Adams: A Biography* (1996).

Alinder, Mary Street, and Andrea Gray Stillman, eds. *Ansel Adams: Letters and Images, 1916–1984* (1988).

Cahn, Robert. ''Ansel Adams, Environmentalist.'' *Sierra* 64 (1979): 31, 33–49.

Gray, Andrea. *Ansel Adams an American Place, 1936* (1982).

Newhall, Nancy. *Ansel Adams: Volume 1, The Eloquent Light* (1963).

Pritzker, Barry. *Ansel Adams* (1991).

Turnage, Robert. ''Ansel Adams, the Role of the Artist in the Environmental Movement.'' *The Living Wilderness* 43 (1980): 4–31.

UNPUBLISHED SOURCES

See Ansel Adams Letters to Alfred Stieglitz, Beinecke Library, Yale University; Ansel Adams Archives, Center for Creative Photography, University of Arizona, Tucson; Ansel Adams Papers, Bancroft Library, University of California, Berkeley; Ansel Adams Letters, Polaroid Corporation Archives, Cambridge, Mass.

Judith Moran Curran

ADAMS, CHARLES CHRISTOPHER. Born Clinton, Illinois, 23 July 1873; died Albany, New York, 22 May 1955. Teacher, museum director, ecologist.

FAMILY AND EDUCATION

Son of William Henry Harrison Adams, clergyman and president of Illinois Wesleyan University, and Hannah Westfall Concklin Adams. Graduated from Illinois Wesleyan in 1895; received M.S. at Harvard in 1898, Ph.D. at University of Chicago in 1908. Honorary Sc.D. conferred by Illinois Wesleyan in 1920. Married Alice Luthera Norton in 1908; a daughter, Harriet Dyer Adams.

POSITIONS

Assistant biologist at Illinois Wesleyan, 1895–1896; assistant entomologist, Illinois State Laboratory of Natural History, 1896–1898; curator, University Museum, Michigan, 1903–1906; director, Cincinnati Society of Natural History, 1906–1907; associate professor of animal ecology at University of Illinois State Laboratory of Natural History, 1908–1914; assistant professor of forest ecology, New York State College of Forestry, Syracuse, in 1914, and professor of forest zoology, New York State College of Forestry, 1916–1926; director, New York State Museum, 1926–1943.

CAREER

While on fellowship at University of Chicago (1900–1903), published papers on base leveling and on Pleistocene climate changes in relation to faunal problems, which contributed to resumption of serious American biogeographic study.

Through museum work at universities of Michigan and Cincinnati, developed idea of the museum as a vital teaching center linked to the community and its natural history. At New York State College of Forestry in 1914, organized Roosevelt Wildlife Experiment Station and directed its activities until named director of New York State Museum in 1923. Here he assembled collection of objects produced and used by Shaker religious sect (which once flourished in northern New York) and built up museum's history and art collection. Continued idea of ecological surveys at the Palisades, Mount Marcy, Allegheny State Park, and in the Rochester region. Also wrote much on wildlife and fisheries in relation to forestry and, after a paper on conservation on 1915, gave increasing emphasis to that problem. Organized symposia on human ecology and insisted that ecology ought to be underwritten on same scale as geology and physical sciences. One of founders, and in 1923 president, of Ecological Society of America (established in 1915). Was also active in Association of American Geographers.

Bibliography includes over 150 titles, mainly journal articles. His book, *Guide to the Study of Animal Ecology* (1913), ranks among the important pioneer studies done in North America. Main emphasis of his published research became relation of human ecology to land use and the problems of interrelationship of human ecology and public policies. Was perhaps at his best in conferences and on committees, where he could work toward having his ideas translated into action.

MAJOR CONTRIBUTIONS

Pioneer in, and one of creators of, the ecological perspective. Never ceased trying to rouse fellow ecologists to appreciate their own importance, warning that unless they did so, their job would be taken over by others.

BIBLIOGRAPHY

''Adams, Charles Christopher.'' *National Cyclopedia of American Biography*. Vol. 46 (1967).

''Charles Adams, Headed Museum.'' *New York Times*, 23 May 1955 (obit.).

''Charles C. Adams, Ecologist.'' *Science* 123 (January–June 1956).

''Dr. Charles C. Adams.'' *Bulletin of the Ecological Society of America* 37. 4 (1956).

UNPUBLISHED SOURCES

See letters in Charles Christopher Adams Papers, University of North Carolina, Southern Historical Collection (correspondence concerning writing and publication of *A Guide to the Winter Birds of North Carolina Sandhills* [1928]); Letters in Charles Christopher Adams Papers, in Archives of American Art, Washington, D.C. (papers relate chiefly to Adams's work as director of New York State Museum at Albany); Correspondence in Jacob Ellsworth Reighard Papers at University of Michigan, Bentley Historical Collection.

Richard Harmond

AGASSIZ, (JEAN) LOUIS (RODOLPHE). Born Motier-en Vuly, Canton Fribourg, Switzerland, 28 May 1807; died Cambridge, Massachusetts, 14 December 1873. Naturalist, zoologist, geologist.

FAMILY AND EDUCATION

Son of Rodolphe Agassiz, Protestant minister, and Rose Mayor Agassiz. At ten, was sent to school at Bienne and at fifteen attended College of Lausanne. Medical training, beginning in 1824, was taken at University of Zurich. In 1826 attended Heidelberg for one year and went on to University of Munich, where he studied embryology with Ignaz von Dollinger. Received Ph.D., University of Erlangen, 1829, M.D. from University of Munich, 1830. Hon. LL.D., Harvard, 1848. Married Cecile Braun, 1833; three children. She remained behind when he went to United States, 1846, died in 1848. Agassiz married Elizabeth Cabot Cary, 1850; no children.

POSITIONS

Professor of natural history at Neuchatel, 1832–1846. Chair of natural history, Lawrence Scientific School of Harvard, from 1848. Lecturer on comparative anatomy at Medical College of Charleston, South Carolina, 1851–1852. Founder of Harvard's Museum of Comparative Zoology, 1859. Founder and lecturer, Anderson School of Natural History, Penikese Island, 1873.

CAREER

Published *The Fishes of Brazil* in 1829. In 1831 was living in Paris and had workshop at the Museum of Natural History of the Jardin des Plantes, where he cultivated the friendship of Cuvier, who was then considered the master of comparative anatomy. Agassiz took over both Cuvier's data and collections, which resulted in his next achievement, The *Recherche sur les Poissons Fossiles* (1833–1844), which was published in five volumes and described the fossil fishes and fauna of Europe. At Neuchatel the Natural History Society was established, and Agassiz was the secretary. He also had his own printing press, with which he published *History of the Fresh Water Fishes of Central Europe* (1839–1842), *Etudes Critiques sur les Mosslusques Fossiles* (1840–1845), and *Nomenclator Zoologicus* (1842–1846), a major systematic reference work. In the 1840s interest in glaciers brought him to Great Britain as well as to glacier wells on the great median moraine of lower Aar glacier in Neuchatel. In 1846 was able to take a trip to America, which was paid for by king of Prussia; he lectured and did fieldwork along the East Coast. In 1847 began survey work with Coast Survey by steamer. Additional Coast Survey work included 1851 exploration of the Florida reefs. Eventually, Agassiz published four-volume *Contributions to the Natural History of the United States*, with a portion of the first volume, *Essay on Classification*, regarded as one of the most important contributions to American natural history. An appointment as professor at Harvard University began in 1848, where he established the Museum of Comparative Zoology in 1859. In 1865 he visited Brazil along with his wife, and they published *A Journey in Brazil* in 1868. Other explorations included trips to Rocky Mountain region, Cuba, and California. Staunch opponent of Darwinian ideas concerning evolution, a position that somewhat reduced his influence in American biological thinking in his final years.

Agassiz played an important role in the metamorphosis of the Association of American Geologists and Naturalists into the American Association for the Advancement of Science (1847), and became its president in 1851–1852. Agassiz was a leader of the Lazzaroni, an informal but determined group of scientists who were actively interested in developing the very highest possible standards for the American scientific community in the mid-19th century. Agassiz wanted Harvard to become an all-graduate institution devoted to pure, as opposed to practical, science and was prepared to let the undergraduate program languish. For a time, he and his friends held sway, but the election of Charles W. Eliot as Harvard president in 1869 thwarted their plans. Agassiz was also instrumental in the organization of the National Academy of Sciences (NAS) in 1863, which he envisioned as an agency through which federal aid might be made available for pure scientific research. He and his associates thought the new agency could be a means of resolving certain national problems through science. The centralization of scientific effort was another objective, as were aspirations to greater prestige for their view of scientific excellence. Little of this was accomplished, however, and it was not until after the death of Agassiz and some of his closest allies that the NAS achieved some prestige for its members and for American science. Unfortunately, Agassiz and his close associates took an elitist view of their craft and were criticized for forming "an illiberal clique." He attempted to block membership in both the Lazzaroni and the NAS to such scientists as S. F. Baird and Asa Gray on grounds that their science was too practical, though some of his opposition was based on personal and professional considerations. Agassiz later admitted that some of the steps taken in the founding of the NAS had gotten it off to a bad start within the scientific community.

MAJOR CONTRIBUTIONS

Was most noted in the United States for his ability to lecture, research, and publish copiously on topics of interest to naturalists. Through almost sheer determination, was able to establish the Museum of Comparative Zoology at Harvard University. The institution provided the site, and through state appropriations, private gifts, and Agassiz's own family money, the museum became a reality. As professor there, Agassiz was able to instruct future generations of American naturalists. His many students were well versed in biology and the direct study of nature. Substantial numbers held influential posts in American colleges and universities into the early years of the twentieth century. Promoted glacial theory.

BIBLIOGRAPHY

Agassiz, Elizabeth Cary. *Louis Agassiz, His Life and Correspondence* (1885).
Bruce, Robert V. *The Launching of American Science* (1987).

Dictionary of American Biography. Vol. 1 (1938).
Lurie, Edward. *Louis Agassiz: A Life in Science* (1960).
Lurie, Edward. *Nature and the American Mind: Louis Agassiz and the Culture of Science* (1974).
Marcou, Jules. *Life, Letters, and the Works of Louis Agassiz* (1896).
New York Times, 15 December 1873 (obit.).
Peare, Catherine Owens. *A Scientist of Two Worlds: Louis Agassiz* (1958).
Teller, James D. *Louis Agassiz, Scientist and Teacher* (1947).
Tharp, Louise Hall. *Adventurous Alliance; The Story of the Agassiz Family of Boston* (1959).
Winsor, Mary P. *Reading the Shape of Nature: Comparative Zoology at the Agassiz Museum* (1991).

UNPUBLISHED SOURCES

See Louis Agassiz Letters, 1867–1874, at Cornell University #14/15/m.82; Louis Agassiz Papers, 1833–1873, at American Philosophical Society Library; Louis Agassiz Papers, 1847–1876, at Harvard University Archives; Louis Agassiz Letters, 1863–1882, to Pedro II, emperor of Brazil, at Houghton Library, Harvard University; Agassiz family papers, 1817–1910, at Museum of Comparative Zoology at Harvard University.

Geri E. Solomon

AKELEY, CARL ETHAN. Born Clarendon, Orleans County, New York, 19 May 1864; died Belgian Congo, 18 November 1926. Taxidermist, sculptor, naturalist, conservationist, inventor.

FAMILY AND EDUCATION

Son of Daniel Webster "Webb" Akeley and Julia Glidden Akeley. Attended Cowles School for several years; later spent six months at Brockport Normal School, New York. Did not attend college; instead embarked on taxidermy study. Married Delia "Mickie" Denning Reiss, December 1922 (divorced March 1923), and Mary Leonore Jobe, October 1924; no children.

POSITIONS

Employed at Ward's Natural History Establishment until November 1886. Left Ward's to accept contract work and part-time position as taxidermist in Milwaukee Public Museum; employed full-time in 1889; resigned September 1892. Opened independent taxidermic and sculpture studio in DeKalb, Illinois. In January 1896 offered position in taxidermy department of Field Museum. Became affiliated with American Museum of Natural History, New York, early 1909 until death.

CAREER

Considered father of modern taxidermy. First to give consideration to musculature, wrinkles, veins, and movement in mounts. Used concept of diorama, or natural settings, in exhibits. Career as sculptor extension of taxidermic work. Subjects generally big-game animals—lions, elephants, gorillas. Invented Akeley camera (portable, wide-range field camera) and a cement gun (version of which still used today); also improved searchlight design for U.S. Army in World War I.

Undertook expeditions to Somaliland (1896), Washington State Olympic Mountain Range (1898), British East Africa (1905–1906, 1909–1911), Belgian Congo (1921), British East Africa and Belgian Congo (1926). Wrote articles for journals and magazines, including *National Geographic*, but only significant work is his autobiography, *In Brightest Africa* (1923).

MAJOR CONTRIBUTIONS

Established Akeley African Hall in American Museum of Natural History, New York, and the Parc National Albert, Belgian Congo. Influenced generations of museumgoers and would-be naturalists and environmentalists. Saw need to preserve some parts of Africa from human development. Played an active part in American conservation movement through his involvement with several conservation societies and groups.

BIBLIOGRAPHY

Akeley, Mary L. Jobe. *Carl Akeley's Africa* (1929).
Akeley Memorial Issue. *National History* 27 (March–April 1927).
Bodry-Sanders, Penelope. *Carl Akeley, Africa's Collector* (1991).
New York Times, 1 December 1926 (obit.).

UNPUBLISHED SOURCES

See American Museum of Natural History, New York, Department of Library Services, Archives and Special Collections; Field Museum of Natural History, Chicago, Archives; University of Rochester, New York, Rush Rhees Library, Department of Rare Books and Special Collections.

Steven Cooper

ALBRIGHT, HORACE MARDEN (MADDEN). Born Bishop, California, 6 January 1890; died Los Angeles, 28 March 1987. Conservationist, lawyer, business executive, consultant.

FAMILY AND EDUCATION

Son of George Langdon, mechanical engineer and millwright, and Mary Clemens (Marden). B.L., 1912, University of California; honorary LL.D., 1951; LL.B., 1914, Georgetown University; honorary LL.D., 1956, University of Montana; honorary LL.D., 1962, University of New Mexico. Admitted to Washington, D.C., and California bars, 1914. Married Grace Marion Noble, 23 December 1915; two children, Robert Mather and Marian Carleen (Schenck).

POSITIONS

Assistant in economics, University of California, 1912–1913; member staff of secretary of the Department of the Interior, Washington, D.C., confidential secretary on staff of Franklin K. Lane, Secretary of the Interior, 1913–1916; assistant attorney of Department of the Interior, assigned to national park affairs, 1915–1917; assistant director National Park Service, Washington, D.C., 1917–1919; acting director, 1917–1918, assistant director field and superintendent Yellowstone National Park, 1919–1929; director of the service and member of National Capitol Park and Planning Commission, 1929–1933; vice president, general manager, U.S. Potash Company, 1933–1946; president, general manager, director, 1946–1956; member National Parks Advisory Council, National Outdoor Recreation Resources Review Committee, 1959–1962; Pacific Tropical Botanical Garden, president, 1964–1971, member of Board of Trustees.

CAREER

Actively involved in the politics and practice of parks, recreation, and related fields for more than a half century. Played a major role in establishment and development of the U.S. National Park Service. While on staff of U.S. Secretary of the Interior Franklin Lane, Albright, who was also assistant attorney for national park affairs, helped draft legislation that created the service in 1916. Shortly thereafter became assistant to its first director, taking over director's duties as needed until he succeeded to the post in 1919. As director Albright lobbied to expand the service, adding national monuments and historic battlefields to its responsibilities and overseeing the creation of additional national parks. Stressed need to keep some areas of the system in an undeveloped, wilderness state.

Left government work in 1936 to work as executive of U.S. Potash Company; continued to be active in issues involving parks and conservation. Member of Board of Directors of Grand Teton Lodge Co., trustee of Colonial Williamsburg Foundation, 1934–1958, and Jackson Hole Preserve, Inc., 1945–1977; member of National Trust for Historic Preservation and Desert Protective Council. Regents Lecturer at University of California, Berkeley, 1961; member of Board of Trustees of Mills College 1939–1942 and 1951–1959. Member of Board of Directors of North American Philips Co. and Philips Trust, 1953–1961, and Arnold Bakers, 1956–1961.

Recipient of many awards and honors, Albright was presented the John Muir Award by the Sierra Club. Received the Order of the Northern Star of Sweden, 1926; Pugley Gold Medal from American Scenic and Historic Preservation Society, 1933; Conservation Award from U.S. Department of the Interior, 1952; Frances K. Hutchinson Medal from Garden Club of America, 1959; Theodore Roosevelt Medal for the Conservation of National Resources from Theodore Roosevelt Association, 1959; University of California established the Horace M.

Albright Scenic Preservation Award; Audubon Medal; Gold Medal from Camp Fire Club of America, 1962; National Park Service established the Horace M. Albright Training Center at Grand Canyon National Park, 1963; Berkeley fellow, 1968; Distinguished Service Award from American Forestry Association, 1968; award from Cosmos Club, 1974; Neasham Medal for Historic Preservation from California Historical Society, 1976; Horace M. Albright Medal, 1979; U.S. Medal of Freedom, 1980; John Muir Award from Sierra Club, 1986.

MAJOR CONTRIBUTIONS

Instrumental in convincing Congress to pass the legislation that established the National Park Service in 1916, was given the Medal of Freedom Award, the nation's highest civilian honor, by President Reagan in 1981. Appointed director of the National Park Service in 1929, he fought successfully for the establishment of three national parks in the area of the Great Smoky Mountains, the Grand Tetons, and Carlsbad Caverns.

His writings include *''Oh, Ranger!'': A Book about the National Parks* (with Frank J. Taylor, 1928, rev. 1946, repr. 1972), *The National Park Service: The Story behind the Scenery* (with Russell E. Dickenson and William Penn Mott, Jr., edited by Mary Lu Moore, 1987), and *The Birth of the National Park Service: The Founding Years, 1913–1933* (with Robert Cahn, 1985). Authored numerous magazine articles on the parks, among them ''Everlasting Wilderness'' (1928) and ''Drift of the Elk'' (with Theodore G. Joslin, 1930) in the *Saturday Evening Post*.

BIBLIOGRAPHY

Audubon, 72. 2 (March 1970).
Contemporary Authors. Vol. 122 (1992) (obit.).
Contemporary Authors. Vol. 124 (1993).
Gendlin, Frances. ''A Talk with Horace Albright.'' *Sierra* 64 (September/October 1979).
National Cyclopaedia of American Biography (current ser.), Vol. H (1967).
New York Times, 29 March 1987 (obit.).
''An Interview with Horace Marden Albright.'' *Parks and Recreation* 5 (December 1970).
Swain, Donald C. *Wilderness Defender: Horace M. Albright and Conservation* (1970).
The Ties That Bind: A Biographical Sketch of Horace M. Albright (1973).
Who Was Who in America. Vol. 9, 1985–1989 (1989).

UNPUBLISHED SOURCES

See Columbia University Oral History Collections; Helen M. MacLachlan Papers, Columbia University; Jackson Hole Preserve Oral History Collection, Columbia University Library; Conservation: State and National Parks Collection, University of California, Berkeley; Hal G. Evarts Papers, University of Oregon; National Park Service Oral History Collection—National Park Service Archives, Harpers' Ferry, West Virginia; Michael Frome Papers, Western Carolina University Library; Lewis C. Cramton Papers, University of Michigan Historical Collections.

Connie Thorsen

ALEXANDER, ANNIE MONTAGUE. Born Honolulu, Hawaii, 29 December 1867; died Berkeley, California, 10 September 1950. Explorer, naturalist, philanthropist.

FAMILY AND EDUCATION

Third of four children of Samuel Thomas Alexander, a sugarcane grower, and Martha Cook Alexander. Early educated by a governess. Attended Punahou School in Honolulu. Moved to Oakland, California, in 1882, attended public primary and secondary schools. Attended La Salle Seminary, Auburndale, Massachusetts, 1886–1888. Studied art in Paris (1888) but had to abandon this field due to poor eyesight. Next attempted career as a nurse but gave this up for the same reason. Attended paleontology classes at the University of California, Berkeley, for several years after 1900; did not take a degree. Never married.

POSITIONS

Active as explorer, naturalist, and paleontologist, 1900–1949.

CAREER

Traveled with parents to Europe, Pacific Islands, and Africa, 1888–1904. First exploring and collecting expedition with friends taken 1899. Began financing paleontological expeditions in northern California and Oregon, 1901–1903, and became an excellent collector of fossils. Met C. Hart Merriam, chief (1885–1910) of the U.S. Biological Survey, 1905. Inspired to begin collecting specimens of bears and other mammals, especially West Coast fauna in danger of extinction. By 1914, she had amassed the best private collection of bear specimens in the country. Continued her exploring and collecting expeditions in the western states and Alaska with friends and professional biologists until the year before her death. In her last decade, her interests centered largely on plant collecting. In 1906, began formulating plans for a museum at Berkeley in furtherance of her original plan of establishing a research collection of western American fauna and encouraging their study. Founder and patron of the Museum of Vertebrate Zoology (MVZ) at Berkeley in 1908 and made possible selection of Joseph Grinnell as first director (1908–1939). Ultimately provided various subventions totaling well over $750,000, together with 20,564 zoological and fossil specimens either collected or purchased by her. She also helped establish and underwrite the Museum of Paleontology at Berkeley (1908–1909) and created several graduate fellowships at the MVZ and the paleontological museum. In addition, she gave 17,851 plant specimens to the university herbarium. Established and maintained a California farm from 1911 until her death. Two mammals, two birds, six fossils, and two plants had been named for her by the end of the 1950s.

MAJOR CONTRIBUTIONS

Intrepid explorer and naturalist in western North America and Hawaii for the first half of the twentieth century. Endowed Museum of Vertebrate Zoology at the University of California at Berkeley. Provided substantial support for it, the university's Museum of Paleontology, and the university's botanical collections. Contributed numerous specimens of animals, plants, and fossils to the university's collections. Actively interested in the operation of the MVZ for the first three decades of its operation.

BIBLIOGRAPHY

Bonta, Marcia Myers. *Women in the Field: America's Pioneering Women Naturalists* (1991).

Grinnell, Hilda W. *Annie Montague Alexander* (1958).

Pfaff, Timothy. "The Evolution of a Giant: Berkeley's Museum of Vertebrate Zoology." *California Monthly* (October 1983).

Sterling, K. B. "Introduction" to *An Account of the Mammals and Birds of the Lower Colorado Valley*, by Joseph Grinnell (1914), 1978.

Zullo, Janet L. "Annie Montague Alexander: Her Work in Paleontology." *Journal of the West* (April 1969).

UNPUBLISHED SOURCES

See Annie Montague Alexander Field Notebooks and Papers, Museum of Vertebrate Zoology and Bancroft Library, University of California, Berkeley. Louise Kellogg Field Notebooks, Museum of Vertebrate Zoology, University of California, Berkeley. A biography of Alexander by Barbara R. Stein is in progress.

Keir B. Sterling

ALLEE, WARDER CLYDE. Born near Bloomingdale, Indiana, 5 June 1885; died Gainesville, Florida, 18 March 1955. Ecologist.

FAMILY AND EDUCATION

Son of John Wesley and Mary Emily Newlin Allee. Attended Friends Academy, Bloomingdale. B.S., Earlham College, 1908; M.S., 1910, Ph.D., 1912, University of Chicago; honorary LL.D., Earlham, 1940. Married Marjorie Hill, 1912 (died 1945); three children. Married Ann Silver, 1953.

POSITIONS

Assistant in zoology, University of Chicago, 1910–1912; instructor in botany, University of Illinois, 1912–1913; instructor in zoology, Williams College, 1913–1914; assistant professor of zoology, University of Oklahoma, 1914–1915; professor of biology, Lake Forest College, 1915–1921; assistant professor of zoology, 1921–1923, associate professor of zoology, 1923–1928, professor of zoology, 1928–1950, dean in the colleges, 1925–1927, University of Chicago; instructor, Marine Biology Laboratory, Woods' Hole, Massachusetts, summers 1914–1921 (invertebrate course, 1918–1921); professor of zoology, National

Summer School, Logan, Utah, 1924–1926; head professor of zoology, University of Florida, 1950–1955.

CAREER

Allee was student of V. E. Shelford at Chicago and was imbued to a considerable extent with the ideas of ecologists there concerning the role of ecological succession. He was also greatly influenced by F. R. Lillie, on faculty at Chicago and also director of Woods' Hole Laboratory, through whom Allee became very much involved in summer research and instructional programs for some years. At Wood's Hole, strong tendency of experimental school of zoologists to study natural inclination of marine organisms toward sources of stimulation as basis for their research was strong influence in his work. Spent some years using results of laboratory experimentation to understand actions and role of animals in nature but did not confine himself to physiological or behaviorist explanations for all animal adaptations. In his *Animal Aggregations: A Study of General Sociology* (1931), sought to demonstrate role of physical factors in animal distribution and discussed nature of biological communities and the role of integration in social communities. Gradually became more concerned with role and function of social hierarchies in nature. Allee's *The Social Life of Animals* (1938) stressed cooperation among animals, as contrasted with competition. He also tried to apply his ideas to human society. His classic text, *Principles of Animal Ecology* (with O. Park, A. E. Emerson, T. Park, and K. P. Schmidt, 1949), was an effort to "start . . . supplying the orientation of which [animal] ecology, a subscience of biology, is in need." Allee drafted major section dealing with an analysis of the environment and chapters on history of ecology and on aggregations. Placed emphasis on capacity of animals to maintain internal equilibrium by making physiological adaptations. Argued that all communities in nature are fundamentally controlled by same principles. Translated into several languages, it continued in use for decades.

MAJOR CONTRIBUTIONS

His *Principles of Animal Ecology* and his exposition of the views of the Chicago school of ecology, together with the work he did with colleagues and through his many students, were the principal ways in which his work had a lasting impact.

BIBLIOGRAPHY

American Men of Science. 9th ed. (1949).

Kimler, William C., *Dictionary of Scientific Biography*. Supplement 2, vol. 18 (1988).

Schmidt, Karl P. "Warder Clyde Allee, 1885–1955." In *Biographical Memoirs, National Academy of Sciences*. Vol. 30 (1957).

Who Was Who in America.

Keir B. Sterling

ALLEN, ARTHUR AUGUSTUS. Born Buffalo, New York, 28 December 1885; died Ithaca, New York, 17 January 1964. Ornithologist.

FAMILY AND EDUCATION

Son of Daniel Williams Allen and Anna Moore. Attended Buffalo High School and Cornell University (A.B., 1907; M.A., 1908; Ph.D., 1911). Married Elsa Guerdrum, 1913; five children. Mrs. Allen (Ph.D., Cornell, 1929) became notable historian of ornithology.

POSITIONS

Leader of American Museum of Natural History expedition to Columbia, 1911–1912; instructor in zoology, Cornell, 1912–1916; assistant professor, 1916–1926; professor, 1926–1953; lecturer for National Audubon Society, 1953–1959.

CAREER

Allen's doctoral dissertation, ''The Red-Winged Blackbird: A Study in the Ecology of a Cattail Marsh,'' was described as ''the best, most significant biography which has thus far been prepared of any American bird.'' It set a pattern for later studies of this character. Inspired teacher of ornithology. More than 10,000 students took his undergraduate and summer school courses. From the 1920s until the early 1940s, Cornell offered the only Ph.D. program in ornithology in the United States. More than 100 graduate students secured advanced degrees under his supervision. Published an account of ornithological education in the United States in *Fifty Years' Progress of American Ornithology, 1883–1933* (1933). Skilled publicist for birds to the public as magazine writer, filmmaker, and lecturer. His *Book of Bird Life* (1930, rev. 1961) was a general introduction to the subject. Forty-seven popular species accounts were published in *American Bird Biographies* (1934) and *The Golden Plover and Other Birds* (1939) and saw frequent use in public schools. Known as North America's leading bird behaviorist. Pioneered in recording birdsong beginning in 1929. Produced first phonograph record of birdsong, 1932. Frequent contributor of articles on birds to *National Geographic Magazine*, 1934–1962. Many of these published in book form as *Stalking Birds with Color Camera* (1951), later twice revised. The Laboratory of Ornithology, long the unofficial name for Allen's ornithological program at Cornell, became a separate department at Cornell in 1955. In 1957, it was housed in separate facilities some distance from the main campus, where Allen continued to observe birds until his death. A founder of the Wildlife Society (1936) and its second president, 1938–1939.

MAJOR CONTRIBUTIONS

Preeminent ornithological educator for aspiring professionals and dedicated amateurs for over half a century. Established outstanding ornithological research and training facility at Cornell. Outstanding photographer of birds and path-breaking recorder of birdsong.

BIBLIOGRAPHY

Pettingill, Olin S. ''In Memoriam: Arthur A. Allen.'' *Auk* (1968).
Sutton, George M. *Bird Student: An Autobiography* (1980).

UNPUBLISHED SOURCES

See A. A. Allen Papers, Ornithological Archives, Cornell University (see *Laboratory of Ornithology Newsletter* [Summer 1978]; Taped interview with Allen in Carolynne H. Cline Papers, Cornell University.

Keir B. Sterling

ALLEN, JOEL ASAPH. Born Springfield, Massachusetts, 19 July 1838; died Croton-on-Hudson, New York, 29 August 1921. Zoologist and museum official.

FAMILY AND EDUCATION

Son of Joel Allen, a house builder and later a farmer, and Harriet Trumbull Allen, a teacher. He was one of five children. Allen attended local public schools and then Wilbraham Academy (Massachusetts) on an intermittent basis (1858–1862). Special student at Harvard (1862–1871) under Louis Agassiz. Received honorary Ph.D., Indiana University, 1886. Married Mary Manning Cleveland, 1874 (died 1879); one son, Cleveland. Married Susan Augusta Taft, 1886; no children.

POSITIONS

Appointed an assistant at Harvard's Museum of Comparative Zoology (1871–1885). Lecturer on ornithology, Harvard, 1870–1875. Special collaborator with the U.S. Geological and Geographical Survey of the Territories, 1876–1882. Curator of ornithology and mammalogy, American Museum of Natural History, New York, 1885–1907; curator of mammals there, 1907–1921; also had charge of invertebrate zoology (1887–1890) and of fishes and reptiles (1887–1901).

CAREER

A student and then a principal assistant to Louis and Alexander Agassiz at Harvard for fourteen years, then spent thirty-six years at the American Museum of Natural History, building, curating, and describing the bird and mammal collections. The American Museum bird collections increased from about 13,000 to in excess of 140,000 specimens, while the mammal collection grew from

1,000 to 40,000 specimens (exclusive of skeletal material) by 1915. Made collecting trips to Brazil [with Louis Agassiz] (1865), the Middle West (1867), east Florida (1868–1869), the Great Plains and Rocky Mountains (1871–1872), Yellowstone (1873), and Colorado (1882). Thereafter largely confined to desk work because of intermittent poor health.

MAJOR CONTRIBUTIONS

Completed several pioneering studies on the biogeography of birds and mammals, including *On the Mammals and Birds of East Florida* (1871) and *The Geographical Distribution of the Mammals, Considered in Relation to the Principal Ontological Regions of the Earth* (1878). Coauthor with Elliott Coues of *Monographs of North American Rodentia* (1877), an important monograph; also authored major studies of other major groups of mammals: *The American Bisons, Living and Extinct* (1876) and *History of North American Pinnipeds* (1880). A leading taxonomist, he described more than 560 new mammals and over forty new species and subspecies of birds. Authored important systematic revisions of various groups of mammals. Also published works on evolution and nomenclatural issues, together with many biographical sketches of ornithologists for *The Auk*. One of three cofounders of the American Ornithologists' Union (AOU) in 1883 and its president for eight years (1883–1891.) One of the five incorporators of the first national Audubon Society, 1886. A founder, director, and member of the Executive Committee of the (second) National Association of Audubon Societies, 1905–1921. Spent a quarter century editing *The Auk* and other AOU publications. Member of the Commission of Zoological Nomenclature of the Internal Congress of Zoology, 1910–1921. Was asked to accept first presidency of the American Society of Mammalogists (1919) but was forced to decline because of poor health.

BIBLIOGRAPHY

Allen, J. A. *Autobiographical Notes and a Bibliography of the Scientific Publications of Joel Asaph Allen*. Special Publication, American Museum of Natural History, New York (1916).

Chapman, Frank. "Biographical Memorial of J. A. Allen." *National Academy of Sciences Memoirs* 21 (1926): 1–20.

Sterling, K. B. ed. *Selected Works of Joel Asaph Allen*. New York: Arno Press (1974).

UNPUBLISHED SOURCES

See J. A. Allen, miscellaneous correspondence in the files of the Ornithology Department, American Museum of Natural History, New York.

Keir B. Sterling

ANABLE, GLORIA ELAINE "GLO" HOLLISTER. Born New York City, 11 June 1901; died Fairfield, Connecticut, 19 February 1988. Conservationist, ichthyologist, zoologist.

FAMILY AND EDUCATION

Daughter of Frank Canfield Hollister, physician and diagnostician, and Elaine Shirley Hollister. Father's encouragement to probe and uncover the mysteries of nature coupled with his educational guidance helped pave the way to Anable's success. Attended Misses Rayson's School, New York, until 1919, followed by one year at Hillside School, Norwalk, Connecticut. B.S. in zoology (studied under Pauline Dederer) from Connecticut College, New London, Connecticut, 1924; M.S. in zoology at Columbia University, 1925 (under tutelage of Florence Lowther, William K. Gregory, and others in department). Married Anthony Anable, chemical and metallurgical engineer, in October 1941; no children.

POSITIONS

Research assistant, Rockefeller Institute for Medical Research (1925–1928); research assistant and expedition leader, Department of Tropical Research at New York Zoological Society (1928–1941); chairman and chairman emeritus, Mianus River Gorge Conservation Committee of the Nature Conservancy (1953–1988); director, Stamford, Connecticut, Museum and Nature Center; trustee, Belle Baruch Conservation Foundation; trustee, Charles Madison Cannon Memorial Fund; national vice-chairman of Conservation Committee, Garden Club of America; chairman of Conservation Committee, Garden Club of Stamford, Connecticut; cofounder, Rockland County, New York, Council of Girl Scouts of America.

CAREER

Anable entered the scientific world as a medical research assistant under Alexis Carrel. Together they experimented on tissues from chicken hearts in the hope of unlocking the cause of cancer. However, Anable left this venue to join William Beebe, head of Department of Tropical Research at the New York Zoological Society; thus began her tenure as an ichthyologist. The goal of the Bermuda Oceanographic Expedition (begun in 1928, her first scientific trip with Beebe) was to discover and categorize the numerous forms of aquatic life found in the waters off the coast of Bermuda. During this undertaking, Anable analyzed deep-sea life from a steel bathysphere. Anable accompanied Beebe and his research team on two other projects: the West Indies Oceanographic Expedition (1932–1933) and the Pearl Islands Oceanographic Expedition (1934), both of which concentrated on the study and collection of fish life.

Anable's research focus changed to zoology when she led three expeditions for the Department of Tropical Research. Two Trinidad Zoological Expeditions (1926, 1936) afforded Anable the opportunity to explore the Arma Gorge and study and photograph the unusual Gaucharo, or "oil bird." Anable's British Guiana Expedition (1936) was an exceptional undertaking; to reach the Kaieteur

Plateau and Falls, the research team traversed a dense, tropical jungle. In addition to the discoveries of jungle fauna, Anable found two treasures: the half-inch golden frog and the Canje pheasant, or Hoatzin (a bird whose physical components combined those of birds and reptiles). Anable wrote prolifically about her expeditions and subsequent scientific findings, chiefly in *Bulletin of the N.Y. Zoological Society, Zoologica*, and *The American Girl.*

During World War II, Anable resigned her job at the Zoological Society to help the American National Red Cross establish the nation's first blood donor center in Brooklyn, New York. She also organized the Speaker's Bureau of the Public Relations Department for the Red Cross's North American region and later was promoted to assistant chief of the National Speaker's Bureau based in Washington, D.C.

Anable once wrote, "Man, whether he realizes it consciously or not, has an urge to run away from what he has made for himself in order to renew his contact with the world as it was before he put his mechanical mark on it." This desire to experience the unadulterated beauty of the natural world motivated her creation of the Mianus River Gorge Conservation Committee, a subdivision of the Mianus Gorge Project. As chairman, she challenged both the building of a dam that would consequently flood the Gorge Valley and the postwar explosion of real estate development. This once-small preserve of 60 acres now encompasses 555 acres of the "finest examples of the habitat of the flora and fauna of the Northeastern United States."

MAJOR CONTRIBUTIONS

On 11 June 1932, Anable set the women's record of 410 feet for the deepest oceanic descent. Her discoveries of the Gaucharo, golden frog, and Hoatzin were impressive contributions to zoology. Gloria Hollister Anable dedicated her life to the exploration of what fascinated her—nature. Her accomplishments in the fields of ichthyology and zoology, along with her considerable contributions to the cause of conservation, were characteristic of Anable's scientific talent and her personal reverence of nature.

BIBLIOGRAPHY

Anable, Anthony. *Biographical Sketch of Gloria Hollister Anable.* Wildlife Conservation Society Library.

Anable, Gloria Hollister. "The Society's Expedition to Kaieteur Fall." *Bulletin of the N.Y. Zoological Society* 39.5 (1936).

New York Times, 24 February 1988 (obit.).

Juliette M. Fernan

ANDERSON, RUDOLPH (MARTIN). Born Winneshiek County, Iowa, 30 June 1876; died Ottawa, Ontario, Canada, 21 June 1961. Zoologist, mammalogist, ornithologist, taxonomist.

FAMILY AND EDUCATION

Son of John Emanuel and Martha Ann Johnson Anderson. After public school, attended high school at Forest City, Iowa. Attended State University of Iowa 1897–1903, received Ph.B. in 1903. Captain of university track team. Honor graduate cadet officer Department of Military Science and Tactics, University of Iowa, 1903. Received Ph.D. (ornithology) in 1906, University of Iowa. Married Mae Belle Allstrand, 1913; three daughters.

POSITIONS

Assistant in zoology and acting curator, Iowa University Museum, 1901–1906. Instructor, Blees Military Academy; captain of the National Guard of Missouri, 1906–1908. Field agent, American Museum of Natural History, 1908–1912 (Alaska, Yukon, and Northwest Territories). Assistant mammalogist, American Museum of Natural History, New York, 1913. Zoologist, Geological Survey of Canada, 1913. Mammalogist, National Museum of Canada, Ottawa, 1913–1920 (arctic Canada, Alberta, Saskatchewan). Chief, Biology Division, National Museum of Canada, Ottawa, 1920–1946 (throughout Canada). Honorary curator of mammals, National Museum of Canada, Ottawa, 1946–1961.

CAREER

Began collecting birds at age twelve in 1888. Became ornithologist as teenager, publishing his first paper at seventeen. Served in Infantry Volunteers in Spanish-American War, 1898. Conducted exploratory and biological work in arctic Alaska, Yukon, and Northwest Territories 1908–1912 as general field agent and mammalogist for American Museum of Natural History, joint sponsor with Geological Survey of Canada, of the Stefansson-Anderson arctic expedition. Appointed zoologist to Geological Survey of Canada in 1913 and became chief of the Southern Party of the Canadian Arctic Expedition (CAE) 1913–1916. Carried out bird and mammal studies along arctic coast of Alaska and from Alaska–Yukon boundary east to Bathurst Inlet. Was in charge of scientific investigations and logistical support, including expedition ships for Southern Party. In 1919 appointed general editor of CAE reports, consisting of sixty-four individual papers in sixteen volumes. Between 1920 and retirement in 1946, carried out research on mammals in all provinces and territories of Canada. Was consulting zoologist to Lands, Parks and Forests Branch of the Department of Mines and Resources. Served on the Wildlife Protection Advisory Board 1917–1946 and Northern Advisory Board 1925–1946. A director of American Society of Mammalogists from 1919 to 1940. Elected corresponding member of Zoological Society of London 1939. Was associate editor (mammalogy) for *The Canadian Field-Naturalist* 1917–1955.

R. M. Anderson published 134 papers and books between 1893 and 1951, including *The Birds of Iowa* (1908), "Report on the Natural History Collections of the Expedition" in V. Stefansson's *My Life with the Eskimo* (1913), *Canadian Arctic Expedition II Report of the Southern Division* (1917), *Methods of Collecting and Preserving Vertebrate Animals* (1932), and *Catalogue of Canadian Recent Mammals* (1946).

As zoologist, he was happiest when in the field collecting specimens. He faced frustrations and privations of year-round Arctic work with notable good humor and excitement. His collecting efforts for over thirty years greatly enriched the collections of the National Museums of Canada and enabled him to publish descriptions of numerous new geographical races of mammals. Several newly recognized subspecies were named in his honor. His contributions to the fields of taxonomy, mammalogy, and Arctic research were recognized by the awarding of honorary membership or fellowship in seven professional societies. Much of his time was devoted to organization and duplication of distributional and other records, editing and report writing, correspondence, and an unfortunate controversy with Vilhjalmur Stefansson concerning the objectives, priorities, and methodology of the Canadian Arctic Expedition and disagreements over its leadership. Some of Anderson's mammalogy research and writing projects suffered from this attention to paperwork and the duties of administration. He was particularly generous with time he gave to correspondence with contacts made across Canada (whaling captains, Inuit hunters, Royal Canadian Mounted police officers, missionaries, young zoologists, and professional colleagues).

MAJOR CONTRIBUTIONS

The records and photographs from his Arctic expeditions provide one of the earliest and most complete documentations of the land, wildlife, and people of the western Arctic. His knowledge of arctic animals played an important role in early action of the Canadian government in drafting the Northwest Game Act to aid in conservation of northern wildlife. The new generation of arctic researchers benefited greatly from his experience and advice. Research on Canadian mammals laid groundwork for the eventual publication of *Mammals of Canada* by National Museums of Canada. Anderson's *Methods of Collecting* has been a standard text since publication in 1932 and is now in the fifth printing of fourth edition (revised in 1965).

BIBLIOGRAPHY

Diubaldo, R. J. *Stefansson and the Canadian Arctic* (1978).

Gray, David P., "R. M. Anderson's 'Camp Robinson Crusoe,' Langton Bay, NWT, 1910," *The Arctic Circular*, 27, nos. 1&2, (1979).

Jenness, D. "Obituary, R. M. Anderson 1877–1961." *Arctic* (1962).

Russell, L. S. "Rudolph Martin Anderson." *Canadian Geographical Journal* 64 (1962) (obit.).

Soper, J. D. "In Memoriam: R. M. Anderson 1876–1961." *Canadian Field-Naturalist* 76 (1962).

UNPUBLISHED SOURCES

See Public Archives of Canada, Ottawa: (1) Manuscript Division, Rudolph Martin Anderson Papers (records, papers, private correspondence); (2) Public Records Division, Transportation Records, Marine Branch RG42 (Canadian Arctic Expedition 1913–1918, Stefansson–Anderson controversy, papers, records, correspondence); (3) National Photography Collection (photographs taken by Anderson on 1908–1912 and 1913–1916 expeditions, later expedition photographs to 1945). Also see National Museums of Canada, Ottawa: (1) Library, R. M. Anderson Collection (personal library, correspondence, expedition notes, manuscripts, material on CAE and CAE reports, maps, card files, newspaper clippings, and collection of photographs); (2) National Museum of Natural Sciences, Vertebrate Zoology Division (scientific correspondence, field notebooks, diaries, CAE records).

David R. Gray

ANDREWS, ROY CHAPMAN. Born Beloit, Wisconsin, 26 January 1884; died Carmel, California, 11 March 1960. Explorer, zoologist, museum executive.

FAMILY AND EDUCATION

Son of Charles Ezra Andrews, wholesale druggist, and Cora May Chapman Andrews. Graduated from Beloit Academy, 1902. B.A., Beloit College, 1906; M.S., Columbia University, 1913; honorary Sc.D., Beloit, 1928, Brown 1935. Married Yvette Borup, 1914 (divorced, 1931); two sons. Married Wilhelmina Anderson Christmas, 1935.

POSITIONS

Joined staff of American Museum of Natural History, New York, as general assistant in Taxidermy Department, 1906; assistant in mammalogy, 1909; assistant curator, Department of Mammals, 1918; curator-in-chief, division of Asiatic exploration and research, 1924–1931. Briefly with U.S. Army Intelligence as a major, 1918, then with U.S. Naval Intelligence, in China, Manchuria, Mongolia, and parts of Siberia, 1918–1919. Participated in American Museum expeditions to British Columbia and Alaska, 1908; southwest Pacific islands and Japanese and Korean whaling stations, 1909–1910; Korea, 1911–1912; Alaska, 1913. Led expeditions to Tibet, southwest China, and Burma, 1916–1917; China and Mongolia, 1919–1920, 1922, 1923, 1925, 1928, 1930. Vice-director, American Museum, 1931–1934, director, 1935–1941; honorary director, 1941–1960.

CAREER

Had ambitions to be an explorer from his elementary school years. Went to New York City directly out of college, begged for any job at the American Museum of Natural History even when told none was available. Began as $40 per month assistant in Department of Preparation, aiding James L. Clark as taxidermist and preparer of exhibits. By 1907, Andrews had begun seven year career of research concerning whales and helped prepare seventy-five-foot blue whale museum model which was on exhibition for more than four decades. Andrews first collected whale specimens in the continental United States, and then in Alaska and British Columbia (1908–1909). He continued his whale research while also collecting other mammals in Borneo, the Celebes, Philippines, China, and Japan with the U.S.S. *Albatross* (1909–1910). He also collected in Northern Korea (1911, 1912), then returned to Alaska with the Borden Alaska Expedition of 1913. By 1914, Andrews had become a well-known authority on the Cetacea, writing a masters' thesis and several papers on the subject.

In 1915, however, Andrews changed emphasis with the enthusiastic approval of Henry Fairfield Osborn, the American Museum's president. Andrews' career as collector and later as expedition leader for the museum coincided with Osborn's presidency (1908–1933) of the institution. During that quarter century, the museum reached the apex of its public influence. Osborn had great interest in supporting work in vertebrate zoology and vertebrate paleontology, his own research specialties. Andrews ended his own active scientific research and for fourteen years (1916–1930) led museum specialists on a series of collecting expeditions to Asia. Osborn posited that central Asia had seen the genesis of reptiles, early mammals, and man, and Andrews was more than willing to seek evidence for this contention. He began with modest scale trips to Yunnan Province, China, and Burma (1916–1917) and to northern China and Outer Mongolia (1918–1919). The first of these yielded 2,100 mammal and numerous other vertebrate specimens; the latter, 1,500 mammals and a number of other vertebrates. His major expeditions to the Gobi Desert region in 1922, 1923, 1925, 1928, and 1930 were underwritten by funds Andrews personally raised from wealthy museum trustees, bankers, and businessmen, together with lecture fees and royalties on his articles and books. Andrews focussed on scientific results, avoiding mere adventuring; his well-planned trips centered on paleontology, geology, and archaeology. Participants were supported by camel caravans and automobiles, the latter Andrews' own contribution to the logistics of desert travel. Major discoveries included a number of duck-billed dinosaur eggs and nests (the first time material of this kind had been uncovered) and many fossils of other dinosaurs. There were also skeletal remains of large extinct mammals including *Baluchitherium (Indicatherium)*, a rhinocerus-like relative, and *Andrewsarchus*, larger than any previously found mammalian predator, which was named in Andrews' honor. In addition, other mammal fossils, including a mastodon jaw, a marsupial, and small extinct insectivores were identified. Geological evidence was found demonstrating that Mongolia had never experienced an ice age. Stone age implements (though no stone age skeletons) dating back 20,000 years were uncovered, and a number of living animals were collected. The total number of specimens brought back to New York exceeded 10,000. Osborn heavily favored Andrews, regarded the younger man's accomplishments as one of the leading achievements of his tenure as museum president, and paid him more than any other curator. This view was not shared by many scientists, who saw Andrews primarily as an administrator and field leader. By the early 1930s, the Great Depression had largely dried up Andrews' sources of funding. Growing anti-western feelings in China, coupled with the beginnings of the Sino-Japanese conflict, rendered further Asian field work impracticable.

From 1931 to 1934, Andrews served as one of several vice-directors of the American Museum while working up the scientific results of the Gobi expeditions and publishing a series of popular accounts of his work. Andrews was made Director of the American Museum in 1935, but he had little taste for administration, and fiscal stringencies left some museum activities underfunded. In 1941, Alexander Ruthven, a herpetologist and president of the University of Michigan, was brought in to assess the museum's structure and functions. He submitted a report critical of its administration and operations and recommended a change in leadership. Andrews was obliged to retire effective 1 January 1942, with title of Honorary Director.

Andrews authored several scientific studies, including ''Mammals Collected in Korea'' (with J. A. Allen, 1913) and ''Monographs of the Pacific Cetacea'' (two volumes, 1914–1916). He oversaw publication of seven volumes of *Central Asiatic Expeditions of the American Museum of Natural History: The Natural History of Central Asia* (1927–1943), and was principal author of volume I of this series, *The New Conquest of Central Asia* (1930). Books that he wrote for the general public included *Whale Hunting with Gun and Camera* (1916); *Camps and Trails in China: A Narrative of Exploration, Adventure, and Sport in little-known China* (with his first wife Yvette, a photographer, 1918); *Across Mongolian Plains* (1921); *On the Trail of Ancient Man* (1926); *Ends of the Earth* (1929); *This Business of Exploring* (1935); *This Amazing Planet* (1940); *Meet Your Ancestors* (1945); *Quest of the Desert* (1950); *Quest of the Snow Leopard* (1951); and *Beyond Adventure: The Lives of Three Explorers* (1952), together with several autobiographical works. His books for children included *All About Dinosaurs* (1953); *All About Whales* (1954); *All About Strange Beasts of the Past* (1956); and *In the Days of the Dinosaurs* (1959). Andrews was a fellow of the American Association for the Advancement of Science, the American Geographic Society, the New York Zoological Society, and the New York Academy of Sciences, and was a member of a numerous other organizations concerned with natural science and exploration.

MAJOR CONTRIBUTIONS

Andrews was considered by some to have been the exemplar for the fictional ''Indiana Jones,'' though Andrews' exploits always had a scientific objective. His accomplishments as leader of expeditions to Asia after World War I were certainly a factor in ''[H.F.] Osborn's successful identification of the [American] Museum with the dominant public values of the nineteen-twenties.'' By 1930, the American Museum's vertebrate paleontology collections ''were certainly the finest in the world,'' and the bird and mammal collections were of world class status. Andrews' Asian specimen collecting was one important factor in these achievements. His ebullience, his many articles and books, and his public relations skills helped to make his name a household word during a period when nature study and outdoor values had greater resonance with Americans than at any other time before or since.

BIBLIOGRAPHY

Andrews, Roy Chapman. *Under a Lucky Star: A Lifetime of Adventure* (1943).

Andrews, Roy Chapman. *An Explorer Comes Home* (1947).

Archer, Jules. *From Whales to Dinosaurs* (1976).

Dictionary of American Biography.

Fairservis, Walter A. *Archaeology of the Southern Gobi of Mongolia* (1993).

Hellman, Geoffrey. *Bankers, Bones and Beetles: The First Century of the American Museum of Natural History* (1969).

National Cyclopaedia of American Biography.

The New York Times, 12 March 1960 (obit.).

Rexer, Lyle, and Rachel Klein. *American Museum of Natural History: 125 Years of Expedition and Discovery* (foreword by Edmund O. Wilson) (1995).

Who Was Who in America.

UNPUBLISHED SOURCES

Kennedy, John Michael, ''Philanthropy and Science in New York City: The American Museum of Natural History, 1868–1968,'' unpublished doctoral dissertation, Yale University, 1968. Journals and correspondence are in the Archives of the American Museum of Natural History and some collected papers are in the Biological Sciences Library, University of California. Other materials are in the Carl L. Hubbs Papers, University of California at San Diego.

Keir B. Sterling

ANTHONY, HAROLD ELMER. Born Beaverton, Oregon, 5 April 1890; died Paradise, California, 29 March 1970. Mammalogist, zoologist, museum curator, author, editor.

FAMILY AND EDUCATION

Son of Alfred Webster Anthony, ornithologist and collector, and Anabel Klink Anthony. Educated at public schools of Portland, Oregon. Attended Pacific University, Forest Grove, Oregon, 1910–1911. Columbia University, B.S., 1915, M.A., 1920. Honorary D.Sc., Pacific University, 1934. Married Edith Irwin Demerell, 5 April 1916 (died 1918); one son. Married Margaret Feldt, 1922; two children.

POSITIONS

Field collector for Biological Survey (U.S. Department of Agriculture), May 1910 (North Dakota and Montana). Additional work with Biological Survey, summer 1911 (Wyoming). Initial employment at American Museum of Natural History in spring of 1910 as a naturalist on albatross expedition to lower California. Regular employment at American Museum began 26 September 1911. Member of staff, 1911–1958; served as associate curator of mammals, 1919–1926; curator, 1926–1958; chair, Department of Mammalogy, 1942–1958; deputy director, museum, 1952–1958; honorary curator, Department of Conservation and General Ecology, 1953–1956; dean of scientific staff, 1942–1947, 1951–1952, 1955–1956; curator, Frick Laboratory, 1958–1966.

CAREER

Was one of the world's premier mammalogists, specializing in mammals of the Western Hemisphere. Entered the army in Plattsburgh, New York, and rose to the rank of 1LT field artillery in late 1917. He became captain in 1918. Following a tour of duty in France during the summer of 1918, he was discharged in March 1919.

From 1910 to 1936 led many expeditions to South and Central America and the West Indies. During these trips he collected numerous fossil specimens, increased the American Museum of Natural History's collections. His leadership and creativity were key factors in the creation of major museum attractions such as the Hall of North American Mammals, the Akeley Hall of African Mammals, and the Hall of South Asiatic Mammals. His work on a pension board from 1926 to 1958 was instrumental in organizing a credit union and pension plan for museum employees.

Published over fifty papers on mammalogy and paleontology from 1913 to 1927, which were primarily published by the American Museum of Natural History. Member of over a dozen scientific societies of ornithology and mammalogy and served as director, chair, or charter member in most of these groups. President, American Society of Mammalogists, 1935–1937. Also served as scientific editor of *Mammals of America* in 1917 and consulting editor of *(Mammals) Book of Knowledge* in 1926. His *Capture and Preservation of Small Mammals for Study* (1925, several subsequent editions) was a widely used practical aid for collectors in the field.

Interests outside natural sciences included membership in wildlife and conservation organizations. Also won awards from the American Orchid Society for his contributions to orchid culture. Was president of the Greater New York Orchid Society.

MAJOR CONTRIBUTIONS

His book *Field Book of North American Mammals* served as the primary authority in the field for over twenty-five years. The authoritative *Mammals of Puerto Rico* was published in two volumes in 1925–1926. Articles appeared in a host of scientific journals. By 1939, Anthony estimated that he had traveled at least 200,000 miles while with the American Museum.

BIBLIOGRAPHY

''Harold Anthony Retires.'' *The Grapevine* 15: 7 (1 April 1958).

Birney, Elmer C. and Jerry R. Choate, eds. *Seventy-Five Years of Mammalogy 1919–1994* (1995).

New York Times, 31 March 1970 (obit.).

''Noted Mammalogist Dies in California.'' American Museum News Release (30 March 1970).

Who Was Who among North American Authors. Vol. 3 (1927–1928).

Who Was Who in America. Vol. 8 (1982–1985).

UNPUBLISHED SOURCES

Correspondence in files of Department of Mammalogy, American Museum of Natural History. Some papers for the years 1913–1929 in Northern California Regional Library Facility, University of California.

Anthony Todman

ANTHONY, JOHN GOULD. Born Providence, Rhode Island, 17 May 1804; died Cambridge, Massachusetts, 16 October 1877. Conchologist.

FAMILY AND EDUCATION

Son of Joseph and Mary (Gould) Anthony. When twelve years of age, parents moved to Cincinnati. Had little formal schooling and went into business, until eye trouble forced his retirement in 1851. Married Anna W. Rhodes, 16 October 1832.

POSITIONS

In 1863, Louis Agassiz placed him in charge of mollusk collections of Museum of Comparative Zoology at Cambridge. In 1865, joined Agassiz on Thayer expedition to Brazil.

CAREER

Despite business career, developed interest in natural history and began collection of freshwater mollusks of the Ohio River. From 1835 on, corresponded with mollusk students in the East and Europe and in 1838 began to publish his findings. Serious eye trouble interrupted his activities, but in 1853 toured Kentucky, Tennessee, and Georgia to collect mollusks. His work drew attention of Agassiz, who offered him a post at Museum of Comparative Zoology. His years at museum were devoted to classification and arrangement of its great collections. Also gathered data for history of Anthony family. Publications include ''Descriptions of Three New Species of Shells'' (1839); ''Descriptions of New Fluviatile Shells of the Genus Melania *Lam.* from the Western States of North

America'' (1854); ''Descriptions of New Species of North American Unionidae'' (1865); ''Descriptions of New American Fresh-water Shells'' (1866); and ''Genealogical Memoranda concerning the Anthony Family'' (n.d.).

MAJOR CONTRIBUTIONS

His published papers were mainly descriptive and had no great scientific influence but supplied material for important work by others.

BIBLIOGRAPHY

Dictionary of American Biography. Vol. 1 (1928–1937).

National Cyclopedia of American Biography. Vol. 10 (1967).

Turner, R. D. ''John Gould Anthony, with a Bibliography and Catalogue of His Species.'' *Occasional Papers on Molluscs, Museum of Comparative Zoology, Harvard University*. Vol. 1 (1946).

UNPUBLISHED SOURCES

See correspondence in Augustus Addison Gould collection at Harvard University. Other letters to be found in Benjamin Tappan collection at Library of Congress (Washington, D.C.).

Richard Harmond

ATWOOD, WALLACE WALTER. Born Chicago, 1 October 1872; died Annisquam, Massachusetts, 24 July 1949. Geologist, geographer, conservationist, university president.

FAMILY AND EDUCATION

Eldest of three children of Thomas Greene and Adelaide Adelia Richards Atwood. Graduate of West Division High School, Chicago. S.B., 1892, Ph.D., University of Chicago, 1902. Honorary D.Sc., Worcester Polytechnic Institute, 1943; honorary LL.D., Clark, 1946. Married Harriet Towle Bradley, 1900; four children.

POSITIONS

Instructor in physiography, Lewis Institute, Chicago, 1897–1899; instructor in geology, Chicago Institute, 1899–1900; assistant in physiography, 1901–1902, associate in physiography, 1902–1903, instructor in geology, 1903–1908, assistant professor of physiography and general geology, 1908–1910, associate professor of geology, University of Chicago, 1910–1913; professor of physiography, Harvard University, 1913–1920; president and professor of physical geography, Clark University, 1920–1946; director, Graduate School of Geography, 1921–1946. Also assistant geologist, 1901–1909, and geologist, U.S. Geological Survey, 1909–1946.

CAREER

Atwood's teaching career began following receipt of his undergraduate degree in 1897. He taught at several institutions in Chicago, then served under his mentor at Chicago, Rollin D. Salisbury, as a junior assistant on the New Jersey Geological Survey (1897) and the Wisconsin Natural History Survey (1898–1899). Following teaching appointments at Chicago and Harvard, he joined the faculty at Clark University as president and professor, remaining there until his retirement. At Chicago and Harvard, became specialist in physiography, now called geomorphology. Explored, studied, and recorded geology and geography of the Rocky Mountains on some twenty-five occasions between 1909 and 1948, becoming preeminent authority on the subject. Atwood established and maintained Graduate School of Geography at Clark despite objections that it received disproportionate share of university budget. He helped found *Economic Geography* (1925), a professional journal with international circulation. He was coauthor of geography texts for elementary and junior high students, and he was an able popularizer of scientific subjects for the public. Atwood was president of the Association of American Geographers, 1933–1934, and of the National Council of Geography Teachers. A dedicated conservationist, he also headed the National Parks Association from 1929 to 1933 and was active in the Sierra Club, Save the Redwoods League, and the National Forestry Association. Following his retirement from Clark, he was a cofounder and first chair of the board of a new institution, Utopia College, in Eureka, Kansas. He authored ''The Evidence of Three Distinct Glacial Epochs in the Pleistocene History of the San Juan Mountains'' (with Kirtley F. Mather, 1912). His principal professional publication was probably *Physiography and Quaternary Geology of the San Juan Mountains, Colorado* (with K. F. Mather, 1932). His best-known popular work was *The Rocky Mountains* (1945). Atwood was a conservative and somewhat controversial president of Clark University, and several contentious disputes during his tenure led some highly rated scholars at Clark to resign and continue their careers elsewhere.

MAJOR CONTRIBUTIONS

Atwood's principal contributions were in the areas of undergraduate and professional training in geography, his development of elementary and junior high school geography curricula, his studies of Rocky Mountain geomorphology, an active concern with environmental issues, and his skill at popularizing geography.

BIBLIOGRAPHY

Cressey, George B. *Annals of the Association of American Geographers* 22 (December 1949).

Dictionary of American Biography.

Ewan, Joseph, and Nesta Dunn Ewan. *Biographical Dictionary of Rocky Mountain Naturalists* (1981).

Kirtley, F. Maher. *Proceedings, Geological Society of America* (1949). *National Cyclopedia of American Biography.*
Van Valkenburg, Samuel. *Geographical Review* (October 1949).
UNPUBLISHED SOURCES
Atwood papers are in the Clark University Archives, the American Geographical Society, New York, the Association of American Geographers, Washington, D.C., Harvard University, and the University of Chicago.

Arthur A. Belonzi

AUDUBON, JOHN JAMES. Born Fougere, Jean Rabin or Jean Jacques Fougere Audubon, Les Cayes Plantation, Santo Domingo (now Haiti), 28 April 1785; died Minnie's Land Farm, Carmansville, New York (now Washington Heights section, New York City), 27 January 1851. Ornithologist, painter, naturalist.

FAMILY AND EDUCATION

Illegitimate son of Jean Audubon, a ship captain, and Jeanne Rabin, French chambermaid in employ of planter in Santo Domingo, who died before Audubon was seven months old. Oldest child of two, who, with younger half-sister, was formally adopted March 1794 by elder Audubon and legal wife, Anne Moynet. No formal education; educated at home by stepmother. Attended Rochefort-Sur-Mer Naval Academy briefly in 1796. Claimed to have studied under painter Jacques Louis David in Paris in 1802–1803, although this now has been discounted. Audubon married Lucy Bakewell, June 1808. Four children, two of whom, Victor Gifford (1809) and John Woodhouse (1812) helped their father with his major book projects. Audubon's two daughters died in infancy.

POSITIONS

Audubon family's bourgeois background allowed young Audubon to live life of "country gentleman" in the United States after first visit in 1803. On second visit to United States in 1806 circumstances changed, and Audubon embarked upon mercantile career in New York, Kentucky, Missouri, and Louisiana. Career in business ended in bankruptcy and brief internment in debtors' prison in 1819. In 1819–1820 worked four months as taxidermist in Cincinnati's Western Museum (only formal scientific position ever held). Thereafter, until departure for Europe in May 1826, lived life of itinerant artist, tutor, and clerk. Effort to enlist subscribers for proposed ornithological works in Europe proved successful and, after return to America in May 1829, was able to devote himself full-time to ornithological and scientific pursuits. Made three more trips to England (1830–1831, 1834–1836, 1837–1839), supervising publication of *The Birds of America* and securing subscriptions for that work, which were also being solicited in America. Was elected to membership in many scientific societies in Europe and and the United States.

CAREER

Audubon's study of nature began at early age, and his first drawings date from 1805. He painted and drew in the media of crayon, water, and oil. Formal publications of works of his art published during his lifetime began in late 1820s and are as follows: *The Birds of America*, 4 vols. (1827–1838). *A Synopsis of the Birds of North America* (summary of species included and index for Folio subscribers) (1839), and a smaller octavo edition of *The Birds of America*, combining pictures and text, in seven volumes (1840–1844). This latter edition of *The Birds* was a runaway best seller, certainly the most successful of any natural history title in the world to that time. Many of the backgrounds for the paintings in *The Birds* were provided by Joseph Mason, George Lehman, and others, and by Robert Havell, Jr., Audubon's English engraver, and his staff artists. The octavo edition of *The Birds* provided Audubon with enough money so that the family could finally build its first permanent home at "Minnie's Land," on northern Manhattan Island, in 1842. The plates of *The Viviparous Quadrupeds of North America* were published in two volumes (1845–1846) and the text in three volumes (1846–1854). Later editions of both *The Birds* and *The Quadrupeds* were published after Audubon and his two sons were dead.

Audubon's eyesight began failing him in 1846 and by 1847 he was completely senile. John Woodhouse Audubon completed 74 of the 150 mammal paintings, his brother Victor painting the backgrounds and handling all editorial and business details. Much of the scientific value of this work lies in the excellent text provided by the Reverend John Bachman, Lutheran minister and amateur mammalogist of Charleston, South Carolina, a long-suffering friend of the Audubons who was also father-in-law to the two Audubon sons. Bachman's second wife, Maria, painted many of the insects and flowers in the backgrounds for both *The Birds* and *The Quadrupeds*. Apart from the *Ornithological Biography*, 5 vols. (1831–1839), which was meant to be the accompanying text to *The Birds of America*, no significant written works of Audubon's are known. The text of *The Birds* was edited, and Audubon's poor command of English cleaned up by, the young Scottish naturalist William MacGillivray, who also supplied the necessary scientific nomenclature. Audubon had a love of the outdoors and feel for nature; however, his knowledge of zoology and ornithological terminology and nomenclature, even by the standards of his time, was weak and developed late in his life. That, along with deficiencies in English, made it necessary for him to collaborate with Bachman and MacGillivray. Only a dozen minor papers and reprints that appeared between 1824 and his death exist. Of his journals only five are extant. The original paintings for the *Birds of America*, save for three studies, were sold by Audubon's widow, Lucy, to the New York Historical Society in 1863 for $4,000. Background details for many of his later paintings, particularly of the *Quadrupeds*, were provided by other hands, notably his son John Woodhouse Audubon. The younger Audubon also completed some of the mammal paintings themselves when his father's mind and eyesight failed in the mid- and late-1840s. John Bachman's second wife Maria Martin was responsible for many of the plants, butterflies, and insects in Audubon's bird paintings.

MAJOR CONTRIBUTIONS

Although Audubon's ornithological studies of American birds had been preceded by Catesby and Wilson (the former by about 100 years, the latter by about 20), Audubon is still considered the preeminent figure in American ornithological art, matched only in the twentieth century by Louis Agassiz Fuertes, due to breadth of Audubon's work and because he was the first to attempt to portray birds in motion. The text of Wilson's earlier work, however, was more substantive and scientifically original. He has been criticized for portraying not a few of his subjects in anatomically impossible positions, adding a certain unscientific romanticism to his work and, in some instances, portraying subjects that he did not observe firsthand. Some critics of his work in the United States favored the earlier *American Ornithology* by Wilson, while others resented Audubon's insensitivity, occasional irresponsibility, and his sometimes proud and difficult personality. Yet his flair for the dramatic, combined with his extensive travels throughout much of the early United States and eastern Canada, brought about the first popular interest in the study and appreciation of nature in America. It should also be noted that in 1803 Audubon was the first individual on record to band birds.

BIBLIOGRAPHY

Adams, A. B. *John James Audubon* (1960).

Audubon, John Woodhouse, ed. *Audubon's Western Journal, 1849–1850* (1906).

Audubon, Maria R. *Audubon and His Journals, with Zoological and Other Notes by Elliot Coues* (1897).

Blaugrund, Annette, and Theodore E. Stebbins, Jr., eds. *John James Audubon. The Watercolors for Birds of America* (1993).

Chancellor, John. *Audubon: A Biography* (1978).

Corning, Howard, ed. *The Letters of John James Audubon, 1826–1840* (1930).

Delatte, Carolyn. *Lucy Audubon: A Biography* (1982).

Ford, Alice. *John James Audubon: A Biography* (1992).

Ford, Alice, ed. *Audubon, by Himself, a Profile of John James Audubon from His Writings* (1969).

Ford, Alice, ed. *The 1826 Journal of John James Audubon* (1967).

Fries, Waldemar. *The Double Elephant Folio: The Story of Audubon's Birds of America* (1973).

Herrick, Francis Hobart. *Audubon the Naturalist.* 2 vols. (1917) (see also 2d ed., 1938).

Mengel, R. M. "How Good Are Audubon's Bird Pictures in the Light of Modern Ornithology?" *Scientific American* 216.5 (May 1967).

Murphy, Robert Cushman. "John James Audubon (1785–1851): An Evaluation of the Man and His Work," *New York Historical Society Quarterly* (October, 1956).

Peterson, Roger Tory, and Virginia Marie Peterson, eds. *The Audubon Society Baby Elephant Folio of Audubon's Birds of America* (1990).

Streshinsky, Shirley. *Audubon: Life and Art in the American Wilderness* (1993).

UNPUBLISHED SOURCES

See American Philosophical Society Library, Philadelphia, John James Audubon Papers; Audubon Museum, Henderson, Ky., Tyler Collection; Filson Club Historical Society, Louisville, Ky., Manuscript Department; New York Historical Society, New York; Princeton University Library, Princeton, N.J., John James Audubon Collection; The Royal Scottish Academy, Edinburgh, Archives; Tulane University, Howard Tilton Memorial Library, La., John James Audubon Papers, Manuscripts, Special Collections.

Welch, Margaret C., ''John Audubon and His American Audience: Art, Science, and Nature, 1830–1860,'' unpublished Ph.D. dissertation, University of Pennsylvania (1988).

Steven Cooper

B

BACHMAN, JOHN. Born Rhinebeck, New York, 4 February 1790; died Charleston, South Carolina, 24 February 1874. Clergyman and naturalist.

FAMILY AND EDUCATION

Son of Jacob Bachman, a farmer, and his wife, Eva. Educated at home until age twelve. Received some brief formal education from a Lutheran minister in Schaghticoke, New York, then attended a school in Philadelphia for several years. Education interrupted about 1806 by illness. About 1808, briefly considered legal career but resumed religious training under Lutheran ministers in Rhinebeck and Philadelphia, 1808–1813. Is said to have attended Williams College for a time, but that institution has no record of it. Honorary M.A., Williams; honorary D.D., Gettysburg College; honorary LL.D., University of South Carolina; honorary Ph.D., University of Berlin, 1838. Married Harriet Martin, 1816 (died 1846); fourteen children, of whom nine lived to adulthood. Married Maria Martin (sister of his first wife), 1848; no children. His daughters Maria and Mary Eliza married the two sons of J. J. Audubon, John and Victor (1837, 1839), but both daughters soon died of tuberculosis (1840 and 1841).

POSITIONS

While still very young, is said by his daughter to have served as secretary to Johannes Knickerbocker during latter's expedition to visit and study Oneida Indians in western New York State, but modern biographers do not mention it. Taught school in Frankfort, Ellwood, and Philadelphia to support himself while completing theological studies. Licensed to preach in Lutheran Church, 1813.

Pastor at three churches in Schaghticoke, Rensselaer County, New York, 1813–1814; recurrence of lung illness led to brief trip to Caribbean. Moved south, in part for health, accepted pastorate at St. John's Lutheran Church, Charleston, South Carolina, 1815–1874; president, Evangelical Lutheran Synod of South Carolina, 1824–1834; trustee, College of Charleston, 1834–1848; professor of natural history there for a number of years from 1848.

CAREER

Began studying wildlife as a child with father's slave George, a competent woodsman and amateur naturalist. As a teenage student in Philadelphia, met Alexander Wilson, Alexander von Humboldt, William Bartram, C. W. Peale, and other scientific luminaries of the day. Met John J. Audubon when latter visited Charleston in 1831. Shared with Audubon his observations on land and water birds of coastal South Carolina, sent him bird skins and eggs; preserved other bird specimens for him in spirits; in the process of doing this enhanced collections of Charleston Museum and University of South Carolina. Traveled to Europe for six months for his health, 1838, where he resumed friendship with von Humboldt, met Louis Agassiz, and spent some time studying educational methods, consulting with naturalists, and studying specimens in natural history museums. Collaborated with Audubon and the latter's sons John and Victor on the *Viviparous Quadrupeds of North America*, 1840–1852, during which time he dissected specimens, studied their nomenclature in the available literature, and completed a carefully researched text. Had assistance from many friends and colleagues in region around Charleston in collecting specimens. The three volumes of the *Quadrupeds* were published 1846–1852. Published *The Doctrine of the Unity of the Human Race* (1850), in which he stated that "man is composed of only one true species." Published *A Defense of Luther and the Reformation* (1853). Offered opening prayer at ratification of Ordinance of Secession, 1860. Worked strenuously to provide support for Charleston hospitals during Civil War.

MAJOR CONTRIBUTIONS

Bachman was primarily responsible for the scientific descriptions of the mammals included in Audubon and Bachman's *Viviparous Quadrupeds of North America*. Made important contributions to the understanding of the mammals of South Carolina and the surrounding region through the collection of specimens, their analysis, and their description. Also assisted Audubon with bird specimens and descriptions when Audubon was completing his *Ornithological Biography* and his *Birds of America*. Bachman's second wife, Maria Martin Bachman, was an able artist, whose drawings of plants, butterflies, and other insects were utilized for background in a number of Audubon's bird paintings.

BIBLIOGRAPHY
Bachman, Catherine L. *John Bachman* (1888).
Herrick, Francis H. *Audubon the Naturalist*. 2 vols. (1938).
Neuffer, Claude H., ed. *The Christopher Happoldt Journal: His European Tour with the Rev. John Bachman (June–December, 1838) . . . with Biographies* (1960) (Bachman biography, 29–118).
Shuler, Jay. *Had I the Wings: The Friendship of Bachman and Audubon*. (1995).

Keir B. Sterling

BAILEY, ALFRED MARSHALL. Born Iowa City, Iowa, 18 February 1894; died Denver, 25 February 1978. Ornithologist, museum curator, author, explorer, National Geographic lecturer.

FAMILY AND EDUCATION

Son of William H. Bailey, an attorney, and Mollie Jelly Bailey. B.A., University of Iowa, 1916; honorary D.Sc., Norwich University, 1944; doctor of public service, University of Denver, 1954. Married Muriel Etta Eggenberg, 1917; two daughters.

POSITIONS

Member, U.S. Biological Survey (USBS) Expedition to Hawaiian Islands, 1912–1913; naturalist and curator of birds and mammals, Louisiana State (Cabildo) Museum, 1916–1919; representative of USBS in Alaska, 1919–1921; leader, Arctic Expedition for Colorado Museum of Natural History, 1921–1922; curator of birds and mammals, Colorado Museum of Natural History, 1922–1926; zoologist for Abyssinian Expedition, Field Museum of Natural History, 1926–1927; director, Chicago Academy of Sciences, 1927–1936; director, Denver Museum of Natural History, 1936–1969; director emeritus, 1970–1978.

CAREER

An ornithologist and mammalogist who, in addition to the expeditions already noted, participated in, or led, expeditions under the aegis of the Denver Museum to Mexico, 1941; Alaska, 1945; Labrador, 1946; the Mid-Pacific, Australia, New Zealand, 1949, 1952–1954, 1957–1958; Ecuador and the Galapagos Islands, 1960; Botswana and South Africa, 1969. Lectured extensively, using own color films, for more than thirty years, including seventeen consecutive years making presentations for the National Geographic Society at Constitution Hall in Washington, D.C. Published a number of volumes, including *Birds of Arctic Alaska* (1948); *Nature Photography with Miniature Cameras* (1951); *Laysan and Black Footed Albatrosses* (1952); *The Red Crossbills of Colorado* (with Robert J. Niedrach and A. Lang Baily, 1953); *Canton Island* (with Niedrach and Robert

Cushman Murphy, 1954); *Birds of New Zealand* (1955); *Birds of Midway and Laysa Islands* (1956); *Birds of Colorado*, 2 vols. (1965); *Pictorial Checklist of Colorado Birds* (1967); *Galapagos Islands* (an account of his 1960 field trip, 1970); *Field Work of a Museum Naturalist, 1919–1922* (1971); and two booklets in the Denver Museum of Natural History, Museum Pictorial Series: *The Hawaiian Monk Seal* (1952) and *Dusky and Swallow-Tailed Gulls of the Galapagos Islands* (1961). Contributed more than 200 articles covering expeditions on six continents to such publications as *The Auk, Condor, National Geographic, Natural History, Nature, Frontier, American Forests Life, Journal of Mammalogy, and Wilson Bulletin*. Member of the American Association for the Advancement of Science and various ornithological and conservation organizations. Received Malcolm Glen Wyer Award for distinguished service in adult education, 1961, and Regis College Civis Princeps Award, 1967.

MAJOR CONTRIBUTIONS

His bird and mammal collections, made over a period of more than half a century, helped make the Denver Museum a major regional natural history facility. A number of his bird skins are also at the Chicago Academy of Science. His collections, lectures, and publications were important contributions to both popular understandings of natural history and the scientific study of ornithology and mammalogy.

BIBLIOGRAPHY

American Men and Women of Science. 13th ed., vol. 1 (1976).
Auk 98 (January 1981) (obit.).
Contemporary Authors. Vol. 41R (1979).
Who Was Who in America. Vol. 7, 1977–1981 (1981).

Connie Thorsen

BAILEY, FLORENCE MERRIAM. Born Locust Grove, Lewis County, New York, 8 August 1863; died Washington, D.C., 22 September 1948. Ornithologist.

FAMILY AND EDUCATION

Daughter of Clinton Levi Merriam, successful businessman and Republican congressman from Lewis County, New York (1871–1875), and Caroline Hart Merriam. Youngest of four children. Only sister died shortly before her birth. Educated at private schools in Lewis County and at Mrs. Pratt's School, Utica, New York. Admitted as special student to Smith College, Northampton, Massachusetts. Attended 1882–1886; later awarded B.A. in 1921 "as of Class of 1886." Awarded honorary LL.D., University of New Mexico, 1933. Married Vernon Bailey, 1899; no children.

POSITIONS

A Victorian lady, she held no paying positions. At Smith College helped organize in March 1886 the Smith College Audubon Society. The year before, became the first woman admitted to the all-male American Ornithologists' Union (AOU), as an associate member. Became a full member in 1929 and in 1931 won the Brewster Medal of the AOU, another first for a woman. In 1887 was a social worker for Grace Dodge, helping Brooklyn poor working girls. She and her family had left Locust Grove for the milder climate of New York City, where her father still had some business and banking interests. Her articles on birds began to appear in the newly formed *Audubon Magazine*, organized by George Bird Grinnell. Earlier (1885–1886) had written a series of articles for the local press condemning the murder of birds for millinery purposes. Was encouraged by her older brother, C. Hart Merriam, employed at the new Ornithological Division of the Department of Agriculture, who often served as first editor for her pieces. At the age of twenty-six published first book, *Birds through an Opera Glass* (1889). Continued to write essays for publication, some for children, others for the general public, in magazines like *St. Nicholas, Our Animal Friends*, the *American Agriculturalist*, and the *Observer*.

CAREER

Aided by Olive Thorne Miller, an established ornithologist, she became a regular contributor to *Auk*, published by the AOU. Traveled to Illinois to teach bird class at the Hull House summer school at Rockford. Also continued her winter work with Grace Dodge's working girls' clubs in New York City. Went to Utah for her health with Miller; reported the trip in her second book, *My Summer in a Mormon Village* (1894). In 1894, traveled to California and visited Twin Oaks in San Diego County, where she observed and took notes on birds, then moved on, making similar observations in San Francisco Mountains in Arizona. These observations were published in 1896 as *A-Birding on a Bronco*. Two years later *Birds of Village and Field* followed; it was meant for beginners in ornithology and proved to be successful, as well as first popular American bird guide.

Returned east to Washington, D.C., to home of her brother Hart, now first chief of U.S. Biological Survey. Here she met Vernon Bailey, a colleague of her brother, whom she married in late 1899. Warm companions, she accompanied him on most of his trips for U.S. Biological Survey. He checked for mammals; she for birds. Articles based on her observations appeared in *Auk*, *Bird-Lore*, and *Condor* magazines. In 1902 published the *Hand Book of Birds of the Western United States*, which went through many editions and became the standard work on the subject. With the help of her husband and support of the Biological Survey, she completed and published the first book on birds of the Southwest: *Birds of New Mexico* (1928), which won her the Brewster Medal of the AOU (1931). Also contributed to her husband's work, especially *Wild Animals of Glacier National Park* (1918) and *Cave Life of Kentucky* (1933). Her home in Washington, D.C., was a mecca for the naturalist fraternity. Founded and promoted Audubon Society of the District of Columbia in 1897. A year later established the society's famous bird classes for teachers. A most productive observer and recorder of bird life, she published some 120 articles in various publications and more than a dozen books including *Birds Through An Opera Glass* (1890), *My Summer in a Mormon Village* (1894), *A-Birding on a Bronco* (1896), and *Birds of Village and Field* (1898). Additionally, continued social work with such groups as the Playground and Recreation Association, the National Child Labor Committee, and the Working Boys Home in Washington, D.C.

MAJOR CONTRIBUTIONS

Observant and rigorous in note taking, she was among first to study live birds in their native habitats. Considered by many as the outstanding female ornithologist of the nineteenth century.

BIBLIOGRAPHY

Ainley, Marianne. "The Involvement of Women in the American Ornithologists' Union," in Keir B. Sterling and M. G. Ainley, *A History of the American Ornithologists' Union* (in progress).

"Florence Merriam Bailey." *Audubon Magazine* (1949) (obit.).

Horner, Elizabeth, and Keir B. Sterling. "Feathers and Feminism in the 'Eighties." *Smith College Alumni Quarterly* (April 1975).

James, Edward T., ed. *Notable American Women, 1607–1950* (1971).

Kofalk, Harriet. *No Women Tenderfoot: Florence Merriam Bailey, Pioneer Naturalist* (1989).

Oehser, Paul H. "In Memoriam: Florence Merriam Bailey." *Auk* (1952).

Storm, Deborah, ed. *Birdwatching with American Women* (1986).

UNPUBLISHED SOURCES

See Florence Merriam Letters, Class of 1886, Smith College; Florence Merriam Bailey autobiographical notes (December 1919), California Academy of Sciences, San Francisco; Florence Merriam Bailey Papers, Bancroft Library, University of California, Berkeley (by far the most extensive collection of her writings; includes some additional autobiographical notes); Florence Bailey Collection, Houghton Library, Harvard University (mostly letters with her publisher, Houghton, Mifflin.)

Judith Moran Curran

BAILEY, LIBERTY HYDE. Born South Haven, Michigan, 15 March 1858; died Ithaca, New York, 15 December 1954. Horticulturalist, botanist, educator, author, philosopher.

FAMILY AND EDUCATION

Attracted to a career of plant collecting by father, Liberty Hyde Bailey, who was pioneer farmer and fruit grower. Mother was Sarah Harrison Bailey. He obtained formal instruction in botany and horticulture at Michigan Agricultural College (MAC) under William Beal, student of famed botanist Asa Gray. He received degrees at MAC: B.S., 1882, and M.S., 1886. Honorary LL.D., Wisconsin, 1907, Alfred University, 1908; honorary Litt. D., Univ. of Vermont, 1919; honorary Sc.D., Univ. of Puerto Rico, 1932. Married Annette Smith, 1883; two children.

POSITIONS

Assistant to Asa Gray at Harvard University Herbarium, 1882–1884. Professor of horticulture and landscape gardening, Michigan Agricultural College, 1885–1888. Professor of practical and experimental horticulture, Cornell University, 1888–1903. Director of State College of Agriculture at Cornell and dean of its faculty, 1903–1913. Chairman, U.S. Commission on Country Life, 1908–1909.

CAREER

As university professor advanced widespread recognition of the new horticulture as an applied science. Pioneered new course methodology with an emphasis upon demonstrations and laboratory experiments. His graduate students staffed universities and government bureaus. He was president of American Society for Horticultural Science from its formation, 1903. His years as educational administrator, 1903–1913, were marked by achievement of state funding for the independent College of Agriculture at Cornell, 1904, and by multiplication of specialized research departments and corresponding physical plant expansion. Concern for social and cultural aspects of rural life reflected in new departments of rural economy and home economics. College program carried to farmers and youth of the state through diversified extension curricula that featured agricultural courses and general education, notably Bailey's nature study. As dean he was credited with creation of foremost agricultural college of the nation. His report of the Country Life Commission reflected his belief in education as the principal ameliorative force in preserving traditional rural values and institutions. His survey technique used in the commission's operations led to development of rural sociology as a discipline. After twenty-five years at Cornell, he took up his plant exploration and botanical taxonomy studies he had earlier pursued. His published classifications of various plant genera based on his own specimen collections from around the world were innovative in their comparisons of native and cultivated species and achieved general acceptance. The scientific world is indebted to the Bailey Hortorium, instituted from his gift of 125,000 specimens to Cornell University, 1935, where studies on systematics of cultivated plants are conducted.

Published some sixty-five books and 700 articles and edited 117 manuscripts over a span of eighty-one years. They include *The Horticulturalist's Rule-Book*

(1889), *Lessons with Plants* (1897), *The Principles of Agriculture* (1898), *Botany* (1900), *The Nature-Study Idea* (1903), *First Course in Biology* (with W. M. Coleman, 1908), *The Training of Farmers* (1909), *The Country-Life Movement in the United States* (1911), *The Amateur's Practical Garden Book* (with C. E. Hunn, 1913), *The Cultivated Evergreens* (1923), *Manual of Cultivated Plants* (1924), *The Cultivated Conifers in North America* (1933), and *Hortus Second* (1941), the latter a revised edition of a dictionary of gardening, cultivated plants, and general horticulture in North America, first published in 1930.

A versatile and prodigious author, Bailey published over 100 scientific papers, which are recognized as substantial contributions to taxonomic botany, including research reports in plant breeding and physiology. He organized and edited new periodical *Gentes Herbarum*, to report on research conducted at the Bailey Hortorium, 1935. Agricultural and horticultural sciences owe him a vast debt for compilation of scientific and practical knowledge found in his series of reference volumes, best typified by *The Standard Cyclopedia of Horticulture* (6 vols., 1914–1917). His reputation as philosopher of country life was reinforced by his widely read volumes of poetry and philosophy, especially *The Holy Earth* (1915), which he prized beyond all others. Bailey visited Europe for research in various botanical collections after World War I. He continued collecting specimens in the field in many parts of the U.S., the Caribbean, and Latin America until the late 1940s while he turned out articles and books on a wide range of botanical subjects.

A fellow of a number of scientific organizations, he was elected President of the American Nature Study Association (1915), American Association for Agricultural Legislation (1919), American Pomological Society (1921), American Association for the Advancement of Science (1925), Botanical Society of America (1925), and the International Congress of Plant Science (1926). Won a number of awards and medals for his accomplishments in botany from organizations in Europe and the United States.

MAJOR CONTRIBUTIONS

Widely respected research scientist, professor, and agricultural college administrator, he sought to maximize agricultural productivity. As conservationist he attempted through nature study and elementary science courses to inculcate reverence for the natural order and need to preserve natural resources, particularly the soil. Paradoxically, he contributed to the accelerating transformation of commercial agriculture while serving as spokesman for the institutional values of a vanishing agrarian tradition.

BIBLIOGRAPHY

Banks, Harlan P. "Liberty Hyde Bailey," *Biographical Memoirs, National Academy of Sciences* 64 (1994).

Bowers, W. L. *The Country Life Movement in America 1900–1920* (1974).

Colman, G. P. *Education and Agriculture: A History of the New York State College of Agriculture at Cornell University* (1963).

Dictionary of American Biography. Supplement (1977).
Dorf, Philip. *Liberty Hyde Bailey: An Informal Biography* (1956).
Ellsworth, C. S. "Theodore Roosevelt's Country Life Commission." *Agricultural History* (October 1960).
Larson, O. F. "Liberty Hyde Bailey's Impact on Rural Life." *Baileya* (March 1958).
Lawrence, George H. M. *Baileya* (March 1955) (obit.).
Oleson, A. and J. Voss. *The Organization of Knowledge in Modern America, 1860–1920* (1979).
Rodgers, A. D. *Liberty Hyde Bailey. A Story of American Plant Sciences* (1949).
UNPUBLISHED SOURCES
See Bailey Papers, Collection of Regional History, Cornell University.

Lawrence B. Lee

BAILEY, LORING WOART. Born West Point, New York, 28 September 1839; died Fredericton, New Brunswick, Canada, 10 January 1925. Geologist, naturalist, chemistry professor.

FAMILY AND EDUCATION

Son of Jacob Whitman Bailey, professor of chemistry at the U.S. Military Academy, and Maria Slaughter Bailey. Mother and sister died in steamboat disaster; one brother trained in science but died in early twenties; one surviving brother, William Whitman, was a professor of botany at Brown University. Attended private schools. At Harvard University (1855–1859), studied science with Cooke, Agassiz, and Gray and received A.B. in 1859. Further chemical studies at Brown University in 1860. Awarded Harvard A.M. in 1862. Honorary Ph.D. from University of New Brunswick in 1873 and honorary LL.D. from Dalhousie University in 1896. Married Laurestine Marie d'Avray in 1863; six children.

POSITIONS

Assistant to Cooke at Harvard in 1860–1861. Professor of chemistry and natural history at the University of New Brunswick, 1861–1907. Geological surveyor for province of New Brunswick, 1863–1865. Geologist with the Geological Survey of Canada (temporary status), 186?–1904. Associated with the Marine Biological Station, St. Andrews-by-the-Sea, New Brunswick, from 1907. Professor emeritus of geology and biology, University of New Brunswick, 1907–1925.

CAREER

Although Bailey's primary studies were chemical, his research interests, like his father's, lay in the description of diatoms, the subject of his earliest papers. On his arrival in New Brunswick, he was immediately drawn into geological surveying, as the province had financed only one official survey (by Abraham Gesner, 1838–1841). Bailey was assisted by paleontologist C. F. Hartt and by G. F. Matthew. The three were the core of the Natural History Society of New Brunswick. Their work was officially sanctioned, 1863–1865. His *Report on the Mines and Minerals of New Brunswick* appeared in 1864. In 1865, Bailey and his collaborators surveyed the geology of southern New Brunswick, publishing *Observations upon the Geology of Southern New Brunswick with a Geological Map* (1865). He continued to work, at a nominal sum, for the Geological Survey of Canada during summers for the next forty years. His expertise was recognized in his election as a founding fellow of the Royal Society of Canada in 1882. Bailey's natural science teaching led to the publication of a school textbook, *Elementary Natural History* (1887), still in use after his death. He returned to his work on diatoms at the Marine Biological Station after his retirement. By 1923, when he ceased active work, he had identified nearly 400 species of diatoms from the Atlantic region, Saskatchewan, and the coast of Vancouver Island. These appeared in his *Annotated Catalogue of the Diatoms of Canada* (1924). Bailey published some seventy articles and reports on geology and biology, along with articles on more general scientific topics. He was a beloved teacher and popular lecturer.

MAJOR CONTRIBUTIONS

Bailey's sound geological work laid the groundwork for later investigations of New Brunswick and parts of Nova Scotia and Quebec by the Geological Survey of Canada. His example as a marine biologist helped establish Canada's excellence in that specialty.

BIBLIOGRAPHY

Bailey, Joseph Whitman. *Loring Woart Bailey. The Story of a Man of Science* (1925).

Ganong, William. "Loring Woart Bailey." *Proceedings and Transactions of the Royal Society of Canada* (1925).

Jarrell, Richard A. "Science Education at the University of New Brunswick in the Nineteenth Century." *Acadiensis* (Spring 1973).

UNPUBLISHED SOURCES

See Bailey family papers (12 linear feet), including scientific materials, University of New Brunswick Archives, Fredericton, N.B.

Richard A. Jarrell

BAILEY, VERNON ORLANDO. Born Manchester, Michigan, 21 June 1864; died Washington, D.C., 20 April 1942. Mammalogist, field naturalist.

FAMILY AND EDUCATION

Son of Hiram and Emily Bailey, pioneer farmers. Family moved to Elk River, Minnesota, when he was six years old. Educated in public schools of Minnesota.

Attended the University of Michigan, 1893, and George Washington University, 1894–1895, but did not receive a degree. Married Florence Augusta Merriam, 1899, sister of C. Hart Merriam and a noted ornithologist; no children.

POSITIONS

Special field naturalist, Division of Ornithology (after 1888, Division of Economic Ornithology and Mammalogy), U.S. Department of Agriculture, 1887–1890; chief field naturalist, Division of Economic Ornithology and Mammalogy (after 1896, Bureau of Biological Survey), U.S. Department of Agriculture, 1890–1933.

CAREER

Childhood on Minnesota frontier provided classroom for self-education in natural history. Received first government appointment in 1887 after impressing C. Hart Merriam with bird and mammal collections sent for identification. Forty-six-year career with the Bureau of Biological Survey marked by exhaustive field research. Conducted investigations in every state of union and in Mexico and Canada. Most important work concentrated in western United States. Member of Biological Survey's Death Valley Expedition, 1891. Executed intensive biological surveys of Texas, New Mexico, North Dakota, and Oregon. Often accompanied on field trips by his wife, who studied birds while he concentrated on mammals. Contributed close to 13,000 specimens to Biological Survey collections, including many new species. Continued fieldwork after his retirement in 1933 and was planning an expedition to Texas at the time of his death.

Primarily recognized as mammalogist, but also had extensive knowledge of birds and plants. Mammalogical research focused on American rodents, particularly voles and gophers. Sixteen mammals named in his honor. Devoted considerable time developing and promoting traps that captured wild animals alive and uninjured. His work in this area was honored by the American Humane Association. An accomplished outdoorsman, he shared his skills as a leader in the Boy Scouts of America. Involved in numerous professional organizations and conservation groups. Was a founder and president of American Society of Mammalogists (1933–1934) and served as president of the Biological Society of Washington and of Audubon Society of the District of Columbia.

Wrote both technical papers and articles for general readers. For many years member of Scientific Consulting Board of *Nature Magazine*. Bibliography totaled 244 titles, including *Spermophiles of the Mississippi Valley* (1893); *Pocket Gophers of the United States* (1895); *Mammals of the District of Columbia* (1900, 1923); *Biological Survey of Texas* (1905); *Life Zones and Crop Zones of New Mexico* (1913); *Wild Animals of Glacier National Park* (with Florence M. Bailey, 1918); *Biological Survey of North Dakota* (1926); *Animal Life of Carlsbad Cavern* (1928); *Animal Life of Yellowstone National Park* (1930);

Mammals of New Mexico (1931); *Cave Life of Kentucky* (with Florence M. Bailey and Leonard Giovannoli, 1933); and *Mammals and Life Zones of Oregon* (1936).

MAJOR CONTRIBUTIONS

Energetic, comprehensive field investigations added much to our understanding of flora and fauna of the western United States. Abiding concern for conservation and the welfare of wild animals. Increased general public's awareness of conservation concerns and biological sciences.

BIBLIOGRAPHY

Bailey, V. O. ''Into Death Valley Fifty Years Ago'' (as told to William Ullman). *Westways* (December 1940).

Birney, Elmer and Jerry R. Choate, eds. *Seventy-Five Years of Mammalogy (1919–1994)* (1995).

Chesnut, A. ''Vernon Bailey. A Nature Nobleman.'' *Nature Magazine* (October 1929).

Palmer, T. S. *Auk* (July 1947) (obit.).

Preble, E. A. *Nature Magazine* (June 1942) (obit.).

Sterling, K. B. *Last of the Naturalists. The Career of C. Hart Merriam*, rev. ed., (1977).

Who Was Who in America. Vol. 1 (1966).

Zahniser, H. *Science* (3 July 1942) (obit.).

UNPUBLISHED SOURCES

See Smithsonian Institution Archives, Record Unit 7267, Vernon Orlando Bailey Papers, 1889–1941 (includes field notes, journals, correspondence, photographs, maps, and newspaper clippings; Smithsonian Institution Archives, Record Unit 7176, U.S. Fish and Wildlife Service, 1860–1961, Field Reports (includes field notes on birds, mammals, and plants taken on his many field trips); University of Wyoming, American Heritage Center, Vernon Bailey Collection (includes correspondence, photographs, journals, biographical material, manuscripts, maps, newspaper clippings, and publications).

William Cox

BAILEY, WILLIAM WHITMAN. Born West Point, New York, 22 February 1843; died Providence, Rhode Island, 20 February 1914. Botanist, educator.

FAMILY AND EDUCATION

Son of Jacob Bailey, a professor of chemistry, mineralogy, and geology at the U.S. Military Academy, and Maria Slaughter Bailey. Younger brother of Loring Woart Bailey. Early education at garrison school, West Point. After father's death in 1858, attended University Grammar School, Providence, Rhode Island. Ph.B., Brown, 1864. Attended Harvard summer school, 1875, 1876, 1879. Honorary LL.D., University of New Brunswick, Canada, 1900; honorary M.A., Brown, 1903. Married Eliza Randall Simmons, educator and author, 1881; two children.

POSITIONS

Assistant librarian, Providence Atheneum, 1861–1871; assistant in Chemistry Department, Brown University, 1864; chemist, Manchester (New Hampshire) Print Works, 1866–1867; assistant in chemistry, Massachusetts Institute of Technology, 1866; botanist, U.S. Survey of the 40th parallel, 1867; teacher in private schools, Providence, 1867–1877; instructor in botany, 1877–1881; professor of botany, Brown University, 1881–1906.

CAREER

Had an interest in plants from boyhood. Lost mother and sister in destruction by fire of Hudson River steamer at age nine, while he himself was rescued by father. This tragedy evidently affected him until end of his life. At Brown, concentrated in chemistry until his junior year, when a fellow student persuaded him to switch to botany, this despite the fact that the university offered no instruction in the subject except as part of certain physiology lectures. Briefly served as private in 10th Regiment of Rhode Island Volunteers, defending Washington, D.C., 1862.

After nearly a year of fieldwork as a botanist with Clarence King and the 40th Parallel Survey in the West, serious illness compelled him to resign. Later, in the early 1870s, Bailey worked with John Torrey at Columbia University, picking up much practical information concerning herbaria. Was a specialist on plant pollination and also malformations and other abnormalities in plants. Authored *The Botanical Collector's Notebook* (1881); *Among Rhode Island Wild Flowers* (1895); *New England Wild Flowers* and *Botanical Notebook* (1897); and *Botanizing* (1898), together with many articles submitted to scientific periodicals and a good deal of poetry published in literary journals. Member and secretary, Board of Visitors, U.S. Military Academy, 1896. President of the Rhode Island Horticultural Society (1902), fellow of the American Association for the Advancement of Science, and active in a variety of scientific, scholarly, and patriotic organizations. Owned substantial botanical library. Left his large herbarium to Brown University.

MAJOR CONTRIBUTIONS

Extremely effective classroom teacher of botany; authority on teratology and pollination; his botanizing work with the King Survey in the West resulted in useful descriptions and collections; author of a number of well-received books and papers on a variety of botanical subjects.

BIBLIOGRAPHY

American Men of Science. 2d ed. (1910).
Bailey, W. W. "Why I Became a Botanist." *The Observer* (1894).
National Cyclopedia of American Biography. Vol. 29.

Who Was Who in America.

Keir B. Sterling

BAIN, FRANCIS. Born on family farm, North River, Prince Edward Island, 25 February 1842; died there, 20 November 1894. Farmer, scientist, author, ecologist.

FAMILY AND EDUCATION

Son of William and Ellen Dockendorff Bain. Educated by local Baptist minister, at a school in the district, in Thomas Leeming's class in Charlottetown for one winter, and later in City's Central Academy for one term. Married Caroline Clark, 1876; nine children.

POSITIONS

Farmer. Columnist for *Charlottetown Daily Examiner*, 1881–1892.

CAREER

Though formal education was limited, read widely. Had to take responsibility for family farm following death of older brother, 1862. Despite family and farm responsibilities, began traveling around Prince Edward Island (PEI) in 1860s; developed expertise about island's shells, insects, birds, plants, rocks, and fossils. Maintained journal from 1865, which demonstrates high degree of skill as observer. Primarily interested in geology; made map of his province's bedrock and collected and described a substantial number of fossils. On basis of preliminary reconnaissance, determined the practicability of tunnel beneath Northumberland Strait between Cape Traverse, PEI, and Cape Tormentine, New Brunswick. Though later (1892) employed by federal government to determine feasibility of this project, government proved reluctant to undertake it. Respected by other local naturalists and became principal contact for scientists and naturalists in other parts of Canada and the United States. When in 1883 Bain identified a fossil dug up by local farmer in New London as *Bathygnathus borealis*, reptile classified since 1867 as dinosaur, J. W. Dawson confirmed his analysis. Bain was credited with having made first recorded identification of dinosaur in Canada until *Bathygnathus* reclassified as nondinosaur in 1905. Dawson named new species of fossil fern, *Tylodendron baini*, in his honor (1890). Bain published some fifty articles on PEI natural history and his natural history investigations over an eleven-year period. Wrote extremely well and had considerable scientific understanding of his subjects, far superior to island contemporaries. In addition, always conveyed sense of infectious wonder. Also published at least twenty scientific papers in Canadian and American journals between 1881 and 1893, many of them lists and records of local fauna, flora, geologic strata, and fossils. Public speaking career dates from at least 1885. Prominent member of Natural History Society founded in Charlottetown, 1889, and spoke before it and other groups on botany, geology, and projected tunnnel to New Brunswick. Author of *The Natural History of Prince Edward Island* (1890), used as textbook in island primary schools, and of *Birds of Prince Edward Island: Their Habits and Characteristics* (1891), which described some 152 species. Despite later work by other authorities, this book still provides best available descriptions for some species.

MAJOR CONTRIBUTIONS

First native of PEI to make substantial contribution to understanding of the island's natural history despite modest means and lack of professional training. Left substantial body of published articles and books and unpublished manuscripts and lecture notes. Later reassessment of some of his work does not minimize significance of his accomplishments. His understanding of interrelationship of plants, animals, and environment marks him as island's first ecologist.

BIBLIOGRAPHY

(Charlottetown, P.E.I.) *Daily Examiner*, 20 November 1894.

(Charlottetown, P.E.I.) *Island Guardian*, 22 November 1894.

Martin, Kathy. ''The First Year of the Prince Edward Island Natural History Society.'' *PEI Natural History Society Newsletter*, no. 30 (March 1978).

Martin, Kathy. ''Francis Bain, Farmer Naturalist.'' *Island Magazine*, no. 6 (1979).

Martin, Kathy. *Dictionary of Canadian Biography*. Vol. 12 (1990).

Watson, L. W. ''Francis Bain, Geologist.'' *Royal Society of Canada Transactions*. 2d series, vol. 9, sect. 4 (1903).

UNPUBLISHED SOURCES

See Journal, lecture notes, unpublished manuscripts, and other items in Public Archives of PEI. Some photocopies of articles from *Daily Examiner* are in PEI Collection at University of PEI Library, Charlottetown.

Kathy Martin

BAIRD, SPENCER FULLERTON. Born Reading, Pennsylvania, 3 February 1823; died Woods Hole, Massachusetts, 19 August 1887. Zoologist, government official, scientific administrator.

FAMILY AND EDUCATION

Son of Samuel Baird, a lawyer, and Lydia Biddle Baird. Upon his father's death in 1833, the family moved to Carlisle, Pennsylvania. Baird graduated from Dickinson College in 1840. After a brief period as a medical student in New York City (1841–1842), he returned to Dickinson, where he earned an M.A. in 1843. Thereafter, Baird's education consisted of self-study and informal instruction from other naturalists. Honorary degrees from Philadelphia College of Medicine (M.D., 1848); Dickinson College (D.Sc., 1856); Columbian University (now George Washington University) (LL.D., 1875); and Harvard University (LL.D., 1886). Married Mary Helen Churchill in 1846; one child.

POSITIONS

Professor of natural history and chemistry, Dickinson College, 1846–1850; assistant secretary, Smithsonian Institution, 1850–1878; director, U.S. National Museum, 1850–1878; secretary, 1878–1887; commissioner, U.S. Commission of Fish and Fisheries, 1871–1887.

CAREER

Baird's scientific work established him as the leading authority in the mid-nineteenth century on North American vertebrate zoology. His own collections as a field naturalist in Pennsylvania soon were eclipsed by the very large collections from governmental expeditions that were deposited in the Smithsonian Institution and by specimens provided by a large network of private collectors. Baird placed a number of young scientists on the Pacific Railway Survey parties in the 1850s, and on other government and private expeditions in the U.S. and Canada in the late 1860s, 1870s, and 1880s. The specimens and observations these individuals brought or sent to Washington provided much of the raw data so masterfully synthesized by Baird in his many publications. In addition to using the materials as the basis for his own writings, these collections served as the basis for the Smithsonian's National Museum, which Baird largely instigated. Baird did not travel extensively from his home in Washington, D.C., except for summer trips to seashore points along the East Coast. During these working vacations, Baird became interested in marine biology. A notable result was Baird's initiative in 1871 in establishing the U.S. Commission of Fish and Fisheries, which, under his direction, conducted basic research in marine biology and also applied scientific knowledge to practical problems by propagating food fishes, encouraging legislation for the conservation of marine resources, and offering direct aid to the commercial fishing industry. After 1881, the fish commission operated a permanent research station at Woods Hole, Massachusetts. Baird also was a notable popularizer of science. Between 1872 and 1879, he edited for Harper Brothers eight annual volumes entitled *Annual Record of Science and Industry*, which disseminated recent advances in science and technology to a general audience.

Baird's bibliography included more than 1,000 entries. Most of these were articles in *Annual Record of Science and Industry* or official reports in his capacity as a Smithsonian and fish commission official, but about ninety titles were basic studies on the taxonomy and life history of birds, mammals, reptiles, amphibians, and fishes. Baird is remembered especially for his work in ornithology, which featured two notable works: *Birds*, vol. 9 of *Reports of Explo-*

rations and Surveys to Ascertain the Most Practicable and Economical Route for a Railroad from the Mississippi River to the Pacific Ocean (1858); and *A History of North American Birds*, 3 vols. (with Thomas M. Brewer and Robert Ridgway, 1874). These works replaced the studies of John James Audubon and remained standard sources up to the time of Baird's death. Six years after the appearance of Darwin's *Origin of Species*, Baird also authored an important commentary on the theory of organic evolution. This was an 1865 paper, "The Distribution and Migrations of North American Birds," read to the National Academy of Sciences, which discussed "general laws" showing "certain influences exerted upon species by their distribution . . . and by their association with each other." Another of Baird's important works was *Mammals*, vol. 8 of *Reports of Explorations and Surveys to Ascertain the Most Practicable and Economical Route for a Railroad from the Mississippi River to the Pacific Ocean* (1857), which was the most comprehensive and reliable taxonomic study of North American species prepared to that point.

Aside from his own investigations and his role as a popularizer of science, Baird's work in organizing governmental scientific institutions was of notable importance. These institutions provided essential support to individual investigators and expanded the professional study of American science as a whole. The most notable of these agencies were the U.S. Commission of Fish and Fisheries and the U.S. National Museum. Unlike the Smithsonian Institution, of which it was part, the National Museum primarily was funded by the U.S. government. Among the scientists who benefited from the patronage of these organizations and from Baird's personal influence were C. Hart Merriam, Robert Ridgway, William H. Dall, and George Brown Goode. Although Baird was noted for his modesty, serenity, and kindness, he also displayed great drive and persuasive skill in obtaining funds from Congress for these growing institutions and in shaping them to meet his standards of professional science.

MAJOR CONTRIBUTIONS

Competence and influence as an investigator of vertebrate zoology, ability to form and expand governmental agencies to support the advancement of professional science in the United States, skill in popularizing science, and application of zoological knowledge for practical ends, including a pioneering interest in the conservation of natural resources.

BIBLIOGRAPHY

Allard, Dean C. *Spencer Fullerton Baird and the U.S. Fish Commission* (1978).

Billings, John Shaw. "Memoir of Spencer Fullerton Baird, 1823–1887." *Biographical Memoirs of the National Academy of Sciences*. Vol. 3 (1895).

Cockerell, T. D. A. "Spencer Fullerton Baird." *Popular Science Monthly* 68 (1906).

Dall, William Healey. *Spencer Fullerton Baird: A Biography* (1915).

Dean, Ruthven, ed. "Unpublished Letters of John James Audubon and Spencer F. Baird." *Auk* 23–24 (1906–1907).

Deiss, William A. "Spencer F. Baird and His Collectors." *Journal of the Society for the Bibliography of Natural History* 9 (1980).
Deiss, William A. "The Making of a Naturalist: Spencer F. Baird, the Early Years." In *From Linnaeus to Darwin: Commentaries on the History of Biology and Geology, Society for the History of Natural History*. Special Publication No. 3 (1985).
Goode, George Brown. *The Published Writings of Spencer Fullerton Baird, 1843–1882* (1883).
Goode, George Brown. "The Three Secretaries." In *The Smithsonian Institution, 1846–1896* (1897).
Herber, Elmer Charles, ed. *Correspondence between Spencer Fullerton Baird and Louis Agassiz, Two Pioneer American Naturalists* (1963).
Jordan, David Starr. "Spencer Fullerton Baird and the United States Fish Commission." *The Scientific Monthly* 17 (1923).
Lindsay, Debra. *Science in the Subarctic* (1983).
Rivinus, E. F., and E. M. Youssef. *Spencer Baird of the Smithsonian* (1992).
UNPUBLISHED SOURCES
See personal papers, Smithsonian Archives, Washington, D.C.; records of the Smithsonian Institution, Smithsonian Archives, Washington, D.C.; records of the U.S. Fish Commission, U.S. National Archives, Washington, D.C.

Dean C. Allard

BAKER, JOHN HOPKINSON. Born Cambridge, Massachusetts, 30 June 1894; died 21 September 1973. Investment banker, conservationist, administrator.

FAMILY AND EDUCATION

Son of George Pierce Baker, professor of dramatic literature at Harvard and Yale, and Christine Hopkinson. Graduated from Harvard in 1915. Married Elizabeth Dabney, 1921; two children.

POSITIONS

Worked in foreign sales department of National Cash Register Company, in Dayton, Ohio, 1915–1917; accounting department of American International Corporation, in New York City, 1919–1921; and with White, Weld and Co., as well as other investment bankers, in New York City, 1921–1934. Board chairman, executive director, and president of National Audubon Society, 1933–1959. Member and chairman of advisory committee on fish and wildlife to Secretary of Interior; board member of National Parks Association; member of advisory committee on conservation to Garden Clubs of America. Also served as president of Linnaean Society of New York.

CAREER

After graduation from Harvard and service in World War I as army pilot overseas, became successful investment counselor in New York. An enthusiastic birder who had proved himself a knowledgeable conservationist as president of Linnaean Society, was recruited for Audubon Association's board in 1933.

Became executive director of Audubon Society in 1934. Was a pioneer in nature center movement; believed that wildlife sanctuaries near cities should offer nature education for children. Also conceived of Audubon summer camps for eighteen-year-olds and older, designed as outdoor training facilities for teachers, youth group leaders, and others. Working with Nature Conservancy and other groups, was leader in establishing Corkscrew Swamp Sanctuary in Florida. Sparked national interest in Everglades area and worked to establish Everglades National Park in 1947.

MAJOR CONTRIBUTIONS

Under his leadership, National Audubon Society acquired number of new wildlife sanctuaries. Membership of Audubon Society increased tenfold during his tenure. Attracted a young and talented staff to Audubon. Campaigned for establishment of Everglades National Park and for preservation of whooping crane.

BIBLIOGRAPHY

Graham, Frank Jr. *The Audubon Ark: A History of the National Audubon Society* (1990).

New York Times, 23 September 1973 (obit.).

Who Was Who in America. Vol. 6 (1976).

UNPUBLISHED SOURCES

There are Baker letters in the Audubon Society Collection at the New York Public Library.

Richard Harmond

BANISTER, JOHN. Born Twigworth, Gloucestershire, England, 1650; died on Roanoke River, Virginia, May 1692. Botanist, entomologist, malacologist.

FAMILY AND EDUCATION

Son of John Banister. Educated Magdalen College, Oxford; matriculated 1667, B.A. 1671, M.A. 1674. Ordained minister of Church of England before 1676. Married Martha (surname unknown) before 1688; one son.

POSITIONS

Clerk, Magdalen College, Oxford, 1674–1676; chaplain, 1676–1678. Church of England missionary in West Indies before 1678. Church of England minister in Virginia from 1678. Planter in Virginia from 1690.

CAREER

Banister's interest in botany can be traced to his years at Oxford, where he compiled a herbarium of the plants in the Oxford Physick Garden. On his arrival in Virginia in 1678, he began to gather materials for a natural history of the region on the model of Robert Plot's *Natural History of Oxfordshire* (1677). Banister sent specimens and descriptions of plants, shells, and insects to Jacob Bobart, Bishop Henry Compton, Samuel Doody, Martin Lister, Robert Morison, John Ray, and Hans Sloane. William Byrd I was his chief patron in Virginia. Although he published nothing himself, his contributions may be seen in Lister's *Historia conchyliorum* (1686–1688); Ray's *Historia plantarum*, vols. 2 and 3 (1688, 1704); Leonard Plukenet's *Phytographia* (1691–1705); and Morison's *Plantarum historiae*, vol. 3 (1699). Lister published "Extracts of four letters from Mr John Banister" in *Philosophical Transactions of the Royal Society* 17 (1693): 667–692. James Petiver published "Herbarium Virginiarum Banisteri" in the *Monthly Miscellany* (December 1707), and "Some Observations concerning insects made by Mr John Banister," *Philosophical Transactions* 22 (1701): 807–814. Banister's *Natural History of Virginia* remained unfinished, in part for financial reasons. His "Account of the Natives" and other materials from his *Natural History* were used (without acknowledgment) by Robert Beverley in his *History and Present State of Virginia* (1708). Banister became a planter near the Appomattox River in 1690 in order to finance his botanizing. He was accidentally shot to death during a collecting expedition along the Roanoke River in Virginia in 1692. Although not widely known, Banister was highly regarded by his contemporaries; Ray referred to him as "*eruditissimus vir et consummatissimus botanicus.*" Banister was a member of the founding committee and an original trustee of William and Mary College, which opened in 1693.

MAJOR CONTRIBUTIONS

The first resident naturalist of the British colonies in America, Banister contributed to several major British works on natural history during this period. His most notable contribution in this regard was his descriptions of some 340 species of plants, used by Ray, Morison, and Plukenet. His catalogs and specimens were later purchased by Hans Sloane, whose collections formed the basis of the British Museum. His botanical work also influenced Linnaeus's accounts of American plants.

BIBLIOGRAPHY

Britten, James. *Dictionary of National Biography*. Vol. 1 (1888).

Elliott, Clark A., ed. *Biographical Dictionary of American Science* (1979).

Ewan, Joseph. *Dictionary of Scientific Biography*. Vol. 1 (1970).

Ewan, Joseph, ed. *Short History of Botany in the United States* (1970).

Ewan, Joseph, and Nesta Ewan. *John Banister and his Natural History of Virginia* (1970).

Gordon, A. C. *Dictionary of American Biography*. Vol. 1 (1928).
Stearns, R. P. *Science in the British Colonies of America* (1970).
UNPUBLISHED SOURCES
Extant manuscript materials on Banister, including correspondence, are reproduced in Ewan and Ewan.

Anita Guerrini

BARTON, BENJAMIN SMITH. Born Lancaster, Pennsylvania, 10 February 1766; died New York City, 19 December 1815. Botanist, zoologist, ethnographer, educator, physician.

FAMILY AND EDUCATION

Son of Thomas and Esther (Rittenhouse) Barton. Father, for twenty years rector of St. James Church (Anglican) in Lancaster, was founder of the Juliana Library Company, a missionary to the Indians, and member of the American Philosophical Society. Mother was sister of celebrated astronomer David Rittenhouse. From 1780 to 1782, Barton was student at respected York Academy in Lancaster, collecting specimens of birds, plants, and insects in his spare time. Exhibited talent for drawing. Attended College of Philadelphia and in 1784 began study of medicine under Thomas Shippen. In 1785, accompanied Rittenhouse and the commissioners appointed to survey the western boundary of Pennsylvania, which aroused interest in Native Americans. Studied medicine at University of Edinburgh, 1786–1788; was awarded Harveian Prize by Royal Medical Society of Edinburgh. Strained relationship with two professors induced him to move to Germany; possibly received M.D. at University of Göttingen, 1789. Married Mary Pennington, 1797; two children.

POSITIONS

Began practicing medicine in Philadelphia in 1789, that same year securing appointment at College of Philadelphia. After merger of that institution with University of Pennsylvania two years later, remained on staff for rest of career. In 1790 appointed professor of natural history and botany; also named to chair of materia medica five years later. In 1813 succeeded Benjamin Rush as professor of theory and practice of medicine, still retaining original position in natural history and botany. Between 1798 and 1815 was physician at Pennsylvania Hospital. From 1802 to 1815 was vice president of the American Philosophical Society; president of Philadelphia Medical Society, 1808–1815. From 1805 to 1808 was editor of his own *Philadelphia Medical and Physical Journal*.

CAREER

First professor of natural history in Philadelphia, if not United States, was also a pioneer professor of botany and built largest collection of native botanic specimens in the country. His *Elements of Botany* (1803) was the first (often reprinted) textbook in that field, while his informative lectures helped to popularize this new discipline. *Collections for an Essay towards a Materia Medica of the United States* (part I, 1798; part II, 1804) contains systematic descriptions of medicinal plants based on his own research, for which he is "still held in memory by the medical profession of America." Although too busy to devote sufficient time to extensive botanical fieldwork, funded expeditions by Frederick Pursh to explore southern states in 1806–1807 and by Thomas Nuttall to explore the Missouri River in 1810–1812. Influenced William Bartram. In principal work on natural history, *New Views of the Origin of the Tribes and Nations of America* (1797), theorized that the "nations of America and those of Asia had a common origin." Other publications included *A Memoir concerning the . . . Rattlesnake and Other American Serpents* (1796); *Fragments of the Natural History of Pennsylvania* (1799); and *Extinct Animals and Vegetables of North-America* (1814). Numerous journal articles covered such diverse topics as North American earthquakes, the hummingbird, and North American alligator. One manuscript, "Notes on the Animals of North America," was originally intended for use by the Englishman Thomas Pennant, who was revising his *Arctic Zoology* in 1793. Barton's manuscript arrived after Pennant's book had gone to press. A copy in the American Philosophical Society Library was published in 1974.

One writer surmises that "there must have been a fire about him that led to his becoming a keenly interesting personality," while another characterizes him as an "industrious collector and a tireless proselytizer for medical and botanical investigation." Plagued by poor health brought about, in part, by overwork, which may account, in part, for his disappointingly modest achievements.

MAJOR CONTRIBUTIONS

A talented and multifaceted person, one of the founders of the study of natural history in the United States and the "father of American botanical and zoological instruction."

BIBLIOGRAPHY

Barton, William Paul Crillion. *A Biographical Sketch . . . of . . . Professor Barton* (1816).

Colbert, Robert. "Benjamin Smith Barton." In James A. Levernier and Douglas R. Wilmes, eds., *American Writers before 1800: A Biographical and Critical Dictionary* (1983).

Dictionary of American Biography.

Dictionary of Scientific Biography.

Elliott, Clark A., ed. *Biographical Dictionary of American Science* (1979).

Ewan, Joseph Andorfer. *Benjamin Smith Barton's Influence on Trans-Allegheny Natural History* (1988).

Graustein, Jeannette E. "The Eminent Benjamin Smith Barton." *Pennsylvania Magazine of History and Biography* (1961).

MacPhail, Ian. *Benjamin Smith Barton and William Paul Crillon Barton* (1986).

Meisel, Max. *Bibliography of American Natural History* (1924–1929).

Pennell, Francis W. "Benjamin Smith Barton as Naturalist." *Proceedings of the American Philosophical Society* (1942).
Sterling, Keir B., ed. *Notes on the Animals of North America*, by Benjamin Smith Barton [1793] (1974).

UNPUBLISHED SOURCES

See Benjamin Smith Barton lecture notes and papers, Historical Society of Pennsylvania (HSP), Philadelphia (5,200 items); Violetta W. Delafield Collection, American Philosophical Society (APS), Philadelphia (contains 5,000 items of Barton's, including letters, scientific notebooks, journals, lectures, papers, medical notes, drawings and plates; Medical School Notes, College of Physicians of Philadelphia, APS.

The following collections contain letters from Barton:

American Philosophical Society Archives; John Bartram papers, Historical Society of Pennsylvania; College of Physicians of Philadelphia, letters, APS; Thomas Jefferson, papers, Library of Congress, Washington, D.C.; North American Indian Languages, APS; Scientists' Letters, APS.

Rolf Swensen

BARTRAM, JOHN. Born Marple, Pennsylvania, 23 May 1699; died Kingsessing, Pennsylvania, 22 September 1777. Botanist.

FAMILY AND EDUCATION

Eldest son of William Bartram, farmer, and Elizabeth Hunt, both Quakers. Educated at local country school; largely self-taught. Married Mary Maris or Morris, 1723 (died 1727); two sons. Married Ann Mendenhall, 1729; five sons, four daughters.

POSITIONS

Farmer at Kingsessing from 1728. King's Botanist, 1765.

CAREER

Expeditions: sources of Schuylkill River, Pennsylvania (1736); Delaware (1737); Virginia and the Blue Ridge (1738); New Jersey and southern Delaware (c.1740); Catskills (1742); Pennsylvania and upstate New York, with Lewis Evans and Conrad Weiser (Onondago expedition) (1743–1744); Catskills, with son William Bartram (1753); Connecticut, with William Bartram (1755); Carolinas (stayed with Alexander Garden) (1760); Ohio River and West Virginia (1761); and South Carolina, Georgia, Florida, with William Bartram (1765–1766).

From 1766 on Bartram remained at Kingsessing on the Schuylkill four miles from Philadelphia, where he worked in his famous botanical garden, established around 1730. He began to correspond with Peter Collinson, English natural philosopher and fellow Quaker, around 1733. They exchanged both ideas and plants. Through Collinson, Bartram became a correspondent of Buffon, Johannes Gronovius, Pehr Kalm, Linnaeus, and Philip Miller. Collinson helped finance Bartram's botanizing by purchasing plants and seeds, and published extracts from his letters in the *Philosophical Transactions of the Royal Society*. Bartram's friendship with James Logan began in the 1720s; Logan loaned him books which helped supplement his meager education. Logan also helped Bartram gain a place on the 1743 Onondago expedition. Bartram's journal of that trip, published in 1751, established his fame. This expedition was funded by a subscription initiated by Benjamin Franklin. Other English patrons of Bartram included Lord Petre and the Quaker physician John Fothergill, to whom he in turn sent specimens. Collinson arranged his appointment as King's Botanist, with a stipend of fifty pounds a year, which enabled Bartram to undertake his last journey, to the southeast. He then retired to Kingsessing. His visitors included Franklin, Washington, John Mitchell, Cadwallader Colden, Garden, Kalm, Andre Michaux, and St.-Jean de Crevecoeur, who published an account of Bartram in his *Letters from an American Farmer* (1782). Bartram published little apart from his Onondago diary. Diaries of other expeditions circulated in manuscript, however, and contributed to his fame.

Bartram was a member of the American Philosophical Society, which he helped Franklin to establish in 1743, and of the Royal Academy of Sciences of Sweden.

MAJOR CONTRIBUTIONS

Bartram was not a systematic botanist, although he carried out hybridization experiments in his garden. Through Collinson and other collectors, he introduced over 100 American species of plants to Europe, and during his expeditions he identified and collected many more. His work and his garden served to inspire the next generation of naturalists.

BIBLIOGRAPHY

Bartram, John. ''Introduction,'' in Thomas Short, *Medicina Britannica* (1751).

Bartram, John. *Observations on the Inhabitants, Climate, Soil, Rivers, Productions, Animals, and Other Matters . . . From Pennsilvania to Onondago, Oswego and the Lake Ontario in Canada* (1751).

Bartram, John. ''Diary of a Journey through the Carolinas, Georgia, and Florida,'' in Francis Harper, ed. *Transactions of the American Philosophical Society* 33, part 1 (1942–1944).

Bell, Whitfield, Jr. *Dictionary of Scientific Biography*. Vol. 1 (1970).

Berkeley, Edmund, and Dorothy Smith Berkeley. *The Life and Travels of John Bartram* (1982).

Berkeley, Edmund, and Dorothy Smith Berkeley, eds. *The Correspondence of John Bartram, 1734–1777* (1992).

Corner, Betsy, and Christopher Booth, eds. *Chain of Friendship. Selected Letters of Dr. John Fothergill* (1971).

Cruickshank, H. G., ed. *John and William Bartram's America* (1957).

Darlington, William. *Memorials of John Bartram and Humphry Marshall* (1849, rpt. ed., 1967).
Earnest, Ernest. *John and William Bartram* (1940).
Elliott, Clark A., ed. *Biographical Dictionary of American Science* (1979).
Peattie, D. C. *Dictionary of American Biography*. Vol. 2 (1929).
Reid, Nina. ''Enlightenment and Piety in the Science of John Bartram.'' *Pennsylvania History* 58 (1991).
Stearns, R. P. *Science in the British Colonies of America* (1970).

Anita Guerrini

BARTRAM, WILLIAM. Born Kingsessing, Pennsylvania, 9 April 1739; died Kingsessing, 22 July 1823. Botanist, ornithologist, artist.

FAMILY AND EDUCATION

Third son of John Bartram and his second wife Ann Mendenhall. Attended Academy of Philadelphia, 1752–1756. Apprenticed to Philadelphia merchant, 1757–1761. Never married.

POSITIONS

Merchant, Cape Fear, North Carolina, 1761–1765. Planter along St. John's River, Florida, ca. 1766–1767. Artist under patronage of Dr. John Fothergill, 1768–1778. With elder brother John, ran their father's botanic garden in Kingsessing near Philadelphia, 1778–1812. Elected professor of botany, University of Pennsylvania, 1782, but declined for reasons of health.

CAREER

While still in school William Bartram began to accompany his father on collecting expeditions and to draw the plants and animals they encountered. His expeditions were: Catskills, with John Bartram (1753); Connecticut, with John Bartram (1755); South Carolina, Georgia, Florida, with John Bartram (1765–1766); and Carolinas, Georgia, Florida (1773–1778).

Through his father's patron, Peter Collinson, William's drawings became known to the Quaker physician John Fothergill, who helped support him by purchasing his drawings. Fothergill financed Bartram's final trip to the southeast between 1773 and 1778. Bartram returned to Kingsessing in January 1778 and remained there until his death in 1823. His account of his last journey was published in 1791 as *Travels through North and South Carolina, Georgia, East and West Florida*. His vivid descriptions of flora, fauna, and Native Americans made the work immediately popular, with multiple editions in the United States and Britain, as well as translations into Dutch, German, and French within the decade.

After the publication of the *Travels*, Bartram continued to write and draw, while also entertaining a stream of visitors to Kingsessing who sought his wisdom. He drew most of the illustrations for Benjamin Smith Barton's *Elements of Botany* (1803) and contributed articles (including a sketch of his father's life) to Barton's *Philadelphia Medical and Physical Journal* and other periodicals. Visitors to Kingsessing included Barton, F. A. Michaux, Thomas Nuttall, Palisot de Beauvois, Thomas Say, and Alexander Wilson. Bartram was elected to the Philadelphia Society for Promoting Agriculture (1785), the American Philosophical Society (1786), and the Academy of Natural Sciences of Philadelphia (1812).

MAJOR CONTRIBUTIONS

For the half-century up to his death, Bartram was the dominant figure in American natural history. His immensely popular *Travels* influenced Romantic poets such as Coleridge and Wordsworth as well as naturalists. He discovered several new plants and his list of 215 native birds in the *Travels* was the most complete before Alexander Wilson's *American Ornithology* (1808–1814). Wilson acknowledged Bartram's influence by naming "Bartram's sandpiper" (the upland plover). Bartram was also the most important American nature artist before Audubon.

BIBLIOGRAPHY

Bartram, William. "Observations on the Creek and Cherokee Indians, 1789." *Transactions of the American Ethnological Society* 3, part 1 (1853).

Bartram, William. "Travels in Georgia and Florida, 1773–4: A Report to Dr. John Fothergill." Francis Harper, ed. *Transactions of the American Philosophical Society* 33, part 2 (1943).

Bartram, William. *Travels through North and South Carolina, Georgia, East and West Florida, the Cherokee Country, the Extensive Territories of the Muscogulges, or Creek Confederacy, and the Country of the Choctaws* (1791).

Bell, Whitfield, Jr. *Dictionary of Scientific Biography*. Vol. 1 (1970).

Clarke, Larry R. "The Quaker Background of William Bartram's View of Nature." *Journal of the History of Ideas* 46 (1985).

Corner, Betsy, and Christopher C. Booth, eds. *Chain of Friendship. Selected Letters of Dr. John Fothergill* (1971).

Cooper, Lane. *Dictionary of American Biography*. Vol. 2 (1929).

Cruickshank, H. G., ed. *John and William Bartram's America* (1957).

Darlington, William. *Memorials of John Bartram and Humphry Marshall* (1849, rpt. ed., 1967).

Earnest, Ernest. *John and William Bartram* (1940).

Elliott, Clark A., ed. *Biographical Dictionary of American Science* (1979).

Ewan, Joseph, ed. *Botanical and Zoological Drawings of William Bartram, 1756–88* (1968).

Kastner, Joseph. *A Species of Eternity* (1977).

Slaughter, Thomas P., ed. *William Bartram: Travels and Other Writings* (1996).

UNPUBLISHED SOURCES

Manuscripts and letters of William Bartram are at the Historical Society of Pennsylvania, the University of Pennsylvania, the Academy of Natural Sciences (Philadelphia), the American Philosophical Society, and the British Museum (Natural History). Some of his drawings may be found at the Victoria and Albert Museum, London.

Anita Guerrini

BARTSCH, PAUL. Born Tuntschendorf, Silesia, Germany, 14 August 1871; naturalized U.S. citizen 1888; died Lorton, Virginia, 24 April 1960. Naturalist, invertebrate zoologist, curator, professor.

FAMILY AND EDUCATION

Born to prosperous merchants, Heinrich and Anna Klein Bartsch, was raised in foothills of the Sudenten range, encouraged in early love of nature by his father. His mother, graduate of University of Breslau in obstetrics, fostered his interest in biology. Family lost money during depression following Franco-Prussian War and emigrated to United States. After year in Missouri, settled in Iowa, where Paul learned taxidermy and collected a menagerie and natural history cabinet. Organized natural history club in high school and became avid ornithologist. Received broad background in science at University of Iowa under the tutelage of geologist Samuel Calvin. Was awarded the B.A. in 1896, the M.A. in 1899 for his thesis, "The Birds of Iowa," and Ph.D. in 1905 for his dissertation, "A Study in Distribution Based upon the Family Pyramidellidae of the West Coast of America." Married Signe Charlotte Gjerdrum in 1902, and they had one son, Henry. Divorced in 1939 and married Elizabeth Parker that same year.

POSITIONS

Division of Mollusks, U.S. National Museum (USNM): museum aide, 1895–1905; assistant curator of mollusks, 1905–1914; curator of marine invertebrates, 1914–1921; curator of mollusks, Cenozoic invertebrates, helminths, and corals, 1921–1946; and associate, 1946–1960. Major expeditions for USNM were 1907 pearl mussel investigation along the Mississippi; 1907–1908 USS *Albatross* Philippine expedition; 1911 Gulf of California; 1912 *Anton Dohrn*/Carnegie Institution expedition to Bahamas; 1914 *Tomas Barrera* expedition to Western Cuba; 1917 Cuba and Haiti; 1923 Cuba, Puerto Rico, and Bahamas; 1928–1930 Walter Rathbone Bacon Travelling Scholarship to West Indies; 1933 Johnson–Smithsonian deep-sea expedition to the Puerto Rico Deep; and 1937 Smithsonian–Roebling exploring expedition to Cuba. Concurrent appointments were Howard University, lecturer, 1901; professor of histology, 1902–1939; and director, Physiological and Histological Laboratory, 1903–1939, and George

Washington University, professor of zoology, 1899–1945; professor emeritus, 1945–1960.

CAREER

With over 450 papers to his credit, Bartsch was prolific, systematic biologist, whose taxonomic work has, for most part, stood the test of time. His first paper, published during his freshman year of college, was on fossil flora of the Sioux quartzite. Despite early devotion to ornithology, his area of expertise shifted to mollusks following his appointment as aide in that division of the National Museum in 1896. First malacological paper, description of California species of *Bittium* with William H. Dall, was published in 1901. Early taxonomic work focused on family Pyramidellidae. During career, worked primarily on marine and land mollusks of the United States, Caribbean, Hawaiian, and Philippine regions. Over years, curatorial responsibilities expanded to include all Cenozoic marine invertebrates, until separate divisions could be established in 1921. Oversaw arrangement of rapidly expanding collections and devised special storage trays for invertebrate specimens. The breadth of his natural history interests was reflected in series of faunal exhibits he designed for the National Museum in the 1910s, notably the "District Room," very popular display of local animal and plant life.

Bartsch's scientific career was notable for its broad approach to biological problems. In 1907 conducted an investigation for the Bureau of Fisheries to determine the status of pearl mussels in Mississippi River Valley. Studied role of ocean currents in distribution of marine fauna, first publishing his theory in 1912. During World War I, determined that common garden snail reacted to the poison gas used in trench warfare, before toxicity levels were hazardous to humans. Troops then utilized these snails as an early warning system for the use of gas masks, thus avoiding untold injury. For Navy Department, conducted pioneering taxonomic research on teredos or shipworms during the 1920s, subsequently patenting process for preserving wood immersed in saltwater. Studies of *Planorbis*, the intermediate host for the blood fluke, *Schistosoma mansoni*, led Bartsch to recommend procedures for preventing infestation of American soldiers during World War II, which proved beneficial to inhabitants of the Orient as well.

Amassed vast biological collections for the USNM during his many trips to the tropics of the Caribbean and the Pacific. Birds, mammals, reptiles, plants, and invertebrates of all varieties, in addition to mollusks, were preserved. Adopted underwater photography as early field tool for exploring reefs and ocean depths. Devised and utilized submarine still and motion photography equipment in the early 1920s. Invented underwater lamps for attracting specimens. Pioneered in use of echo sound recordings to determine ocean depths, notably on 1933 Johnson–Smithsonian expedition to the Puerto Rico Deep.

Bartsch initiated a systematic program for banding black-crowned night herons in 1902, first time this method had been used since Audubon's attempts. His popular accounts of these life cycle studies sparked interest and led to massive bird-banding programs of today.

Assumed teaching responsibilities in zoology at George Washington University shortly after his arrival in Washington and directed graduate studies until 1945. Taught medical students at Howard University, and for thirty-seven years directed the Physiology and Histology Laboratories. Was active leader in local Boy and Girl Scouts, guiding field trips and teaching conservation.

Bartsch's concern for conservation of wildlife began early, as evidenced in his second publication, at age twenty-four, entitled, ''Birds Extinct in Iowa and Those Becoming So.'' Rapid destruction of American flora and fauna led him to take a public role in their preservation. In addition to work with groups such as the Audubon Society, the Wild Flower Preservation Society, and the Emergency Conservation Committee, began publishing articles on conservation in the 1930s. Purchased some 458 acres along Mason's Neck, Virginia, in 1942 to establish a wildlife sanctuary, Lebanon, where he hosted many conservation and Scouting groups.

MAJOR CONTRIBUTIONS

Although maintained lifelong interest in ornithology and botany, his museum career was devoted to systematic and ecological research on mollusks. Bartsch's field research was distinguished by reliance on his broad biological background and innovative use of technology. Pioneered in environmental studies of animal behavior and interactions with surroundings. Applied his malacological expertise to important medical problems. Introduced several generations of Scouts and students to nature study and conservation. Developed his home into wildlife refuge and publicly encouraged conservation of habitats, wildflowers, and animals.

BIBLIOGRAPHY

Barnes, I. ''Paul Bartsch'': pt. 1, ''Biologist and Naturalist''; pt. 2, ''Smithsonian Curator''; pt. 3, ''The Lebanon Years.'' *Atlantic Naturalist* 7–9 (1952–1954).

Rehder, Harald A. *Journal of Conchology* (July 1961) (obit.).

Ruhoff, F. A. *Bibliography and Zoological Taxa of Paul Bartsch, with a Biographical Sketch by Harald A. Rehder.* (1973).

Seidenschnur, C. E., and S. G. Shetler. ''The Botanical Activities of Paul Bartsch (1871–1960).'' *Proceedings of the Biological Society of Washington* 78 (1965).

UNPUBLISHED SOURCES

See Letters, field notes, manuscripts, and photographs in the Paul Bartsch Papers, Smithsonian Archives; Letters and reports in Division of Mollusks Records, Smithsonian Archives; Letters in the Tyron-Pilsbry Correspondence and Witmer Stone Papers, Academy of Natural Sciences, Philadelphia.

Pamela M. Henson

BATES, MARSTON. Born Grand Rapids, Michigan, 23 July 1906; died Ann Arbor, Michigan, 3 April 1974. Zoologist, cultural biologist, university professor, author.

FAMILY AND EDUCATION

Only son of Glenn F. and Amy Mabel Button Bates. Glenn Bates was a farmer and horticulturist in Florida. Marston attended public schools in Fort Lauderdale. Majored in biology at University of Florida, Gainesville. B.S. degree in 1927. Graduate studies in zoology at Harvard University; A.M. degree in 1933 and Ph.D. in 1934. Sheldon traveling fellow, Harvard, 1934–1935. Postdoctoral work at Johns Hopkins University. Honorary Sc.D., Kalamazoo College, 1956. Married Nancy Bell Fairchild, daughter of botanist David Fairchild, 11 January 1939; one son, three daughters.

POSITIONS

From 1928 to 1931, worked for Servicio Técnico de Cooperación Agricola of the United Fruit Company in Honduras and Guatemala, first as a research assistant and then as director. On staff of Harvard Museum of Comparative Zoology in 1935–1937. Joined staff of Rockefeller Foundation, studying mosquito biology and malaria in Albania, Egypt, and Colombia, 1937–1950. Served as consultant, Office of Naval Research, 1950–1951. Professor of zoology at University of Michigan, 1952–1972. Director of research, University of Puerto Rico, 1956–1957. Timothy Hopkins lecturer, Stanford University, 1954. Fellow, Center for Advanced Studies, Wesleyan University, 1961. Trustee, Cranbrook Institute of Science, 1955–1962.

CAREER

Bates' doctoral thesis at Harvard concerned "The Butterflies of Cuba." He spent most of his two years at Harvard's Museum of Comparative Zoology (MCZ) on leave with the Rockefeller Foundation, looking into the biology of mosquitoes in Albania. He resigned from the MCZ in 1937 and continued his work in Albania until 1939. He remained with the Rockefeller Foundation until 1952, except for one year of study at Johns Hopkins University, investigating malaria in Egypt and then, with the coming of World War II, yellow fever in Colombia. Although new methods of prevention were developed, this study ended with the conclusion that the "elimination of yellow fever from America—was impossible." During his postdoctoral year at Johns Hopkins (1948–1949) Bates wrote *The Natural History of Mosquitoes* (1949), described by one authority as "a masterpiece of entomology for mosquitoes, which had to be known before control of mosquito-borne diseases was possible." Bates returned to the Rockefeller Foundation and remained with it in several capacities before joining the zoology faculty of the University of Michigan (1952). Bates insisted that scientists should address a larger audience and he turned out a series of popular volumes with that objective in mind. They included *The Nature of Natural History* (1950), widely praised by specialists and lay reviewers; *Where Winter Never Comes* (1952), a discussion of life in the tropics; *The Prevalence of People* (1955), a study of populations and problems in demography; *Coral Island* (1958); and *The Forest and the Sea: A Look at the Economy of Nature and Ecology of Man* (1960), perhaps his most influential book. His other volumes included *Man in Nature* (1960); *Animal Worlds* (1963); *The Land and Wildlife of South America* (1964); *Gluttons and Libertines* (1968); and *A Jungle in the House* (1970). A number of Bates' books were translated into several other languages, but some critics felt that he was too prone to accept the conclusions of authorities in fields where he himself had not specialized. Some professionals suggested that his later books popularizing environmental science had little to offer to students of ecology.

In 1953, received Centennial Certificate of Merit from University of Florida. Recipient of Daly Medal of the American Geographical Society, 1967. University of Michigan gave him its Distinguished Faculty Achievement Award, 1968.

His wife described their personal and professional experiences in her 1947 book, *East of the Andes and West of Nowhere*, and in her 1948 *National Geographic Magazine* article, ''Keeping House for a Biologist in Colombia.''

Fellow of American Association for the Advancement of Science; Phi Beta Kappa; Sigma Xi. Member National Science Foundation; Pacific Science Board; American Society of Naturalists (president 1961); Council on Foreign Relations; American Academy of Arts and Sciences; *American Scholar*, editorial board, 1955–1958.

MAJOR CONTRIBUTIONS

Research on mosquitoes, malaria, and yellow fever. Explored relations between public health and the problem of human population. Major works acclaimed for readability as well as scholarship.

BIBLIOGRAPHY

American Men and Women of Science. 12th ed.
Current Biography (1956), 34–35.
Facts on File (1974) (obit.).
Madden, Charles T., ed. *Talks with Social Scientists* (1968).
New York Times, 5 April 1974 (obit.).
Who Was Who in America. Vol. 6 (1974–1976), 23–24.

UNPUBLISHED SOURCES

See Marston Bates Papers, 1913–1914, in University of Michigan Bentley Historical Library, Michigan Historical Collections, Ann Arbor, Michigan, and Paul Lester Errington Papers, 1930–1962, in Iowa State University Library, Ames, Iowa (correspondence from Bates).

Carmela Tino

BAYNES, ERNEST HAROLD. Born Calcutta, India, 1 May, 1868; died Meriden, New Hampshire, 21 January 1925. Naturalist, author, lecturer.

FAMILY AND EDUCATION

Eldest of four children of John Baynes, an English inventor of photographic modeling, and Helen Augusta Norwill Baynes. Placed in English boarding school when parents went to the United States, 1874, and did not arrive there himself until age eleven. Attended high school in the Bronx, New York, and attended City College of New York, 1887–1891. Married Louise Birt O'Connell, 1901, no children.

POSITIONS

Reporter for the *New York Times* from 1891 to 1892. Assistant to his father, 1893–1900. Member Company K, 3d Connecticut Volunteers, during Spanish-American War, 1898. Writer and lecturer on natural history topics from 1900 until his death. Special correspondent for *Harper's Magazine* while traveling in Europe, Egypt, and Palestine, describing the part played by animals in World War I.

CAREER

Baynes' articles appeared in a wide range of periodicals including Scribner's *Country Life in America, St. Nicholas, Forest and Stream*, and *Munsey's* during his quarter-century writing career. Was critical of ''Nature Fakers,'' particularly Reverend William Long. Lived in and near Blue Forest Reservation in New Hampshire, 1904–1915, studying, photographing, and writing about animals. Was major force behind creation of American Bison Society, 1905. Served as treasurer and later as vice-president, securing support of Theodore Roosevelt, William Hornaday, and other leading conservationists of the time. Instrumental in raising buffalo on Blue Forest Reservation, a number of which were later shipped to Oklahoma and other locations in the west to swell growing federally protected herds. Participated with Percy MacKaye and others in ''Sanctuary,'' a masque about the preservation of birds performed before President Woodrow Wilson, which many audiences found as an effective means of dramatizing the issue of bird protection. An effective lecturer, Baynes was active on the Chautauqua Lecture circuit for Redpath Agency, promoting conservation issues, including conservation of birds. Organized the Meriden, New Hampshire, Bird Club, which expanded to include some 114 local bird clubs in 30 states, many of which subsequently became part of the Audubon Society movement. Published a number of books, including *Wild Bird Guests* (1915) with a preface by Theodore Roosevelt; *The Book of Dogs* (c.1919); *Vivisection and Modern Miracles* (1923); *War Whoop and Tomahawk* (1929); *Polaris, the Story of an Es-*

kimo Dog (1922); *Jimmie: The Story of an Eskimo Dog* (1923); *The Sprite* (1924); *Animal Heroes of the Great War* (1925); and *Three Young Crows and Other Bird Stories* (c.1927).

MAJOR CONTRIBUTIONS

Was tireless exponent of a wide range of conservation issues, having mainly to do with animals and birds. Believed in the necessity of vivisection to aid in the scientific study for prolonging life of both mankind and animals. Actively concerned with preservation of the buffalo.

BIBLIOGRAPHY

Gorges, Raymond. *Ernest Harold Baynes* (1928).
Who Was Who in America. Vol. 1 (1897–1942).

UNPUBLISHED SOURCES

See correspondence of Baynes and the office of the president of the American Bison Society, specifically about bison conservation, New York Zoological Society, Archives, Bronx, N.Y.

Geri E. Solomon

BEEBE, (CHARLES) WILLIAM. Born Brooklyn, New York, 29 July 1877; died Simla, Trinidad, 4 June 1962. Naturalist, oceanographer, zoological society executive.

FAMILY AND EDUCATION

Son of Charles Beebe, a paper company executive, and Henrietta Marie Younglove Beebe. Family moved to East Orange, New Jersey, when Beebe was a small child. One younger brother died in infancy. Graduated from East Orange High School in 1896. Attended Columbia University as a special student in zoology, 1896–1899, but did not receive a degree. Honorary Sc.D. degrees from Colgate and Tufts, 1928. Married Mary Blair Rice, 1902 (divorced 1913), and Elswyth Thane Ricker, 1927; no children.

POSITIONS

Assistant curator of birds, New York Zoological Society (NYZS), 1899–1902, curator, 1902–1916, honorary curator, 1916–1952. Founder-director, Department of Tropical Research, NYZS, Bartica and Kartabo, British Guiana, 1916–1926; Nonsuch Island, Bermuda, 1928–1938; various locations in Venezuela, 1942–1948, and at Simla, Trinidad, 1950–1952. Director emeritus, 1952–1962.

CAREER

Offered his first professional position with the NYZS at age twenty-two at the suggestion of Henry Fairfield Osborn (q.v.), his professor at Columbia. Began travels abroad 1900. Published *A Monograph of the Pheasants*, 4 vols. (1918–1922), termed by one authority "perhaps the greatest ornithological monograph of the present century." With French Aviation Service, 1917–1918. Visited the Galapagos, 1923, 1925. First undertook helmet diving to study fish and other marine species during visits to the Galapagos and Sargasso Sea, 1925. Studied fish and coral life in Haitian waters, 1927. First used spherical diving device, the bathysphere (with Otis Barton, who financed and developed it), off Nonsuch Island, Bermuda, 1930. Made more than thirty descents and achieved record depth of 3,028 feet, 1934. This record not broken until 1949. Other oceanographic research conducted off Baja California (1936) and the Pacific coast of Central America (1937 and 1938). Purchased land in the Arima Valley of Trinidad near Simla, 1949; established research station there, which he later presented to the NYZS.

Published some 800 articles and twenty-four books, 1905–1953, including *Two Bird Lovers in Mexico* (with Mary Rice Beebe, 1905), *The Bird, Its form and function* (1906), *Tropical Wildlife in British Guiana* (with C. I. Hartley and P. G. Howes—only one of two vols. published, 1917), *Jungle Peace* (1918), *Edge of the Jungle* (1921), *Beneath Tropical Seas* (1928), *Nonsuch: Land of Water* (1932), *Field Book of the Shore Fishes of Bermuda* (1933), *Half Mile Down* (1934), *Zaca Adventure* (1938), *High Jungle* (1949) and *Unseen Life of New York* (1953).

Wrote well on both technical and popular subjects, but not recognized as a major scientific figure by most professional biologists, who were probably reluctant to accord high standing to a successful popularizer. Named eighty-seven species of fish and one form of bird, but most of his systematic work has been superseded, due to his avoidance of the necessary formal training when a young man and to his inability, because of heavy professional commitments, to make a comprehensive study of the fields in which he was later active. Maintained high standards in his research work, but balanced this with a good sense of humor in his personal dealings.

MAJOR CONTRIBUTIONS

Breadth and detail of his field observations, emphasis upon interrelationship of living forms, abiding concern with conservation, felicity of expression in his published writings. Did much to close gap between the world of science and the general public.

BIBLIOGRAPHY

Berra, T. H. *William Beebe: An Annotated Bibliography* (1977).

Bridges, W. *Gathering of Animals: An Unconventional History of the New York Zoological Society* (1974).

Crandall, Lee S. *Auk* (January 1964) (obit.).

Dictionary of American Biography. Supplement 7 (1981).

Goddard, Donald, ed. *Saving Wildlife: A Century of Conservation* (1995).

New York Times, 6 June 1962 (obit.).

Welker, R. H. *Natural Man: The Life of William Beebe* (1975).

UNPUBLISHED SOURCES

See Letters in Alpheus Hyatt Correspondence, Princeton University.

Keir B. Sterling

BELANEY, ARCHIBALD STANSFELD, alias Grey Owl (Wa-Sha-Quon-Asin). Born Hastings, England, 18 September 1888; died Prince Albert, Saskatchewan, 13 April 1938. Writer, lecturer, conservationist.

FAMILY AND EDUCATION

Family included great-uncle who wrote theological animal rights tracts, and another, charged with murdering his wife, who wrote on falconry. Alcoholic father, George Belaney, abandoned his daughter by his wife's sister. Mother Kittie (Cox) Belaney. Removed and raised by two aunts. Fascinated by writings on Indians (Fenimore Cooper, Longfellow). Attended Hastings Grammar School, 1899–1904. Good student but did not complete school. Married Angele Egwuna 1910; two daughters. One son by trapline partner Marie Girard. She died, 1915. Married Ivy Holmes, 1917, who divorced him for bigamy, 1922. Pronounced married to Gertrude Bernard (Anahareo) by Algonquin Chief Ignace (Nias Papaté), Lac Simon 1926; one daughter. Married Yvonne Perrier, using the name "Archie McNeil," 1936.

POSITIONS

Chore boy, Temagami Inn, summers 1907, 1908. Guide, Camp Keewaydin boys' camp, summers 1910–1911. Fire ranger, Mississaga Forest Reserve, 1912–1914. Enlisted Canadian Expeditionary Force. Units included Black Watch, 13th Battalion, 1915–1916. Deputy chief ranger, Mississaga Forest Reserve, 1920–1921. Thereafter, various ranger, trapping, and seasonal work, Ontario and Quebec, to mid-1920s.

CAREER

Left school 1904 and left for Canada after two troubled years. Traveled from Halifax to Toronto, then to Lake Temiskaming in northern Ontario. Learned trapping from Bill Guppy. Canoed Lake Temagami, 1907. Claimed to be part Scot, part Apache, and grew his hair. Briefly in England. Returned to work at Temagami Inn, 1908, where he met Angele Egwuna. Joined her Ojibwa family for winter camp, Austin Bay, 1909–1910. Learned languages, trapping, and bushcraft. Summer guide at boys' camp, married Angele in August. Left her and his new daughter for traplines, fall 1911. Fire ranger, summers 1912–1914, winters as trapper. Trapped with Métis woman Marie Girard, winter 1913–1914. She died of tuberculosis after giving birth to his son, fall 1915. Joined Canadian Expeditionary Force, spring 1915, wounded 1916. Convalescence in England.

Married Ivy Holmes before returning to Canada alone, 1917. Severely troubled individual and alcoholic. Increasingly took refuge in being an Indian, even dyeing his skin. Summer work as forest ranger. Two years with another Ojibwa family, the Espaniels. Began writing.

Met Gertrude Bernard (Anahareo), town-raised Iroquois waitress, Camp Wabikon, Lake Temagami 1925. They moved to the bush. She convinced him to abandon trapping and to espouse conservation. Adopted and began to study two young beaver kits, winter 1928–1929. Sold first article, written as a trapper, to *Country Life* in 1929. Survivor of second pair of beavers named "Jelly Roll." Moved to cabin on Hay Lake, Témiscouta region, Quebec. Winter 1929–1930 Anahareo prospecting while he wrote *The Men of the Last Frontier* (1931). Published articles, now as an Indian conservationist, in *Canadian Forest and Outdoors*. Canadian Parks Commissioner J. B. Harkin suggested a "living argument for conservation," a film on Grey Owl, Anahareo, "Jelly Roll," and "Rawhide." Filmmaker Bill Oliver made several Parks-funded documentaries. Lecture career began with address and film for Canadian Forestry Association at Montreal. Position created for him and the beavers, to promote conservation and tourism in the national parks. Briefly at Riding Mountain National Park, Manitoba. Moved after six months to Prince Albert National Park, Saskatchewan (1931). Wrote autobiography, *Pilgrims of the Wild* (1934). Daughter born. Wrote childrens' novel, *Adventures of Sajo and Her Beaver People* (1935). Four-month British lecture tour (1935–1936) with Lovat Dickson. Severe drinking binges on his return. Anahareo absent. Completed short story collection, *Tales of an Empty Cabin* (1936). Married Yvonne Perrier in 1936, and they wintered at Prince Albert National Park. Next two films funded by Macmillan, Lovat Dickson, and himself, as Parks refused support. Second tour of Britain, included Buckingham Palace (10 December 1937). Tour continued for three months in United States and eastern Canada, leaving Yvonne hospitalized with exhaustion. Grey Owl died shortly afterward. North Bay *Nugget* then revealed to the startled public that he was not an Indian, something they had deliberately withheld. Buried behind his now restored cabin. Writings underwent revival during 1970–1972 environmental movement.

MAJOR CONTRIBUTIONS

Grey Owl was a cultural icon, combining the themes of wilderness conservation, wildlife preservation, and North American Indian cultural survival. Proposed Indians as official stewards of Canada's wilderness. Pioneered nature education on film. Important educator of children. Promoted role of parks as places for active nature education, not merely scenic vistas.

BIBLIOGRAPHY

Anahareo. *My Life with Grey Owl* (1940).

Anahareo. *Devil in Deerskins. My Life with Grey Owl* (1972).

Dickson, Lovat. *Wilderness Man: The Strange Story of Grey Owl* (1973).

Grey Owl. *Pilgrims of the Wild* (1934).
Smith, Donald B. *From the Land of Shadows: The Making of Grey Owl* (1990).
Smith, Donald B. "Archibald Stansfeld Belaney." *Canadian Encyclopedia.*
Waiser, Bill. *Saskatchewan's Playground: A History of Prince Albert National Park* (1989).

UNPUBLISHED SOURCES

See Smith and Waiser. National Archives of Canada, Ottawa, has the Lovat Dickson, Grey Owl Collections, Indian Affairs files (RG 10), and the Prince Albert Park files (RG 84). See also Archives of Ontario Grey Owl Collection; the Macmillan Archive, William Ready Archives, McMaster University, Hamilton, Ontario, and Thomas Randall's Grey Owl Collection, Dalhousie University Archives, Halifax, Nova Scotia.

Lorne Hammond

BENDIRE, CHARLES EMIL (KARL EMIL BENDER). Born König im Odenwald, Hesse-Darmstadt, 27 April 1836; died Jacksonville, Florida, 4 February 1897. Ornithologist, oölogist, curator of birds' eggs, cavalry officer.

FAMILY AND EDUCATION

The eldest of a family of two sons and three daughters. Privately educated in the home until age twelve. Five years at a theological seminary at Passy, France. Informal training in natural history in 1850s from Bernard J. D. Irwin. Apparently a lifelong bachelor.

POSITIONS

Enlisted man, U.S. Army, 1854–1864; rose from 2d lieutenant to captain, 1864–1873, retired due to disability 1886, brevetted major on retired list, 1890. Honorary curator of birds' eggs, U.S. National Museum, Washington, D.C., 1883–1897.

CAREER

Soon after immigrating to United States, he enlisted in the 1st Dragoons, 10 June 1854. Rose through the noncommissioned ranks to hospital steward, 4th Cavalry. Commissioned 2d lieutenant, 2d Infantry, 1864; transferred to 1st Cavalry, September 1864, and promoted to 1st lieutenant, November 1864; captain, 1873. Brevet 1st lieutenant, 1864, "for gallant and meritorious service in battle of Trevillian Station, Virginia," and brevet major, 1890, "for gallant service in action against Indians at Canyon Creek, Montana, 13 September 1877." Retired from army, 1886. Returned to Germany for brief visits in 1853 and 1867.

Serious collecting of eggs and nests began about 1868. Correspondence with Elliott Coues, J. A. Allen, and others resulted in articles by them based on Bendire's notes. By the late 1870s he was writing his own papers, which appeared in *American Naturalist, Bulletin of the Nuttall Ornithological Club, The*

Auk, Ornithologist and Oologist, Proceedings of the U.S. National Museum, and *Forest and Stream*. Worked mostly on birds, but contributed to other branches of natural history; for example, he demonstrated that the "red fish" of Idaho was actually the breeding young male of *Oncorhynchus nerka*. Articles, monographs, and notes total about seventy.

According to C. Hart Merriam, "No other American naturalist in modern times has spent half so much time in the field as Bendire." Most of his army years were spent in the trans-Mississippi West, his main duty stations being Cantonment Burgwin, New Mexico; Forts Bowie, McDowell, Lowell, and Whipple, Arizona; Fort Bidwell and Camp Independence, California; Forts Klamath and Harney, Oregon; Vancouver Barracks and Fort Walla Walla, Washington; and Boise Barracks and Fort Lapwai, Idaho. James C. Merrill wrote that Bendire "first obtained in the United States several Mexican species and discovered certain new ones."

His "honorary" (i.e. unpaid) curatorship of birds' eggs at the National Museum began three years before his retirement from the army. The collection was entirely revamped by him and grew spectacularly through the donation of his personal collection and specimens solicited by him. The two volumes of his magnum opus, *Life Histories of North American Birds*, were published in 1892 and 1896. A founder of the American Ornithologists' Union.

MAJOR CONTRIBUTIONS

The leading figure in American oology and nidiology in his day. A widely traveled field naturalist with a dedication to accuracy. His *Life Histories* were the bases of A. C. Bent's *Life Histories* series.

BIBLIOGRAPHY

Hume, E. E. *Ornithologists of the U.S. Army Medical Corps* (1942).
Knowlton, F. H. *The Osprey* (March 1897) (obit.).
Merriam, C. H. *Science* (12 February 1897) (obit.).
Merrill, J. C. "In Memoriam: Charles Emil Bendire." *Auk* (January 1898).

UNPUBLISHED SOURCES

See Charles E. Bendire Correspondence, Library of Congress, Manuscripts Division.

Michael J. Brodhead

BENT, ARTHUR CLEVELAND. Born Taunton, Massachusetts, 25 November 1866; died Taunton, Massachusetts, 30 December 1954. Businessman, ornithologist, oologist.

FAMILY AND EDUCATION

Son of William Henry Bent, business executive, and Harriet Fellowes Hendee Bent. His mother died when he was about six years old. Attended local public schools in Taunton and Bristol Academy through high school. Received the A.B.

from Harvard, 1889. Married Rosalba Peale Smith, 1895 (divorced 1911); no children, and Madeline Vincent Godfrey, 1914; one son, three daughters.

POSITIONS

"Took charge of" Safety Seamless Pocket Co., 1891–1895. Treasurer, Plymouth Electric Light Co., 1892–1931, president, 1900–1931. Treasurer, Atlantic Covering Co., 1891(?)–1893. General manager, Mason Machine Works, 1900–1914. Honorary appointments included associate in ornithology, Museum of Comparative Zoology, 1911–1916; collaborator, Division of Birds, U.S. National Museum, 1927–1954; and associate in ornithology, 1927–1942, research fellow in ornithology, 1942–1948, associate in ornithology, 1948–1954, Harvard University.

CAREER

Spent most of his life as successful businessman in Plymouth and Taunton, Massachusetts. Also involved at various times during his business career as vice president and manager, Campbell Printing Press and Manufacturing Co., New York; vice president, Autoplate Co. of America; director, Jager Engine Co.; director, Corr Manufacturing Co. As boy, developed interest in natural history, particularly ornithology and oology. Began to develop large collection of birds and birds' eggs, the former later donated to Harvard, and the latter to Smithsonian Institution. From 1901 to 1930 conducted extensive fieldwork throughout North America, with several trips to Saskatchewan, Newfoundland, and Florida. In 1910 reached agreement with Smithsonian to continue the work of life histories of North American birds begun by Charles E. Bendire (1836–1897) but left unfinished at his death. Between 1910 and his death completed twenty volumes, one of which was published posthumously, and had accumulated notes for three other volumes, which were published in 1968 under editorship of Oliver L. Austin, Jr. The *Life Histories* were published as series of *Bulletins* of the U.S. National Museum. The *Life Histories* were regarded as collaborative work; and Bent solicited data, notes, and photographs from hundreds of professional and amateur ornithologists throughout North America. Most species accounts were written by Bent, but some were prepared by other ornithologists. Although never formally trained as ornithologist, he served as president, American Ornithologists' Union, 1935–1937 and member, Nuttall Ornithological Club, Cooper Ornithological Club, Boston Society of Natural History, and the Academy of Natural Sciences of Philadelphia.

Possessed of enormous energy, devoted long hours to business and civic affairs, as well as to ornithology. Was active in Taunton civic affairs and served as a vestryman, St. Thomas Episcopal Church in Taunton. Received the Daniel Giraud Elliot Medal of National Academy of Sciences, the William Brewster

Award of American Ornithologists' Union, and (a source of great pride to Bent) the Silver Beaver Award for distinguished service to Boy Scouts of America.

MAJOR CONTRIBUTIONS

Life Histories of North American Birds, published between 1919 and 1968, remains a standard reference work and a starting point for most research dealing with avian biology. Although somewhat dated, most species accounts have not been superseded, remain the standard reference, and are frequently consulted by both amateur and professional ornithologists. A 1969 review of Bent's work concluded that "if a question about food, courtship, or some other aspect of avian biology arises, a common reaction is to look to see what Bent said about it. The comments made by Bent, or the references given by him, have saved countless searches through the older literature. The series is cited with great frequency and regularity in present day ornithological writing." Entire series, reprinted by Dover Publications in 1960s, remains in print.

Organizing large network of correspondents to produce *Life Histories*, major achievement in itself.

BIBLIOGRAPHY

Banks, Richard C. *Auk* 86 (1969).

Mason, C. Russell. "Arthur Cleveland Bent, 1866–1954." *The Bulletin of the Massachusetts Audubon Society* 39 (1955).

New York Times, 1 January 1955 (obit.).

Taber, Wendell. "In Memoriam: Arthur Cleveland Bent." *Auk* 72 (1955).

Teale, Edwin Way. "A. C. Bent: Plutarch of the Birds." *Audubon Magazine* January–February 1946.

Tyler, Winsor M. "Ornithologists Alive! 1. Arthur Cleveland Bent." *The Bulletin of the Massachusetts Audubon Society* 28 (1944).

Who's Who in the East. 4th ed. (1953).

UNPUBLISHED SOURCES

See Smithsonian Institution Archives, Record Unit 7120, Arthur Cleveland Bent Papers, c.1910–1954 (includes notes, correspondence, information, and photographs accumulated while writing the *Life Histories*; Smithsonian Institution Archives. Additional Bent correspondence may be found in the Alexander Wetmore Papers (Record Unit 7006) and the Division of Birds Records (Record Unit 105) as well as other record units.

William A. Deiss

BETHUNE, CHARLES JAMES STEWART. Born West Flamborough, Upper Canada, 11 August 1838; died Toronto, 17 April 1932. Entomologist, clergyman.

FAMILY AND EDUCATION

Third son of Rt. Rev. A. N. Bethune, later bishop of Niagara and of Toronto. Attended Upper Canada College; B.A., 1859, M.A., 1861, Trinity University;

honorary D.C.L., Trinity, 1883. Married Harriet Alice Mary Furlong; four children.

POSITIONS

Ordained Anglican priest, 1861–1870; headmaster, Trinity College School, Fort Hope, 1870–1899; professor of entomology and zoology, Ontario Agricultural College, Guelph, 1906–1920.

CAREER

Following nine years as an Anglican priest, made Trinity one of the leading preparatory schools in Canada. One of his students there was William Osler, later leader in medical profession on faculty of Oxford University. Was joint founder with William Saunders of Entomological Society of Canada (1863), which later became Entomological Society of Ontario. Established *Canadian Entomologist* (1868), served as its editor until the mid-1890s. Also edited the society's *Annual Reports* and was considered leading entomological systematist in Canada for most of his life. Served as president of the society from 1871 to 1876, from 1890 to 1893, and again in 1912. Known for his "prodigious memory for references to literature, entomological history and scientific names." Also served as president of Entomological Society of America, 1913. Headed program in entomology and zoology at Ontario Agricultural College until his retirement in 1920 and trained many of the entomologists of the next generation in Canada.

MAJOR CONTRIBUTIONS

Leading Canadian entomological systematist and educator at both the secondary and college levels in latter part of nineteenth and early twentieth centuries.

BIBLIOGRAPHY

Dearness, J. "Reminiscences of the Early Days of the Society." *Canadian Entomologist* 71 (1939).

Mallis, Arnold. *American Entomologists* (1971).

Spencer, G. J. "A Century of Entomology in Canada." *Canadian Entomologist* 96 (1964).

Wallace, W. Stewart, ed. *The Macmillan Dictionary of Canadian Biography*. 4th ed. (1978).

Who Was Who in Canada. Vol. 1.

Keir B. Sterling

BICKMORE, ALBERT SMITH. Born Tenant's Harbor, St. George, Maine, 1 March 1839; died Nonquit, Massachusetts, 12 August 1914. Naturalist, museum founder and executive, educator.

FAMILY AND EDUCATION

Son of John and Jane (Seavey) Bickmore; father was sea captain and shipbuilder. Attended New London Academy in New Hampshire (graduated 1856) and Dartmouth (A.B., 1860, A.M., 1863); studied under Louis Agassiz at the Lawrence Scientific School, Harvard (B.S., 1864); later awarded honorary Ph.D.s at Hamilton, 1869, and Dartmouth, 1889; and honorary LL.D., Colgate, 1905. Married Charlotte A. Bruce, 1873.

POSITIONS

Professor of natural history in Madison (Colgate) University, 1865–1869; founder and superintendent, American Museum of Natural History, 1869–1884; curator, Department of Public Instruction, American Museum of Natural History, 1884–1904 (became curator emeritus in 1885); trustee of American Museum of Natural History, 1885–1914.

CAREER

Served in Civil War with nine-month regiment, 44th Massachusetts Volunteers, 1862–1863. While at Harvard was appointed assistant at Museum of Comparative Zoology under Louis Agassiz. In 1862 traveled to Bermuda with P. T. Barnum expedition to collect marine specimens for the museum; while on special duty from regiment he collected specimens for Agassiz in North Carolina and recorded meteorological data for U.S. Army Medical Department. Upon graduation from Harvard, with assistance of "generous friends of science," he traveled to the Dutch East Indies, Malay Archipelago, China, Siberia, and Japan (1865–1867). He collected birds and shells and was credited with discovering the Ainu of Yezo. He published an account of journey, *Travels in the East Indian Archipelago* (1868), which was translated into German and published in Jena in 1869. He also published several papers on Ainu and on China: "Sketch of a Journey from Canton to Hankow," *Journal of the Royal Geographical Society*, no. 38 (1868), and *American Journal of Science* (July 1868), "A Description of the Banda Islands," *Proceedings of the Royal Geographical Society* (1867–1868): 324 ff., "Some Remarks on the Recent Geological Changes in China and Japan," *American Journal of Science* (March 1868), and "The Ainos, or Hairy Men of Yesso" and "The Ainos, or Hairy Men of Saghalien and the Kurile Islands," *American Journal of Science* (May 1868). While still student at Harvard he conceived an idea for great natural history museum to be housed on Manhattan Island. During his travels he corresponded with William E. Dodge of New York City regarding his idea. With the aid of Theodore Roosevelt, Sr., Joseph H. Choate, and a number of prominent New Yorkers, young Bickmore succeeded in founding American Museum of Natural History and became its first superintendent in 1869. He devoted his time to perfecting plans for museum.

He actively pursued acquisition and installation of early collections and establishment of the library. In 1880 he organized the museum's Department of Public Instruction to instruct public school teachers in the use of the exhibitions in school curricula by means of lectures illustrated with specimens and lantern slides. He traveled throughout world to obtain lantern slides and to personally explore all the areas that were subjects of his lectures. His lecture series became popular New York events. In 1885 he was elected to Board of Trustees after he had resigned as superintendent and had become curator of public instruction. Ill health forced him to retire in 1904, but he continued as trustee until his death in 1914.

MAJOR CONTRIBUTIONS

He is best known as founder of the American Museum of Natural History and as teacher and public lecturer on natural history. His concept of public instruction was copied by many museums and is considered major contribution to the education of working people in New York and other cities.

BIBLIOGRAPHY

Gratacap, L. P. ''Memories of Professor Bickmore.'' *American Museum Journal* (February 1915).

Hellman, G. *Bankers, Bones and Beetles: The First Century of the American Museum of Natural History* (1968).

Hovey, E. O. ''Professor Albert S. Bickmore: Educator.'' *American Museum Journal* (November 1911).

Kennedy, John M. ''Philanthropy and Science in New York City: The American Museum of Natural History, 1868–1968.'' Ph.D. diss., Yale University (1968).

Preston, Douglas J. *Dinosaurs in the Attic* (1986).

''Resolutions to Professor Bickmore on the Occasion of His Seventy-fifth Birthday.'' *American Museum Journal* (April 1914).

Sherwood, G. S. ''A Gift of Peculiar Value.'' *American Museum Journal* (October 1911).

UNPUBLISHED SOURCES

See Archives of the American Museum of Natural History; A. S. Bickmore, *An Autobiography* (1908).

Nina J. Root

BIGG, MICHAEL ANDREW. Born London, England, 26 December 1939; died Duncan, British Columbia 18 October 1990. Naturalist, marine mammalogist.

FAMILY AND EDUCATION

Son of Andrew B. Bigg, newspaperman and Irene Lilian (Sones). Mother partially blinded in the London Blitz. Two sisters. Family arrived in British Columbia, February 1948. Lived at Bamberton, attended Cowichan District Senior Secondary, Duncan (1955–1958). Undergraduate, Victoria College, 1958–1961. University of British Columbia, B.Sc. (1962). Spent 1963 backpacking Southeast Asia, India, Australia, and New Zealand. Graduate studies, Department of Zoology, U.B.C. 1964–1972 (M.Sc. 1967, Ph.D. 1972). Married Silke Margaret Hansen (13 May 1967). Two children. Divorced 1982.

POSITIONS

Head, Marine Mammal Research Program in British Columbia, Pacific Biological Station, Canadian Department of Fisheries and Oceans, Nanaimo, B.C. from 1970 until his death. Arctic Biological Station, Ste. Anne de Bellevue, Quebec (1970–1978). Canadian scientific representative, North Pacific Fur Seal Commission (1970–1984). Advisor, International Whaling Commission, and North Pacific Fisheries Commission. Co-chair, Federal/Provincial Johnstone Straits Killer Whale Committee (1990). Member Society for Marine Biology.

CAREER

Dissertation on harbor seal (*Phoca vitulina*) supervised by H. D. Fisher. Continued studies of same at Pacific Biological Station after Ph.D. Studied life history of Pribilof and B.C. fur seals (*Callorhinus ursinus*). Maintained fur seal colony at Nanaimo for observation and first successful breeding in captivity. Organized 1972 translocation and monitoring of sea otters (*Enhydra lutris*) from Alaska to Bunsby Islands, B.C. Also long-term studies of Steller sea lions (*Eumetopias jubata*) and California sea lions (*Zalophus californianus*).

Finest work involved killer whales (*Orcinus orca*), previously the least understood cetacean. Involvement began in 1964 with death of "Moby Doll" at the Vancouver Aquarium. Prepared skeleton. In 1970s pioneered method for reliable photo-identification and systematic assessment of cetacean populations in the wild, using individual markings. By 1976, with a network of hundreds of volunteers, identified virtually every individual orca in B.C. coastal waters. Began fourteen years of detailed studies of population dynamics, and social organization, often conducted outside official duties. Studies revealed maternal social organizations, and distinctions between fish-eating residents and mammal-hunting transient orcas. With research colleagues Kenneth C. Balcomb III, Graeme Ellis, and John J. B. Ford published *Killer Whales: A Study of Their Identification, Geneology, and Natural History* (1987). This group, including others, such as Peter F. Olesiuk, delineated the natural history of the orca.

Diagnosed with cancer in 1984, he moved philosophically toward a deep ecology position on wildlife. Numerous public education engagements. Escorted H.R.H. Prince Philip whale-watching (14 October 1987). Two television specials, "The Island of Whales," (N.B.C. *Nova* 1989), and "Beautiful Killers," (A.B.C. TV 1989). Events of his last year included publications in the International Whaling Commission Reports (Special Issue 12), accolades in the 56 papers presented at the Third International Orca Symposium and Workshop (University of Victoria, 1990), and appointment to Johnstone Straits Killer Whale Committee. Recommended rules to protect orcas from commercial fisheries, and well-meaning whale watchers, and to protect ocean habitat from silt runoff caused by clearcut logging. World's first killer whale reserve renamed the Robson Bight (Michael Bigg) Ecological Reserve after his death. Like his friend Robin Morton, his ashes were scattered in Robson Bight.

MAJOR CONTRIBUTIONS

Mentor to a generation of marine mammalogists, beneficiaries of his insistence that knowledge be given freely. World's foremost authority on *Orcinus orca*. Pioneer in creative and unorthodox methods of field observation, and the marine application of gender sensitive geneology. Cetacean classification system now in world wide use. General approach to marine mammals notable for its holistic ecological unity, such as his last area of focus on the impact of irresponsible forestry on inhabitants of world's oceans.

BIBLIOGRAPHY

Ford, J. K. B. ''Michael Andrew Bigg 1939–1990,'' *Marine Mammal Science* 7 (3) (July 1991).

Hoyt, E., *Orca: The Whale Called Killer* (1981).

Obee, B, and G. Ellis. *Guardians of the Whales: The Quest to Study Whales in the Wild* (1992).

UNPUBLISHED SOURCES

Extensive working papers and field books held by the Marine Mammal Unit, Pacific Biological Station, Nanaimo, B. C., Gordon Miller, station librarian. See also Vancouver Public Aquarium.

Lorne Hammond

BINNEY, AMOS. Born Boston, 18 October 1803; died Rome, Italy, 18 February 1847. Naturalist, zoologist, businessman.

FAMILY AND EDUCATION

Son of Col. Amos, businessman and shipping merchant, and Hannah Binney. Family was prominent in shipping, commerce, and New England public affairs. Entered Derby Academy, Hingham, Massachusetts, at age ten. Entered Brown University four years later and graduated in 1821. M.D., Harvard 1826. Married cousin, Mary Ann Binney, 20 December 1827. Of their five children, the third, William Greene Binney, continued father's work and became major authority on American land mollusks.

POSITIONS

Helped to found Boston Society of Natural History in 1830; curator, 1830–1832; treasurer, 1832–1834; corresponding secretary, 1834–1837; vice president, 1837–1843; president, 1843. Became one of the society's chief financial, scientific patrons and contributors. Member of Massachusetts legislature, 1836–1837. Established American Society of Geologists and Naturalists.

CAREER

Studied natural sciences and began collecting shells while in college. Initially, medical study served as a means to secure scientific training, not to practice the profession. After receipt of M.D. worked as a merchant; later joined father in real estate and mining businesses. Scientific studies pursued as hobby in leisure time after business pursuits.

After founding of Boston Society of Natural History, contributed papers on zoological and paleontological topics. Most works published in *Boston Journal of Natural History*. Led effort to acquire permanent space for society. Formed and contributed to its museum by donations of extensive personal collection of shells, fossils, and birds. After election to Massachusetts state legislature obtained funds for zoological, geological, and botanical surveys. In 1835 began studies for chief book, *Terrestrial Air-Breathing Mollusks of the United States and the Adjacent Territories of North America* (published posthumously in three volumes from 1851 to 1857). Was a serious patron of the arts.

MAJOR CONTRIBUTIONS

Terrestrial . . . Mollusks established Binney as an authority on the subject. Preparations included expeditions to Texas and Florida, combined with hiring of the best artists for illustrations. Also contributed many years of hard labor and much expense. Son, William Greene, became a conchologist and succeeded father as authority on American land mollusks; he also continued father's work in two supplementary volumes.

BIBLIOGRAPHY

Dictionary of American Biography.

Elliott, Clark A., ed. *Biographical Dictionary of American Science* (1979).

Gould, A. A. "Memoir." In A. A. Gould, ed., *The Terrestrial Air-Breathing Mollusks of the United States and the Adjacent Territories of North America* (1851).

National Cyclopaedia of American Biography. Vol. 17 (1967).

Who Was Who in America. Historical vol. (1607–1896).

Anthony Todman

BIRGE, EDWARD ASAHEL. Born Troy, New York, 7 September 1851; died 9 June 1950. Limnologist and university administrator.

FAMILY AND EDUCATION

Son of Edward White Birge, carpenter, dairy farmer, and partner in a bakery, and Anna (Stevens) Birge. Received A.B. from Williams College in 1869; A.M.,

1876, Ph.D., 1878, from Harvard. Also spent year (1880–1881) in Leipzig studying histology and physiology. Married Anna Wilhelmina Grant, 15 July 1880 (died 14 December 1919); a son and daughter.

POSITIONS

Instructor (1875–1879), professor of zoology (1879–1911), dean of College of Letters and Sciences (1891–1918), acting president (1900–1903), and president (1918–1925) at University of Wisconsin. Director of Wisconsin Geological and Natural History Survey (1897–1919). President of American Microscopical Society (1902) and American Fisheries Society (1907). Member and president (1890–1891, 1918–1921) of Wisconsin Academy of Sciences, Arts and Letters. Also member of Wisconsin Board of Forestry Commissioners (1905–1915) and Wisconsin Conservation Commission (1908–1915). President (1919–1922) and life senator of Phi Beta Kappa.

CAREER

At Harvard, began study of Cladocera (minute crustaceans known as water fleas); continued this research at Wisconsin University, where he became authority on their taxonomy. Subsequently, enlarged scope of his work to include Cladocera's lake habitat, physical and chemical factors that conditioned their movements, their biological productivity, and life history. As director of Wisconsin Geological and Natural History Survey, supervised variety of research projects and carried on his own studies of Wisconsin lakes as individual, integrated entities. College and university administrative chores cut into research time, but after retirement continued research on lakes. Authored or coauthored over seventy papers, the last appearing when he was nearly ninety. Among these were "The Thermocline and Its Biological Significance," *Transactions of the American Microscopical Society* 25 (1904); coauthor, "Inland Lakes of Wisconsin: The Dissolved Gases of the Water and Their Biological Significance," *Bulletin of the Wisconsin Geological and Natural History Survey*, no. 22, Sci. Ser. No. 7 (1911); coauthor, "A Limnological Study of the Finger Lakes of New York," *Bulletin U.S. Bureau of Fisheries* 32 (1914); coauthor, "The Organic Content of the Water of Small Lakes," *Proceedings of the American Philosophical Society* 66 (1927); coauthor, "Hydrography and Morphometry of Some Northeastern Wisconsin Lakes," *Transactions of the Wisconsin Academy of Sciences, Arts, and Letters* 33 (1941).

MAJOR CONTRIBUTIONS

As limnologist, did important work on revealing the mechanics of stratification and showing how photosynthesis, respiration, and decay combine to produce concurrent stratification of the dissolved gases. Also pioneer in explaining why some lakes are more productive and support larger populations of plankton than do other lakes. His teaching was marked by great breadth of knowledge and clarity of presentation, as well as ability to stimulate students' investigative interests. Won enthusiastic regard of large number of future physicians and scientists.

BIBLIOGRAPHY

Dictionary of American Biography. Supplement 4 (1974).
Sellery, G. C. *E. A. Birge: A Memoir* (1956).
Who Was Who in America. Vol. 3 (1960).

UNPUBLISHED SOURCES

Birge's papers, including autobiographical material, are in the State Historical Society of Wisconsin.

Richard Harmond

BLAKISTON, THOMAS WRIGHT. Born Lymington, Hampshire, England, 27 December 1832; died San Diego, California, 15 October 1891. Naturalist, magnetic observer, explorer.

FAMILY AND EDUCATION

The second son of John Blakiston (1785–1867) and Jane, daughter of Rev. Thomas Wright of Market Harborough, Leceistershire. Brothers were Matthew, Lawrence, and John. Grandson of Matthew Blakiston, second baronet. Attended St. Paul's School, Swansea, then entered the Royal Military Academy at Woolwich. In 1885, at age fifty-three, married Anne Mary Dun of London, Ohio, and settled in New Mexico. One son and one daughter.

POSITIONS

Commissioned as lieutenant in the Royal Artillery in December 1851. Served in England, Ireland, Nova Scotia, and the Crimea, where his brother Lawrence was killed in action. Appointed as magnetic observer to the Palliser expedition, sailing from Thames on 21 June 1857 and arriving at Fort Carlton on Saskatchewan River on 23 October. Here he kept daily magnetic and meteorological observations from 12 November through 16 April 1858 and made collections of birds and mammals. In 1858 he was first surveyor of two passes through the Rocky Mountains—Kootenai Pass and Boundary Pass (the latter crosses the 49th parallel obliquely)—naming the Waterton Lakes. Mount Blakiston, the highest peak on Waterton Lakes National Park, bears his name. He left Red River in early March 1859.

After exploring the upper reaches of the Yangtze-Kiang River in China, for which he was awarded the Patron's Medal of the Royal Geographical Society, settled in 1861 as a merchant in Hakodate on the island of Hokkaido, Japan. He left Japan in 1884 and the next year settled in New Mexico.

CAREER

Blakiston published observations of 100 bird species collected and another 29 species observed near Fort Carlton. He made even more important collections in Japan.

MAJOR CONTRIBUTIONS

First to describe the nest and eggs of the ferruginous hawk, taken near present Dinsmore, Saskatchewan, on 30 April 1858. Blakiston's observations at Fort Carlton complement those of John Richardson and Thomas Drummond in 1820 and 1827, providing unsurpassed knowledge of presettlement bird life. In Japan, he first recognized the Tsugaru Straits as an important boundary in bird and mammal distribution—now known as Blakiston's line.

BIBLIOGRAPHY

Blakiston, T. "On Birds Collected in the Interior of British North America." *Ibis* vols. 3, 4, 5 (1861–1863).

Holmgren, E. J. "Thomas Blakiston, Explorer." *Alberta History* 24 (1976).

Houston, C. S. "A Bird-Watcher's Outing in 1858." *Saskatchewan History* 29 (1976).

Houston, C. S. and M. G. Street. *The Birds of the Saskatchewan River, Carlton to Cumberland* (1959).

C. Stuart Houston

BODMER, (JOHANN) KARL. Born Zurich, Switzerland, 11 February 1809; died Paris, France, 30 October 1893. Artist, ethnologist.

FAMILY AND EDUCATION

Third child of cotton merchant Heinrich Bodmer. After his formal education, was trained to draw, paint, and engrave by his maternal uncle, well-known landscape painter Johann Jacob Meyer. Married Anna Maria Magdalena Pfeiffer, 1876; three sons.

POSITIONS

Artist, expedition into the interior of North America (under contract to Maximilian, prince of Wied-Neuwied), 1832–1834. Artist of the Barbizon school, associated with Jean François Millet, Theodore Rousseau. Awarded the ribbon of the Légion d'Honneur, 1876.

CAREER

At age nineteen, went to Rhineland, where work as an artist was acclaimed in *Rhine and Mosel Gazette* by Jacob Hoelscher, Coblenz publisher (1828).

Under contract to Maximilian, prince of Wied-Neuwied, joined two-year scientific investigation, 1832–1834, to interior of North America, one of the most important explorations ever made of the upper Missouri region. Bodmer was to provide precise pictorial records of flora, fauna, and ethnology.

Prince Maximilian and his party arrived in Boston on 4 July 1832. They traveled to New York, then to Philadelphia, and reached Bethlehem, Pennsylvania, where they remained for six weeks. On 17 September, they proceeded to New Harmony, Indiana, where party stayed for five months, engaging in scientific studies with Thomas Say and Charles-Alexandre Lesueur, from 19 October 1832 to 16 March 1833. This also allowed the prince time to recover from a serious illness. While based at New Harmony, Maximilian and his associates traveled down the Ohio and Mississippi Rivers to New Orleans (January–February 1833). The party then journeyed down the Ohio River and up the Mississippi to St. Louis; on 10 April they traveled up Missouri River on American Fur Company steamer, *Yellowstone*.

At Ft. Pierre (Pierre, South Dakota) they switched to *Assiniboine*. On 18 June they reached Ft. Union (Williston, North Dakota) and on twenty-fourth went on keel boat to Ft. McKenzie on Maria's River (Great Falls, Montana), where they remained two months. On the return, the party spent the winter at Ft. Clark (Bismarck, North Dakota) among the Mandan Indians. They returned to St. Louis in May 1834 and retraced their route to New Harmony, where they visited 9–13 June. They traveled back to New York via the Ohio River, the Ohio and Erie Canals, and the Hudson River and sailed for Europe on 16 July 1834.

Published account of journey *Reise in das Innere Nord-America in den Jahren 1832 bis 1834*, 2 vols. (Coblenz, 1840, 1841), French edition (Paris, 3 vols., 1840, 1841, 1843), English edition (London, 1 vol., 1843). The same *Atlas* of eighty-one prints was published with all of the editions.

Bodmer supervised artists and engravers in Paris. Some editions issued with black and white plates only, some with few color plates, and some with all plates colored by hand. Besides the *Atlas*, a table edition took up his time, *Die Mosel und ihre nächste Umgebungen von Metz bis Coblenz* (Coblenz, 1841).

In 1849 moved to Barbizon. In following fruitful years, among others, illustrations to Jean de La Fontaine's *Fables* (Paris, 1873) and Victor Hugo's *Quatre-vingt-treize* (Paris, 1873). Also regular contributions to *L'illustration*, magazine with forest and animal pictures.

MAJOR CONTRIBUTIONS

The expedition up Missouri River was timely; permanent settlement disrupted nomadic life and led to dispersal of tribes. In the case of the Mandan Indians, the people most comprehensively documented by Maximilian and Bodmer, the timing was uncanny. The Mandans were almost obliterated by smallpox within a few years.

While the major contribution of the work lies in its ethnological significance, it also provides highly important historical description of natural conditions west of Mississippi and in Bethlehem, Pennsylvania, and New Harmony, Indiana, as well. Maximilian praises the work of Bodmer in the author's preface to *Travels in the Interior*, saying that Karl Bodmer "represented the Indian nations with great truth, and . . . his drawings will prove an important addition to our knowledge of this race of men."

Bodmer's pictorial record of the upper Missouri inhabitants is, both artistically and ethnologically, among the finest records that exist. Exacting attention to detail reflects skill of close observation gained from training as draftsman and makes Bodmer's works reliable historical documents and remarkable art.

BIBLIOGRAPHY

Davidson, Marshall B. "Carl Bodmer's Unspoiled West." *American Heritage* 14, 3 (April 1963).

Goetzmann, William H., and Joseph C. Porter. *The West as Romantic Horizon* (1981).

Läng, Hans. *Indianer Waren Meine Freunde* (1976).

Thomas, Davis, and Karen Ronnefeldt. *People of the First Man* (1976).

Thwaites, Reuben. *Early Western Travels*. Vols. 22, 23, 24 (1966).

UNPUBLISHED SOURCES

See Historic New Harmony, Inc.: Maximilian-Bodmer Collection, and InterNorth Art Foundation: Center for Western Studies, Joslyn Art Museum (Omaha, Nebr.); Maximilian–Bodmer Collection; more than 400 of Bodmer's original sketches and paintings; diaries, journals, account books, correspondence.

Ralph G. Schwarz

BONAPARTE, CHARLES LUCIEN JULES LAURENT, Prince of Canino and Musignano. Born Paris, France, 24 May 1803; died Paris, 29 July 1857. Naturalist, systematist.

FAMILY AND EDUCATION

Son of Napoleon's brother Lucien and Alexandrine de Bleschamp. Most of his childhood and youth spent in the Papal States, where he began his investigations of natural history. Married his cousin Zéhälde, 1822; twelve children, four of whom did not survive him.

POSITIONS

Vice president, Maclurian Lyceum of Philadelphia, 1826–1827(?).

CAREER

Sailed to the United States, 1823, and lived with his uncle and father-in-law, Joseph Bonaparte, at Bordentown, New Jersey. Active in the Academy of Natural Sciences of Philadelphia and published several ornithological papers in its *Journal*. Also a member of other American scientific societies. Described in

1825 as "a little set, blackeyed fellow, quite talkative, and withal an interesting and companionable fellow." Principal works on American birds include *American Ornithology; or, the Natural History of Birds Inhabiting the United States, Not Given by Wilson*, 4 vols., (1825–1833, ostensibly a continuation of, or supplement to, Alexander Wilson's *American Ornithology*, but generally considered a separate work; and *A Geographical and Comparative List of the Birds of Europe and North America* (1838). Helped to launch the career of John J. Audubon (who provided one of the plates for Bonaparte's *American Ornithology*). Returned to Europe, 1826; visited the United States again only briefly, 1827.

Residing chiefly in Rome, he divided his time among writing several important books and papers, notably on systematic ornithology, visiting several European museums, organizing international scientific congresses, and participating in the movement to create a united, republican Italy. Resided in London and Leiden after the fall of the Roman republic in 1849. In 1850 his cousin Louis Napoleon (later, Napoleon III) permitted him to live in France. Resided in Paris until his death. Among his more significant later publications was *Conspectus generum avium*, 2 vols. (1850–1857). According to Erwin Stresemann, "Bonaparte's achievements in ornithological systematics ensure him an outstanding place in history. His work left deep and indelible marks; but there has never been any lack of ornithologists who disliked them." Casey A. Wood described him as "a devoted and efficient student of nature fit to rank with the best systematic writers on zoology."

MAJOR CONTRIBUTIONS

His *American Ornithology* was an important pioneering work that described several new species collected in the American West and South. His several classifications of birds and other fauna, although controversial, are respected as milestones in systematics and evolutionary theory. Bonaparte's comparative listing of the birds of Europe and North America was an important contribution to zoogeography.

BIBLIOGRAPHY

Élie de Beaumont, L. *Notice sur les travaux scientifiques de S. A. le Prince Charles Lucien Bonaparte* (1866).

Brodhead, J. "The Work of Charles Lucien Bonaparte in America." *Proceedings of the American Philosophical Society* 122 (August 1978).

Petit, Georges. *Dictionary of Scientific Biography*. Vol. 2.

Stresemann, E. *Ornithology: From Aristotle to the Present* (1975).

UNPUBLISHED SOURCES

See C. L. Bonaparte Correspondence, Museum d'Histoire Naturelle, Paris; C. L. Bonaparte Correspondence, American Philosophical Society (APS) (and other manuscript collections in the APS).

Michael J. Brodhead

de BONNÉCAMPS, JOSEPH-PIERRE. Born Vannes, France, 3 September 1707; died Gourin, France, 28 May 1790. Teacher of hydrography, astronomer, cartographer.

FAMILY AND EDUCATION

Son of Nicolas and Anne Muerel de Bonnécamps. Became a Jesuit novice in 1727, studied philosophy at the Collège de La Flèche (1729–1732) and theology at the Collège Louis-le-Grand (1739–1742). Took vows as priest in 1746.

POSITIONS

Taught in Jesuit colleges of Caen (1732–1736) and Vannes (1736–1739). Named professor of hydrography at the Collège de Québec in 1744. Returned to France (1759) on fall of Quebec. Taught Caen 1761–1762; in Saint-Pierre and Miquelon about 1765; prison chaplain in Brest about 1770. At time of death, a private tutor to aristocratic family.

CAREER

Began teaching sciences (mathematics, hydrography, astronomy, and cartography) at Québec. Made meteorological observations in 1746, published next year in *Journal de Treroux*. Collected astronomical instruments with hopes of erecting an observatory. Made astronomical observations at Fort Frontenac (modern Kingston, Ontario) in 1752. His best-known work was his description of the Ohio River country as part of the expedition of Céloron de Blainville in 1749. His report included geographical, natural history, and astronomical information. He also made the first map of the territory. His report is reprinted in Thwaites's *Jesuit Relations* (vol. 49).

MAJOR CONTRIBUTIONS

Provided first in-depth description and map of the Ohio River country and one of the first to make serious astronomical observations in Canada.

BIBLIOGRAPHY

Cossette, Joseph. ''Joseph-Pierre de Bonnécamps.'' *Dictionary of Canadian Biography*. Vol. 4 (1979).

Jarrell, Richard. *The Cold Light of Dawn: A History of Canadian Astronomy* (1988).

Richard A. Jarrell

de BONNEVILLE, BENJAMIN LOUIS EULALIE. Born Paris, France, 14 April 1796; died Fort Smith, Arkansas, 12 June 1878. Army officer, explorer.

FAMILY AND EDUCATION

Son of Nicholas and Margaret Brazier de Bonneville. Came to United States at age seven. Evidently received most early education from mother. Entered U.S. Military Academy at West Point, 1813; graduated, 1815. Married Ann Lewis at a date not known; one daughter (both died c.1827–1830). Married Susan Neis, 1870.

POSITIONS

Filled variety of garrison and recruiting posts, 1815–1820. Assigned to troops building military road in Mississippi, 1820–1821 (?). Assigned to garrison duty, Fort Smith, Arkansas, and Fort Gibson, Oklahoma, 1821–1825. Detached to serve as aide to Marquis de Lafayette during latter's visit to the United States, 1825; briefly visited France as guest of the marquis, 1825–1826. Promoted captain, 1826, and returned to Fort Gibson, remained there four years except for a period of months in St. Louis. Received eight-month leave of absence to investigate possibility of independent foray in western fur trade and raise funds therefore, 1830–1831. On leave in western mountains for more than three years, 1832–1835. Had been dropped from army for overstaying leave, 1834, but restored to rank by President Andrew Jackson, 1836. Sent once again to Fort Gibson, but managed detached service elsewhere. Promoted major, 1845. Served in Mexican War with gallantry, promoted lieutenant colonel. Held variety of posts thereafter, was commandant at Fort Kearny in late 1840s and at Fort Vancouver, 1853–1855. Also served in Wisconsin, New York, California, New Mexico, Texas, and Missouri. Promoted colonel, 1855. Retired, 1861, but immediately reentered service. Commanded Benton Barracks, St. Louis, 1862–1865. Promoted brigadier general, 1865. Retired 1866. Oldest officer on retired list at his death in 1878.

CAREER

Bonneville is probably best known for his thirty-nine month trip to the West between May 1832 and August 1835, which was privately financed through moneys given him by several New York financiers, including Alfred Seton. Bonneville's objective was to explore what was then known of the northern Great Plains and intermontane region while investigating possibilities for successful foray into fur-trading business. Many Americans were, of course, actively trapping and trading furs in the area. Bonneville was also made aware of Hudson's Bay Company's (HBC) growing power along Columbia River following its absorption of Northwest Company in 1821. Bonneville therefore reluctantly concluded that he could not successfully compete with men who knew fur industry and the West far better than he did. What he did accomplish was to explore and map much of the northern basin and range not seen by Lewis and Clark. He investigated the sources of the Yellowstone and visited portion of the Columbia River despite strong disapproval of HBC officials. He produced a map of region that, despite amateurish inaccuracies, was widely publicized for many years. Some of his men contributed greatly. In 1833, Bonneville dispatched Joseph Walker, a trapper, on separate mission. Walker took smaller party from northern Utah west through the Rockies to the Sierras and back to Salt Lake, identifying what would become the principal route used by later overland travelers to California. Walker also evidently explored a portion of the western end of Great Salt Lake at Bonneville's direction, while Bonneville's route passed to the north of it, and he did not see or visit it. Great Salt Lake had, in fact, been discovered in 1825 by James Bridger and had been circumnavigated by four men belonging to a party led by W. H. Ashley in 1826. Nevertheless, the much larger ancient inland sea that once existed in that same region was named in Bonneville's honor by the geologist Grove C. Gilbert in 1890. Bonneville met novelist Washington Irving in 1835, sold him copy of his original report and notes for $1,000. Irving published *The Rocky Mountains, or, Scenes, Incidents, and Adventures in the Far West; Digested from the Journal of Captain B. L. E. Bonneville* . . . in two volumes (1837). Well received for its accuracy, which depended on Bonneville's journal and partly on Irving's own western travels. Bonneville expedition also subject of critical account by Gen. Hiram M. Chittenden (1902), who pronounced it a commercial and scientific failure. Yet Chittenden called Bonneville highly successful as leader of men, "for he lost not a single life in any case where the men were under his personal control". He was also noted for his "exceptional humaneness" toward Indians.

MAJOR CONTRIBUTIONS

Was the first western explorer to complete major reconnaissance of the northern and western basin and range region and map it. As James P. Ronda has stated, "Bonneville did not make history. Rather, he was present to record happenings of signal importance." Irving's account captured the great changes Bonneville observed in the West during his trip, particularly the rivalries among the great fur-trading enterprises. Bonneville's diary was, in fact, "the precious record of a western trapping culture poised on extinction."

BIBLIOGRAPHY

Bartlett, Richard A. *Great Surveys of the American West* (1962).
Chittenden, Hiram M. *The American Fur Trade of the Far West.* 2 vols. (1902).
Cullum, G. W. *Biographical Register of the Officers and Graduates of the United States Military Academy at West Point, New York.* 3d ed., 3 vols. (1891).
Dictionary of American Biography.
Goetzmann, William. *Army Exploration in the American West, 1803–1866* (1959).
Irving, Washington. *The Rocky Mountains; or Scenes, Incidents, and Adventures of the Far West; Digested from the Journal of Capt. B. L. E. Bonneville, of the Army of the United States* . . . 2 vols. (1837).
Morgan, Dale L. *The Great Salt Lake* (1947).

Todd, Edgeley W., ed. *The Adventures of Captain Bonneville, USA, in the Rocky Mountains and the Far West, Digested from His Journal by Washington Irving* (1986).
Wheat, Carl I. "Mapping the American West, 1540–1857." *Proceedings of the American Antiquarian Society* 64 (1954).
Wheeler, George M. "Memoir upon the Voyages, Discoveries, Explorations, and Surveys to and at the West Coast of North America and Interior of the United States West of the Mississippi River between 1500 and 1880." *U.S. Geographical Survey West of the 100th Meridian.* Vol. 1, Appendix F (1889).
Who Was Who in America.

Keir B. Sterling

BOURGEAU, EUGENE. Born Brizon, Hautes-Alpes, France, 20 April 1813; died Paris, France, February 1877. Botanist.

FAMILY AND EDUCATION

Son of Jacques Bourgeaux and Françoise Missilier. While tending his father's herds high in the Alps, Bourgeau became interested in flowers. He had little, if any, schooling. His grammar and spelling were atrocious. Botany was his passion.

POSITIONS

Employed first in the botanic gardens at Lyons, France, where he was taught by Nicolas-Charles Seringe and Alex Jordan. He next worked in Philip Barker Webb's herbarium in Paris in 1843, collecting for him in the Canary Islands in 1845–1846. From 1847 through 1856 he was a botanical collector for the Association Botanique Française d'Exploration in Spain, Algeria, and southern France. Botanist to the Palliser Exploring Expedition in western Canada, 1857–1860, then a collector in Lycia, Asia Minor (1860), Pontic Alps in Turkey (1861), Spain (1863–1864), Mexico (1865–1866), Island of Rhodes (1870). At the time of his death he was employed in the Museum d'Histoire Naturelle in Paris.

CAREER

At the request of William Jackson Hooker, who considered him "the prince of botanical collectors," Bourgeau was appointed botanist to the Palliser expedition, which surveyed the Canadian prairies and the Rocky Mountains, 1857–1860. Although he spoke no English, and his legs were too short to ride a horse comfortably, Bourgeau was described by Palliser as a "most active energetic and excellent companion . . . the most sociable jovial disposition . . . he not only prosecuted his researches indefatigably in the field, but was also most careful and successful in preserving the specimens in the evenings and during night.

. . . Little Bourgeau is a Brick.'' Bourgeau was of immense assistance to Lt. Blakiston, the magnetic observer, in recording temperatures of the air and of the soils below ground. His large collection of pressed plants was delivered to Kew Gardens in England, where William Jackson Hooker said: ''He performed his duty and more than his duty . . . most thoroughly. A better botanical collector never accompanied any British scientific mission.''

MAJOR CONTRIBUTIONS

The third serious botanist, after Thomas Drummond and David Douglas, to visit the Canadian Plains and Rocky Mountains, where he collected thousands of specimens of at least 819 species of plants. A genus of Compositae, *Bourgaea*, two species of plants, *Astragalus bourqovii* and *Rosa bourqeauiana*, and a mountain range southwest of Banff (51° 05' north, 115° 37' west) have been named for him. The French government named him knight of the Légion d'Honneur. Contrary to his chief, Palliser, who spoke of the ''Great American desert,'' Bourgeau correctly recognized the land between the North and South Saskatchewan Rivers as particularly suited to grain farming and stockbreeding.

BIBLIOGRAPHY

Bourgeau, E. ''Letter to Sir J. Hooker.'' *Journal of the Proceedings of the Linnaean Society of London* 4 (1860).

Raymond, M. ''Eugene Bourgeaux.'' *Dictionary of Canadian Biography,* (1972).

Spry, I. ed. *The Palliser Papers, 1857–60.* (Toronto: Champlain Society, 1968).

C. Stuart Houston

BOWERS, EDWARD AUGUSTUS. Born Hartford, Connecticut, 2 August 1857; died New Haven, Connecticut, 8 December 1924. Lawyer, government employee, conservationist.

FAMILY AND EDUCATION

Son of Connecticut state senator Caleb Bailey Bowers and Fanny M. Cutler. Father moved family from Hartford to New Haven in 1867 for educational purposes. A.B., 1879, LL.B., 1881, Yale University. Received Townsend Prize in law school. Never married.

POSITIONS

Lawyer, 1882–1886, judge, 1884–1885, Dakota Territory. Inspector of public lands, U.S. Department of the Interior, 1886–1889. Lawyer, 1889–1893, Washington, D.C. Assistant commissioner, General Land Office, 1893–1895. Assistant comptroller, U.S. Treasury, 1895–1898. Lawyer, 1898–1907, New Haven,

Connecticut. Lecturer on forest law and administration, Yale University, 1901–1917.

CAREER

For health reasons, moved to Dakota Territory in 1882, settling in Brown County. Made several remote trips, including one to Devil's Lake. Worked as lawyer and wheat farmer, with interest also in banking and politics.

Received commission to be inspector of public lands, with headquarters in Cheyenne, Wyoming, and was charged to curtail illegal fencing on public lands by large-stock men. Succeeded in opening up millions of acres and stopping practice for number of years. Acting as personal representative of Secretary of Interior Lamar, traveled throughout West trying land cases and supervising local land officers. Along with other Democrats, went out of office in 1889 and returned in 1893. Was elected vice president of New Haven Trust Company in 1899.

Prepared special report for secretary of interior in 1888 calling for conservation of public timberlands. Another report, "The Present Condition of the Forests on the Public Lands," presented in December 1890 at a joint meeting of the American Forestry Association and the American Economic Association (and published in the *American Economic Association Publications* [1891]), led to demands for their protection and conservation.

Said to have drafted act of 3 March 1891, called the "most important legislation in the history of forestry," which created forest reserves. Contributed further to national forest system through act of 4 June 1897, which is credited to his influence and which made it possible to practice forestry on the reserves and provided for their administration.

MAJOR CONTRIBUTIONS

His work in government service brought about the creation of national forests that would be managed in the public interest. Wrote numerous papers and made many speeches criticizing governmental neglect of public lands.

BIBLIOGRAPHY

Stroud, R. H. *National Leaders of American Conservation* (1985).

Who Was Who in America. Vol. 4 (1968).

Williams, F. W. *A History of the Class of Seventy-Nine Yale College* (1906).

John A. Drobnicki

BOYD, LOUISE ARNER. Born San Rafael, California, 16 September 1887; died San Francisco, 14 September 1972. Arctic explorer, geographer.

FAMILY AND EDUCATION

Daughter of mining magnate John Franklin Boyd, who headed his own investment company, and Louise Cook Arner Boyd. Two younger brothers died of rheumatic fever in their mid-teens. Mother died in 1919, father died in 1920. Attended Miss Stewart's School, San Rafael, and Miss Murison's School, San Francisco. Did not attend college. Honorary LL.D., University of California, 1939, Mills College, 1939; Honorary D.Sc., University of Alaska, 1969. Never married, no children.

POSITIONS

Born into socially prominent northern California family. Did some early travel with family, but much of her young womanhood spent as active socialite. Left in charge of her father's business affairs at age thirty-two following his death in 1920. Became interested in Arctic exploration following tourist visit to Spitzbergen, 1924. Organized and went on seven Arctic exploratory trips at her own expense between 1926 and 1941, making possible considerable increase in knowledge of Arctic geography, especially of east-central Greenland and lands in waters of Greenland and Norwegian Seas. Led her final Arctic expedition at request of federal government (1941) to facilitate study of magnetic radio phenomena in Arctic region and served as consultant to federal government on mapping and geography of the Arctic during World War II. Flew over North Pole, 1955. Visited Far East as part of round the world trip, 1958. Active in civic affairs in San Francisco Bay area throughout much of her life. Two brief final trips made to Alaska, 1964 and 1969.

CAREER

Led first Arctic voyage, a polar bear hunt for group of her friends, to Franz Josef Land on chartered Norwegian sealing vessel, 1926, and career-long practice of photographing Arctic terrain and wildlife. Gave up plans for 1928 Arctic exploratory trip and instead volunteered her charted vessel and crew to help search for Roald Amundsen's missing plane, the *Latham*, travelling 10,000 miles in region between Norway, Greenland, Spitzbergen, and Franz Josef Land during three month effort. Though Amundsen was not found, Boyd received order of St. Olaf of Norway for her services (1928), the first foreign woman to be so honored. Thereafter began focusing on serious geographic exploration and photography. Made increasing use of highly sophisticated photographic equipment. Visited Franz Josef Land and northern Scandinavia, 1930; made first detailed reconnaissance of inner reaches of fjord region of east Greenland aboard chartered vessel *Veslekari* during four expeditions undertaken in 1931, 1933, 1937, and 1938. Also visited Jan Mayen Island, Spitzbergen, and northwest coast of Norway. During the last three of these trips, employed photogrammetric mapping procedures recommended to her by Isaiah Bowman of the American Geographical Society (AGS). A number of her findings necessitated correction of existing maps and hydrographic charts. One of the east Greenland islands she charted and photographed was named Weisboydlund (Miss Boyd Land) by Danish Geodetic Institute in her honor (1932), as was a local glacier. The AGS sponsored and helped her staff her 1933–1938 trips, undertaken at her expense, with accompanying geologists, physiographers, paleontologists, hydrographers, and biologists, most of them salaried. During each expedition she continued her practice of collecting botanical specimens while also observing and photographing birds and mammals. During the winter of 1937–1938, she sought some instruction in plant science from the American botanist Alice Eastwood. All of Boyd's voyages were successfully completed despite considerable risk to Boyd, her associates, and her vessel, principally from harsh climatic conditions in Arctic waters. All land travel and exploration was accomplished by Boyd and her associates on foot, without the use of sled dogs.

The AGS published all three of Boyd's books. The first was *The Fiord Region of East Greenland* (1935). She attended the Fourteenth International Geographical Congress in Warsaw (1934) with Bowman and others as an official delegate of the AGS, the National Academy of Sciences, and the U.S. State Department. Prior to the congress, Boyd organized an extensive automobile tour of northern Poland and neighboring areas with Bowman and Polish geographers. Her book *Polish Countrysides*, accompanied by hundreds of her photographs, was published in 1937 (there is also a recent Polish edition, published 1991, with an introduction by Susan Gibson Mikos). Discovery of new higher profile of sea bottom off Jan Mayen Island during her 1937 trip resulted in this feature being named Louise A. Boyd Bank (1938). Publication of her later geographic findings in Greenland was deferred for national security reasons until after World War II. It appeared as her third book, *The Coast of Northeast Greenland*, in 1948. She also published articles on her photographs of the east Greenland fjords in *The Geographical Review* (October, 1932) and on her seven Arctic expeditions in *Photogrammetric Engineering* (1950). During the summer and early fall of 1941, at request and under the aegis of National Bureau of Standards, Boyd led a final self-financed expedition to waters off western Greenland and northeastern Canada to enable scientists to study transmission of radio waves. The results were never published, although the Bureau of Standards was highly complimentary about her contribution. Boyd also served as War Department consultant in 1942 and 1943.

In 1962, Boyd sold the family mansion and lived in reduced circumstances during her final years. Sources disagree as to whether this was the result of the heavy expenses of her Arctic expeditions or was due to some other factor, given her careful earlier management of her resources. She battled cancer toward the end of her life, and friends underwrote her final years in a nursing home. Her excellent library on the Arctic and Scandinavia went to the Universities of California and Alaska, and a local science museum in San Rafael named in her honor was given her books on horticulture and on California. Boyd was a member of a number of organizations concerned with geography, exploration, photography, and horticulture, and was made a councilor of the American Geographic Society (1960) and a director and honorary member of the American

Polar Society. She received many honors and awards, including the Andree Plaque from the Swedish Anthropological and Geographical Society (1932), the medal of King Christian X of Denmark, the Cullum Medal from the American Geographic Society for distinguished geographic discoveries (1938), and decoration as Chevalier of the French Legion of Honor.

MAJOR CONTRIBUTIONS

Though not formally trained as a scientist, Boyd's enormous interest, organizational abilities, financial support, and active leadership greatly facilitated scientific studies of Arctic geography, hydrography, photography, and natural history, particularly in the area of plant collecting. Her contributions to Arctic exploration were among the last underwritten by private philanthropy prior to World War II. Such enterprises have since been mounted on a much larger scale by government and private agencies from the standpoint both of personnel and equipment. Her use of highly sophisticated photographic equipment made it possible for geographers to develop valuable maps of portions of the east Greenland coast and subterranean land masses in the Arctic region.

BIBLIOGRAPHY

Gillis, Anna M. "A Socialite Conquers An Arctic Wilderness." *Washington Post*, 10 April 1996.

The New York Times, 17 September 1972 (obit.).

Olds, Elizabeth F. *Women of the Four Winds* (1985).

Who Was Who in America.

UNPUBLISHED SOURCES

L. A. Boyd papers in Marin County (California) Historical Association. Photographs, copies of cables, and reports in files of the Society of Women Geographers. Photographs in American Geographical Society collections, University of Wisconsin-Milwaukee. Information concerning the 1941 expedition for National Bureau of Standards in National Archives. See also bibliographical notes in Olds, above.

Keir B. Sterling

BRADBURY, JOHN. Born near Stalybridge, England, 20 August 1768; died Middletown, Kentucky, 16 March 1823. Naturalist, botanical collector, western traveler.

FAMILY AND EDUCATION

Attended a local school taught by John Taylor, who encouraged his interest in mathematics and botany. Nothing is known of family except that poverty forced Bradbury to leave school at an early age to work in a cotton mill. Continued his studies independently, learning to read and write French, and, at age eighteen, organized night school for young men. Serving as a teacher at the school, bought a microscope and other scientific apparatus, by which he helped to further his own education. His growing knowledge of natural history brought him to attention of Joseph Banks, and at age of twenty-four was elected a member of the Linnaean Society. His wife, Elizabeth, and their eight children returned with him on his second trip to the United States in 1817, where all remained.

POSITIONS

Did landscape gardening for wealthy English patrons but, unable to support his large family, proposed a three-year expedition to Louisiana to collect natural history specimens for the Liverpool Museum, intending that one of his sons transship from garden in New Orleans living specimens of such economic plants as Bradbury would discover in the wild. The Lancashire men who underwrote the scheme, in part motivated to find a better source of cotton than that coming from the West Indies, were unwilling to finance venture at level greater than £100 a year, with the result that Bradbury had to go alone. After his western travels, tried unsuccessfully to obtain government appointments (to establish a botanical garden in Washington, D.C., and to survey a road through the Louisiana Territory). At end of his life, was again a laborer in a textile mill in Louisville.

CAREER

Sailed in the spring of 1809 and, with a letter of introduction by William Roscoe, president of the Liverpool Botanic Garden, visited Monticello to see Thomas Jefferson, who persuaded him to explore the northern part of Louisiana Territory rather than base himself in New Orleans. This decision prevented Bradbury's making any contribution to Manchester's cotton needs and may have led to cooling of his sponsors' interest. On the other hand, it led to his being one of the first naturalists to explore the region recently opened up by Lewis and Clark.

During the winter of 1809–1810, Bradbury laid out a garden near St. Louis and obtained sixteen species of undescribed plants; also collected nearly sixty species of birds, eight or nine of which he believed were undescribed, as well as unidentified insects and mammals. Along with Thomas Nuttall, left St. Louis, 12 March 1811, to join Astoria Expedition. Proceeding up Missouri River under the leadership of Wilson Hunt, the Astorians' boats reached present northern border of South Dakota by 12 June. Here, at the villages of the Arikara Indians, the Astorians struck out overland for the Pacific, and Bradbury and some others traveled by horseback another 200 miles northward to Fort Mandan near present Bismarck, North Dakota. On journey upriver they had been overtaken by the fur trader Manuel Lisa, accompanied by Henry Marie Brackenridge. Bradbury and Brackenridge, who became close friends, returned to St. Louis during latter part of July, while Nuttall apparently remained in Mandan country until fall. Perhaps because of professional rivalry, the two botanists had little to do with each other; however, the presence of the two of them had made this the best-manned natural history exploration so far undertaken in North America. Brackenridge reported that during the journey of 2,000 miles Bradbury discovered nearly 100 undescribed plants.

Bradbury dispatched seven parcels of plants and seeds collected near St. Louis, which netted the Liverpool Botanic Garden more than 1,000 living specimens. From Missouri Valley trip he brought back more than 4,000 roots to transplant in St. Louis, but most of these died. Meanwhile, rankled by the parsimony of Liverpool sponsors, Bradbury resigned his commission in August 1811. Traveling via New Orleans in December, he reached New York but could not cross the Atlantic before War of 1812 detained him there. His living plants, seeds, and herbaria specimens reached Liverpool ahead of him. Duplicates of these were distributed by Roscoe, some of them going to A. B. Lambert, in whose herbarium Frederick Pursh examined the thirty-nine new species he published in his *Flora Americae Septentrionalis* (1814), giving Bradbury credit only as collector.

Bradbury returned to England in 1816 to see his only book through the press, *Travels in the Interior of America* (Liverpool, 1817), which included merely a list of ninety-eight "rare or valuable" plants, not discoveries, but was first description of the Missouri Valley to appear after Lewis and Clark preliminary reports. Earlier, he had contributed information included in Dawson Turner and L. W. Dillwyn's *Botanist's Guide through England and Wales* (London, 1805). C. S. Rafinesque wrote a "Florula Missurica" based on specimens obtained from Bradbury, but this apparently was lost when sent to William Swainson in 1817.

MAJOR CONTRIBUTIONS

Though Bradbury had little credit for his discoveries, his book is a thrilling adventure story, replete with observations on flora, fauna, and mineral productions of the region through which he traveled. Also has many incisive observations on life and customs of Plains Indians, whom Bradbury viewed realistically though he arrived in America imbued with the European romantic concept of "noble red man." Was one of the first books to deplore the wanton slaughter of the bison—yet Bradbury was willing to contract with Indians to supply "buffalo wool" if that commodity would please the Manchester mill owners. Washington Irving based eight chapters of *Astoria* (1836) on Bradbury's book and Brackenridge's parallel account. Reprinted as Vol. 5 in Thwaites's *Early Western Travels*.

BIBLIOGRAPHY

Rickett, H. W. "John Bradbury's Explorations in Missouri Territory." *Proceedings of the American Philosophical Society* 94 (1950).

Stansfield, H. "Plant Collecting in Missouri: A Liverpool Expedition, 1809–11." *Liverpool Libraries, Museums and Arts Committee Bulletin* 1 (1951).

True, R. H. "A Sketch of the Life of John Bradbury . . ." *Proceedings of the American Philosophical Society* 68 (1929).

UNPUBLISHED SOURCES

See Letters in Thomas Jefferson Correspondence, Library of Congress and Massachusetts Historical Society; City of Liverpool Museums. These letters were published by Rickett and True.

Charles Boewe

BRANDEGEE, MARY KATHERINE LAYNE CURRAN ("KATE"). Born western Tennessee, 28 October 1844; died Berkeley, California, 3 April 1920. Botanist.

FAMILY AND EDUCATION

Daughter of a miller-farmer. Family moved often from her birth until she was nine, when they settled near Folsom, California. The second of ten children. Early schooling sporadic, possibly including some time at short-lived female seminary at Folsom. Attended medical school at the University of California at San Francisco, 1875–1878, receiving M.D. Married Hugh Curran, 1866 (died 1874), and Townshend Stith Brandegee, 1889; no children.

POSITIONS

Probably taught school in Folsom prior to first marriage. Practiced medicine, though not very successfully from a medical point of view, 1883–1894. Curator of herbarium, California Academy of Sciences, 1883–1894. Continued botanical studies with husband in San Diego, 1894–1906. Worked voluntarily at University of California, Berkeley Herbarium, where her husband was honorary curator.

CAREER

Began study of plants while medical student but serious collecting not undertaken until 1882. Left floundering medical practice to become curator of the herbarium of California Academy of Sciences and remained active in botany until shortly before her death. With T. S. Brandegee collected extensively in California (including their 1889 honeymoon walk from San Diego to San Francisco) and in Nevada and Baja (1893). Established and edited *Bulletin* series of the California Academy of Sciences (1880). With Townshend Stith Brandegee and Harvey W. Harkness founded and edited the journal *Zoe*, where many of her articles appeared (1890–1908). With husband established personal herbarium and botanical library. Deposited herbarium and library at Berkeley in 1906 when they moved there.

Brandegee's reviews, articles, and notes published in the *Bulletin* and *Pro-*

ceedings of the California Academy of Sciences, especially *Zoe*. Her writing was often extremely critical, especially when the subject was Edward Lee Groene, whose rejection of Eastern authorities and Darwinism and tendency to split species were at odds with her embrace of Asa Gray's classification and evolutionary thought.

Dedicated collector of California flora, especially its variation and the stages of development of plants. Several plants were named after her (Curran). Her position as curator of academy herbarium gave her an unusual seat of power for a woman, and she used it to further her study of California flora. Somewhat casual in matter of recording data on specimen labels, but nonetheless her collections are still considered very useful.

MAJOR CONTRIBUTIONS

Contributed greatly to the knowledge of flora of California. Helped establish the Bay Area botanical institutions.

BIBLIOGRAPHY

Bonta, Marcia M. *Women in the Field: America's Pioneering Women Naturalists* (1991).

Crosswhite, Frank S., and Carol D. Crosswhite. "The Plant Collecting Brandegees, with Emphasis on Katherine Brandegee as a Liberated Woman Scientist of Early California." *Desert Plants* 7 (1985).

Dupree, Hunter, and Marian L. Gale. *Notable American Women*. Vol. 1 (1971).

Elliott, Clark A. *Biographical Dictionary of American Science* (1979).

Ewan, Joseph. "Bibliographical Miscellany—IV: A Bibliographical Guide to the Brandegee Botanical Collections." *American Midland Naturalist* 27 (1942).

Jones, Marcus E. "Katherine Brandegee, A Biography." *Cactus and Succulent Journal* 41 (1969).

Setchell, William Albert. "Townshend Stith Brandegee and Mary Katherine (Layne) (Curran) Brandegee." *University of California Publications in Botany* 13 (1924–1927).

UNPUBLISHED SOURCES

See herbarium, library, and manuscripts at University of California herbarium, Berkeley; also, material at Bancroft Library, California Academy of Sciences, and Harvard University herbarium.

Liz Barnaby Keeney

BRAUN, (EMMA) LUCY. Born Cincinnati, Ohio, 19 April 1889; died Cincinnati, Ohio, 5 March 1971. Botanist, ecologist, conservationist, educator.

FAMILY AND EDUCATION

Daughter of George Frederick and Emma Moriah Wright Braun. Father was a public school principal and mother had strong interest in botany. Paternal ancestors were of German-French descent; maternal ancestors were of English origin. Only one sister, Annette (1884–1978), an entomologist, became internationally known as an authority on Microlepidoptera (moths) and was first woman to obtain a Ph.D. from the University of Cincinnati. Early interest in the natural world and environment was fostered by parents, who took the girls to the woods and identified wildflowers. E. Lucy received primary and secondary education in Cincinnati public schools. Earned all college degrees from the University of Cincinnati: B.A. (1910), M.A. in geology (1912), Ph.D. in botany (1914). One summer (1912) of study with Henry C. Cowles, pioneer plant ecologist, at the University of Chicago. Never married, no children. Lived with her sister, who shared in all of her scientific work, and both retained the strict Victorian lifestyle of their parents. A portion of their home and garden was used as a science laboratory, where many rare and unusual plants were transplanted for close observation and study.

POSITIONS

Entire career was at the University of Cincinnati: assistant in geology (1910–1913), assistant in botany (1914–1917); instructor in botany (1917–1923), assistant professor of botany (1923–1927), associate professor of botany (1927–1946), and professor of plant ecology (1946–1948), professor emerita of plant ecology (1948–1971).

CAREER

Early scientific publications in plant ecology concerned the physiographic ecology and vegetation of the Cincinnati region, especially the unique vegetation of the unglaciated limestone in Adams County, Ohio; the forests of the Illinoian Till Plain of southwestern Ohio; and the forests of the Cumberland Plateau and Mountains of Kentucky. By 1950 had become recognized authority on the deciduous forests of eastern North America based on twenty-five years of concentrated field study and over 65,000 miles of travel. First person to use the term *mixed mesophytic forest* to distinguish a complex segregate of the eastern deciduous forest. Expanded on the theory that the southern Appalachians were the center of survival of plants during the continental glaciation, that from there the species and communities spread, and that the prairie flora had a southwestern origin and migrated into southern unglaciated Ohio long before Wisconsinian glaciation and the onset of the Xerothermic Period. Believed that during continental glaciation the deciduous forest was maintained south of the glacial boundary, rather than a zone of tundra and boreal forest. Contributed extensively to the floristics and taxonomy of vascular plants by comparing the flora of Cincinnati in the 1920s and 1930s with the known flora of 100 years earlier based on the records of Thomas G. Lea. Named and described nine new taxa from the region and prepared an annotated catalog of the seed plants of Kentucky. Organized an Ohio Flora Committee in 1951 and contributed two authoritative books on the woody plants and monocots of the state. Stressed saving natural areas by founding the Cincinnati chapter of the Wildflower Preservation Society of North America (1917), editing the Society's national magazine, *Wild Flower*

(1928–1933), and contributing signed and unsigned articles to the magazine; published first inventory of Ohio's natural areas in the *Naturalist's Guide to the Americas* (1926); and exerted leadership roles in the founding of the Ohio chapter of the Nature Conservancy and in the fight to save prairie remnants and establish nature preserves, especially in Adams County, Ohio. Herbarium of over 11,000 specimens was presented to the Smithsonian Institution; many specimens representing the flora of Cincinnati deposited at the University of Cincinnati.

Published four books and over 180 articles in twenty scientific and popular journals; *The Physiographic Ecology of the Cincinnati Region* (1916), *The Vegetation of the Mineral Springs Region of Adams County, Ohio* (1928), *The Lea Herbarium and the Flora of Cincinnati* (1934), *Forests of the Illinoian Till Plain of Southwestern Ohio* (1936), *Forests of the Cumberland Mountains* (1942), *An Annotated Catalogue of Spermatophytes of Kentucky* (1943), *Deciduous Forests of Eastern North America* (1950), *The Phytogeography of Unglaciated Eastern United States and its Interpretation* (1955), *The Woody Plants of Ohio* (1961), *The Monocotyledoneae* [*of Ohio*] (1967), *An Ecological Survey of the Vegetation of Fort Hill State Memorial Highland County, Ohio* (1969).

First woman elected president of the Ohio Academy of Science (1933–1934) and the Ecological Society of America (1950) and first woman to be listed in the Ohio Conservation Hall of Fame (1971). Awarded many honors, including Mary Soper Pope Medal for achievement in the field of botany (1952), a Certificate of Merit from the Botanical Society of America (1956), honorary doctor of science degree from the University of Cincinnati (1964), and Eloise Payne Luquer Medal for special achievement in the field of botany by the Garden Clubs of America (1966).

MAJOR CONTRIBUTIONS

One of the most original thinkers in the developing field of plant ecology during the first half of the twentieth century in North America. A faithful and dedicated worker in the field of conservation and the preservation of natural areas. An independent investigator with strong beliefs who allowed little deviation for argument and disagreement. A perfectionist striving for thoroughness and accuracy with meticulous attention to detail in her scientific work. Her book, *Deciduous Forests*, was acclaimed by F. Fosberg as a definitive work that "reached a level of excellence seldom or never before attained in American ecology or vegetation science, at least in any work of comparable importance." It remains her most remembered and lasting achievement. Her legacy continues in the Lucy Braun Association for the Mixed Mesophytic Forest dedicated to the research, understanding, and preservation of the deciduous forest.

BIBLIOGRAPHY

Bonta, M. M., ed. "E. Lucy Braun 1889–1971." In *American Women Afield, Writings by Pioneering Women Naturalists* (1995).

Durrell, L. "Memories of E. Lucy Braun," In R. L. Stuckey and K. J. Reese, eds. *The Prairie Peninsula—In the Shadow of Transeau: Proceedings of the Sixth North*

American Prairie Conference. Ohio Biological Survey Biological Notes 15 (1981).
Lampe, L. "E(mma) Lucy Braun." *Ohio Journal of Science* 71 (1971) (obit.).
Peskin, P. K. "A Walk Through Lucy Braun's Prairie." *The Explorer* 20, 4 (1978).
Stuckey, R. L. "E. Lucy Braun (1889–1971), Outstanding Botanist and Conservationist: A Biographical Sketch with Bibliography." *Michigan Botanist* 12 (1973).
Stuckey, R. L. "Emma Lucy Braun, April 19, 1998–March 5, 1971, Botanist, Conservationist," In B. Sicherman and C. H. Green, eds. *Notable American Women: The Modern Period. A Biographical Dictionary* (1980).
Stuckey, R. L. "E. Lucy Braun, Foremost Woman Botanist of Ohio," In *Women Botanists of Ohio Born Before 1900* (1992).
Stuckey, R. L. *E. Lucy Braun: Ohio's Foremost Woman Botanist. A Collection of Biographical Accounts, Maps, and Photos* (1994).
UNPUBLISHED SOURCES
Field notes, photographs, and some correspondence in the Cincinnati Museum of Natural History.

Ronald L. Stuckey

BREITUNG, AUGUST JULIUS. Born Muenster, Saskatchewan, 9 May 1913; died Lakewood, California, 27 September 1987. Taxonomic botanist.

FAMILY AND EDUCATION

Son of Heronimus and Veronica (née Fuller) Breitung, recent immigrants from Germany. Attended public school at Loring and Beech Grove rural schools, south of Tisdale, Saskatchewan, and three years of high school at Tisdale. In the later 1940s he took night classes in English literature and composition at Carleton University, Ottawa. Married Mathilde Presch 4 May 1949 in Ottawa (died 20 February 1984); no children.

POSITIONS

Assistant to botanist A. E. Porsild, Canol Road collecting expeditions in Yukon Territory, 1944, and in Banff and Jasper National Parks, 1945 and 1946. Assistant technician, Herbarium, Division of Botany and Plant Pathology, Canada Department of Agriculture, Ottawa, November 1946 to late 1952. Botanized privately in Waterton Lakes National Park, 1953. Took a course in applied aerodynamics, apparently at the California Institute of Technology, Pasadena. Draftsman in aerospace industry, California, 1954–c.1982.

CAREER

A self-taught taxonomic botanist, he collected about 10,000 carefully prepared and numbered plant specimens (25,000 if duplicates are included) in east-central Saskatchewan, 1,500 in Cypress Hills, and 2,000 in Waterton Lakes National Park.

MAJOR CONTRIBUTIONS

He named one new taxonomic variety, ten new forms, and twenty-seven new name combinations. Named for him were the genus *Breitungia* Love and Love, and two species *Antennaria breitungia* Porsild and *Thalictrum breitungii* Boivin. He donated his central-eastern Saskatchewan plants to the herbarium in Ottawa and left the proceeds of his estate to the August J. Breitung and Mathilde K. Breitung Memorial Trust Fund to support taxonomic research at the W. P. Fraser Herbarium at the University of Saskatchewan. His taxonomic papers remain the standard references for Saskatchewan, not yet superseded.

BIBLIOGRAPHY

Breitung, A. J. "Catalogue of the Vascular Plants of Central-Eastern Saskatchewan." *Canadian Field-Naturalist* 61 (1947).

Breitung, A. J. "A Botanical Survey of the Cypress Hills." *Canadian Field-Naturalist* 68 (1954).

Breitung, A. J. "Annotated Catalogue of the Vascular Flora of Saskatchewan." *American Midland Naturalist* 58 (1957).

Breitung, A. J. "Plants of Waterton Lakes National Park." *Canadian Field-Naturalist* 71 (1957).

Harms, V. L. "August Julius Breitung, Noted Saskatchewan Amateur Botanist." *Blue Jay* 46 (1988).

C. Stuart Houston

BREWER, THOMAS MAYO. Born Boston, 21 November 1814; died Boston, 23 January 1880. Ornithologist, oologist, publisher.

FAMILY AND EDUCATION

Son of Col. James Brewer, Revolutionary patriot who took part in Boston Tea Party. A.B., 1835, M.D., 1838, Harvard. Married Sally R. Coffin, 1849; two children.

POSITIONS

Medical doctor, Boston, 1838–1840. Contributor, then editor, *Boston Atlas*, 1840–1857. Partner, then president, Swan and Tileston publishing firm (later, Brewer and Tileston), 1857–1875.

CAREER

After brief career as dispensary physician in Boston, became a correspondent for *Boston Atlas*. Published first paper, on native birds of Massachusetts, in *Boston Journal of Natural History* (1837) as a supplement to Hitchcock's *Catalogue of the Birds of Massachusetts*, adding forty-five species. Active in cause of public education, serving on Boston School Committee, 1844 until his death.

Collaborated with Spencer F. Baird and Robert Ridgway on *A History of North American Birds*, 3 vols. (1874), for which he wrote all of the life histories (about two-thirds of material), and the posthumously published *The Water Birds of North America*, 2 vols. (1884).

Joined Boston Society of Natural History in 1835 and contributed papers to its *Proceedings* for over forty years. Prepared and published new edition of *Wilson's American Ornithology* (1840) and contributed to it his own synopsis of all known birds of North America. His journal, *North American Oology*, published by the Smithsonian, appeared only once (1856) due to expenses of color plates. Translated Francis E. Sumichrast's *The Geographical Distribution of the Native Birds of the Department of Vera Cruz, with a List of the Migratory Species* from French into English (1869). Toured principal museums of Europe, 1875–1876, inspecting oological collections; wrote description of trip in August 1877 in *Popular Science Monthly*. Left own large collection of birds' eggs to Museum of Comparative Zoology at Harvard.

Although an amateur, enjoyed international reputation as ornithologist. Financed and publicized the Swede Thure Kumlien as a naturalist and brought him worldwide recognition. An enthusiastic and indefatigable worker given to argument and criticism, Brewer regarded field of Massachusetts birds as his private domain and did not think highly of younger generation of ornithologists. Quarrel with Elliot Coues over English sparrow was carried on in pages of *American Naturalist*. Was close friend of Audubon's during later years of Audubon's life and received acknowledgment in *Birds of America*.

MAJOR CONTRIBUTIONS

Considered America's first oologist, in its broadest sense, since he was concerned with breeding habits and life histories, as well as eggshells. Work on geographical distribution of birds.

BIBLIOGRAPHY

Appleton's Cyclopaedia of American Biography. Vol. 1 (1887).

Cutright, Paul R., and Michael J. Brodhead. *Elliott Coues: Naturalist and Frontier Historian* (1981).

Dictionary of American Biography. Vol. 2 (1929).

Elliott, C. A. *Biographical Dictionary of American Science: The Seventeenth through the Nineteenth Centuries* (1979).

Kinnell, S. K., ed. *People in History* (1988).

McAtee, W. L. *Nature Magazine* (November 1953).

National Cyclopaedia of American Biography. Vol. 22 (1932).

Russell, F. W. *Mount Auburn Biographies* (1953).

The Twentieth Century Biographical Dictionary of Notable Americans. Vol. 1 (1904).

Who Was Who in America. Historical vol. (1967).

John A. Drobnicki

BREWER, WILLIAM HENRY. Born Poughkeepsie, New York, 14 September 1828; died 2 November 1910. Chemist, geologist, botanist.

FAMILY AND EDUCATION

Son of Henry and Rebecca (Du Bois) Brewer. Graduated Sheffield Scientific School, Yale College, in 1852 with Ph.B. Continued studies abroad at Heidelberg, Munich, and Paris, furthering his knowledge of chemistry, botany, and geology. Received LL.D. from Yale in 1903. Married Angelina Jameson, 15 August 1858 (died 1860), and Georgiana Robinson, 1 September 1868 (died 1889).

POSITIONS

Professor of chemistry and geology, Washington College, Pennsylvania, 1858–1860; first assistant, Geological Survey of California, 1860–1864; professor of chemistry, University of California, 1863–1864; and professor of agriculture, Sheffield Scientific School, Yale College, 1864–1903. Also, president, Connecticut Board of Health, 1892–1909; secretary and treasurer of Connecticut experiment station, 1877–1902; and member of U.S. Forestry Commission, 1896. Was also member of National Academy of Sciences and, in 1902, chairman of committee organized at request of President Theodore Roosevelt to draw up statement and plan for scientific survey of Philippine Islands.

CAREER

Man of great versatility and explorer of all sciences underlying agriculture. As first assistant on Geological Survey of California gained intimate knowledge of state's geography, geology, and botany. Collected about 2,000 species of plants, basis of his work on flora of California, published as "Polypetalae," in vol. 1 of *Geological Survey of California.*

As professor of agriculture at Sheffield Scientific School, improvement of agriculture was chief goal. Helped to establish first agriculture experiment station in nation. For tenth census prepared careful study of cereal production of the United States. Contributed frequently to annual reports of Connecticut Board of Agriculture on variety of topics of interest to farmers (such as origin and constitution of soils and pollution of streams). Also pioneer in public health work. Published about 140 papers in scientific journals on chemical, geological, sanitary, agricultural, and other subjects.

MAJOR CONTRIBUTIONS

Was influential in establishment of Yale Forest School (1900). One of most important pieces of his scientific work concerned development of the American trotting horse as a contribution to evolution of breeds. Also carried on experiments on conditions influencing suspension of clays in river water with reference to the bearing of river deposits on delta formation.

BIBLIOGRAPHY
Dictionary of American Biography. Vol. 2 (1928–1937).
National Cyclopedia of American Biography. Vol. 13 (1967).
Who Was Who in America. Vol. 1 (1943).

Richard Harmond

BREWSTER, WILLIAM. Born Wakefield, Massachusetts, 5 July 1851, died Cambridge, Massachusetts, 11 July 1919. Ornithologist.

FAMILY AND EDUCATION

Son of John Brewster, a Boston banker, and Rebecca Parker Noyes Brewster. Attended public schools in Cambridge; did not attend college because of poor health. Honorary M.A., Amherst, 1880; Harvard, 1889. Married Caroline F. Kettell, 1878. No children.

POSITIONS

Was clerk in father's banking house, 1869–1870; assistant in charge of collection of birds and mammals at Boston Society of Natural History, 1880–1889; in charge, similar departments at Cambridge Museum of Natural History, Harvard University, 1885–1900; curator, Department of Birds, Museum of Comparative Zoology, Harvard, 1900–1919. Director of State Fish and Game Commission for a number of years, elected president in 1906, served two terms. Led in formation of American Game Protective and Propagation Association, 1911; was appointed member of its advisory board for remainder of his life.

CAREER

A frail lad with eyes that tended to overstrain, precluding intensive formal study at Harvard. Interest in birds stimulated at about age ten by Daniel C. French, older neighbor and taxidermist. This led young Brewster to begin collection of bird specimens deemed "finest . . . in existence." One of eight young men who founded Nuttall Ornithological Club in Brewster's home, 1873. Brewster, then age twenty-two, was president for forty-six years until his death. Maintained private ornithological museum in his Cambridge home, which was given after his death to Museum of Comparative Zoology at Harvard. One of principal organizers of *Bulletin of the Nuttall Ornithological Club*, which began publication, 1876, and which offered "good will and subscription list" when *The Auk*, organ of American Ornithologists' Union (AOU), began publication, 1883. Brewster was also one of three cofounders of the AOU in the fall of 1883. Was president of the AOU, 1895–1898. Nuttall Club has continued operations until the present, though always primarily a regional organization. Maintained property with rustic cabin near Concord, Massachusetts, and a camp on Lake Umbagog, Maine, where he carried on observations of bird habits. Organized one of the first state Audubon Societies in Massachusetts; later a director of the National Association of Audubon Societies and long president of the Massachusetts Audubon Society. Took collecting trips to various parts of the country and to Trinidad, and skilled collectors in his employ secured specimens for him elsewhere. Author of more than 250 scientific and literary papers, as well as books and reviews. Published *Bird Migration* (1886); *Descriptions of First Plumage in Various North American Birds* (1878–1879); *Birds of the Cape Region of Lower California* (1902); *Birds of the Cambridge Region of Massachusetts* (1906); and *Birds of the Lake Umbagog Region, Maine*, not completed at his death, which was published posthumously (1924–1925). Edited H. D. Minot's *Land Birds and Game Birds of New England* (1894). His writings comprised important additions to literature of American ornithology, development of which he notably influenced. The William Brewster Memorial Medal created by the AOU in his honor, 1919, is awarded periodically to authors of works on birds of Western Hemisphere. Fellow of the American Association for the Advancement of Science and of American Academy of Arts and Sciences. Longtime trustee of Brewster Free Academy, Wolfboro, New Hampshire, founded by his father.

MAJOR CONTRIBUTIONS

His outstanding collection of North American birds now at Museum of Comparative Zoology, Harvard. His publications comprise important additions to literature of American ornithology, the development of which he notably influenced. Major figure in founding of Nuttall Ornithological Club, 1873, and of American Ornithologists' Union, 1883.

BIBLIOGRAPHY

Appleton's Cyclopaedia of American Biography. Vol. 1 (1886).

Batchelder, Charles F. *An Account of the Nuttall Ornithological Club, 1873 to 1919* (1937).

Batchelder, Charles F. *A Bibliography of the Published Writings of William Brewster* (1951).

Boston Transcript, 12 July 1919 (obit.).

Brooks, Paul. "Birds and Men." *Audubon* 82 (July 1980).

Davis, William E., Jr. *History of the Nuttall Ornithological Club, 1873–1986* (1987).

Dictionary of American Biography. Vol. 3 (1929).

Henshaw, H. W. *Auk* (January 1920) (obit.).

National Cyclopaedia of American Biography. Vol. 12 (1904); Vol. 22 (1932).

Stroud, Richard, ed. *National Leaders of American Conservation* (1985).

Who Was Who in America, 1897–1942. Vol. I (1981).

Connie Thorsen

BRITTON, ELIZABETH GERTRUDE (KNIGHT). Born New York, 9 January 1858; died New York, 25 February 1934. Teacher, botanist, editor.

FAMILY AND EDUCATION

Daughter of James and Sophie Ann (Compton) Knight. Spent much of her childhood in Cuba, as her father and grandfather manufactured mahogany furniture and operated a sugar plantation there. Graduated from the Normal (later Hunter) College, New York, in 1875. Married Nathaniel Lord Britton in New York City on 27 August 1885.

POSITIONS

Teacher in training department, Normal College, 1875–1882, assistant in botany, 1882–1885. Edited *Bulletin of the Torrey Botanical Club*, 1886–1888. Life member of the Botanical Society of America and president of the Sullivant Moss Society (1916–1919), which later became the American Bryological Society.

CAREER

In 1878, she joined the Torrey Botanical Club and gained an international reputation for her studies of mosses. After marriage to Nathaniel Lord Britton, a geology professor at Columbia College who shared her interest in botany, she had unofficial charge of the moss collection at this school. Was also interested in ferns and published an enumeration of Henry Hurd Rusby's collection of South American ferns. In later life, she worked to conserve wildflowers and participated in the founding of the Wild Flower Preservation Society of America.

Published extensively on mosses, including a book, *The Life History of Vittaria Lineata* (with Alexandrina Taylor, 1902), and numerous articles later reprinted in the *Bulletin of the Torrey Botanical Club* and in *Torreya*.

MAJOR CONTRIBUTIONS

In 1889 she began in the *Bulletin* her ''Contributions to American Bryology,'' which, with a subsequent series, ''How to Study the Mosses'' in *The Observer*, ranked her as a leader in the bryological field in America. She has been credited with inspiring the establishment of the New York Botanical Garden, with her husband serving as its first director in 1891. Also interested in ferns, she published a revision of the North American species of *Ophioglossum* in 1897 and wrote life histories of the curly grass fern and the tropical *Vittaria lineata*, but it was mainly as a specialist on mosses that she was known to the botanical world. The moss genus *Brytobrittonia* and fifteen species of plants were named for her. The Brittons' will established an endowment fund at the New York Botanical Garden that was stipulated to be used for ''botanical and horticultural research, exploration and publication, and increase of collections of plants, books, and specimens, but not for maintenance or construction.''

BIBLIOGRAPHY

Bailey-Ogilvie, Marilyn. *Women in Science: Antiquity through the Nineteenth Century: A Biographical Dictionary with Annotated Bibliography* (1986).

Elliot, Clark A. *Biographical Dictionary of American Science: The Seventeenth through the Nineteenth Centuries* (1979).

Howe, Marshall A. ''Elizabeth Gertrude Britton.'' *Journal of the New York Botanical Garden* 35 (May 1934).

James, Edward T., ed. *Notable American Women, 1607–1950; A Biographical Dictionary* (1971).

Karen M. Venturella

BRITTON, NATHANIEL LORD. Born New Dorp, Staten Island, New York, 15 January 1859; died Bronx, New York 25 June 1934. Botanist, author, educator, administrator.

FAMILY AND EDUCATION

Son of (Jasper) Alexander Hamilton and Harriet Lord (Turner) Britton and a descendant of James Britton, who came from London, England, in 1635. Received degree in engineering of mines in 1879 from Columbia College, and in 1881 earned Ph.D. from Columbia University. Married Elizabeth Gertrude Knight, 27 August 1885; no children.

POSITIONS

Assistant in geology at Columbia and botanist and assistant geologist for Geological Survey of New Jersey. In 1887, instructor at Columbia University in botany and geology; in 1890 became adjunct professor of botany, in 1891 professor of botany, and in 1896 professor emeritus. Assumed directorship of newly established New York Botanical Garden. Retired from post in 1929.

Instrumental in organizing Botanical Society of America and served as its vice president in 1894 and 1896 and president in 1898 and 1920. Also vice president of American Association for the Advancement of Science in 1896 and president of New York Academy of Sciences in 1907. Member of American Philosophical Society and National Academy of Sciences.

CAREER

Best known for his role in establishment and development of New York Botanical Garden, which he guided for thirty-three years, with enthusiasm, drive, and organizing ability. Under his direction it became ''one of the world's great botanical institutions.'' Also founded *Bulletin of the New York Botanical Garden* in 1896 and the garden's *Journal* and *Memoirs* in 1900 and *Addisonia* in 1916.

For over forty years was actively engaged in research on the floras of the United States, West Indies, and parts of South America, resulting in a number of books and articles. Among former were *Illustrated Flora of the Northern United States, Canada, and the British Possessions*, 3 vols. (with Addison Brown, 1896–1898; 2d ed. 1913); *Manual of the Flora of the Northern States and Canada* (1901); *North American Flora* (1905–1930); *North American Trees* (with J. A. Safer, 1908); *Flora of Bermuda* (1918); *The Bahama Flora* (with C. F. Millspaugh, 1920); and *Botany of Porto Rico and the Virgin Islands*, 2 vols. (with P. Wilson, 1923–1930).

MAJOR CONTRIBUTIONS

Greatest achievement was leadership of New York Botanical Garden during its formative years. His written works retain value ''as comprehensive descriptive surveys.''

BIBLIOGRAPHY

American Men of Science. Eds. 1–5.
Dictionary of American Biography. Vol. 1 (1928–1937).
Dictionary of Scientific Biography (1970).
Merrill, E. D. ''Nathaniel Lord Britton.'' *National Academy of Sciences Biographical Memoirs* 19 (1938) (also has references to Elizabeth Britton).
New York Times, 26 June 1934 (obit.).

UNPUBLISHED SOURCES

Family papers are located at Staten Island Institute of Arts and Sciences Archives (New York). Correspondence may also be found in George F. Atkinson and Liberty Hyde Bailey Collections at Cornell University Library. Other letters are located in Charles V. Piper Papers at Washington State University Library (Pullman).

Richard Harmond

BRODIE, WILLIAM. Born Peterhead, Aberdeenshire, Scotland, 1831; died Toronto, Ontario, 6 August 1909. Naturalist, entomologist, botanist, dentist.

FAMILY AND EDUCATION

Son of George Brodie and Jane Milne, who immigrated to Whitchurch, York Township, Ontario, from Scotland in spring of 1835. Obtained primary education at Whitchurch and was one of the earliest graduates of the Royal College of Dental Surgeons at Toronto. Married Jane Anna McPherson (also recorded as Jean MacPherson). Four daughters and a son, William G. A. Brodie, who was a cohort of Ernest Thomson Seton until the former drowned in the Upper Assiniboine River, 14 May 1883.

POSITIONS

Founding president of the Toronto Entomological Association, 26 February 1877 (became the Natural History Society of Toronto, 14 May 1878). Provincial biologist of Ontario, 1903–1909, and curator of the natural history collection for the Education Department (Provincial Museum).

CAREER

Obtained his early education in Whitchurch, York Township, Ontario, where he subsequently taught school for three years. He then became a doctor of dentistry and practiced his profession in Markham for two years before moving to Toronto in 1865. Studied natural history from the time he was a child and became well versed in all facets of the field. Recognized the ornithological importance of Point Pelee as early as 1879. Catalyst for the formation of the Toronto Entomological Society, which became the Natural History Society of Toronto in 1878 and later affiliated with the (Royal) Canadian Institute as the Biological Section in 1886. This organization eventually became autonomous as the Biological Society of Ontario and published *Ontario Biological Review* in 1894. Provincial government of Ontario purchased Brodie's private natural history collection in 1903 (90,000+ specimens), making it the core of a natural history museum, and appointed him as the first provincial biologist and natural history curator of the Provincial Museum. He published very little but delivered many lectures and public talks.

MAJOR CONTRIBUTIONS

First provincial biologist of Ontario and the region's best all-round naturalist of his time. Founded first natural history society in Toronto in 1878. Significant influence and mentorship extended to a host of leading naturalists, including Ernest Thompson Seton, J. H. Fleming, and the Group of Seven painter Tom Thomson. World leader in the study of gall-forming insect. Described at least twenty new species and forms. Massive private collection of these insects (20,000 specimens) now in the Smithsonian Institution. One of the earliest to recognize the importance of biological control of agriculturally harmful insects. The Brodie Club (founded 1921, Toronto) is a prestigious natural history group named in his honor.

BIBLIOGRAPHY

Morris. F. "Dr. William Brodie." *Canadian Entomologist* 41 (1909).

McNicholl, M. K. and J. L. Cranmer-Byng. *Ornithology in Ontario* (1994).

UNPUBLISHED SOURCES

The primary extant papers of William Brodie are held by the Royal Ontario Museum, Archives, Special Collections: Brodie Papers (SC20A) and J. H. Fleming Papers (SC29). A memorial evening was held by his namesake organization, the Brodie Club (Toronto), 22 December 1925. The club minutes for this event provide useful background on activity and character.

Michael S. Quinn

BROLEY, CHARLES (LAVELLE). Born Gorrie, near Goderich, Ontario, 7 December 1879; died Delta, Ontario, 4 May 1959. Naturalist, ornithologist, bander, banker.

FAMILY AND EDUCATION

Son of strict Methodist minister. Family moved several times to various villages and towns in general vicinity of Elora, Ontario, settling in Islet Rock in 1886. Attended public schools with his brother and two sisters but did not undertake further formal education. Married Ruby Stevens, about 1908 (died 1921) and Myrtle McCarthy, 1923 (died 1958). One daughter.

POSITIONS

Manager, Merchant Bank, Delta, Ontario 1905–1918. Manager, Croydon Avenue branch, Bank of Montreal, Winnipeg, Manitoba 1918–1938. Secretary, Ornithological Section, Natural History Society of Manitoba, (NHSM) 1924–1925, 1928–1929; chairman 1925–1928, 1929–1930, [vice president, NHSM, 1931–1934.] Appointed to Wildlife Committee, Florida Audubon Society, shortly before his death, 1959.

CAREER

Professionally, banker, first in southern Ontario, then in Winnipeg, Manitoba. Interest in nature, apparently stirred at his cabin in Ontario, did not develop strongly until after the death of his first wife. In Manitoba, became very active and contributed regularly to A. G. Lawrence's bird column, "Chickadee Notes," in which Broley reported number of significant observations. Applied his thorough, careful banker's discipline to observing and reporting and quickly became known for his reliability. These traits, combined with his good humor and active nature, led to participation and popularity in the Natural History Society of Manitoba, with the enthusiastic support of his second wife, Myrtle. Among his many observations there was the mysterious death of large number of Franklin's gulls, attributed by him to their eating poisoned grasshoppers.

Broley retired from bank in 1938 at the age of fifty-eight, and after brief period of European travel, he and Myrtle started annual pattern of wintering in Florida and summering at the cabin in Delta, Ontario. In Florida, he started an extensive program of banding bald eagles, of which only 166 had been banded previously. Soon became adept at climbing to nests, eventually banding over 1,200 eagles, primarily in Florida, with about 100 also in Ontario. Recoveries showed that "resident" eagles of Florida underwent extensive wanderings to the north after nesting and completely revised previous thinking on the life history of the species. These significant results earned him elective membership in the American Ornithologists' Union (1959) but more significantly his popular lectures on subject brought much publicity and popularity to the eagles, and he became popular subject of writers, notably in J. J. Hickey's classic bird-watching guide and an *Audubon* article by R. T. Peterson. This work and his earlier observations in Manitoba also resulted in honorary life membership in the Natural History Society of Manitoba in 1944.

Broley's careful records at large numbers of nests took on new significance in 1947, when hatching success suddenly declined. The coincidence of this decline with the start of DDT spraying in the area was not unnoticed, and after carefully considering other possibilities, Broley concluded that DDT was the culprit, and his several articles and lectures on the topic were among the early warnings of the dangers of chemical pollution in the environment. Rachel Carson's *Silent Spring* effectively used Broley's data in combination with others to sound this warning.

Manitoba observations were published in Lawrence's column. Broley wrote about ten papers on his eagle work, and Myrtle published popular book on his work (1952).

MAJOR CONTRIBUTIONS

Studies elucidated many details of nesting and post-nesting biology of Bald Eagles, and lectures helped popularize the eagle. Investigations were among the first to implicate DDT in raptor declines and thus pesticides as environmental threats.

BIBLIOGRAPHY

Bodsworth, F. "How to Catch an Eagle." *Maclean's Magazine* 65, 3 (1952).

Broley, C. L. "The Eagle and Me." *Canadian Banker* 60 (1953).

Broley, M. J. *Eagle Man* (1952).

Gerrard, J. *Charles Broley: An Extraordinary Naturalist* (1983).

Hickey, J. J. *A Guide to Bird Watching* (1943).

Mager, D "Charles Broley in Florida." In J. M. Gerrard and T. M. Ingram, eds., *The Bald Eagle in Canada* (1985).

Mason, C. R. *Auk* 77 (1960) (obit.).

McKeating, G. "Charles Broley: Eagles Then and Now in Southern Ontario." In J. M. Gerrard and T. M. Ingram, eds., *The Bald Eagle in Canada* (1985).

McMaster, A. "Charles Broley in Manitoba." In J. M. Gerrard and T. M. Ingram, eds. *The Bald Eagle in Canada* (1985).

McNicholl, M. K. *The Canadian Encyclopedia* (1985, 1988).

Peterson, R. T. "Eagle Man." *Audubon* 50 (1948).

UNPUBLISHED SOURCES

Eagle records and films at Cornell Laboratory of Ornithology, Ithaca, New York.

Martin K. McNicholl

BROOKS, ALLAN (CYRIL). Born Etawah, Uttar Pradesh, India, 15 February 1869; died Comox, British Columbia, 3 January 1946. Ornithologist, artist, illustrator, naturalist.

FAMILY AND EDUCATION

Son of William Brooks, British engineer and ornithologist, and Mary Jane Renwick Brooks. Sent to England age five, where he attended day school in Northumberland, his only formal education. In 1881 the family of three boys and two girls moved to Canada and settled on farm near Milton, Ontario. In 1887 the family moved to Chilliwack, British Columbia, where Brooks became avid student of the natural history of the West. Apart from several short periods and First World War, Brooks lived the rest of his life in British Columbia. Married Marjorie Holmes 1926; one son.

POSITIONS

Freelance artist, illustrator, natural history collector. First collected for William Brewster in 1891. Soon other East Coast naturalists wanted his services. From 1894 he had many advance orders for small birds and mammals of British Columbia from Brewster, Outram Bangs, and Gerrit Miller. Later he collected parasites for British Museum. Discovered several species of mammals and fleas for science. Studied natural history of the Okanagan, 1897–1899. Cariboo Region, 1901, Vancouver Island, 1903–1904. Settled at Okanagan Landing, 1905. Illustrator for natural history and recreation journals from end of the nineteenth century, for books on ornithology and mammalogy from 1906. His first show was at the International Sportsmen Exhibition, Vienna, Austria, in 1911.

CAREER

Began studying North American birds while farming in Ontario in early 1880s. Met, and was greatly influenced by, ornithologists Henry Seebohm and Thomas McIlwraith. His first contributions to outdoor magazines started in 1897 (*Recreation, St. Nicholas, Forest and Stream*). A freelance naturalist by choice, Brooks turned down several offers for museum positions, such as those of Provincial Museum of British Columbia, Victoria, 1916, Museum of Vertebrate Zoology, Berkeley, 1920, State of Georgia Museum 1923, and American Museum of Natural History and Cleveland Museum joint expedition in 1923. His first major commissions to illustrate works on ornithology were W. B. Mershon, *The Passenger Pigeon* (1907) and W. L. Dawson's two-volume *The Birds of Washington* (1909). He undertook many field trips in connection with his commissions to collect and illustrate. First trip to California in 1910–1911 as preamble for his illustration of W. L. Dawson, *Birds of California* (1923). In 1913 traveled to Arizona. Spent war years in Europe. Promoted captain and major 1915, awarded DSO in 1919. In 1920–1921 studied birds of Florida, Texas, and California and spent most of following year in California. In 1930 Brooks traveled to Maritime Provinces of Canada, and in 1931 he took his family to New Zealand. In order to study pelagic birds he took round-the-world trip in 1934–

1935. He successfully contributed to E. H. Forbush, *Birds of Massachusetts*, in which he finished the set of plates started by the late Louis A. Fuertes, in 1927, and to J. C. Phillips, *Natural History of Ducks* (1923–1926). According to a fellow artist-ornithologist, ''Major Brooks' interest in waterfowl makes him particularly successful as a painter of ducks.'' In 1926 his illustrations appeared in P. A. Taverner, *Birds of Western Canada*, and in 1928 in F. M. Bailey, *Birds of New Mexico*. In the 1930s he contributed plates and drawings to T. S. Roberts, *Birds of Minnesota* (1932), J. B. May, *Hawks of North America* (1935), and M. B. Grosvenor and A. Wetmore, *The Book of Birds* (1932–1937).

Illustrated about a dozen major works on ornithology and several on mammalogy and was considered by another bird painter as a ''master of *accessories* . . . the twigs, rocks, moss, leaves and grass . . . enhancing rather than detracting from the birds.'' First and foremost an ornithologist, it was said of him that ''he has established for himself an enviable reputation in the New World not alone as a painter of birds but as a scientist and sportsman as well.'' Brooks the scientist is particularly well represented in A. Brooks and H. Swarth, *Distributional List of the Birds of Aitlin, British Columbia* (1925). His last published paper (*Auk* [1945])disclosed his discovery of ''the almost unbelievable underwater action of the alula of the wings of the White-winged and Surf Scoters.'' His many illustrations are found in *Auk, Condor, Wilson Bulletin*, and *National Geographic Magazine* over period of fifty years. In spite of his lack of formal education he was well read in literature and well versed in anatomy and systematics of birds. He was often critical of ''experts'' of systematics and tended to be opinionated. He had an excellent memory, yet he kept detailed field journals, which contain valuable natural history information on British Columbia during six decades of fieldwork.

MAJOR CONTRIBUTIONS

Greatly added to knowledge of ornithology, entomology, and mammalogy of British Columbia. In his paintings and illustrations his thorough knowledge of species and appropriate background made his work outstanding. He was first-class taxidermist. His early collections are now in the Museum of Vertebrate Zoology, Berkeley.

BIBLIOGRAPHY

Brooks, M. ''Allan Brooks, a Biography.'' *Condor* (1938).

Dawson, W. L. ''Allan Brooks—an Appreciation.'' *Condor* (1913).

Harris, H. ''An Appreciation of Allan Brooks, Zoologist and Artist, 1869–1946.'' *Condor* (1946).

Laing, H. M. ''Allan Brooks, 1869–1946.'' *Auk* (1947).

Laing, H. M. *Allan Brooks, Artist-Naturalist*. B.C. Victoria, Provincial Museum (1979).

UNPUBLISHED SOURCES

See letters in P. A. Taverner Correspondence, National Museum of Natural Sciences, Ottawa.

Marianne Gosztonyi Ainley

BUCHHEISTER, CARL WILLIAM. Born Baltimore, Maryland, 20 January 1901; died Chapel Hill, North Carolina, 25 July 1986. Conservationist, administrator, field naturalist.

FAMILY AND EDUCATION

Son of George Albrecht and Mary Hermine (Koch) Buchheister. Father was tobacco exporter who took his six children on nature walks. Omnivorous reader of Ernest Thompson Seton and other writers of natural history. B.A., Johns Hopkins University, 1923; postgraduate work in Latin and Greek, 1923–1925. Honorary degrees included LL.D., Pace College, and H.H.D., Bowdoin College. Married Harriet Nettleton Gillilan 1924; three children.

POSITIONS

Taught Latin, Park School, Baltimore, 1925–1926; Lawrence School, Hewlitt, Long Island, New York, 1927–1936, where he took students on nature walks at five in the morning. ''We'd comb the countryside looking for birds and birds' nests before settling down to Caesar,'' he later remarked. In 1927, founded Camp Moccasin, private boys' camp in Lochmere, New Hampshire, which he directed until 1934. Director of Audubon Camp of Maine in 1936–1957. Executive director of Massachusetts Audubon Society, 1936–1939. Assistant director of National Audubon Society, 1940–1944; senior vice president, 1944–1959; president, 1959–1967; president emeritus, 1967–1986.

CAREER

Activities at Lawrence School included highly successful nature club, which attracted the attention of John Hopkinson Baker, president of the National Association of Audubon Societies. In 1935, Baker invited him to become first director of the Audubon Nature Camp, soon renamed Audubon Camp in Maine. Consisting of series of two-week sessions offered by Roger Tory Peterson and others, the camp trained adults in ''ecological concept of nature and in methods of conservation education.'' As direct result of the first highly successful season, became executive secretary and treasurer of Massachusetts Audubon Society in 1936. Baker was so impressed with ability to inspire people and his gift for public speaking, he brought Buchheister to New York in 1940 to newly created post of assistant director of subsequently renamed National Audubon Society. Named senior vice president in 1944, traveled widely to maintain touch with local chapters and soothe feathers ruffled by his imperious superior.

Succeeding Baker in 1959, served as president until his retirement in 1967. During his tenure, built stronger bonds with state and local chapters and societies, ironed out organizational rifts, encouraged children to become active in conservation and natural history through creation of Nature Centers for Young

America, revamped society's magazine, *Audubon*, and applied modern methods of public relations and fund-raising. Society also became increasingly involved with national causes, including protection of bald eagle and its retention as the national symbol of the United States, the fight to keep the California condor from extinction, support of Wilderness Act, and opposition to indiscriminate use of pesticides. Provided strong Audubon backing of Rachel Carson's *Silent Spring* (1962). Roger Tory Peterson wrote that Buchheister's "influence in America can be observed wherever we find naturalists working with people in the out-of-doors."

Honorary president, Alice Rich Northrup Association, 1950–1960; honorary vice president, American Forestry Association, 1963; chairman, Natural Resources Council of America, 1964–1966. Served on Board of Directors of Chewonki Foundation, Canadian Audubon Society, 1960–1967; National Parks and Conservation Association; Wildfowl Foundation, Inc.; World Wildlife Fund; and American Ornithologists' Union. Recipient of Distinguished Service Award of American Forestry Association, 1973; Award of Honor of National Resources Council of America, 1979; and Frances K. Hutchinson Medal of Garden Club of America, 1980. The National Arboretum in Washington, D.C., opened the Carl W. Buchheister National Bird Garden in 1982.

MAJOR CONTRIBUTIONS

One of the foremost American conservationists and conservation educators in the twentieth century. Transformed National Audubon Society from an "organization with limited focus into one that is truly national."

BIBLIOGRAPHY

Berle, Peter A. A. "Carl Buchheister's Legacy." *Audubon* (1986).
Buchheister, Carl W. "Four of Six Conservation Goals Reached." *Audubon* (1962).
Graham, Frank, Jr., with Carl Buchheister. *The Audubon Ark: A History of the National Audubon Society* (1990).
"New President." *New Yorker* (1960).
Peterson, Roger Tory. "Carl W. Buchheister." *Audubon* (1967).
Stroud, Richard H., ed. *National Leaders of American Conservation* (1985).
Who Was Who in America (1989).

UNPUBLISHED SOURCES

See National Audubon Society, Records. Manuscripts Division, Research Library, New York Public Library (contains extensive correspondence and other materials pertaining to Buchheister's tenure with the society, 1935–1970).

Rolf Swensen

BUMPUS, HERMON CAREY. Born Buckfield, Maine, 5 May 1862; died Pasadena, California, 21 June 1943. Zoologist, evolutionist, biometrist, natural history educator.

FAMILY AND EDUCATION

Second of three sons born to Laurin Bumpus, cabinetmaker, missionary, and social worker, and Abbie Eaton Bumpus. Ph.B., Brown, 1884; Ph.D., Clark, 1891; Sc.D., Tufts, 1905, Brown, 1905; LL.D., Clark, 1909. Married Lucy Ella Nightingdale, 1886; two sons.

POSITIONS

Assistant in zoology, Brown, 1884–1886; professor of zoology, Olivet College, 1886–1889; fellowship, Clark University, 1889–1890; assistant professor of zoology, 1890–1891, associate professor of biology, 1891–1892, professor of comparative anatomy, 1892–1901; established summer school for biology students, Marine Biological Laboratory, Woods' Hole, 1889. Assistant director, 1893–1895, scientific director, Marine Biological Laboratory, 1895–1900, member Board of Trustees, 1895–1942. Assistant to the president and curator of invertebrate zoology, 1900–1902. Director, American Museum of Natural History, 1902–1911. Member, Faculty of Pure Science, Columbia University, 1905–1911; business manager, University of Wisconsin, 1911–1914. President, Tufts College, 1914–1919. Member, Board of Fellows, 1905–1942, secretary of the corporation, Brown University, 1924–1939.

CAREER

Bumpus evinced an early love of nature and was an eager collector of specimens. When he entered Brown, enjoyed reputation as one "who had shot, skinned, stuffed, and eaten every living animal." Was illustrator of animals as an undergraduate, and some of his pictures appeared in print. His dissertation dealt with embryology of American lobster, and at Wood's Hole, he both energized research program and published a number of papers on various forms of vertebrate and invertebrate sea life. In recognition of his work on scientific study of fisheries, was elected president of Fourth International Fisheries Congress (1908). Pioneer in employing biometry in analysis of biological information and encouraged scholars at Woods' Hole to utilize this methodology. Utilized this approach in his study of various organisms, including snails, mud puppies, and birds. Following a severe storm in Providence in February 1898, assessed morphological traits of sparrows that survived and those that did not. Concluded that those killed in storm demonstrated "process of selective elimination." This classic 1899 study, "The Elimination of the Unfit as Illustrated by the Introduced Sparrow, *Passer domesticus*," appeared in summarized form in many textbooks. At American Museum, stressed that research and teaching both obligatory parts of overall program. Established education department there. Following his retirement from Tufts, promoted outdoor education—known today as environmental education—while chair of National Park Service

Advisory Board (1924–1931). Received several awards for having inspired and popularized trailside museum concept. Consulting director, Buffalo Museum of Science, 1925–1930. President of American Morphological Society, 1902; American Society of Zoologists, 1903; Rhode Island Audubon Society, 1899, 1902; American Association of Museums, 1906. Active in a number of other scientific, civic, and cultural organizations. Published a number of scientific papers, also articles concerning museum education, museum work in national and state parks, and other subjects.

MAJOR CONTRIBUTIONS

Leader in fisheries science and education, museum and parks education, museum administration, and as college president. A pioneer in statistical measurement of biological processes.

BIBLIOGRAPHY

Bumpus, Hermon C., Jr. *Hermon Carey Bumpus, Yankee Naturalist* (1947).
Bajema, C. K., *Dictionary of Scientific Biography*. Supplement 2, vol. 17 (1990).
Science 99 (14 January 1944).
Who Was Who in America.

Keir B. Sterling

BURGESS, THORNTON WALDO. Born Sandwich, Massachusetts, 14 January 1874; died Hampden, Massachusetts, 5 June 1965. Author, naturalist, conservationist.

FAMILY AND EDUCATION

Son of Thornton W. and Caroline Hayward Burgess. Educated at Sandwich High School; also spent term at a business college in Boston. Received honorary Litt.D., Northeastern University, 1938. Married Nina Osborne, 30 June 1905 (died in 1906), one son; Fannie P. Johnson, 30 April 1940 (died in 1950).

POSITIONS

After leaving Boston business school, worked as assistant bookkeeper in Boston shoe factory. In 1895 went to work as office boy and then reporter for Phelps Company, in Springfield, Massachusetts, publisher of *Good Housekeeping* and *Springfield Homestead*. Was reporter and editorial "utility man" for *Homestead*. From 1901 to 1911 was literary and household editor of group of agricultural newspapers published in association with Phelps. Was also editor of *Good Housekeeping*, 1904–1911. In latter year *Good Housekeeping* was sold, and Burgess was discharged. Then left editorial work to become full-time author.

CAREER

First book, *Bride's Primer*, a collection of his pieces from *Good Housekeeping*, appeared in 1905. In 1910 published *Old Mother West Wind*, first of his children's nature books. Between 1910 and 1965 published over eighty books for children, featuring such characters as Peter Rabbit, Sammy Jay, Reddy Fox, and Danny Meadowmouse. From 1912 to 1955, also wrote some 15,000 nature stories, first for Associated Newspapers, a newspaper syndicate, and later for *New York Tribune* (later *New York Herald Tribune*), where his stories appeared six days a week. By the time he retired in 1960, his books had sold over 7.5 million copies, including editions in French, German, Spanish, and Chinese. Was host of radio program, and in 1924 began Burgess Radio Nature League to interest children in wildlife. At one point league had roughly 50,000 members. Recipient of gold medals from National Conservation Society and from Wild Life Protection Fund of New York Zoological Society, for distinguished service to wildlife. His autobiography, *Now I Remember*, was published in 1960.

MAJOR CONTRIBUTIONS

Influenced generations of children with his nature stories. (Over thirty of his books remain in print.) Besides imparting moral lessons, his stories cultivated respect for nature among young readers. Although not fully realistic (Reddy Fox never captures Peter Rabbit or Danny Meadowmouse), Burgess, who did not hesitate to consult authorities like Frank Chapman, filled his stories with accurate information about nature. His animals and birds ate, nested, and raised young like their real-life counterparts.

BIBLIOGRAPHY

Audubon 66 (September–October, 1964): 312–13; 85 (September 1983): 99–101.
Dictionary of American Biography. Supplement 7, 1961–1965 (1981).
New York Times, 6 June 1965 (obit.).
Wright, Wayne W. *Thornton W. Burgess: A Descriptive Book Bibliography* (1979).

Richard Harmond

BURROUGHS, JOHN. Born Roxbury, New York, 3 April 1837; died Ohio (while on a train), 29 March 1921. Essayist, naturalist, literary critic.

FAMILY AND EDUCATION

Son of Chauncey A. and Amy Kelly Burroughs. Raised on a farm, the seventh of ten children. Educated in local schools. Began teaching in rural schools in New York and Illinois. Attended Hedding Literary Institute and Cooperstown Seminary. Received honorary doctorate from Yale University in 1910 and the University of Georgia in 1915, among others. Married Ursula North in 1857; one son.

POSITIONS

Taught in country schools from 1854 to 1863. Clerk and, later, chief in the Organization Division in the Bureau of National Banks, Treasury Department, Washington, D.C., 1864–1872. Bank examiner in New York, 1873–1885. Returned to active farming, 1875.

CAREER

Began publishing philosophical essays in 1856 under the literary influence of Emerson. Began publishing nature essays in 1861 in the *New York Leader*, ''From the Back Country.'' Wrote his famous poem, ''Waiting,'' in 1862. Published *Notes on Walt Whitman as Poet and Person* in 1867. From 1871 to 1922 his literary output, mainly essays on nature, philosophy, science, and literary criticism, consisted of twenty-seven volumes. In addition many of his more popular articles were arranged in special editions during his lifetime and since his death. His lifetime (since 1863) friendship with Walt Whitman was important for both writers. Voyaged to Alaska with John Muir in 1899 and with other naturalists and scientists as a member of the Harriman Expedition. Journeyed to Yellowstone National Park with President Theodore Roosevelt in 1901. Traveled extensively in North America and Britain, observing and writing about natural history subjects. With Theodore Roosevelt, C. H. Merriam, and other writers, was a fierce critic of the early twentieth century ''Nature Faker'' school of nature writing, as exemplified by William Long and several other authors. Later changed his mind about at least one popular writer, Ernest Thompson Seton, after meeting Seton and learning something of his methods of research.

MAJOR CONTRIBUTIONS

John Burroughs was the chief nature interpreter to the American public in his later years and is called by some ''the father of the modern nature essay.'' Had an important influence on President Theodore Roosevelt's conservation policies. He popularized the biological science of his day, including the theory of evolution. Highly regarded as a literary critic, especially of Whitman, Emerson, and Thoreau.

BIBLIOGRAPHY

Barrus, Clara. *The Life and Letters of John Burroughs*. 2 vols. (1925).
Barrus, Clara, ed. *The Heart of Burroughs' Journals* (1928).
Kelley, Elizabeth Burroughs. *John Burroughs, Naturalist* (1959).
Payne, Daniel G. *Voices in the Wilderness: American Nature Writing and Environmental Politics* (1996).
Renehan, Edward. *John Burroughs: An American Naturalist* (1992).
Renehan, Edward. *John Burroughs' Slabsides* (1974).
Westbrook, Perry. *John Burroughs* (1974).
Wiley, Farida, ed. *John Burroughs' America* (1951).

Jim Stapleton

BURT, WILLIAM HENRY. Born Haddam, Kansas, 22 January 1903; died Boulder, Colorado, 4 December 1987. Zoologist, mammalogist.

FAMILY AND EDUCATION

Son of Frank B. and Hattie Carlson Burt. B.A., 1926, teaching fellow, 1926–1927, M.A., 1927, University of Kansas. Doctoral program, University of California, Berkeley, 1927–1930. Research fellow, California Institute of Technology, 1928–1935. Ph.D., University of California, 1930. Married Leona Suzan Galutia, 1928 (died 1973).

POSITIONS

Remained at California Institute of Technology as teaching fellow until 1935. Instructor to professor of zoology, University of Michigan, 1935–1949. Associate curator of mammals, Museum of Zoology, University of Michigan, 1935–1938, curator, 1938–1969, curator emeritus, 1969–1987. Visiting lecturer and honorary curator, University of Colorado Museum, Boulder, Colorado, 1969–1987.

CAREER

Developed early interest in mammals from observing activities of prairie dogs on family farm. Early article in *Journal of Mammalogy*, "A Simple Live Trap for Small Mammals" (1927), led to development of live trap, since then widely used by mammalogists worldwide. Doctoral dissertation, "Adaptive Modifications in the Woodpeckers," published 1930. Taught courses in a wide range of subjects at Michigan; chaired twenty completed doctoral committees and served on doctoral committees of another 100 graduate students. Exacting taskmaster with students, but gave considerable time to them. His "Territorial Behavior and Populations of Some Small Mammals in Southern Michigan" (1940) set forth concepts of territoriality and home range that have undergone little modification since. It was for several decades one of the more widely cited articles on the subject. His *Mammals of Michigan* (1946, revised 1948) was a landmark regional study, extended to cover a larger territory in *Mammals of the Great Lakes Region* (1957, revised by Allen Kurta, 1995). In 1952, Burt published *A Field Guide to the Mammals* (with illustrations by R. P. Grossenheider), which has since been twice revised and is the one book in mammalogy that has sold more copies than any other. His comprehensive "Bacula of North American Mammals" appeared in 1960. Burt also gave some attention to the mammals of Middle America, with papers concerning species to be found in Sonora (1938) and Sonora and Chihuahua (with Emmet J. Hooper, 1941). He also examined the effects of the new Mexican volcano Paricutín on the vertebrates in its vicinity (1961) and completed *The Mammals of El Salvador* (with R. A. Stirton) in 1961. Burt served on the National Research Council from 1963, was a member of the Council of the Society of Systematic Zoology (1957–1960), and was active in a number of organizations concerned with mammals, birds, ecology, wildlife, and related subjects. He was successively secretary (1935–1938), editor of the *Journal of Mammalogy*, (1947–1952), vice president (1951–1953), and president (1953–1955) of the American Society of Mammalogists (ASM). He was named an honorary member of the ASM (1968) and received the H.H.T. Jackson Award (1979) for his long service to the society.

MAJOR CONTRIBUTIONS

Burt's work in various aspects of mammalogy, including home range, territoriality, morphology, behavior, and evolution, were of considerable value. His regional studies (Michigan, the Great Lakes, Sonora, El Salvador) were excellent syntheses, and his *Field Guide to the Mammals* has greatly stimulated mammal observation among the general public.

BIBLIOGRAPHY

Birney, E. C., and J. R. Choate, eds. *Seventy-Five Years of Mammalogy (1919–1994)* 1994.

Muul, Illar. *Journal of Mammalogy* 71 (1990) (obit.).

Karen M. Venturella

C

CAHALANE, VICTOR HARRISON. Born Charlestown, New Hampshire, 17 October 1901; died Dormansville, New York, 6 May 1993. Naturalist, conservationist.

FAMILY AND EDUCATION

Son of David Victor and Elizabeth Harrison Cahalane. B.S. (landscape gardening), Massachusetts Agricultural College, 1924; M.F[orestry]., Yale University, 1927, completed course work for Ph.D., at University of Michigan but left without completing dissertation, 1927–1929. Married Isabelle Porter, 1928; one daughter.

POSITIONS

Field naturalist, Roosevelt Wildlife Forest Experimental Station, Syracuse University, 1926; instructor in School of Forestry and Conservation, University of Michigan, 1927–1929; deer mammal investigator, Michigan Department of Conservation, 1929–1930; director, Cranbrook Institute of Science, Bloomfield Hills, Michigan, 1931–1934; wildlife technician, National Park Service (NPS), 1934–1939, chief of wildlife division, NPS, 1939–1940; in charge of Section on National Park Wildlife, U.S. Fish and Wildlife Service (USF&WS), 1940–1944; chief of Biology Branch, NPS, 1944–1955; collaborator with NPS, 1955–1970; assistant director, New York State Museum, Albany, 1955–1967. President, Wildlife Society (1940); Defenders of Wildlife, 1963–1971. Vice chairman, American Committee for International Wildlife Protection, 1960–1970, historian from 1971. Consultant, National Parks Board of South Africa, 1950–1951, and U.S. National Park Service. Active in a number of professional, conservation, and wildlife advocacy organizations.

CAREER

In the course of his career with the NPS and USF&WS, Cahalane pioneered in the development of a philosophy of habitat protection and management while controlling the populations of ungulates. Moved from NPS to USF&WS in 1940 when the wildlife work he had been doing was transferred to the new agency. Returned to NPS with his fellow wildlife researchers in 1944, but frustrated by continuing lack of support for scientific program, resigned 1955. Spent the next dozen years in Albany as assistant director of the New York State Museum. Subsequently active in a number of conservation advocacy organizations. Through his prolific writing and other activities, he made important contributions to the conservation of wildlife worldwide. His other professional concerns included mammalian life histories and distribution, also life zones of the southwestern United States. Was an able nature photographer. Author of a number of books, including *Fading Trails: The Story of Vanishing American Wildlife* (with others, 1942); *Meeting the Mammals* (1943); *Mammals of North America* (1947); *A Biological Survey of Katmai National Monument* (1959); editor of *National Parks: A World Need* (1962); editor of J. J. Audubon and John Bachman, *The Imperial Collection of Audubon Animals* (1967); editor of *America's Natural Resources* (rev. ed. 1967); editor of Audubon and Bachman, *A Selected Treasury for Sportsmen: Audubon Game Animals* (1968); author (with C. C. Johnson) of *Alive in the Wild* (1969); *Man and Wildlife in Missouri* (with Charles Callison, 1953).

MAJOR CONTRIBUTIONS

Indefatigable researcher on wildlife and conservation issues, principally concerning mammalogy, in the United States, Europe, and Africa. Leader in government and private conservation advocacy process.

BIBLIOGRAPHY

Auk (January 1994) (obit.).

Cahalane, V.H., C. Presnall, and D.B. Beard. "Wildlife Conservation in Areas Administered by the National Park Service, 1930–1939: A Report to the Special Committee of the U.S. Senate on the Conservation of Wildlife Resources," in *The Status of Wildlife in the United States* (1940).

Cahalane, V.H. "Conserving Wildlife in the National Parks," *Atlantic Naturalist,* 16, no. 3 (1961).

Contemporary Authors (permanent ser.). Vol. 2.

Coronet Magazine (July 1955).

Stroud, Richard H., ed. *National Leaders of American Conservation* (1985).

Who's Who in America (1982–1983).

Wright, R. Gerald. *Wildlife Research and Management in the National Parks* (1992).

Keir B. Sterling

CAIN, STANLEY ADAIR. Born Jefferson County, Indiana, 19 June 1902; died Santa Cruz, California, 1 April 1995. Botanist, ecologist.

FAMILY AND EDUCATION

Son of Oliver Ezra and Lillian Florence (Whitsitt) Cain. Received B.S. from Butler University, 1924; M.S., 1927, and Ph.D. (plant ecology), 1930, from University of Chicago. Married Louise Gilbert, 1940; one son.

POSITIONS

Instructor, assistant professor, and associate professor, Butler University, 1925–1931; assistant professor at Indiana University, 1931–1933, and Waterman Institute research associate, 1933–1935. At the University of Tennessee, rose from assistant professor to professor, 1935–1946. In 1940 was chosen Guggenheim Fellow. In 1945–1946, was chief of science section at the U.S. Army University, France, and from 1946 to 1950 was botanist at Cranbrook Institute of Science. Served as professor of conservation and chairman of the Department of Conservation, University of Michigan, 1950–1961. From 1965 to 1968 was assistant secretary of the U.S. Department of the Interior for Fish, Wildlife, and Parks. Subsequently taught at University of Michigan, where he was head of the Institute for Environmental Quality, and after 1977 at the University of California, Santa Cruz.

CAREER

Began his academic career as botanist but broadened his scientific interests well beyond world of plants. Focused increasingly on the relationship between people and environment. As early as 1948 warned of dangers posed by overpopulation and expanding consumption to the world's store of resources. Served as vice president and president of Ecological Society of America; member of advisory board of the Conservation Foundation; and chairman of the panel on environmental biology for National Science Foundation. Was member of Michigan Conservatory and served as its chairman. Also member of advisory board of U.S. Department of Interior and chairman of ad hoc committee of International Biological Program of National Academy of Science (to which he was elected a member in 1970). Secretary and vice president of American Association for the Advancement of Science. Author of more than 100 articles and two books: *Foundations of Plant Geography* (1944) and, with G. M. de O. Castro, *A Manual* of *Vegetation Analysis* (1959).

MAJOR CONTRIBUTIONS

Pioneering conservationist who helped develop the science of ecology. Partly because of his work, conservation became a national concern. Founded Department of Conservation at University of Michigan.

BIBLIOGRAPHY

New York Times, 15 September 1948.
New York Times, 2 April 1995 (obit.).

Stroud, Richard H., ed. *National Leaders of American Conservation* (1985).
Who's Who in America. 40th ed. (1978–1979).

Richard Harmond

CALL, RICHARD ELLSWORTH. Born Brooklyn, New York, 13 May 1856; died New York City, 14 March 1917. Teacher, geologist, malacologist.

FAMILY AND EDUCATION

Son of William A. and Sarah (Ellsworth) Call, who moved to Des Moines when son was young man. After public school, Call graduated from Cazenovia Seminary in 1875; attended Syracuse University but dropped out to teach in country schools. Continued his education at Indiana University, taking his B.A. in 1890, his M.A. in 1891. Took M.D. at Hospital College of Medicine in Louisville in 1893 but never practiced. Honorary Ph.D. from Ohio University, Athens, 1895. Married Ida L. Long, 1884; no children.

POSITIONS

Assistant, U.S. Geological Survey, 1884. Assistant geologist, Arkansas Geological Survey, 1887–1901 (summers). Science teacher, Des Moines High School, 1886–1892. Teacher of physics and chemistry, Manual Training High School, Louisville, 1892–1895. Superintendent of schools, Lawrenceburg, Indiana, 1895–1897. Teacher of physiography, Erasmus Hall High School, Brooklyn, 1897. Curator, Children's Museum, Brooklyn Institute of Arts and Sciences, 1898–1905. Nature study lecturer, Pratt Institute, Brooklyn, 1903–1905. Teacher of biology, DeWitt Clinton High School, New York City, 1906.

CAREER

Considered superb classroom teacher, Call made scientific investigation his avocation. However, his ''usual lack of mental equipoise,'' as Keyes characterized his notorious absent-mindedness, also caused most of his teaching appointments to be of short duration. When secretary of Iowa Academy of Science he neglected to keep any record of meetings; he was habitually confused about the ownership of books—his own and others'.

Call's most substantial book, published as the annual report of the Arkansas Geological Survey, was *The Geology of Crowley's Ridge* (1891). Somewhat slighter, but more influential, was his *Descriptive Illustrated Catalogue of the Mollusca of Indiana* (1900), which appeared in annual report of that state's Department of Geology and Natural Resources. His work with the U.S. Geological Survey (USGS), though of brief duration, led to study of molluscan shells in the ancient Lake Bonneville. This resulted in publication of a sixty-six-page paper in the USGS *Bulletin*, No. 11 (1884), ''On the Quaternary and Recent Mollusca of the Great Basin, with Descriptions of New Forms,'' but also made Call, for a time, a leading authority on loess. Unfortunately, his experience on shores of the Great Salt Lake caused him to conclude that loess is deposited by water rather than by air. With H. C. Hovey he published *Mammoth Cave* (1902), which long remained the standard guidebook to that geological spectacle. He also published descriptive papers in ichthyology and herpetology and series of papers on artesian wells.

His interest in fishes and shells led him to review the work of C. S. Rafinesque, with the result that he published first and so far only book-length biography of that colorful emigrant naturalist, *The Life and Writings of Rafinesque* (1895). This book also was first to try to order the confused Rafinesque bibliography.

MAJOR CONTRIBUTIONS

Remembered today as the first to try to "rehabilitate" the reputation of Rafinesque, Call is also worthy of remembrance for his modest contributions in geology, while his descriptive work on shells is still fundamental.

BIBLIOGRAPHY

Johnson, R. I. "R. Ellsworth Call with a Bibliography of His Work on Mollusks and a Catalogue of His Taxa." *Occasional Papers: Mollusks, Museum of Comparative Zoology* 4 (1975).

Keyes, C. "R. Ellsworth Call and Iowa Geology." *Proceedings of the Iowa Academy of Science* 27 (1920).

Who Was Who in America. Vol. 4 (1961–1968).

Charles Boewe

CALLIN, ERIC MANLEY. Born Whitewood, Saskatchewan, 9 March 1911; died Fort Qu'Appelle, Saskatchewan, 6 November 1985. Accountant, birder.

FAMILY AND EDUCATION

Son of John Callin (1864–1966) of Blyth, Ontario, and Anna Vickberg from Jamtland, Sweden. Educated in Park School and Whitewood High School, Saskatchewan. Married Margaret Electa Fyke in November 1944 (died 19 January 1976); no children.

POSITIONS

Bank clerk, Royal Bank at Punnichy and Battleford, Saskatchewan, bookkeeper for small businesses at Whitewood and Kipling, 1929–1942. Assistant accountant, Fort Qu'Appelle Sanatorium, 1943–1947. Chief accountant, Saskatchewan Anti-Tuberculosis League, Fort Qu'Appelle, 1947–1974.

CAREER

A superb observer with a good ear, Callin kept careful bird records for fifty-nine years, culminating in one of the best regional annotated bird lists in North America. He studied intensively the unidentified "kicker" song in the marsh, which proved to be made by the Virginia rail. Eighteen articles in *Blue Jay* included four firsts for Saskatchewan: yellow-crowned night heron, eastern wood pewee, parula warbler, and orchard oriole, in addition to the first specimen record of the western tanager in the southern third of the province and the farthest-east nesting of the yellow-breasted chat.

MAJOR CONTRIBUTIONS

Notable Saskatchewan birder. Maintained valuable bird records for nearly six decades.

BIBLIOGRAPHY

Brazier, F. H. ''In Memoriam: Eric Manley Callin, 1911–1985.'' *Blue Jay* 44 (1986).

Callin, E. M. ''Vocalization of the Virginia Rail: A Mystery Solved.'' *Blue Jay* 26 (1968).

Callin, E. M. *Birds of the Qu'Appelle, 1857–1979* (1980).

C. Stuart Houston

CALLISON, CHARLES HUGH. Born Lousana, Alberta, Canada, 6 November 1913; died Columbia, Missouri, 23 February 1993. Conservation society executive writer.

FAMILY AND EDUCATION

Son of Guy A. and Dorinda Stuart Callison. Moved with family to Holliday, Missouri, at age five. Naturalized U.S. citizen 1937. B.Jour., 1937, honorary D.Sc., 1979, University of Missouri. Married Amelia D. Ferguson, 1951; four children.

POSITIONS

Editor, *Garnett Review* (Kansas), 1937–1938, *Boonville Advertiser* (Missouri), 1938–1941. Education and information specialist, Missouri Conservation Commission, 1941–1946. Executive secretary, Conservation Federation of Missouri, 1946–1951. Assistant conservation director, 1951–1952, secretary and director of conservation, 1953–1960, National Wildlife Federation. Chairman, Natural Resources Council of America, 1957–1959. Assistant to president, 1960–1966, executive vice president, 1966–1977, National Audubon Society. Founder and director, Public Lands Institute, 1978–1986. Founder and director, Missouri Parks Association, 1982–1991.

CAREER

Combined interests in both conservation and journalism, serving as executive while also editing various societies'/agencies' journals, including *Missouri Conservationist, Missouri Wildlife*, and *Legislative News Service*. Gained recognition as authority on laws affecting natural resources, testifying over 100 times before congressional committees. Appointed by Presidents Eisenhower, Kennedy, Nixon, and Carter to various committees involved in conservation issues.

Wrote widely on conservation subjects, including *Man and Wildlife in Missouri* (1952), *America's Natural Resources* (1957; rev. ed. 1967), *Areas of Critical Environmental Concern on the Public Lands*, 2 vols. (1984–1986), and *Overlooked in America* (1991). From 1961 to 1977, contributed ''National Outlook'' column to *Audubon* magazine. Received Frances K. Hutchinson Medal, Garden Club of America, 1974; Audubon Medal, National Audubon Society, 1978; Conservation Award, Department of the Interior, 1980; Distinguished Service Award, American Forestry Association, 1986; Charles H. Callison Award, Missouri Audubon Council, 1991.

MAJOR CONTRIBUTIONS

Turned National Audubon Society into a significant wildlife advocacy organization with a political presence and built a network of local Audubon groups. Strong and vocal critic of Bureau of Land Management.

BIBLIOGRAPHY

Graham, Frank, Jr. *Audubon* (May–June 1993) (obit.).
New York Times, 26 February 1993 (obit.).
Stroud, R. H., ed. *National Leaders of American Conservation* (1985).
Who's Who in America. 1992–1993 ed. (1992).

John A. Drobnicki

CARR, ARCHIE FAIRLY. Born Mobile, Alabama, 16 June 1909; died Micopony, Florida, 21 May 1987. Taxonomist, evolutionary biologist, educator, author, conservationist.

FAMILY AND EDUCATION

Son of Archibald Fairly and Louise Gordon (Deaderick). His father was a Presbyterian minister, and his mother a teacher of piano. Family moved to Fort Worth, Texas, then to Georgia, where in 1928 he entered Davidson College as an English major. He switched his major to biology and attended University of Florida at Gainesville. He received his bachelor's degree in 1932, his master's degree in 1934, and the University of Florida's first doctorate in biology in 1937. Married Majorie Harris, 1 January 1937; five children.

POSITIONS

Assistant biologist, University of Florida at Gainesville, 1933–1937, fellowship at Harvard's Museum of Comparative Zoology, 1937–1943, instructor of biological sciences at University of Florida, 1938–1940, assistant professor, 1940–1944, associate professor, 1945–1949, professor of biology, 1949–1959, graduate research professor, from 1959. Professor of biology at Escuela Agricola

Panamerica, Honduras, 1945–1949. Biologist, United Fruit Company, Honduras, 1949. Research associate, American Museum of Natural History, from 1949. Associate, Florida State Museum, from 1953. Technical adviser to faculty at University of Costa Rica, 1956–1957. Technical director, Caribbean Conservation Corporation, 1959–1968, Executive vice president, Caribbean Conservation Corporation, from 1968.

CAREER

Following completion of his doctorate at Florida, Carr was offered and accepted a fellowship at Harvard from 1937 to 1943. He then returned to Florida and held a succession of positions within the university as educator and biologist. Coauthored monograph entitled *Antillian Terrapins*. It consisted of descriptions of new turtles in Cuba, Haiti, and the Bahamas. Taught biology for Escuela Agricola Panamerica in Honduras, 1945–1949. Full professor at University of Florida, 1949. Published *The Handbook of Turtles* (1952), classified seventy-nine species and subspecies in the United States, Canada, and Baja, California. William Harecrow invited Carr to join numerous expeditions, Africa, 1952, Panama, Costa Rica, Trinidad, 1953. National Science Foundation's principal investigator of marine turtle migration project on expeditions to Central America, 1955; Brazil, French West Africa, Portugal, and Azores, 1956; Union South Africa, Argentina, Chile, 1958; Africa, Madagascar, 1963; yearly expeditions to Caribbean from 1960. Published *The Windward Road* (1955), won O. Henry Award for "Black Beach" chapter. Grants brought him to Tortuguero in 1955 to study the breeding grounds of the turtles. Invented the "five dollar tag" to tag turtles. Joshua Powers, Ben Phipps, Alberto Gainza, Spruille Braden, Ricardo Castro, James Oliver, and Mrs. Muñoz Marin formed the Brotherhood of the Green Turtle, an organization that funded Carr's projects. It later changed its name to the Caribbean Conservation Corporation. The Office of Naval Research was interested in the turtle's navigation prowess and decided to lend Carr a hand. He had the use of Military Air Transport to relocate the turtles. In 1978 he was an assistant in Green Turtle Expedition of the R.V. Alpha Helix project in Costa Rica.

Published several hundred articles and Time Life Books series on natural history of Africa. Authored ten books from 1937 to 1986, including *The Handbook of Turtles* (1952), *High Jungles and Low* (1952), *The Windward Road* (with Coleman Goin, 1955), *Amphibians and Freshwater Fishes of Florida* (1956), *Guideposts of Animal Navigation* (1962), *The Reptiles* (1963), *Ullendo* (1964), *Africa Wildlife* (1964), *So Excellent a Fishe: A Natural History of Sea Turtles* (1967), *The Everglades* (1973).

Well versed in the English language, as well as Gullah, an African dialect. He was recognized as the foremost authority on turtles. He classified seventy-nine species and subspecies of turtles. His extensive work at Tortuguero and Ascension Island earned him the title of "Turtle Man." Maintained a systemic organized standard when researching turtles in various parts of the world. He would aggressively pursue the opportunity to educate his students, personnel, and local people about turtles and his studies. His peaceful nature and sense of humor helped to break the barrier between the locals and his crew. This enabled him to employ everyone in his efforts of research and conservation.

Carr received numerous awards for his work in the field. Received decoration as officer, Order of the Golden Ark, Netherlands; recipient of the Daniel Giraud Elliot Medal, National Academy of Science; O. Henry Award for best short story (1956); Edward H. Browning Award for outstanding achievement in conservation, Smithsonian Institution (1975); John Burroughs Medal for exemplary nature writing; first annual research award Sigma Xi, University of Florida; Merit Award, Florida Audubon Society (1961); Gold Medal World Wildlife Fund, New York Zoological Society, (1973); Archie F. Carr Medal established in his honor at Florida State Museum (1979); Archie F. Carr postdoctoral fellowship established, University of Florida Department of Zoology (1983); fellow, Linnaean Society of London; Phi Beta Kappa, Sigma Xi.

MAJOR CONTRIBUTIONS

Carr helped dispel the myths and folklore about turtles by classifying seventy-nine species. His extensive studies of migratory, nesting, mating, and nutritional habits of turtles enabled him to locate the optimal areas for turtles to live and breed. His consistent effort for the conservation of turtles helped to increase their population throughout the world.

BIBLIOGRAPHY

Bowen, J. David. ''To Save the Green Turtle'' *Americas* (1960).

Carr, Archie. *The Windward Road* (1955).

Ehrenfield, David. *Copeia 4* (1987).

Graham, Frank, Jr. ''What Matters Most: The Many Worlds of Archie and Majorie Carr'' *Audubon* (1982).

New York Times, 21 May 1987 (obit.).

Who's Who in America with World Notables. Vol. 9 (1985–1989).

Richard Mangeri

CARSON, RACHEL (LOUISE). Born Springdale, Pennsylvania, 27 May 1907; died Silver Spring, Maryland, 14 April 1964. Marine biologist, ecologist, writer.

FAMILY AND EDUCATION

Rachel was the third and youngest child of Robert Warden and Maria (McLean) Carson. Father sold insurance and real estate; the family lived outside Springdale on sixty-five acres, which carried a few farm animals but were not operated commercially. After attending local public schools, she became an English major at Pennsylvania College for Women (later Chatham College) in nearby Pittsburgh. During her junior year became so fascinated by a required course in biology, she changed her major. She graduated magna cum laude in

1929, feeling that, while she lacked the imagination to be a creative writer, biology gave her interesting material to write about. Honorary D.Litt., Pennsylvania College for Women, 1952, Drexel, 1952, Smith, 1953; honorary D.Sc., Oberlin, 1952.

Received a scholarship at Johns Hopkins University, where she pursued experimental zoology under H. S. Jennings and biological statistics under Raymond Pearl. Completed her M.S. in 1932. Never married; following her father's death in 1935 and her sister's death a year later, she provided a home for her mother and two orphaned nieces. In 1957, after the death of one of these nieces, she adopted her five-year-old son, Roger Christie, as well as continuing to care for her aged mother.

POSITIONS

In 1931 she took a teaching position in zoology at the University of Maryland. Also began writing feature articles on biology for the *Baltimore Sunday Sun*. In 1936 joined the U.S. Bureau of Fisheries, the first woman in the agency to hold a scientific rather than a secretarial position. Rose quickly in responsibility to staff biologist and editor of the agency's publications; in 1949 she became editor in chief. Also continued independent writing; in 1951 success of her second book, *The Sea around Us*, gave her sufficient income to resign her position and become freelance writer and to purchase modest piece of property in Maine.

CAREER

Her mother, who had taught school before marriage, had a profound influence on Rachel, developing in her a deep interest in music, books, and nature. She was a solitary, self-sufficient child who loved books and aspired to be a writer. At ten, one of her stories was published in *St. Nicholas Magazine*. Also spent much time outdoors, exploring the nearby woods and streams and identifying plants, insects, and birds.

Although she carried out investigations in marine biology, her principal contributions were as a writer. While in government service, wrote and edited numerous publications, of particular importance being the series ''Conservation in Action,'' a layman's guide to some of the national wildlife refuges. In spare time was establishing herself as a nature writer. At irregular intervals her pieces appeared in *Nature Magazine, Yale Review, Science Digest, Reader's Digest, New Yorker, Life*, and other magazines. First book, *Under the Sea Wind* (1941), received good reviews but was not a publishing success. Her second, *The Sea around Us* (1951), received excellent reviews and was on *New York Times* best-seller list for eighty-six weeks, thirty-nine of them in first place. Its appeal lay in the author's talent for combining careful marine research with sensitive literary style. Reprint of *Under the Sea Wind* now also had good acceptance, as did a third book, *The Edge of the Sea* (1956).

Carson's most influential book, *Silent Spring*, appeared in September 1962. Before the end of the 1950s she recognized, as did many biologists, that a number of wild species were in a state of decline, possibly in response to increasing use of insecticides (like DDT) that had been introduced during World War II. She made a careful study of the problem, sought aid of working biologists, and produced a book that focused attention on, and raised violent controversy between, ecologists and the agribusiness community and its associated agricultural scientists.

Opponents of her position began their attacks as soon as parts of the book were published in June in the *New Yorker*, three months before the completed volume appeared. Once published, the book came under barrage of critical reviews from scientists and industrial firms. Carson was depicted as a popular writer, gifted with words, but a nonscientist who sought to protect birds and fish by forcing farmers to manage without chemical pesticides. However, her major theme was opposition to *uniform* and *irresponsible* use of chemicals, as in mass spray programs set up by government agencies for "extermination" of target pests (gypsy moth, fire ant) without concern for damage to wildlife, streams, and even people on farms and in cities. Her opponents were particularly incensed by the suggestion that a cozy relationship existed between academic scientists and industrial firms that supported their research. The industry charged her with a one-sided advocacy position, while closing its eyes to one-sided, commercially biased presentations that had been commonplace in agricultural, forestry, commercial, and even general circulation periodicals ever since promotion of agricultural chemicals began. They charged her with failure to document her charges—a charge that ignored or distorted the fifty-five page list of her principal sources.

Environmentalists were generally ecstatic about publication of *Silent Spring*. She had said, in forceful, readable prose, what they had been saying ineffectively for a decade. While confessing certain shortcomings in the book, they pushed ahead to stimulate public discussion and action on a problem that had been smoldering for years: introduction of massive amounts of new, frequently nonnatural chemicals into the environment with no attention to possible harm.

In the decade that followed, soundness of her position was fully established. Public attack on use of persistent insecticides was mounting in state and federal agencies and in the courts of law. In the early 1970s, U.S. Environmental Protection Agency placed an almost total ban on the use of DDT in United States; and in years following, similar bans were placed on dieldrin, adrin, and other chlorinated hydrocarbon insecticides. Carson was not around to participate in the success of the movement she had generated, for she died of cancer less than two years after *Silent Spring* was published.

Her last book, published posthumously in 1965, was *A Sense of Wonder*, a nature book for children. Beautifully illustrated, the book was an outgrowth of her experiences with her adopted grandnephew, Roger, at the Maine seashore.

She was usually more comfortable with children, who responded readily to her teachings, than with adults; the book is a fitting memorial.

MAJOR CONTRIBUTIONS

Most notable contribution is *Silent Spring*, which brought public awareness of the environmental damage caused by persistent organic pesticides and generated a corrective movement. Earlier books on natural history of the sea were important in developing a broad interest in the subject. Her grace in the use of language was effective in generating interest in her subjects.

BIBLIOGRAPHY

Brooks, Paul. *The House of Life: Rachel Carson at Work* (1972).

Chemical and Engineering News (20 April 1964) (obit.).

Dictionary of American Biography. Supplement 7 (1981).

Dunlap, Thomas R. *DDT: Scientists, Citizens, and Public Policy* (1981).

Freeman, Martha, ed. *Always Rachel: The Letters of Rachel Carson and Dorothy Freeman, 1952–1964* (1995).

Notable American Women, The Modern Period (1980).

Payne, Daniel G. *Voices in the Wilderness: American Nature Writing and Environmental Politics* (1996).

Whorton, James. *Before Silent Spring* (1974).

UNPUBLISHED SOURCES

Letters, speeches, articles on the public debate over *Silent Spring*, and miscellaneous papers are in the Rachel Carson Papers in the Yale University Library.

Aaron J. Ihde

CARTIER, JACQUES. Born St. Malo, France, 7 June or 23 December 1491; died St. Malo, 1 September 1557. Navigator, explorer, commander of three French expeditions to Canada.

FAMILY AND EDUCATION

Son of Jamet and Geseline Jansart Cartier (some sources give Jean and Guillemette Baudoin Cartier). Married Catherine des Granches, 1520, from an influential family of St. Malo; no children.

POSITIONS

Received commission (not yet located) in 1534 to command an expedition to the northern New World. Commissioned 30 October 1534 by admiral of France to return with three vessels ''to complete the navigation already begun . . . of the lands to be discovered beyond Newfoundland.'' Received a commission (17 October 1540) from Francis I, king of France, as captain and pilot general of another exploratory expedition.

CAREER

Appears to have made voyages to Newfoundland and Brazil before 1532. In 1534 commanded a French expedition to the lands beyond Newfoundland in search of gold, spices, and a western route to Asia; explored the coast of Labrador ("the land God gave to Cain") and the Gulf of St. Lawrence without discovering St. Lawrence River. Drafted a brief relation of the voyage for Francis I, who decided to sponsor another exploration. In 1535 Cartier's three ships explored the St. Lawrence River from Sept-Iles to Montreal Island. Visited Laurentian Iroquoian villages at Stadaconé (Quebec), Achelacy (near Portneuf), and Hochelaga (Montreal). Wintered near Stadaconé. By spring a quarter of the crew had died from scurvy; the rest were saved by a native medicinal remedy made from the bark and leaves of a local conifer (probably *Thuja occidentalis*). Left the St. Lawrence Valley in May after kidnapping several Stadaconans, whose testimony regarding the presence of precious minerals and spices in adjacent lands prodded Francis I to commission the lord of Roberval to colonize the St. Lawrence Valley. Cartier led the first part of the expedition (five ships) in May 1541. Revisited Stadaconé and Hochelaga; established a base at Cap-Rouge, where "Diamonts, and a quantite of Golde ore" (actually quartz and iron pyrites) were discovered. Hostilities with the people of Stadaconé prompted abandonment of the settlement in spring of 1542. Cartier's ships met Roberval's fleet off St. John's, Newfoundland, but ignored Roberval's order to return to Canada and stole away at night to France. This ignominious return, Roberval's failed attempt at colonization (1542–1543), and the dictates of European diplomacy ended French attempts to colonize Canada for over a half century. Versions of Cartier's *relations* were published in Paris in 1545 (first relation); in Italian by Giovanni Battista Ramusio in 1556 (first and second relations); in English (1580); and in French (first relation only, 1598). Cartier's written and oral testimony marked André Thevet's *Singularitez de la France Antarctique* (Paris, 1558). Richard Hakluyt published an account of the third voyage in his *Principall Navigations*, vol. 3 (London, 1600), drawn from an as yet undiscovered manuscript.

MAJOR CONTRIBUTIONS

Made an unprecedented contribution to sixteenth-century European cartography and toponymy of North America. His relations provided the first written descriptions of the geography, flora, and fauna of the Gulf of St. Lawrence and the St. Lawrence Valley. Named forty plants and approximately sixty animals, almost always through association with similar European species. This predilection for the familiar was matched by a relative absence of the marvelous: Cartier inferred from native testimony the existence of unipeds "and other marvels too long to relate" (for which he was mocked by Rabelais) but did not claim to have seen these himself. Described a handful of unfamiliar species:

maize, tobacco, the walrus, and beluga. His impressions of native peoples were cast in the Renaissance mold of the "wild man" but nevertheless provide unique ethnographic data for the period.

BIBLIOGRAPHY
Bideaux, M., ed. *Jacques Cartier. Relations* (1990).
Biggar, H. P., ed. *The Voyages of Jacques Cartier* (1924).
Biggar, H. P., ed. *A Collection of Documents relating to Jacques Cartier and the Sieur de Roberval* (1930).
Gagnon, F.-M., and D. Petel. *Hommes effarables et bestes sauvaiges. Images du Nouveau-Monde d'après les voyages de Jacques Cartier* (1986).
Rousseau, J. "La botanique canadienne à l'époque de Jacques Cartier." *Annales de l'ACFAS* (1937).
Rousseau, J. "L'annedda ou l'arbre de vie." *Revue d'histoire de l'Amérique française* (1954).
Trudel, M. "Jacques Cartier." *Dictionary of Canadian Biography*. Vol. 1 (1966).
UNPUBLISHED SOURCES
All important manuscript sources have been published; see Bideaux (1990) and Biggar (1924, 1930).

Peter L. Cook

CARTWRIGHT, BERTRAM WILLIAM. Born Malvern, Worcestershire, England, 15 September 1890; died Richmond, British Columbia, 16 June 1967. Printer, naturalist.

FAMILY AND EDUCATION

Son of Thomas William Cartwright and Emily Florence Wheeler. Attended school at Bolton, Lancashire. Declined a scholarship to the University of Manchester to emigrate to western Canada in 1911. Married Hilda Walker in 1915; three children.

POSITIONS

Employed in Winnipeg by Bishop Printing, then became a partner in Gregory-Cartwright, a stationery and office supply business.

CAREER

Member of the Bolton Field-Naturalists Club, England, 1907–1911. With A. G. Lawrence and H. M. Speechly, founded the Manitoba Natural History Society in Winnipeg in 1921. Author of weekly nature column, "Wild Wings," in *Winnipeg Tribune*, 1928–1950. Chief naturalist for Ducks Unlimited, 1938–1960 (based in Winnipeg, covering chiefly the prairie provinces).

MAJOR CONTRIBUTIONS

Developed methods for aerial censusing of waterfowl, subsequently refined and carried on by governments. Fostered interest of western farmers in habitat, chiefly wetlands conservation, by enrolling them as unpaid "kee-man" observers to report on duck numbers and marsh water levels. These individuals, members of Ducks Unlimited, contributed yearly or twice yearly observations, and occasionally mentioned marshes worth saving.

BIBLIOGRAPHY

Cartwright, B. W., T. M. Shortt, and R. D. Harris. *Baird's Sparrow* (1937).
Shortt, A. H., and B. W. Cartwright. *Know Your Ducks and Geese* (1948).

C. Stuart Houston

CARVER, GEORGE WASHINGTON. Born on farm near Diamond Grove, Missouri, c.1861 (some sources give 1864 or 1865); died Tuskegee, Alabama, 5 January 1943. Botanist, mycologist, agricultural experimentalist.

FAMILY AND EDUCATION

Last of three children born to slave parents. Father killed in accident when his son was an infant. Shortly thereafter, Carver, his mother, Mary, and only sister kidnapped by raiders and taken to Arkansas. Carver returned to owner, Moses Carver, by bushwhacker in return for $300 horse and raised by Carver's wife, Susan. Left Carver family at age fourteen to seek formal education. Attended grade schools in several Missouri and Kansas communities. Completed high school in Minneapolis, Kansas. Denied admission to Highland College in northeastern Kansas because of his race, 1885. Admitted to Simpson College, Indianola, Iowa, 1890; studied art, music, and natural sciences. Transferred to Iowa State College, 1891. B.S. (1894) and M.S. (1896) in agriculture. Honorary D.Sc., Simpson College, 1928, University of Rochester, 1941. Never married.

POSITIONS

While pursuing his M.S. from 1894 to 1896, worked as an assistant botanist to L. H. Pammell, botanist on Iowa State faculty focusing on mycology (study of fungi). In charge of college greenhouse and carried on work in propagation and cross-fertilization of plants. Invited to teach and head new agriculture program and Experimental Institute at Tuskegee Institute (first all-black agricultural experiment station) by Booker T. Washington, 1896. Greatly preferred experimental work and gave most of his attention to it from 1910 until his death. Appointed collaborator, Mycology and Plant Disease Survey, Bureau of Plant Industry, U.S. Department of Agriculture, 1935. Tested soils and fertilizers in effort to change one-crop system employed by southern black farmers.

CAREER

As graduate assistant at Iowa State, Carver met and was influenced by two faculty members and future secretaries of agriculture: James Wilson and Henry C. Wallace. Carver's experimental work won him high praise from Wilson. In area of mycology, Carver discovered new species of fungi, collected over 200,000 specimens, and identified those types that caused diseases in plants. Published list of newly discovered Cerocosporae, "Some Cerocosporae of Macon County Alabama" (1899). Prior research on *C. cannescens metaspheria* on leaves of frog plant led Carver to discover another fungus on soybeans. This was named *Taphrina carveri* in his honor (1901). Published some forty-four bulletins concerning cultivation and utilization of crops and other agricultural issues between 1898 and 1943. Stressed crop diversification; urged planting of peanuts, sweet potatoes, cowpeas, and similar products in place of cotton. Suggested early cultivation and harvesting as one means of preventing crop loss. His bulletin on cowpeas (1903) indicated that this legume was useful as animal feed but also replenished nitrogen in soil. Wrote on the many uses of sweet potatoes in rubber making, dyes, and soil replenishment. Published *How to Build Up Worn Out Soils* (1905). Spent much of his career extolling peanut as another alternative in soil enrichment and developed hundreds of other uses for it, some of which had commercial potential. His Bulletin 31, *How to Grow the Peanut and 105 Ways of Preparing It for Human Consumption*, was published in 1916. Most of his publications were aimed at consumers, rather than scientists. Elected a fellow of the Royal Society of Arts of London, 1916. Presented the Spingarn Medal for his work in agricultural chemistry, 1923. His other awards included one from the Theodore Roosevelt Association in 1939.

MAJOR CONTRIBUTIONS

Applied findings in mycology to efforts to improve crop yield of southern farmers. By identifying damaging fungi, Carver was able to suggest corrective measures that limited crop loss. Through promotion of peanuts, sweet potatoes, and cowpeas, demonstrated how farmers might shift from their emphasis on cotton or tobacco, which depleted soil. The soil-enhancing crops that he recommended not only increased yields but served as source of needed cash for poor farmers.

BIBLIOGRAPHY

Adair, Gene. *George Washington Carver* (1989).
Dictionary of American Biography.
Dictionary of American Negro Biography (1982).
Holt, Rackham. *George Washington Carver* (1943).
Journal of Southern History 42 (November 1976).
Kremer, Gary R., ed. *George Washington Carver in His Own Words* (1987).
Mackintosh, Barry. "George Washington Carver and the Peanut." *American Heritage* (August 1977).

McMurry, Linda O. *George Washington Carver: Scientist and Symbol* (1981).
Whitman, Alden. *American Reformers* (1985).
UNPUBLISHED SOURCE MATERIALS
See personal papers at Tuskegee Institute.

Cherie Parsons

CASSELS, ELSIE (ELIZA MCALISTER). Born at Megget, Scotland, 16 February 1864; died Red Deer, Alberta, 12 November 1938. Naturalist, ornithologist, conservationist, popularizer of science.

FAMILY AND EDUCATION

Daughter of Janet Reid and Archibald McAlister, schoolmaster. Interested in nature from an early age. Particularly fascinated by birds. Married William Anderson Cassels in Scotland, in April 1889, then immigrated to Canada the same year. Homesteaded on the Albertan prairie. No children.

POSITIONS

Vice president of the Alberta Natural History Society (ANHS) 1917–1924. Member 1906–1926 (the society was inactive 1926–1937). One of several game officers for the Canada Bird Protection Service of the Dominion Parks Branch, Ottawa (1921). One of first active women to hold an official position in a Canadian natural history society.

CAREER

Lived and carried out fieldwork on flora and fauna in Red Deer and Sylvan Lake regions, Alberta. Friend and correspondent with William Rowan and Percy A. Taverner and respected amateur naturalists/ornithologists Frank Farley, Tom Randall, and Charles Snell. In conjunction with the ANHS, helped establish Purple Martin colonies at Sylvan Lake, campaigned from 1906 (unsuccessfully) for the creation of a provincial park to encompass the Red Deer River Canyon, and formulated plans for the Gaetz Lake Sanctuary in Red Deer (known as the Red Deer Bird Sanctuary), which became a Dominion Wildlife Refuge in 1924. Studied the pigmy nuthatch. Contributed ornithological notes to *Canadian Field-Naturalist* (1920–1935) on the Hudsonian chickadee, Jaeger, White gryfalcon, Rufous hummingbird, and Red-breasted nuthatch. Wrote articles and lectured about birds for naturalist societies and the general public. Kept a field journal for many years, including the migratory bird account for the ANHS. Strongly opposed to shooting wildlife for the purpose of collection or identification. A sensitive and energetic woman who lived an almost penniless existence her entire adult life. Those who knew her well called her "a woman of charm and spirit."

MAJOR CONTRIBUTIONS

One of the earliest women to work as a game officer for the Canadian Bird Protection Service of the Dominion Parks Branch, Ottawa (1921). Promoted conservation through education. Popularizer of science. A keen observer of birds whose expertise was often sought and freely given.

BIBLIOGRAPHY

The Canadian Field-Naturalist. Vol. 53 (February 1939) (obit.).

Cassels, Mrs. W. A. "Birds of Alberta." *Red Deer Advocate* (15 July 1921).

UNPUBLISHED SOURCES

See Elsie Cassels and William Rowan, selected correspondence 1922–1934, Rowan Papers, University of Alberta Archives; Elsie Cassels and Percy A. Taverner, selected correspondence 1920–1921, Taverner Papers, Canadian Museum of Nature Archives, Ottawa; E. Tina Crossfield, "A Natural Adaptation: Eliza McAlister Cassels (1863–1938), Scottish Immigrant and Naturalist on the Albertan Prairie" (1995).

E. Tina Crossfield

CASSIN, JOHN. Born Upper Providence Township. Delaware County, Pennsylvania, 6 September 1813; died Philadelphia, 10 January 1869. Ornithologist, author.

FAMILY AND EDUCATION

Born on a farm in Upper Providence Township, Delaware County, Pennsylvania, about a mile from the present town of Media. Son of Thomas Cassin, farmer, and Rachel (Sharpless) Cassin, grandson of Luke and Ann (Worrell) Cassin, and great-grandson of Joseph Cassin, who emigrated from Queen's County, Ireland, in 1724 and settled in Philadelphia. Family was Quaker, and Cassin was educated at a Friend's school in Westtown, Pennsylvania, and under private tutors. Was married and had several children.

POSITIONS

Engaged in mercantile pursuits in Philadelphia and was employed in the U.S. Custom House. Upon death of J. T. Bowen, then principal engraver and lithographer in Philadelphia, took over management of business and continued it throughout his life. An ongoing interest in the study of ornithology led him to join the Academy of Natural Sciences of Philadelphia in 1842 and begin to study the society's collection of birds. Served on Philadelphia City Council. Vice president of Academy of Natural Sciences, Philadelphia, and curator 1842–1869. Army enlisted man during Civil War, much of which he spent in Confederate prisoner-of-war camp.

CAREER

Cassin had early interests in ornithology from the natural history emphasis in the Quaker school he attended. During the period 1846–1850, the assemblage of a large collection of 26,000 birds and a related library by Thomas B. Wilson in the Academy of Natural Sciences in Philadelphia led Cassin to assume responsibility for the collections' proper arrangement and identification. Work on the Wilson collection and resultant publications established Cassin as leading American ornithologist of the time. Emphasized physical description and classification; dealt with problems of synonymy and nomenclature. Authored revision of Titian Peale's controversial account of the mammals and birds of the Wilkes Expedition. Effects of his long captivity in Confederate prison during Civil War shortened Cassin's life.

MAJOR CONTRIBUTIONS

Published descriptions of 194 new bird species, mainly from West African specimens collected by P. B. Du Chaillu. Most papers appeared in *Proceedings of Academy of Natural Sciences, Philadelphia*; also prepared ornithological sections for reports of several government-sponsored expeditions, including Wilkes Expedition and Pacific Railroad Survey, and of expeditions to China seas, Japan, and Southern Hemisphere. Ornithological work for government surveys included his *Mammalogy and Ornithology*, 2d ed., vol. 8 of U.S. Exploring Expedition report (Philadelphia, 1858); contributions to the Report of Pacific Railroad Survey, vol. 9, pt. 2 (*Senate Executive Document no. 78* and *House Executive Documents no. 91*, 33d Congress, 2d session); and *Illustrated Birds of California, Texas, Oregon, British and Russian America* (Philadelphia, 1856), covering the species discovered since the appearance of Audubon's *Birds of America* (1827–1838).

BIBLIOGRAPHY

Dictionary of American Biography. Vol. 3 (1929).

Elliott, Clark A., ed. *Biographical Dictionary of American Science: The Seventeenth through the Nineteenth Centuries* (1979).

National Cyclopaedia of American Biography. Vol. 22 (1967).

Stone, Witmer. "John Cassin." *Cassinia: Proceedings of the Delaware Valley Ornithological Club* 5 (1901).

Connie Thorsen

CATESBY, MARK. Born supposedly at Sudbury in Suffolk, England, c.1679 (some authorities cite 1682 as year of birth); died London, England, 23 December 1749. Naturalist, botanist, ornithologist.

FAMILY AND EDUCATION

Little is known. In his *Natural History* told of move to London to improve access to a scientific education. Indicated that he had had some formal training; also mentions his sister was married to secretary of the colony of Virginia. Both factors indicate a middle-class family background.

POSITIONS

In 1720 commissioned by Royal Society of London for Improving Natural Knowledge to observe and provide samples of plants and species found in America that would be useful to the society.

CAREER

Sailed to America in 1712; lived with sister and brother-in-law. Met William Byrd II and other leaders in Virginia. Spent next seven years in Virginia mountains and tidewater regions studying plants and their habitats. Also traveled to Bermuda and Jamaica. Returned to England in 1719 with specimens. Provided a box of plants to Samuel Dale, an important naturalist. In 1720 at behest of Royal Society returned to America for further study. As William Goetzmann has pointed out, Catesby may have been America's first ecologist.

Spent the next three years traveling in the Carolinas, Georgia, and Florida, taking careful notes. Was particularly interested in the impact of the seasons on plant and animal life. Recorded precise descriptions of the plants, animals, birds, fishes, and Indians that he observed. Sent back to England many specimens new to members of the Royal Society. In 1725 in the Bahamas studied its fish life. Wished to publish a natural history of North America upon his return to England but could not raise the money.

With the help of the French artisan Joseph Goupey, Catesby learned engraving. Later published in English and French *A Natural History of Carolina, Florida, and the Bahama Island*, vol. 1 (1731), vol. 2 (1743), and an Appendix (1748). Elected a member of the Royal Society in 1733. In the following year *Philosophical Transactions* published his paper repudiating the notion that birds hibernate underwater. Also wrote *Hortus Britanno-Americanus*; *or, A Currious* [*sic*] *Collection of Trees and Shrubs*, illustrated and published posthumously, 1763–1767.

MAJOR CONTRIBUTIONS

Recognized and respected as a man of accomplishments who was able, modest, and upright and whose volumes presented the most impressive illustrations of eighteenth-century America then in existence. Demonstrated that the Americas were lands of plenty. Showed the animals, trees, and flowers in all their splendor. He can be adulated as America's first ecologist and one having the distinction of having a frog named after him.

BIBLIOGRAPHY

Darling, William, ed. *Memorials of John Bartram and Humphrey Marshall, with Notices of Their Botanical Contemporaries* (1849).

Elman, Robert. *First in the Field* (1977).

Frick, George Frederick, and Raymond Phineas Stearns. *Mark Catesby: The Colonial Audubon* (1961).
Goetzmann, William H. *New Lands, New Men: America and the Second Great Age of Discovery* (1986).
Kastner, Joseph. *A Species of Eternity* (1977).
McAtee, W. L. ''Mark Catesby.'' *Nature Magazine* 52 (1959).
McBurney, Henrietta. *Mark Catesby's Natural History of America: The Watercolors from the Royal Library Windsor Castle* (1997).
Park, Nellie Wolcott. ''Catesby—Man of Many Talents.'' *Antiques* 61 (1952).
Stearns, R. P. *Science in the British Colonies of America* (1970).
UNPUBLISHED SOURCES
See Mark Catesby Papers, Royal Society of London and the American Philosophical Society (Philadelphia); Elsa G. Allen Papers, Cornell University Library (contains number of research notes on Catesby, part of a study of early ornithologists).

Thomas J. Curran

CHAMPLAIN, SAMUEL de. Born Brouage, France, c.1570; died Quebec, 25 December 1635. Geographer, navigator, draftsman, explorer, colonizer.

FAMILY AND EDUCATION

Champlain's father, Antoine, and an uncle were sea captains. His mother was Marguerite LeRoy Champlain. Before 1599 he had been schooled in navigation, draftsmanship, and cartography. Married Hélène Boullé, daughter of a secretary of the King's Chamber, in 1610; no children.

POSITIONS

Served in the army of Henri IV, king of France, until 1598. Named lieutenant to a succession of lieutenant generals (1608–1612) and viceroys (1612–1627) of New France. From 1627 to his death, commanded in New France ''in the absence of my Lord the Cardinal de Richelieu.'' Member of the Compagnie des Cent-Associés.

CAREER

Traveled to Spain and thence to the West Indies and Mexico, 1599–1601. Drafted an illustrated account of this voyage (the *Brief discours*). Received a royal pension in 1601. Accompanied a 1603 fur-trading expedition to the St. Lawrence Valley as observer; his *Des Sauvages* (Paris, 1603) described the region from Tadoussac to Montreal, the customs of the native peoples, and the latter's reports of western waterways. From 1604 to 1607, at the invitation of the lieutenant general of New France, explored and mapped Atlantic Coast from LaHave, Nova Scotia, to Martha's Vineyard, wintering on Dochet Island (St. Croix River) and at Annapolis Royal (Nova Scotia). Commanded an expedition (1608) to the St. Lawrence Valley, where he established a habitation at Quebec on behalf of a fur-trading company. Over the winter, sixteen of twenty-five men died of scurvy; although aware that a native medicinal remedy had saved Jacques Cartier in similar circumstances, Champlain was unable to locate the prescribed ingredients. Between 1609 and 1615, Algonkian and Huron allies of the French allowed Champlain to explore interior waterways (Ottawa, Richelieu, and Trent Rivers; Georgian Bay) in return for aid against their enemies. Champlain published accounts of these voyages in 1613 and 1619. After submitting a program of colonization to Louis XIII, king of France, in 1618, Champlain concentrated on establishing the colony at Quebec. In France during the English occupation of Quebec (1629–1632), published the *Voyages de la Nouvelle-France*, dedicated to Richelieu, along with a *Traitté de la marine* (Paris, 1632). Following restitution of the colony to France, sent explorer Jean Nicollet to the Great Lakes and established a new habitation at Trois-Rivières.

MAJOR CONTRIBUTIONS

Produced outstanding maps of northeastern North America (especially large map of 1632, described by one historian as "long unequalled in 17th-century cartography"), which incorporated native geographical knowledge. Was primarily interested in the economic potential of flora, fauna, and minerals; a quarter of the *Brief discours* manuscript is devoted to description of useful plants. His drawings of plants and cartographic details exhibit the influence of a new empiricism in botanical classification. Probably brought to France many of the Canadian plants described in the *Canadensium plantarum historia* (Paris, 1635) of Jacques Cornut. His colonization program of 1618 reveals how seventeenth-century European eyes viewed the natural resources of eastern North America; his descriptions of Algonkian and Iroquoian societies provide valuable information on how natives utilized the same resources.

BIBLIOGRAPHY

Biggar, H. P., ed. *The Works of Samuel de Champlain*. 6 vols. (1922–1936).

Doyon, P.-S. "L'iconographie botanique en Amérique française." Diss., Université de Montréal, 1993.

Heidenreich, C. E. "Explorations and Mapping of Samuel de Champlain, 1603–1632." *Cartographica*, no. 17 (1976).

Le Blant R., and R. Baudry, eds. *Nouveaux documents sur Champlain et son époque*. Vol. 1: *1560–1622* (1967).

Lescarbot, M. *History of New France*. 3 vols. (1907–1914; orig. pub. Paris, 1609).

Rousseau, J. "Samuel de Champlain, botaniste mexicain et antillais." *Cahiers des Dix* (1951).

Trudel, M. *Histoire de la Nouvelle France*. Vols. 1–3 (1963–1979).

Trudel, M. "Samuel de Champlain." *Dictionary of Canadian Biography*. Vol. 1 (1966).

UNPUBLISHED SOURCES

The most important source material relating to Champlain has been published; see Biggar (1922–1936) and Le Blant and Baudry (1967).

Peter L. Cook

CHAPMAN, FRANK MICHLER. Born Englewood Township, New Jersey, 12 June 1864; died New York City, 15 November 1945. Ornithologist.

FAMILY AND EDUCATION

Son of Lebbeus and Mary Augusta (Parker) Chapman. His only formal education consisted of ten years at Englewood (New Jersey) academy, graduating in 1880. He married Fannie (Bates) Embury in 1898. A widow with four children, her union with Chapman produced one son.

POSITIONS

Starting in 1887 as a part-time cataloger (classifying his own as well as other collections) for the American Museum of Natural History, Chapman began an association with the museum that would last until his retirement in 1942. In 1888 became a full-time staff member as assistant to Joel A. Allen, who was the department head of the museum's Mammalogy and Ornithology Division. In 1901 became associate curator of that same department. In 1908 was named curator of birds. From 1920 until his retirement he held the position of chairman for the Department of Birds. In 1888 Chapman was elected as a fellow of the American Ornithologists' Union, eventually serving as its president in 1911. Also served as president of Linnaean Society of New York, receiving in 1912 that society's first medal.

CAREER

Chapman is most popularly known as one of the originators of "habitat groups." This method of exhibition displayed the mounted birds in poses that they would normally strike in their natural environs. Further, the entire display case would be a re-creation of each animal's natural habitat. This technique is still the norm of museums worldwide. Even before the "habitat group" innovation, Chapman, in effort to make the exhibits more appealing to the public, created separate displays for birds that could be found in and around New York City and other geographic areas.

In 1899 he founded and was editor until 1934 of *Bird-Lore* (now *Audubon Magazine*), "a magazine for the study and protection of birds." He was very instrumental in the battle against the practice of destroying millions of birds for the sole purpose of decorating women's hats.

Chapman made several expeditions to South America, discovering many new species. His essays classifying the birds by ecological association are considered pioneering in that they advanced "a bold new hypothesis on the history of South American birds." Chapman's research established altitudinal boundaries in explaining the vast differences that existed in bird populations in different geographical areas. The research concentrated on how fauna in higher altitudes evolved and sought their connection with fauna in lower altitudes.

In addition to his numerous technical papers Chapman published about 200 popular articles. Of the eighteen books he wrote, some of the more important included *Handbook of Birds of Eastern North America* (1895), *The Distribution of Bird Life in Colombia* (1917), *and The Distribution of Bird Life in Ecuador* (1926). Chapman also published two autobiographies: *Autobiography of a Bird-Lover* (1933) and *Life in an Air Castle* (1938).

MAJOR CONTRIBUTIONS

Chapman was looked upon as pioneer in biogeography. He was a leader in the conservation movement in the United States, helping to push through protective legislation and creating bird sanctuaries. His efforts to show the bird as a living entity in its own environment played major role in popularizing the study of birds. He developed the staff and collection of the Department of Birds at the American Museum of Natural History. He was one of the major contributors in placing America in the forefront of ornithology.

BIBLIOGRAPHY

Chapman, F. M. *Autobiography of a Bird Lover* (1933).

Dictionary of American Biography. Supplement 3 (1941–1945).

Dictionary of Scientific Biography. Vol. 17, supplement 2 (1990).

Gregory, W. K., "Frank Michler Chapman, 1864–1945." *National Academy of Sciences, Biographical Memoirs* 25 (1948).

Murphy, R. C. *Auk* (July 1950) (obit.).

New York Times, 17 November 1945 (obit.).

UNPUBLISHED SOURCES

Primary source materials in the form of notebooks and letters can be found in the Department of Ornithology of the American Museum of Natural History, New York City.

Rodney Marve

CLARK, WILLIAM. Born Caroline County Virginia, 1 August 1770; died St. Louis, Missouri, 1 September 1838. Explorer, army officer, territorial governor.

FAMILY AND EDUCATION

Ninth child of John and Ann Rogers Clark. His elder brother was General George Rogers Clark. Grew up on family plantation. Family moved to plantation near Louisville, Kentucky, 1785. Had very little formal education. Married Julia Hancock, 1808 (died 1820); four children. Married Harriet Kennerly, widowed cousin of first wife, 1821; one son.

POSITIONS

Militiaman at intervals from either 1786 or 1789. Commissioned lieutenant of infantry, 4th Sub Legion, served under Gen. Anthony Wayne for most of four years, including Battle of Fallen Timbers (August 1794). Resigned commission, 1796. Returned to family plantation near Louisville, Kentucky, attended to family business affairs, and, after 1799, ran plantation left to him by deceased parents. Joined Meriwether Lewis on latter's invitation as co-captain of expedition to West, 1803–1806. Resigned from army 1807, appointed brigadier general of militia and superintendent of Indian affairs, Missouri Territory, 1807–1813. Appointed governor of Missouri Territory, 1813–1821. U.S. superintendent for Indian affairs, St. Louis, 1821–1838. Surveyor general for Illinois, 1824–1825. Laid out town of Paducah, Kentucky, 1828.

CAREER

Lived life of well-to-do plantation owner's son in Virginia and later in Kentucky. Probably active as militiaman from age sixteen, but certainly from age nineteen. Entered regular army in 1792, distinguished himself in several campaigns, but not promoted. During seven years of private life (1796–1803), attempted to straighten out tangled business affairs of elder brother Gen. George Rogers Clark (1752–1818) while operating family plantation, which had become his after both parents died (1799). Invited to serve as cocaptain of Lewis and Clark Expedition, but President Jefferson was unable to secure higher substantive rank for him than 1st lieutenant. Was the more experienced frontiersman, maintained own diary (as did Lewis), and was cartographer and artist for expedition. Clark also had responsibility for preparing expedition diaries for publication following Lewis's death in 1809, and these appeared in edited form in 1814. The edition of 1893, prepared by Elliott Coues, had invaluable notes. Complete diaries did not appear until 1904–1905 (ed. R. G. Thwaites). Clark spent most of the last three decades of his life in St. Louis. Felt he must decline President Jefferson's invitation to accept governorship of Louisiana (later Missouri) Territory in succession to Lewis, 1809, but accepted when offer was again made by President Madison in 1813. Held that office for eight years; undertook military expedition into upper Northwest Territories in 1814. First to raise American flag in what is now Wisconsin. Had responsibilities for Indian affairs in Missouri, 1807–1813 and 1821–1838. Successfully negotiated treaties with Indians involving territorial and other concessions to U.S. government at close of War of 1812. Also dealt with disputes between tribes and spent most of his final years strongly defending their interests to authorities in Washington, D.C. Indian relics that he displayed in a home museum in St. Louis were largely lost after his death.

MAJOR CONTRIBUTIONS

Clark was notable for the major role he played in success of Lewis and Clark Expedition and in overseeing the publication of edited journals of the expedition leaders. Respected by both whites and Indians, he had some success as intervenor in Indian disputes and as defender of Indian interests. Expedition did not find overland water connection between Missouri and Columbia Rivers, and was considered by some contemporaries a failure for that reason.

BIBLIOGRAPHY

Ambrose, Stephen A. *Undaunted Courage: Meriwether Lewis, Thomas Jefferson, and the Opening of the American West* (1996).

Allen, John L. *Passage Through the Garden: Lewis and Clark and the Image of the American Northwest* (1975).

Botkin, Daniel. *Our Natural History: The Lessons of Lewis and Clark* (1996).

Burroughs, Raymond D. *Exploration Unlimited: The Story of the Lewis and Clark Expedition* (1953).

Coues, Elliott, ed. *History of the Expedition under the Command of Lewis and Clark*, 4 vols. (1893).

Cutright, Paul R. *Lewis and Clark: Pioneering Naturalists* (1969).

Cutright, Paul R. *A History of the Lewis and Clark Journals* (1976).

Ewan, Joseph, and Nesta Dunn Ewan. *Biographical Dictionary of Rocky Mountain Naturalists* (1981).

Goetzmann, William H. *Army Exploration in the American West, 1803–1863* (1959).

Moulton, Gary E., ed. *The Journal of the Lewis and Clark Expeditions*. 8 vols. to date (1983–).

Osgood, Ernest. *The Field Notes of Lewis and Clark, 1803–1805* (1964).

Thwaites, G. "William Clark: Soldier, Explorer, Statesman." *Missouri Historical Society Publications* 2. 7 (1906).

Wheat, Carl I. *Mapping the Transmississippi West*. Vol. 2. *From Lewis and Clark to Fremont, 1804–1845* (1958).

UNPUBLISHED SOURCES

See original manuscript journals, American Philosophical Society, Philadelphia; Lewis and Clark Papers, National Archives and Records Administration, Washington, D.C.; Clark's field notes and manuscript maps, Beinecke Library, Yale University; some of Clark's journals and other papers, Missouri Historical Society, St. Louis.

Keir B. Sterling

CLAYTON, JOHN. Born Fulham, England, fall of 1694; died Gloucester County, Virginia, first week of January 1774 (*Dictionary of National Biography* gives dates of 1693–1773; *Dictionary of American Biography*, c.1685–15 December 1773). Botanist.

FAMILY AND EDUCATION

Son and grandson of barristers. His father, John, came to Virginia c.1705 and was appointed secretary to the lieutenant governor and subsequently (from 1713) attorney general for the Virginia colony while also holding several other posts. His mother was Lucy Clayton. The younger John Clayton may have attended Eton and Cambridge University. Married Elizabeth Whiting, 1723; eight children.

POSITIONS

Appointed assistant to the clerk of Gloster (Gloucester) County, Virginia, sometime after 1720 (some authorities suggest 1715). Appointed first clerk of Gloucester County, 1722. Held this position for half a century until his death. Independent plant collector.

CAREER

Came to Virginia about 1720 (though some authorities suggest he may have arrived as early as 1705, then returned to England for further schooling). He may have had formal training in law (or possibly medicine). He became a prosperous planter and landowner (his estate was called Windsor) and spent much of his time as a plant collector while creating his own botanical garden. Both Clayton and his father were good friends of William Byrd II. Corresponded with European and colonial botanists, including Mark Catesby, Linnaeus, John Frederick Gronovius, and Alexander Garden, and was also in touch with Peter Collinson and Pehr (Peter) Kalm. Sent seeds and plants to John Bartram and, from c.1729, to various European correspondents. Most of Clayton's collecting was confined to a modest-sized region south of the Rappahannock and north of the James Rivers and west to the Blue Ridge. Papers based on his botanical observations and other subjects were published by the Royal Society in its *Philosophical Transactions*. His plant collections formed the basis for his ''Catalogue of Plants, Fruits, and Trees Native to Virginia,'' completed in the late 1730s. In it, Clayton used the classification scheme developed by John Ray. He sent this to Gronovius, who revised it in line with the Linnean system and published it under his own name as *Flora Virginica Exhibens Plantas Quas V. C. Johannes Clayton in Virginia Observarit atque Collegit* (with some assistance from Linnaeus) in two parts (Leiden, 1739, 1743). The work covered some 900 species. Clayton seems not to have been consulted about this arrangement. A second edition, edited by Gronovius but seen through the publication process by Gronovius's son Laurans, was published in 1762. Clayton had prepared his own, much superior second edition in 1757, which was sent to Collinson but was not printed because the younger Gronovius completed his work first. A copy of Clayton's manuscript was preserved in England. Clayton traveled in parts of Canada in 1746 and in the region west of the Appalachians in 1748. Many of the plants he collected are held in British and Swedish herbaria. Several other handwritten volumes by Clayton, together with drawings and plants, were destroyed toward the close of the Revolution. Copies of his botanical letters were extant until at least the end of the nineteenth century but may since have been lost. Gronovius said of Clayton to Linnaeus in 1739 that ''Clayton, if not unmatched, is at least the most distinguished foreign resident [among his contemporary naturalists].'' Linnaeus named the ''spring beauty,'' an American flower, *Claytonia virginica*.

MAJOR CONTRIBUTIONS

Clayton's collections provided the basis for the first important work on British North American plants, though much of the credit for this has been attributed to the elder Gronovius.

BIBLIOGRAPHY

Berkeley, Edmund, and Dorothy Berkeley. *John Clayton, Pioneer of American Botany* (1963).

Davis, Richard Beale. *Intellectual Life in the Colonial South, 1585–1763* (1978).

Dictionary of American Biography.

Dictionary of National Biography.

Ewan, Joseph, and Nesta Ewan. *John Banister and His Natural History of Virginia, 1678–1692* (1970).

Stearns, Raymond Phineas. *Science in the British Colonies of America* (1970).

Keir B. Sterling

CLEMENTS, FREDERIC EDWARD. Born Lincoln, Nebraska, 15 September 1874; died Santa Barbara, California, 26 July 1945. Plant ecologist, botanist.

FAMILY AND EDUCATION

Son of Ephraim G. Clements, photographer, and Mary Scoggin Clements. Two younger sisters. B.Sc., 1894, M.A., 1896, Ph.D., 1898, LL.D., 1940, University of Nebraska. Married Edith Gertrude Schwartz, 1899; no children.

POSITIONS

Instructor in botany, University of Nebraska, 1897, adjunct professor, 1899, professor of plant physiology, 1906; professor of botany and head of the Botany Department, University of Minnesota, 1907–1917; research associate in ecology, Carnegie Institution, 1917–1941.

CAREER

Educated at University of Nebraska under Charles E. Bessey, one of the leaders in movement to modernize botanical instruction and research in America. Early work concerned classification of fungi. Participated in the Botanical Survey of Nebraska under direction of Roscoe Pound. Publication, with Pound, of results of survey constituted Clements's doctoral dissertation and represents one of the pioneering formal ecological studies in North America. Pound and Clements devised quantitative research methods and introduced concepts from European phytogeography into study of American vegetation. Clements's subsequent work dealt with dynamics of vegetation change and development of an elaborate system of terminology to characterize plant communities in various stages of development. Clements taught botany for ten years at Nebraska and another ten years at University of Minnesota before taking position in 1917 as research associate with Carnegie Institution of Washington. Worked in winters at Carnegie Institution's Desert Laboratory at Tucson, Arizona, 1917–1925, and later at institution's Coastal Laboratory at Santa Barbara, California, 1925–1941. Did summer research at Alpine Laboratory below Pikes Peak in Colorado, established by him and his wife in 1900 and taken over by Carnegie Institution in 1917. Did not engage in formal teaching after 1917 but worked with numerous graduate students as assistants at various Carnegie laboratories.

His published works include numerous articles and several lengthy monographs on botanical and ecological subjects (1893–1942), among them *The Phytogeography of Nebraska* (with Roscoe Pound, 1898), *The Development and Structure of Vegetation* (1904), *Research Methods in Ecology* (1905), *Plant Physiology and Ecology* (1907), *Plant Succession: An Analysis of the Development of Vegetation* (1916), *Experimental Vegetation: The Relation of Climaxes to Climates* (1924), *Plant Competition: An Analysis of Community Functions* (1929), and *Bio-Ecology* (with Victor Shelford, 1939).

Considered "by far the greatest individual creator of the modern science of vegetation" by British plant ecologist Arthur G. Tansley. Along with Chicago botanist Henry Chandler Cowles, Clements helped introduce and establish the study of plant ecology as scientific discipline in America. Emphasized ecological succession, classification of plant communities, and, especially, concept of the climax formation, the collection of plants optimally suited to climate of a particular region. Often criticized by biologists for his treatment of the plant community as superorganism and for his neo-Lamarckian evolutionary views, but his work became the model for ecological research and writing for generation of American botanists.

MAJOR CONTRIBUTIONS

Development of formal theoretical framework for plant ecology, creation of system of ecological nomenclature, introduction of quantitative methods to the study of vegetation, extensive field research in plant ecology. His concept of climax formation constituted first successful paradigm in American ecology and became subject of controversy during the dust bowl incidents of the 1930s.

BIBLIOGRAPHY

Clements, Edith S. *Adventures in Ecology* (1960).

Ewan, Joseph. *Dictionary of Scientific Biography*. Vol. 3 (1973).

Pound, Roscoe. "Frederic Clements as I Knew Him." *Ecology* (1954).

Sears, Paul B. *Dictionary of American Biography*. Supplement 3 (1973).

Shantz, H. L. *Ecology* 26 (1945) (obit.).

Tobey, Ronald C. *Saving the Prairies: The Life Cycle of the Founding School of American Plant Ecology, 1895–1955* (1981).

Worster, Donald. *Nature's Economy: The Roots of Ecology* (1977).

UNPUBLISHED SOURCES

See Edith G. Clements and Frederic E. Clements Collection, Division of Rare Books and Special Collections, University of Wyoming, Laramie.

Eugene Cittadino

CLEPPER, HENRY EDWARD. Born Columbia, Pennsylvania, 21 March 1901; died Washington, D.C., 28 March 1987. Forester, administrator, historian.

FAMILY AND EDUCATION

Son of Martin Neal and Charolette (Keech) Clepper. Received B.F. in 1921 from Pennsylvania State Forest Academy in Mont Alto. Married Clorinda McFerren (died in 1974); had son and daughter (died in 1986).

POSITIONS

From 1921 to 1936 employed by Pennsylvania Department of Forests and Waters, first as field forester and subsequently as assistant chief of Bureau of Research and Education. From 1936 to 1937 was information specialist in Washington, D.C. Between 1937 and 1966 (save for two years with the War Production Board) served as executive secretary of Society of American Foresters; was also managing editor of *Journal of Forestry*. From 1957 to 1965 was adviser to Forestry Committee of Food and Agricultural Organization of United Nations at organization's biennial conferences in Rome. Between 1970 and 1971, was consulting editor for American Fisheries Society; from 1974 to 1981 was consulting editor of Sport Fishing Institute.

CAREER

Helped to establish Society of American Foresters quarterly journal, *Forest Science*, in 1955, and research series, "Forest Science Monographs," in 1955. Prolific author who wrote over 100 articles and bulletins on forestry and related resources, many of historical nature. Books include *Forestry Education in Pennsylvania* (editor and coauthor, 1957); *America's Natural Resources* (coeditor and coauthor, 1967); *American Forestry—Six Decades of Growth* (author, 1960); *Origins of American Conservation* (editor and coauthor, 1966); *Leaders in American Conservation* (author, 1971); *Crusade for Conservation* (author, 1975); *Famous and Historic Trees* (coauthor, 1976); *Marine Recreational Fisheries* (coeditor, 1976–1981).

MAJOR CONTRIBUTIONS

Accomplishments in forestry conservation and service to Society of American Foresters were recognized by Gifford Pinchot Medal (1957), American Forestry

Association's Distinguished Service Medal (1970), and John Aston Warder Medal (1977).

BIBLIOGRAPHY

American Forests 93 (May/June 1987) (obit.).
Journal of Forestry 85 (June 1987) (obit.).
Stroud, Richard H., ed. *National Leaders of American Conservation* (1985).
Who's Who in America. 42d ed. (1982–1983).

Richard Harmond

COLDEN, CADWALLADER. Born Ireland, 7 February 1688; died Long Island, New York, 28 September 1776. Lieutenant governor of New York, scientist, botanist.

FAMILY AND EDUCATION

Son of Alexander Colden, a Scottish minister. Born while mother visiting in Ireland. B.A., University of Edinburgh, 1705, later received M.A. Studied medicine in London but did not take a degree. Married Alice Christie, 1715; ten children.

POSITIONS

Practiced medicine and in business in Philadelphia, 1710–1718. Surveyor general of New York Colony, 1720. Appointed to Governor's Council, New York, 1721–1776. Appointed lieutenant governor of New York, 1761–1776, and served as acting governor for much of that period.

CAREER

Attempted without success to create medical lectureship in Philadelphia. Gave up medical practice there and moved to New York Colony on the suggestion of Governor Robert Hunter, who met him and promised post of surveyor general of colony if Colden moved to New York. For most of his life thereafter, Colden used income from political sinecures to underwrite work on his scientific interests. Actual duties often performed by deputies. These interests were many, including botany, medicine, and physics but also history, mathematics, and philosophy. His first work, *The History of the Five Indian Nations Depending on the Province of New York* (1727, 2d ed., 2 vols., 1745, 1755), subsequently reprinted several times, was a useful account, though turgid and disordered. As trained physician, wrote a number of treatises in that field. In one (1751), discussed supposed treatment of cancer using pokeweed, based on reports from a Connecticut doctor. Admitted he had not tested this theory and suggested research might be done. Had studied plants at Edinburgh University, worked extensively on them following move to Coldengham in 1728. Linnaeus, who once referred to him as "Summus Perfectus," published catalog of plants sent to him by Colden (1743) and named a flower after him. Colden participated for a time in the Natural History Circle, exchanging seeds, plants, and corespondence with other plant collectors in colonies and in Europe. Also made occasional observations concerning insects, fossils, and other fauna. Colden mastered Linnaean system and later suggested it could be made more "natural," without attempting to follow through. Within about a dozen years, had tired of botany and turned most collecting over to his daughter Jane. Then turned his attention to physics and by 1745 claimed to have uncovered basis of gravitation. Pursued his theory of this subject until his death, despite biting criticism from better-informed contemporaries. This experience demonstrated that Colden's ideas were not always well ordered and not based on complete understanding of the literature that was available to him. In addition, while a good observer, he did little experimental work, nor did he utilize experimental methods. One speculative paper, "An Inquiry into the Principles of Vital Motion" (1761), was an attempt to apply his ideas concerning gravity to microscopic life, but without the benefit of firsthand scientific observation. Eagerly sought, but did not win, scientific reputation comparable to that of leading scientists of his time. For last fifteen years of life, was de facto head of New York colonial government save for occasional periods when colonial governors were in residence. Followed strongly pro-British policy. Retired shortly before death.

MAJOR CONTRIBUTIONS

Able and well trained, eager to pursue scientific and philosophical inquiry. Made most useful contributions as collector of botanical specimens and in exchanges of correspondence, seeds, and plants within Natural History Circle.

BIBLIOGRAPHY

Dictionary of American Biography.

Hindle, B. *The Pursuit of Science in Revolutionary America, 1735–1789* (1956).

Hindle, B. "A Colonial Governor's Family: The Coldens of Coldengham." *New York Historical Society Quarterly* 45 (1961).

Hindle, B. *Dictionary of Scientific Biography.*

Keys, Alice M. *Cadwallader Colden* (1906).

Stearns, R. P. *Science in the British Colonies of America* (1970).

UNPUBLISHED SOURCES

See papers in New York Historical Society; smaller collections at Newberry Library, Chicago, Huntington Library, San Marino, California, Rosenbach Foundation, Philadelphia, University of Edinburgh Library.

Keir B. Sterling

COLDEN, JANE. Born New York City, 27 March 1724; died New York City (?), 10 March 1766. Botanist.

FAMILY AND EDUCATION

Second daughter and fifth child (of ten) born to Cadwallader and Alice Christie Colden. Father a physician trained at University of Edinburgh and in London, who served as surveyor general of the New York Colony, lieutenant governor, and several times acting governor. He was also a well-read observer and philosopher of science. Jane educated at home. Married William Farquhar, a widower, 1759; one child.

POSITIONS

A largely self-trained botanist active from about age twenty-five (c.1749) to the time of her marriage (1759).

CAREER

Jane moved with her family to Coldengham, her father's estate several miles west of Newburgh, New York, in 1739 (some sources suggest they may have moved at least ten years earlier). Some sources suggest that her father, wearying of "the endless round of collecting plants, seeds and the like with members of the natural history circle" in the American colonies and Britain, encouraged his daughter to take up the study of botany. She had demonstrated considerable interest in learning generally and in natural history. In addition, her father faced growing problems with his vision (due to increasing age), making it harder for him to make accurate observations. He later (1755) wrote J. F. Gronovius, a Dutch colleague, "that botany is an amusement which may be made greater to the Ladies who are often at a loss to fill up their time" and that "[t]heir natural curiosity and the pleasure they take in beauty" made them better fitted for botanical study than men. The elder Colden secured working library of reference works from London for his daughter's use. Also taught his daughter the Linnaean system. She did not, however, learn Latin. She took up much of her father's botanical correspondence, exchanged seeds with some of the collectors with whom her father had worked, and drafted a "Flora Nov-Eboracensis." This manuscript, with simple and limited descriptions of 341 plants, also included hundreds of her clear but unsophisticated drawings and is now in British Museum. Cadwallader was enthusiastic about Jane's work, and this elicited favorable comments from Alexander Garden, Peter Collinson, and others. Collinson wrote Linnaeus in May 1756 that "[Colden] is well; but, what is marvellous, his daughter is perhaps the first lady that has so perfectly studied your system. She deserves to be celebrated." Linnaeus was reportedly pleased. Jane's gender seems to have elicited more interest than what she actually accomplished. She married at age thirty-five, had one child, gave up most of her botanical work, and died at age forty-one. Only one of her descriptions of a plant was published (by Alexander Garden), included in his "The Description of a New Plant" in

Essays and Observations, Physical and Literary (1756). This was of the flower known today as the gardenia, a specimen of which Garden had picked near New York City and which she named in his honor.

MAJOR CONTRIBUTIONS

Jane Colden was a thorough and competent observer and describer of the plants in the vicinity of her home in southern New York Colony. Her "Flora Nov-Eboracensis" is noteworthy as the major botanical work compiled by an American woman scientist during the colonial era. She is also remembered for having named the gardenia for Alexander Garden.

BIBLIOGRAPHY

Berkeley, Edmund, and Dorothy Smith Berkeley. *Dr. Alexander Garden of Charles Town* (1969).

Britten, James. "Jane Colden and the Flora of New York." *Journal of Botany, British and Foreign* 33 (1895).

Rickett, H. W., and Elizabeth Hall, eds. *The Botanic Manuscript of Jane Colden* (1963).

Smith, James E., ed. *A Selection of the Correspondence of Linnaeus and Other Naturalists.* Vol. 1 (1821).

Vail, Anne Murray. "Jane Colden, an Early New York Botanist." *Contributions from the New York Botanical Garden* 4 (1966–1967).

UNPUBLISHED SOURCES

See manuscripts in Colden Family Papers, New York Historical Society.

Barbara Lovitts and Keir B. Sterling

COMSTOCK, ANNA BOTSFORD. Born Cattaraugus County, New York, 1 September 1854; died Ithaca, New York, 24 August 1930. Naturalist, educator, lecturer, illustrator.

FAMILY AND EDUCATION

Only daughter of Quaker mother and farmer father. After local schools, attended Cornell University, 1874 to 1876, and completed B.S. degree after intermittent study in 1885. Studied wood engraving with John P. Davis of Cooper Union in late 1880s. Married Cornell University professor of entomology, John Henry Comstock in 1878; no children.

POSITIONS

Collaborator with husband in reports for Department of Agriculture, 1879–1881, helped illustrate his lectures at Cornell and his textbooks on entomology, sometimes credited as joint author; also prepared wood engravings for other Cornell faculty, including Liberty Hyde Bailey and Burton G. Wilder, in the 1880s and 1890s. Used position as member of Committee for the Promotion of Agriculture to identify rural problems and to formulate program to prepare teachers to present nature in aesthetic, practical, and experiential terms. Appointed first woman assistant professor at Cornell in 1898 for the Extension Division, but protest of Board of Trustees reduced her rank to lecturer by 1900. Taught nature study until retirement in 1922 (with rank of assistant professor after 1913 and professor after 1920).

CAREER

Early career was collaborative with husband, working as collector, laboratory analyst, editor, and illustrator, especially for *Manual for the Study of Insects* (1893) and *Insect Life* (1897). Illustration activity led also to freelance assignments for other scientists and to prizewinning submissions at art and engraving expositions. In late 1890s her involvement with extension work in rural New York led her to promote the concept of nature study, term used to suggest the injunction of earlier naturalists to "study nature, not books." Taught approach at Cornell, at summer institutes in nearby normal schools, at Chautauqua, and through numerous publications over the next thirty years. Her pamphlets for students, guidelines for teachers, and home-study courses were written and distributed by state, while other essays appeared in popular magazines. She served on editorial board of movement's central magazine, *Nature Study Review*, throughout its existence, 1905 to 1923, and was chief editor after 1917. Most widely used was her comprehensive *Handbook of Nature Study* (1911), which went through twenty-four editions by 1939. Her *Handbook* was published by the Comstock Publishing Company, whose motto was "through books to nature."

Liberty Hyde Bailey, her colleague at Cornell, called her a leader of the movement because of her "accurate and extensive knowledge of her subject matter, clear knowledge of the situation in the schools, and warm personality." Her approach to nature had romantic overtones, and she used poetry and literature along with practical descriptions of plants and animals. Much of her time and energy was absorbed by local claims for her attention as collaborator and faculty wife and later as educator, editor, and publisher. Nonetheless, she was nationally known spokesperson for nature study, invited to educational conferences, and named one of twelve outstanding women of America in 1923 by the League of Women Voters.

MAJOR CONTRIBUTIONS

Artistic work on woodcuts undoubtedly made more attractive and understandable the entomological texts of her husband and the extension service pamphlets on which Comstock worked. The nature study movement had perhaps its greatest success in New York and surrounding areas, in part because of her energy and initiative. Her avid followers presented additional dimensions to nature study, "an outlook over all the forms of life and their relationship to one another,'' and she applauded their efforts to take this theme outside the schools into urban gardens and parks, wildlife trails, nature clubs, and scouting.

BIBLIOGRAPHY

Champagne, Audrey B., and Leopold E. Klopfer. ''Pioneers of Elementary-School Science: II. Anna Botsford Comstock.'' *Science Education* 163, 3 (1979).

Comstock, Anna Botsford. *The Comstocks of Cornell*. Ed. R. G. Smith (1953).

Jacklin, Kathleen. *Notable American Women* (1971).

Nature Magazine (October 1930).

Needham, James G. ''The Lengthened Shadow of a Man and His Wife.'' *Scientific Monthly* (February, March 1946).

UNPUBLISHED SOURCES

Manuscripts and related sources are in the Cornell Collection of Regional History and University Archives.

Sally Gregory Kohlstedt

COMSTOCK, JOHN HENRY. Born near Janesville, Wisconsin, 24 February 1849; died Ithaca, New York, 20 March 1931. Entomologist.

FAMILY AND EDUCATION

Son of Ebenezer and Susan Allen Comstock. Father left for California goldfields soon after his son's birth and died of cholera in vicinity of Platte River. Mother lost farm, returned to New York, worked as housekeeper. Comstock in orphan asylum for about a year, then farmed out to relatives. Worked as cook on Great Lakes vessels from age sixteen. Studied entomology privately. Entered Cornell, 1869, became ill, withdrew, entered again, 1870. B.S., Cornell, 1874. Student at Museum of Comparative Zoology, Harvard, summer 1872. Studied with Addison Verrill at Yale, 1874–1875, and student at Leipzig under K. G. F. R. Leuckart, 1888–1889. Had no time to complete graduate degrees. Married Anna Botsford, former student, 1878; no children.

POSITIONS

Sailor cook on Great Lakes vessels for several years after 1865. Assistant in entomology, 1872–1873, instructor in entomology, 1873–1876, assistant professor of entomology, Cornell, 1876–1879. At Yale with A. Verrill, 1874–1875; lecturer on zoology, Vassar College, 1877, field agent for U.S. Department of Agriculture (USDA), summer 1878. Chief entomologist, U.S. Department of Agriculture, Washington, D.C., 1879–1882; professor of entomology and invertebrate zoology, Cornell, 1882–1914. Spent parts of years 1891–1893 as nonresident entomologist at Stanford University, organizing and teaching courses in entomology. Also lectured at State Normal School, San Jose, California, 1892.

CAREER

Comstock's interest in insects was triggered by his purchase in 1870 of T. W. Harris's *Treatise on Insects Injurious to Vegetation*, published in the late 1850s. Comstock began his teaching career at Cornell in the spring of his sophomore year at Cornell at request of thirteen classmates. On strength of his considerable knowledge of entomology (self-taught from Harris text and others), university trustees appointed him assistant, later instructor in entomology prior to his graduation. During his tenure as chief entomologist with USDA, Comstock completed report on cotton worm and began study of scale insects in Florida and California. With aid of his wife, developed system of classification of these organisms based on configuration of their pygidia (posterior parts). Published *Report on Cotton Insects* (1880) and *Report of the Entomologist of the United States Department of Agriculture for the Year 1880* (1881), which contained his classification of scale insects. Began summer courses at Cornell, 1885, and instruction on bee-keeping, 1886. Established insectary (in effect, an insect greenhouse) at Cornell, 1887, which made possible study of living insects under laboratory conditions. Published first edition of his *An Introduction to Entomology* (1888). Declined full-time post at Stanford from D. S. Jordan, Stanford president and former student, but developed entomology program there, 1891–1893, turned over in latter year to Comstock's former student Vernon L. Kellogg on Comstock's recommendation. Published *Manual for the Study of Insects* (1894, 5th ed., 1904) and *Insect Life*, for elementary teachers (1897), both with assistance of Mrs. Comstock. Purchased father-in-law's dairy farm, 1894, operated it successfully for two decades. Published *The Wings of Insects* (1899, rev. ed., 1918) and *Wings of the Sesiidae* (1901). Did fieldwork on spiders in several southern states, 1903. Published *How to Know the Butterflies* (with wife, 1904) and *The Spider Book* (1912). Published revised version of his earlier *Introduction to Entomology* (1920), a massive tome of over 1,000 pages, which by 1940 had undergone a total of nine revisions. Students presented check to him for $2,500 to establish library of entomology on the occasion of his retirement, which he turned over to Cornell. University established John Henry Comstock Memorial Library of Entomology. Continued active research until suffering brain hemorrhage in 1926. Inactive last five years of life. One of his students, V. L. Kellogg, headed entomology program at Stanford after 1893, and another, L. O. Howard, would later serve as chief of Bureau of Entomology in USDA. Described and named a number of insect species during his career (some California species listed in Essig).

MAJOR CONTRIBUTIONS

Comstock established Department of Entomology at Cornell, first to be established anywhere in the world. He is said to have taught 5,000 students during his career. His textbooks enjoyed wide use in classrooms for over half a century, and their usefulness was much enhanced by his wife's illustrations. Success of their books led to establishment of Comstock Publishing Company, which also published notable texts in other areas of zoology. Comstock's reports on cotton and scale insects were well done and highly thought of.

BIBLIOGRAPHY

Comstock, A. B. *The Comstocks of Cornell: John Henry Comstock and Anna Botsford Comstock: An Autobiography*. Ed. G. W. Herrick and R. G. Smith (1953).

Dictionary of American Biography.

Essig, E. O. *A History of Entomology* (1931).

Herrick, G. W. "Professor John Henry Comstock." *Annals of the Entomological Society of America* 24 (1931).

Mallis, Arnold. *American Entomologists* (1971).

Needham, J. G. "John Henry Comstock." *Science* 73 (April 1931).

Needham, J. G. "The Lengthened Shadow of a Man and His Wife." *Scientific Monthly* 62 (1946).

New York Times, 21 March 1931 (obit.).

Who Was Who in America.

UNPUBLISHED SOURCES

See John Henry Comstock Correspondence, Cornell University Libraries, Collection of Regional History and University Archives; also, correspondence, manuscript proofs, and illustrations in Cornell University Press and Comstock Publishing Company records in Cornell University Libraries, Department of Manuscripts and University Archives.

Keir B. Sterling

COOLIDGE, HAROLD JEFFERSON. Born Boston, 15 January 1904; died Beverly, Massachusetts, 15 February 1985. Conservationist, mammalogist, curator, author.

FAMILY AND EDUCATION

Son of Harold and Edith (Lawrence) Coolidge. Graduate Milton Academy, 1922; student, University of Arizona, 1922–1923; B.S., Harvard, 1924; student at Cambridge University, England, 1927–1928. Hon. D.Sc., George Washington University, 1959, Seoul National University, 1965, Brandeis University, 1970. Married Helen Carpenter Isaacs, 1931 (divorced 1972); three children. Married Martha Thayer Henderson, 1972.

POSITIONS

Assistant mammalogist, Harvard African Expedition to Liberia and Belgian Congo, 1926–1927. Leader, Indochina Division, Kelley–Roosevelt Asian Expedition (on behalf of Field Museum, Chicago), 1928–1929. Assistant curator of mammals, Museum of Comparative Zoology (MCZ), Harvard, 1929–1946. Director, Film Service, Harvard, 1934–1936; organizer and leader, Asiatic Primate Expedition, Northern Siam, Burma, and Sumatra, 1936–1937. Associate in mammalogy, MCZ, 1946–1970. Executive director, Pacific Science Board, National Academy of Sciences and National Research Council, 1946–1970. President, 1966–1972, then honorary president, International Union for the Conservation of Nature and Natural Resources, 1972–1985.

CAREER

Was specialist on African and Asian primates, publishing, among other papers, "A Revision of the Genus Gorilla" (1929); "Notes on the Gorilla: The African Republic of Liberia and the Belgian Congo. Based on the Observations Made and Material Collected during the Harvard African Expedition, 1926–1927" (1930); "Pigmy Chimpanzee from South of the Congo River," (1933); "Notes on a Family of Breeding Gibbons" (1933); "Zoological Results of the George Vanderbilt African Expedition of 1934. Part 4: Notes on Four Gorillas from the Sanga River Region" (1936); and "The Living Asiatic Apes" (1938). In 1933 completed book, *Three Kingdoms of Indo-China* (with Theodore Roosevelt, Jr.), a popular account of the Field Museum Expedition in 1928–1929, which dealt with the work of the division of the expedition that specialized in collecting birds, small mammals, and reptiles in the northern part of the country. During World War II (1943–1945), was uniformed member of Office of Strategic Services (OSS) with rank of major, U.S. Army. Among other projects, was involved in attempts to develop shark repellent for airmen downed at sea. Vice president, International Commission for National Parks, 1948–1954 and 1963–1966 (also chairman, 1958–1963, member, 1963–1976, honorary member, 1976–1985); collaborator, U.S. Park Service, 1948–1985; secretary, American Committee for International Wildlife protection (later American Committee for International Conservation), 1930–1951, chairman, 1951–1971, honorary chairman, 1971–1985. Chairman, Species Survival Service Commission, 1949–1958, member, 1958–1976, honorary member, 1976–1985; also member of a large number of other committees and commissions concerned with conservation, wildlife research, and related fields, including World Wildlife Fund (founding director) and Charles Darwin Foundation for the Galapagos Islands (founder, director, and trustee). Secretary general, Tenth Pacific Science Congress, Honolulu, 1961. Chairman, First World Conference on National Parks, Seattle, 1962. Awarded Silver Medal, U.S. National Parks Centennial Commission; J. Paul Getty Medal for Wildlife Conservation, 1979; Browning Medal from the Smithsonian Institution, 1978; Seventy-Fifth Anniversary Medal of Merit, University of Arizona, 1960; and others. Decorated by the French, Laotian, Cambodian, Belgian, Ecuadorian, and Annamite governments.

MAJOR CONTRIBUTIONS

Coolidge's efforts in the field of international conservation and his expertise in primate and bovid classification won him an international reputation. His service as founder, adviser, trustee, and executive furthered the objectives of many scientific, national park, and conservation organizations in Africa, the Pacific, and the United States.

BIBLIOGRAPHY

American Men of Science. 12th ed.
New York Times, 16 February 1985 (obit.).
Stroud, Richard H., ed. *National Leaders of American Conservation* (1985).
Washington Post, 16 February 1985 (obit.).
Who Was Who in America.
Who's Who in the World (1975–1976).

UNPUBLISHED SOURCES

Coolidge's papers are mainly at Harvard University Archives and at Archives of the National Academy of Sciences in Washington. Papers at Harvard are primarily personal and professional from approximately the time of his entry into preparatory school (Milton Academy) until c.1985. Coverage includes expeditions to Africa and Southeast Asia and work at Harvard. There is also transcript of extended oral history interview done with K. B. Sterling in 1983. Papers at National Academy of Sciences are primarily professional in nature, covering period c.1946–1970, but with some personal correspondence and Harvard-related materials.

George A. Cevasco

COOPER, JAMES GRAHAM. Born New York, 19 June 1830; died Hayward, California, 19 July 1902. Naturalist, ornithologist, malacologist, physician, biogeographer.

FAMILY AND EDUCATION

Son of an amateur naturalist, William Cooper, and Frances Graham Cooper, who died in 1835. Family moved to a farm near Slogha, New Jersey, in 1837. Graduated from College of Physicians and Surgeons in New York City in 1851. Married Rosa Wells, 9 January 1866; three children.

POSITIONS

Physician, New York City Hospital, 1851–1853. Physician/naturalist, Pacific Railway Survey Northern Route (western end), 1853–1855. Physician, Orange, New Jersey, 1856. Physician/naturalist, Ft. Kearny-Honey Lake Wagon Road, 1857. Collected in Florida, 1859. Physician/naturalist, Military Expedition to the Columbia River, 1860. Physician, Ft. Mojave, California, 1860. Zoologist, California Geological Survey, 1860–1873 (on and off). Army physician, 1865. Private physician, 1866–1890.

CAREER

Collected extensively in Washington Territory for the Pacific Railway Survey and the Smithsonian Institution. Published several sections in the Railroad Reports—botany, mammals, birds, reptiles, crustacea. Published popular articles on travels in Florida and Military Expedition to Columbia River. Published early articles on plant biogeography and on need for national parks. Authored an early essay (1859) calling for a national forest system; essay was circulated but never published. Published extensively on work for California Geological Survey, particularly on birds, mollusks, botany, and paleontology.

In all, published 154 titles, the most important of which are *The Natural History of Washington Territory* (with George Suckley, the private edition of a Railroad Report volume, 1860), *Geographical Catalogue of the Mollusca Found West of the Rocky Mountains* (1867), *Ornithology. Land Birds* (by Spencer Baird, from "manuscript and notes of Cooper," 1870), and *Catalogue of Californian Fossils* (1894).

Collected extensively in several parts of the country, and his material was used by many scientists describing new species of plants and animals. His most important personal contributions were on the birds, mollusks (both living and fossil), and biogeography. He was one of the first to divide North America into biogeographical provinces, and his early calls for conservation—both published and privately circulated—helped set the stage for more serious later efforts.

MAJOR CONTRIBUTIONS

Provided extensive, well-collected materials for workers on the East Coast and became one of the first formally trained naturalists to move to the West Coast. Played a major role in elucidating the fauna and flora of Washington and California.

BIBLIOGRAPHY

Coan, E. *James Graham Cooper, Pioneer Western Naturalist* (1982).
Dall, W. H., and J. G. Cooper. *Science* 16 (1902).
Emerson, W. O., and James G. Cooper. *Cooper Ornithological Club Bulletin* 1 (1899).
Emerson, W. O. "In Memoriam: Dr. James G. Cooper." *The Condor* 4 (1902).

UNPUBLISHED SOURCES

See California Academy of Sciences (journals, manuscripts, photographs, correspondence); Bancroft Library, University of California, Berkeley (journal, records of the California Geological Survey, correspondence); Smithsonian Institution Archives (journals, letters, notes); National Archives (Wagon Road and army records).

Eugene Coan

COPE, EDWARD DRINKER. Born Philadelphia, 28 July 1840; died Philadelphia, 12 April 1897. Zoologist, paleontologist.

FAMILY AND EDUCATION

Son of Alfred and Hannah Edge Cope, wealthy Quakers. Attended Friend's School, Wettown, Pennsylvania, Friend's Select School, Philadelphia. Studied at University of Pennsylvania (with Joseph Leidy); private study at Academy of Natural Sciences of Philadelphia and also at Smithsonian Institution (with S. F. Baird). Did not take undergraduate degree. Honorary A.M., Haverford, 1870, honorary Ph.D., Heidelberg, 1885. Married Annie Pim, 1865; one child.

POSITIONS

Professor of comparative zoology and botany, Haverford College, 1864–1867; paleontologist with Wheeler Survey, 1874; with Hayden Survey, 1870–1879; professor of geology and mineralogy, 1889–1895, professor of zoology and comparative anatomy, University of Pennsylvania, 1895–1897. Owner and senior editor, *The American Naturalist*, 1878–1897.

CAREER

As a youth, strong interest in nature demonstrated in many notebooks of sketches he made. Described as having been almost "dangerously precocious." At age nineteen, published first paper, "On the Primary Divisions of the Salamandridae, with a Description of Two Species." In 1863–1864, spent several months in study abroad. By age twenty-two, recognized as one of leading specialists in vertebrate paleontology. From 1866, devoted remaining three decades of his life to exploration and research. His studies in vertebrate paleontology began with fossil reptiles found in New Jersey, Maryland, and Virginia. In 1868, studied remains of air-breathing vertebrates of the upper Mississippi River. Two years later, described fossils collected by Hayden Expedition in Wyoming with Leidy. Save for some fieldwork in Mexico, spent most of active research career in American West. Throughout his career, was active and tireless worker. Entered into ferocious professional competition with Othniel C. Marsh of Yale, which exhausted personal resources of both men and damaged both their reputations. Controversy disgusted Leidy, who retired altogether from paleontological activity. Part of this had to do with struggle for support of paleontological work by U.S. Geological Survey (USGS) after 1879, which Marsh won, forcing Cope's retirement from government service. Cope often tended to rush to conclusions. Former mentor and associate Leidy observed that "haste led him to superficiality." In one instance, Leidy demonstrated that dinosaur skeleton reconstructed by Cope was assembled "wrong end to." Cope sometimes intuited conclusions unsupported by materials with which he was working. On other hand, another authority, William Berryman Scott, wrote that "it was in the unravelling of the complexities of the freshwater tertiaries that Cope's most splendid services to geology were rendered."

Had some 1,390 publications to his credit, including books, papers, notes, editorials, and reviews on living and fossil vertebrates, evolution, heredity, psychology, sociology, and education. Among his most important monographs were *The Vertebrata of the Cretacious Formations of the West* (1875) and *The Ver-*

tebrata of the Tertiary Formations of the West, vol. I (1884), a massive, 1,009–page tome that was known as "Cope's bible." Marsh's control of publication program at USGS precluded publication of volume 2 of this work, although some of the plates (but not the text) were published after his death (in 1915). Adverse outcome of unseemly row concerning ownership of specimens collected for USGS and for Cope's private collections (most at his expense) deeply embittered Cope. Exhaustion of silver mines in which he had invested compelled him to take teaching position at University of Pennsylvania, and Cope was later compelled to sell his remaining collections to University of Pennsylvania to meet living expenses. Often placed professional interests ahead of family ties. "His house," wrote Scott, "was a museum with almost no furniture in it." Also published *The Origin of the Fittest* (1886) and *The Primary Factors of Organic Evolution* (1896), which confirmed his position as leading neo-Lamarckian of his time. Advocated belief in evolution of different genera through what was known as parallelism. Was influenced by English evolutionist Herbert Spencer. Elected to membership in Academy of Natural Sciences of Philadelphia at unusually early age of twenty-one. Later a curator (1865) there and member (1879–1880) of Academy Council. Member, National Academy of Sciences, president of American Academy for Advancement of Science, 1896.

MAJOR CONTRIBUTIONS

As noted by W. B. Scott after Cope's death at age fifty-seven, "The greatest and most enduring monument to his fame will prove to be the gigantic work which he accomplished among the extinct vertebrates of the far West." One of handful of leading American vertebrate paleontologists of his generation, along with Leidy and Marsh.

BIBLIOGRAPHY

Colbert, Edwin H. *The Great Dinosaur Hunters and Their Discoveries* (revision of his *Men and Dinosaurs*) (1984).

Frazer, Persifor. "Life and Letters of Edward Drinker Cope." *American Geologist* (August 1900).

Gill, Theodore, H. F. Osborn, and W. B. Scott. *Addresses in Memory of Edward Drinker Cope* (1897).

Howard, Robert W. *The Dawnseekers: The First History of American Paleontology* (1975).

Osborn, H. F. "Edward Drinker Cope." *Biographical Memoirs of the National Academy of Sciences* 13 (1930).

Osborn, H. F. *Cope: Master Naturalist* (1931).

Scott, W. B. *Some Memories of a Paleontologist* (1939).

Shor, Elizabeth Noble. *The Fossil Feud between E. D. Cope and O. C. Marsh* (1974).

Simpson, George G. "Hayden, Cope, and the Eocene of New Mexico." *Proceedings of the Academy of Natural Sciences of Philadelphia* 103 (1951).

Stocking, George. "Lamarckianism in American Social Science, 1890–1915." *Journal of the History of Ideas* 23 (1962).

Arthur A. Belonzi

COTTAM, CLARENCE. Born St. George, Utah, 1 January 1899; died Corpus Christi, Texas, 30 March 1974. Biologist, conservationist.

FAMILY AND EDUCATION

Son of Thomas Punter and Emmaline (Jarvis) Cottam and grandson of Thomas Cottam, who came to the United States from England in 1852 and settled in St. George. Father was farmer and public official. One brother, Walter P., conservationist. Educated at public schools in St. George and attended Dixie College, St. George, in 1919–1920 and the University of Utah in the summer of 1923. Graduated, A.B., 1926, M.S. (biology), 1927, Brigham Young University; graduate study at American University, 1931; Ph.D., 1936, George Washington University. Married Margery Brown, 20 May 1920; four daughters.

POSITIONS

Principal, Consolidated Schools, and teacher in the high school, Alamo, Nevada, 1922–1925; instructor in chemistry and biology, Brigham Young University, Provo, Utah, 1925–1929; junior biologist, 1929–1931, assistant biologist, 1931–1935, senior biologist in charge of Food Habits Research Section, U.S. Biological Survey, 1935–1940; also held same position with U.S. Fish and Wildlife Service (USF&WS), 1940–1942; in charge of economic wildlife investigation, USF&WS, 1942–1944, assistant to director, USF&WS, 1944–1946; chief, Division of Wildlife Research, USF&WS, 1945–1946; assistant director, USF&WS, 1946–1954; professor of biology and founding dean, College of Biological and Agricultural Sciences, Brigham Young University, 1954–1958; first director, Rob and Bessie Welder Wildlife Foundation, Sinton, Texas, 1955–1974. Consultant to Department of the Interior and other federal agencies.

CAREER

Active career staff member with U.S. Biological Survey and later with U.S. Fish and Wildlife Service for a quarter century. His lifelong commitment to biological research, conservation, and wise game management involved him in several important causes. Cottam began inveighing against insecticides in 1946, published a government circular outlining their effects on fish and wildlife (1946), and opposed government-subsidized pest-control programs. In 1969, he campaigned against the use of a highly toxic pesticide, Dieldrin, in Texas. Active in research on long-range effects of DDT, a similar pesticide. Instrumental in passage of federal law that largely eliminated use of pesticides on lands controlled by the Interior Department. Supported creation of Padre Island National Seashore, 1967–1969, and also expansion of Arkansas Wildlife Refuge. As director of the Rob and Bessie Welder Wildlife Foundation, fostered the founder's wish that conservation education and appreciation be fostered. Some 145 fellowships in ecology, totaling $575,000, were made available to students from thirty-three colleges and universities.

Published some 250 articles and contributed to many books, including *Food Habits of North American Diving Ducks* (1939); *Insects* (with Herbert Zim, 1951), a pocket-sized guide to 225 species still in print and later translated into French; *Ecological Study of Birds in Utah* (with Angus M. Woodbury, 1962); and *Whitewings: The Life History, Status, and Management of the White-Winged Dove* (1968). Also contributed chapters and sections to a number of books on wildlife, conservation, pesticides, and other subjects. Received many awards and honors, including the Aldo Leopold Award of the Wildlife Society, 1955, the Conservation Distinguished Service Medal of the National Audubon Society, 1961, the Paul Bartsch Award from the Audubon Naturalist Society, 1962, and, following his death, the Alban-Heiser Award from the Zoological Society of Houston. A series of annual lectures in his name and that of his brother Walter P. Cottam were established at the University of Utah. Cottam was president of the Wildlife Society, 1949–1950, the Texas Ornithological Society, 1957, and the National Parks Association, 1960–1970, and was a member of at least twenty other conservation, biological, service, and honorary organizations. He was also extremely active in the Church of Latter-Day Saints, rising to the presidency of the Corpus Christi Stake, with ecclesiastical authority over some 3,000 coreligionists.

MAJOR CONTRIBUTIONS

Staunch supporter of wildlife research, conservation, wise use of national parks, effective wildlife management policies; opposition to widespread use of pesticides.

BIBLIOGRAPHY

Commire, Anne, in *Something about the Author*. Vol. 25 (1981).
Contemporary Authors. Vols. 97–100 (1981).
National Cyclopedia of American Biography. Vol. 58 (1979).
National Parks Magazine 40 (June 1974).
New York Times, 3 April 1974 (obit.).
Stroud, R. H., ed. *National Leaders of American Conservation* (1985).
Who Was Who in America.

Connie Thorsen

COUES, ELLIOTT. Born Portsmouth, New Hampshire, 9 September 1842; died Baltimore, 25 December 1899. Naturalist, bibliographer, anatomist, historian, surgeon.

FAMILY AND EDUCATION

Son of Samuel Elliott Coues, a merchant, and Charlotte Haven Ladd. Family moved to Washington, D.C., in 1854 after his father received clerkship in Patent

Office. One sister, one brother, one adopted sister, and half-brother who lived to adulthood, and brother, sister, and half-sister who did not. Attended Washington Seminary (Gonzaga College, Washington, D.C.), 1855–1857. A.B. (1861), A.M. (1862), and M.D. (1863) from Columbian College (George Washington University). Married Sarah A. Richardson, 1864 (divorced 1864), Jeannie Augusta McKinney, 1867 (divorced 1886), and Mary Emily Bates, 1887. Five children by second wife, two of whom died in childhood.

POSITIONS

Medical cadet, 1862–1863, assistant surgeon, U.S. Army, 1864–1881. Surgeon and naturalist, U.S. Northern Boundary Commission, 1873–1876. Resident collaborator in mammalogy and ornithology, Smithsonian Institution, 1874–1875. Secretary and naturalist, U.S. Geological and Geographical Survey of the Territories, 1876–1880. Curator of mammalogy, U.S. National Museum, 1879. Lecturer on anatomy, 1877–1880, professor of anatomy, 1881–1887, National Medical College, Washington, D.C.

CAREER

One of the many protégés of Spencer F. Baird. Began publishing articles and monographs on natural history, chiefly ornithology, in 1861. Major works written mostly during his nineteen-year army career. Traveled, collected, and observed in the West: Arizona (1864–1865, 1880–1881), Dakota and Montana (1872–1874), and Colorado (1876). Elected in 1877 to National Academy of Sciences. A founder of the American Ornithologists' Union (AOU) (1883), vice president (1883–1890), and president (1892–1895). Associate editor of *Bulletin of the Nuttall Ornithological Club* (1876–1883) and *The Auk* (1884–1887). Chairman of AOU Committee on Classification and Nomenclature (1883–1895). Natural history editor for *Century Dictionary* (1883–1891). Active in Theosophical Society and psychic research in the 1880s. Supported women's rights movement. Edited fifteen volumes, in the 1890s, of western travel journals of Lewis and Clark, Z. M. Pike, and others. Provided original descriptions of six genera, two subgenera, eight species, and twenty subspecies of birds and four genera, two species, and twelve subspecies of mammals. Described by Alfred Russel Wallace as "a man of brilliant talents, wide culture, and delightful personality." According to Casey A. Wood, he was "in many respects . . . the most brilliant writer on vertebrate zoology America has so far produced."

Among his more important works are *Key to North American Birds* (1872, 1884, 1887, 1890, 1903, 1927), *Check List of North American Birds* (1873, 1882), *Birds of the Northwest* (1874), *Field Ornithology* (1874), *Fur-Bearing Animals* (1877), *Monographs of North American Rodentia* (with J. A. Allen, 1877), *Birds of the Colorado Valley* (1878), and "Universal Bibliography of Ornithology" (four installments, 1878–1880). Coues was probably the best-known and most widely published American ornithologist of the late nineteenth century. Various editions of the *Key* were especially appreciated by both laymen and professionals. Published in journals of learned societies as well as in sportsmen's periodicals. Frequently involved in controversy, for example, the "Sparrow War," and had a stormy private life; yet was capable of kindness and was helpful to young students of natural history, notably Louis Agassiz Fuertes.

MAJOR CONTRIBUTIONS

A prolific, lucid writer and an energetic field naturalist. Orinthologists continue to consult his works with profit.

BIBLIOGRAPHY

Allen, J. A. "Elliot Coues." *The Nation* (4 January 1900) (obit.).

Brodhead, M. J. *A Soldier-Scientist in the American Southwest: Being a Narrative of the Travels of Brevet Captain Elliott Coues* (1973).

Cutright, P. R., and M. J. Brodhead. *Elliott Coues: Naturalist and Frontier Historian* (1981).

Elliott, D. G. "In Memoriam: Elliott Coues." *Auk* (January 1901).

Hume, E. E. *Orinthologists of the United States Army Medical Corps* (1942).

UNPUBLISHED SOURCES

See letters in Spencer F. Baird Correspondence, Smithsonian Institution Archives, and J. A. Allen Correspondence, American Museum of Natural History, N.Y.

Michael J. Brodhead

COUPER, WILLIAM. He may have been Scottish, but nothing is known of his origins or birthdate. Arrived in Canada c.1843, left in 1885, and probably died in Troy, New York, c.1890. Naturalist, entomologist, ornithologist, taxidermist.

FAMILY AND EDUCATION

Couper seems to have been self-taught in natural history. He worked as a typographer when he arrived in Canada. He married at least once; only one son known, William (also a naturalist) of Troy, New York.

POSITIONS

Worked and lived in Toronto until 1860, moving to lower Canada (Trois-Rivières in 1860, Quebec City, 1860–1869), to Ottawa (1869–1871) and Montreal (1871–1885). He seems to have worked as both typographer and taxidermist (also teaching taxidermy) before 1871. In Montreal, he listed his occupation as "naturalist," collecting and selling specimens. Was proprietor of *The Canadian Sportsman and Naturalist* (1881–1884). Probably in retirement upon moving to son's home in Troy, New York, in 1885.

CAREER

Couper was active in the scientific circles of Toronto, contributing several papers on collecting *coleoptera* to the *Canadian Journal* and exhibiting insect collections at the Provincial Exhibition in 1852 and 1856. After moving to lower Canada, he concentrated upon the entomology of the Quebec City region. Several contributions to the *Canadian Naturalist and Geologist* during this period include general notes on insect collecting and preservation, *hymenoptera, lepidoptera,* and special topics such as nest architecture, larvae, and parasites. He was instrumental in the creation of one of the first two branches of the Entomological Society of Canada (ESC) in Quebec in 1864. When Couper worked in Ottawa, he read papers to the Ottawa Natural History Society. During this period, he made the first of several collecting trips financed by collectors on speculation; his published report, *Investigations of a Naturalist between Mingan and Watchicouti, Labrador* (1868), was a pioneering effort. During the summer of 1872, he surveyed the insects of Anticosti Island—the entomology of which was virtually unknown—and the coast of Labrador. His collections, mostly earmarked for American collectors, were destroyed by local Indians in revenge for Couper's complaints about their salmon-fishing methods during his earlier visit. He returned to Anticosti in 1873.

He published a number of articles on insects, particularly on economic entomology, in the transactions of the Literary and Historical Society of Quebec, the Entomological Society of Philadelphia, the Fruit Growers' Association of Quebec, and Montreal Horticultural Society. Couper was a regular contributor to the ESC's journal, *Canadian Entomologist*, from its inception in 1869. In 1874, he joined several others in forming the Montreal branch of the reorganized ESC—the Entomological Society of Ontario—in November 1873. Couper was first branch president but ceased active participation in the late 1870s. His last contribution to the *Canadian Entomologist* was in 1880. Couper's magazine, *The Canadian Sportsman and Naturalist*, was an excellent combination of natural science with the avid Canadian pursuits of hunting and fishing. Couper maintained close relations with American as well as Canadian collectors and naturalists. He was a corresponding member of the Entomological Society of Philadelphia, and his knowledge of bird distribution and habits in eastern Canada made a contribution to E. A. Samuels's important *Our Northern and Eastern Birds* (1883).

MAJOR CONTRIBUTIONS

He was meticulous and systematic in his collecting, mounting and describing of insects, teaching these techniques to Canadians at a time when scientific entomology was just beginning. His explorations of Anticosti and Labrador, especially his studies of the distribution of species, paved the way for further explorations by both Canadian and American naturalists.

BIBLIOGRAPHY
Paradis, Rodolphe-O. "William Couper." *Dictionary of Canadian Biography*. Vol. 12 (1982).
Paradis, Rodolphe-O. "Etude biographique et bibliographique de William Couper, membre fondateur et premier président de la Société entomologique du Québec." *Annales de la Société entomologique du Québec* 19 (1974).
UNPUBLISHED SOURCES
Egg-collecting book (c.1879) is in University of Toronto Library.

Richard A. Jarrell

COVILLE, FREDERICK VERNON. Born Preston, New York, 23 March 1867; died Washington, D.C., 9 January 1937. Botanist, naturalist, plant breeder.

FAMILY AND EDUCATION

Son of Joseph Addison Coville, bank director, and Lydia More Coville. Grew up in in small town of Oxford, New York, and graduated from Oxford Academy. Enrolled at Cornell University, where he graduated with honors and received a B.A. in 1887. Honorary Sc.D. from George Washington University in 1921. Married Elizabeth Harwood Boynton, 1890; five children.

POSITIONS

Instructor in botany, Cornell University, 1887–1888. Assistant botanist, Geological Survey of Arkansas, 1888. Assistant botanist at U.S. Department of Agriculture (USDA), 1888–1893, botanist, 1893–1937. Botanist of the Death Valley Expedition, 1891. Botanist of the Harriman Alaska Expedition, 1899. Honorary Curator of the U.S. National Herbarium, 1893–1937. Acting director of the U.S. National Arboretum, 1929–1937.

CAREER

First fieldwork was on Geological Survey of Arkansas, 1888, which resulted in his publication of *A List of the Plants of Arkansas* (with John C. Branner, 1891). Most important fieldwork was the Death Valley Expedition, 1891, and his *Botany of the Death Valley Expedition* (1893) was considered classic in study of desert vegetation. Produced articles as result of the Harriman Alaska Expedition, but "Flora of Alaska" was never completed or published. Travels to western United States created in him continuing interest in botanical studies of the desert, in medicinal and other plants used by Indians, and in the effect of grazing on the land. His reports on sheep grazing and the forest were called by Gifford Pinchot "bold and masterly" and "the essentials of sound and far-sighted grazing policy." Helped bring about establishment of the Carnegie Institution's Desert Botanical Laboratory in 1903 in Tucson, Arizona. As chairman of National Geographic's Research Committee from 1920 to 1937, he had great influence on choice of areas for exploration.

Helped establish the Seed Laboratory of USDA and contributed approximately 170 titles to botanical literature. Was long acknowledged the American authority on rushes (*Juncaceae*) and currants and gooseberries (*Grossulariaceae*). Helped write and rewrite botanical definitions of the revised *Century Dictionary* and, with Frederick Law Olmstead and Harlan P. Kelsey, produced *Standardized Plant Names* (1923), guide for American nurserymen.

Around 1910 became primarily interested in blueberries. First to discover the necessity of acid soil and effects of cold on blueberries and other plants. Improved strains of wild blueberries and succeeded in raising diameter of the blueberry from 12mm to 25.9mm. Blueberries thus became a valuable commercial crop in Northeast. Received for this work the George Roberts White Medal of Honor from the Massachusetts Horticultural Society in 1931.

Led a long campaign to establish the National Arboretum in Washington, D.C., by enlisting sponsors, arranging soil surveys, examining proposed sites, testifying before Congress, and drafting the enabling legislation finally passed in 1927. Last years were spent on a revision of the Death Valley plants.

MAJOR CONTRIBUTIONS

With forty-nine years of government service in the USDA, helped provide direction for government's plant and commodity investigations and researches. Did significant work in economic botany, especially concerning the West, and in taxonomic studies, especially the rushes and berries. Best remembered for "taming the wild blueberry" and as father of the National Arboretum.

BIBLIOGRAPHY

Barnett, Claribel. "Frederick Vernon Coville: Friend of the Library." *Agriculture Library Notes* 12, 2 (February 1937).

Brown, F. C., and Arthur W. Palmer. "Frederick Vernon Coville." *The Cosmos Club Bulletin* (January 1967).

Journal of the Washington Academy of Sciences 27.

Maxon, William R. "Frederick Vernon Coville." *Science* 85 (1937).

UNPUBLISHED SOURCES

See office files, in the records of the Bureau of Plant Industry, Record Group 54, National Archives and Records Service, Washington, D.C.; letters, notes, and surveys on medicinal plants, in Frederick Vernon Coville Papers, Record Unit 7272, Smithsonian Archives, Washington, D.C. (primarily c.1893–1900); letters and unprocessed materials on the history of the National Arboretum, National Arboretum Herbarium, Washington, D.C.; letterpress book and reprints, the Frederick V. Coville Collection, University of Wyoming, Division of Rare Books and Special Collections, Laramie, Wyoming. Coville Field Diary, 1891, Bancroft Library, University of California, Berkeley.

Susan W. Glenn

CRIDDLE, NORMAN. Born Addlestone, Surrey, England, 14 May 1875; died, Brandon, Manitoba, Canada, 4 May 1933. Naturalist, entomologist, botanist, artist.

FAMILY AND EDUCATION

Oldest son of Percy Criddle, a merchant in Addlestone, and Alice Nicol. Elder brother of Stuart Criddle. Family emigrated to Canada in 1882, settled on a homestead at Aweme, Manitoba. No formal schooling, taught by highly educated parents. Awarded an honorary diploma in agriculture by the University of Manitoba, 1933. Never married.

POSITIONS

Worked on his father's farm at Aweme, Manitoba, till 1910. Winters of 1910–1913 was a seed analyst, Calgary, Alberta. Held temporary entomological field officer position in 1913, Treesbank, Manitoba. Entomologist, Dominion Entomological Laboratory, Treesbank, Manitoba, 1914–1933.

CAREER

In his preprofessional days he observed, studied, recorded, and described the flora and fauna on the farm and its environs. Submitted annual ecological notes of the region to the National Museum of Canada. Originated (1901) the ''Criddle mixture,'' a horse manure-arsenical sawdust bait for the control of grasshoppers. Published thirty papers on the biology of insects, birds, and mammals prior to 1914. Although expert in the identification, life history, habits, and control of dozens of species of field crop, garden, and household insects, his most comprehensive work involved grasshoppers. Life history and habits of seventy species were described and figured; the ecology and food habits were determined for another ten. Basic biological research on the wheat stem sawfly, Hessian fly, crickets, and lesser insect pests laid a firm foundation for future entomological work on the prairies. He was a recognized authority on the control of field crop insect pests; his advice was sought and respected by all levels of government. As a representative at Dominion-Provincial conferences of game officials he, for many years, influenced the adoption of adequate conservation measures regarding prairie wildlife.

Published 130 scientific and popular articles on insects, including tiger beetles (1910), wheat stem sawfly (1917), locust control (1920), grasshopper control (1931), biology of Acrididae (1933); and on birds: bluebird (1904), snowbird (1911), crow (1914), magpie (1923), grouse (1930); and on mammals: skunk (1913), rat (1918), coyote (1923), weasel (1925). His artistic excellence is displayed in his paintings of seventy-one weed species (*Farm Weeds* [1909]) and twenty-seven colored plates (*Fodder and Pasture Plants* [1913]). Membership and active participation in eight Canadian and American scientific societies of natural history, entomology, and ornithology gave credence to his work and gained the respect of his peers.

Endowed with an artistic temperament and a love of music, his lifelong study of animals and plants was nurtured by exceptional powers of observation, love of nature, and a preservation of life. Although possessed of a kind, modest, and quiet-spoken demeanor, a dedicated thoroughness of investigation and a zealous defense of any wild creature under attack lent authority and respect for his actions and public works. No one individual has matched or replaced his authoritative impact on economic entomology of the Canadian prairies.

MAJOR CONTRIBUTIONS

Devised simple and effective control measures, such as the "Criddle mixture," for pest insects; developed a basic survey method of grasshopper population density to forecast outbreaks; showed that successful economic insect pest control was possible following the judicious analysis of the basic biology, bionomics, and ecology of a pest species. By word and action he popularized the conviction that lowly and oft denigrated plants and animals deserved attentive consideration, they being an integral and necessary part of nature.

BIBLIOGRAPHY

Criddle, Alma. *Criddle-de-diddle-ensis* (1973).
Gibson, A., and H. G. Crawford. *The Canadian Entomologist* (September 1933) (obit.).
Lloyd, Hoyes. *The Canadian Field-Naturalist* (November 1933) (obit.).
Riegert, Paul W. *From Arsenic to DDT, A History of Entomology in Western Canada* (1980).

Paul W. Riegert

CRIDDLE, STUART. Born Addlestone, Surrey, 4 December 1877; died Sidney, British Columbia, 23 October 1971. Farmer, mammalogist, plant breeder.

FAMILY AND EDUCATION

Son of Percy and Alice Nicol Criddle and brother of naturalist Norman Criddle. Emigrated with family to Aweme, Manitoba, 1882. Taught by his well-educated mother at home. Married Ruth Watkins, 1919, (died 1960), one son and one daughter. Married Kathleen Haynes, 1963. Although he had no formal schooling whatever, he was awarded an honorary D.Sc. at the first convocation of Brandon University in 1968.

POSITIONS

Farmer for most of his life. Moved to British Columbia, 1960. A charter member of the American Society of Mammalogists, a founding member of the Manitoba Museum Association, serving on its executive council for twenty-five years, and an honorary game guardian in the 1930s. Served on the Manitoba Game Advisory Committee.

CAREER

He published two papers in *Ottawa Naturalist*, fifteen in *Canadian Field-Naturalist*, one in *Copeia*, and two in *Journal of Mammalogy*. Moved to Sidney, British Columbia, in November 1960.

MAJOR CONTRIBUTIONS

He kept meticulous field notes of mammal (and bird) life. He carried out many breeding experiments with sunflowers, corn, and lilies and had a new variety of lily named after him.

BIBLIOGRAPHY

Bird, C. D. ''Stuart Criddle, 1877–1971.'' *Canadian Field-Naturalist* 87 (1973).

Criddle, Alma. *Criddle-de-diddle-ensis, A Biographical History of the Criddles of Aweme, Manitoba Pioneers of the 1880's* (1973).

C. Stuart Houston

CROFT, HENRY HOLMES. Born London, England, 6 March 1820; died San Diego, Texas, 1 March 1883. Chemist, entomologist.

FAMILY AND EDUCATION

Son of William Croft, deputy paymaster general of the Ordnance. Croft attended school in London and developed interest in chemistry privately. On Faraday's advice, was sent to University of Berlin in 1838 to study with Mitscherlich but did not take the Ph.D.

POSITIONS

Took a clerkship in the Ordnance Office before entering Berlin. Appointed professor of chemistry at King's College, Toronto, in 1842. Vice chancellor of University of Toronto 1850–1853. Professor of chemistry and chairman of the board of the School of Practical Science, Toronto, 1877–1880. Retired 1880.

CAREER

Primary teaching interest was experimental chemistry. Built up the University of Toronto's chemical laboratory in late 1850s. Highly regarded as a teacher and published *Course of Practical Chemistry, as Adopted at University College, Toronto* in 1860. Published little research. Was an experienced toxicologist and employed regularly for forensic purposes. Active entomologist. Founding member and later president of the Canadian (later Royal Canadian) Institute. Active in the Toronto Mechanics' Institute and Entomological Society of Canada.

MAJOR CONTRIBUTIONS

Foremost academic chemist in midcentury Canada; created the basis for the Department of Chemistry at Toronto. Active promoter of scientific studies in local organizations.

BIBLIOGRAPHY

Craig, G. M. ''Henry Holmes Croft.'' *Dictionary of Canadian Biography*. Vol. 11 (1982).
[Ellis, W. H.] *University of Toronto Monthly* (1901–1902).
King, John. *McCaul, Croft, Forneri: Personalities of Early University Days* (1914).

Richard A. Jarrell

CUSTIS, PETER. Born Deep Creek, Accomack County, Virginia, 1781; died New Bern, North Carolina, 1842. Explorer, naturalist, botanist.

FAMILY AND EDUCATION

Son of a landowner and mill owner on Eastern Shore peninsula of Virginia, and member of famous Custis dynasty of that state, related in marriage to the Randolphs, Lees, and Byrds. A distant relative to Martha Washington's first husband, Daniel Parke Custis. Details of early life sketchy, but it is known that he asked his father to sell his share of the family estate that he might receive a ''Latin education'' and ''be brought up to one of the learned professions.'' A solid secondary education prepared him to enter the medical and natural history program at the University of Pennsylvania in 1804. There he became a pupil, friend, and protégé of Benjamin Smith Barton, the ranking U.S. academic in the field. Took his doctorate at Pennsylvania in 1807, writing as a thesis ''Bilious Fever of Abemarle County [Virginia].'' Following graduation returned to Accomack and started a medical practice at Ocancock, Virginia. By October 1808 had relocated to New Bern, North Carolina, and there apparently practiced medicine until his death at the age of sixty-one.

POSITIONS

Naturalist on President Thomas Jefferson's government exploration of the Red River of the South in 1806. Botanical collector in Virginia for the Barton Herbarium, 1807.

CAREER

At age twenty-five appointed naturalist on Jefferson's second most ambitious exploring probe into the Louisiana Territory. Chosen by expedition leader Thomas Freeman at the suggestion of Custis's mentor, Barton. The twenty-four-man Grand Excursion or Exploring Expedition of Red River, as it was variously termed, set out from Natchez 19 April 1806. Instructed to ascend the Red River to its headwaters, believed to lie near Santa Fe, the party successfully traversed the Great Swamp, a 100–mile maze of waterways created by the immense log-jam known as the Great Raft. Some 635 miles above the mouth of the river, the explorers were blocked and turned back by a superior Spanish force under Capt. Don Francisco Viana. As naturalist of the exploration, Custis assembled one of earliest meteorological charts of Red River Valley, with detailed wind, water, temperature, and storm data. From observations and talks with hunters and Indians, compiled a chart of Red River landmarks. Examined historic villages and sites of the Alabama-Coushatta and Caddo Indians, whose culture he studied and described. Most important work: the natural history catalogs he kept from 2 May to 8 September 1806 and the specimen collections he made, the first such work done in the trans-Mississippi West north of present Mexico. Cataloged twenty-two mammals, thirty-six birds, seventeen fishes, reptiles, and amphibians, three insects, forty-four trees, and 130 vegetables. Subsequent taxonomical work indicates that as many as forty-four of these were new species, and another twenty-eight new subspecies, although Custis recognized no more than a dozen as such. Collected twenty-six botanical specimens and eleven fossil and mineral specimens. Made ecological references respecting early invasion of Red River Valley by some fifteen European plants, as well as advance of several tropical invaders. Wrote of disruption of river valley by advance of the Great Raft. Credited Indian fires with creating the Southern Plains. Following return finished degree at Pennsylvania and briefly collected plants in Virginia for Barton. Apparently abandoned natural history for medicine after 1808.

Coauthor of *An Account of the Red River in Louisiana, Drawn Up from the Returns of Messrs. Freeman and Custis to the War Office of the United States Who Explored the Same, In the Year 1806* (U.S. government) and author of "Observations Relative to the Geography, Natural History, etc., of the Country along the Red River, in Louisiana," *The Philadelphia Medical and Physical Journal* (1806). A remnant of his botanical collection exists in Barton Herbarium of American Philosophical Society collection in the Academy of Natural Sciences, Philadelphia.

Youth and inexperience prevented Custis from publishing and capitalizing on his Red River work. His name is now attached to but a single mammal, while Wilson, Rafinesque, Nuttall, and Pursh are now credited with many species he first encountered and described in his government report. Apparently allowed or encouraged Barton to publish his *Siren quadrupeda* in 1808, a rendering superseded by Cuvier in 1827. Most of his observations too subtle for this early period of discovery. Failure of Samuel Mitchill to publish any notice of his work in *The [New York] Medical Repository*, while doing so in the cases of Lewis and Clark, Dunbar and Hunter, and Pike, must have discouraged him. Failure to promote his report and gain recognition from contemporaries sealed his obscurity.

MAJOR CONTRIBUTIONS

First trained scientist to do a comprehensive natural history survey in trans-Mississippi West north of present Mexico. Made first scientific plant collection in Southwest. Provides modern ecologists with one of the best early descriptions of a major American river system on eve of Anglo occupation. Appointment set a precedent for attaching trained naturalists to American exploring expeditions. Made early ecological references that dispel notions of a static "climax" and paradisiacal continent.

BIBLIOGRAPHY

Barnhart, J. H. "Brief Sketches of Some Collectors of Specimens in the Barton Herbarium." *Proceedings, Philadelphia Botanical Club* (1926).

Barton, Benjamin Smith. "A Discourse on Some of the Principal Desiderata in Natural History, and the Best Means of Promoting the Study of This Science in the United States" (1807).

Barton, Benjamin Smith. "Some Accounts of the *Siren lacertina* and Other Species of the Same Genus." *Philadelphia Medical and Physical Journal Supplement* (1808).

Cox, Isaac J. *The Early Exploration of Louisiana* (1906).

Flores, Dan. *Rendezvous at Spanish Bluff: The Chronicle of Jefferson's Red River Exploration* (1982).

Jackson, Donald. *Jefferson and the Stony Mountains: Exploring the West from Monticello* (1981).

Lowery, George H. *Mammals of Louisiana and Its Adjacent Waters* (1974).

Rowland, Eron Dunbar. *Life, Letters and Papers of William Dunbar* (1930).

UNPUBLISHED SOURCES

See Custis Family Papers in the University of North Carolina Library (Chapel Hill); letters in the Benjamin Smith Barton Papers in the Library of the American Philosophical Society (Philadelphia); Peter Force Collection, Division of Archives and Manuscripts, Library of Congress (Washington, D.C.); Thomas Jefferson Papers, Library of Congress; Accomack County Deeds (Ocancock, Va.); Wills and Administrations of Accomack County, 1663–1800; Craven County, N.C., Superior Court, Will Book D (New Bern).

Dan Flores

CUTLER, MANASSEH. Born Killingly, Connecticut, 13 May 1742; died Hamilton, Massachusetts, 28 July 1823. Congregationalist clergyman, botanist, scientist, colonizer.

FAMILY AND EDUCATION

Son of Hezekiah and Susanna (Clark) Cutler. Graduated Yale University in 1765 with a bachelor of arts degree and in 1768 with a master of arts degree. Later studied law, theology, and medicine. In 1791, received an honorary LL.D. from Yale. Married Mary Balch on 7 September 1766; four children. Was admitted to the Massachusetts bar in 1767. Licensed to preach in 1770, was ordained a Congregationalist clergyman in 1771.

POSITIONS

Pastor of Ipswich Hamlet (now Hamilton), Massachusetts, 1771–1823. Served as a chaplain in the Revolutionary War. In 1782, opened a private boarding school that operated for more than twenty-five years. An organizer of the Ohio Company. A colonizer of the Ohio River Valley (1786) and one of the founders of Marietta, Ohio (1788). Instrumental in drafting the Ordinance of 1787 for the administration of the Northwest Territory and was elected a director of the Ohio River Company in 1789. Declined position as a judge on the Supreme Court of Ohio Territory (1795). Representative from Ipswich to Massachusetts General Court in 1800. Elected to the U.S. House of Representatives (1800–1804).

CAREER

Pursued many careers and fields of study in his lifetime. Taught in Dedham, Massachusetts, in 1766. From 1766 to 1769, operated a store on Martha's Vineyard. During this period, was admitted to the Massachusetts bar. Due to the influence of his father-in-law, Rev. Thomas Balch of Dedham, he decided to study theology from 1769 to 1770. To supplement his income, began practicing medicine on a part-time basis in 1779.

Involved in numerous activities, found time to pursue his scientific interests in astronomy, meteorology, physical sciences, and botany. Measured the distance of some of the stars with a sextant and telescope; entered the positions of Jupiter's moons in his journal; observed hairs and other objects through a microscope; described an aurora borealis; and inoculated people for smallpox.

Due to his knowledge of astronomy taught seamen the art of navigation, particularly lunar observations. In company with six others, made the first ascent to the summit of Mt. Washington in 1784 and overestimated its height by about 3,000 feet. Is best remembered for his botanical studies. Was a member of the American Academy of Arts and Sciences and contributed papers to its *Proceedings*. Also, his scientific articles appeared in *Memoirs of American Academy of Sciences* (1785). Was a member of the American Philosophical Society (1784), Philadelphia Linnaean Society (1809), American Antiquarian Society (1813), and New England Linnaean Society (1815) and honorary member of the Massachusetts Medical Society.

MAJOR CONTRIBUTIONS

Chief scientific contribution was his systematic cataloging of over 350 species of the flora of New England according to the Linnaean method of botanical classification.

BIBLIOGRAPHY

Barnhart, Clarence L., ed. *New Century Cyclopedia of Names* (1954).

Cutler, William Parker, and Julia Perkins Cutler. *Life, Journals, and Correspondence of Rev. Manasseh Cutler, LL.D.*. 2 vols. (1987).

Dictionary of American Biography. (1930)
Elliot, Clark A. *Biographical Dictionary of American Science: The Seventeenth through the Nineteenth Centuries* (1979).
Van Doren, Charles, and Robert McHenry, eds. *Webster's American Biographies* (1974).

UNPUBLISHED SOURCES

See Manasseh Cutler Papers, Essex Institute, Salem, Massachusetts (collection of some of his botanical notebooks and seven letters); Manasseh Cutler Papers, Northwestern University Library (extensive collection of his diaries, sermons, letters, amd miscellaneous writings); Manasseh Cutler Papers, Ohio University Archives, Athens, Ohio (seventy items, including correspondence, sermons, and botanical notebooks); Manasseh Cutler Papers, Yale University Library; Ohio travel journals, Western Reserve Historical Society Collections, Cleveland, Ohio (accounts of travels in, or to, Ohio that include the writings of Manasseh Cutler).

Karen M. Venturella

D

DALL, WILLIAM HEALEY. Born Boston, 21 August 1845; died Washington, D.C., 27 March 1927. Naturalist, malacologist, paleontologist.

FAMILY AND EDUCATION

Son of Charles Henry Appleton Dall, Unitarian minister, and Catherine Wells Healey Dall. Educated in Boston public schools and at Boston Latin School. Studied for several years at Harvard (1861–1863) under Louis Agassiz and Jeffries Wyman, concentrating on anatomy, medicine, and mollusks. Studies interrupted by military service, 1863. Did not take a degree. Honorary M.A., Wesleyan, 1888; Honorary D.Sc., University of Pennsylvania, 1904; Honorary LL.D., George Washington University, 1915. Married Annette Whitney, 1880; four children.

POSITIONS

With Col. J. S. Foster in Geological Survey of Lake Superior area, 1863–1864. Clerk, Illinois Central Railroad, Chicago, 1864–1865. Lieutenant and naturalist, International Telegraph Expedition to Alaska, 1865–1868 (became leader of expedition on death of Robert Kennicott in Alaska, 1866); studied his collections at Smithsonian, 1868–1870; with U.S. Coastal Survey, Alaska, 1871–1884. (In command, four cruises among Aleutians and north to point just below Point Barrow, 1871–1880); assistant with Coast and Geodetic Survey, 1881–1884; paleontologist, U.S. Geological Survey, 1884–1925 (made six trips to northwest coast, 1890–1910); honorary curator, U.S. National Museum, 1880–

1927; honorary professor of invertebrate paleontology, Wagner Institute of Philadelphia, for many years after 1893.

CAREER

Displayed interest in collecting natural history specimens and shells as a child. Studied zoology with Agassiz and anatomy with Jeffries Wyman at Harvard, but studies interrupted by brief Civil War militia service and not resumed (1863). Owing to father's extensive missionary work in India, was encouraged to enter tea business there but declined. Clerked for Illinois Central Railroad while studying natural history at Chicago Academy of Sciences at night under William Stimpson and Robert Kennicott. Invited by Kennicott to be part of Western Union expedition to Alaska, 1865, took charge of expedition party following Kennicott's death in 1866. Explored parts of Alaska, Aleutian Islands, and Pacific Coast after Western Union abandoned plans for overland telegraph following laying down of second Atlantic cable that same year. Published authoritative text *Alaska and Its Resources* (1870); several editions followed, last in 1897. Spencer F. Baird was instrumental in Dall's receiving appointment to U.S. Coast Survey, 1871. Had charge of four cruises to Aleutians and more northerly waters; published *Pacific Coast Pilot, Coasts and Islands of Alaska* (1883). Transferred to U.S. Geological Survey as paleontologist, 1884, where he concentrated on Cenozoic mollusks. Led six expeditions to Pacific Northwest between 1890 and 1910. He ultimately published his findings on mollusks of the Northwest in *Summary of the Marine Shellbearing Mollusks of the Northwest Coast of America from San Diego, California, to the Polar Sea* (1921). Spent much of his life at Smithsonian after 1884 when not in the field. Honorary curator there from 1881, also held honorary curatorship at Bishop Museum, Honolulu. Was a participant in the Harriman–Alaska Expedition of 1899–1900. Other publications included "Contributions to the Tertiary Fauna of Florida" in six parts (1890–1903). His bibliography, compiled in 1946, listed 1,607 books, monographs, articles, reviews, and other items, including papers on vertebrates, climate, and anthropology of Alaska, mollusks collected by the Coast Survey steamer *Blake* in the Caribbean and Gulf of Mexico, land shells of the Galapagos, and other subjects. Dall also authored first full-length biography of his friend and mentor Spencer F. Baird in 1915. During his career, named and described more than 5,400 "genera, subgenera, and species of molluscs and brachipods, both recent and fossil" from various parts of the world. Dall was a member of the American Association for the Advancement of Science, National Academy of Sciences, Phi Beta Kappa, American Academy of Arts and Sciences, California Academy of Sciences, and a number of other organizations.

MAJOR CONTRIBUTIONS

Dall's Alaskan specimens and his writings on mollusks there, in the Caribbean, and elsewhere, continue as a basic source of information for malacologists and paleontologists. The Smithsonian and the National Museum of Natural History share his collections and studies.

BIBLIOGRAPHY

Boss, K. J., J. Rosewater, and F. A. Ruhoff. *The Zoological Taxa of William Healey Dall* (1968).

Dictionary of American Biography. Vol. 5 (1936).

Dictionary of Scientific Biography. Vol. 3 (1971).

National Cyclopaedia of American Biography. Vol. 27 (1967).

Who Was Who in America. Vol. 1 (1897–1942).

Woodring, W. P., "William Healey Dall," *Biographical memoirs, National Academy of Sciences*. Vol. 31 (1958).

UNPUBLISHED SOURCES

Papers are in following collections: Smithsonian Institution Archives, William Dall Papers; Division of Mollusks, U.S. National Museum; Baird Ornithological Club of Washington, D.C.; Western Union Telegraph Expedition Collection.

Anthony Todman

DANA, JAMES DWIGHT. Born Oneida County, New York, 12 February 1813; died New Haven, Connecticut, 14 April 1895. Geologist, educator, zoologist, professor.

FAMILY AND EDUCATION

Son of James Dana, successful businessman in Oneida County, and Harriet Dwight Dana. Eldest of ten children. Attended high school at Bartlett Academy, Utica, New York, 1827–1830. Attended Yale University, received S.B., 1833. Received honorary Ph.D., University of Munich, 1871; honorary LL.D.s from Harvard, 1886, and from the University of Edinburgh, 1890. Married Henrietta Frances Silliman, daughter of Benjamin Silliman of Yale, 1844; four children. Eldest son, Edward Salisbury Dana, became leading American mineralogist of his day.

POSITIONS

Instructor in mathematics, U.S. Navy, 1833–1836. Assistant to Benjamin Silliman, Yale University, 1836–1837. Geologist and mineralogist accompanying Capt. Charles Wilkes's naval expedition that explored the South Seas, 1838–1842. Engaged in writing official expedition reports, Washington, D.C., 1842–1856. Professor of natural history, Yale University, 1849–1864; professor of geology and mineralogy, Yale University, 1864–1890.

CAREER

As student at Yale, Dana was encouraged by future father-in-law, Benjamin Silliman, to pursue a career in the field of science. Dana's first book, *A System of Mineralogy*, appeared in 1837. His participation in the Great United States Exploring Expedition (commonly known as the Wilkes Expedition in honor of its commander, Charles Wilkes), was probably the making of his scientific career. He initially served as geologist but added marine biology to his responsibilities when Joseph P. Couthouy left the expedition in Australia. Following his return in 1842, spent fourteen years completing three volumes based on observations made and specimens collected on expedition.

Zoophytes (1846), *Geology* (1849), and *Crustacea* (1852–1854) considerably heightened his standing in the scientific community, as did his experience with the Wilkes Expedition. His account of zoophytes (since renamed coelenterates) was a pioneering one, since so little had previously been known about the subject. Many of the classification decisions made by Dana have remained standard for 150 years. His account of the corals was separately printed and underwent three editions during his lifetime. His study of coral reefs essentially substantiated Darwin's earlier argument that subsidence was the critical element in their formation. Dana also spent much time in the study of volcanoes. Originally a catastrophist with respect to biology, Dana resisted Darwin's views on evolution for many years and did not modify his views until the mid-1870s. Dana could not accept his full-time professorship at Yale until 1856 because of his work on the Wilkes reports. The majority of his subsequent career was spent at Yale, teaching, researching, and writing. In 1859, suffered a nervous breakdown, and for the rest of his life, he continued with his work only by carefully marshaling his strength. Nevertheless, he produced twenty-four books and over 200 articles and papers, an output unequaled by many of his contemporaries. His most important books included *Manual of Geology* (1862), *Corals and Coral Islands* (1872), and *Characteristics of Volcanoes* (1890). President, American Association for the Advancement of Science, 1854–1855.

MAJOR CONTRIBUTIONS

Dana's work was characterized by both his breadth of vision and devotion to scientific detail. His report of the Wilkes Expedition broke new ground in the study of zoophytes and crustacea. For many years his books on mineralogy and geology were the standard works in the field. Considered one of the greatest American geologists of the nineteenth century.

BIBLIOGRAPHY

Bartlett, H. H. "The Reports of the Wilkes Expedition, and the Work of the Specialists in Science." *Proceedings of the American Philosophical Society* 82 (1940).

Dana, E. S. "James Dwight Dana." *American Journal of Science* 49 (1895).

Dictionary of American Biography.

Dictionary of Scientific Biography.

Gilman, Daniel C. *The Life of James Dwight Dana, Scientific Explorer, Mineralogist, Zoologist, Professor in Yale University* (1899).

Sanford, W. F., Jr. "Dana and Darwinism." *Journal of the History of Ideas* 26 (1965).

Stanton, William. *The Great United States Exploring Expedition of 1838–1842* (1975).
Viola, Herman, and Carolyn Margolis, eds. *Magnificent Voyagers: The U.S. Exploring Expedition, 1838–1842* (1985).

UNPUBLISHED SOURCES

Letters of J. D. Dana, letters in Dana Family Papers, letters in Miscellaneous Manuscripts Collections, letters in Silliman Family Papers, and Gilman Family Papers are all at Yale University. Other letters are at American Philosophical Society, Harvard University, University of Rochester, and University of Illinois.

Patrick J. McNamara

DARLING, JAY NORWOOD ("DING"). Born Norwood, Michigan, 21 October 1876; died Des Moines, Iowa, 12 February 1962. Pulitzer Prize-winning cartoonist, conservationist, chief of U.S. Biological Survey, founder of National Wildlife Federation.

FAMILY AND EDUCATION

Jay Darling's father, Rev. Marcellus Warner Darling, was a school superintendent and college professor turned preacher. His mother was Clara Woolson. Both parents were graduates of Albion (Michigan) College. The family moved several times and finally settled in Sioux City, Iowa, when Jay was nearly ten years old. He graduated from high school there and went to Yankton (South Dakota) College in 1894. He attended Beloit (Wisconsin) College from 1895 to 1898, when he was expelled. He worked for year and returned to graduate from Beloit with a bachelor's degree in 1900. He received honorary Litt.D. from Beloit in 1925 and honorary doctorate from Drake University in 1926. Married Genevieve Pendleton, 31 October 1906; three children:

POSITIONS

Reporter and cartoonist, *Sioux City Journal*, 1900–1906; cartoonist, *Des Moines Register and Leader*, 1906–1911; cartoonist, *New York Globe*, 1911–1913; cartoonist, *Des Moines Register and Tribune*, 1913–1949; syndicated by *New York Herald Tribune*, 1916–1949; chief, U.S. Biological Survey, 1934–1935.

CAREER

Upon his graduation from Beloit College, Darling planned to seek medical degree, but his family could not afford tuition. He hoped to save for his education while working at the *Sioux City Journal* as reporter and photographer. He once submitted original sketch, when he failed to get the photo his editor ordered, and his cartooning career was launched. The *Journal*'s circulation brought him to attention of several other editors, who offered him positions. He finally joined Gardner Cowles, Sr., in 1906, in rebuilding faltering *Des Moines Register and Leader*. Darling's work as cartoonist became extremely popular, and his signature—a contraction of his last name—became a household word. In 1911, accepted offer from the *New York Globe*, in part because paper offered to syndicate his cartoons. He disliked living and working in New York, however, and returned to the *Register* and newly renamed *Tribune* in Des Moines in 1913. In 1916, his work was syndicated nationally by the *New York Herald Tribune*, whose management did not insist that Darling reside in New York. In 1924, he received his first Pulitzer Prize for editorial cartooning. The second was awarded for cartoon published in 1942. A conservative Republican, Darling was sworn in as chief of the U.S. Biological Survey 10 March 1934 and served one and a half years as the nation's conservation officer in Franklin Roosevelt's New Deal administration. While head of survey, Darling cracked down on game law violators, greatly expanded national wildlife refuge system, and began the Duck Stamp program to raise funds for land acquisition. He designed first Duck Stamp and flying goose symbol that today marks every federal wildlife refuge. Following resignation as chief of the Biological Survey, Darling concentrated on creating federation of conservation organizations to help combat abuse and waste of soil, wildlife, and water. He was founder and first president of National Wildlife Federation. By the time he retired in 1949, Darling had fashioned approximately 15,000 cartoons for the *Des Moines Register and Tribune*, the *New York Globe*, the *New York Herald Tribune* and its syndicate, *Collier's*, *The Saturday Evening Post*, *Better Homes and Gardens*, and other publications. Anthologies of his work were printed by the *Register and Leader* and the *Register and Tribune*. He wrote and illustrated several popular books, including *Ding Goes to Russia* (1932), *The Cruise of the Bouncing Betsy* (1937), and *It Seems Like Only Yesterday* (1960). With John Henry, he compiled *Ding's Half Century* (1962).

MAJOR CONTRIBUTIONS

In words and pictures, Darling was able to explain complex conservation issues in easily understood and dramatic terms. He was articulate exponent of the principles of conservation for sake of humankind's survival. Many of his cartoons are as timely now as when they were drawn, and they continue to appear in contemporary texts and cartoon anthologies. The national wildlife refuge system is perhaps his greatest legacy. The J. N. "Ding" Darling National Wildlife Refuge on Sanibel Island, Florida, was dedicated in his memory in 1978.

BIBLIOGRAPHY

Laycock, George. *The Sign of the Flying Goose: The Story of Our National Wildlife Refuges* (1973).

Lendt, David L. *Ding: The Life of Jay Norwood Darling* (1978).

Mills, George. *Harvey Ingham and Gardner Cowles, Sr.: Things Don't Just Happen* (1977).

Trefethen, James B. *Crusade for Wildlife: Highlights in Conservation Progress* (1961).

David L. Lendt

DARLINGTON, WILLIAM. Born Dilworthtown, Pennsylvania, 28 April 1782; died West Chester, Pennsylvania, 23 April 1863. Physician, botanist, author, congressman, architect, engineer, bank president.

FAMILY AND EDUCATION

Son of Edward Darlington and Hannah Townsend. Father was a farmer, state representative, and road surveyor. Received M.D. from University of Pennsylvania in 1802. Married Katherine Cacey and had five children who survived into adulthood.

POSITIONS

Worked in mapping local botanical populations and in charting weed introduction, 1826–1853; also worked in assessing the therapeutic values of medicinal botanicals (1854). Published five books at his own expense on botany and botanical scientists, two of which are well known for their botanical-historical facts. Designed and built the local West Chester Academy. Founded (1809–1828) two medical societies (president of one 1828–1852). Helped to organize state medical society; delegate to American Medical Association convention in 1851. Three-term congressman, 1815–1822; three-time candidate for governor of Pennsylvania. Served on Pennsylvania Canal Commission (1826–1827) and helped to design and engineer the project; designed and engineered a state road from the Delaware River to the Susquehanna River (1830–1832). President of the Bank of Chester County (1830–1863) and delegate to banking conventions in the interim. Founder and president of Chester County Cabinet of Natural Science, 1828–1863 (now West Chester State College). Founder of Teacher's Institute, 1855, and of West Chester Library and the West Chester Athenaeum. Founder and president of West Chester Rail Road, 1830–1844. Organizer and officer of the West Chester Grays, an official militia organization that metamorphosed into the Republican Artillerists, 1814–1828. Vocal advocate of abolition of capital punishment. Newspaper editor, 1808, and author of innumerable professional articles.

CAREER

Renowned on a local, state, national, and international level as a working botanical scientist deeply interested in the development of educational resources of scientific farming. Practiced medicine for thirty-five years, although often absent during a ten-year span when active on national political scene. Founder of medical and scientific societies and member of over forty professional or scientific societies. Among his works are *Florula Cestrica* (1826); *Flora Cestrica* (1837, 1853); *Agricultural Botany* (1847; 1859 editor: George Thurber, who took over the book and reissued it in the 1880s); *Reliquiae Balwinianiae* (1843); *Memorials of Bartram and Marshall* (1849); *Natae Cestriensis* (with J. Smith Futhey, 1862); "Medical Botany," *Chester, Delaware Medical Recorder* (1854).

MAJOR CONTRIBUTIONS

Notable botanist and agricultural educator, who also found time for careers in banking, railroads, and politics.

BIBLIOGRAPHY

James, T. P. *Memorial of William Darlington, M.D.* (1853).
Kelly H. A., and W. Burrage. *American Medical Biographies* (1920).
Lansing, D. I. *William Darlington* (1965).
Pleasants, H. S. *Three Scientists from Chester County, Pa* (1936).
Townsend, W. *Memoir of Dr. William Darlington* (1853).

Dorothy I. Lansing

DAVIDSON, FLORENCE EDENSHAW, alias Jadał q' egəngá (Story Maid). Born in Haida village of Masset, Queen Charlotte Islands, British Columbia, 15 September 1896; died Masset, British Columbia, 13 December 1993. First Nation Elder, ethnobotanist, artist.

FAMILY AND EDUCATION

Daughter of Charles (1839–1920) and Isabella Edenshaw (1858–1926). Father was a major Haida artist, whose creations were collected or studied by Marius Barbeau, Franz Boas, C. F. Newcombe, and John Swanton. Received a traditional Haida education from her mother. Attended Anglican mission school in Masset until fourth level. Non-traditional formal education ceased with her arranged marriage to widower Robert Davidson, 1911. Thirteen children. Widowed, 1969.

POSITIONS HELD

Mother. Member of Anglican Church's White Cross ladies, later Women's Auxiliary (W.A.). Elected W.A. President (1923). Responsibilities included preparation of the community's dead for funeral services. Life service in W.A. in various positions: president, vice-president, secretary, and treasurer. Waged positions included work as a cannery worker (Masset, Naden, Tow Hill, and Rose Inlet) baker, cook, operator of a small home-based restaurant and boarding house. Artist renowned for her button blankets and woven items such as baskets and hats. Cultural ambassador on many formal occasions, including dinner with Prime Minister Pierre Elliot Trudeau and welcoming Queen Elizabeth II to Sandspit (1971).

CAREER

Grew up in a household that combined old traditions and new values. Family's annual cycle consisted of traditional seasonal movement, modified by father's work as a successful Haida artist. From February to March, drying halibut at Kung, return to Masset in April, then to Yatz where women prepared potato gardens and men fished halibut. Brief return to Masset before women and children went to Yatz in early May to pick and dry seaweed. Late May, Florence and her mother gathered spruce roots for weaving hats and baskets. In June family went to the mainland, selling hats and baskets and buying winter clothes. Father carved while mother worked at the Inverness cannery, Skeena River, or less often, Kasaan cannery, Alaska. Late summer, family returned. Women gathered huckleberries, crabapples, and salal berries. Mission school began for children. Late fall, women and children gathered cranberries and harvested potatoes. Family purchased or traded for fall salmon. Later as an adult Florence spent late fall upriver with husband, slicing, smoking, and canning sockeye and dog salmon. After her traditional puberty seclusion, a marriage was arranged with Robert Davidson. Couple lived with her parents. After birth of two children they moved into their own home. In 1929 the couple moved into a two-story house. In 1952 they lost everything in a house fire and began again. Florence's life was dominated by childrearing. Husband worked as carpenter, commercial fisherman, handlogger, and as a trapper in the Yakoun River valley. New house paid for by argillite carvings, a skill her husband learned from Charles Edenshaw. Husband also carved canoes, and Florence painted one in a traditional pattern in 1937. His grandson and namesake raised first totem pole of the twentieth century on the Queen Charlotte Islands a month before his death in September 1969.

After a period of mourning she began to take in boarders: visitors, education and medical professionals, and researchers. As Nani (''grandmother''), her kitchen became a meeting place for those involved in the exploration and celebration of Haida culture. From 1977–1981 worked with anthropologist Margaret B. Blackman on the publication of her life history (1982), later revised. A distinguished Elder and the oldest member of the Masset Haida Band, she acted as ceremonial advisor and custodian of her lineage's chieftainship. Died at her place of birth.

MAJOR CONTRIBUTIONS

''Respect for self'' is an essential concept for Haida culture. Her life stands as a model for her ability to build community from that statement. First Haida woman to leave a detailed life history, her story marks a major shift in a century of anthropological literature dominated by stories recorded by males about males. As Nani she became a major source for information on Northwest coast ethnobotany. Sharing her extensive knowledge of plants and their usage with researchers such as Nancy Turner, she expanded understanding of the traditional interaction of First Nation women with the natural world.

BIBLIOGRAPHY
Blackman, M. B. *During My Time: Florence Edenshaw Davidson: A Haida Woman* (1982, Revised 1992).
Turner, N. "Plant Taxonomic Systems and Ethnobotany of Three Contemporary Indian Groups of the Pacific Northwest (Haida, Bella Coola, and Lillooet)." *Syesis* (7) supplement 1 (1974).
UNPUBLISHED SOURCES
For information concerning the interviews and the family's wishes concerning the same, contact Dr. Blackman, Chair, Anthropology, State University of New York at Brockport.

Lorne Hammond

DAWSON, GEORGE MERCER. Born Pictou, Nova Scotia, Canada, 1 August 1849; died Ottawa, Ontario, 2 March 1901. Geologist, explorer, surveyor, ethnologist, botanist.

FAMILY AND EDUCATION

Son of J. William Dawson (principal of McGill University 1855–1899), and Margaret Ann Young Mercer. Studied at Montreal High School until illness resulted in permanent spinal curvature, making tutoring by parents and private teachers necessary. Graduated in 1872 from Royal School of Mines, London, winning several awards for natural history and paleontology, including the Edward Forbes Medal and the Murchison Medal in geology.

Received honorary doctorates from Princeton, Queens, McGill, and Toronto; Royal Society of London, fellow; Zoological Society of London, elected member; American Society for the Advancement of Science, fellow; Ethnological Survey of Canada, member; Companion of Order of St. Michael and St. George received from Queen Victoria, Bigsby Gold Medal from the London Geological Society and Gold Medals from the Geological Society of America and the Royal Canadian Geographical Society.

POSITIONS

Royal School of Mines, associate; Morrin College, Quebec, lecturer 1872. British-American International Boundary Commission, geologist and botanist 1873–1875; Geological Survey of Canada (GSC), geologist 1875, director 1895–1901; Royal Society of Canada (RSC), charter member, president, 1893; British commissioner on Bering Sea Natural Resources Arbitration 1892 and on Northwestern Tribes 1894; British Association for the Advancement of Science, Geological Section president 1897; Geological Society of America, president 1900.

CAREER

Surveyed the 49th parallel from Lake of the Woods to the Rockies 1873–1875; published important Boundary Commission Report on the geology and resources of the region. Geologist for British Columbia with the Geological Survey of Canada, then director. Fieldwork in America included areas adjacent to the boundary, the Pacific West Coast and Alaska. Organized Canadian exhibit for Philadelphia International Exhibition, 1876. Published *Comparative Vocabularies of the Indian Tribes of British Columbia* (with W. F. Tolmie, 1884).

Survey of natural resources of British Columbia influenced government decisions on route of Canadian Pacific Railway (CPR). First comprehensive work on Canada's physiography, *A Descriptive Sketch of the Physical Geography and Geology of the Dominion of Canada* (with A. Selwyn, 1884). In 1887 he surveyed the Alaska boundary at the Yukon River and investigated gold discoveries, leading to the gold rush of 1898. Dawson City was named after him. In 1889 the GSC and CPR published his authoritative *Mineral Wealth of British Columbia* and in 1890 *On Some of the Larger Unexplored Regions of Canada.* His RSC 1893 Presidential Address was entitled "Progress and State of Scientific Investigation in Canada." In 1900 *Economic Minerals of Canada* was published for the Paris International Exhibition. He made original contributions to geology, natural history, geography, and anthropology.

MAJOR CONTRIBUTIONS

Fundamental studies of the geology and natural resources of the Western Provinces, British Columbia, and Yukon; pioneering glaciation work (Cordilleran glacier; Laurentide ice sheet, Missouri Couteau); first scientific exploration of the Queen Charlotte Islands and Yukon. Ethnographic observations. Photographs of Haida. Important contributions of artifacts, geological specimens, photographs, paintings, and written works.

BIBLIOGRAPHY

Avrith-Wakeam, Gale, and Suzanne Zeller. "Dawson, George Mercer." *Dictionary of Canadian Biography* Vol. 13 (1994).

Barkhouse, Joyce C. *George Mercer Dawson, The Little Giant* (1974).

Cole, Douglas, and Bradley Lockner. *The Journals of George Mercer Dawson: British Columbia, 1875–1878.* 2 vols. (1989).

Cole, Douglas, and Bradley Lockner. *To the Charlottes: George Mercer Dawson's 1878 Survey of the Queen Charlotte Islands* (1993).

Lockner, Bradley, ed. *No Ordinary Man* (1993).

Winslow-Spragge, Lois. *The Life and Letters of George Mercer Dawson 1849–1901* (1962).

UNPUBLISHED SOURCES

See National Archives of Canada, Ottawa; McGill University Archives, Montreal; Canadian Museum of Civilization, Ottawa; Geological Survey of Canada, Ottawa; McCord Museum, Montreal.

Peter Geldart

DAWSON, JOHN WILLIAM. Born Pictou, Nova Scotia, 13 October 1820; died Montreal, Quebec, 19 November 1899. Paleobotanist, geologist, naturalist, educator, principal, McGill University.

FAMILY AND EDUCATION

Son of Scottish-born Mary Rankine and James Dawson, who eventually became printer in Pictou, Nova Scotia. Young Dawson's enthusiasm for natural history was fanned by attending Pictou Academy, where principal Thomas McCulloch had assembled small museum of local and British specimens. After graduation, Dawson spent several terms in Edinburgh and received an M.A. from the university in 1856. LL.D., Edinburgh (1884); D.L., Columbia. Married Margaret A. Y. Mercer of Edinburgh, 1847. Of six children, five survived to adulthood: three sons and two daughters.

POSITIONS

Superintendent of education for Nova Scotia, 1850 to 1853. Principal and professor of natural science, McGill University, 1855 to 1893, which he transformed from a languishing provincial college to a major university of international stature by recruiting influential patrons and enterprising professors. Principal, McGill Normal School (which trained teachers for Protestant schools in Quebec), 1857 to 1870. From 1893 until his death, Dawson continued to serve McGill as principal emeritus and professor, governor's fellow, and honorary curator of the Peter Redpath Museum.

CAREER

Because of his knowledge of Nova Scotia geology, in early 1840s Dawson met William Logan, first director of the Geological Survey of Canada, and accompanied Charles Lyell on his geological tours of exploration. During this period, government of Nova Scotia employed Dawson to survey coalfields. He also lectured at local educational institutions on geology and natural history.

At age twenty-one, Dawson wrote his first scientific paper (on Nova Scotian field mice), which he contributed to Edinburgh's Wernerian Society. From the 1860s, he published an average of ten scientific papers a year, in addition to articles on educational, social, and religious matters. His scientific writings appear in a wide range of journals and society proceedings, but the *Canadian Naturalist, Quarterly Journal of the Geological Society*, and *American Journal of Science* each published more than thirty papers. Dawson collected data for *Acadian Geology* of 1855 (2d ed., 1868; 3d ed., 1880; 4th ed., 1891) during his travels through Nova Scotia as superintendent of education. This work, the most complete treatment of maritime geology of its day and but slightly modified by the findings of the Geological Survey years later, made his reputation as a geologist of the first rank.

Dawson's contributions to paleontology and geology extend beyond publications. His unstinting fieldwork increased the number of known species of post-Pliocene fossils from around 30 to over 200. His own collection of

Canadian rocks and fossils formed the nucleus of holdings of the Peter Redpath Museum, which was donated to McGill in Dawson's honor in 1882. Dawson discovered the puzzling fossil *Eozoon canadense* in 1864. For the rest of his life he argued that specimen proved presence of animal life in the Laurentian rocks, despite mounting evidence of its inorganic composition.

Dawson also wrote numerous popular scientific works. Some of these served polemical purposes in his arguments against the glacial theory and evolution or to promote his views on eozoon. Examples include *The Story of the Earth and Man* (1872), *The Dawn of Life* (1875), and *Modern Ideas of Evolution* (1890). Almost all, however, sought to reconcile scientific truths with religious doctrine, such as *Archaia* (1857), *Nature and the Bible* (1875), and *The Meeting Place of Geology and History* (1894). Because of Dawson's devout Presbyterianism, intense antievolutionism, and firm commitment to letter of biblical Scripture, he was in great demand as lecturer at religiously affiliated educational institutions across United States and Canada. Reflecting his abiding interest in education and lucid scientific exposition, also wrote textbooks geared toward Canadian students: *Agriculture for Schools* (1864); *Hand-book of Canadian Zoology* (1871); *Hand-book of Canadian Geology* (1889).

In Canada and abroad, Dawson accumulated scientific honors and awards. He was first president of the Royal Society of Canada in 1882 and active member of Montreal Natural History Society, over which he also presided for many years. He became fellow of Geological Society in 1854 and of Royal Society in 1862. An enthusiastic supporter and onetime president of the British (1886) and American (1882) Associations for the Advancement of Science, he brought both organizations to meet in Montreal in the early 1880s. He was knighted in 1884. In 1893 he became president of American Geological Society. He also received honorary or corresponding memberships in many other learned societies.

MAJOR CONTRIBUTIONS

Apart from his important role as a popularizer and institution builder, Dawson's scientific reputation rests upon his work in paleobotany. He investigated formations from the maritimes to western Canada (culminating in the *Geological History of Plants* [1888]) and published several papers on the subject every year. His more controversial contributions to paleozoology treated variety of organisms, ranging from the lowest forms of life to prehistoric man (in *Fossil Men* [1880]). In addition, Dawson's legacy includes pioneering work in Canadian geology, particularly for eastern provinces and St. Lawrence River Valley.

BIBLIOGRAPHY

Adams, F. D. "Memoir of Sir J. William Dawson." *Bulletin of the Geological Society of America* 11 (1899).

Ami, H. M. "Sir John William Dawson: A Brief Biographical Sketch." *The American Geologist* 26 (July 1900).

Clark, T. H. *Dictionary of Scientific Biography*. Vol. 3 (1971).
Collard, E. A. "Lyell and Dawson: A Centenary." *The Dalhousie Review* 22 (July 1942).
"John William Dawson." *The Canadian Biographical Dictionary* (1881).
UNPUBLISHED SOURCES
Major repository is the J. W. Dawson Papers, McGill University Archives (Montreal): letters, papers (published and manuscripts), scrapbooks, and so on.

Susan Sheets-Pyenson

DAY, ALBERT M. Born Humbolt Nebraska, 2 April 1897; died Harrisburg, Pennsylvania, 21 January 1979. Biologist, environmentalist, U.S. government official.

FAMILY AND EDUCATION

Youngest of five children of John Breece and Laura Thayer Day. Family were farmers and ranchers. Family moved to 2,000–acre ranch in eastern Wyoming when Albert was three. There he attended Goldendale High School and Jireh Junior College, graduating in 1916. Corporal in Army Signal Corps during World War I. B.S., University of Wyoming, 1923 (some sources suggest 1921 or 1922). Married Gertrude Wichmann, 1923; two children, and Eva Kendall, 1944; one child.

POSITIONS

Field biologist, U.S. Biological Survey (USBS), 1918; With USBS in Wyoming on predator and rodent control work, 1920–1928. Leader of Division of Predator and Rodent Control in Laramie, Wyoming, 1928–1930. Assistant chief, Division of Predator and Rodent Control (with responsibilities throughout the United States), USBS, Washington, D.C., 1930–1935; chief of that division, 1935–1938. Chief, Division of Federal Aid in Wildlife Restoration (to promote game management throughout the United States, Puerto Rico, and the Virgin Islands), USBS (and, after 1940, U.S. Fish and Wildlife Service [USF&WS]), 1938–1942. Liaison officer among Fish and Wildlife Service, War and Navy Departments, and other government agencies, Washington, D.C., 1942–1943. Assistant director, USF&WS, with wartime offices in Chicago and, later, Washington, D.C., 1943–1946; director, USF&WS, 1946–1953; assistant to the director, USF&WS, 1953–1955; director, Wildlife Research, Arctic Institute of North America, 1955–1958. Director, Oregon Fish Commission, 1958–1960; executive director, Pennsylvania Fish Commission, 1960–1964; conservation consultant, 1964–1979.

CAREER

Major early influence on his later career was childhood in eastern Wyoming. Intrigued by damage done to land and family grain supplies by rodents and had some experience of coyotes, wolves, and bobcats. Studied agriculture, biological sciences, and animal husbandry at University of Wyoming following World War I. Earnings from early work with USBS helped fund his studies. Early publications (under university auspices) included *Magpie Control in Wyoming* (1927) and *Cooperative Wildlife Conservation and Control in Wyoming* (1928). Moving to Washington, D.C., in 1930, had national responsibilities, first for predator and rodent control and later for restoration of wildlife and game in cooperation with the states and territories. Transferred to new U.S. Fish and Wildlife Service, 1940, when the old Biological Survey (from Agriculture) and Bureau of Fisheries (from Commerce) were combined in the Interior Department. As assistant director of USF&WS during World War II, was instrumental in ensuring that certain wildlife products, such as goosedown-filled sleeping bags, deerskin gloves, and deer meat, were utilized to offset decline of hunting during wartime. As director, USF&WS, was responsible for national and international game law enforcement, licenses, protection, and restoration of wildlife, and wildlife research. Was adviser to United Nations Conference on Law of the Sea and to North Pacific Fisheries Commission. Was member, Pacific Marine Fisheries Commission and also International Pacific Salmon Fisheries Commission, 1947–1954. Published *Hunter's Handbook* (1932), *North American Wildfowl* (1949), and *Making a Living: A Guide to Outdoor Careers* (1971) and contributed articles to various publications. National director of Izaak Walton League. Member, American Association for the Advancement of Science; International Association of Game, Fish, and Conservation Commissioners; Wildlife Society; and many other conservation and wildlife organizations.

MAJOR CONTRIBUTIONS

Worked consistently to foster cooperation in wildlife, fisheries, and conservation issues with the States, Canada, Mexico, and other nations. Following his retirement from government service, active conservation consultant and in advisory capacity on international fisheries matters, especially in the Pacific region.

BIBLIOGRAPHY

American Men and Women of Science: The Physical and Biological Sciences (1971–1973).

Current Biography (1948).

Stroud, Richard H., ed. *National Leaders of American Conservation* (1985).

Washington Post, 24 January 1979 (obit.).

Who Was Who in America. Vol. 7 (1977–1981).

Anthony Todman

DEKAY, JAMES ELLSWORTH. Born Lisbon, Portugal, 12 October 1792; died Oyster Bay, New York, 21 November 1851. Zoologist.

FAMILY AND EDUCATION

Son of George, a sea captain, and Catherine Colman DeKay. Attended Yale University, 1807–1812. Began reading medicine in Guilford, Connecticut, around 1811. Attended University of Edinburgh, 1818–1819, received M.D., 1819. Married Janet Eckford, 1821; at least three children.

POSITIONS

Active as editor and librarian, Lyceum of Natural History, New York, 1819–1830. Traveled with father-in-law, Henry Eckford, to Turkey for several months, 1831–1832, and while there undertook thorough study of Asiatic cholera. Commissioned by state of New York to complete zoological portion of State Natural History Survey, 1836–1844.

CAREER

A lover of books and writers from an early age. Also had an interest in plants and animals and put both to good use as librarian of the Lyceum of Natural History, where he edited the first several volumes of its *Transactions*, developed one of the leading collections of scientific literature in the nation for that period, and assisted in the creation of a museum. Published his *Anniversary Address on the Progress of the Natural Sciences in the United States, Delivered before the Lyceum of Natural History of New York* (February 1826). This was an excellent summary of what had been accomplished by American naturalists to that point. He included some discussion of the several federal exploring expeditions to the West. He does not seem to have given much attention to a medical career, although he quickly responded when asked to help treat patients during several cholera outbreaks following his return from Turkey. In 1833, DeKay anonymously published an account of his sojourn and observations there in *Sketches of Turkey by an American*. The New York State Geological and Natural History Survey, "the oldest continuously functioning geological survey in the New World," was established by the state legislature in 1836. It came into being, in part, as a result of petitions submitted by the Lyceum of Natural History and the Albany Institute. This resulted from a growing interest in a survey of the state's natural resources, particularly coal. Offices were established in Albany, and DeKay was appointed zoologist to the survey following the very brief tenure of another man, who had resigned to go into business. His *Zoology of New York*, 5 vol. (1842–1844) was a pioneering work that dealt with fossil, as well as recent, species. The work embraced mammals, birds, reptiles and amphibians, fish, mollusks, and crustacea. This was the second state faunal account to appear in this country, preceded only by several less elaborate volumes in a series put out by Massachusetts. DeKay's work was the most comprehensive effort for any state for some years and a model for its time.

MAJOR CONTRIBUTIONS

In his *Zoology of New York*, DeKay identified over 1,600 species of state fauna and used common names wherever possible to facilitate their use by the general public.

BIBLIOGRAPHY

American Journal of Science (second series) 13 (1852).
Dictionary of American Biography.
Elliott, Clark A., *Biographical Dictionary of American Science* (1979).
National Cyclopedia of American Biography. Vol. 5.
New York Herald, 23 November 1851 (obit.).
Socolow, Arthur A., ed. *The State Geological Surveys: A History* (1988).

Rodney Marve and Keir B. Sterling

DENYS, NICOLAS. Born Tours, France, 1598; died Nepisiguit, Acadia, (near Bathurst, N.B.), 1688. Trader, colonist, writer.

FAMILY AND EDUCATION

Son of Jacques Denys de la Thibaudière, from an engineering family, and Marie Cosnier. Had several brothers: Simon, ennobled by Louis XIV, Jacques, and perhaps Henri. Left school early to learn the fishery. Married Marguerite Lafite in 1642; seven or eight children.

POSITIONS

Agent and representative of la Compagnie de la Nouvelle-France at La Rochelle, 1632 to 1640s; lieutenant governor general of Acadia with Letters Patent for territory from Newfoundland to Virginia, 1653–1660's.

CAREER

Learned the cod fishery early in life. Hired by la Compagnie de la Nouvelle-France in 1632 to outfit and accompany expeditions to Acadia. Made several subsequent trips as he attempted unsuccessfully to establish fishing, trading, and lumbering businesses. Financial problems plagued his many ventures. Had powerful enemy in Charles de Menou d'Aulnay, who assumed control of Acadia in 1635. Made several voyages back to France to petition the French government to restore and protect his possessions in Acadia from d'Aulnay. His lifelong goal of establishing permanent settlement in Acadia, motivated by obligations of his land grant and his optimism and love of the country, ended in failure.

MAJOR CONTRIBUTIONS

One of the founders and the most persistent colonist of Acadia. Final and most notable contribution was his reminiscences entitled "Description géographique et historique des costes de l'Amerique septentrionale: avec l'Histoire naturelle du Païs'' and ''Histoire naturelle des Peuples, des animaux, des arbres et plantes de l'Amerique Septentrionale, & de ses divers climats.'' Although flawed by errors, they are the most extensive source of historical, geographical, and ethnological information on seventeenth century.

BIBLIOGRAPHY

Bélisle, Louis-Alexandre ''Nicolas Denys.'' *Références Biographiques: Canada-Québec.* Vol. 2 (1978).

d'Entremont, Clarence-Joseph. *Nicolas Denys: sa vie et son oeuvre* (1982).

Ganong, William F., ed. *The Description and Natural History of the Coasts of North America*, by Nicolas Denys (1908).

Hansen, Denise. ''Les poteries du XVIIe siècle écouvertes au fort Saint-Pierre de Nicolas Denys à St. Peter's, Nouvelle-Écosse'' [Environment Canada: Service Canadien des Parcs.] *Bulletin de Recherches* 271 (March 1989).

MacBeath, George. ''Nicolas Denys.'' *Dictionary of Canadian Biography.* Vol. 1 (1966).

Société historique Nicolas Denys. ''Catalogue des fonds d'archives conservés au Centre de documentation de la SHND.'' Shippegan, N.B.: Centre de documentation, Centre universitaire de Shippegan (1986).

Theresa Redmond

DERICK, CARRIE (MATILDA). Born Clarenceville, Quebec, 14 January 1862; died Montreal, Quebec, 10 November 1941. Natural scientist, botanist, professor.

FAMILY AND EDUCATION

Daughter of Edna Colton Derick from the United States and Frederick Derick of Quebec's Eastern Townships. Early education at Clarenceville Academy, where she taught at the age of fifteen. Teacher's training at McGill Normal School in Montreal, awarded the J. C. Weston Prize and Prince of Wales Medal. McGill University, B.A., 1890, M.A., 1896. Never married; no children.

POSITIONS

Demonstrator of botany in 1891. Lecturer and demonstrator in botany in 1896. Assistant professor in 1904. Acting chair of botany 1909 for three years. In 1912 was appointed professor of morphological botany, McGill University. McGill's first woman professor emeritus.

CAREER

Principal of Clarenceville Academy in 1881, then two years later, returned to Montreal to teach at the Trafalgar Institute, a private school. Enrolled in Faculty of Arts at McGill in 1887, earned her B.A., achieving first-class honors. Awarded prizes in classics, zoology, and botany and won the Logan Gold Medal in natural science. Invited by David Penhallow, professor of botany, to be his part-time assistant. One year later, he arranged for her to be a part-time demonstrator in botany, a position she held for four years while working on her M.A. Academic advancement and pay increases were slow. Appointed lecturer and demonstrator, then finally promoted to assistant professor in 1904. Studied at Harvard during three summers and at the Marine Biological Laboratory at Woods' Hole, Massachusetts, for seven. In 1898, attended Royal College of Science in London, England, then spent eighteen months at the University of Bonn, Germany. Completed the work required for a Ph.D, but the University of Bonn did not grant degrees to women at that time. Acted as chair of botany in 1909 after Penhallow became ill, a job she did well for three years. When a new chair was sought, Francis E. Lloyd from Alabama was chosen, and Derick was given the "courtesy title" of professor of morphological botany, a position that included neither a seat on the faculty nor an increase in salary. Gave many public lectures and wrote many scientific as well as popular articles, such as "Flowers of the Field and Forest." Gave scientific testimony on a variety of topics to commissions, government departments, and professional organizations. As a geneticist, she was influenced by contemporary Marie Stopes, led public debates on eugenics, and advocated birth control. At a time when women's participation in professional organizations was limited, Derick was a fellow of the American Association for the Advancement of Science, vice president of the Natural History Society of Montreal, and member of the Botanical Society of America, the American Genetics Association, Montreal Philosophical Club, the Canadian Public Health Association, the Executive Committee of the National Council of Education, and first woman on the Protestant Committee of Public Instruction (Quebec), an appointment she held from 1920 to 1937. She resigned from McGill in 1929 due to poor health.

Publications include "The Early Development of the Florideae," *The Botanical Gazette* (Chicago), which was abstracted in *The Journal of the Royal Microscopic Society*. Lectures included "The Origin of New Forms of Plant Life" and "Nature Study and City Life." Other papers on evolution, heredity, genetics, and environment were published in *Science*, *The Canadian Record of Science*, and *Transactions of the Royal Society*. She was listed in *American Men of Science* (1910). A respected scholar, gifted teacher, and concerned public citizen who helped break new ground for women in science.

MAJOR CONTRIBUTIONS

Introduced the teaching of genetics at McGill. Her research on heredity received national and international recognition. First woman appointed on McGill's instructional staff. First woman in Canada to be made full professor.

BIBLIOGRAPHY
Gillett, Margaret. "Carrie Derick (1862–1941) and the Chair of Botany at McGill." *Despite the Odds: Essays on Canadian Women and Science*. Ed. Marianne G. Ainley. (1990), 74–87.
Gillett, Margaret. *We Walked Very Warily: A History of Women at McGill* (1981).
Gillett, Margaret. *The Canadian Encyclopedia* (1988).
Morgan, Henry James. *The Canadian Men and Women of the Time* (1912).

E. Tina Crossfield

DICE, LEE RAYMOND. Born Savannah, Georgia, 15 July 1887; died Ann Arbor, Michigan, 31 January 1977. Zoologist, geneticist, professor.

FAMILY AND EDUCATION

Father a farmer. Before his second birthday, family moved to York, Pennsylvania, and then to Prescott, Washington. Working on family ranch helped develop his love of nature. Supplemented his education with correspondence courses in electrical engineering, drawing, and ornithology from the University of Chicago. Began studies at the Washington State Agricultural College at Pullman in 1908. In 1909, went to the University of Chicago, and next year to Stanford University. He received his B.A. there in 1911. Received M.S. from University of California at Berkeley in 1914 and his Ph.D. in 1915. Married Doris S. Lemon, 1918; three children.

POSITIONS

Deputy fur warden, Alaska Fisheries Service, 1911–1913. Instructor of zoology, Kansas State Agricultural College (Manhattan), 1916–1917. Assistant professor of zoology, University of Montana, 1917–1918. Field assistant, U.S. Biological Survey, 1918. Field assistant, Michigan Geological and Biological Survey, 1919. Curator of mammals, Museum of Zoology, University of Michigan, (UM) 1919–1938; instructor, 1919–1928; assistant professor, 1928–1934; associate professor, 1934–1942; professor, 1942–1957. Research assistant, Laboratory of Vertebrate Genetics (UM) (later renamed the Laboratory of Vertebrate Biology), 1927–1933; director, 1934–1949. Director, Institute of Human Biology (UM), 1950–1957. Consultant, Federal Bureau of Fisheries. Consultant, Carnegie Institution. Trustee, Cranbrook Institute of Science. Board of Directors, American Eugenics Society. Member, Scientific Advisory Committee, Animal Welfare Institute. President, Ecologists Union, 1948. Charter member, American Society of Mammalogists; vice president, 1947–1951. President, Ecological Society of America, 1952–1953. President, Society of Systematic Zoology, 1949–1950. Vice president, Society for the Study of Evolution, 1947–1948, 1953–1954. President, American Society of Human Genetics, 1950–1951. President, Michigan Academy of Science, Arts and Letters, 1936–1937.

CAREER

As deputy fur Warden, main duty was to enforce trapping laws, but also asked to collect specimens and gather information on their distribution. In 1912 traveled by dogsled into the interior of Alaska to collect and after several months traveled with a friend 1,000 miles down the Kuskokwin River. Upon returning to Washington, D.C. eye strain forced him to give up scholarship at Columbia with T. H. Morgan. During summer of 1913 worked with C. C. Adams at University of Montana's Biological Station at Flathead Lake. Next summer collected data for his dissertation near Wallula on Columbia River and in the Blue Mountains. As field assistant in the U.S. Biological Survey he and his new wife spent summer of 1918 in Washington, D.C., working on rat control along the Potomac. After brief tour of duty as army laboratory technician in 1919, spent several months as part-time graduate assistant at the University of Illinois. That summer studied mammals of Warren Woods (Berrien County, Michigan) for the Michigan Geological and Biological Survey.

In 1919 was offered position at University of Michigan and was employed there until his retirement. By 1922 had begun maintaining various species of *Peromyscus* to study their variation, ecology, behaviors, and genetics. When UM Laboratory of Mammalian Genetics was established in 1925 to study heredity of house mice, Dice was made a research associate and later its director. Stock from Scripps Institute was added in 1930, making the laboratory the country's principal center for study of *Peromyscus* genetics. This study suggested implications for human genetics, and in 1940 Dice received university grant to begin this research. In 1941 established Heredity Clinic, and in 1946 J. V. Neel became its director. As more studies were added, Dice's supervision of these projects was centralized by including them all in the Institute of Human Biology, which he directed from its formation in 1950 until his retirement.

Maintained interest in field studies and went on collecting expeditions every summer, 1920–1940. These included several trips to New Mexico, Arizona, and Chihuahua (Mexico), which resulted in report coauthored by P. M. Blossom, ''Studies of Mammalian Ecology in Southwestern North America, with Special Attention to the Colors of Desert Animals'' (1937). Trips to North Dakota and Nebraska provided information on populations of *Peromyscus maniculatus*. Recognized importance of careful measurement and documentation and wrote several papers dealing with methodology. Early training in electronics enabled him to modify the existing equipment so as to obtain more precise measurements. Was interested in problems of geographical and ecological distribution pertaining to plants and animals. As he studied this, he came to the conclusion that concepts of life zone and formation were inadequate. Developed the concept of the biotic province, a large, continuous area characterized by a unique combination of ecologic communities, climate, soil, and physiography. His *The Biotic Provinces of North America* (1943) is an important contribution on biogeography. His other books include *Natural Communities* (1952) and *Man's Nature and Nature's Man* (1955). Also authored over 200 scientific papers.

Taught courses in ecology and human biology and supervised work of graduate students. In 1957, University of Michigan recognized his exceptional career by establishing the Lee R. Dice Distinguished Professorship. A man who enjoyed being active, he rarely took vacations from his work. Still, he found time to listen to music and to maintain a garden, the produce of which he generously shared with friends.

MAJOR CONTRIBUTIONS

Research in genetics and directorship of the genetics program at the University of Michigan; concept of biotic province.

BIBLIOGRAPHY

Evan, Francis C. "Lee Raymond Dice (1887–1977)." *Journal of Mammalogy* (1978) (obit.).

New York Times, 3 February 1977 (obit.).

UNPUBLISHED SOURCES

See Lee R. Dice Papers, Historical Collections, University of Michigan (chiefly correspondence).

Cynthia D. Chambers

DILG, WILL H. Born in fall of 1867, Milwaukee, Wisconsin; died 27 March 1927, Washington, D.C. Publicist, advertising salesman, nature writer, conservationist.

FAMILY AND EDUCATION

Little is known about background or personal life. Was a publicist and advertising salesman. Married Marguerite Ives (divorced 1925); one son and one daughter.

POSITIONS

A founder and first president (1922–1926) of Izaak Walton League of America.

CAREER

With fifty-three other sportsmen, established Izaak Walton League in Chicago, on 14 January 1922. Was central figure in early years of league, which had grown to 100,000 members by 1925. Launched *Izaak Walton League Monthly* in 1922 (after 1923 called *Outdoor America*). Devised, and helped to shepherd through Congress in 1924, Upper Mississippi River Wild Life and Fish Refuge—the first major such refuge to be funded by Congress, and to be under supervision of federal government. Set up Izaak Walton League Fund to save starving elk in Jackson Hole, Wyoming. Last months of life spent campaigning in Washington, D.C., for Conservation Department in president's cabinet. Edited *Tragic Fishing Moments* (1922).

MAJOR CONTRIBUTIONS

Helped to organize Izaak Walton League of America, first mainstream conservation organization with mass appeal. Zeal and drive crucial to league's early development. More than any other single person, was responsible for passage of Upper Mississippi River Wild Life and Fish Refuge. Honored by granite monument on charterhouse grounds at Winona, Minnesota, near Upper Mississippi Refuge.

BIBLIOGRAPHY

New York Times, 29 March 1927 (obit.).
"A 60–Year Battle to Preserve the Outdoors." *Outdoor America* 47 (March/April 1982).
Voigt, William, Jr. *Born with Fists Doubled: Defending Outdoor America* (1992).

Richard Harmond

DIONNE, CHARLES-EUSÈBE. Born Saint-Denis-de-Kamouraska, Quebec, 20 July 1846; died Quebec City, Quebec, 25 January 1925. Naturalist, taxidermist, curator.

FAMILY AND EDUCATION

Son of Eusèbe, shoemaker and Emilie Lavoie Dionne. Attended elementary school in Saint-Denis (1853–1859). His interest in natural sciences was already apparent at this age. Carved wooden sculptures of common birds, which he also attracted by the imitation of their song. At age eleven mounted the first specimen of his collection. On account of great natural curiosity, was permitted to attend lectures at the local private school (1862–1864). Moved to Quebec City in 1865. Married Marie-Emélie Pelletier in 1879; no children. Received honorary degrees from Laval University in 1902 (maîtrise ès lettres) and 1924 (docteur ès sciences).

POSITIONS

Handyman, Quebec City Seminary, 1865–1866. Laboratory attendant, Laval University, 1866–1873, assistant librarian, 1873–1882. Curator, Zoological Museum of Laval University, 1882–1925. Conducted his field investigations near Quebec City and, occasionally, at Saint-Denis.

CAREER

At Laval University found ideal surroundings necessary to become an accomplished autodidact. Soon mastered taxidermy and became fully knowledgeable in all branches of natural history. His sole scientific expedition was conducted on board research vessel cruising the Gulf of Saint Lawrence in 1882. Published *Les oiseaux du Canada* in 1883, book he soon admitted had many flaws not apparent to local reviewers. The presentation of his personal collection of mounted birds at Provincial Exhibition of 1888 established his reputation with the public, and accuracy of the *Catalogue des oiseaux de la province de Québec* (1889) won him the respect of American ornithologists. Published over fifty short papers, mainly on birds. Planned and prepared several private collections. At the height of his career, wrote *Les mammifères de la province de Québec* (1902) and *Les oiseaux de la province de Québec* (1906). Quebec thus became one of the very few (if not the only) political divisions of North America covered by ''state books'' on both mammals and birds. Was one of founding members of *La société Provancher d'histoire naturelle du Canada* (1919).

MAJOR CONTRIBUTIONS

Did not undertake original research but was meticulous compiler of life histories, descriptions, and distributional occurrences. As writer and naturalist as well, studied birds above all. Both learned naturalists and laymen found him always ready and willing to answer their queries.

For many decades, Dionne's French-speaking compatriots turned to *Les oiseaux de la province de Québec* (1906) as the standard reference on the birds of Quebec. It contains original notes on status of birds at the turn of the century that remain an invaluable source of information.

BIBLIOGRAPHY

Déry, D. A. *The Canadian Field-Naturalist* (March 1925) (obit.).

Gaboriault, V. *Bio-bibliographie de Charles-Eusèbe Dionne*. Montreal: Ecole des bibliothécaires de l'Université de Montréal (1947).

UNPUBLISHED SOURCES

Shortly after his death, Dionne's papers were destroyed by his wife.

Normand David

DIXON, JOSEPH SCATTERGOOD. Born Galena, Cherokee County, Kansas, 5 March 1884; died Escondido, California, 23 June 1952. Ornithologist, mammalogist, naturalist.

FAMILY AND EDUCATION

Son of Benjamin Franklin Dixon, he was also the nephew of Joseph M. Dixon, former Republican senator from Montana and first assistant secretary of the interior in the Hoover administration. At an early age, his parents moved to

Escondido, and he graduated from the public high school there. Graduate of Throop Polytechnic Institute in Pasadena and from Stanford University, 1910. Married 1914, four children.

POSITIONS

Became staff member at Museum of Vertebrate Zoology (MVZ), University of California–Berkeley on its opening in 1910. Assistant curator of Mammals, MVZ, 1915–1918; economic mammalogist, MVZ, 1918–1931. Taught at Yosemite School of Field Natural History for several years beginning in 1931. Associated with George Melendez Wright in founding Wildlife Branch, National Park Service, for which he was field naturalist, 1931–1946.

CAREER

Met Joseph Grinnell at Throop Polytechnic Institute, and they became good friends. Dixon was invited to join Grinnell at Berkeley when the Museum of Vertebrate Zoology was created there through the liberality of Annie Montague Alexander in 1910. Dixon was described as "an active, careful, and accurate field worker," who made many extended field trips throughout California and Alaska. In 1907–1908, he located a nest of the surfbird in the vicinity of Mt. McKinley in the course of the Alexander Expedition to Alaska. He was a member of the Harvard Expedition that visited parts of Alaska and Siberia in 1913–1914. His researches included work on the mammals of Sequoia National Park, food habits of the mule deer in California, and ecological research in the Mt. Lassen region of California. His articles included "Contribution to the Life History of the Alaska Willow Ptarmigan in *Condor* (July–August 1927). His major publications included "Vertebrate Natural History of a Section of Northern California through the Lassen Park Region" (with Grinnell and J. M. Linsdale, 1930); *Fur-Bearing Animals of California* (with Grinnell and Linsdale, 1937); and *Birds and Mammals of the Sierra Nevada* (with Lowell Sumner, 1953). In addition, he was responsible for an elegant portfolio focusing on the wildlife of the national parks of the western United States. Dixon was active in a number of organizations, notably the American Ornithologists' Union and the Cooper Ornithological Club. He was vice president, California Academy of Sciences, board member, American Society of Mammalogists, and regional director of the First North American Wildlife Conference in 1936.

MAJOR CONTRIBUTIONS

Made important contributions to understandings of western wildlife and exploration that did much to enhance the reputation of the Museum of Vertebrate Zoology at the University of California–Berkeley as a center of wildlife research during the first three decades of its existence. His major publications, particularly the *Fur-Bearing Mammals of California* and the Mt. Lassen study, have become minor classics of western natural history and zoology.

BIBLIOGRAPHY
American Men of Science. 8th ed.
Palmer, Theodore S. *Auk* (January 1954) (obit.).
Palmer, Theodore S. "Joseph Scattergood Dixon." *Zoologist* 71 (1954).
Stroud, Richard H., ed. *National Leaders of American Conservation* (1985).

Arthur A. Belonzi

DOBIE J(AMES) FRANK. Born on a ranch in Live Oak County, Texas, 26 September 1888; died Austin, Texas, 18 September 1964. Educator, folklorist, writer.

FAMILY AND EDUCATION

Son of Richard Jonathan and Ella Byler Dobie. A.B., Southwestern University, Georgetown, Texas, 1910; M.A., Columbia University, 1914; honorary M.A., Cambridge University (England), 1944. Married Bertha McKee, 1916.

POSITIONS

Principal, public schools, Alpine Texas, 1910–1911; teacher of English and secretary to the president, Southwestern University, 1911–1913; instructor in English, University of Texas, 1914–1917, 1919–1920, 1921–1923; 1st Lieutenant, Field Artillery, U.S. Army, 1917–1919; manager of a ranch in southwestern Texas, 1920–1921; head of English Department, Oklahoma A&M College, 1923–1925; adjunct professor of English, 1925–1926, associate professor of English, 1926–1933, professor of English, 1933–1947, University of Texas (with leaves of absence at intervals); research fellow, Laura Spelman Rockefeller Foundation, 1930–1931 and 1934–1935; Guggenheim Fellowship, 1932–1933; visiting professor of American history, Cambridge University, England, 1943–1944; with information and education Division of U.S. Army, lecturer at Shrivenham American University in England and to troops at various locations in Europe, 1945–1946. Member, U.S. National Commission to UNESCO, 1945–1948. Research grant, Huntington Library, San Marino, California, 1948–1949. Published Sunday newspaper column in several Texas papers from 1939 until his death.

CAREER

Distinguished teacher; author or editor of more than thirty books. Dobie was the leading authority on the culture of the Southwest. Joined Texas Folklore Society, 1916, edited society's annual publications as secretary-editor for twenty-one years. His books included *A Vaquero of the Brush Country* (1929);

Coronado's Children (1931; winner of Literary Guild Award); *On the Open Range* (1931). Several research grants in early 1930s made possible 2,000–mile travel by muleback through various parts of Southwest, resulting in four collections of folktales: *Tongues of the Monte* (1935); *Tales of the Mustang* (1936); *The Flavor of Texas* (1936); *Apache Gold and Yaqui Silver* (1939). Also published *John C. Duval: First Texas Man of Letters* (1939). The animals of the Southwest were covered in a series of books, including *The Longhorns* (1941), story about this breed of cattle and the men who herded them; *The Voice of the Coyote* (1949), which followed thirty years of observation and research and gave character and personality to this species; *The Mustangs* (1952), history of the wild horses of the Old West, with theories of their origin and use by Native Americans and cowboys; and (posthumously) *The Rattlesnakes* (1965), modest-sized volume with observations, legends, actions, and attitudes of this animal. Following return from visiting professorship at Cambridge, published critically acclaimed *A Texan in England* (1944). Longtime and often stormy relationship with University of Texas ended 1947, when, after refusing to cut back on his many leaves of absences, he was dismissed. After leaving the university, published seven more books, including *Guide to Life and Literature of the Southwest* (1943. rev. ed. 1952); *The Ben Lilly Legend* (1950); *Up the Trail from Texas* (1955); *Tales of Old-Time Texas* (1955); *I'll Sell You a Tale* (1960); *Cow People* (1964).

MAJOR CONTRIBUTIONS

The feeling for nature he was able to convey to readers of his books shaped a sympathetic attitude toward animals and the natural order of things. His writings combined folktale, history, and natural history into stories that described how all life forms are relevant and interdependent.

BIBLIOGRAPHY

Dictionary of American Biography. Supplement 7 (1961–1965).
Dobie, J. Frank. *Some Part of Myself.* Ed. Bertha McKee Dobie. (1967).
New York Times, 19 September 1964 (obit.).
Tinkle, Lon. *J. Frank Dobie: The Making of an Ample Mind* (1968).

UNPUBLISHED SOURCES

Primary source materials including letters and manuscripts, and his personal library is housed in the Harry Huntt Ransom Humanities Research Center at the University of Texas at Austin. Other collections can be found at the Texas A&M (College Station) and Southwestern University (Georgetown, Texas).

Rodney Marve

DORNEY, ROBERT STARBIRD. Born Milwaukee, Wisconsin, 1 April 1928; died Waterloo, Ontario, 28 July 1987. Wildlife ecologist, environmental planner, university professor.

FAMILY AND EDUCATION

Son of John Alexander and Abigail Andrews Starbird Dorney. Received all degrees from the University of Wisconsin: a B.S. in education and biology, an M.S. in wildlife management, and a Ph.D. in veterinary science and wildlife management. Married twice; three children from the first marriage. Survived by Lindsay Dorney, second spouse, a specialist in literature and women's studies.

POSITIONS

Conservation biologist, Wisconsin Conservation Department, Madison, Horicon, and Ladysmith, Wisconsin, 1949–1957. Instructor, Department of Veterinary Science, University of Wisconsin, 1958–1959. Assistant professor, Department of Botany and Zoology, University of Wisconsin Center, Green Bay, 1959–1962, associate professor, 1963–1964. Program specialist, Department of Scientific Affairs, Pan American Union (Organization of American States), Washington, D.C. Associate professor and professor, School of Urban Planning, University of Waterloo, Ontario, 1967–1987.

CAREER

Influenced significantly while an undergraduate student by Aldo Leopold and others in conservation studies and wildlife management at the University of Wisconsin. His early scholarly work focused on the ecology and parasitology of the ruffed grouse. During graduate studies he developed an abiding affection for field studies and conservation. Was one of the first faculty members hired in the School of Urban and Regional Planning in the newly created (1967) Faculty of Environmental Studies at the University of Waterloo. Work quickly spread from strictly ecological and wildlife research and management to interdisciplinary ecological planning by 1970. Became one of the first ecologists to be certified as a member of the Canadian Institute of Planners. He formed, along with several colleagues and students, one of the first exclusively ecological consulting firms in North America, Ecoplans. Dorney found it much easier to develop new approaches to planning and design when working outside the academy. He maintained this connection to professional consulting for the remainder of his career. Briefly, he operated a nursery to supply native plant stock for various consulting contracts, but demand was too small in the mid-1970s. Was a founder of the Ontario Society for Environmental Management, a professional organization, and an award was named in his honor in 1979. An accomplished naturalist, he worked with colleagues at Waterloo to establish the first network of regional environmentally sensitive areas in Ontario (1969).

Published sixty articles, dozens of technical reports, and one book, *The Professional Practise of Environmental Management* (1989; published posthumously).

Dorney was known by colleagues, clients, and students as an innovator. His ideas were generally well ahead of conventional wisdom, for example, his development of the ''mini-ecosystem'' approach to urban ecological restoration, the ''landscape evolution model'' for charting changes in a region, and techniques for large-scale ecological restoration. Heterodox training and experience were beginning to provide rewards in the form of offers of major teaching and research positions and international recognition at the time of his death of a heart attack at age fifty-nine. The Robert Starbird Dorney Ecology Garden was created in his honor at the University of Waterloo.

MAJOR CONTRIBUTIONS

Ability to fuse management, planning, and ecology; formative role in the development of environmental professionalism in Canada; innovations in ecological restoration and urban ecology.

BIBLIOGRAPHY

Higgs, E. S. ''A Life in Restoration: Robert Starbird Dorney, 1928–87.'' *Restoration and Management Notes* 11.2 (Winter 1993).

Eric S. Higgs

DOUBLEDAY, NELTJE DE GRAFF (NELTJE BLANCHAN). Born Chicago, 23 October 1865; died Canton, China, 18 February 1918. Author.

FAMILY AND EDUCATION

Daughter of Liverius and Alice (Fair) de Graff. Was educated at St. John's School, New York City, and at the Misses Masters' School at Dobbs Ferry, New York. On 9 June 1886, she married Frank Nelson Doubleday at Plainfield, New Jersey, by whom she had two sons and a daughter. Her husband was a magazine and later book publisher.

POSITIONS

Author and literary critic, 1889–1917.

CAREER

Little is known of her early life other than she received most of her secondary education in the east. She was an enthusiastic bird watcher. Took to writing in large part because of her husband's encouragement and support. Frank N. Doubleday, known from his initials to intimates as ''Effendi,'' was in the magazine publishing business, and after 1897, was a leading figure in the publishing firm known successively as Doubleday McClure, Doubleday Page, Doubleday Doran, and finally Doubleday and Company. All of her books, save the first, were published by her husband's firm, which issued them in oversize, well-illustrated format. Most remained in print for decades, and some were reissued in smaller reprint editions. She assumed the pen name of Neltje Blanchan with her first book, *The Piegan Indians* (1889). Though she later published a few shorter pieces concerning the education and handicrafts of Native Americans, she soon began focusing on nature, especially birds and flowers. Her *Bird Neighbors* (1897) and *Birds that Hunt and are Hunted* (1898), later republished as *Game Birds: Life Histories of One Hundred and Seventy Birds of Prey, Game Birds, and Water Fowls* (1917), quickly established her reputation as a competent natural history writer, and both books sold widely. Such errors as they contained were attributed by some critics to the sometimes faulty sources of information at her disposal. Blanchan published two books on flowers and gardening, *Nature's Garden: Our Wild Flowers and Their Insect Visitors* (1900) and *The American Flower Garden* (1909). These were well written and illustrated, but their author wrote from the point of view of one with a fairly large home and grounds comparable to the one she and her husband enjoyed on Long Island, and this may not have been as useful to readers with smaller gardens. *How to Attract the Birds and Other Talks About Bird Neighbors* (1902) and *Birds Every Child Should Know* (1907), both primarily written for children, were followed by *Birds Worth Knowing* (1917), the latter with nearly fifty illustrations by Louis Agassiz Fuertes. By 1917, Blanchan, writing "The Audubon Societies: A Note of Appreciation," in her *Game Birds*, felt that the American public had developed a great interest in the conservation of birds, which she attributed to the work of the "persistent and highly intelligent educative work by the Audubon Societies, directed by scientific and altruistic men and women, in reaching school children, clubs of many kinds, granges, editors, and legislators." She also concluded that the "vast economic value [of birds] to the country" had been demonstrated. *Birds*, a posthumous selection of previously printed pieces, was published in 1926. Blanchan also published some articles, literary criticism, and book reviews on other subjects in such publications as *Country Life*. A highly moralistic person, Blanchan was said to have urged her husband not to publish Theodore Dreiser's controversial novel *Sister Carrie*. For contractual reasons, Mr. Doubleday had no choice but to proceed, but he issued a small printing in 1900 and gave the book little publicity. Dreiser was furious, and never forgave Blanchan, though firm evidence of her involvement in the matter was never established. A widely read author for over thirty years, Blanchan's nature writings were sometimes anthropomorphic, and she was criticized for suggesting in her books that certain predatory species should be killed, rather than protected. The Doubledays were active in the American Red Cross and early in December, 1917, sailed to China as special Commissioners on a special mission for that organization. Two months later, Blanchan died very suddenly in Canton of unannounced causes, and was buried near her Long Island home.

MAJOR CONTRIBUTIONS

Widely read in her day, she helped to create a popular interest in birds, which, in turn, facilitated passage of protective legislation for them. Robert H. Welker has suggested that Blanchan, along with Olive Thorne Miller, Florence Merriam Bailey, and Mabel Osgood Wright, the outstanding quartet of late nineteenth century women bird writers, were important more as educational than literary figures.

BIBLIOGRAPHY

James, Edward T., ed. *Notable American Women, 1607–1950; A Biographical Dictionary* (1971).

Johnson, Allen, and Dumas Malone. *Dictionary of American Biography* (1930).

Mainiero, Lina, ed. *American Women Writers: A Critical Reference Guide from Colonial Times to the Present* (1980).

New York Times, 23 February 1918 (obit.).

Strom, Deborah, ed. *Birdwatching with American Women: A Selection of Nature Writings* (1986).

Swanberg, W. A. *Dreiser* (1965).

Welker, Robert H. *Birds and Men: American Birds in Science, Art, Literature, and Conservation, 1800–1900* (1955).

Karen M. Venturella

DOUGLAS, DAVID. Born Scone, Perthshire, Scotland, 25 July 1799; died near Laupahoehoe, Hawaiian Islands, 12 July 1834. Botanist.

FAMILY AND EDUCATION

Second of six children in the family of stonemason John Douglas and Jean Drummond, educated at schools at Scone and Kinnoul, apprenticed at the age of eleven to head gardener at Scone Palace, seat of the earl of Mansfield; never married.

POSITIONS

From 1817 to 1820, undergardener to Robert Preston of Valleyfield; 1820–1823 gardener at the Botanic Gardens at Glasgow University under William James Hooker; sent by Joseph Sabine in 1823 as plant collector for the Royal Horticultural Society to America; and botanized in eastern United States and Canada. Success of initial venture led to return in 1824 under the sponsorship of the Royal Horticultural Society and the Hudson's Bay Company to explore and collect along the Columbia River in Pacific Northwest; reached Columbia River in April 1825 and botanized in this region until early 1827. Then crossed the Rocky Mountains to reach Hudson's Bay to take ship for London. Returned to Pacific Northwest for further collecting in 1830, but Indian unrest led to his devoting much of the next three years to botanizing in California and Hawaii, and after 1832 along the Fraser River. Returned to botanize in Hawaii while enroute back to England shortly before his death.

CAREER

Became familiar with natural history of his native region as a child and soon began collecting plants. Developed interest in books dealing with plants while apprenticed to Earl of Mansfield at about the age of twelve. Further intensified interest in literature of botany while in employ of Sir Robert Preston, and was taken under the wing of William Hooker in Glasgow. Sponsored by Hooker, was sent to North America to collect plants for Horticultural Society of London; spent most of the year 1823 in what is now Ontario and New York State. His efforts were much praised, and after taking time to improve his understanding of natural history and collecting procedures, was next sent by Horticultural Society to the Pacific Northwest region. Spent a bit more than three years traveling from northern California and what are now Oregon and Washington into present day British Columbia, Alberta, Saskatchewan, and Manitoba. Skilled at mountain climbing; named several mountains in the vicinity of Athabaska Pass, one of them for his mentor William Hooker. Encountered John Richardson and Thomas Drummond, both with Sir John Franklin's Arctic Expedition of 1825–1827, and traveled with Franklin across Lake Winnipeg and thence to Red River settlement in vicinity of present-day Winnipeg, where he stopped and collected plants. Nearly lost his life during storm while in small boat in Hudson's Bay with a dozen other men; survived, but suffered effects for some time. The plants and seeds brought back to England by Douglas late in 1827 were so numerous that the Horticultural Society was obliged to share the largesse with several museums, Cambridge University, and some nurseries. William Hooker's *Flora Boreali Americana* (1840) included descriptions of specimens collected by Douglas, Richardson, and Drummond, and various animals collected by Douglas were discussed in several volumes of Richardson's *Fauna Boreali Americana* (1829–1837). His geographical and astronomical notes and field sketches were also much prized. His eyesight had deteriorated during his second expedition and this tended to limit what he was able to accomplish on his third and final expedition, which began in the summer of 1830. For several years (1830–1833) he again ranged from southern California north to Washington and western Canada, and spent some time in Hawaii. Now blind in one eye, he decided to return to England via Alaska across Russia, but a canoe accident near Prince George, British Columbia, at the juncture of the Fraser and Nechako rivers terminated that effort. He lost his plant collection, plant notes, and canoe, but retained some maps, other notes, and equipment. He was obliged to rest for a time at Fort Vancouver. Recovering, he returned to Hawaii in late December 1833 and lost his life while hiking in the northern part of that island in mid-

July, 1834. His badly mangled body was found in a cattle trap also occupied by an angry bull, but whether he fell by accident or met his end under other circumstances has never been determined. A dedicated, persistent, and highly skilled naturalist with a shy and often appealing personality, Douglas was sometimes insensitive to the feelings of traders who met or helped him on his travels. He added approximately 7,000 species of plants to the 92,000 then known worldwide, a great number of them from the Pacific coast of the United States and the region between Alaska and British Columbia and present-day Manitoba. Douglas was made a member of the Geological, Linnaean, and Zoological Societies of London following his second North American trip. The Douglas Fir of the Pacific northwest, some smaller plants, a tree squirrel native to the Pacific northwest coast, and a mountain near Lake Louise, Alberta, have all been named in his honor.

MAJOR CONTRIBUTIONS

Douglas was not only energetic field botanist and collector but also a skilled ornithologist and field naturalist. Published in one of his eight papers is the first account of the California condor. As the first professional botanist to work in the Pacific Northwest, he discovered many new species of plants and introduced more than 250 plants into England, including the great tree that bears his name, the Douglas fir.

BIBLIOGRAPHY

Aethelstan, George Harvey. *Douglas of the Fir* (1947).

Davies, John. *Douglas of the Forests, the North American Journals of David Douglas.* (1979).

Dictionary of Canadian Biography.

Dictionary of National Biography. Vol. 5 (1949–1950).

Douglas, David. *Journal Kept by David Douglas During his Travels in North America, 1823–1827* (1914) (contains list of the eight scientific papers published by Douglas during his lifetime).

McKelvey, Susan. *Botanical Explorations of the Trans-Mississippi West* (1955).

Morwood, William. *Traveller in a Vanished Landscape* (1973).

UNPUBLISHED SOURCES

Journals and letters are in Royal Horticultural Society Archives, London, England. Plant specimens at British Museum; Royal Botanic Gardens, Kew; and Cambridge University.

Phillip D. Thomas

DOWNS, ANDREW. Born New Brunswick, New Jersey, 27 September 1811; died Halifax, Nova Scotia, 26 August 1892. Zoo owner, animal breeder, naturalist.

FAMILY AND EDUCATION

Son of Robert Downs, a tinsmith and Elizabeth Plum Downs. Largely self-educated. Married Mary Elizabeth (Eliza?) Matthews (died 1858), four daughters, and Mathilda E. Muhlig, 1859, one daughter.

POSITIONS

Assistant curator, Halifax Mechanics Institute Collection, 1845; curator 1846. Proprietor, Down's Zoological Garden, 1847–1868 and 1869–1872.

CAREER

Started work as a tinsmith in Halifax. Met J. J. Audubon on his visit to Nova Scotia in 1833 and subsequently corresponded with him. Became member of the governing committee of the Halifax Mechanics Institute in 1835. By 1838 developed plans for a provincial museum and started to assemble a collection of animals from Nova Scotia for export to England. In 1845–1846 worked as assistant curator and then curator of the Halifax Mechanics Institute. Main claim to fame is his 1847 establishment of the first zoological garden in North America, north of Mexico City. The zoo was located at Dutch Village, Halifax, Nova Scotia, and housed a wide variety of animals and birds, obtained from donations, purchases, and exchange agreements with foreign zoos interested in the North American animals that he bred. The animals were displayed in near-natural conditions, in contrast to most European zoos, which displayed animals in small cages. The zoo steadily expanded from 5 acres to 100 acres, with improvements partially supported by grants from the Nova Scotia government and from the city of Halifax. In addition to his animal breeding activities, was a supporter of agriculture in the region and trained carrier pigeons.

Sold his zoo in 1868 with the intention of taking charge of the Central Park Zoo in New York. However, the position was not offered to him, and he returned to Halifax and reopened his zoo in 1869. The second zoo was less successful than the first, and he sold out in 1872. Later years were partly spent in compilation of a catalog of the birds of Nova Scotia, but he continued his interest in natural history exhibitions until his death. An agnostic, he said, "My religion is that of humanity. The woods and lakes are my church."

MAJOR CONTRIBUTIONS

Downs moved far beyond the role of showman. After he died, it was said, "While Audubon killed and stuffed and painted, he preserved and propagated." His work on animal breeding and display of animals in near-natural surroundings preceded similar efforts in modern zoos throughout the world.

BIBLIOGRAPHY
Dictionary of Canadian Biography. Vol. 12 (1990).
Fergusson, C. B. *Dalhousie Review* 27 (1947–1948).
Hallock, Charles. *Forest and Stream* (2 September 1899).
Piers, Harry. *Proceedings and Transactions, Nova Scotia Institute of Science* 10 (1898–1902).
UNPUBLISHED SOURCES
Documents of various government departments are in Public Archives of Nova Scotia.

Terence Day

DRAKE, DANIEL. Born Bound Brook, New Jersey, 20 October 1785; died Cincinnati, Ohio, 5 November 1852. Physician, educator, scientist, civic leader.

FAMILY AND EDUCATION

Son of Isaac and Elizabeth Shotwell Drake, illiterate, hardworking pioneer farming parents, Drake's boyhood was spent in Mays Lick, Kentucky, until age fifteen when taken to Cincinnati to study medicine under Dr. William Goforth. Studied medicine at University of Pennsylvania (1805–1806, 1815–1816) and came under the influence of Drs. Benjamin Rush, Benjamin Smith Barton, and Caspar Wistar; took medical degree (1816). Married Harriet Sisson (1807); had five children of which three (one son and two daughters) lived to maturity.

POSITIONS

Physician at Mays Lick (1806) and at Cincinnati (1807–1852), except when in other cities. Professorships or lectureships held at Cincinnati Lancasterian Seminary (1814, 1818–1819); Medical Department of Transylvania University (1817–1818, 1823–1827), served as Dean (1824–1827); Medical College of Ohio (1819–1822, 1831–1832, 1849–1850, 1852); Jefferson Medical College (1830–1831); Ohio Mechanics Institute (1832–1833); Medical Department of Cincinnati College (1835–1839), served as Dean; Louisville Medical Institute (1839–1849, 1850–1852).

CAREER

As Cincinnati's foremost citizen of the nineteenth century, Drake organized and founded colleges, hospitals, clinics, libraries, and literary, scientific, and professional societies, including Cincinnati Lyceum (1807); Cincinnati Circulating Library Association (1807); First District Medical Society (1812); School of Literature and the Arts (1813); Cincinnati Manufacturing Company (1813); Cincinnati Lancasterian Seminary (1814); Cincinnati College (1819); Medical College of Ohio (1819), which developed into the College of Medicine of the University of Cincinnati; Cincinnati Society for the Promotion of Agriculture,

Manufactures, and Domestic Economy (1819); Western Museum Society (1819); Medico-Chirurgical Society of Cincinnati (1820); Commercial Hospital and Lunatic Asylum (1821), which was the first hospital for medical instruction in the United States staffed exclusively by professors of a medical college; Cincinnati Eye Infirmary (1826); Historical and Philosophical Society of Ohio (1831); Medical Department of Miami University, which was consolidated with the Medical College of Ohio (1831); Medical Department of Cincinnati College (1835), in which Drake assembled one of the most distinguished medical faculties in the country; Western Academy of Natural Sciences (1835); Physiological Temperance Society of the Louisville Medical Institute (1841); and Kentucky School for the Blind (1842).

As a medical educator, taught materia medica, botany, anatomy and physiology, theory and practice of medicine, clinical medicine, and pathological anatomy. As a scientist, most interested in the natural sciences, especially botany, in which he was the first resident of the West to publish a list of trees, as well as a phenological record and medical properties of indigenous plants (*Notices*, 1810; *Picture*, 1815), and to give public lectures on botany (1818). Promoted botany by reviewing publications on the flora of eastern North America, urging the writing of local and regional floras of the Mississippi valley, pointing out the necessity for young physicians to study the specialized sciences, especially botany, and arguing for more professorships in natural history and botany in the medical schools. As a public-minded citizen, was active in local affairs and politics, worked for establishment of industry, promoted development of canals and railroads, and supported projects for improved social conditions, including freeing of slaves, education for the workingman, and the cause of temperance.

As an author, Drake's writings were voluminous but principally on a medicogeographic theme that linked health and disease to the environment and topography; his contributions based on original observations involving over 30,000 miles of travel in the Mississippi Valley and written in a skillful, clear, and vigorous style with great descriptive powers, including the following selections: *Some Account of the Epidemic Diseases which Prevail at Mays-Lick, in Kentucky* (1808), *Notices concerning Cincinnati* (1810), *Natural and Statistical View, or Picture of Cincinnati and the Miami Country* (1815), *Geological Account of the Valley of the Ohio* (1825), *A Practical Treatise on the History, Prevention, and Treatment of Epidemic Cholera, . . .* (1832), *Practical Essays on Medical Education, and the Medical Profession, in the United States* (1832), *Rail-road from the Banks of the Ohio River to the Tide Waters of the Carolinas and Georgia* (1835), *A Memoir on . . . Milk-Sickness . . . in the State of Ohio* (1841), *The Northern Lakes a Summer Residence for Invalids of the South* (1842), *A Systematic Treatise, . . . on the Principal Diseases of the Interior Valley of North America, . . .*, 2 vols. (1850, 1854), *Pioneer Life in Kentucky . . .*, edited by his son, Charles D. Drake (1870, reprinted 1907; new edition edited by E. F. Horine 1948). For twenty-two years edited the *Western Journal of the Medical and Physical Sciences* (1827–1838) at Cincinnati and its successor, the *Western Journal of Medicine and Surgery* (1840–1849) at Louisville, the most influential medical journals of the West. Elected to membership in a number of professional societies, including American Antiquarian Society, American Medical Association, American Philosophical Society, Academy of Natural Sciences of Philadelphia, and Wernerian Natural History Society of Edinburgh.

MAJOR CONTRIBUTIONS

Drake's brilliant career profoundly influenced the scientific and cultural development of the Mississippi Valley; his early writings on climate, geology, and natural history were pioneering attempts to access the habitability of the trans-Applachian West and revealed an awareness of the problems and importance of urban development. As evaluated by his contemporaries, his *Systematic Treatise* was stated by A. Stille as belonging "to the very highest rank of our medical literature, and may very probably come to be regarded as the most valuable original work yet published in America"; F. J. Garrison noted, "There is nothing like this book in literature." His *Pioneer Life* has been acclaimed as "the greatest of all Kentucky books," by J. C. Bay, and considered to provide one of the best and most vivid portrayals of pioneer family life in the United States. A. E. Waller considered Drake's published original observations on the interactions of organisms and their environment as overlooked contributions to modern ecology that were far ahead of their time.

BIBLIOGRAPHY

Shapiro, H. D., and Z. I. Miller, eds. *Physician to the West: Selected Writings of Daniel Drake on Science and Society* (1970).

Stuckey, R. L. "Medical Botany in the Ohio Valley (1800–1850)." *Transactions and Studies, College of Physicians of Philadelphia* 45 (1978).

Waller, A. E. "Daniel Drake, as a Pioneer in Modern Ecology." *Ohio State Archaeology and Historical Society Quarterly* 56 (1947).

UNPUBLISHED SOURCES

Most significant collections are at the Cincinnati Historical Society and the Medical Library of the Cincinnati General Hospital.

Ronald L. Stuckey

DRUMMOND, THOMAS. Born Scotland, c.1780; died while botanizing in Havana, Cuba, in early March 1835. Botanist, collector of birds and mammals.

FAMILY AND EDUCATION

The youngest brother of James Drummond (director of the Botanical Garden at Cork and later an important Australian botanical explorer).

POSITIONS

He succeeded George Don in the nursery at Forfar, Scotland. Assistant naturalist with the second arctic exploring expedition led by John Franklin, 1825–1827. Curator of the Belfast Botanical Garden, 1828–1829. Botanical collector in southern United States, including Missouri and Louisiana, but chiefly Texas, 1831–1835.

CAREER

Before leaving Scotland, Drummond distributed individual sets of his collection of mosses, "musci Scotici." In North America, he explored the Rocky Mountains near Jasper while John Franklin and John Richardson went north to explore the arctic coastline. He was at Cumberland House, Saskatchewan, from 27 June through 20 August 1825, then spent the winter and the following summer near Jasper House in the Canadian Rockies. Here he collected 1,500 specimens of plants, 150 birds, and 50 mammals, including type specimens of the black-backed woodpecker, *Picoides arcticus*, and the white-tailed ptarmigan, *Laqopus leucurus*, as well as three new subspecies of grouse. Fifty mammals included type specimens of the pika, *Ochotona princeps*, and of three new subspecies. He had a narrow escape from the jaws of a grizzly bear and once went seven days without food. At Carlton House on the Saskatchewan River, where Drummond rejoined Richardson on 5 April 1827, the two men collected 103 specimens of birds. Drummond shot the type specimen of Swainson's hawk, *Buteo swainsoni*, and of two new subspecies and then on 18 July, just before reaching Cumberland House, the type specimen of Forster's tern, *Sterna forsteri*. His greatest contribution was his botanical collecting—he collected five new species of mosses in the mountains; ten additional species of mountain plants alone and the genus *Drummondia* has been named for him.

Drummond's trip to the United States began in New York in 1831. He walked twenty-five miles a day over the Allegheny Mountains from Frederickstown to Wheeling, but severe and prolonged illness while at St. Louis allowed him to collect only 250 species of plants. His health improved at New Orleans, where he again collected plants with enthusiasm. He did not live to publish information concerning the 750 species of plants he collected in Texas between March 1833 and December 1834.

MAJOR CONTRIBUTIONS

His collections of many important new species of plants, mammals, and birds in the Rocky Mountains and in present Saskatchewan. An important pioneer botanist in Texas.

BIBLIOGRAPHY
Bird, C. D. "The Mosses Collected by Thomas Drummond in Western Canada, 1825–1827." *Bryologist* 70 (1967).
Drummond, T. "Sketch of a Journey to the Rocky Mountains and to the Columbia River in North America." *Botanical Miscellany* (1830).
Geiser, S. W. "Naturalists of the Frontier: VII: Thomas Drummond." *Southwest Review* 15 (1930).
Hooker, W. J. "Notice concerning Mr. Drummond's Collections Made in the Southern and Western Parts of the United States." *Journal of Botany* 1 (1834).
Houston, C. S., and M. G. Street. *Birds of the Saskatchewan River, Carlton to Cumberland.* Regina: Saskatchewan Natural History Society (1959).

C. Stuart Houston

DRURY, NEWTON BISHOP. Born San Francisco, 9 April 1889; died Berkeley, California, 14 December 1978. Conservationist.

FAMILY AND EDUCATION

Son of Wells, prominent newspaper editor and columnist, and Ella Lorraine Bishop Drury. B.L., University of California, Berkeley, 1912. Honorary LL.D., University of California, 1947. Married Elizabeth Frances Schilling, 1918; three children.

POSITIONS

Reporter for several San Francisco and Oakland newspapers, 1906–1911; instructor in English, assistant professor of forensics, and secretary to the president, University of California, 1912–1918; 2d lieutenant, Army Air Service (Balloon Corps), 1918–1919, 1st lieutenant, Air Service Rescue, 1919–1920; Partner, Drury Co. Advertising Agency, also public relations counsel, 1919–1940; secretary, 1919–1940, 1959–1971, president, 1971–1975, chairman, Board of Directors, Save the Redwoods League, 1975–1978; land acquisitions officer, California Park Commission, 1929–1940; director, National Park Service, 1940–1951; chief, California Division of State Parks, 1951–1959.

CAREER

Established Save the Redwoods League (1919), spent time developing that organization. Over fifty-year period, raised over $25 million for organization as executive secretary. League enlarged Redwood preserves in California by some 135,000 acres and established thirty state parks. In 1927, assisted in drafting legislation creating first California State Parks Commission, whose mandate extended to preservation of coastal and forest lands within the state. Directly responsible for establishment of fifty-six state parks within twenty-year period. Secured many private donations to Redwoods preservation effort, including

$3 million from Rockefeller family. First offered directorship of national parks in 1933, but declined, in part because of responsibilities with Save the Redwoods League. Appointed director of national parks by President Franklin D. Roosevelt, Drury was known for his uncompromising attitude toward anticonservationist and commercial interest groups, particularly during difficult World War II period. Sharp budget restrictions, loss of key personnel to the armed forces, and wartime relocation of Park Service headquarters to Chicago among challenges with which he coped. Had strong support from Interior Secretary Ickes for much of his tenure. Led Civilian Public Service Corps, designed to utilize civilian conscientious objectors in park work. Several important new parks and monuments added during his tenure: Jackson Hole National Monument (Wyoming) and Big Bend (Texas), also several national historic sites: Independence Hall, T. Roosevelt and F. D. Roosevelt homes. Was research associate with Carnegie Institution of Washington, 1938–1943. Criticized by Ickes and other conservationists in 1940s when he tried to deal with demands for logging in Olympic National Park in diplomatic fashion. Due to serious policy differences with Oscar Chapman, Secretary of Interior under Truman (over Echo Park Dam in Dinosaur National Monument—which Drury opposed—and other issues), resigned and returned to California to head State Division of Parks and Beaches. Returned to Save the Redwoods League following retirement from state position in 1959 and fought for creation of Redwood National Park, achieved in 1968. Faced challenges from lumber industry and with more militant Sierra Club over this issue. Two redwood groves designated as living memorials to Drury, Drury Brothers Grove in Prairie State Park and Newton B. Drury Grove in Humboldt Redwoods State Park. Honorary vice president of Sierra Club. Regional trustee of Mills College, 1941–1943. Won many awards from federal government, conservation, and professional organizations.

MAJOR CONTRIBUTIONS

Through his organizational, state, and federal positions, contributed greatly to conservation movement in the United States. Strong defender of National Parks during extremely difficult period of its history. Long a vital figure in defense of California redwoods and in creation of Redwood National Park.

BIBLIOGRAPHY

Albright, Horace M., with Robert Cahn. *The Birth of the National Park Service: The Founding Years, 1913–1933* (1985).

Davis, Richard C., ed. *Encyclopedia of American Forest and Conservation History* (1983).

New York Times, 15 December 1978 (obit.).

Shankland, Robert. *Steve Mather of the National Parks* (1951).

Stroud, Richard H., ed. *National Leaders of American Conservation* (1985).

Swain, Donald C. *Wilderness Defender: Horace M. Albright and Conservation* (1970).

Who Was Who in America. Vol. 7.

Who's Who in the West (1969).

UNPUBLISHED SOURCES

See Newton Drury, ''Parks and Redwoods, 1917 to 1971,'' unpublished oral history by Ameria Roberts Fry and Susan R. Schrepfer, University of California Regional Oral History Office, 1972.

Joseph Bongiorno

DUNN, EMMETT REID. Born Alexandria, Virginia, 21 November 1894; died Bryn Mawr, Pennsylvania, 13 February 1956. Herpetologist, professor, museum curator.

FAMILY AND EDUCATION

Son of Emmett Clarke and Mary Cassandra (Reid) Dunn. Father was civil engineer. Received B.A. degree in 1915 and M.A. degree in 1916 from Haverford College. In 1921 graduated from Harvard University with Ph.D. Married Alta Merle, 1930; no children.

POSITIONS

Assistant zoologist, Smith College, 1916; instructor in zoology, 1917–1921; assistant professor, 1921–1927; associate professor, 1927. Received Guggenheim Fellowship, 1928. Associate professor of biology, Haverford College, 1929–1935; David Scuff Professor of Biology, 1935–1956. Aide, U.S. National Museum, 1919. Field naturalist, American Museum of Natural History, 1916, 1926; research associate, 1946–1956. Curator of reptiles and amphibians, Philadelphia Academy of Natural Sciences, 1937–1956. Editor, *Copeia* (journal of the American Society of Ichthyologists and Herpetologists [ASIH]), 1924–1929; President, ASIH, 1930–1931. Member American Society of Mammalogists; Ecological Society of America; Society for the Study of Evolution; Reptile Study Society; Phi Beta Kappa; Academia Colombiana de Ciencias; Biological Society of Washington, D.C.; Harvard Travellers' Club; Explorers Club of New York City.

CAREER

In 1915 published *Some Amphibians and Reptiles of Delaware County, Pa* and the next year began career as educator. During World War II served in the Naval Reserve as an ensign on submarine patrol and mine-sweeping duty. Resigned from Smith College in 1928 upon receipt of Guggenheim Fellowship for fieldwork in the American tropics and research in European museums. Joined faculty of Haverford College in 1929 and remained there until his death.

His wife, another teacher at Smith College, frequently accompanied him on expeditions and also coauthored some of his writings. Early expeditions were to the mountains of Virginia and the southern Blue Ridge Mountains. In 1926 was member of the Douglas Burden Expedition to Komodo in the East Indies. Bulk of his fieldwork was conducted in the American tropics. Collected in Cuba and Jamaica, spent a year in Colombia, traveled in Mexico, and was part of more than a dozen expeditions to Panama and Costa Rica. In 1944, he and his wife spent year doing fieldwork in South America as part of the Inter-American Cultural Exchange program of the Nelson Rockefeller Committee.

His writings include *Salamanders of the Family Plethodontidae* (1926), considered a classic addition to the literature; *Lower Categories in Herpetology* (1943); and some 200 scholarly papers. He held positions at several museums and was also associated with the Museum of Comparative Zoology, Harvard University. His work at Philadelphia Academy of Natural Sciences was one of his important contributions to the discipline and included identifying and cataloging large number of specimens.

Dunn, known as "Dixie" to his friends, was considered excellent field researcher and museum curator. Specialized in work with amphibians and reptiles of the eastern United States, West Indies, and Central and South America and was an outstanding authority on the herpetology of Panama. His phenomenal memory and vast knowledge of herpetology made him valuable source of information and guidance to both his students and colleagues.

MAJOR CONTRIBUTIONS

Discovered some forty species of frogs, salamanders, snakes, lizards, and turtles. His ordering of the salamander fauna of eastern North America made proper scientific understanding of these amphibians possible. Other outstanding achievements include revision of the classification and phylogeny of salamander identification and work on comparative anatomy and zoogeography, distribution, and ecology.

BIBLIOGRAPHY

Conant, Roger. "E. R. Dunn, Herpetologist." *Science* (1956) (obit.).
National Cyclopaedia of American Biography. Vol. 43.
New York Times, 14 February 1956 (obit.).
Schmidt, Karl P. "Emmett Reid Dunn, 1894–1956." *Copeia* (1957) (obit.).

Cynthia D. Chambers

DU PRATZ, ANTOINE SIMON LE PAGE. Born Holland (?), 1689; died La Rochelle, France (?), 1775. New World traveler, explorer, plant collector, historian.

FAMILY AND EDUCATION

Virtually nothing known of early life or background. Apparently received a considerable scientific education in Europe. Well versed in practical engineering and astronomy and had some familiarity with the natural sciences, particularly botany, and to a lesser degree zoology. Du Pratz also possessed expertise in mineralogy.

POSITIONS

''Overseer of Public Plantations,'' for John Law's *La Compagnie des Indes*, or ''Mississippi Company,'' and later for the imperial French Crown from 1718 to 1734, in which capacity he became an explorer and the first historian of Louisiana and the lower Mississippi Valley.

CAREER

Du Pratz's known career begins at age twenty-nine, with his migration from France to New Orleans as an official in the effort to colonize the Mississippi Valley. Settled briefly on Bayou St. Jean in 1718 but soon transferred to a woodland site between Fort Rosalie on the eastern bluffs of the Mississippi and the principal village of the Natchez Indians. Here spent most of the ensuing eight years and became accepted and esteemed among the Natchez, whom he studied. In 1720 suffered a variety of physical afflictions, which European medical practices seemed incapable of remedying. But the Natchez shamans, ''who are both surgeons, devines and sorcerers,'' affected cures with incantations and poultices made from native plants. So impressed were Du Pratz's superiors that he was ordered to begin collections of Louisiana plants deemed to have medicinal properties. In sixteen years spent in the Mississippi Valley cataloged and collected more than 300 native American plants, mostly those used medicinally by the tribal peoples. His position necessitating extensive travel throughout Louisiana, Du Pratz explored and kept notes on the topography and natural history of the Mississippi lands as far north as the mouth of the Illinois, eastward to the Chickasaw country of present Tennessee, and westward along the St. Francis River into today's Missouri and Arkansas. Also collected the journal of Etienne Veniard, Sieur de Bourgmond, a French officer who made an early foray (1723) into the plains country along the Missouri River. In addition to botanical collections, Du Pratz cataloged and described numerous mammals, birds, and reptiles and listed the varied natural resources, including minerals, with notations on climate, soil, topography. Upon appointment of a new governor in 1734, relieved of duties and returned to La Rochelle. Died in France at age eighty-six.

Author *Histoire de la Louisiane. Contenant la Découverte de ce vaste Pays, sa Description géographique: un Voyage dans les Terres, l'Histoire Naturelle, les Moeurs, coûtumes, et Religion des Naturels avec leur Origines; deux Voyages dans le Nord du nouveau Mexique dont un jusqu'à la Mer du Sud*, published in three volumes in Paris in 1758, with forty copperplate engravings and two maps. English translations appeared in 1763 (2 vol.) and 1774 (1 vol.);

latter was utilized in *An Account of Louisiana, Exhibiting a Compendious Sketch of Its Political and Natural History and Topography*, published in America in 1804, following the Louisiana Purchase. The 1774 version is also available in edited form, by Stanley Clisby Arthur.

Enthusiasm and energy, rather than precision are Du Pratz's distinguishing characteristics. Many of his natural history descriptions are flawed or inaccurate, and the illustrations crude. He compiled in his *Histoire* a sympathetic and remarkably thorough account of the Indians of the Mississippi Valley. Believed most Amerindians were descendants of the Tartars; but from close familiarity with the Natchez, came to believe they were the remnants of a Phoenician colony arrived from Africa via Brazil centuries before. His *Histoire* apparently brought him fame, if not wealth.

MAJOR CONTRIBUTIONS

Ranks with John Lawson and Robert Beverly as an early American traveler who provided Europeans with factual intimacy respecting natural history and Indian life of North America. Provided first fairly intensive natural history and resources catalogs of lower Mississippi Valley. Promoted respect for effectiveness of plant lore integral to traditional medicine. Enthusiasm for Louisiana country, which he regarded as superior to any other part of the continent, probably contributed to imperial rivalry for its control. Sympathetic to, and interested in, Indians; his descriptions of the religion, language, customs, ceremonies, and history of the Natchez remain the best firsthand accounts of the great mound-building complex of the Mississippi Valley. Du Pratz's *Histoire* stimulated the noble savage aspect of the emerging romantic movement. François Rene Chateaubriand's romantic epic, *The Natchez* (published in *Oeuvres Completes*, 1826–1831), drew heavily from Du Pratz.

BIBLIOGRAPHY

An Account of Louisiana, Exhibiting a Compendious Sketch of Its Political and Natural History and Topography (1804).

Brebner, John B. *The Explorers of North America, 1492–1806* (1933).

DeVoto, Bernard. *The Course of Empire* (1952).

Du Pratz, Antoine Simon Le Page. *The History of Louisiana or of the Western Parts of Virginia and Carolina* (1774).

Michaud, Andre, ed. *Biographie Universelle* (1842–1865).

Phelps, Albert. *Louisiana: A Record of Expansion* (1905).

Savage, Henry. *Discovering America, 1700–1875* (1979).

Dan Flores

DUTCHER, WILLIAM. Born Stelton, New Jersey, 20 January 1846; died Plainfield, New Jersey, 2 July 1920. Ornithologist, businessman.

FAMILY AND EDUCATION

Son of Rev. Jacob Conklin and Margaretta Ayres Dutcher. Educated in public schools of Owasco, New York. Married Catherine Oliver Price, 1870; two children.

POSITIONS

Employed in a bank in New York for several years from age thirteen; later worked on a farm near Springfield, Massachusetts. Around 1866, employed by Brooklyn Life Insurance Company as clerk, cashier, and secretary until the winter of 1894–1895, then became agent for Prudential Insurance Company of New Jersey for approximately fifteen years.

CAREER

A recreational hunter, Dutcher became interested in birds as a subject of scientific study in the late 1870s. Became active in the Linnaean Society of New York and the Nuttall Ornithological Club and joined the fledgling American Ornithologists' Union (AOU) when founded in 1883. From c.1879 until the mid-1890s, assembled information concerning Long Island birds for a projected book on the subject, which was not completed. Became member of AOU Committee on Protection of Birds, 1884 (chair, 1896, 1897), and was one of its most active members. Helped to organize first Audubon Society with George Bird Grinnell, George Sennett, and others; began work on model law (also known as AOU law) that later served as framework for legislation protecting birds in the various states. Treasurer of AOU, 1887–1903. In 1899, began lobbying state legislatures and the Congress for passage of protective legislation with aid of T. S. Palmer and others. The Lacey Act (1900), first federal legislation in this area, passed with strong support from Dutcher. Audubon laws had been enacted in thirty-three states by 1905. With others, persuaded President Theodore Roosevelt to set aside first bird refuge in nation (Pelican Island, Florida) in 1903. With moneys left to Audubon Society, hired first game wardens, a responsibility later assumed by federal government. Dutcher chaired AOU committee set up in 1901 to explore organization of umbrella Audubon Committee. In 1905, with support of the thirty-six state Audubon societies organized to that point, Dutcher founded and became first president (1905–1920) of the National Association of Audubon Societies. Pressed New York state legislature to pass Shea–White bill banning sale of bird plumes for decorative purposes (1910). Dutcher faced opposition not only from the millinery industry but from some professional ornithologists who thought the Audubon group wanted to stop scientific collecting. Dutcher was dropped from AOU Bird Protection Committee, which left the drive for the model law to the Audubon Societies. Dedicated and driven, Dutcher's exertions led to a stroke in 1910, which left him unable to speak and essentially incapacitated him. T. Gilbert Pearson, Audubon secretary since 1905, was made executive officer to assist Dutcher and ultimately succeeded him at helm of national Audubon organization. Dutcher published about 100 papers dealing with Long Island birds; annual reports and notes on bird protection in his various capacities with AOU and Audubon groups; popular leaflets for educational purposes; and general ornithological articles.

MAJOR CONTRIBUTIONS

Principal creator of the modern Audubon movement, fought hard for protective legislation at the state and national level. Inveighed against use of bird parts for ladies' hats and other wear.

BIBLIOGRAPHY

Auk (October 1921) (obit.).
Fox, Stephen. *John Muir and His Legacy: The American Conservation Movement* (1981).
Graham, Frank, Jr. *Man's Dominion: The History of Conservation in America* (1971).
Graham, Frank, Jr. *The Audubon Ark: A History of the National Audubon Society* (1990).
New York Times, 4 July 1920 (obit.).
Pearson, T. Gilbert. *Adventures in Bird Protection* (1937).
Stroud, Richard H. *National Leaders of American Conservation* (1985).

UNPUBLISHED SOURCES

See Dutcher Correspondence in National Audubon Collection, New York Public Library.

Arthur A. Belonzi

DUTTON, CLARENCE EDWARD. Born Wallingford, Connecticut, 15 May 1841; died Englewood, New Jersey, 4 January 1912. Geologist, soldier.

FAMILY AND EDUCATION

Son of Samuel and Emily (Curtis) Dutton. His father was shoemaker and postmaster. Dutton attended school at Ellington, Connecticut. Entered Yale in 1856, at the age of fifteen. Graduated in 1860; pursued graduate study at Yale for two more years. Married Emiline Babcock in 1864; two children, daughter and son.

POSITIONS

Appointed adjutant, 21st Connecticut Volunteer Infantry, with rank of 1st lieutenant, September 1862. Promoted captain, March 1863. Appointed 2d lieutenant, Ordnance Corps, Regular Army, January 1864. Assigned to Ordnance Depot, Army of the Potomac, 1865. Stationed at Watervliet Arsenal, New York, West Troy, New York, 1865–1870. Transferred to Frankford Arsenal, Philadelphia, 1870. Transferred to Office Chief of Ordnance, Washington, D.C., 1871.

Promoted captain, 1873. Detailed to Survey of the Rocky Mountain Region as geologist, 1875. Remained with this survey until 1879 and with U.S. Geological Survey, 1879–1890.

In 1890, Dutton rejoined military service. The following year was promoted major and became head of San Antonio Arsenal. He held this position until 1899. He retired in 1901 but continued with scientific research until his death.

CAREER

Dutton's career as a geologist flourished once he began working with John Wesley Powell and Grove K. Gilbert in the 1870s. Was associated with Gilbert on latter's study of the Henry Mountains and contributed report on igneous rocks for Gilbert's *Geology of the Henry Mountains* (1877). Dutton's fieldwork over the next decade entailed study of plateau region in Arizona and Utah. During this period, wrote a number of reports for the Geological Survey. Three of his more notable titles were *Report on the Geology of the High Plateaus of Utah* (1880); *Tertiary History of the Grand Canyon District* (1882); and *Mount Taylor and the Zuni Plateau* (1885). In 1882, spent six months in Hawaii, studying volcanoes, subsequently drafting a number of short papers for the Geological Survey based on this fieldwork. In 1886, studied effects of an earthquake in Charleston, South Carolina. Later published *Earthquakes in the Light of the New Seismology* (1904). Dutton was a member of the Philosophical, Geological and Seismological Societies of Washington. He was elected to the National Academy of Sciences in 1884.

MAJOR CONTRIBUTIONS

Dutton's geologic studies of Arizona and Utah region are noted for their clarity and accuracy. The reports provided detailed geologic information and portrayed the scenic beauty of the area. He also made scientific contributions to volcanism and seismology.

BIBLIOGRAPHY

American Men of Science. 1st, 2d eds.

Dictionary of American Biography.

Dictionary of Scientific Biography.

Diller, Joseph S. "Memoir." *Bulletin of the Geological Society of America* (March 1913).

Longwell, Chester. "Clarence Edward Dutton." In *National Academy of Sciences Biographical Memoirs*. Vol. 32 (1958).

Stegner, Wallace E. *Clarence Edward Dutton: Geologist and Man of Letters* (1935).

UNPUBLISHED SOURCES

See Clarence Dutton's Letters, 1891–1907, University of Texas.

Paul Cammarata

E

EASTWOOD, ALICE. Born Toronto, Canada, 19 January 1859; died San Francisco, 30 October 1953. Botanist.

FAMILY AND EDUCATION

Oldest daughter of Colin Skinner Eastwood, a shopkeeper, and Eliza Jane Gowdey Eastwood. Mother died when Eastwood was six, leaving also younger brother and sister. Lived for several years with an uncle, who introduced her to botany. Returned to father and attended local school. At age eight she entered convent, where she continued her interest in botany. In 1873 she rejoined her father in Denver, where she entered public school. Graduated from East Denver High School in 1879, ending her formal education. Did not marry.

POSITIONS

High school teacher, 1879–1889. Traveled to East Coast 1890 and California, 1890–1891. Succeeded Katherine Brandegee as curator of the herbarium of the California Academy of Sciences, 1892–1949.

CAREER

From first introduction to botany by uncle, her interest seems to have been unwavering. First met Asa Gray in 1881. Spent summers exploring the Colorado flora, 1879–1889. Guided Alfred Russel Wallace in the Rockies, summer 1881. In 1890–1891 traveled to California, where she met Katherine and Townshend Brandegee. In winter 1891–1892 returned to California to help organize Academy of Sciences herbarium. Moved to San Francisco in late 1892 to become curator and acting editor of *Zoe*. In April 1906 rescued the type specimens of the academy during the San Francisco earthquake and fire, having had foresight to store them separately, a practice not then common. Spent following six years visiting American and European herbaria while the academy was rebuilt. Her botanical explorations by rail, stage, horse and muleback, foot, and eventually car covered western America from Alaska to Baja and inland to New Mexico and Utah.

Published over 300 articles and reviews, 1891–1949 in *Zoe, Proceedings of the California Academy of Sciences, Erythra, Bulletin of the Torrey Botanical Club, Muhlenbergia*, and especially the journal she helped found, edit, and subsidize, *Leaflets of Western Botany*. Several articles and floras were printed as monographs. Most renowned were her critical studies of western Liliaceae, which culminated with three articles in the *Leaflets of Western Botany* 5 (1948): 103–4, 120–23, 133–38.

Wrote extensively for both colleagues and public. Recognized as leading authority on California flora, receiving many honors from the scientific community. Two American genera, *Eastwoodia* Brandegee and *Aliciella* Brand, and many species and varieties were named to honor her. Noted for her candor, disregard for social convention—especially in matters of dress—and preference for simplicity whether in art or food, was considered staunch feminist as well. A much-loved and highly regarded member of scientific community.

MAJOR CONTRIBUTIONS

Built the herbarium of the California Academy of Sciences twice (before and after 1906). Contributed to knowledge of western American flora.

BIBLIOGRAPHY

Abrams, Leroy. "Alice Eastwood—Western Botanist." *Pacific Discovery* 11 (1949).

Dakin, Susanna Bryant. *The Perennial Adventure: A Tribute to Alice Eastwood, 1859–1953* (1956).

Howell, John Thomas. "Alice Eastwood, 1859–1953." *Taxon* 3 (1954).

Howell, John Thomas. "I Remember, When I Think . . ." *Leaflets of Western Botany* 7 (1954).

MacFarland, F. M., R. C. Miller, and Veronica J. Sexton. "Biographical Sketch of Alice Eastwood" and "Bibliography of the Writings of Alice Eastwood." *Proceedings of the California Academy of Sciences* 25 (1943–1949).

Notable American Women: The Modern Period.

Wilson, Carol Green. *Alice Eastwood's Wonderland: The Adventures of a Botanist* (1955).

UNPUBLISHED SOURCES

Some material is at California Academy of Sciences.

Liz Barnaby Keeney

EATON, AMOS. Born Chatham, New York, 17 May 1776; died Troy, New York, 10 May 1842. Botanist, geologist, educator.

FAMILY AND EDUCATION

Son of Capt. Abel Eaton, farmer and Revolutionary War soldier, and Azuba Hurd Eaton. Graduated from Williams College, B.A., 1797. Attended Yale between 1815 and 1817 and studied science. Married Polly Thomas 1779, Sally Cady 1803, Alice Bradley 1816, and Alice Johnson 1827. Had ten children. Admitted to the New York bar 1802.

POSITIONS

Lawyer, land agent, and surveyor, Catskill, New York, 1802–1810; incarcerated in Greenwich Jail, New York, convicted of alleged forgery in a land dispute; while there, became interested in science with assistance of John Torrey, son of prison warden; student of Benjamin Silliman and Eli Ives at Yale, 1815–1817. Began offering series of scientific lectures at Williams College, 1817. Traveling lecturer in region between West Point, New York, and Castleton, Vermont, where he was professor of natural history in Castleton Medical Academy, 1820–1821. Senior professor, Rensselaer School (later Institute), 1824–1835; Senior professor, Rensselaer Institute, 1835–1842.

CAREER

Briefly a teacher following college, then an attorney. Published *Art without Science*, text on surveying (1802). Forgery conviction ended career as lawyer; turned to study of science in prison and thereafter. Successful teacher of popular science. Became interested in geology, in part because of belief that valuable mineral resources could be found, especially in western New York State. Undertook surveys in vicinity of Erie Canal and considered himself something of an authority on the geology of the region. Made certain assumptions concerning relationships between geological strata of the United States and the European continent unsupported by available data. Asa Gray at first depended heavily on Eaton's *Manual of Botany* but was later very critical of its shortcomings. Played important role in persuading Stephen Van Rensselaer to found school, which was forerunner of Rensselaer Polytechnic Institute, first scientific school in the United States after U.S. Military Academy at West Point. There many modern methods of science teaching had their inception. Published *A Botanical Dictionary* (1817); first edition of *Manual of Botany* (1817 and seven subsequent editions, last one in 1840); *An Index to the Geology of the Northern States* (1818); *Chemical Instructor* (1822; and subsequent editions); *A Geological and Agricultural Survey of the District Adjoining the Erie Canal* (1824); *Zoological Text-Book* (1826); *Geological Nomenclature for North America* (1828); *Prodromus of a Practical Treatise on the Mathematical Arts* (1838).

MAJOR CONTRIBUTIONS

His popular lectures, crossing entire spectrum of science, aroused considerable public interest. His emphasis on learning by doing rather than by rote, as was standard practice at the time, was significant innovation in science education. Was for some years virtually the only member of teaching faculty at Rensselaer School, later Polytechnic Institute, in the founding of which he was largely instrumental.

BIBLIOGRAPHY

Daniels, George H. *American Science in the Age of Jackson* (1968).

Dictionary of American Biography.

Dictionary of Scientific Biography.

Elliott, Clark A. *Biographical Dictionary of American Science: The Seventeenth through the Nineteenth Centuries* (1979).

McAllister, Ethel M. *Amos Eaton, Scientist and Educator* (1941).

Merrill, G. P. *The First One Hundred Years of American Geology* (1924).

Nickles, J. M. *Geological Literature on North America, 1785–1918* (1923).

Rezneck, Samuel. *Education for a Technological Society: A Sesquicentennial History of Rensselaer Polytechnic Institute* (1968).

UNPUBLISHED SOURCES

Correspondence and other papers of Eaton and other botanists relating to teaching of natural history in the United States are located in the Syracuse University Library. Correspondence and other material, chiefly relating to Eaton's professorship at Rensselaer School (later Rensselaer Polytechnic Institute) under the patronage of Stephen Van Rensselaer and to his geological tours along route of the Erie Canal, are located in New York State Library, Manuscripts and Special Collections (Albany).

Roberta Pessah

ECKSTORM, FANNIE PEARSON HARDY. Born Brewer, Maine, 18 June 1865; died Brewer, Maine, 31 December 1946. Ornithologist, authority on history, folksongs, and Indians of Maine.

FAMILY AND EDUCATION

Eldest of six children (five girls and one boy) born to Manley and Emeline Freeman (Wheeler) Hardy. Father was fur trader and naturalist. Attended high school in Bangor, then spent a year at Abbott Academy, Andover, Massachusetts. Graduated from Smith College, 1888. Received honorary M.A. from University of Maine, 1929. Married Rev. Jacob A. Eckstorm in 1893; two children.

POSITIONS

Superintendent, Brewer schools, 1889–1891. Reader of scientific manuscripts, D. C. Heath, Boston, 1891–1892. Conducted research in ornithology, 1893–

1901. Researched and wrote about the history, folksongs, and Penobscot Indians of Maine, 1902–1945.

CAREER

Her father was known as authority on the birds and mammals of Maine and was regular contributor to *Field and Stream*. He also maintained a close relationship with the Penobscot Indians of the area and knew their language. From childhood, she spent much time with her father and learned to know and respect the Indians, as well as how to observe animals and take notes on her findings. While at Smith College she founded a college Audubon Society and became interested in further study of the area Indians. Summer after graduation, she accompanied her father on the first of several long canoe trips into Maine wilderness.

One of first women to serve as a school superintendent in Maine, but town of Brewer's failure to appropriate necessary funding caused her to resign. During this time, worked with her father in crusade to preserve Maine's big game and control out-of-state hunters. As part of this effort, wrote two series of articles for *Forest and Stream* in 1891. Spent year in Boston working for D. C. Heath. Then after serving parishes in Oregon City, Oregon, and Eastport, Maine, the family moved to Providence, Rhode Island, in 1898. After husband's death the next year, she and children returned to Brewer.

Her time in Oregon is reflected in early ornithological writing for *Auk* and *Bird-Lore*. After returning to Brewer, she wrote two books on birds, *The Bird Book* (1901), a children's book, and *The Woodpeckers* (1901), which introduced techniques of bird-watching. Both books were praised for their scientific accuracy. Concern over the disappearing traditional culture of northern Maine can be seen in her next two books, *The Penobscot Man* (1904) and *David Libbey: Penobscot Woodsman and River Driver* (1907). In these she attacked paper companies that were taking over the forests and rivers of her state.

During 1920s, turned to another area of Maine's culture—ballads and folksongs. In collaboration with Mary Winslow Smyth, wrote *Minstrelsy of Maine* (1927) and with third collaborator, Phillips Barry, *British Ballads from Maine* (1929). After this, concentrated most of her energies on Indian history and language, writing *The Handicrafts of the Modern Indians of Maine* (1932); *Indian Brother* (1934) and *The Scalp Hunters* (1936), written with Hubert Vansant Coryell; *Indian Place-Names of the Penobscot Valley and Maine Coast* (1941), which established her as the leading authority on the Penobscot Indians; and *Old John Neptune and Other Maine Indian Shamans* (1945).

Actively involved in community affairs during her life, she founded a public library in Brewer in 1908, did Red Cross work during World War I, and was first chair of the local Women's Republican Committee in 1920. Also cofounded the Folk-Song Society of the Northeast in 1930. Throughout her life identified deeply with her home state and its people. Was known for her generous help to other scholars as well as sharp criticism of those with only a superficial interest in the areas of knowledge she held dear. Her books and more than thirty published articles reflect her scientific training and methodical, precise research.

Member of American Ornithologists' Union, 1887 (one of first women elected); Society of Mayflower Descendants; Maine State Historical Society; Piscataquis County Historical Society; Bangor Historical Society. Honorary member, National Bird Museum of Canada.

MAJOR CONTRIBUTIONS

She was the leading authority on Penobscot Indians of Maine. Work in ornithology, Maine history, and folksongs continues to be important to researchers.

BIBLIOGRAPHY

Dictionary of American Biography. Supplement 4.

James, Edward T., ed. *Notable American Women, 1607–1950* (1971).

Mainiero, Lina, ed. *American Women Writers* (1979).

Ring, Elizabeth. "Fannie Hardy Eckstorm: Maine Woods Historian." *New England Quarterly* (1953).

UNPUBLISHED SOURCES

See Eckstorm-Hardy Manuscripts, Bangor Public Library (notes, diaries, miscellaneous papers); Fannie Hardy Eckstorm Papers, Raymond H. Fogler Library, University of Maine (letters, notes, texts, ballads); Fannie Hardy Eckstorm Letters, Class of 1888, Smith College Archives; Houghton Mifflin Co. Correspondence, Houghton Library, Harvard University (more than 100 items dealing with Eckstorm, mostly letters with her publisher and manuscripts).

Cynthia D. Chambers

EDGE, MABEL ROSALIE. Born New York City, 3 November 1877; died New York City, 30 November 1962. Conservationist, reformer.

FAMILY AND EDUCATION

Daughter of John Wylie Barrow (a first cousin of Charles Dickens) and Harriet Bowen Barrow. Had private school education. Honorary Litt.D., Wagner College, Staten Island, New York, 1948. Married Charles Noel Edge, 1909; two children.

POSITIONS

Creator and chair, Emergency Conservation Committee (ECC), 1929–1962. Organizer, director, and president, Hawk Mountain Sanctuary Association, 1936–1962.

CAREER

Lived in Europe for several years following her marriage. Involved in womens' suffrage movement. Became interested in birds c.1915. Became conservation activist following separation from husband, a mining engineer. As head of ECC, attacked inertia of National Audubon Society Board, its finances, and its association with sportsmen hunting interests at 1929 meeting. Ultimately, in 1934, she helped bring about reorganization of Audubon Society and rejuvenation of its conservation ethos. Was described by former reform colleague (Willard Van Name) as "the only honest, unselfish, indomitable hellcat in the history of conservation." Edge purchased option on Hawk Mountain, near Drehersville, Pennsylvania, and turned it into refuge for hawks, which previously had been targets for hunters. Beginning in 1931, Edge, who was a dedicated preservationist, inveighed against the U.S. Biological Survey for its practice of trapping and poisoning predators and small mammals to protect the interests of ranchers and livestock growers. Terming it "the Bureau of Destruction and Extermination," she sought to change its policy, but without success. In the West, Edge overcame bitter opposition of lumber companies to serve as a midwife to the creation of Olympic National Park (1938) and played major role in creation of Kings Canyon National Park (1940). She also spearheaded effort to place 6,000 additional acres of old-growth sugar pines within borders of Yosemite National Park (1937). The ECC was small, basically kept alive by Edge's dwindling fortune and modest contributions, and continued its advocacy of her conservation causes on a more limited basis in later years. Hawk Mountain was always a priority. By 1947, the ECC budget was less than $3,000, and membership hovered around 900. The defeat of the Echo Canyon dam project in the mid-1950s was Edge's last great success, though hers was but one voice in the chorus of opposition to that undertaking. When Edge died in 1962, the ECC died with her. Author of *White Pelicans of Great Salt Lake* (1935); *Finishing the Mammals* (1936); *Migratory Bird Treaty with Mexico* (1936); *Roads and More Roads in the National Parks and the National Forests* (1936); and "Motor Power," poem in *Nature Magazine* (December 1940). A photcopied edition of the typescript of her autobiography, with penciled corrections (see Unpublished Sources) was released under the title of *An Implacable Widow* in 1978.

MAJOR CONTRIBUTIONS

Persistent and sometimes strident advocate of a range of conservation causes over a third of a century. Brought about some reforms in the management of the National Audubon Societies in the 1930s. Her creation and sustained support of the Hawk Mountain preserve were noteworthy. While her effort to end the Biological Survey's poisoning and trapping practices in the 1930s was unsuccessful, later reformers brought about changes in federal wildlife management procedures in the 1960s and 1970s.

BIBLIOGRAPHY
Broun, Maurice. *Hawks Aloft: The Story of Hawk Mountain* (1949).
Davis, R. C., ed. *Encyclopedia of American Forest and Conservation History*. Vol. 1 (1983).
Fox, Stephen. *John Muir and His Legacy: The American Conservation Movement* (1981).
Graham, Frank, Jr. *Man's Dominion: The Story of Conservation in America* (1971).
Taylor, Robert Lewis. *The New Yorker* (17 April 1948).
Time Magazine (3 November 1930).
UNPUBLISHED SOURCES
See Irving Brant Papers, Library of Congress, including his unpublished manuscript "Adventures in Conservation," and Rosalie Edge, "Good Companions in Conservation: Annals of an Implacable Widow," manuscript in possession of family.

Roberta Pessah and Keir B. Sterling

EIGENMANN, CARL H. (did not stand for a name). Born Flehingen, Germany, 9 March 1863; died Chula Vista, California, 24 April 1927. Ichthyologist, zoologist, educator.

FAMILY AND EDUCATION

Son of Philip and Margaretha Lieb Eigenmann. Brought up in Rockport, Indiana, from age fourteen by an uncle. Graduated Indiana State University with a bachelor's degree in zoology, 1886; master's degree, 1887; Ph.D., 1889. Married Rosa Smith, a West Coast ichthyologist, 1887.

POSITIONS

Curator, San Diego Natural History Society, 1888. Acting curator of fishes, California Academy of Sciences, 1889–1891. Professor of zoology, Indiana University, 1891–1908. Dean of Graduate School, Indiana University, 1908–1927. Founder-director, Indiana University's Biological Station at Turkey Lake (later removed to Winona Lake), 1895–1920. Honorary curator of fishes, Carnegie Museum, Pittsburgh, 1909–1918.

CAREER

Encouraged in zoology by David Starr Jordan, his professor of natural history at Indiana University, who later appointed him as professor of zoology. With his wife, soon after their marriage, studied collections of South American fish at the Museum of Comparative Zoology, under Alexander Agassiz. In 1892 collected fishes in the Canadian and American Northwest for the British Museum. Studied the origin and differentiation of sex cells in fish collected from the Pacific Coast. In the 1890s studied blind fish and other creatures inhabiting caves of southern Indiana and Kentucky, explaining this phenomenon as resulting from degenerative evolution. In 1909 published his best-known work, *Cave*

Vertebrates of North America, A Study in Degenerative Evolution, described as a "splendid" volume. In 1903 resumed work on South American freshwater fish, particularly the Brazilian *Characidae*. With financial aid from the Carnegie Museum of Pittsburgh, made an expedition to British Guiana in 1908 to study the geographical distribution of freshwater fishes there. The trip yielded twenty-eight new genera and 128 new species. In 1912 issued *The Fresh-Water Fishes of British Guiana*, regarded by one authority as "the most important single work in South American ichthyology." In 1912 he also traveled to Colombia and organized similar expeditions for his students upon his return. These expeditions resulted in *The Fresh-Water Fishes of Northwestern South America*, published in 1922 by the Carnegie Museum. In 1918, led an expedition to high Andes, Peru, and coast of Chile. In meantime, he had put together his magnum opus, *The American Characidae*, concerning the largest and most diversified family of the immense neotropical fish fauna. First three volumes appeared between 1917 and 1925; the fourth and fifth volumes were published posthumously.

Published 221 articles and reports and seven monographs (1885–1929), including *The Cheirodontinae* (1915); *The Pygiidae, a Family of South American Cat-Fishes* (1918); and *The Fresh-Water Fishes of Chile* (1927). Among his important articles were; with R. S. Eigenmann, "Preliminary Notes of South American Nematognathi" (1888) and "A Catalogue of the Fresh-Water Fishes of South America" (1891); "Fresh-Water Fishes of Patagonia and an Examination of the Archiplata-Archhelenis Theory" (1909); and a suggestive summary of his South American work, "The Fishes of the Pacific Slope of South America and the Bearing of Their Distribution on the History of the Development of the Topography of Peru, Ecuador, and Western Columbia" (1923).

Had a phenomenal capacity for work, his monographs tending to monumental lengths, in addition to his numerous articles, only twenty of which were intended for popular consumption. His research publications were characterized as "critical, painstaking, detailed," and important. He was kindly, jolly, and sympathetic, yet was stubborn and iron-willed with respect to his work. He believed in letting his students learn for themselves and was beloved by his students and colleagues alike.

MAJOR CONTRIBUTIONS

Considered one of the four greatest ichthyologists of his time. Best remembered for his evolutionary studies of cave-dwelling species, but he made his reputation with developmental studies of Pacific fish. His biogeographic and faunistic studies of South American freshwater fish were the most complete and extensive ever undertaken to that time. Named numerous new species and genera and noted faunistic similarities with African species, lending support to Von Ihering's theory of the former existence of a landmass connecting South America and Africa. Also "a great and inspiring teacher" who trained many prominent ichthyologists.

BIBLIOGRAPHY
Myers, George S. ''Carl H. Eigenmann—Ichthyologist 1963–1927.'' *Natural History* (January, February 1928).
Myers, George S. ''Eigenmann, Carl H.'' *Dictionary of American Biography.*
Stejneger, Leonhard. ''Biographical Memoir of Carl H. Eigenmann 1863–1927.'' *National Academy Biographical Memoirs* 18 (1937).

Jennifer M. Hubbard

EISELEY, LOREN C. Born Lincoln, Nebraska, 3 September 1907; died Philadelphia, 9 July 1977. Anthropologist, educator, essayist, poet.

FAMILY AND EDUCATION

Son of Clyde Edwin and Daisy (Corey) Eiseley. His grandfather was a native of Germany who homesteaded in the Nebraska Territory. After public schools in Lincoln, entered University of Nebraska, earning a B.A. in 1933. In 1935 completed an M.A. and in 1937 a Ph.D. at the University of Pennsylvania. Later awarded several honorary degress, among them an L.H.D. from Western Reserve University in 1959, an Litt.D. from University of Nebraska in 1960, a D.Sc. from Franklin and Marshall College in 1960, and an LL.D. from Alfred University in 1963. Married Mabel Langdon, 1938.

POSITIONS

Appointed assistant professor of anthropology and sociology at the University of Kansas, 1937, associate professor, 1942. Appointed department chairman and professor of anthropology and sociology, Oberlin College, 1944; then to same position at University of Pennsylvania, 1947–1962. President of American Institute of Human Paleontology, 1949–1952. Named provost of the University of Pennsylvania, 1959; served until 1961 and in that year named first Benjamin Franklin Professor of Anthropology and the History of Science at the University of Pennsylvania.

CAREER

An eminent scientist and eloquent prose stylist, Eiseley's eighteen books, more than 200 essays, and some 150 poems are difficult to categorize. To a large extent, his literary work recalls that of Francis Bacon and Henry David Thoreau. Eiseley thought of himself as a ''naturalist'' in the Victorian sense of a man of science with a questioning mind and gift of expression. Additionally, he was a poet who wrote of the dust of the Nebraska landscape.

Initiated his career with essays on natural history for popular periodicals. Among his more noteworthy books are *The Immense Journey* (1957), *The Firmament of Time* (1960), *The Mind as Nature* (1962), *The Unexpected Universe*

(1964), *The Brown Wasps* (1969), *Darwin and the Mysterious Mr. X: New Light on the Evolutionists* (1970), *The Invisible Pyramid* (1970), *The Man Who Saw Through Time* (1973), and *The Star Thrower* (1978). There were also four volumes of poetry. With Thoreau's gift for finding metaphors in "natural facts," Eiseley frequently wrote of man's existence. Lacking the genius of a Darwin to assemble and sift through facts, Eiseley failed to develop into an original scientist as understood within the confines of the profession; but he did become an outstanding scholar and critic of science. Unable to abide the separation of fact and value, he questioned many of the orthodoxies of his profession. In his own writings are found rich statements of what it means to be a human creature within evolutionary time.

MAJOR CONTRIBUTIONS

Even if his books and essays did little more than create interest in natural history, his works hit their mark. His poetry spans geological and biological time. As scientist and poet, he wrote eloquently of man's "immense journey." A harbinger of a new humanism based on connections forged between man and nature, he is in the tradition of such luminaries as Emerson, Thoreau, John Muir, John Burroughs, and Rachel Carson.

In his *Darwin's Century* (1958) he stressed two approaches to the reconstruction of the history of life: the study of living organisms and the examination of fossil records. Critical of Darwin, Eiseley maintained that the British scientist overemphasized the brute struggle in the course of human evolution and that Darwin's determinism failed to distinguish between physical and cultural determinants of human nature.

More than anyone else, Eiseley should be adulated for originating a style so popular with writers on scientific matters such as Carl Sagan, Stephen Jay Gould, and other humanistic scientists.

BIBLIOGRAPHY

Angyl, Andrew J. *Loren Eiseley* (1983).

Auden, W. H. "Considering the Unpredictable." *New Yorker* (21 February 1970).

Carlisle, E. Fred. *Loren Eiseley: The Development of a Writer* (1983).

Christianson, Gale E. *Fox at the Wood's Edge: A Biography of Loren Eiseley* (1990).

Gerber, Leslie E., and Margaret McFadden. *Loren Eiseley* (1983).

Heuer, Kenneth, ed. *The Lost Notebooks of Loren Eiseley* (1987).

Nemerov, Howard. "Loren Eiseley: 1907–1977." *Proceedings of the American Academy of Arts and Letters* 2 (1978).

Schwartz, James M. "Loren Eiseley: The Scientist as Literary Artist." *Georgia Review* 31 (1977).

UNPUBLISHED SOURCES

Eiseley manuscripts and papers are in American Museum of Natural History, American Philosophical Society, Columbia University Archives, Harvard University Archives, Library of Congress, National Archives, National Museum of Natural History, University of Kansas Archives, University of Pennsylvania Archives, and others. See list in Christianson. For complete list consult Jeanne de Palma Gallagher, *Bibliographic Card Index for the Writings of Loren Eiseley*, copies of which may be secured from University of Pennylvania Archives, Philadelphia.

George A. Cevasco

EISENMANN, EUGENE. Born Panama City, Panama, 19 February 1906; died New York City, 16 October 1981. Ornithologist, lawyer.

FAMILY AND EDUCATION

Son of an American father, Gustave Eisenmann, and a Panamanian mother, Esther Ethel Brandon Eisenmann. One of four children. Received primary and secondary education in New York City. Received S.B. degree, magna cum laude, Harvard University, 1927; received LL.B. degree, magna cum laude, Harvard Law School, 1930. Remained a lifelong bachelor.

POSITIONS

Partner and trial lawyer, Proskauer and Mendelsohn, later Proskauer, Rose, Goetz, and Paskus, New York City, 1930–1957. Research associate, Department of Ornithology, American Museum of Natural History, 1957–1981.

CAREER

Although he had a lifelong interest in Middle and South American birds, Eisenmann did not pursue ornithology on full-time basis until his retirement from the legal profession at age fifty-one. Prior to 1957 he had been particularly active in New York's Linnaean Society and the American Ornithologists' Union (AOU).

Published some thirty articles and six books, the most important being *Annotated List of Birds of Barro Colorado Island, Panama Canal Zone* (1952) and *The Species of Middle American Birds* (1955). Collaborated with Rodolphe Meyer De Schauense on *The Species of South American Birds and Their Distribution* (1966), although he declined the offer of a coauthorship. Eisenmann's research and writing on neotropical birds were particularly valued by the American ornithological community. Gregarious by nature, was always helpful to researchers and students of the subject. As member of the Pan-American Section of the International Council for Bird Preservation (ICBP), his financial and legal astuteness was strongly relied upon.

President, Linnaean Society of New York, 1951–1953. Editor, *The Auk*, AOU, 1957–1959. Vice president, AOU, 1967–1969. Chairman, AOU Check List Committee (6th ed.), 1966–1981. At the time of his death, Eisenmann was chairman, AOU Standing Committee on Nomenclature; vice chairman, Pan-American Section, ICBP; member, International Commission on Zoological Nomenclature.

Considered authority on neotropical birds, with focus on Panama but extending north into Middle America and south into western South America. Able editor and coadjutor to those working on birds of these regions. Played major role in completing sixth edition of AOU *Checklist of American Birds*. His extensive notes on systematics, behavior, and distribution of Middle American birds an invaluable resource for anyone studying the subject.

MAJOR CONTRIBUTIONS

Considered an expert on neotropical birds, centered in Panama and radiating north and south therefrom. Although he did not publish many papers, his behind-the-scenes research on their systematics, behavior, and distribution is still an invaluable source.

BIBLIOGRAPHY

American Men and Women of Science. 14th ed. (1979).
Bull, John, and Dean Amadon. *Auk* (January 1983) (obit.).
Graham, Frank, Jr. ''Eisenmann's Monument.'' *Audubon Magazine* (January 1986).
New York Times, 17 October 1981 (obit.).
Who's Who in New York (City and State). 13th ed. (1960).

UNPUBLISHED SOURCES

See American Museum of Natural History, Department of Ornithology, Historic Correspondence, 1864–present. This collection contains Eisenmann's correspondence pertaining to his involvement with the International Council for Bird Preservation and also to his chairmanship of the Check List Committee for the AOU.

Patrick J. McNamara

ELLIOTT, STEPHEN. Born Beaufort, South Carolina, 11 November 1771; died Charleston, South Carolina, 28 March 1830. Naturalist, humanist, bank president, professor, educator.

FAMILY AND EDUCATION

Son of rice planter William Elliott (died 1778) and Mary Barnwell Elliott (died 1774). Grew up in home of older brother William. Attended school in Beaufort. Graduated from Yale College (1791) and was elected to Phi Beta Kappa (1791). Honorary LL.D., Yale, 1819; Harvard, 1822. Married Esther Wylly Habersham, 1796; twelve children.

POSITIONS

Planter, coastal South Carolina and Georgia, 1791–1830. President of Bank of the State of South Carolina, 1812–1830. Botanized extensively in South, especially Georgia and South Carolina, 1791–1830.

CAREER

Member of lower house of state legislature, 1794–1800. Member of South Carolina Senate, 1808–1812. Wrote law creating public school system and bank. Active as layman in Episcopal Church. Enthusiastic collector of specimens in fields of botany, mineralogy, entomology, conchology, and ornithology. Visited Charles Wilson Peale in Philadelphia and Henry Muhlenberg of Lancaster, Pennsylvania, in 1808 and later. Assisted Alexander Wilson with bird collections. Corresponded and exchanged specimens with many persons prominent in natural history. Was elected vice president of American Geological Society at its founding in 1819 and served until his death. Was corresponding member of Academy of Natural Sciences at Philadelphia (1815) and member of American Philosophical Society (1819). Was honorary member of Lyceum of Natural History of New York, fellow of the American Academy of Sciences, and member of Linnaean Society in Paris. Petitioned state legislature for charter for the Literary and Philosophical Society of South Carolina and was its president, 1814–1830. Was president of Charleston Library Society in 1816 and undertook task of cataloging its 14,000 volumes. Was cofounder in 1821 of South Carolina Academy of Fine Arts and active supporter until his death. Served as commissioner for the United States in dispute over damages between South Carolina and federal government in 1822 and made strenuous efforts to secure a Naval Depot at Charleston in 1824–1825. Received honorary LL.D. from Columbia (1825). Was founding trustee of Beaufort College in 1795, trustee of South Carolina College from 1820 and of the College of Charleston from 1826. Assisted in founding of Medical College of Charleston in 1824 and taught as professor of natural history without pay. Was publisher, editor, and contributor to the *Southern Review* from 1828 to 1830.

Published *A Sketch of the Botany of South-Carolina and Georgia* in thirteen parts issued between 1816 and 1824, bound in two volumes dated 1821 and 1824. An article, "Observations on the Genus Glycine, and Some of its Kindred Genera," was published in 1818, and posthumous article on the Cherokee rose as hedge appeared in 1831. Other publications include speeches, memorials, and detailed report on his bank. He wrote seventeen essays for *Southern Review* and reputedly enough poetry to fill small volume. The poetry is lost. He projected a *Prodromus* to be added to his *Sketch* and intended special treatment of native orchids, including drawings similar to those of the grasses in his *Sketch.*

Named more than 150 new species of plants, including the *Elliottia racemosa* out of Muhlenberg's *Catalogue*. Described plants in parallel columns of English and Latin in his *Sketch*, which is still indispensable to botanical taxonomist. Was self-trained botanist who crossed that invisible threshold from amateur naturalist to careful, analytical scientist.

MAJOR CONTRIBUTIONS

Precise study and detailed observation of plant specimens set high standards of exactness and accuracy. Literary essays and unpublished letters have a grace and felicity of expression that render them delight to read. Efforts to enhance education of children, college students, and adults benefited many.

BIBLIOGRAPHY

Barnwell, Stephen B. *The Story of an American Family* (1969).
Childs, Arney R. *Dictionary of American Biography* Vol. 6 (1931).
Dexter, F. B. *Biographical Sketches of the Graduates of Yale College* (1907).
Ewan, Joseph. "Introduction" to facsimile edition of *A Sketch of the Botany of South-Carolina and Georgia* by Stephen Elliott (1971).
Gee, Wilson. *Bulletin of the University of South Carolina*, no. 72 (1918).
Hoch, John Hampton. *Annals of Medical History* (n.s.) Vol. 7, no. 68, 1935.
Moultrie, James. *An Eulogium on Stephen Elliott* (1830).
Puckette, Clara Childs. *Charleston News and Courier*, 27 February 1933.
Ravenel, H. W. *Botanical Gazette* Vol. 8, no. 7, July 1883.
Sargent, C. S. *Garden and Forest*, 23 May 1894.

UNPUBLISHED SOURCES

See Elliott's Herbarium and a few letters in Charleston Museum, Charleston, S.C.; Letters in Elliott-Gonzales Collection, Southern Historical Collection, University of North Carolina; Letters and Manuscripts, Arnold Arboretum Library; Letters and herbarium specimens, Academy of Natural Sciences at Philadelphia.

George A. Rogers

EMMONS, EBENEZER. Born Middlefield, Massachusetts, 16 May 1799; died Brunswick County, North Carolina, 1 October 1863. Geologist, physician.

FAMILY AND EDUCATION

Son of Ebenezer Emmons, a farmer, and Mary Mack Emmons. Received early education in Middlefield and in Plainfield, Massachusetts. A.B., Williams College, 1818. Studied at Albany Medical College and Renssalaer School (now Polytechnic Institute), graduating in 1826 (some authorities state that Emmons attended Renssalaer before going to medical school). Married Maria Cone, 1818, had at least three children, including one son.

POSITIONS

Lecturer in chemistry, Albany Medical College, 1826; began medical practice in Chester and later in Williamstown, Massachusetts, starting in 1827; lecturer in chemistry, Williams College, 1828–1834; junior professor of chemistry at Renssalaer Institute, 1830, and lecturer at Castleton (Vermont) Medical School. Professor of natural history, Williams, 1833–1859; geologist on staff of New York Natural History Survey, 1836–1842; professor of chemistry, Albany Medical College, 1838–1851; and custodian of New York State collections, 1842–1851. Appointed North Carolina state geologist, 1851–1859; returned to Williams as professor of geology and mineralogy, 1859–1863 (but was caught in North Carolina by the Civil War, 1861–1863, and died there).

CAREER

Strongly influenced by Amos Eaton, one of his teachers at Renssalaer Institute, and became interested in geology. Made sufficient income from his obstetrical practice to underwrite much of his research in geology. Best known for having established a sequence of geological strata for North America that was independent of European approach to the subject. State of New York had been divided into districts for purposes of natural science research, and Emmons and his assistant, James Hall (1811–1898), had responsibility for the second of these. Emmons named the Adirondack Mountains and proposed a new Taconic system for the rocks in the vicinity of that range of mountains. Though he was able to advance paleontological evidence for his theory, it was bitterly attacked by a majority of other geologists, both here and abroad, including his former assistant Hall, Louis Agassiz, James D. Dana, J. D. Whitney, Charles Lyell, and many others. The outcome led to Emmon's "excommunication . . . from the ranks of American science" and his embittered departure from the state for a new position as state geologist in North Carolina. There Emmons did work of a high order, including the uncovering of fossils in the Deep and Dan River regions. Returning to Massachusetts on the eve of the Civil War, he resumed teaching at Williams College for several years and, by the time of his death, had managed to persuade some geological colleagues of the correctness of his views concerning the Taconic strata. This controversy continued to rage until nearly the end of the century, and there have been some reverberations since. When, in 1888, the American Committee of the International Geological Congress proposed to accept the Taconic system, several of Emmons's professional opponents, notably J. D. Dana, managed to quash the idea. In later years, Emmon's classification of rock strata and his terminology achieved more general acceptance and set the stage for much of the work that has since been done in the United States. Emmons published several major studies, including *Manual of Mineralogy and Geology* (1826, 2d ed., 1832); "Geology of the Second District," in *Geology, Natural History of the State of New York*, part 4 (1842); *Agriculture of New York*, 5 vols. (1846–1854); and his *American Geology*, 3 vols. (1855–1857), which was very controversial. While in North Carolina, he published three volumes of geological reports, 1856–1860. His *Manual of Geology* (1860), on the other hand, was highly influential among students, for whom he was "an underground favorite." He is said to have had strong support from junior geology faculty "in the strongholds of his opponents," (C. J. Schneer, in *Dictionary of Scientific Biography*) and the geology text of his staunchest opponent, James D. Dana, closely approximated Emmons's from the point of view of both organization and basic essentials. The Association of American Geologists (later the American Association for the Advancement of Science) was projected by a group meeting in Emmons's Albany home in 1838. Emmons's own papers were probably destroyed following his death in North Carolina during the Civil War.

MAJOR CONTRIBUTIONS

Emmons's development of his Taconic system laid the basic framework for American geological studies in the twentieth century, and he is also remembered for having named the Adirondack Mountains. He also was a highly effective college instructor. Emmons also published the first modern account of a state's mammalian fauna in *Zoology of Massachusetts* (1840).

BIBLIOGRAPHY

Clarke, J. M. *James Hall of Albany* (1923).
Dictionary of American Biography.
Elliott, Clark A. *Biographical Dictionary of American Science* (1979).
Merrill, George P. *The First One Hundred Years of American Geology* (1924).
National Cyclopedia of American Biography.
Schneer, C. J. *Dictionary of Scientific Biography.*
Schneer, C. J. "Ebenezer Emmons and the Foundations of American Geology." *Isis* 60, part 4 (1970).

Keir B. Sterling

EMORY, WILLIAM HEMSLEY. Born Queen Anne County, Maryland, 7 September 1811; died Washington, D.C., 1 December 1887. Army officer, explorer.

FAMILY AND EDUCATION

Son of Thomas and Maria Hemsley Emory. Graduate of the U.S. Military Academy, West Point, 1831. Married Matilda Wilkins Bache, 1831.

POSITIONS

Second lieutenant, 4th Artillery, 1831, resigned September 1836. Appointed 1st lieutenant, Topographical Engineers, 1838, principal assistant on Northeastern Boundary Survey between the United States and Canada, 1844–1846. Chief engineer officer and acting assistant adjutant-general, Army of the West during Mexican War. Captain, 1851, chief astronomer, Boundary Survey between the United States and Mexico, 1854–1857 (commissioner and astronomer to survey boundaries set forth by Gadsden Treaty, 1855); major (cavalry), 1855, lieutenant colonel, cavalry, 1861, brigadier general of volunteers, March 1862, major general of volunteers, September 1865–January 1866; colonel cavalry, 1863, brigadier general (retired) 1876, seven brevets for gallantry and meritorious service between 1846 and 1865.

CAREER

Emory is perhaps best known for his explorations made while with the advance guard of the Army of the West. He and his force departed Fort Leavenworth, Kansas, on 24 July 1846, and arrived at Raton Pass, near present Raton, New Mexico, early in August. They spent some time in New Mexico, not crossing the Colorado River into California until 22 November, and they reached the San Diego area early in January 1847, having been delayed by several military actions with Mexican forces in the region. They departed San Diego on 25 January 1847 and returned to the East by water. Emory made detailed comments concerning the Native Americans encountered, the topography, mammals, birds, fish, vegetation, native and Spanish architecture, and other matters of interest. Several detailed botanical appendixes (by John Torrey and George Engelmann, with thirteen plates) accompanied Emory's Report, *Notes of a Military Reconnaissance, from Fort Leavenworth, in Missouri, to San Diego, in California, including Parts of the Arkansas, Del Norte, and Gila Rivers*, which was published as Senate Document No. 7, 30th Congress, 1st Session, in 1848. Emory later, in the course of his boundary commission work (1848–1853), made notes on the fauna and flora observed, and his plant and mammal specimens are in the Museum of Natural History in Washington, D.C. He published "Notes . . . of the Boundary . . ." in 1851 and *Report on the U.S. and Mexican Boundary Survey* as a U.S. House of Representatives Document, 1857–1859. In addition, Emory published *Observations, Astronomical, Magnetic, and Meteorological, Made at Chagres and Gorgona, Isthmus of Darien and at the City of Panama, New Granada* in 1850. He was commanding officer of several forts in Indian Territory at the onset of the Civil War and successfully brought his men back to Fort Leavenworth, where they helped prevent Missouri from going into secession. He served with gallantry during a number of actions during the Civil War, rising to a major generalcy of U.S. Volunteers. He was subsequently commander of the Department of Washington, D.C. (1865–1869), District of the Republican, 1869–1871, and Department of the Gulf (1871–1875) before his retirement in 1876.

MAJOR CONTRIBUTIONS

Primarily known for his military reconnaissance from Fort Leavenworth to California, 1846–1847, and for his boundary survey work, 1854–1857, both of which resulted in detailed reports of the Southwest, together with botanical, zoological, and ethnological observations. His primary responsibilities during the boundary surveys were for astronomical observations, but his plant and mammal specimens from the boundary region were among the first from that region to be placed in the U.S. National Museum (now Museum of Natural History).

BIBLIOGRAPHY

Coville, Frederick V. "Three Editions of Emory's Report, 1848." *Bulletin of the Torrey Botanical Club* 23 (1896).
Dictionary of American Biography.
Elliott, Clark A. *Biographical Dictionary of American Science* (1979).
Emory, William H. "Extracts from His Journal While with the Army of the West." *Niles National Register*, no. 71 (31 October 1846).
Ewan, Joseph, and Nesta Dunn Ewan. *Biographical Dictionary of Rocky Mountain Naturalists* (1981).
National Cyclopedia of American Biography.
Warner, Ezra J. *Generals in Blue: Lives of the Union Commanders* (1964).

UNPUBLISHED SOURCES

Papers relating to Emory's Boundary Survey work are in Bancroft Library, University of California–Berkeley.

Keir B. Sterling

ENGELMANN, GEORGE. Born Frankfort-am-Main, Germany, 2 February 1809; died St. Louis, Missouri, 4 February 1884. Botanist, meteorologist, physician.

FAMILY AND EDUCATION

Eldest of thirteen children and son of George Engelmann, a professor at the University of Halle, and Julia May Engelmann. Entered University of Heidelberg in 1827 but left after student uprising, spent two years at University of Berlin, received M.D. from University of Würzburg in 1831. Married Dorothea Horstmann in 1840; one son.

POSITIONS

Began practice of medicine at St. Louis in 1835, which lasted until his death, and became one of the city's most distinguished physicians. From 1856 to 1858 studied botany briefly at Cambridge, Massachusetts, and then journeyed to Europe for further study and travel. Botanized after 1880 in the Rocky Mountains, Colorado, and forests of the Pacific Northwest.

CAREER

After studying medicine and other scientific subjects at Paris in 1832 with Alexander Braun and Louis Agassiz, he began a career that combined the practice of medicine with the study of botany and meteorology. After meeting Asa Gray in 1840, Engelmann began to unofficially coordinate the activities of many of the prominent western plant collectors, including Ferdinand Lindheimer and August Fendler, and sent many examples of western flora to eastern scientists for study and classification. With the aid of Henry Shaw, he established in 1859 the Missouri Botanical Garden—a major center for the study of systematic botany. A careful and meticulous student, he produced fundamental classification studies of the genus *Cuscuta* and *Cactaceae* as well as the *Yucca* and *Agave*. He also authored important studies on American oaks, conifers, grapes, mistletoes, and quillworts. His ''Diseases of Grapes'' (1873) was one of the nineteenth century's most significant studies of plant diseases.

Authored more than 100 scientific studies, the titles of which may be found in C. S. Sargent's ''Botanical Studies of George Engelmann,'' *Botanical Gazette* 9 (1884): 69–74 and *Botanical Works of the Late George Engelmann Collected for Henry Shaw* (1887).

MAJOR CONTRIBUTIONS

Meticulous examinations of some of the more difficult floral groups of the American West, preparation of detailed reports on plants collected on government expeditions west of the Mississippi River, development of taxonomic clarity on the conifers, grapes, yuccas, and rushes.

BIBLIOGRAPHY

Dictionary of American Biography. Vol. 6 (1931).
Elliott, Clark A., ed. *Biographical Dictionary of American Science*. Vol. 88 (1979).
Ewan, Joseph. *Dictionary of Scientific Biography* ''George Engelmann.''15 (1978).
Gray, Asa. *Proceedings of the American Academy of Arts and Sciences* 19 (1884).
White, Charles A. ''George Engelmann.'' *National Academy of Sciences, Biographical Memoirs* 4 (1902).

UNPUBLISHED SOURCES

Extensive manuscript materials, letters, notes, and drawings are in the archives and library of the Missouri Botanical Garden, St. Louis.

Phillip D. Thomas

ERRINGTON, PAUL L(ESTER). Born Bruce, South Dakota, 14 June 1902; died 5 November 1962. Wildlife ecologist, naturalist, writer, conservationist.

FAMILY AND EDUCATION

Grew up with mother and stepfather in Brookings and at Lake Tetonkaha, South Dakota. One leg crippled by polio at age eight, a handicap eventually largely overcome by strenuous hiking and hunting. Inspired partly by E. T. Seton's books, trapped for thirteen winters, including college years, when tuition largely paid for by fur sales. B.S., 1930, South Dakota State College. Ph.D., 1932, University of Wisconsin. Married Carolyn Storm, 1934; two sons.

POSITIONS

Industrial fellow, University of Wisconsin, 1929–1932. Research assistant professor (zoology), Iowa State College, 1932–1938; research associate professor, 1938–1948; research professor, from 1948. Guggenheim Fellow and visiting professor at Lund University, Sweden, 1958–1959, when he also worked at Erken Limnological Laboratory and Boda Research Station in Sweden.

CAREER

Associated with Aldo Leopold, who arranged financial support for graduate study. Published on bobwhite quail 1930–1945, culminating in "The Northern Bob-White's Winter Territory" (with F. N. Hamerstrom, Jr., 1936); on food habits of predatory birds 1930–1940, including "The Great Horned Owl and Its Prey in North-Central United States" (with Frances Hamerstrom and F. N. Hamerstrom Jr., 1940); and on muskrats from 1937 on. Muskrat studies spanned twenty-five years and more than 30,000 hours in the field. His most important work, *Muskrat Populations* (1963, 665 pp.) was published posthumously, as was his *Of Predation and Life* (1967). His publications totaled more than 200 technical papers, popular articles, books, and book reviews on over thirty-five species of vertebrates (see bibliographies in Carlander and Weller 1964 and Schorger 1966).

Received the Wildlife Society's Aldo Leopold Medal (1962); twice honored by Wildlife Society for outstanding publications ("The Great Horned Owl and Its Prey . . ." in 1941 and "Predation and Vertebrate Populations" in 1947). An Alfred Eisenstaedt photograph of him appeared in *Life*, 22 December 1961, canoeing in one of his long-term Iowa study areas. He explored widely in North America, visiting and studying wetlands in Minnesota, North Dakota, Montana, Wyoming, Idaho, Oregon, Utah, Nebraska, Georgia, Florida, Manitoba, Saskatchewan, and other states and provinces, as well as Sweden, Finland, Norway, and Denmark. His work was well known in Europe.

He was an accomplished writer and had a distinctive style, with little jargon and an almost quaint turn of phrase. His popular book *Of Men and Marshes* (1957) was one of first on wetlands ecology. He was criticized after his death for insufficient statistical analysis of his muskrat population data. Few biologists have been as adept at reading wildlife signs and as insightful in the ways of wild animals as Errington; these consummate field naturalists clear the necessary ground for biomathematicians and quantitative ecologists.

MAJOR CONTRIBUTIONS

His studies of predator–prey relationships and mortality from disease and climatic crises pointed to the idea that population regulation in the wild is usually a result of a complex of interacting factors rather than a single factor. Before Errington, many game species were thought to be limited in numbers by predation alone. He developed the ideas of compensating adjustments in mortality that affect the "biological surplus" of a population. He was profoundly interested in the sharp fluctuations and sometimes periodicity of population density.

He was also known for epidemiological studies of wildlife and was the namesake of Errington's disease of muskrats, a hemorrhagic syndrome the causes of which are not fully understood.

Errington's love of wildness and "untampered nature" resulted in his strong and reasoned pleas for conservation of wildlands and for not overmanaging parks and reserves by excessive manipulation of habitats or killing of "vermin." A wetlands reserve in Iowa has been dedicated to his memory.

BIBLIOGRAPHY

American Men of Science. 12th ed.

Carlander, K. D., and M. W. Weller. "Survey of a Life's Writings: Paul L. Errington's Bibliography." *Iowa State Journal of Science* 38 (1964).

Dunlap, Thomas. *Saving America's Wildlife* (1988).

Scott, T. G. "Paul L. Errington, 1902–1962." *Journal of Wildlife Management* 27 (1963).

Schorger, A. W. *Auk* 83 (1966) (obit.).

Stroud, Richard H., ed. *National Leaders of American Conservation* (1985).

UNPUBLISHED SOURCES

See Errington Papers, Iowa State University. Extensive collection includes correspondence, articles, papers and stories, manuscripts and galleys of books, bibliography of writings, and research and lecture notes pertaining to vertebrate ecology, population dynamics, and professional organizations. Unpublished manuscripts include "Of Birds and a Marsh and 30 Years."

Erik Kiviat

ESCHSCHOLTZ, JOHANN FRIEDRICH. Born Dorpat, Estonia (then part of Russia), 1 November 1793; died Dorpat, Estonia, 7 May 1831. Physician, naturalist.

FAMILY AND EDUCATION

Son of Johann Gottfried and Katherine Hedwig Ziegler Eschscholtz. M.D., University of Dorpat, 1815. Married Christine Friedrike Ledebour, 1819; two sons.

POSITIONS

Physician-naturalist on vessel *Rurik*, under command of Otto von Kotzebue, which circumnavigated globe from 1815 to 1818. Visited Unalaska and other sites in the Aleutian Islands, August–September 1816; visited San Francisco Bay, October–November 1816; visited Hawaiian Islands, November 1816; also made stops in the Philippines, Brazil, Chile, and other countries. Associate professor, University of Dorpat, 1819; director of the Zoological Cabinet there, 1822. Made second voyage with Kotzebue on vessel *Predpriaetie (Enterprise)*, 1823–1826. Again visited San Francisco Bay, September–November 1824. Professor extraordinary of medicine, professor of zoology, and director of the Zoological Museum, University of Dorpat, 1828–1831.

CAREER

During his two visits to California, Eschscholtz, who has been described as ''the ablest of the early Russian entomologists'' and ''the outstanding entomological figure during the Russian occupation of California,'' collected various insects, primarily Coleoptera, Orthoptera, Hymenoptera, and Hemiptera. Other animal specimens were reportedly scarce during his first visit. Also visited Unalaska and Sitka during his second visit in 1825. While in California, visited Santa Clara, San Rafael, Ross (site of the principal Russian settlement in Northern California), and Bodega Bay. On his first visit in 1816, also collected plants with Adelbert von Chamisso (1781–1837), the expedition naturalist, a French-born botanist exiled in Germany, who later described many of the species both men found. Eschscholtz was largely responsible for the 78 species brought back from his first visit to San Francisco Bay (1816). Eschscholtz collected mammals and amphibians, in addition to insects and plants, and picked up sand dollars at Unalaska. Additional specimens were collected in 1824. Probably 100 butterflies and twenty beetles among the insects collected represented new species. Many of these he described prior to his death, but some ninety-two others were described by Carl Gustav von Mannerheim (1804–1854), Swedish entomologist and sometime governor of Finland following Eschscholtz's early death at age thirty-eight. Among others who later described Eschscholtz material were Pierre François Marie Auguste Dejean (1780–1845), a French coleopterist, and Gotthelf Fisher von Waldheim (1771–1853), a Russian entomologist. Eschscholtz's own published papers on the animals and plants he collected along the northwest Pacific Coast included *Entomographien* Berlin: (Erste Liefurung, 1822); *Zoologischer Atlas, enhaltend Abbildungen neuer Thierarten wahrend des Flottcapitans von Kotzebue zweiter Reise um die Welt, auf der Russich-Kaiserlichen Kriegeschlupp Predpriaetie in den Jaren 1823–1826*, Berlin, five pts., 1829–1833; and ''Description es plantarum novae Californiae, adjectis florum exoticorum analysibus,'' *Memoires de l'Academie de St. Petersburg* 10 (1826) (includes discussion of insects, salamanders, mammals, birds, reptiles, and invertebrates collected and described; final part published by M. A. Rathke and includes brief outline of Eschscholtz's life). Most of his collections were deposited in the University of Dorpat Museum and in the collections of the Society of Natural History, at the Imperial Museum of Moscow. Von Chamisso described many of the plants both men collected in his *Works* (6 vols., 1836, 2nd ed., 1842), and in ''Remarks and opinions of the naturalist of the expedition'' in von Kotzebue's account, listed in the bibliography below. Named in his honor was Eschscholtz Bay on Kotzebue Sound, a large inlet in northwestern Alaska near Bering Strait, and the California poppy, *Eschscholtzia californica*.

MAJOR CONTRIBUTIONS

Made substantial collections of various insects, other fauna, and plants during two visits to Hawaii, Alaska, and California, 1816 and 1824, most of which ended up in Estonian and Russian museums. Responsible for descriptions of a number of new species. Considered to have been the ablest Russian entomologist to have visited North America during his lifetime.

BIBLIOGRAPHY

Eastwood, Alice. "The Botanical Collections of Chamisso and Eschscholtz in California." *Leaflets in Western Botany* 4 (1944).

Essig, E. O. *A History of Entomology* (1931).

Linsley, E. Gorton, ed. *Beetles from the Early Russian Explorations of the West Coast of North America, 1815–1857* (1978).

Mahr, August C. "The visit of the 'Rurik' to San Francisco in 1816." *Stanford University Publications, University Series: History, Economics and Political Science* 2, 2 (1932).

von Kotzebue, Otto. *A Voyage of Discovery into the South Sea and Beering's [sic] Straits, for the purpose of Exploring a north-east passage, undertaken in the years 1815–1818 . . . in the ship Rurick.* Trans. by H. E. Lloyd, 3 vols. (1821).

World Who Was Who in Science (1968).

Keir B. Sterling

EVERMANN, BARTON WARREN. Born Albia, Monroe County, Iowa, 24 October 1853; died Berkeley, California, 27 September 1932. Ichthyologist, educator, conservationist.

FAMILY AND EDUCATION

Son of Andrew Evermann and Nety Gardner Evermann. Received early education at local schools in Carroll County, Indiana, and at Howard College, Kokomo, Indiana. Received S.B., Indiana State University, 1886; A.M. in 1888 and Ph.D. in 1891 from Indiana State University. Received honorary LL.D. degrees from the University of Utah in 1922 and Indiana State University in 1927. Married Meadie Hawkins, 1875; two children.

POSITIONS

Teacher in public schools in Indiana and California, 1871–1881. Professor of biology, Indiana State Normal School, 1886–1891. Ichthyologist, U.S. Commission of Fish and Fisheries (later the Bureau of Fisheries), 1891–1914; U.S. Fur Seal Commissioner, 1892; chief of the Division of Statistics and Methods of Fisheries, 1902–1903; assistant in charge of scientific inquiry, 1903–1910; chief of the Alaska Fisheries Service, 1910–1914. Director of the California Academy of Sciences, 1914–1932; while there he founded and was director of the Steinhart Aquarium, 1921–1932.

CAREER

After teaching in public schools for ten years, returned to school to pursue a biology degree on a full-time basis, 1881–1886. While teaching biology at Indiana State Normal School, studied for the doctoral degree. While there he became acquainted with professor of zoology David Starr Jordan, with whom he later coauthored several books in the field of ichthyology. Upon receiving that degree, entered the federal service, where he served in a variety of capacities for twenty-three years. As fur seal commissioner in 1892, he advocated management policies for the Alaska seal herd that helped conserve a species once on the verge of extinction.

The number of his published books and articles reached 387. Several of his most important books were coauthored with David Starr Jordan, including *The Fishes of North and Middle America*, 4 vols. (1896–1900), *American Food and Game Fishes* (1902), and *A Checklist of the Fishes and Fishlike Vertebrates of North and Middle America* (1896). By himself, he authored *The Golden Trout of the High Sierras* (1906), *The Fishes of Alaska* (1907), and *The Fishes of Peru* (1915). Although major field of endeavor was ichthyology, about one-half of his publications cover zoological and ornithological topics, among others.

MAJOR CONTRIBUTIONS

Although he did not write rapidly, Evermann's works were strong on scientific detail and thorough research. As a scientist, was especially noted for dogged determination to complete any task he started. As fur seal commissioner, he pursued conservation policies that were years ahead of their time. At the time of his death was hailed as one of the foremost authorities on ichthyology in the United States.

BIBLIOGRAPHY

Dictionary of American Biography. Supplement 1 (1944).
New York Times, 28 September 1932 (obit.).
Stroud, Richard H., ed. *National Leaders of American Conservation* (1985).
Who's Who in America. Vol. 16 (1930–1931).

UNPUBLISHED SOURCES

Barton Warren Evermann Papers, California Academy of Sciences, San Francisco.

Patrick J. McNamara

F

FASSETT, NORMAN CARTER. Born Ware, Massachusetts, 27 March 1900; died Boothbay Harbor, Maine, 16 September 1954. Botanist.

FAMILY AND EDUCATION

Son of Joseph Lorenzo and Helen Stearns Carter Fassett. B.S., 1922, A.M., 1923, Ph.D., 1925, Harvard University. Married Katherine Hill Knight, 1925 (divorced 1950); three children.

POSITIONS

Teaching assistant in botany, Harvard, 1922–1924; teaching assistant in mycology, Radcliffe College, 1923–1924; biologist, U.S. Biological Survey, 1926; instructor in botany, 1925–1930, assistant professor, 1930–1937, associate professor, 1937–1945, professor, 1945–1954, University of Wisconsin–Madison. Chairman of the Botany Department, University of Wisconsin, 1948–1949; curator of the herbarium there, 1937–1954. Member, Colombian Cinchona Mission, 1944.

CAREER

Fassett was on the faculty of the University of Wisconsin for twenty-nine years. His primary interests had to do with the taxonomy and ecology of the plants of the Middle West. He was author of *Spring Flora of Wisconsin* (1932); *Leguminous Plants of Wisconsin* (1939); and *Manual of Aquatic Plants* (1940). He also published a number of shorter papers on the taxonomy, ecology, and phytogeography of plants of the Middle West and elsewhere. He was a member of the American Society of Naturalists, Society for the Study of Evolution, Botanical Society of America, American Society of Taxonomists (of which he was secretary and treasurer, 1936–1944, and president, 1953–1954), Wisconsin Academy of Science, and Sigma Xi. In addition, Fassett was a member of the Torrey Botanical and New England Botanical Societies.

MAJOR CONTRIBUTIONS

Fassett was an able teacher, researcher, and departmental administrator. His published works were important contributions to the botany of the upper Midwest.

BIBLIOGRAPHY

New York Times, 17 September 1954 (obit.).
Who Was Who in America. Vol. 3 (1963).
Who's Who in America. Vol. 26 (1950).

Susan Ignaciuk

FEATHERSTONHAUGH, GEORGE WILLIAM. Born London, England, 9 April 1780; died Le Havre, France, 28 September 1866. Geologist, explorer, boundary surveyor.

FAMILY AND EDUCATION

Son of George and Dorothy Simpson Featherstonhaugh. His widowed mother moved to Scarborough, Yorkshire, during Featherstonhaugh's infancy. He was educated at Yorkshire Academy and attended Oxford University, where he specialized in classical literature, languages, and science. Following graduation he toured Europe, visiting France, Spain, Italy, and various German principalities. He moved to United States in 1806 and soon settled in upstate New York. Married Sarah Duane, 1808 (died 1828); five children, and Charlotte Carter, 1831; four children.

POSITIONS

Corresponding secretary, New York State Board of Agriculture, 1820–c.1823. Vice president and director, Mohawk and Hudson Railroad Company, 1826–1829. Founder and editor of *The Monthly American Journal of Geology and Natural Science*, 1831–1832. Geologist for the U.S. War Department, 1834–1837. Boundary surveyor for Great Britain, 1839. British consul for the Department of the Seine at Le Havre, France, 1844–1866.

CAREER

From time of his first marriage to about 1825 Featherstonhaugh was gentleman farmer in Mohawk Valley. His advocacy of scientific agriculture stimulated his interest in both railroads and geology. In association with Stephen Van Rensselaer, scion of an old patroon family, he organized Mohawk and Hudson Railroad, precursor of the New York Central. It was formed to help Schenectady merchants, who had been bypassed by recently completed Erie Canal. Wishing to study British railroad developments, Featherstonhaugh returned to England in fall of 1826. During his stay, which lasted to March 1828, he renewed acquaintances with leading geologists. His association with Impey Murchison and Adam Sedgwick led to his election in 1827 as fellow of the Geological Society of London. As disciple of modern geology, which had been formulated by the Englishmen James Hutton and William Smith, Featherstonhaugh stressed uniformitarianism, which held that those natural forces that had shaped the earth were still operative.

Following death of his first wife and the destruction by fire of his Mohawk Valley home, Featherstonhaugh embarked on literary and scientific career in Philadelphia. Rekindling his youthful interest in classics, he translated Cicero's *Republic* (1829) and wrote drama, *The Death of Ugolino: A Tragedy* (1830) about a Pisan politician who had been immortalized in Dante's *Inferno*. While engaged with these works he also gave public lectures on the new geology. Their popularity caused him to establish the short-lived *Monthly American Journal of Geology and Natural Science*, one of earliest scientific journals published in the United States.

Featherstonhaugh's periodical brought him to the attention of Lt. Col. John J. Abert, chief of the Corps of Topographical Engineers in U.S. Army. Abert, who was interested in mineralogical surveys of public lands, commissioned Featherstonhaugh to act as principal government geologist. In that capacity Featherstonhaugh, from 1834 to 1837, conducted three major surveys. The first was principally of Ozark region, the second of Minnesota River Valley, and third of the lead mines of Wisconsin and Missouri and Cherokee Indian lands in eastern Tennessee and western North Carolina.

After leaving U.S. government service, Featherstonhaugh moved to England, where he was soon employed as cosurveyor of the controversial Maine–New Brunswick boundary. As a reward for this service he was appointed British consul at Le Havre, a position he held for last twenty-two years of his life.

Featherstonhaugh's geological surveys provided basis for both official reports and popularly written travel accounts. His Ozark expedition was detailed in *Report of Geological and Mineralogical Survey of the Elevated Country between the Missouri and Red Rivers*, in 23d Congress, 2d session, *House Executive Documents*, no. 151 (serial 274). It was also separately printed by Gales and Seaton of Washington, D.C. Its literary counterpart, *Excursion through the Slave States* (1844), was published by John Murray in London. Featherstonhaugh's entertaining and lively *Canoe Voyage up the Minnay Sotor* about his surveys of 1835 and 1837 was published three years later. Although it includes many geological descriptions, Featherstonhaugh's scientific observations had earlier been covered in *Report of a Geological Reconnaissance Made in 1835, from the Seat of Government, by the Way of Green Bay and the Wisconsin Territory to the Coteau de Prairie* (24th Congress, 1st session, *Senate Executive Documents*, no. 333, serial 282; also separately printed by Gales and Seaton in 1836.) Featherstonhaugh's boundary duties led to his authorship of *Historical Sketch of the Negotiations at Paris in 1782* (1842) and *Observations upon the Treaty of Washington* (Webster–Ashburton Treaty) (1843). The first was intended to bolster British claims in New England, and the second to defend the reasonableness of the treaty negotiated by Lord Ashburton and Daniel Webster.

Featherstonhaugh's geological reports were lucid and concise but, because they were very general and somewhat superficial and contained errors, were soon outdated. Consequently, his scientific works have only historical value. His popular writings were vehemently condemned by American frontiersmen, who reacted to his characterization of them as crude, materialistic, and uncultured commoners. His unpopularity among westerners was partially attributable to his haughtiness, Tory inclinations, fastidiousness, and insistence on abstinence from use of liquor and tobacco. However, he was well accepted by his English peers, who honored him with membership in the Royal Society.

MAJOR CONTRIBUTIONS

Dissemination of information about modern geology through his journal and lectures. Apparently played a key role in persuading the War Department to use geologists. He was first scientist to hold the unofficial position of geologist of the United States. His reports were first in which geological formations of large frontier areas were identified by stratigraphical classifications. His observations of the Minnesota River Valley finally dispelled the long-standing myth that it held valuable copper ore deposits.

BIBLIOGRAPHY

Armytage, W. H. G. "G. W. Featherstonhaugh, F.R.S. 1780–1866, Anglo-American Scientist." In *Notes and Records of the Royal Society of London* (March 1955).

Berkeley, Edmund, and Dorothy Smith Berkeley. *George William Featherstonhaugh: The First U.S. Government Geologist* (1988).

Featherstonhaugh, J. D. "Memoir of Mr. G. W. Featherstonhaugh." *American Geologist* (April 1889).

Lass, William E. "Introduction to the Reprint Edition" of Featherstonhaugh's *Canoe Voyage up the Minnay Sotor* (1970).

White, George W. "Editor's Introduction." *Monthly American Journal of Geology and Natural Science* (repr. 1969).

UNPUBLISHED SOURCES
See George William Featherstonhaugh and Family, Papers, 1771–1856. Microfilm copy (eleven rolls) is in Minnesota Historical Society, St. Paul. Originals are owned by James Duane Featherstonhaugh, Duanesburg, N.Y.

William E. Lass

FERNALD, MERRITT LYNDON. Born Orono, Maine, 5 October 1873; died Cambridge, Massachusetts, 22 September 1950. Botanist.

FAMILY AND EDUCATION

Son of Merritt Caldwell Fernald, twice president of Maine State College of Agriculture and Mechanic Arts, and Mary Lovejoy Heywood Fernald. Graduated from Orono High School. B.S., magna cum laude, 1897, Harvard. Honorary D.C.L., 1933, Acadia University, and honorary D.Sc., 1938, University of Montreal. Married Margaret Howard Grant, 1907; three children.

POSITIONS

Assistant, Gray Herbarium, Harvard, 1891–1902, curator, 1935–1937, director, 1937–1947. Instructor of botany, Harvard, 1902–1905, assistant professor, 1905–1915, Fisher Professor of Natural History, 1915–1947, professor emeritus, 1947–1950. A founder of Alstead (New Hampshire) School of Natural History, where he taught summers 1899–1901.

CAREER

Early interest in botany led to correspondence with Sereno Watson, director of Gray Herbarium. After publication of first paper (*Bulletin of the Torrey Botanical Club* [1890]), invited by Watson to work at the herbarium and continue studies at Harvard; served at herbarium for next fifty-six years.

Specialized in the study of plants of temperate North America (the "Gray's Manual range"): North America east of the Missouri and Mississippi Rivers and north of the Carolinas. Majority of work concerned the identification, description, definition, and geographical distribution of plants of this area. Did fieldwork on Gaspe Peninsula in youth and in tidewater regions of Virginia in later years.

Prolific writer of approximately 900 items, spread among forty different periodicals. Author of *Edible Wild Plants of Eastern North America* (with B. L. Robinson, 1943), the 7th edition of *Gray's Manual of Botany* (with B. L. Robinson, 1908), and the 8th (centennial) edition of *Gray's Manual* (1950), which was called "the most critical and searching study in descriptive botany that has ever appeared on the flora of any part of North America." His paper "Persistence of Plants in Unglaciated Areas of Boreal America'' (*Memoirs of the American Academy of Arts and Sciences* [1925]) rebutted the prevailing theory that glaciers had obliterated all plant and animal life in their path.

Served as associate editor, 1899–1928, and editor in chief, 1929–1950, of *Rhodora*. Received Leidy Gold Medal, Academy of Natural Sciences, 1940; Gold Medal, Massachusetts Horticultural Society, 1944; and the Marie-Victorin Medal, 1950, for outstanding services to botany in Canada. President, New England Botanical Club, 1911–1914; vice president, American Association for the Advancement of Science, 1941; president, American Society of Plant Taxonomist, 1938; president, Botanical Society of America, 1942; honorary president, International Botanical Congress, Stockholm, 1950. Member or honorary member of a number of American and foreign scientific societies. A tireless worker and exacting critic, he was remembered as an outstanding teacher. It was said he could remember years later exact spots where plants had been collected.

MAJOR CONTRIBUTIONS

Trained the future leaders in taxonomy and descriptive botany and helped maintain a high level in botanical writing through his work as editor of *Rhodora* and through his published critiques of the work of colleagues. Eighth edition of *Gray's* remains a landmark in botany.

BIBLIOGRAPHY

Dictionary of American Biography. Supplement 4 (1974)
Dictionary of Scientific Biography. Vol. 4 (1971).
Humphrey, H. B. *Makers of North American Botany* (1961).
Merrill, Elmer. *Biographical Memoirs National Academy of Sciences*. Vol. 28 (1954) (obit.).
National Cyclopaedia of American Biography. Vol. 38 (1953).
New York Times, 24 September 1950 (obit.).
Rollins, R. C. *Bulletin of the Torrey Botanical Club* (May 1951) (obit.).
Russell, F. W. *Mount Auburn Biographies* (1953).
Who Was Who in America. Vol. 3 (1960).

John A. Drobnicki

FERNOW, BERNHARD EDUARD. Born Inowrazlaw (now Inowroclaw), Prussia (now Poland), 7 January 1851; died Toronto, Ontario, Canada, 6 February 1923. Forester, university professor and dean, conservationist.

FAMILY AND EDUCATION

Son of Eduard Fernow, estate owner and lawyer, and his second wife, Clara Nordman Fernow. His father held distinguished position in the service of the Prussian government. Young Fernow graduated from the gymnasium at Bromberg, 1869. Study at Prussian Forest Academy, Munden, interrupted by Franco-Prussian War. Spent one year at Law School of University of Königsberg, returned to Forest Academy. Secured forestry license, 1873. Emigrated to United States, 1876. Naturalized, 1883. Honorary LL.D., University of Wisconsin, 1896, Queen's University, Kingston Ontario, 1903, University of Toronto, 1923. Married Olivia Reynolds, 1879; five children.

POSITIONS

Lieutenant in Prussian army, 1870–1871. Member of Prussian Forest Service, 1874–1876, rising to rank of Forstkandidat. After arrival in United States was clerk in New York law firm and did private tutoring in German, 1876–1878. Manager of Cooper-Hewitt and Company's forest preserve in Pennsylvania, 1878–1883. Returned to New York but continued to have charge of forest property until 1887. Chief, Division of Forestry, U.S. Department of Agriculture, 1886–1898. Organized first U.S. school of forestry, Cornell, 1898–1903. Private consultant in United States, Cuba, and Mexico, 1903–1907; lecturer in forestry, Yale, 1904; initiated forestry instruction at Pennsylvania State College, 1906–1907; professor and dean, Department of Forestry, University of Toronto, 1907–1919.

CAREER

Fernow was slated for career in Prussian Forestry Department until he met, and became engaged to, an American girl who was living with her family in Germany. Became an associate of the American Institute of Mining Engineers, 1878; discovered electrical procedure for removing tin from "tin cans" but could not market it because of poor business conditions, 1878. Worked with Cooper-Hewitt on several projects, particularly charcoal making. Began his intensive study of American forests and management practices. As chief of Division of Forestry, undertook research in full range of reforestation, management, and technology. Began assessing forest resources and state policies in relation thereto. Stressed practical aspects in forest management, together with concept of sustained yield. Advocated legislation protecting and managing forests and drafted model legislation to that end. Urged that forestry instruction be expanded in American higher educational institutions. Assisted in organizing the American Forestry Congress (1882), which later became American Forestry Association. After considerable effort, first federal forest reservation policy enacted by Congress, 1897, inaugurating rational plan of management. Fernow began New York State College of Forestry at Cornell, 1898, which was not funded after 1903 due to opposition from Adirondack landowners to the Cornell Forest, a 30,000–acre, university-owned demonstration area where management practices, including logging, were undertaken. State financing for Cornell School vetoed by Governor Odell, resulting in its closure. Fernhow then went into consulting, while teaching at Yale and for one term at Pennsylvania State College. While at the University of Toronto, Fernow also provided counsel to dominion and provincial governments concerning forest management, forest products, and related matters. He promoted scientific forestry, urged creation of more parks, and was actively involved in Canadian conservation issues. Was for thirteen years (1910–1923) member of Canadian Conservation Commission.

Authored some 200 articles and addresses on forestry issues, not including fifty U.S. government bulletins and circulars. Founding editor of *Forest Quarterly* for a number of years from 1902. Published *The Economics of Forestry* (1902); *A Brief History of Forestry in Europe, the United States, and Other Countries* (1907); *The Care of Trees in Lawn, Street, and Park* (1910). Co-organizer and first president, Canadian Society of Forest Engineers, 1908–1916; president, Society of American Foresters, 1914, 1916.

MAJOR CONTRIBUTIONS

A pioneer in the study of scientific forestry and forestry education in the United States and Canada. Supported conservation methods that would enhance natural beauty of forests and yet not destroy the economic interests of lumber and other industries. Raised consciousness of Americans and Canadians regarding environmental issues and gave dignity to this movement. Supplied leadership and expertise in legislative efforts.

BIBLIOGRAPHY

Dictionary of American Biography.

Encyclopedia of American Forest and Conservation History. Vol. 1 (1983).

Globe [Toronto], February 1923 (obit.).

Journal of Forestry (April 1923).

Randall, C. E. ''Fernow, the Man Who Brought Forestry to America.'' *American Forests* (April 1964).

Rodgers, Andrew Denny, III. *Bernhard Eduard Fernow, a Story of North American Forestry* (1951).

Steen, Harold K. *The U.S. Forest Service: A History* (1976).

Arthur Belonzi

FIDLER, PETER. Born Bolsover, Derbyshire, England, 16 August 1769; died Fort Dauphin, Hudson's Bay Territories (Manitoba, Canada), 17 December 1822. Fur trader, explorer, surveyor.

FAMILY AND EDUCATION

Family and education details not known. Fidler was well read. He married Mary, a Cree woman, at York Factory in late 1794. Between 1795 and 1822, they had fourteen children, of whom eleven survived childhood. Arrival of the first ordained minister made formal marriage possible on 14 August 1821.

POSITIONS

Signed on as laborer with the Hudson's Bay Company in London on 19 April 1788 and served the first winter at York Factory on Hudson Bay. In June 1790 studied astronomy and surveying under Philip Turnor at Cumberland House, with the title of assistant surveyor, and from 1795 carried the title of surveyor. Fidler was in charge of Chesterfield House, 1800–1802, of Nottingham House, 1802–1806, and of Ile-a-la-Crosse, 1809–1811. At Brandon House he was surveyor and district master from 1814 to 1819. His last years, 1819–1822, were spent at Fort Dauphin, as district master.

CAREER

During his life as a fur trader-surveyor-mapmaker with the Hudson's Bay Company, Fidler built new trading forts at Carlton House on Assiniboine (1795), Bolsover House at Meadow Lake and Greenwich House on Lac la Biche (1799), Chesterfield House on the South Saskatchewan (1800), Nottingham House on Lake Athabasca (1802) and Halkett House at Red River forks (1816–1817).

Fidler surveyed and mapped 4,700 miles of river and lakeshore and named Wollaston Lake. Was also the first to survey residential lots in western Canada (Red River, May 1813, summer 1817, and September 1818). In 1792–1793 he was the first white man to survey the Battle, Bow, and Red Deer Rivers and a small Canadian portion of the Rocky Mountains and the first to write of tar sands (15–21 June 1791, near present McMurray), coal (12 February 1793, near present Drumheller), and cactus (February 1793, near Trochu). He was a major supporter of the new Selkirk colonists on Red River, Manitoba, 1813–1819.

MAJOR CONTRIBUTIONS

Fidler's major contributions to natural history were contained in two official reports to the Hudson's Bay Company: a general report of the Red River district (May 1819) and "Report of District" from Fort Dauphin in the spring of 1820. The latter document contains important wildlife observations, including the first evidence of synchronous eight-to ten-year cycles in snowshoe hare and lynx. Other observations include mention of the vast flights of the passenger pigeon and declining number of swans. Natural history items in his routine journal entries include an important, detailed, unmistakable description of the channel catfish at Cumberland House in 1797. He was prolific reader and purchaser of books, many of which dealt with natural history.

BIBLIOGRAPHY

Atton, F. M. "Early Records of the Channel Catfish, Ictalurus punctatus, in Cumberland Lake, Saskatchewan." *Canadian Field-Naturalist* 99 (1985).

Macgregor, J. G. *Peter Fidler: Canada's Forgotten Surveyor, 1769–1822* (1966).

C. Stuart Houston

FITCH, ASA. Born 24 February 1809, Fitch's Point, Salem, New York; died Salem, New York, 8 April 1879. Entomologist.

FAMILY AND EDUCATION

Son of Asa, doctor and farmer, and Abigail Martin Fitch. Graduate of Rensselaer School (now Rensselaer Polytechnic Institute), 1827; M.D., Vermont Academy of Medicine, 1829. Additional study at Rutgers Medical College, New York City, and also with a Dr. Mach in Albany, New York. Married Elizabeth McNeil, 1832; at least one child.

POSITIONS

Assistant professor of natural history, Rensselaer School, 1830; medical pactice, Greenville, Illinois, 1830–1831; in private medical practice, Fort Miller and Stillwater, New York, 1832–1838. Managed father's business in Salem, New York, for a time from 1838. Collector of insects for New York State Cabinet of Natural History, 1845–1854 (?). State entomologist in New York, 1854–1870.

CAREER

First collected plants as a child. Encouraged by Amos Eaton, his teacher at Rensselaer, and joined Eaton and other students on collecting trips to Lake Erie. Entered medical career at father's request because training and careers in entomology not then available. Read all entomological texts he could find in Albany libraries, including state collections. Reputed to be a leading American entomologist by age twenty-one. Continued with his collecting during desultory medical career, published first paper in 1845. Began collecting insects for state of New York in 1847. Maintained small frame "bughouse" near his residence in Salem. His daughter illustrated many of his papers. Given courtesy title of state entomologist in 1854, the first to be appointed to such a post in the United States. Primarily concerned with relationship of insects to agriculture. Published fourteen annual "Reports on the Noxious, Beneficial, and Other Insects of the State of New York," which appeared in *Transactions of the New York State Agricultural Society* (1855–1872). Eleven of these were also separately printed. In his third report (1856), Fitch wrote, "I sometimes think there is no kind of mischief going on in the world of nature around us but that there is some insect at the bottom of it." Felt that "their performances are . . . tending probably in an equal degree to our benefit in one direction as to our detriment in another." Proposed useful control methods, which were soon supplanted by better ones. Worked by himself for many years, lacking needed works of reference, though he carried on wide correspondence with other entomologists. Following his death, his collections of more than 120,000 specimens, mostly Coleoptera, were sold, some going to private collectors, some to the State Museum in Albany, and some to the U.S. National Museum. Accumulated some 148 notebooks with carefully collected data and a substantial library. Most notebooks went to U.S. National Museum. Many of his specimens were found to be misnamed, but Fitch evidently received little aid from other specialists to whom he had written and hence did the best he could on his own. Contributed technical articles to several agricultural publications.

MAJOR CONTRIBUTIONS

A leading economic ornithologist of his day. Very conversant with insect ethology due to close personal observation; hence, his annual reports were respected and widely used for some years.

BIBLIOGRAPHY

Collins, D. L. "The Bug Catcher of Salem." *Bulletin of the Schools* (New York) 40 (1954).

Dictionary of American Biography.

Howard, L. O. *A History of Applied Entomology* (1930).

Mallis, A. "The Diaries of Asa Fitch, M.D." *Bulletin of the Entomological Society of America* 9 (1963).

Mallis, A. *American Entomologists* (1971).

Rezneck, S. "Diary of a New York Doctor in Illinois, 1830–1831." *Journal of the Illinois State Historical Society* 54 (1961).

Rezneck, S. *Dictionary of Scientific Biography.*

Riley, C. V. "Dr. Asa Fitch." *American Entomologist* 3 (1880).

UNPUBLISHED SOURCES

Diaries are in Collection of Regional History and University Archives, Cornell University; in Historical Manuscript Collections, Yale University; and in Smithsonian Institution.

Keir B. Sterling

FLEMING, J. H. "HARRY" (JAMES HENRY). Born Toronto, Ontario, 5 July 1872; died Toronto, Ontario, 27 June 1940. Naturalist, ornithologist, collector, curator.

FAMILY AND EDUCATION

Only son of James Fleming, a prominent Scottish immigrant and seedsman, and Mary Elizabeth Wade. Educated at the Saint James Provincial Model School and Upper Canada College (graduated 1889). Attended the School of Mines in London, England, 1890–1891. Self-taught as an ornithologist. Married Christine MacKay Keefer (died 1903), 1897, and Caroline Toovey, 1908. Two children from the first marriage, Annie Elizabeth (1899–1946) and Thomas Keefer (1901–1987).

POSITIONS

British Empire member of the British Ornithological Union, corresponding member of the Zoological Society of London, 1913. Member d'Honneur Etranger Société Ornitholigique et Mammalogie de France, 1931. American Ornithologists' Union: associate, 1893; member, 1901; fellow, 1916; vice president, 1926–1932; president, 1932–1935. Honorary curator of birds, National Museum, 1913; Royal Ontario Museum, 1927. Several honorary member positions in leading Toronto natural history and ornithological groups.

CAREER

Began collecting bird specimens at the age of twelve. Visited the British Museum of Natural History in 1886 and set himself the life task of assembling a worldwide representative collection of birds. Associate member of the (Royal) Canadian Institute at age sixteen. Traveled in Europe 1889–1891 and to the West Indies in 1892. Established as a naturalist in Toronto by 1892 and pursued his interest in collecting through a partnership in a taxidermy studio. The inheritance of his father's estate allowed him to make amateur natural history a full-time avocation for the rest of his life. Returned to Europe in 1895 to visit the British Museum of Natural History and Walter Rothschild's collection at Tring. The personal contacts he made on this trip established a lifelong association with leading European ornithologists. Attended International Ornithological Congress in London, 1905, as the only Canadian representative. Actively involved in assembling his collection and participating in local, regional, national, and international ornithology throughout the first quarter of the twentieth century. Built a three-level museum annex to his residence in 1925 to house an expanding collection and library. Elected as the first Canadian president of the American Ornithologists' Union, October 1932. Published at least eighty-four papers and notes, including "The Birds of Toronto, Pt. I & II" (1906–1907), several papers on new species, as well as historical notes. Eventually amassed a collection of over 32,000 bird specimens containing representatives of nearly all living bird families, 75 percent of all genera and more than 6,300 species as well as a comprehensive ornithological library of over 2,000 bound volumes.

MAJOR CONTRIBUTIONS

Acknowledged as the "dean of Canadian ornithology." Distinguished amateur ornithologist recognized internationally for his comprehensive private collection and knowledge of systematic ornithology. Bequeathed the world's largest and most comprehensive private ornithological collection to the Royal Ontario Museum, Toronto, Ontario. Instrumental in the establishment of a National Museum of Canada and a representative of his country at several International Ornithological Congresses and the International Council for the Preservation of

Birds. Compiled detailed historical records of ornithological (and other natural history) notes for Ontario.

BIBLIOGRAPHY

James, R. D. "James Henry Fleming (1872–1940)." In M. K. McNicholl and J. L. Cranmer-Byng, eds., *Ornithology in Ontario* (1994).

Quinn, M. S. "Natural History of a Collector: J. H. Fleming (1872–1940), Naturalists, Ornithologists and Birds. Diss., York University, 1995.

Snyder, L. L. "In Memoriam: James Henry Fleming." *Auk* 58 (1941).

Taverner, P. A. "James Henry Fleming 1872–1940: An Appreciation." *Canadian Field-Naturalist* 55, 5 (1941).

UNPUBLISHED SOURCES

J. H. Fleming left his entire collection of birds, books, and papers to the Royal Ontario Museum. The Archives, Special Collection (SC29), contains manuscripts, photographs, correspondence, notes, journals, and scrapbooks. "The Fleming Memorial Papers," a collection of personal reminiscences and biographical details, was produced by the Brodie Club in 1940.

Michael S. Quinn

FLETCHER, JAMES. Born Ashe, Kent, England, 28 March 1852; died Montreal, Canada, 8 November 1908. Entomologist, Canadian government official.

FAMILY AND EDUCATION

Second son of Joseph Flitcroft and Mary Ann Hayward Fletcher. Educated at King's School, Rochester, England. Honorary LL.D. Queen's University, 1896. Married Eleanor Gertrude Schreiber, 1879; two daughters.

POSITIONS

Clerk, Bank of British North America, London, 1871–1874. Transferred to branch bank in Montreal, 1874, posted to Ottawa, 1875–1876. Resigned to accept position as library assistant, Library of Parliament, Ottawa, 1876. Named honorary dominion entomologist and botanist, 1884–1887. Position made permanent, 1887–1908.

CAREER

As young librarian, had access to material on science and natural history. Self-taught in botany and entomology, joined Entomological Society of Canada, 1876, and contributed annually to the *Canadian Entomologist* from 1880 until his death. An originator of the Ottawa Field-Naturalists' Club (1879) and of the Association of Economic Entomologists of North America (1889). He was the "father of economic entomology" in Canada. His position was originally made honorary so that he might demonstrate the importance and value of insect control and entomological research in the dominion. Until 1892, solely responsible for identification of insects and weeds and for recommendations for the control of pest species. Thereafter, had the help of assistants. One, Arthur Gibson, later served as third dominion entomologist (1920–1942). Instituted nationwide network of correspondents, comprising farmers and gardeners, to report on noxious pests and remedies for their control. By rigorous self-improvement, observation, and discussion, he became an expert taxonomist and a recognized specialist on moths and butterflies. Traveled extensively throughout Canada, collecting weeds and insects, investigating insect outbreaks and damage, lecturing to students, addressing farmers and growers, and advising on practical weed and insect control. His pictorial book *Farm Weeds in Canada* (1906) was an outstanding contribution to science. Established National Herbarium on Central Experimental Farm, Ottawa. Began collection of the insects of Canada that burgeoned into today's prestigious Canadian National Collection of Insects. Responsible in large measure for enactment of first federal legislation in Canada pertaining to insects—the San José Scale Act (1898). This permitted imposition of quarantine restrictions on imported insect-infested plant material. Under his direction, first federal insect fumigation stations were built at ports of entry to treat nursery stock entering Canada. He also promoted some of the first experimental work using arsenical and plant-extracted insecticides to control pest insects.

The *Transactions of the Royal Society* contain many of his early scientific papers on practical entomology (1895), lists of injurious insects (1899), and descriptions of new butterflies (1903). Most of the dozens of other papers in other journals defined the taxonomy and systematics of diurnal moths and butterflies. Seventeen species of butterflies were named after him. His best and most valued publications are the twenty-two *Annual Reports of the Dominion Entomologist and Botanist* (1886–1908) and the large number of *Bulletins of the Dominion Department of Agriculture*. In these are outlined not only life histories of, damage by, and control of, a vast number of noxious insects in Canada but also detailed descriptions and accurate drawings of their life stages. In great demand as public speaker, captivating audiences of agriculturists, scientists, or students with lively humor and artful language. Students were eager to be with him on nature excursions.

MAJOR CONTRIBUTIONS

Imparted knowledge of noxious insects to, and instructed, Canadian public in most practical methods of controlling them; demonstrated continuing need for entomological research by governments; established economic entomology, interrelated with productivity of man, as succinct, scientific discipline; emphasized absolute need of determining precise taxonomic and bionomic status of an insect before proceeding with applied control measures; was committed to philosophy that noxious weeds and insects must be controlled with minimum of environmental disarray so as to maximize national agricultural productivity.

BIBLIOGRAPHY
Bethune, C. J. S. *Canadian Entomologist* 30, 1 (1898).
Bethune, C. J. S. *Canadian Entomologist* 40, 12 (1908) (obit.).
The Canadian Who Was Who. Vol. 1 (1934).
Derraugh, Rita. *Entomological Newsletter* 33, 2 (1955).
Dictionary of National Biography. Supplement 2.
Estey, R. H. "James Fletcher (1852–1908) and the Genesis of Plant Pathology in Canada." *Canadian Journal of Plant Pathology* 5 (1983).
Gibson, Arthur, and Herbert Groh. *Ottawa Naturalist* (January 1909).
Howard, Leland O. *A History of Applied Entomology* (1930).
Le Naturaliste Canadien 35, 1 (1908) (obit.).
Proceedings of the Royal Society of Canada (1909) (obit.).
Riegert, P. W. *From Arsenic to DDT: A History of Entomology in Western Canada* (1980).
Riegert, P. W. *The Canadian Encyclopedia*. Vol. 2 (1989).
UNPUBLISHED SOURCES
See letters, correspondence, and papers of James Fletcher, Entomology Division, Dominion Department of Agriculture, Public Archives of Canada, Ottawa.

Paul W. Riegert

FORBUSH, EDWARD HOWE. Born Quincy, Massachusetts, 24 April 1858; died Westboro, Massachusetts, 8 March 1929. Ornithologist.

FAMILY AND EDUCATION

Son of Leander Pomeroy and Ruth Hudson (Carr) Forbush. Father was principal of Coddington School (Quincy, Massachusetts). Received primary education in West Roxbury and Worcester but left school at fifteen. Never attended college. Married Etta L. Hill, 1882; four children.

POSITIONS

Curator, Worcester Natural History Society Museum, from 1874. Director, Massachusetts Commission on the Gypsy Moth, 1890–1900. Director, Division of Ornithology, Massachusetts Department of Agriculture, 1893–1908, 1921. State ornithologist, Massachusetts, 1908–1928.

CAREER

From childhood was interested in outdoors and in watching and hunting birds and animals. At fourteen taught himself taxidermy and soon became skilled in preparing specimens. Left school at fifteen to assist father in latter's business, made ornithology his avocation until age twenty-two. Was determined to equip himself through reading and experience for work in that field. Was appointed curator of Worcester Natural History Society Museum at age sixteen. Two years later went to Florida on collecting expedition and brought back large variety of specimens. Established Naturalists' Exchange with William S. Perry to provide specimens and taxidermists' supplies.

Under his leadership, the Worcester Natural History Society expanded and improved its collections and conducted many classes. In 1885, the society established summer natural history camp for boys at Lake Quinsigamond, Massachusetts, near Worcester, first such enterprise since Louis Agassiz's Penekese Island school of 1873. Some also regard this effort as forerunner of summer camp movement.

In 1888, conducted another extensive collecting expedition, this time to Alaska, western Canada, and Washington Territory. However, as years passed, he began to recognize an important shift in emphasis occurring in his own thinking and that of other naturalists—(in his words) ''that it was more essential to preserve the living than the dead.'' Spent years from 1890 to 1900 studying the gypsy moth and developing plans for controlling this pest in Massachusetts. Wrote *The Gypsy Moth* (1896), an important work on the subject.

During this time, continued to be interested in ornithology and wrote on importance of birds in destroying harmful insects. These reports resulted in his appointment as ornithologist to the Massachusetts State Board of Agriculture in 1893. Educated the public on relationship between birds and agriculture and stressed the economic value of bird preservation and conservation. Lectured throughout the state and also published series of reports on birdlife and bird protection. Was New England field agent for National Association of Audubon Societies, 1905–1927. Published a number of annual reports outlining his efforts to secure protective legislation from state legislatures.

Also published three extensive works on birds: *Useful Birds and Their Protection* (1907), which he illustrated; *A History of the Game Birds, Wild Fowl and Shore Birds* (1912); and three-volume *Birds of Massachusetts and Other New England States* (1925–1929). This set was his most important contribution to ornithology and included summation of his life's study of birds as well as important observations of other ornithologists. Published some 172 titles in all.

A companionable man, with a good sense of humor, was known for the joy he brought to his work with birds. Fellow and council member, American Ornithologists' Union. Founder and, for twelve years, president, Massachusetts Audubon Society. President, Northeastern Bird Banding Association. President, Federation of Bird Clubs of New England. President, Worcester Natural History Society.

MAJOR CONTRIBUTIONS

One of America's greatest ornithologists and a pioneer in economic ornithology. Inspired game and conservation laws of Massachusetts and other New England states, which served as models for bird legislation elsewhere. His three books continue to be considered valuable contributions to the field.

BIBLIOGRAPHY
Brooks, Paul. "Birds and Men." *Audubon* (1980).
Dictionary of American Biography. Vol. 6.
May, John B. "Edward Howe Forbush: A Biographical Sketch." *Proceedings of the Boston Society of Natural History* (1928).
National Cyclopaedia of American Biography. Vol. 21.
Pearson, T. Gilbert. *Auk* (April 1930) (obit.).

Cynthia D. Chambers

FORSTER, JOHANN REINHOLD. Born Dirschau (Tczew), nineteen miles south of Danzig, Poland, 22 October 1729; died Halle (Saale), Germany, 9 December 1798. Cleric, explorer, naturalist, translator.

FAMILY AND EDUCATION

Only son of a scholarly mayor of Dirschau, descended from Yorkshire immigrants of probable Scottish Forrester origin. Both grandfathers and paternal great-grandfather also mayors of Dirschau. Early education limited. Attended Latin School of Rektor Swiderski at age eleven in Dirschau. From 1745 to 1748 attended Joachimsted Gymnasium near Berlin, excelling in classics and biblical languages, with strong interest in cosmology and ethnology. From 1748 to 1751, studied theology at Friedrichs University in Halle, where he informally studied medicine and natural history. Largely self-taught thereafter. Received honorary doctor of civil laws from Oxford (1775) and honorary doctorates of philosophy (1780) and medicine (1781) from Friedrichs University of Halle. Married Justina Elisabeth Nicolai, 1754; seven children.

POSITIONS

Ordinand Reformed Congregation of St. Peter and St. Paul, Danzig, 1751–1753. Pastor, Reformed country parish of Hochzeit-Nasschuben (Nassenhof), West Prussia, 1753–1765. Commissioned by Czarina Catherine II to survey Saratov-Tsaritsyn region of the lower Volga River, 1765–1766. Tutor in modern languages and natural history, Dissenters' Academy, Warrington, Lancashire, England, 1767–1769. Teacher of French, Winwick Grammar School (near Warrington), 1769. Naturalist, James Cook's second voyage around the world, 1772–1775. Professor of natural history and minerology, Friedrichs University of Halle, 1779–1798, professor of medicine, 1781–1798, pro-rector, 1790–1791.

CAREER

Showed early oratory abilities in sermons as ordinand in Danzig. Survey of German settlements along the Volga for eighteen months, in which he was accompanied by oldest son, George, was notable for observations on numerous facets of nature and culture. Frustration over lack of response by Russian government to his report led Forster to move to England. In England, reports on the Volga brought him scientific recognition, leading to election as fellow of the Royal Society and other honors. *Specimen historiae naturalis Volgensis* (1767) was among the first major works to apply Linnaean system to botany, and black lark (*Alauda yeltoniensis*) was first of numerous bird species named by Forster.

On arrival in London in 1766, barely able to speak English, but soon learned and after his friend Joseph Priestly left Warrington, Forster partially replaced him, teaching one of first natural history courses in England. Although lectures well received, disagreements with administration, partly over extracurricular teaching at Winwick, led to dismissal in 1769.

While still at Warrington, published *Introduction to Mineralogy* (1768), then started to use fluency in seventeen languages to translate scientific works into English and from English. At urging of Thomas Pennant, translated Pehr Kalm's *Travels into North America* from Swedish (translation published 1770–1771). *A Catalogue of the Animals of North America* and *Flora Americana Septentrionalis* (both 1771), prepared as "appendices" to Kalm translation, were first attempts at compiling all species then known in North America. Also translated Jean Bernard Bossu's account of travels through Louisiana Territory from French (1771). Commissioned by Hudson's Bay Company (HBC) to publish paper on quadrupeds, birds, and fish sent to England from HBC territories (1772–1773). Bird paper included formal descriptions of nine new species and comments on a tern later formally described by Thomas Nuttall as Forster's tern (*Sterna forsteri*).

Abrupt withdrawal of Joseph Banks as naturalist of second major James Cook expedition brought Forster to most significant work as Banks's replacement. Assisted by son, George, and Swedish botanist Anders Sparrman, Forster took numerous collections and copious notes back to Europe, which laid foundations of antarctic and New Zealand ornithology; indicated the phytogeographical relations of New Zealand with the East Indies, Melanesia, and Polynesia and (with Cook) pioneered Melanesian anthropology. Preceded Charles Darwin in theories on coral formation and theorized on ice formation at sea and on volcanology. At least 220 plant species, 114 bird species, and seventy-four fish species named from expedition collections, including new bird family, Chionidae (sheathbills). *Aptenodytes forsteri* (emperor penguin), numerous plant species, and several geographic features commemorate contributions to South Pacific science.

On return to England, prevented by Admiralty from publishing official account of expedition (finally edited by Heintich Lichenstein as *Descriptionales Animalis* in 1844), but many observations from the journey appeared in *Observations Made during a Voyage Round the World* (1787), in monographs on plants (with George, 1776), penguins (1781) and albatrosses (1785), in numerous scientific papers, and in footnotes and annotations to his translations. George also published a major work on the journey. Forster continued to publish prolifically until his death and to prepare numerous annotated translations on the works of others, including a translation into German of Samuel Hearne's accounts of travels in arctic Canada.

MAJOR CONTRIBUTIONS

Contributed substantially to development of biogeography, botany, ethnology, ichthyology, geology, linguistics, ornithology, and other sciences. Among first to use and promote Linnaean system of nomenclature. Numerous annotated translations promoted scientific dialogue at time of extensive exploration and numerous discoveries. More direct contributions to North America through translations, early compilations of fauna and flora, and descriptions of several new species. Never visited North America.

BIBLIOGRAPHY

Dictionary of Scientific Biography.

Gruson, E. S. *Words for Birds* (1972).

Hoare M. E. ''Johann Reinhold Forster.'' *Journal of Pacific History* 2 (1967).

''Johann Reinhold Forster (1729–98): Problems and Sources of Biography.'' *Journal of the Society of Bibliography of Natural History* 6 (1971).

Lysaght, A. M. *Joseph Banks in Newfoundland and Labrador 1766. His Diary, Manuscripts and Collections* (1971).

Mearns, B., and R. Mearns. *Biographies for Birdwatchers. The Lives of Those Commemorated in Western Palearctic Bird Names* (1988).

Medway, D. G. *The Significance of Captain Cook's Voyages for New Zealand Ornithology.* In B. J. Gill and B. D. Heather (eds.), *A Flying Start/Commemorating Fifty Years of the Ornithological Society of New Zealand 1940–1990* (1990).

Stresemann, E. *Ornithology from Aristotle to the Present* (1975).

UNPUBLISHED SOURCES

Extensive collections of letters, manuscripts and other materials are located in numerous museums, universities, and other institutions, mostly in Europe, but also in Australia, New Zealand, and the United States. Current locations of botanical collections of both Forsters (not always readily distinguished) are discussed by R. C. Carolin in *Proceedings of the Linnean Society of New South Wales* 88 (1963). Manuscripts in Berlin are detailed by M. E. Hoare in *Journal of Pacific History* 7 (1972).

Martin K. McNicholl

FORTIN, PIERRE (ETIENNE). Born Verchères, Quebec, December 1823; died Laprairie, Quebec, 15 June 1888. Fisheries officer, magistrate, naturalist, politician, physician.

FAMILY AND EDUCATION

Son of Pierre Fortin, a carpenter, and Marie-Anne-Julie Crevier Duvernay Fortin. Grew up at Laprairie, near Montreal. Attended Collège Saint-Suplice and then McGill University, where received M.D. in 1845. Never married.

POSITIONS

Practiced medicine between 1845 and 1852. Came to attention of colonial authorities when he organized cavalry troop in 1849 to quell disturbances in Montreal. In 1852 was appointed "stipendiary magistrate for the protection of the fisheries in the Gulf of St. Lawrence" and made annual voyages there until 1867. Served terms as member of Quebec Legislative Assembly and Canadian House of Commons between 1867 and 1887 and was chairman of the Committee on Navigation and the Fisheries of the latter body, 1867–1874. Founding president of the Quebec Geographical Society, 1878–1879. Appointed to Canadian Senate, 1887.

CAREER

When Fortin began his patrols in 1852, the northern part of Gulf of St. Lawrence lacked any effective legal presence, and fishermen carried on their trade without regard to conservation or to each other's life and property. Fortin immediately imposed rule of law in the area and began to gather detailed statistics and observations, which he used as basis for rational regulation of the fishery. This early start in data collection has given northern Gulf the longest continuous sequence of comprehensive fishery statistics anywhere in the world (Huntsman 1943). In 1858 Fortin imposed licensing system for salmon fishermen, and he rigidly enforced regulations that prevented blocking of rivers by nets. In his *List of Fishes Taken in the Gulf and River St. Lawrence*, published from 1862 to 1865 in the Sessional Papers of the Parliament of Canada, he described anatomy and distribution of the region's fresh-and saltwater fishes and named one new species.

From beginning of his expeditions Fortin attempted to suppress intensive commercial egging of the seabird colonies along north shore of the Gulf. In 1858, at Fortin's instigation, Canadian game laws were amended to prohibit removal or destruction of bird eggs on the Gulf islands, and in the same year he stationed seven-man party on Mecatina Murre Rocks to ward off egg hunters. Although Fortin was unable to apprehend many eggers because of the multiplicity of nesting islands, his severe reputation was effective in discouraging their activities. This reputation was vindicated in 1865, when he summarily confiscated Nova Scotian schooner and about 29,000 eggs and sentenced vessel's crew to jail terms.

After Fortin's patrols in the Gulf ended in 1867, he continued to push for rational fisheries management and conservation as member of the Quebec and Canadian Parliaments. He was widely known and respected in Quebec for his work in the Gulf and because of newspaper articles he wrote on marine affairs.

MAJOR CONTRIBUTIONS

His promulgation of fisheries regulations based on careful and objective observation presaged the development of scientific fisheries management and the modern concept of government's responsibility as guarantor of the conservation of natural resources. In his efforts for the protection of birds for their intrinsic worth and not for any economic benefit, he was far ahead of his time.

BIBLIOGRAPHY

Gemmill, J. A., ed. *Canadian Parliamentary Companion 1883* (1883).

Huntsman, A. G. "Fisheries Research in Canada." *Science* 98, 2536 (1943).

Potvin, Damase. *Le roi du golfe* (1952).

Préfontaine, Georges. "Le développement des connaissances scientifiques sur les pêcheries maritimes et intérieures de l'est du Canada." *Actualité économique* 2.3 (1946).

David K. Cairns

FOTHERGILL, CHARLES. Born York, England, 23 May 1782; died Toronto, Ontario, 22 May 1840. Naturalist, writer, journalist, publisher, politician.

FAMILY AND EDUCATION

Born to a prominent Quaker family of Yorkshire, England. Son of John Fothergill, a maker of ivory brushes and combs, and Mary Anne Forbes. Grandnephew of naturalist and philanthropist John Fothergill and nephew of James Forbes, F.R.S. Married Charlotte Nevins, 1 December 1811 (died 1822), and Eliza Richardson, 19 March 1825. At least three sons from the first marriage and at least four sons and two daughters from the second.

POSITIONS

Postmaster for Port Hope, 1817–1820, King's Printer, 1822–1826 at York (Toronto), elected to the House of Assembly for the County of Durham, 1824–1830, justice of the peace, member of the Land Board, naturalist in the environs of Pickering Township, 1830–1840.

CAREER

Trained in his father's business but preferred natural history and artistic pursuits. Published *Ornitholigia Britannica*, a list of 301 species of British birds, when he was seventeen. Labored at writing natural and civil histories for Yorkshire and the Northern Isles and then enrolled in medicine at Edinburgh in 1813. Published "Essay on the Philosophy, Study, and Use of Natural History" in 1813. Possessed a marked propensity to live beyond his means. Forced to leave his academic pursuits to avoid arrest for his debts. Began work on a comprehensive natural history of the British Empire, a lifelong project that was never completed. Emigrated from England in July 1816, arriving at Quebec in August. Traveled to upper Canada in February 1817 and settled at Smith's Creek (Port Hope), east of York (Toronto). Published the *Upper Canada Gazette* (1822–1826, including *Weekly Advertiser* 1822–1825) and *York Almanac* (1823–1826) as the King's Printer. Cofounder of the Literary and Philosophical Society of Upper Canada in 1831 and primary advocate for a Lyceum of Natural History and the Fine Arts, which was to comprise a museum, art gallery, botanic garden, and zoo. His manuscript "An Essay Descriptive of the Quadrupeds of British North America" won the Natural History Society of Montreal Silver Medal for 1830, and his paper on protecting the Lake Ontario salmon fishery was read at the Literary and Historical Society of Quebec in 1835. Entered the publishing scene again with *Palladium of British America and Upper Canada Mercantile Advertiser* (1837–1839) and *Toronto Almanac* (1839). His success as a naturalist far surpassed that of his political and financial ventures. Tragically, many of his papers and his entire natural history collection were destroyed shortly after his death.

MAJOR CONTRIBUTIONS

One of the most important pioneering naturalists in Canada and one of the best of the period anywhere. His views with regard to the formation of a lyceum were decades ahead of his time. In 1830 he attempted to initiate a natural history expedition to the Pacific Ocean with the governments of upper and lower Canada and several natural history associations. Originated a bill to establish agricultural societies in upper Canada. Principal contributions are the detailed and precise natural history depictions in his papers and paintings. Wrote the first nature columns published in an upper Canadian newspaper, wherein he espoused a conservation ethic. His natural history notes promoted observation and protection of fauna and flora, and he was a strong advocate of protection for the Lake Ontario salmon fishery.

BIBLIOGRAPHY

Baillie, J. L. "Charles Fothergill 1782–1840." *Canadian Historical Review* 25 (1944).

Black, R. D. "Charles Fothergill's Notes on the Natural History of Eastern Canada, 1816–1837." *Transactions of the Royal Canadian Institute* 20 (1935).

Gladstone, H. "The Fothergill Family as Ornithologists." *The Naturalist* 785 (1922).

Romney, P. "A Man Out of Place: The Life of Charles Fothergill; Naturalist, Businessman, Journalist, Politician, 1782–1840." Diss., University of Toronto, 1981.

Romney, P. "Fothergill, Charles." *Dictionary of Canadian Biography*. Vol 7 (1988).

Theberge, E. "Fothergill: Canada's Pioneer Naturalist Emerges from Oblivion." *Beaver* 68 (1988).

UNPUBLISHED SOURCES

Most of Fothergill's notebooks, letter books, diaries, rough manuscripts, and scrapbooks are held at the University of Toronto, Fisher Library, in the Fothergill Papers (MS coll. 140) and the James Little Baillie Papers (MS coll. 126). Further manuscript material and paintings are located at the Public Archives of Canada and the Royal Ontario Museum. "An Essay Descriptive of the Quadrupeds of North America" is at McGill University Libraries, Blacker-Wood Library.

Michael S. Quinn

FOX (OR FOXE), LUKE. Born Kingston-Upon-Hull, Yorkshire, 20 October 1586; died c.15 July 1635. Navigator, arctic explorer.

FAMILY AND EDUCATION

Son of Richard Fox, a master mariner from Hull. No information on education, but it was limited. Married Anne Barnard, 1613. No information as to children.

POSITIONS

Frequently sailed in European waters as youth and young man. Captain of Pinnace HMS *Charles*, April or May 1631 to October 1631.

CAREER

Grew interested in seeking Northwest Passage from age twenty. Was well versed in arctic history and navigation, though he had little formal education. Through intercession of Henry Briggs, mathematician, and John Brooke, secured loan of vessel of seventy or eighty tons from King Charles I (late 1629) and sailed as captain and pilot for Hudson Bay region either at the end of April or the beginning of May 1631. Moving through Hudson Strait, he arrived off western shore of Hudson Bay in late June and sailed north, traveling through Foxe Channel (so named two centuries later), past Foxe Peninsula (southeastern extremity of Baffin Island) into Foxe Basin, reaching point he named Cape Dorchester (just south of Arctic Circle) in late September. He then returned to England. Found no passage but noted that tide flowed through Foxe Channel from southeast, thus effectively ruling out Hudson Bay as a means of reaching waters farther to the northeast. In his *North-West Fox* (1635), he reviewed some twenty kinds of birds seen, with useful descriptive comments. These included what were probably the little brown crane, duck, swans, geese, cormorant, sea mew (gull), plover, stint (a sandpiper), blackbird, crow, eagle, hawk, jay, owl, partridge, pheasant, ptarmigan, raven, and thrush. Discussed the eggs of several species and was also one of the first to mention the whooping crane. He described it as "longheaded, long neckt, and a body almost answerable." Some attention was also given to other fauna and to plant life. Some of the geographical designations he gave physical features on his voyage are still in use. Fox's book has been described as egotistical, and one authority considered it "the quaintest and most amusing narrative in the whole range of Polar literature." The circumpolar map he included in the book has been termed "one of the most interesting and important documents in the history of Arctic exploration." Fox himself has been characterized as a clever and conceited person. Because he returned after only six months and did not find Northwest Passage that he and others so eagerly sought then and later, he may have felt defensive about what he had accomplished. Poverty-stricken following his return, he died in relative obscurity in his 48th year. Not until the middle of the eighteenth century was enthusiasm for arctic exploration rekindled to any degree.

MAJOR CONTRIBUTIONS

Expanded what was known concerning area north of Hudson Bay; brought back navigational notes and information concerning fauna and flora observed.

BIBLIOGRAPHY

Allen, Elsa G. ''American Ornithology before Audubon.'' *Transactions of the American Philosophical Society* 41, pt. 3 (1951).

Coote, C. H. *Dictionary of National Biography.*

Fox, Luke. *North-west Fox.* Repr. in *Voyages of Foxe and James.* Vol. 2 (Hakluyt Society, 1894).

Morley, William F. E. *Dictionary of Canadian Biography.* Vol. 1 (1967).

Keir B. Sterling

FRANKLIN, JOHN. Born Spilsby, Lincolnshire, England, 16 April 1786; died off King William Sound, Northwest Territories, 11 June 1847. Naval officer, explorer.

FAMILY AND EDUCATION

Twelfth and youngest son of Willingham Franklin, mercer, and Hannah Weekes. Educated at preparatory school in St. Ives, Huntingdon, and, from age twelve, at Louth Grammar School in Lincolnshire. Received honorary D.C.L. from University of Oxford in July 1829. Married Eleanor Anne Porden on 18 August 1823 (died February 1825) and Jane Griffen on 5 November 1828. One daughter, by Eleanor Porden.

POSITIONS

First-class volunteer on the *Polyphemus*, 1800–1801; midshipman on the *Investigator*, 1801–1804; midshipman on the *Bellerophon*, 1805–1807; midshipman on the *Bedford*, 1807–1814; promoted to lieutenant 11 February 1808; commanded the *Trent*, 1818; commanded the *Prince of Wales* and led team to explore north coast of American continent, 1819–1821; promoted to commander 1 January 1821; promoted to post-captain 20 November 1822; elected a fellow of the Royal Society, 1822; led a second overland venture to arctic coast, 1825–1827; awarded gold medal of the Societe de Geographie de Paris, 1828; commanded the *Rainbow*, 1830–1833; served as lieutenant governor of Van Diemen's Land (Tasmania), 1836–1843; given command of arctic expedition to find Northwest Passage 7 February 1845; promoted to rear admiral 26 October 1852. He was carried on the navy's active list until death confirmed in 1854.

CAREER

Joined Royal Navy in 1801 and served on several vessels. Fought in several battles, including the Battle of Copenhagen (1801), the Battle of Trafalgar (1805), and the Battle of New Orleans (1814). Discharged on half-pay in 1815, reinstated in 1818 upon joining Royal Navy explorations of the Arctic. Joined Comr. David Buchanan's 1818 arctic expedition, commanding the brig *Trent*; expedition was a failure due to its inability to penetrate the pack ice. Chosen in 1819 to lead expedition to chart the north coast of North America east of the Coppermine River. Traveled overland, through unexplored land, up the Yellowknife River to the Coppermine. Surveyed 340 kilometers of coast, reaching Kent Peninsula, before turning back due to dwindling supplies; nine men died of starvation during return journey. Chosen to lead expedition in 1823 to explore the north coast east and west of the Mackenzie River delta. Franklin's group charted approximately 440 kilometers of coast west of the delta, while second party surveyed east as far as the mouth of the Coppermine River. Commanded the frigate *Rainbow* in the Mediterranean in 1830–1835; received the Order of the Redeemer of Greece and the Royal Hanoverian Order as a result of his peacekeeping diplomacy. Appointed lieutenant governor of Van Diemen's Land (Tasmania) in 1836; was dismissed in 1843. Appointed (7 February 1845) to command arctic expedition to explore coast along Barrow Strait with the most equipped expedition yet sent, with steam-driven screw propellers, steam heat for the berths, and instruments for botanical, geological, zoological, and magnetic experiments. Set sail 19 May 1845. Franklin was last spotted by a group of whalers in northern Baffin Bay on 26 July 1845. The demise of the Franklin expedition was not discerned until 1854.

Franklin, author of three books, wrote two that dealt with arctic exploration. His expeditions allowed men such as John Richardson to examine and document observations of birds, mammals, fish, and plants. Franklin's personality was that of a British naval officer, not a seasoned fur trader. This deficiency in Franklin's character is evidenced by his refusal to turn back in 1819, which resulted in the death of nine men, and his eventual demise in 1847. Despite these catastrophes, he was viewed as a hero during his era. The Royal Geographical Society stated in 1845, in regard to Franklin's chances of success on his final expedition: "The name Franklin alone is, indeed, a national guarantee."

MAJOR CONTRIBUTIONS

Charted the unknown arctic coast from approximately longitude 105 to 150 degrees west; this area comprised approximately half of Canada's arctic coast and part of the Alaskan seaboard. His explorations resulted in an expansion of geographical and climactic knowledge of an area unknown to Europeans. His death led to increased arctic exploration from 1847 to 1859 as numerous search teams scoured the Arctic searching for him and members of his crew.

BIBLIOGRAPHY

Beattie, Owen, and John Geiger. *Frozen in Time: Unlocking the Secret of the Franklin Expedition* (1987).

Cooke, Allan, and Clive Holland. *The Exploration of Northern Canada 500–1920: A Chronology* (1978).

Franklin, John. *Narrative of a Journey to the Shores of the Polar Sea, in the Years 1819, 20, 21, and 22* (1823; repr. 1969).

Franklin, John. *Narrative of a Second Expedition to the Shore of the Polar Sea, in the Years 1825, 1826, and 1827* (1828; repr. 1971).

Holland, Clive. ''Sir John Franklin.'' *Dictionary of Canadian Biography*. Vol. 7 (1988).

Neatby, Leslie H. *In Quest of the North West Passage* (1958).

Owen, Roderic. *The Fate of Franklin* (1978).

Richardson, John. *Arctic Ordeal: The Journal of John Richardson, Surgeon-Naturalist with Franklin, 1820–1822*. Ed. C. Stuart Houston (1984).

Sutherland, P. D. *The Franklin Era In Canadian Arctic History, 1845–1850* (1985).

David Calverly

FRASER, SIMON. Born Mapleton, Vermont, 1776; died near St. Andrews, Stormont County, Canada West (now Ontario), 18 August 1862. Fur Trader, explorer.

FAMILY AND EDUCATION

Youngest of eight children of Simon Fraser, Scottish immigrant to the American colonies and a loyalist during the Revolution who died when his son was two, and Isabella Grant Fraser. Mother moved to Coteau-du-Lac, west of Montreal, 1784. Simon received brief schooling in Montreal and was apprenticed to North West Company at age sixteen. Married Catherine Macdonell, 1820, eight children reached adulthood.

POSITIONS

Was serving as clerk for North West Company in Athabaska department, 1799. Made partner in North West Company, 1802. Given responsibility for North West Company operations west of the Rocky Mountains, 1805. Undertook several western trips including major expedition to explore route to west, 1805–1808. Again assigned to Athabaska department, 1810–1814. At Fort William (Northwest Company headquarters, now in Ontario), 1815–1817. Tried for treason, conspiracy, and as accessory to murder with five fellow partners, 1818, following Northwest Company dispute with Thomas Douglas, Lord Selkirk, at latter's Red River Settlement, Fort Douglas, Assiniboia (now Manitoba), acquitted. Farmer on Raisin River, near St. Andrews, Upper, later West Canada, 1818–1862. Captain, First Regiment, Stormont Militia, 1837–1838. Involved in various unsuccessful business initiatives in vicinity of Saint Andrews, notably sawmill and gristmill.

CAREER

Fraser had very little formal education before being apprenticed to North West Company in 1792, most likely because several of his mother's relatives were involved in the fur trade. Was a successful fur trader, but available information concerning his first decade with North West Company is scanty. Became partner in North West Company at age twenty-five. One of his tasks was to look for practical travel routes to the west and ultimately the Pacific Ocean, while constructing trading posts. The Columbia River (which he did not find) made up a part of the route the company wanted him to follow. At Trout Lake (McLeod Lake), Fraser built what became Fort McLeod, initial white settlement in what is now British Columbia, and found what is now the Fraser River (discovered by Alexander Mackenzie in 1793, but given its present name by David Thompson in 1813) and Stuart Lake in 1806. While on his way west in 1806 and 1807, built trading posts later named Forts St. James, Fraser, and George. His final effort came in 1808, when, with a party of two dozen men, he attempted, though without sufficient information about what lay before him, to move down the Fraser River (which he thought was the Columbia), from Fort George to the present Georgia Strait south of the present city of Vancouver. The round trip of well over 1,000 miles was made in 71 days. Fraser was dissuaded from making a thorough reconnaissance of the river's mouth by the presence of hostile natives. The area through which he traveled was first named New Caledonia, probably by Fraser. Fraser considered this expedition a failure, because he had not found the Columbia and had not advanced the interests of the North West Company. Five years later John Stuart, Fraser's able assistant of 1808, worked out a route combining the Fraser south to Alexandria, British Columbia, thence overland in southeasterly direction to Kamloops, British Columbia, and then southwest via the Thompson, Okanagan, and Columbia Rivers to the Pacific. This route (from the coast east to the New Caledonia settlements) was utilized for supply purposes by the North West Company and later the Hudson's Bay Companies from 1814 until the mid-1820s, when it was abandoned. This was due in large part to uncertainty about long-term British territorial prospects in the region vis-à-vis the United States.

Fraser spent the years 1810 to 1814 in the Athabaska department and was at Fort William, on northwest shore of Lake Superior, in 1815–1816. There his time was mainly taken up with the violent quarrels between the North West Company and colonists led by Thomas Douglas, fifth Earl of Selkirk and a Hudson's Bay Company stockholder. Lord Selkirk had been granted 116,000 acres in Assiniboia (now southern Manitoba, northern and eastern North Dakota, and northwestern Minnesota) by the Hudson's Bay Company, where he proposed to create a settlement for Scots Highlanders and Irishmen emigrating to Canada. The colonists arrived in 1811 and built Fort Douglas at the juncture of the Red and Assiniboine Rivers in 1812. Fraser was not involved in various North West Company efforts to destroy this settlement, 1812–1816, but was arrested with five other North West Company partners following the shooting of twenty settlers and their governor at Seven Oaks, northwest of present-day Winnipeg, in June 1816. Was tried in York (now Toronto), 1818, and acquitted. Spent much of 1817 at Fort William while on bail. Left fur trade, spent remaining forty-three years of his life on his farm on Raisin River in vicinity of St. Andrews, Stormont County, West Canada (now southeastern corner of Ontario). Various business enterprises with which he was involved, including a saw mill, were not notably successful. After sustaining a knee injury while captain of his militia regiment during the rebellion of 1837–1838, he was partially incapacitated and spent the rest of his life in very modest circumstances.

MAJOR CONTRIBUTIONS

Fraser was a courageous and persistent man, who dealt calmly with many disappointments and setbacks during his career. He is best known for the expeditions he undertook in western Canada during the years from 1805 to 1808 while seeking a river route to the Pacific, and he was in large part responsible for construction of the first permanent settlements in what is now British Columbia.

BIBLIOGRAPHY

Lamb, W. Kaye, ed. *The Letters and Journals of Simon Fraser, 1806–1808* (1960).

Lamb, W. Kaye. *Dictionary of Canadian Biography.*

Scholefield, E. O. S. ''Simon Fraser.'' *Westward Ho! Magazine* (October, 1908–March, 1909).

Scholefield, E. O. S., and F. W. Howay. *British Columbia From the Earliest Times to the Present.* Vol. 1 (1914)

UNPUBLISHED SOURCES

Transcripts of Fraser's 1806 journal, a portion of his 1808 journal, and some of his letters are at Bancroft Library, University of California, Berkeley. His manuscript journal for the 1808 trip is in Metropolitan Toronto Central Library. Copies of a few letters are in Provincial Archives of British Columbia. Copies of a number of the aforementioned and other documents are in National Archives of Canada.

Keir B. Sterling

FRÉMONT, JOHN CHARLES. Born Savannah, Georgia, 21 January 1813; died New York City, 13 July 1890. Army officer, explorer.

FAMILY AND EDUCATION

Son of Jean Charles Frémon, French teacher, dancing master, sometime painter, and Mrs. Ann Beverley Whiting Pryor (couple was not married in 1813 but married sometime prior to 1818, when father died). John Charles changed name to Frémont c.1838. Attended Charles Robertson's School, Charleston; entered College of Charleston as a junior, 1829, expelled for excessive absences, 1831. Married Jessie Benton, 1841; four children.

POSITIONS

Secured position as teacher of mathematics aboard U.S. naval vessel, 1833–1835; offered appointment as seagoing professor of mathematics for U.S. Navy but declined, 1835. Joined Capt. W. G. Williams, U.S. Army, on survey of railroad route from Charleston to Cincinnati, 1835–1837, and another contingent of topographical engineers in survey of Cherokee country in Georgia. Commissioned 2d lieutenant, U.S. Army Corps of Topographical Engineers, 1838. Accompanied French scientist Joseph Nicollet on several expeditions to what is now Minnesota and the Dakotas, 1838 and 1839; spent much of 1840 assisting Nicollet in drafting report. Led his own first expedition to Des Moines River, 1841. Directed five expeditions to the West, the first three of them under government auspices, in 1842, 1843–1844, 1845–1846, 1848–1849, and 1853–1854. Brevet captain, 1844. As major, California Rifles, was acting military governor of California Territory for two months under auspices of Commod. Robert F. Stockton and Comdr. John D. Sloat, 1846. Shortly thereafter, appointed lieutenant colonel, California Rifles. Court-martialed for alleged insubordination to Brig. Gen. Stephen W. Kearny, 1847–1848. Convicted, but penalty remitted by President James K. Polk. Frémont resigned from army, March 1848. Elected to short term, U.S. Senate from California, 1850–1851. Nominated for president of United States on Republican ticket, 1856. Considered for post of minister to France by President Abraham Lincoln, but this opposed by Secretary of State William Seward, and appointment not made. Major general, U.S. Volunteers, 1861–1864. Commanded at St. Louis and later Mountain Department (West Virginia). Nominated for president by radical Republicans and war Democrats, 1864, but withdrew. Lost his Mariposa Ranch in California due to financial reverses, 1864. Involved in railway speculation, 1870, lost fortune. Territorial governor of Arizona, 1878–1883. Restored to army as major general on retired list less than three months before death.

CAREER

He battled implications of his illegitimate birth much of his life, though his parents married soon after his birth after Mrs. Frémon's first husband died. Frémont learned his craft as wilderness traveler on early expeditions as civilian under auspices of Army Corps of Topographical Engineers and as assistant to French scientist Joseph Nicollet. His 1841 marriage to Jessie Benton, daughter of Senator Thomas Benton of Missouri, began a forty-nine-year partnership that ended only with his death. Benton initially opposed the match but later gave his son-in-law valuable backing. Frémont's first western expedition (1842) was taken with (among others) German-born cartographer Charles Preuss (1803–1854), who was also part of Frémont's next three expeditions and whose maps added materially to Frémont's reports. Frémont's first expedition (1842) took him from Independence, Missouri, through present-day northeastern Kansas and

Nebraska into Wind River Range of modern Wyoming. Second expedition (1843–1844) followed more southerly route into Wyoming, thence through parts of present-day Idaho, Utah, Oregon, southeastern Washington, California as far south as the Mojave Desert, and back through southern Nevada, Utah, Colorado, and Nebraska. Frémont's well-written *Report of the Exploring Expedition to the Rocky Mountains in the Year 1842, and to Oregon and North California in 1843–1844*, combining accounts of the first two expeditions, was among the most popular government documents of the era. Jessie Benton Frémont was her husband's able collaborator and deserves much credit for their success. His third expedition (1845), about which he later provided a full discussion in his *Memoirs of My Life* (1887), took him through Kansas, Colorado, Utah, northern Nevada, and then central and northern California north to Klamath Lake. This was Frémont's most controversial mission. He played a role in American takeover of California before and after onset of Mexican War.

His privately financed fourth expedition (1848), through Kansas and Colorado, was a disaster. His guides led him astray, he failed to turn back with the onset of poor winter weather in the San Juan Mountains, he lost ten men, and imputations of cannibalism among the surviving men of his party followed him for years. He and his men had to winter in Taos, New Mexico Territory. His fifth and last expedition (1853–1854), which took him through previously unexplored parts of southern and eastern Utah, was contemporary with other Pacific Railway surveys and was designed to identify a central railroad route. Frémont did not write formal reports of these last two expeditions, though LeRoy and Ann W. Hafen have published documents concerning the fourth trip, and S. N. Carvalho an account of the last one. Frémont's 1845 report was particularly noteworthy for his detailed observations, including much information about plants and animals. Many plant specimens were noted in the reports and were brought back from most of his trips and can be found today in such repositories as the U.S. National Herbarium, Washington, D.C., Gray Herbarium, Harvard University, Missouri Botanical Garden, and the Museum d'Histoire Naturelle, Paris, among others. Ownership of his ranch in California came about because gold was fortuitously found on property that had been bought for him. His wealth was later dissipated because of his preoccupation with Civil War duties, poor business judgment, and later reverses growing out of poor management of several railways with which he was involved. Frémont was an effective wartime administrator but a poor battlefield commander, and his premature freeing of the slaves in Missouri in the spring of 1861 did not help his standing with Lincoln. His later years were difficult, and only his wife's writing kept them afloat financially.

MAJOR CONTRIBUTIONS

Frémont is, of course, best known for his five expeditions to the West, which captured the public imagination. From these, much valuable information concerning geography, transportation routes, plants, and animals was derived, distinguished more for its breadth than its depth. His *Report* (1845) was among the very best written by any western explorer. The objectives of his expeditions were, for the most part, practically oriented. His international reputation was enhanced owing to fulsome praise from Alexander von Humboldt in the latter's *Aspects of Nature* (1849).

BIBLIOGRAPHY

Carvalho, S. N. *Incidents of Travel and Adventure in the Far West with Frémont's Last Expedition* (1857).

Dictionary of American Biography.

Egan, Ferol. *Frémont: Explorer for a Restless Nation* (1977).

Frémont, Jessie Benton. *Souvenirs of My Time* (1887).

Frémont, John Charles. *Memoirs of My Life* (1887).

Goetzmann, William H. *Army Exploration in the American West, 1803–1863* (1959).

Goetzmann, William H. *Exploration and Empire: The Explorer and the Scientist in the Winning of the American West* (1966).

Hafen, LeRoy R., and Ann W. Hafen, eds. *Frémont's Fourth Expedition: A Documentary Account of the Disaster of 1848–1849* (1960).

Heitman, Francis. *Historical Register of the U.S. Army* (1903).

Herr, Pamela. *Jessie Benton Frémont, a Biography* (1987).

Jackson, Donald, and Mary Lee Spence, eds. *The Expeditions of John C. Frémont.* 5 vols. to date (1970–).

Nevins, Allan. *Frémont, the West's Greatest Adventurer.* 2 vols. (1928).

Nevins, Allan. *Frémont, Pathfinder of the West* (1962).

Preuss, Charles. *Exploring with Frémont: The Private Diaries of Charles Preuss, Cartographer for John C. Frémont on His First, Second, Third, and Fourth Expeditions to the Far West.* Ed. Erwin G. and Elisabeth K. Gudde (1958).

UNPUBLISHED SOURCES

Most surviving Frémont materials (others were destroyed by fire) are in the Bancroft Library, University of California–Berkeley, including the manuscripts for the unpublished second volume of his autobiography and his wife's unpublished "Memoirs." There are also materials in the Bancroft Collections by, or having reference to, a number of individuals who had some involvement with Frémont.

Keir B. Sterling

FUERTES, LOUIS AGASSIZ. Born Ithaca, New York, 7 February 1874; died Potter's Crossing, Unadilla, New York, 22 August 1927. Artist, naturalist, explorer.

FAMILY AND EDUCATION

Father Estevan Antonio Fuertes, professor of civil engineering at Cornell University; mother, Mary Stone Perry, a musician. Attended public schools in Ithaca, New York, and prep school in Zurich, Switzerland, 1892. Received degree in architecture from Cornell University, 1897. Studied with Abbott H. Thayer, artist, for one year following college. Married Margaret F. Sumner, 1904; two children.

POSITIONS

Illustrated yearbooks for the Department of Agriculture and articles for magazine *Outing*. Illustrated *Citizen Bird* (1897); *Handbook of Birds of the Western*

U.S. (1902); *Birds of the Rockies* (1902); *The Water Fowl Family* (1903); *Key to North American Birds* (1903); illustrator for *Bird-Lore* (1904–1927); *Birds of New York* (1910); a series for *National Geographic* magazine, 1914–1919; *Burgess's Bird Book for Children* (1919); *Burgess's Animal Book for Children* (1920); and lecturer in ornithology, Cornell University, 1923–1927.

CAREER

Began exhibiting considerable artistic talent in youth. Became associate member of American Ornithologists Union at age seventeen. From age twenty his career was strongly supported by the leading American ornithologist Elliott Coues, whose protégé he became, and later Frank M. Chapman of the American Museum of Natural History. Became student of Abbott Thayer, prominent artist-naturalist, and accompanied him on collecting trip to Florida, 1898. By his early twenties he had been commissioned to illustrate books by Coues and others, including the leading American women ornithologists Florence M. Bailey, Olive Thorne Miller, and Mabel O. Wright. In addition, illustrated articles for the periodicals *Osprey, Bird Lore, The Condor, St. Nicholas*, and *National Geographic Magazine*. At invitation of C. Hart Merriam, Chief of the U.S. Biological Survey, joined Harriman Alaska expedition, 1899–1900. Between 1900 and 1926 provided illustrations for several of the *North American Fauna* series and other Biological Survey publications. Provided some illustrations for Abbott and Gerald Thayers' *Concealing Coloration in the Animal Kingdom* (1909), which proved to be a controversial work. Chapman and Theodore Roosevelt were critical of some of the Thayers' conclusions; Fuertes tried to play the role of mediator, but no resolution was achieved. The Thayer volume is today regarded as a classic, and Fuertes later developed his views of the problems of concealing coloration in his lectures at Cornell (1923–1926). In 1901 went with the U.S. Biological Survey to Texas and in 1902 went to the Bahamas with Frank Chapman. Other expeditions included Jamaica, 1904; Saskatchewan and Alberta, 1907; Cape Sable and Cuthbert Rookery, Florida, 1908; Magdalen Islands and Bird Rock, 1909; Mexico, 1910; and Colombia, 1911 and 1913. Some of these trips were taken with Chapman and other American Museum staff members; several were under other auspices. During each of these expeditions he made copious notes and sketches, which are still useful to ornithologists. In addition, he collected over 3,500 bird skins, which he used as models. After 1913, spent most of his time in Ithaca, where he gave lectures and bird walks and worked on commissioned paintings. One commission was series of twenty-four oils for home of Frederick F. Brewster in New Haven, Connecticut. In 1925 went to Wyoming with his daughter Mary and met James Baum, a sportsman, who urged him to explore Ethiopia. The trip, which took place in 1927, was sponsored by such organizations as the Field Museum of Natural History in Chicago and *Chicago Daily News*. Fuertes, Baum, and Wilfred H. Osgood, curator of zoology at the museum in Chicago, explored Ethiopia's wildlife, customs, and countryside. During the trip back to Ithaca Fuertes was killed at a railroad crossing.

MAJOR CONTRIBUTIONS

Noted for his rigorous fieldwork, artistic ability, and exceptional memory for exact poses of birds in their natural habitat. His powers of concentration and the ability to see minute details gave his illustrations their exacting, lifelike quality. He was more than an artist as he explored unknown territories and more than a naturalist as he painted thousands of images that conveyed the knowledge and practice of a man dedicated to his art. Regarded by most authorities as the most able and sensitive bird artist since Audubon.

BIBLIOGRAPHY

Allen, Arthur A. "The Passing of a Great Teacher." *Bird Lore* 29 (1927).

Boynton, Mary Fuertes. *Louis Agassiz Fuertes, His Life Briefly Told and His Correspondence Edited* (1956).

Chapman, Frank. *Autobiography of a Bird Lover* (1933).

Dictionary of American Biography. Vol. 4 (1958).

Marcham, Frederick G. *Louis Agassiz Fuertes and the Singular Beauty of Birds* (1971).

Peck, Robert McCracken. *A Celebration of Birds: The Life and Art of Louis Agassiz Fuertes* (1982).

Who Was Who in America. Vol. 1 (1897–1942).

UNPUBLISHED SOURCES

See Cayuga Bird Club Records, 1913–1975, Department of Manuscripts and University Archives, Cornell University; Harriett Mathilda Davidson, ornithology notebooks, c.1931, Department of Manuscripts and University Archives, Cornell University; Louis Agassiz Fuertes Papers, 1892–1954, Department of Manuscripts and University Archives, Cornell University; Harriman Alaska Expedition, arctic field photographs, 1899, American Museum of Natural History, Department of Library Services, New York. Major collections of Fuertes' paintings are at the American Museum of Natural History, Cornell University, Museum of the State of New York at Albany, New York Zoological Society, Department of Interior, National Audubon Society, and in private collections.

Geri E. Solomon

FURBISH, CATHERINE (KATE). Born Exeter, New Hampshire, 19 May 1834; died Brunswick, Maine, 6 December 1931. Botanist, artist.

FAMILY AND EDUCATION

Oldest of six and only daughter of Benjamin Furbish, hardware store owner, manufacturer of stoves and tinware, civic leader, and amateur botanist, and Mary A. Lane Furbish. Family moved to Brunswick, Maine, in 1835. Began study of botany at age twelve under father. Attended local schools with supplemental instruction in drawing in Portland, Maine, and Boston. Attended botanical lectures at Harvard University by George L. Goodale in 1860. Studied French literature in Paris. Never married.

POSITIONS

Practicing botanist and artist, 1851–1931.

CAREER

Began serious study of botany in 1870. In 1873 inherited sufficient sum from her father to be independent for life. Took many extended trips to collect and paint throughout Maine, especially wild and unsettled north, notably Aroostook County and the St. John River in 1880 and 1881, described in the *American Naturalist.* Discovered a new species of Tousewort in 1880, *Pedicularis furbishiae*, which in 1977 threatened, perhaps successfully, the Dickey–Lincoln Hydroelectric Project because the dam threatened the only known habitat of the species. In 1895 helped found the Josselyn Botanical Society of Maine, president 1911–1912.

Her publications were few: articles describing field research appeared in the *American Naturalist* in 1881, 1882, and 1901. More significant perhaps are sixteen folio volumes of botanically accurate watercolors of Maine plants, presented to Bowdoin College's library in 1908 and still on display.

For eighty years Furbish was first-rate amateur botanist and illustrator, winning the respect of Harvard botanists Gray, Watson, and Fernald. Regarded by some as eccentric, crazy, and certainly unladylike for her travels through wild country and exploits in pursuit of plants, she was regarded by fellow botanists as having "undaunted pluck" and a "faithful brush" and as having done "more than any other to make known the wonderful flora of the 'Garden of Maine,' " as well as being a "delightful" field companion.

MAJOR CONTRIBUTIONS

One of best of the nineteenth-century amateur botanists, refusing to allow her lack of formal education or her gender to hold her back. Pioneer collector, cataloger, and painter of Maine flora, including at least two new species.

BIBLIOGRAPHY

Adams, Sally Aldrich. "The Rare Flower That Challenged a Dam: It All Started with Plucky Kate Furbish." *Christian Science Monitor* 13 (April 1977).

Brunswick (Maine) *Record*, 10 December 1931 (obit.).

Coburn, Louise H. "Kate Furbish, Botanist." *Maine Naturalist* 4 (1924).

Graham, Ada, and Frank Graham, Jr. *Kate Furbish and the Flora of Maine* (1995).

Saltonstall, Richard, Jr. "Of Dams and Kate Furbish." *Living Wilderness* 40 (1977).

Schwarten, Lazella. *Notable American Women*. Vol. 1 (1971). (sketch).

UNPUBLISHED SOURCES

See letter in the Historical Letter File, Gray Herbarium, Harvard University and Bowdoin College Library.

Liz Barnaby Keeney

G

GABRIELSON, IRA NOEL. Born Sioux Rapids, Iowa, 27 September 1889; died Washington, D.C., 7 September 1977. Conservationist.

FAMILY AND EDUCATION

Son of Frank August and Ida Jansen Gabrielson. Attended public schools in Sioux Rapids. B.A., Morningside College; 1912, advanced study at Lakeside Laboratory, Iowa State University, summers 1911, 1912. Honorary LL.D., Morningside, 1941, honorary D.Sc., Oregon State College, 1936, Middlebury, 1959, Colby, 1969. Married Clara Speer, 1912; four daughters.

POSITIONS

Taught biology, Marshalltown, Iowa, High School, 1912–1915. Employed by U.S. Biological Survey (USBS), 1915. Investigated economic ornithology and food habits of wildlife for the USBS in Washington, D.C., 1915–1918. From 1918 to 1930, headed rodent control programs for USBS in Oregon, the Dakotas, and Iowa. Pacific Coast regional supervisor, Rodent and Predator Control, USBS, 1930–1934. Head of Division of Wildlife Research, USBS, Washington, D.C., 1935. Chief of the Biological Survey, 1935–1940. Director, U.S. Fish and Wildlife Service, 1940–1946. President, Wildlife Management Institute (WMI), 1946–1970. Chairman of the board, WMI, 1970–1977.

CAREER

For the first twenty years of his career with the Biological Survey, Gabrielson became an authority on the control of noxious rodents and predatory species.

As head of the Biological Survey and its successor organization, the U.S. Fish and Wildlife Service (USF&WS), he gave attention to the cooperative wildlife research unit programs established by his predecessor (J. N. "Ding" Darling), administered a rapidly expanding national wildlife refuge program, created the Patuxent Wildlife Research Refuge (Maryland), and spearheaded effective wildlife law enforcement force. After 1940, assumed responsibility for fish propagation and conservation programs. He organized the first North American Wildlife and Natural Resources Conference in 1936. Supplemented his modest depression-era salary with raising and selling rock-garden plants and by writing articles for gardening and sporting periodicals, sometimes using pseudonyms. During World War II, he took on additional responsibilities as deputy coordinator of fisheries for the federal government, with the task of ensuring essential seafood production. Toward the end of his tenure with the USF&WS, he became increasingly involved in international fishery programs and was a delegate to the International Whaling Conference of 1946. Retiring from federal service to enter the private sector, Gabrielson became head of the Wildlife Management Institute. In that post, he supported a variety of wildlife, park, and wilderness conservation measures and was active at annual North American Wildlife Conferences. With leaders of other conservation groups, he blocked planned dam construction in Echo Park, Colorado. As member of a special advisory board on wildlife management appointed by Interior Secretary Stewart Udall in 1962, Gabrielson and others urged that the highest possible priority be placed on the maintenance and restoration of national parks. He was co-organizer of the International Union for the Conservation of Nature (IUCN) in 1948 and the World Wildlife Fund (United States) in 1961. He had also directed staff studies of the operations of wildlife departments in thirty-one states and two Canadian provinces by the time of his death. Joint author of *Birds of Oregon* (1940); *Birds: A Guide to the Most Familiar American Birds* (1949); and *Birds of Alaska* (1959). He was author of *Western American Alpines* (1932); *Wildlife Conservation* (1941); *Wildlife Refuges* (1943); and *Wildlife Management* (1951). He also edited *Fisherman's Encyclopedia* in 1951 and *New Fish Encyclopedia in 1964*. Authored some 500 papers and reports, which appeared in over 120 technical and popular periodicals and other publications. Gabrielson's primary research interests were in the field of ornithology. His collection of nearly 9,000 skins have now been deposited at the National Fish and Wildlife Laboratory in Washington, D.C. Active in a wide range of conservation, wildlife management, and parks advocacy organizations. Recipient of a number of awards and honors for his services to the conservation movement.

MAJOR CONTRIBUTIONS

A man with a wide-ranging understanding of preservation and management issues. Gabrielson was known to the general public as "Mr. Conservation." During his sixty-two years of activity in the field, half of it in federal service and half as head of a private conservation agency, Gabrielson became an increasingly influential promoter of sound wildlife and parks management policies. An authority on game, range, salmon, and waterfowl management, conservation agencies and legislation, fish and fisheries, forestry, marine mammals, federal–state relationships, wilderness, and a number of related fields.

BIBLIOGRAPHY

Auk (October 1985) (obit.).

Cart, Theodore W. "New Deal for Wildlife: A Perspective on Federal Conservation Policy, 1933–1940." *Pacific Northwest Quarterly* (July 1972).

Dictionary of American Biography. Supplement 10 (1995).

Dunlap, Thomas. *Saving America's Wildlife* (1988).

Stroud, Richard H., ed. *National Leaders of American Conservation* (1985).

Swain, Donald C. *Federal Conservation Policy, 1921–1933* (1963).

Trefethen, James B. *An American Crusade for Wildlife* (1975).

UNPUBLISHED SOURCES

See memoirs (for the years 1912–1970, transcribed by Wildlife Management Institute, Washington D.C.) and papers in Denver, Colorado, Conservation Library.

Keir B. Sterling

GANNETT, HENRY. Born Bath, Maine, 24 August 1846; died Washington, D.C., 5 November 1914. Geographer, topographer, editor, geographical society executive, conservationist.

FAMILY AND EDUCATION

Son of Michael Farley and Hannah Church Gannett. Attended local schools before entering Harvard in 1865. Received a bachelor of science from the Lawrence Scientific School, Harvard, in 1869 and a mining engineering degree from the Hooper Mining School, Harvard, in 1870. Awarded an honorary LL.D. from Bowdoin College, 1899. Married Mary E. Chase of Waterville, Maine, 1874. His wife, two daughters, and a son survived him.

POSITIONS

Topographer, Western Division of the F. V. Hayden Survey (Rocky Mountain region), U.S. Geological and Geographical Survey of the Territories, 1872–1879. Geographer in charge, 10th U.S. census, 1879–1882; 11th U.S. census, 1890–1891; 12th U.S. census, 1900–1901. Chief geographer and topographer, U.S. Geological Survey, 1882–1914 (geographer in charge, Division of Geography and Forestry, U.S. Geological Survey, 1899–1914). Cofounder and member, National Geographic Society, 1888–1906, chairman of the Research Committee, National Geographic Society, 1906–1914, president, National Geographic Society, 1910–1914. Chairman, U.S. Board of Geographic Names (later, U.S. Geographic Board), 1894–1914. Assistant director, Philippine census, 1903; Cuban census, 1908. Cofounder and member, Association of American Geographers,

1904–1914. Secretary, Eighth International Geographic Congress, Washington, D.C., 1904–1905. Geographer and editor, National Conservation Commission, 1908–1909.

CAREER

Worked as an assistant at the Harvard Astronomical Observatory, 1870–1871, accompanying Edward Pickering to Spain in 1871 to observe a total solar eclipse. Topographer with Hayden of the U.S. Geological and Geographical Survey of the Territories, 1872. Mapped, named, and explored large areas of then-unfamiliar Wyoming, Colorado, and Utah, including the newly designated Yellowstone National Park. Returned to Washington, D.C., in 1879 to serve as geographer and statistician for the 10th U.S. census, a role he also held during the 11th and 12th U.S. censuses. Established an efficient and precisely designed system for census enumeration and produced comprehensive statistical atlases of the nation after each census. Began working for the U.S. Geological Survey in July 1882, under Director John Wesley Powell. Assigned task of coordinating the effort to produce a topographic map of the United States. Under his direction the U.S. Geological Survey mapped large portions of the United States, particularly the Appalachians, Great Plains, and Rocky and Cascade Mountains, occasionally visiting field parties at the various sites. Translated field data into useful maps. His keen cartographic skill and thorough attention to detail and organization earned him recognition as "father of American mapmaking." As chairman of the U.S. Board of Geographic Names, brought order to the chaotic process of establishing place-names in the United States. Role at U.S. Geological Survey reduced, first in 1890 due to budget cuts and political necessity, then again in 1896 as a result of survey reorganization by Director Charles Walcott. Traveled to Oregon and California, 1896, to supervise topographic work in those regions. Commissioned by Walcott, 1897, to prepare topographic surveys and resource inventories of the newly established national forest reserves. Prepared three extensive and comprehensive U.S. Geological Survey forest reports, 1897–1899, which included his own studies of forests of Washington state. (Reports hailed by Gifford Pinchot as "by far the best statement yet made of the forest condition of the United States.") Drew attention to devastation by forest fires and need for forest fire prevention, using economic arguments. Performed topographic work with the Harriman Alaska Expedition, 1899. Served in Philippines as geographer and assistant director of the census of the former Spanish possession, 1903, and was geographer and assistant director of the Cuban census, 1908. Edited *Proceedings of the Eighth International Geographic Congress*, 1904, and the *Report of the National Conservation Commission*, 1908–1909.

Published numerous articles and books containing geographical and statistical information about the United States and its possessions in the late-nineteenth and early-twentieth centuries. Noteworthy publications include *The Building of a Nation* (1895), "The United States," *Stanford's Compendium of Geography and Travel* 2 (1898), *Physiographic Types*, 2 vols. (1898–1900), *Commercial Geography* (with C. L. Garrison and E. J. Houston, 1905), and *Topographic Maps of the U.S. Showing Physiographic Types* (1907). Many of his articles published by the U.S. Geological Survey in thirty bulletins (1884–1906) and eighteen annual reports (1883–1900), which include dictionaries on elevations, boundaries, and place-names in the United States, gazetteers on various states and territories, and manuals on topographic methods. Surveys of forests of state of Washington found in *U.S. Geological Survey: Annual Reports 19–21* (1898–1900). Published several articles and statistical atlases describing demographic characteristics of the United States and its territories. Wrote several other articles on American forests and the need for conservation.

Publications marked by a high degree of thoroughness and clarity, though they emphasize compilation and organization of data rather than geographical theory. Incorporated much historical information into his writings. After 1897 became fairly active in forest conservation, though not as forceful politically as Pinchot. Always interested in making geography more comprehensible for the public while retaining high quality in his work.

MAJOR CONTRIBUTIONS

Raised quality of American cartography to a high level comparable to the level of work done in Europe. Brought attention to the extent of American forest resources and the need for forest conservation. Established comprehensive and efficient systems for census enumeration and designation of geographic place-names. Helped make physical and human geography more accessible to the general public. Helped establish forums for the interchange of geographical information and for interaction among professional geographers.

BIBLIOGRAPHY

American Men of Science. 1st, 2d eds.

Brown, Robert M. "Henry Gannett." *Dictionary of American Biography*. 1st ed. (1931).

Bulletin of the American Geographic Society 47 (January 1915).

Darton, N. H. "Memoir of Henry Gannett." *Annals of the Association of American Geographers* 7 (1917).

"Gannett, Henry." *Who's Who in America* (1912–1913).

Harvard University Quinquennial Catalog (1925).

"Henry Gannett." *National Geographic Magazine* 26 (December 1914).

Manning, Thomas. *Government in Science* (1967), Chapter 5, "Advancements in Topography: The National Map."

Michael J. Boersma

GANONG, WILLIAM FRANCIS. Born St. John, New Brunswick, 19 February 1864; died St. John, New Brunswick, 7 September 1941. Botanist, naturalist, historian.

FAMILY AND EDUCATION

Eldest of two sons of James H. Ganong, Canadian of loyalist descent, and Susan E. Brittain Ganong. Attended schools in St. John and nearby St. Stephen. B.A., 1884, M.A., 1886, Ph.D., Ad eundem gradum, 1898, and honorary LL.D., 1920, University of New Brunswick; A.B., Harvard, 1887; Ph.D., University of Munich, 1894. Married Jean Murray Garman, 1888 (died 1920); no children. Married Anna Hobbet, 1923; two children.

POSITIONS

Assistant and later instructor in botany, Harvard, 1887–1893; professor of botany and director of the Botanic Garden, Smith College, 1894–1932; professor emeritus, 1932–1941.

CAREER

Studied zoology at University of New Brunswick and at Harvard, where his interests shifted to botany. After teaching briefly at Harvard, studied botany at the University of Munich, 1893–1894. His dissertation, under Karl Goebel, concerned adaptations of cacti to extremes of heat and aridity. Established Department of Botany at Smith College in 1894. One of founders and first secretary of the Society for Plant Morphology and Physiology, later to merge with the Botanical Society of America. President, Botanical Society of America, 1908. In the 1890s and early 1900s devoted much attention to plant ecology, undertaking extensive research project on the salt marshes of the Bay of Fundy. Spent summers in New Brunswick woods, publishing notes concerning over 100 such excursions in the *Bulletin of the Natural History Society of New Brunswick*. Contributed over a third of material published in the *Bulletin* from 1884 to 1914. Became recognized authority on the natural history, geography, place-names, history, and cartography of New Brunswick and eastern Canada. Reproduced numerous maps by early Canadian explorers, particularly Jacques Cartier and Samuel de Champlain.

Published over 400 articles and five books, 1884–1937, including *A Laboratory Course in Plant Physiology, Especially as a Basis for Ecology* (1901), *The Living Plant: A Description and Interpretation of Its Functions and Structure* (1913), and "The Vegetation of the Bay of Fundy Salt and Diked Marshes; An Ecological Study," in *Botanical Gazette* (1903). Translated and edited "Book I. Acadia and New England," in *The Works of Samuel de Champlain*, vol. 1, ed. H. P. Biggar (1922). Many of Ganong's cartographic works were compiled by T. E. Laying in *Crucial Maps in the Early Cartography and Place-Nomenclature of the Atlantic Coast of Canada* (1964).

Although primarily plant physiologist, his ecological work was regarded highly by contemporaries such as Henry Chandler Cowles. An enthusiastic, popular, and demanding teacher. An experienced woodsman, his numerous articles on natural history of New Brunswick successfully combined a love of nature with extensive knowledge of botany, zoology, geography, and local history. A conservationist, particularly regarding forests of New Brunswick. Disliked the killing of animals in the woods; corresponded with Theodore Roosevelt concerning the misrepresentation of the habits of animals in popular fiction.

MAJOR CONTRIBUTIONS

Natural history and cartography of the Atlantic provinces of Canada, field research and publication in plant ecology.

BIBLIOGRAPHY

Smith, F. G., and H. A. Choate. *Science* 94 (1941) (obit.).
Webster, J. C. *Transactions of the Royal Society of Canada* (3d ser.) 36 (1942) (obit.).
Webster, J. C. ed. *William Francis Ganong Memorial* (1942).

UNPUBLISHED SOURCES

See J. A. Rayburn, "Bibliography of Books, Papers, Notes, and Reviews Written and Edited by W. F. Ganong (1864–1941)," Canadian Permanent Committee on Geographical Names, Fredericton, New Brunswick; letters, field notes, documents at the New Brunswick Museum Archives, St. John.

Eugene Cittadino

GARDEN, ALEXANDER. Born Birse, Scotland, January 1730; died London, England, 15 April 1791. Physician, naturalist, botanist.

FAMILY AND EDUCATION

Son of Scottish minister. Mother died in 1743, at which time Garden was apprenticed to James Gordon and studied medicine at Marischal College, Aberdeen, c.1743–c.1746. Continued medical studies at University of Edinburgh, where he developed love for botany while studying under Charles Alston. Received M.D. and A.M. from Marischal College, 1753. Moved to South Carolina, 1752. Married Elizabeth Peronneau, 1755; three children.

POSITIONS

Qualified as surgeon's second mate in British navy, 1746, but did not receive appointment. Probably returned to Aberdeen to continue apprenticeship with Gordon, 1746. Surgeon's first mate in British navy, 1748–1750. Physician, Prince William Parish, South Carolina, 1752–1754. Physician, Charlestown, South Carolina, 1755–1783. Actively engaged in natural history of South Carolina, particularly botany, but including zoology, mineralogy, and meteorology, 1752–1783. Member Royal Society of Arts, 1755; American Philosophical Society, 1768; Royal Society of London, 1773; and others. Was loyalist during Revolution and forced to return to England, 1782.

CAREER

From outset showed interest in natural history of South Carolina, partly as aid to medical practice. Traveled north to New York State, where he met botanist Cadwallader Colden, and to Philadelphia, where he visited botanist John Bartram, 1754. Entered international scientific circle, exchanging specimens and descriptions with American and European naturalists, including Linnaeus. Joined an expedition to Cherokee country west of Charles Town, 1755. Carried out medical experiments on anthelminthic properties of Indian pink (*Spigelia marilandica*) (work completed in 1757, although not published until 1771). Identified the mud iguana (*Siren lacertina*) as new genus, requiring establishment of new class of amphibia, *Sirendiae*, 1765. Discovered the two-toed Congo eel (*Amphiuma means means*), 1771.

Published three articles: ''An Account of the *Gymnotus Electricus*, or Electrical Eel . . . ,'' *Transactions of the Royal Society of London* 65 (1775): 102–10; ''An Account of the Indian Pink . . . ,'' *Essays and Observations, Physical and Literary* 3 (1771): 145–53 and plate; ''The Description of a new Plant . . . ,'' *Essays and Observations, Physical and Literary* 2 (1756): 1–7. Also had several descriptions published in other naturalists' articles.

Involved in discovery and description of some new species and genera, but given little credit since he published few articles. Major strength was in skill as collector. Respected by naturalists at home and abroad for diligence. Collections were always accompanied by catalog with location and habitat where specimens were found and descriptions based on dissection and microscopical studies.

MAJOR CONTRIBUTIONS

Some original work in taxonomical classification. Major contribution was in establishing contacts between European and New World naturalists and in supplying Europeans, most notably Linnaeus and British naturalist John Ellis, with descriptions and specimens of New World fauna and flora. Acquired ''a Species of Eternity'' when Ellis named gardenia after him, 1762.

BIBLIOGRAPHY

Berkeley, E., and D. S. Berkeley. *Dr. Alexander Garden of Charles Town* (1969).

Denny, M. ''Linnaeus and His Disciple in Carolina.'' *Isis* 38 (1948): 161–74.

Gee, W. ''S. Carolina Botanists . . .'' *Bulletin of the University of South Carolina* 72 (1918): 14–16.

Goodnight, C., and M. Goodnight. ''Alexander Garden: Physician and Naturalist.'' *Nature Magazine* 40 (1947): 525–26, 552.

Jenkins, P. G. ''Alexander Garden, M.D., F.R.S. (1728–1791).'' *Annals of Medical History* 10 (1928): 149–58.

Pitts, T. A. "The Life of Alexander Garden." *The Recorder* 22 (1959): 24–28.
Rees, A. "Garden." *The Cyclopedia* (1819).
Smith, J. E. *A Selection of the Correspondence of Linnaeus and Other Naturalists*. 2 vols. (1821).

Arleen M. Tuchman

GILCHRIST, FREDERICK CHARLES. Born Port Hope, Ontario, 20 April 1859; died Fort Qu'Appelle, Assiniboia (now Saskatchewan), 20 March 1896. Fisheries inspector, naturalist, taxidermist, farmer.

FAMILY AND EDUCATION

Son of Charles Gilchrist, a naturalist involved in the wild-rice business in Ontario, and his wife, Belle (maiden name unknown). A good scholar, Frederick planned to study medicine like his United Empire loyalist grandfather, but the death of his mother when he was eighteen and his father's remarriage changed these plans. Married Harriet Newbegin, a Methodist missionary teacher working among the Indians, in 1882. Had four daughters and two sons.

POSITIONS

Operated a store for three years (1880–1883) at Rice Lake, Ontario. Homesteaded as a farmer in 1883 near the present town of Dysart, Saskatchewan. In 1886 moved to Simpson's Point (now B-Say-Tah) on Echo Lake. He became fisheries overseer for Qu'Appelle River and Lakes on 15 September 1884. In May 1891 he was promoted to inspector of fisheries for the Northwest Territories, a federal appointment he held until his death.

CAREER

As Inspector of Fisheries, Gilchrist covered much of his territory by canoe. Interested in public affairs, he was also a justice of the peace, municipal councillor, school trustee, and Agricultural Society director.

Published two articles in *Forest and Stream* (18 March 1890) on "Sawdust in Ontario Streams" and (7 April 1892), "The Tullibee." Reports on fish were sent to the Smithsonian Institution in Washington, D.C. In a diary kept 1883–1893 and 1895–1896, he carefully recorded life on the prairies, including many observations about hunting and natural history. He died from either pneumonia or appendicitis.

MAJOR CONTRIBUTIONS

One of the earliest western Canadian settlers to keep a diary that recorded natural history observations and the first full-time fisheries inspector on the Canadian prairies.

BIBLIOGRAPHY
Bocking, D. H. "D. H. Gilchrist Diaries." *Saskatchewan History* 20 (1967).
Houston, Mary, and C. Stuart Houston. "F. C. Gilchrist's Diary—Fort Qu'Appelle 1883–1896." *Blue Jay* 24 (1966).

Mary I. Houston

GILL, THEODORE NICHOLAS. Born New York, 21 March 1837; died Washington, D.C., 25 September 1914. Ichthyologist, taxonomist, zoologist.

FAMILY AND EDUCATION

Son of James Darrell Gill and Elizabeth Vosburgh Gill. Mother died when he was nine. Educated by private tutors and attended private schools with classical curricula strong in Latin and Greek. Briefly studied law in firm of his uncle, R. A. Gaines, but never applied for admission to the bar. Later attended Wagner Free Institute of Science in Philadelphia. Received four honorary degrees from Columbian College (now George Washington University), A.M., 1865, M.D., 1866, Ph.D., 1870, LL.D., 1895. Never married.

POSITIONS

Adjunct professor of physics and natural history, Columbian College, 1860–1861, lecturer, 1864–1866, 1873–1884, professor of zoology, 1884–1910, professor emeritus, 1910–1914; librarian, Smithsonian Institution, 1862–1866; assistant librarian, Library of Congress, 1866–1875; associate in zoology, Smithsonian Institution, 1894–1914.

CAREER

Initial interest in ichthyology nurtured by daily visits to Fulton Fish Market on his way to and from school. In the mid-1850s met William Stimpson, who introduced him to Spencer F. Baird, who published a paper by Gill in Smithsonian annual report in 1857. Sent by Baird on expedition to the West Indies in 1858, where he made extensive collections of fishes. Collected in Newfoundland in 1859. Returned to Smithsonian to help prepare the report on collections of the Northwest Boundary Survey under Archibald Campbell. There he worked with other Baird protégés such as Stimpson, Robert Kennicott, Fielding B. Meek, and Ferdinand V. Hayden. Served as librarian of Smithsonian, 1862–1866, and assistant librarian of Congress, 1866–1875. Taught at Columbian College occasionally during the 1860s and 1873–1910. Associated with Smithsonian, where he maintained an office, for most of his life.

Published over 500 scientific papers, 1856–1908, mainly on ichthyology. Considered master in the art of taxonomy and classification, particularly of the higher groups. While still in his twenties he was characterized by Robert Kennicott as "a man of decidedly wonderful talent in the matter of classification and study of the higher groups—that is in investigation of genera, families, and orders, in contradistinction to the study of species. And if he would work harder and more patiently at the details he would be one of the finest zoologists of the age. As it is he ranks as about the best Ichthyologist of America." A few years later David Starr Jordan acknowledged Gill as his "master in fish taxonomy" and the "keenest interpreter of taxonomic facts yet known in the history of ichthyology." Showed little talent for classification at the species level but concentrated on the classification of families and orders of fishes on the basis of osteology. Influenced Jordan's classification of fishes. After about 1870, began publishing on mammals, amphibians, birds, and mollusks and general taxonomic issues, as well as fishes.

As a member of American Ornithologists' Union (AOU) worked closely with the Committee on Nomenclature and added greatly to the AOU *Code of Nomenclature* and *Check-List of North American Birds* in 1866 and revision of the code in 1906. Acquired *The Osprey* in 1898 and published it until its demise in 1902.

MAJOR CONTRIBUTIONS

System of classification of fishes was basis for the generally accepted scheme of classification used by most European and American ichthyologists for many years. Many of his conclusions still accepted. Synthesized the literature on the natural history of many fishes.

BIBLIOGRAPHY

Dall, William H. "Biographical Memoir of Theodore Nicholas Gill, 1837–1914." *National Academy of Sciences Biographical Memoirs* 8 (1914).

Dictionary of American Biography.

Hobbs, C. L. "Ichthyology in the U.S. after 1850." *Copeia* (1964).

Palmer, T. S. *Auk* 32 (1915) (obit.).

Shor, Elizabeth Noble. *Dictionary of Scientific Biography.*

Stone, Witmer. In T. S. Palmer, ed., *Biographies of Members of the American Ornithologists' Union* (1954) (obit.).

Who Was Who in America.

William A. Deiss

GILLISS, JAMES MELVILLE. Born Georgetown, District of Columbia, 6 September 1811; died Washington, D.C., 9 February 1865. Naval officer, astronomer, natural history collector.

FAMILY AND EDUCATION

Third child, first son of George Gilliss, federal government employee, and Mary Melville Gilliss. Details of early education not known. Student at University of Virginia, 1833, and in Paris, 1835. Married Rebecca Roberts, 1837. Had at least one son, but total number of children not known.

POSITIONS

Midshipman, U.S. Navy, 1826; Passed Midshipman, 1833 (some sources say 1831). Promoted lieutenant, 1838, commander, 1861, captain, 1862. With Naval Depot of Charts and Measurements, Washington, 1836–1842, in charge of depot, 1837–1842. In Paris securing equipment for new U.S. Naval Observatory, 1842–1843. Assigned to Naval Observatory, 1843–1846. With Coast survey, 1846–1848. Leader of naval astronomical expedition to the southern hemisphere, 1848–1852, in Chile, 1849–1852; with Naval Observatory, Washington, 1852–1865; studied solar eclipses in Peru, 1858, and in Washington Territory, 1860. Director of Naval Observatory, 1861–1865.

CAREER

Gilliss entered the navy at age fifteen, participated in the usual training while completing four cruises, 1826–1835. Studied briefly at University of Virginia while on special leave, 1833, but obliged to leave because of eye trouble. Studied in Paris, 1835, but recalled to Washington for duty with Office of Charts and Measurements, 1836, which he directed from 1837 until 1842. Spent much of this time making astronomical observations in support of the Wilkes Expedition of 1838–1842; also observed some 1,248 stars for the purpose of placing them as ''landmarks of the universe.'' Sent to Paris late in 1842 to secure equipment and books for the new Naval Observatory, returned 1843, but was not given anticipated directorship, which went to Lieutenant Matthew Fontaine Maury, who had been his successor as head of Charts and Measurements. Worked loyally under Maury, 1843–1846, spending most of his time preparing his *Astronomical Observations made at the Naval Observatory* (1846), first publication of this type to appear in the United States. With Coast Survey, 1846–1848. On basis of correspondence with Professor C. L. Gerling of Germany, proposed that more precise measurement of solar parallax could be made. Secured support of scientific organizations and scientists for this effort, and with the endorsement of Navy Secretary J. Y. Mason, $5,000 was appropriated in 1848 for the expedition to Chile which began in 1849. Additional funds were required for travel and other purposes, and Gilliss spent the better part of four years trying to meet the often conflicting requirements of three secretaries of the navy serving both Whig and Democratic administrations. Maury did not support the project and later failed to make arrangements for the necessary parallel observations at the Naval Observatory. Just why Maury chose not to cooperate is unclear, though he opposed ''sending . . . civilian professionals on overseas missions,'' and entertained certain anti-foreign views. In addition, he was interested in hydrography and wind patterns, not astronomy. Gilliss secured necessary equipment, including first astronomical lens of any size made in the United States, and spent three years (1849–1852) in Chile successfully completing the planned astronomical observations there, and recording information concerning earthquakes and magnetic force. A young man named W. Edmund Smith (some sources give his name as Edward R. Smith), was appointed naval captain's clerk. Gilliss sent him to Harvard and Yale to receive training in biology from Agassiz, James Dana, and Benjamin Silliman. He also was instructed in mineralogy and Spanish. While in Chile, Smith and three young Chileans appointed for the purpose by the Chilean government collected substantial numbers of birds, mammals, fish, reptiles, dried and growing plants, tree and flower seeds, fossils, shells, and minerals. Gilliss took some part in this activity as well and most of these specimens eventually ended up at the Smithsonian, with some plants, seeds, and bulbs going to William D. Brackenridge, who maintained a greenhouse filled with Wilkes Expedition plants for the Patent Office in Washington. With Brackenridge's aid, Gilliss arranged to have American tree and flower seeds sent to Chile in exchange for comparable Chilean plant material. In 1852, with the approval of authorities in Washington, Gilliss sold most of his equipment and the pre-fabricated observatory building he had brought with him from the United States to the Chilean government for a sum equal to their actual cost. The Chileans continued to operate the observatory he had established.

Returning to Washington, Gilliss spent several years drafting and overseeing an account of the work accomplished by members of his team. For the next several years, Gilliss and his colleagues completed six manuscript volumes of reports, of which only four were initially published. The first volume of *U.S. Astronomical Expedition to the Southern Hemisphere, During the Years 1849–'50–'51–'52* consisted of a report on Chilean geography, climate, natural resources, and society by Gilliss, and featured his personal and somewhat biased views on Chilean society. Volume II (1855) contained a travel account by another expedition member, together with the natural history information. The latter was made up of sections on dried plants by Asa Gray, the American botanist; living plants, seeds, and bulbs by William Brackenridge; and mammals by Spencer F. Baird. Unfortunately, information concerning the habitat of the mammals collected was often lacking. Volume III (1856) concerned an account of the astronomical work. Volume VI, which was actually the fourth to be published (1857), recorded magnetic and meteorological data. Publication of the remaining two volumes was delayed for nearly forty years, finally appearing as an appendix to a House Document (no. 219, in the 54th Congress, 1st Session, 1896).

Gilliss continued to correspond with his Chilean colleagues, sent books to them for some years, and finally arranged that they be included in regular Smithsonian exchanges. In 1855 Gilliss was placed on a "reserved list" of officers by a naval retiring board, on grounds that he had seen no sea duty for two decades, but he was continued on full pay to complete his Chilean reports. He traveled to Peru in 1858 and to Washington Territory in 1860, in both cases to observe solar eclipses, and reports of these experiences were published by the Smithsonian and Coast and Geodetic Survey, respectively. On the latter trip, he was accompanied by his son, a young Army officer. When Maury resigned in 1861 to join the Confederate Navy, Gilliss was elevated to the directorship of the Naval Observatory. During his four year tenure, he did much to reduce the backlog of work that had not been published, provided support for the manufacture of American-made astronomical equipment, and instituted cooperative ventures with other American observatories.

MAJOR CONTRIBUTIONS

In addition to his outstanding accomplishments as a pioneering American astronomer and director of the Naval Observatory, Gilliss was responsible for the collection of a number of natural history specimens from Chile which were among the first from Latin America to become part of the Smithsonian Institution collections.

BIBLIOGRAPHY

Dictionary of American Biography.

Gould, Benjamin A. Biography of Gilliss in *National Academy of Sciences Biographical Memoirs.* Vol. 1 (1877).

Harrison, John P. "Science and Politics: Origins and Objectives of Mid-Nineteenth Century Government Expeditions to Latin America." *Hispanic–American Historical Review* 35 (May 1955).

Kazar, John D. "The United States Navy and Scientific Exploration, 1837–1860." Unpublished doctoral dissertation, University of Massachusetts (1973).

National Intelligencer (Washington) 10 February 1865 (obit.).

Ponko, Vincent, Jr. *Ships, Seas, and Scientists: U.S. Naval Exploration and Discovery in the Nineteenth Century* (1974).

Rasmussen, Wayne D. "The United States Astronomical Expedition to Chile, 1849–1852." *Hispanic–American Historical Review* 34 (February 1954).

UNPUBLISHED SOURCES

Gilliss' correspondence is in the Navy Department Archives and in the Records of the U.S. Naval Astronomical Expedition to the Southern Hemisphere, National Archives, Washington. Gilliss' letters in L.A.J. Quetelet correspondence, American Philosophical Society (microfilm originals in Bibliotheque Royale, Brussels).

Keir B. Sterling

GLOVER, TOWNEND. Born Rio de Janiero, Brazil, 20 February 1813; died Baltimore, 7 September 1883. Entomologist, illustrator, federal government official.

FAMILY AND EDUCATION

Son of Henry Glover, an English businessman in Brazil, and Mary Townend Glover, originally from Yorkshire. His mother died when he was six weeks old, and his father six years later. Glover was then raised by relatives in Leeds, England, and was formally taken into the custody of his paternal grandfather, Samuel Glover, and a maiden aunt. Attended private school in Leeds. Worked for a year from age twenty as apprentice to local wool merchants but left when he received inheritance from father at twenty-one. Studied art in Munich with Mattenheimer, inspector of Munich Art Gallery, 1834–1836. Married Sarah Byrnes in Fishkill, New York, 1840 (later separated); one adopted daughter.

POSITIONS

Briefly operated painting studio in England. Traveled to United States, 1836, much taken with varied scenery, did some fishing and hunting, and decided to remain. Spent much of his time traveling and exploring various parts of the country. Managed father-in-law's estate for six years, purchased own farm in 1846. Ran it as showplace for fruits he raised; became adept at creating models of fruit that exhibited effects of insect depredation. These he displayed at agricultural fairs. Sold his collection to federal government for $10,000, 1853, went to work for nascent Bureau of Agriculture in U.S. Patent Office, 1854. He was given task of gathering information on fruits, seeds, and insects. His appointment marked onset of professional work in American entomology. He remained with bureau until 1859 and in that year accepted position as professor of natural sciences, Maryland Agricultural College (later University of Maryland). Was faculty member there for just under four years. Appointed first U.S. entomologist in Department of Agriculture, 1863. Held this post until retirement in 1878.

CAREER

Was much interested in floriculture, natural history, and taxidermy from the time of his arrival in the United States. Studied pomology after beginning career as farmer in 1840s, and this led to his idea of developing fruit models. At Maryland Agricultural College, began project of engraving insects on copper plates, a project that was continued for much of the rest of his life. While with the Bureau of Agriculture, traveled throughout southern United States and parts of South America, studying and reporting on insects. His papers on insects observed appeared in annual reports of the commissioner of patents. Continued fieldwork and collecting while at Maryland Agricultural College. Glover's tenure as first U.S. entomologist was taken up with numerous scientific studies, administrative duties, and development of museum within Department of Agriculture. His accounts of these activities were published in annual reports of Department of Agriculture. In 1860, he began work on lifetime project cataloging and illustrating all insects of North America. Unfortunately, he gave preference to illustrations rather than to collection of actual specimens, thus delaying assemblage of much-needed nationwide federal insect collection for a number of years. He also vacillated between addressing the needs of professionals and those of the general public. His great work was never completed, nor was it officially published. Between 1872 and 1878, five volumes of proofs in limited editions of no more than fifty copies (and sometimes fewer) were printed. These were entitled *Illustrations of North American Entomology in the Orders of Coleoptera, Orthoptera, Neuroptera, Hymenoptera, Lepidoptera, and Diptera*, and *Manuscript Notes from My Journal*. Illustrations were highly praised, but accompanying notes were deemed somewhat out of date.

MAJOR CONTRIBUTIONS

Glover was one of the first professional entomologists in the United States (along with Asa Fitch, the New York State entomologist) and was first to serve as U.S. entomologist. One of the first entomologists to discuss damage done to American agriculture by insects arriving in the United States on plants imported from abroad. Failed to spearhead federal efforts to counter insect depredations in the American West during the 1870s, something that awaited appointment of his successor, C. V. Riley, in 1878. Generally avoided formal publication of his findings; hence, others were often credited with new species he had first uncovered.

BIBLIOGRAPHY

Dictionary of American Biography.

Dodge, Charles R. *Life and Entomological Work of the Late Townend Glover* (1888).

Essig, E. O. *History of Entomology* (1931).

Howard, L. O. ''A History of Applied Entomology.'' *Smithsonian Miscellaneous Collections* 84 (1930).

Mallis, Arnold. *American Entomologists* (1971).

Sorensen, W. Conner. *Brethren of the Net: American Entomology, 1840–1880* (1995).

Walton, W. R. *Proceedings of the Entomological Society of Washington* (1921) (discussion of Glover's engravings).

UNPUBLISHED SOURCES

See papers and illustrations, Smithsonian Institution Archives.

Paul Cammarata

GODMAN, JOHN DAVIDSON. Born Annapolis, Maryland, 20 December 1794; died Philadelphia, 17 April 1830. Naturalist, editor, anatomist.

FAMILY AND EDUCATION

Son of Capt. Samuel Godman, Revolutionary officer, and Anna Henderson Godman. Mother died before he was two, and father before he was five. Cared for by aunt, who died when he was around six. Lived with sister in Baltimore. Apprenticed to Baltimore newspaper printer, winter 1811–1812. (Had first symptoms of tuberculosis during this period.) Joined the navy, 1814, and was present at Battle of Ft. McHenry. Began medical studies 1815 with William N. Luckey in Elizabethtown, Pennsylvania; completed studies with John B. Davidge in

Baltimore; graduated March 1818 from University of Maryland. While student served as a demonstrator in anatomy and gave some lectures in anatomy. He was proficient in Latin, Greek, French, German, and Italian, mostly self-taught. Married Angelica Kauffman Peale, daughter of Rembrandt Peale, 6 October 1821.

POSITIONS

First practiced medicine in New Holland, Pennsylvania; moved to village near Baltimore after a few months, anticipating an appointment as professor of anatomy at the University of Maryland. When that did not materialize, he moved to Philadelphia, where he lectured in anatomy and physiology. Joined Medical College of Ohio in Cincinnati as professor of surgery in October 1821 and resigned after one term because of dissent within the faculty. Then edited new journal, the *Western Quarterly Reporter of Medical, Surgical and Natural Science*, the first medical journal published west of the Alleghenies. Returned to Philadelphia in October 1822 and became private lecturer on anatomy and physiology. Wrote for, and became working editor in 1824 of, the *Philadelphia Journal of the Medical and Physical Sciences*. (Through his efforts, the name was changed four years later to the *American Journal of the Medical Sciences*.) He continued to write for it until shortly before his death. In 1826 he joined faculty of Rutgers Medical College in New York City and occupied the chair of anatomy but resigned in midsession because of worsening tuberculosis. In early 1828 he and his wife went to West Indies in vain attempt to arrest his illness. Returned to Germantown, Pennsylvania, where he spent the remainder of his life.

CAREER

Godman was a prolific writer and translator. Published *Anatomical Investigations* in 1824. His major work was three-volume *American Natural History* (1825–1828), in which several species of fossils were mentioned. He wrote natural history articles for the *Encyclopaedia Americana*, 1829–1833. In 1829 he published collection of his papers as *Addresses Delivered on Various Public Occasions*. In 1830, described a supposed new mastodon genus and species, *Tetracaulodon mastodontoideum*, later determined to be invalid, which began a bitter feud between Richard Harlan and Isaac Hays. Translated works include Levasseur's two-volume account of Lafayette's tour in America in 1824, 1825, 1829 and *Travels of the Duke of Saxe-Weimar*. He wrote a series of natural history articles for the *Friend*, weekly Philadelphia religious journal. They were based on observations made when he had lived in New Holland and near Baltimore and when he had walked in Germantown through Turner's Lane and along banks of the Frankford and Wissahickon Creeks. The articles were published in book form in 1833 as *Rambles of a Naturalist*.

MAJOR CONTRIBUTIONS

Accurate observations on natural history subjects. His *American Natural History* was first descriptive work on mammals by an American. His charming and descriptive writing about animals such as crayfish, shrew-moles, and crows makes him one of America's earliest environmentalist writers.

BIBLIOGRAPHY

Drake, Daniel. *Western Journal of the Medical and Physical Sciences* (January, February, March 1831).

Godman, John Davidson. *American Natural History*. 3 vols. Philadelphia (1825, 1826, 1828).

Godman, John Davidson. "Description of a New Genus and Species of Extinct Mammiferous Quadruped." *Transactions of the American Philosophical Society* 2, Part III (1830).

Godman, John Davidson. "Rambles of a Naturalist." (1833). (contains reprint of biography by Drake, 1831).

Gross, Samuel D. *Lives of Eminent American Physicians and Surgeons of the Nineteenth Century* (1861).

Gross, Samuel D. *Autobiography* (1887) (edited by his sons).

Miller, William Snow. "John Davidson Godman." *Dictionary of American Biography*. Vol. 7 (1931).

Morris, Stephanie. "John Davidson Godman (1794–1830): Physician and Naturalist." *Transactions and Studies of the College of Physicians of Philadelphia* (4th ser.) 41 (April 1974).

Carol Faul

GOLDMAN, EDWARD ALPHONSO. Born Mount Carmel, Illinois, 7 July 1873; died Washington, D.C., 2 September 1946. Field naturalist, mammalogist.

FAMILY AND EDUCATION

Son of Jacob Henry and Laura Carrie Nicodemus Goldman. Parents were former Pennsylvania farmers who had migrated to Illinois and Nebraska before settling in Tulare County, California. Father changed family name from Goltman to Goldman due to pronunciation difficulties encountered by neighboring farmers. Educated in public schools and later took college courses to aid career but never received degree. Married Emma May Chase, 1902. Three children, including Luther C. Goldman, an ornithologist with the U.S. Fish and Wildlife Service.

POSITIONS

Field naturalist, Division of Economic Ornithology and Mammalogy (after 1896, Bureau of Biological Survey), U.S. Department of Agriculture, 1892–1917; biologist in charge, Division of Biological Investigations, Bureau of Biological Survey, 1919–1925; chief, Division of Game and Bird Reservations, Bureau of Biological Survey, 1925–1928; senior biologist, Division of Wildlife Research, Bureau of Biological Survey (after 1939, U.S. Fish and Wildlife Service, U.S. Department of Interior), 1928–1943; associate in zoology, Smithsonian Institution, 1943–1946.

CAREER

His father, an amateur naturalist, influenced early interest in natural history. Began collecting birds and mammals as young boy. At age seventeen, began working as foreman in a vineyard near Fresno, California. Received first appointment with Bureau of Biological Survey on the recommendation of Edward William Nelson, who had hired him as his personal assistant in 1891. Conducted extensive biological surveys of Mexico with Nelson, 1892–1906, which have been described as among most important ever achieved by two workers in any single country. Collected and studied fauna in every state and territory of Mexico. Added over 30,000 specimens of Mexican animals to the Biological Survey's collections, including 354 species and subspecies of mammals described as new. A summary of his Mexican fieldwork with Nelson was published posthumously under the title *Biological Investigations in Mexico* (1951). During construction of the Panama Canal he worked on biological survey of the Isthmus sponsored by Smithsonian and Departments of War and Agriculture, 1911–1912. Results were published in *Mammals of Panama* (1920). Conducted field investigations in Arizona, 1913–1917. Served with American Expeditionary Forces in France, 1918–1919, as major in the Sanitary Corps in charge of rodent control work. Major, Sanitary Reserve Corps, U.S. Army, 1922–1937. Administrative duties superseded major fieldwork and research, 1919–1928. Relieved of administrative work in 1928. Assisted in negotiations between United States and Mexico for the protection of migratory birds and game mammals, 1936. As a result, Mexican officials stated that he was ''an authority on Mexican fauna . . . more familiar with the subject than were the Mexicans themselves.''

Published 206 books and papers, 1902–1946, including *Revision of the Wood Rats of the Genus Neotoma* (1910); *Revision of the Spiny Pocket Mice (Genera Heteromys and Liomys)* (1911); *Rice Rats of North America (Genus Oryzomys)* (1918); *Revision of the Jaguars* (with E. W. Nelson, 1933); *Revision of the Pocket Gophers of the Genus Orthogeomys* (with E. W. Nelson, 1934); *Review of the Spider Monkeys* (with A. Remington Kellogg, 1944); *The Wolves of North America* (with Stanley P. Young, 1944); and *The Puma, Mysterious American Cat* (with Stanley P. Young, 1946).

Skilled at writing for the general reader as well as specialist. Was outstanding outdoor photographer. Described as the ''Noah'' of the Biological Survey due to large number of mammals, birds, plants, reptiles, and invertebrates named for him—a total of more than fifty. Described over 300 forms of mammals, mostly subspecies. Goldman Peak, in Baja California, named in his honor.

MAJOR CONTRIBUTIONS

Considered a leading authority on mammals of middle and North America. Contributed substantially to our understanding of the fauna of Mexico. Influenced legislation protecting migratory birds and mammals.

BIBLIOGRAPHY

Palmer, Theodore S. *Auk* (July 1947) (obit.).
Young, Stanley P. *Journal of Mammalogy* (May 1947) (obit.).
Young, Stanley P. *Journal of the Washington Academy of Sciences* (January 1947) (obit.).
Who Was Who in America. Vol. 2 (1966).

UNPUBLISHED SOURCES

See Smithsonian Institution Archives, Record Unit 7176; U.S. Fish and Wildlife Service, 1860–1961, Field Reports (includes field notes on mammals, birds, and plants submitted by Goldman; especially well represented are notes documenting his Mexican field investigations with E. W. Nelson).

William Cox

GOODE, GEORGE BROWN. Born New Albany, Indiana, 13 February 1851; died Washington, D.C., 6 September 1896. Museologist, ichthyologist, naturalist, government administrator.

FAMILY AND EDUCATION

Son of successful merchant and amateur naturalist, Francis Collier Goode, and Sarah Woodruff Crane Goode; latter died when George Brown Goode was eighteen months old. Father then married Sally Ann Jackson of Amenia, New York, where Francis Collier Goode retired in 1857, suffering from poor health. George Brown Goode was educated by private tutors in Amenia until he entered Wesleyan University, 1866. Graduated from Wesleyan, 1870, later awarded honorary LL.D., 1893. Attended Harvard as special student, 1870–1871; studied under Louis Agassiz, who was also a regent of the Smithsonian Institution. While working as volunteer for U.S. Fish Commission in Eastport, Maine, in 1872, Goode became protégé of Spencer F. Baird. Goode married Sarah Ford Judd, daughter of Orange Judd, publisher of *The American Agriculturist* and the *Prairie Farmer*, 1877; four children.

POSITIONS

Curator of the Museum, Wesleyan University, 1871–1877. Assistant curator and subsequently curator, Smithsonian Institution, 1873–1881. Assistant director, U.S. National Museum, 1881–1887. Assistant secretary, Smithsonian Institution, 1887–1896. U.S. commissioner, International Fisheries Exposition, Berlin, Germany, 1880. U.S. commissioner, London Fisheries Exposition, London, England, 1883. Representative of the Smithsonian Institution at the following exhibitions: Louisville, Kentucky, 1884; New Orleans, 1885; Cincinnati, 1888; Chicago, 1893; Atlanta, Georgia, 1895. U.S. commissioner to the Columbian Exposition, Madrid, Spain, 1892. Member National Academy of Sciences (1888–1896), American Philosophical Society, American Society of Naturalists; fellow of the American Academy of Arts and Sciences; elected to membership in Société des Amis des Sciences Naturelles de Moscou, Société Zoologique de France, Zoological Society of London, and Société Scientifique du Chile. Member of Council of the American Historical Association, vice president general of Sons of the American Revolution and commander of the order of Isabella the Catholic (Spanish government award). Served as president of both the Biological Society and Philosophical Society of Washington, D.C.

CAREER

Working career began in 1871 as curator of the natural history collection of Wesleyan University. In 1872, volunteer for the U.S. Fish Commission. In 1873 was appointed assistant curator at the Smithsonian Institution, where he was employed for remainder of his life. After being promoted to the position of curator, he became the first assistant director of U.S. National Museum when it was organized in 1881. In 1887 became assistant secretary of Smithsonian Institution in charge of U.S. National Museum. Goode's first publication was as an undergraduate in 1869, when he prepared brief article for the *College Argus II.* His first descriptive paper was "The Catalogue of the Fishes of the Bermudas" (1876). Most distinguished scientific work, *Oceanic Ichthyology* (1896). The bibliography of Goode's works, published in the annual report of the Smithsonian Institution for 1897, lists 196 papers, sixteen reports of the U.S. National Museum, fifty-one papers by Goode and others, and twenty-three papers edited by Goode. Of these last, the largest was "The Fisheries and Fishery Industries of the United States (1884–1887)," part of the report of the tenth census. This work ran to seven volumes containing 3,931 pages and 581 plates and charts. Also, Goode reviewed six papers by others. While great majority of these writings were on scientific subjects, particularly ichthyology, others were on such diverse subjects as photography, the history of the Smithsonian Institution, museology, the history of science in America, historiography, and genealogy. Goode supervised preparation and installations of the Smithsonian's large display at the Centennial Exposition held in Philadelphia in 1876 and performed similar work for Smithsonian participation in all the expositions previously noted. Demonstrated extraordinary administrative skills at this work and was one of world's leading figures in the theory of museum administration and management, as represented in his publications, "Museum History and Museums of History" (1889), "First Draft of a System of Classification for the World's Columbian Exposition" (1890), "The Museums of the Future" (1891), and "The Principles of Museum Administration" (1895). The most insightful and seminal of Goode's writings on museology is, however, the earliest. His "Plan of Organization and Regulations, Circular No. 1, U.S. National Museum'' (1881) was written when he was but thirty years of age. Goode's reputation as historian of science rests upon his ''The Beginnings of Natural History in America'' (1886), ''The Beginnings of American Science'' (1886), ''The Beginnings of American Science—The Third Century'' (1887), ''The Origins of the National Scientific and Educational Institutions of the United States'' (1890), and ''The Genesis of the National Museum'' (1893). He also edited and contributed to *The Smithsonian Institution, 1846–1896* (1897). Interested in his family's history since boyhood, Goode in 1887 published his 526–page *Virginia Cousins*, a Goode family genealogy that stands as a masterpiece of its kind. Typically, Goode wrote of it, ''I am sorry that the book is not better and more accurate, but time for this work has been stolen from leisure hours.''

Goode was described in 1891 as ''a slender, dark complexioned man of unassuming manner and wearing a full beard. He is nervous in his actions and smokes cigarettes continually and does not cease even to converse.'' Associates were struck by his highly developed moral and ethical character and his unusually sensitive understanding of colleagues and subordinates. Tremendous drive and capacity for work exceeded his physical stamina. Suffered physical collapse following his work on the Centennial of 1876, from which he never fully recovered. His work and actions were characterized by the most minute attention to detail. Regarded by his peers as an unusually capable administrator and as near-genius for his organizational ability.

MAJOR CONTRIBUTIONS

The development of a philosophy and rationale for functions of the research museum and the use of museum collections and exhibitions for study of history as well as science and for the purposes of public education.

BIBLIOGRAPHY

Annual Report of the U.S. National Museum for 1897 (1901).

Chicago News, 1 October 1891.

Lindsay, G. Carroll. ''George Brown Goode.'' *Keepers of the Past* (1965).

Oesher, Paul H. *The Scientific Monthly* (March 1948).

UNPUBLISHED SOURCES

Author interviewed Charles G. Abbott, late secretary of the Smithsonian Institution and contemporary of Goode.

G. Carroll Lindsay

GOSSE, PHILIP HENRY. Born Worcester, England, 6 April 1810; died St. Marychurch, Devon, 23 August 1888. Naturalist, writer, artist.

FAMILY AND EDUCATION

Son of Thomas Gosse, an impoverished miniature-painter, and Hannah Best, his practical, strong-willed wife. Family settled in Poole, Dorset, in 1812, where he had a fair education in classics and mathematics and developed interest in natural history. Went to Newfoundland at seventeen as clerk to shipping firm and in his eight-year stay ''suddenly and conscientiously became a Naturalist and a Christian.'' Married Emily Bowes, 1848 (died 1857), one son; Eliza Brightwen, 1860.

POSITIONS

Through most of his career was self-employed, as author, illustrator, collector, teacher, and lecturer; at various times held part-time appointments at the British Museum.

CAREER

Farmed and taught school near Compton, Quebec, 1835–1838; was corresponding member of Natural History Society of Montreal and Literary and Historical Society of Quebec; accumulated material for his first book. Traveled in United States and taught in backwoods school in Alabama 1838–1839; completed script of *The Canadian Naturalist* during return to England. Its publication in 1840 rescued him from extreme poverty and introduced him to fellow naturalists.

Went to Jamaica on collecting trip October 1844–June 1846, where his own observations were supplemented by notes and drawings of Anthony Robinson and by close collaboration with Richard Hill. His resulting books, *The Birds of Jamaica* (1848), *Illustrations of the Birds of Jamaica* (1849), and *A Naturalist's Sojourn in Jamaica* (1851), broke new ground and confirmed his reputation. According to David Lack (1976), the first of these ''was far ahead of its time and remained one of the best bird books on any part of the world for at least half a century.'' His subsequent work, entirely in Britain, included scholarly and detailed studies on rotifers and on marine life, especially of the intertidal zone. He was elected fellow of the Royal Society in 1856 for this and his development of seawater aquaria. He also wrote many ''popular'' books on various scientific topics for the Society for Promoting Christian Knowledge (SPCK). In a profound depression after his first wife's death, he tried in *Omphalos: An Attempt to Untie the Geological Knot* (1857) to resolve the conflict between the new sciences and traditional religious beliefs. This paradoxical behavior on the part of ''the finest descriptive naturalist of his day'' has been sympathetically analyzed by S. J. Gould (1985). Some of his books reach the same high standard as his Jamaican ones: *A Naturalist's Rambles on the Devonshire Coast* (1853), *The Aquarium* (1854), *The Ocean* (1860), *A Year at the Shore* (1865). Though he published little after 1865 and enjoyed semiretirement with his orchids and his cats, his continuing work on rotifers made up a large part of a two-volume work, *The Rotifera*, published jointly with C. T. Hudson (1886). Gosse's life alternated between periods of intense productivity and morbid introspection. His son thought of him as essentially an artist, extraordinarily talented with pen and brush but thwarted by his rigid theological preoccupations from giving full rein to his innate sensitivity. But he was far more than "an honest hodman of Science," as Huxley termed him. Today we recognize him as one of first descriptive zoologists to penetrate beyond taxonomy and morphology into the field of animal behavior.

MAJOR CONTRIBUTIONS

Gosse's books on Canada and Jamaica set the pattern for much of his later work and established value of field observations as contrasted with custom of the time, when "Natural History [was] far too much a science of dead things: a necrology" (Preface to *A Naturalist's Sojourn in Jamaica).* Stephen Jay Gould calls him "the David Attenborough of his day, Britain's finest popular narrator of nature's fascination."

BIBLIOGRAPHY

Gosse, Edmund. *The Life of Philip Henry Gosse, F.R.S.* (1890).
Gosse, Edmund. *Father and Son, a Study of Two Temperaments* (1907).
Gould, S. J. *The Flamingo's Smile, Reflections in Natural History* (1985), Chapter 6.
Lack, David. *Island Biology Illustrated by the Land Birds of Jamaica* (1976).
Stageman, P. *A Bibliography of the First Editions of Philip Henry Gosse, F.R.S.* (1955).
Stewart, D. B., Philip Henry Gosse, and his collaborators. *Gosse's Jamaica 1844–45* (1984), Appendix 2.

UNPUBLISHED SOURCES

See P. H. Gosse, *A Voyage to and Residence in Jamaica 1844–46*, manuscript in National Library of Jamaica, East Street, Kingston, Jamaica.

David B. Stewart

GOULD, AUGUSTUS ADDISON. Born New Ipswich, New Hampshire, 23 April 1805; died Boston, 15 September 1866. Naturalist, conchologist, physician.

FAMILY AND EDUCATION

Son of Nathaniel Duren and Sally Prichard Gould. Father was a farmer, musician, music teacher, and engraver of modest circumstances. As a child, Gould worked on his father's farm and attended common school. At age fifteen, he took charge of the farm. Attended New Ipswich Appleton Academy. Graduated Harvard, 1825, having to work his way through school. Employed for a year as a private tutor in Maryland. Studied medicine with James Jackson and Walter Channing at Massachusetts General Hospital and graduated Harvard Medical School, 1830. Married Harriet Cushing Sheafe, 1833, and had ten children, of whom seven lived to maturity.

POSITIONS

Earned living as a practicing physician in Boston. Physician at Massachusetts General Hospital, 1857–1866. Was active in medical societies. Taught botany and zoology at Harvard College in 1835 and 1836.

CAREER

Became interested in natural history as a student and devoted his leisure to this science, specializing in conchology. Was an original member of Boston Society of Natural History, founded 1830, and served as a curator, recording secretary, 1838–1839, and second vice president, 1860–1866. Fellow of the American Academy of Arts and Sciences, 1841. Charter member of the National Academy of Sciences, 1863.

Prepared report on mollusks for the Massachusetts Zoological and Botanical Survey, 1837, published as *Report on the Invertebrata of Massachusetts, Comprising the Mollusca, Crustacea, Annelida, and Radiata* (1841). Described about 275 mollusks and 100 crustacea and radiates and drew the illustrations himself. Report long remained a standard work and did much to promote interest in conchology. According to Jeffries Wyman, it "gave him an honorable name among the naturalists of Europe and America and so he attained to eminence." A preliminary report on this work, presented to the Boston Society of Natural History (1841), is noteworthy for its consideration of geographical distribution. Showed that Cape Cod acted as a natural barrier to many species and genera of mollusks. W. H. Dall wrote that publication of the report initiated the "Gouldian Period" of American conchology, characterized by concern with the anatomy of soft parts, geographical distribution, and the precise discrimination of forms. Edited and completed *The Terrestrial Air-Breathing Molluscs of the United States*, 2 vols. (1851–1857), left unfinished by his friend, Amos Binney.

His work on the mollusks collected by Joseph P. Couthouy during the U.S. Exploring (Wilkes) Expedition, 1838–1842, was considered his greatest contribution to natural history. Shells of the expedition were sent to Boston for him to describe. To his regret, he was not allowed to dissect specimens and had to limit himself to description of genera and species based on the shell. Described 443 new species. The introduction to the volume *United States Exploring Expedition, vol. 12, Mollusca and Shells* (1852) contained original observations on geographical distribution. Showed that shells were distributed according to regions and that each region had a distinctive style of shell. Described several collections of shells made by army officers during the Mexican War period. Described the collections of the North Pacific Exploring Expedition, 1853–1855, traveling to Europe in 1857 to compare shells with those in European collections. His *Otia Conchologica: Description of Shells and Mollusks, from 1839 to 1862* (1862) reprinted all of his descriptions of new species of shells. In his final years, was working on a revision of his 1841 report for Massachusetts. Other works include a translation from the French, *Lamarck's Genera of Shells* (1833), his first work; *Principles of Zoology* (1848), a college textbook written in collaboration with Louis Agassiz (q.v.); and *The Naturalist's Library* (1851), a popular compilation. Published in addition over 100 papers, many in the *Proceedings of the Boston Society of Natural History*. Was remembered for his industriousness, geniality, and simplicity of manner. Was a religious man and an active member of the Baptist Church.

MAJOR CONTRIBUTIONS

Foremost American conchologist of his day. Described all known mollusks in Massachusetts and described and classified the shells of two major government exploring expeditions. In all, described some 1,100 new species of mollusks. Took a philosophical view of natural history, recognizing the importance of geographical distribution, comparative anatomy, and paleontology to the study of mollusks. One of the earliest naturalists in America to be occupied with problems of geographical distribution.

BIBLIOGRAPHY

Baker, Frank C. *Dictionary of American Biography.*

Gifford, G. E., Jr. *Dictionary of Scientific Biography.*

Gifford, G. E., Jr. "The Forgotten Man of the Ether Controversy." *Harvard Medical Alumni Bulletin* (Christmas 1965).

Johnson, R. I. *The Recent Mollusca of Augustus Addison Gould.* Bulletin 239, *Smithsonian Institution, Museum of Natural History* (1964).

Wyman, Jeffries. "Augustus Addison Gould," *Biographical Memoirs of the National Academy of Sciences* (1905) (obit.; additions by William Healey Dall, with bibliography).

UNPUBLISHED SOURCES

Manuscripts are in Boston Museum of Science; Houghton Library, Harvard University; and Countway Library, Harvard Medical School.

Toby A. Appel

GRAHAM, ANDREW. Born, probably near Edinburgh, Scotland, c.1733; died Prestonpans, Scotland, 8 September 1815. Fur trader, naturalist.

FAMILY AND EDUCATION

Early education unknown. Married Patricia Sherer 6 May 1770 in Edinburgh, while home on his first furlough, and spent four weeks with bride before returning to Hudson Bay. He also had a native wife at Hudson Bay; he took his daughter, born about 1769, back to England on second furlough in 1775–1776. A son was born, presumably at Severn House, about 1773. The date of Graham's Edinburgh marriage to Barbara Bowie is not known, but she died in 1812 or 1813. His will left £700 in trust for "Andrew Graham, the son of Miss Menzie Waterstone," perhaps a natural or even an adopted son.

POSITIONS

A servant of the Hudson's Bay Company as follows: servant to the master of the sloop *Churchill*, sailing up and down Hudson Bay, 1749–1752; assistant writer at York Factory, 1753–1759; bookkeeper, trader, and second-in-command at York Factory, 1759–1761. Master at Severn House, 1761–1764; at York Factory, 1765–1766; at Severn, 1766–1769 and 1770–1771; at York Factory, 1771–1772, the only time he had a fellow scientist (Thomas Hutchins) with him to share his interests. Master at Severn, 1772–1774, and at Churchill from February 1775 until relieved by Samuel Hearne. Graham returned permanently to Edinburgh in late August 1775. From late 1786 through early 1791 Graham acted as agent for the Hudson's Bay Company in paying, from Edinburgh, the company's servants in the Orkneys.

CAREER

Among the sixty-four skins of thirty-nine bird species sent in 1771 by Graham from Severn were the great gray owl, *Strix nebulosa*; boreal chickadee, *Parus hudsonicus*; blackpoll warbler, *Dendroica striata*; and white-crowned sparrow, *Zonotrichia leucophrys*. These and one fish, the longnose sucker, *Catostomus catostomus*, were given their definitive Latin names by Johann Reinhold Forster in 1772. At York Factory in 1771–1772, Graham wrote important accounts of native peoples, birds, and mammals of Hudson Bay, including the first description of the "plunge-holes" made by the great gray owl in catching mice beneath deep snow. His natural history observations included forty-one mammals and seventeen fish, as well as 111 birds, to which Hutchins added another 12. The contributions of the two men, a genuine collaboration, have now been sorted out.

MAJOR CONTRIBUTIONS

One of the first naturalists in North America to collect specimens (including five new species) and provide observations on bird behavior.

BIBLIOGRAPHY

Houston, C. S. "Birds First Described from Hudson Bay." *Canadian Field-Naturalist* 97 (1983).

Houston, C. S. *18th-Century Naturalists of Hudson Bay*. Toronto: University of Toronto Press (in press).

Williams, G. *Andrew Graham's Observations on Hudson's Bay, 1767–91* (1969).

Williams, G. "Andrew Graham and Thomas Hutchins: Collaboration and Plagiarism in 18th-Century Natural History." *The Beaver* 308, 4 (1978).

C. Stuart Houston

GRANT, MADISON. Born New York City, 19 November 1865; died New York City, 30 May 1937. Lawyer, zoologist, genealogist, eugenicist.

FAMILY AND EDUCATION

Oldest of four children, his father, Gabriel Grant, as military surgeon, won Congressional Medal of Honor at Battle of Fair Oaks in 1862. Mother, Caroline Amelia Manice of Jamaica, New York, came from well-to-do background. Was educated in private schools in New York City; then with brother, De Forest, was tutored for several years in Dresden, Germany. Attended Yale University and received bachelor of arts in 1887; and LL.B. from Columbia University in 1890. Never married.

POSITIONS

Practiced law in New York City, but private means allowed him to shift his major priorities to animal and natural world. In 1893, joined the Boone and Crockett Club. With Theodore Roosevelt as president and one of founders, Grant became secretary. At time of his death, was president of this conservationist organization. Played leading role in organization of New York Zoological Society in 1895, along with Theodore Roosevelt, Henry Fairfield Osborn, Elihu Root, C. Grant La Farge, and others. Served as its secretary (1895–1924), chairman of Executive Committee (1908–1936), and president (1925–until death). In 1905, also helped establish American Bison Society. In 1919, along with Henry Fairfield Osborn and John C. Merriam, founded the Save the Redwoods League to preserve remaining groves in California. In 1922, became a vice president of Immigration Restriction League and retained position until his death.

CAREER

As friend and colleague, Henry Fairfield Osborn suggested law was not major preoccupation with Grant. Major concerns were animal life, the natural world, and changing nature of American people. Early in life, had participated as hunter and explorer in forests of North America. Not interested in hunting game, but rather in discovering new species. Name has been given to a unique form of caribou on Alaskan peninsula, now called in his honor *Rangifer granti*. Principal goal was to create wilderness preserve in the Bronx, so that future generations would have available live specimens of animals beginning to become extinct. In writings, tried to warn readers of need to protect and preserve natural world from ravages of progress and to recall what he viewed as America's traditional values being endangered by new immigrant groups then inundating America's shores. His publication ''Distribution of the Moose,'' originally part of *Seventh Annual Report of the Forest, Fish and Game Commission of the State of New York*, was given circulation in George Bird Grinnell's *American Big Game in Its Haunts: The Book of the Boone and Crockett Club* (1904). In it, Grant discussed history of relationship between the elk and moose from Pliny through medieval period, to colonial Vermont and New Hampshire, on to this day. Theme was decline of the moose and its habitat. In another article that appeared in George Bird Grinnell and Charles Sheldon's *Hunting and Conservation: The Book of the Boone and Crockett Club* (1925), Grant provided article, ''Saving the Redwoods.'' In article Grant saw Boone and Crockett Club as an agency that would help preserve natural habitat for game and maintain, as part of America's heritage, natural resources such as redwoods for future generations. Earlier had written article on ''Conditions of Wild Life in Alaska'' (1913) in another of Boone and Crockett Club publications. In 1894 with brother, De Forest, actively engaged in effort to elect William Lafayette Strong as reform mayor of New York City. Strong's election made possible creation of New York Zoological Park in the Bronx. To make Bronx Zoo accessible, Grant became leader in planning construction of Bronx River Parkway to reach the park. Zoo was first of its kind in any large eastern city. Became president of the Bronx Parkway Commission. Also became a member, in 1924, of Taconic Park Commission. In 1929, was presented gold medal by the Society of Arts and Sciences for his active role in providing many delightful bridges in Bronx Park. Was the author of *The Origins and Relationships of the Large Mammals of America.* Another of his interests, one that has overshadowed role as naturalist, was in field of eugenics. Was one of eight members of International Committee of Eugenics and member of the American Defense Society. Published *The Passing of the Great Race* (1916), which sold 16,000 copies. Also wrote introduction to Lothorp Stoddard's *The Rising Tide of Color* (1920). A member of the Society of Colonial Wars, helped mark colonial battlefields and came to be considered expert genealogist. In will, he left $25,000 to the New York Zoological Society to continue its work of wildlife protection. In addition, provided $5,000 to the American Museum of Natural History and additional $5,000 to Boone and Crocket Club. Published dozens of articles on large mammals. Also wrote *The Founders of the Republic on Immigration* (1928), edited by C. S. Davison, and *The Alien in Our Midst* (1930), also a compilation edited with help of Davison. Final book, *The Conquest of the Continent* (1933), traced role of racial elements in the settlement of America.

MAJOR CONTRIBUTION

While most historians stress his nativistic and anti-immigrant activities, major contribution to the environment was his role in organizing, establishing and maintaining Bronx Park Zoo. It provided example that was followed by growing number of cities.

BIBLIOGRAPHY

Bridges, William. *Gathering of Animals: An Unconventional History of the New York Zoological Society* (1974).

''Grant Will Aids Zoo.'' New York Times, 5 June 1937.

Horowitz, Helen. ''Animal and Man in the New York Zoological Park.'' *New York History* 56 (1975).

Ludmerer, M. *Genetics and American Society: An Historical Appraisal* (1972).

''Madison Grant.'' *New York Times*, 3 June 1937 (obit.).

UNPUBLISHED SOURCES

See Ellsworth Huntington Papers (1876–1947), Yale University Library (includes a number of letters from Grant to Huntington, one of the founders of the American Eugenics Society); William Temple Hornaday Papers (1876–1960), Library of Congress.

Thomas J. Curran

GRAY, ASA. Born Sauquoit, New York, 18 November 1810; died Cambridge, Massachusetts, 30 January 1888. Botanist, educator.

FAMILY AND EDUCATION

Son of Moses Gray, tanner-farmer, and Roxana Howard Gray. Oldest of five children. Family moved to Paris Furnace, New York, shortly after Gray's birth and then back to Sauquoit in 1823. Attended Clinton (New York) Grammar School and Fairfield (New York) Academy. Graduated Medical College of the Western District of New York, M.D., 1831. Honorary degrees from Oxford (D.C.L., 1887), Cambridge (D.Sc., 1887), Endinburgh (LL.D., 1887). Married Jane Lathrop Loring, 1848; no children.

POSITIONS

Practiced medicine, Bridgewater, New York, 1831–1832. Science teacher in various New York academies, 1832–1835. Curator, New York Lyceum of Natural History, and coworker with John Torrey. Botanist with U.S. Exploring Expedition (Wilkes Expedition), 1836–1838, resigned before sailing. Professor of botany, University of Michigan, traveling in Europe to collect books for library, 1838–1842. Worked with Torrey on *Flora of North America*, 1839–1842. Fisher Professor of Natural History, Harvard University, 1842–1888. Retired from teaching but kept title, 1873.

CAREER

From 1832 until his acceptance of Fisher Professorship, worked closely with Torrey, collecting on East Coast, examining collections in Europe, 1838–1839, and writing their *Flora.* At Harvard as "the only adequately supported professional botanist in the United States," taught undergraduate and occasionally advanced students, supervised botanic garden, and described, classified, and published on American specimens, primarily collected by others. *Manual of the Botany of the Northern United States* (1848 and five more editions in his lifetime) filled need for one-volume flora of flowering and some lower plants using natural system of classification. Study on species and genera common to both eastern Asia and eastern North America concluded they were not separate creations but descendants of common stock that moved south and separated during glaciation. Led the fight to guarantee a fair scientific hearing for the *Origin of Species.* Was principal professional opponent of his Harvard colleague Louis Agassiz in the debate that followed publication of the *Origin* (1859). Traveled in Europe to see European colleagues and their collections of North American plants in 1850–1851, 1855, 1868–1869, 1880–1881, and 1887. Traveled and collected in the trans-Mississippi West (1877 with Joseph Hooker) and Mexico (1885), as well as frequent eastern trips.

Published some 780 articles and reviews, many in the *American Journal of Science*, and books including *Flora of North America*, 2 vols. (with John Torrey, 1838–1843); *Manual of the Botany of the Northern United States* (1848); and *Darwiniana*. Also see the *Scientific Papers of Asa Gray*, 2 vols. (Charles S. Sargent, ed., 1889) and *The Letters of Asa Gray* 2 vols. (Jane Loring Gray, ed., 1893).

Effectively used specimens collected by others to study taxonomy and distribution of North American plants, doing very little fieldwork himself. Emphasized his research at expense, perhaps, of his teaching. Few of his students became professional botanists, though did train scores of collectors, and his texts were dominant until turn of the century. Gave professional scientific support to Darwin and to evolution, and personal support to the reconciliation of theism and Darwinism. Regarded as "our greatest botanist" and a "trump in all senses" by contemporaries, is recalled as America's foremost botanist of the century and a leader of the emergent scientific community.

MAJOR CONTRIBUTIONS

Instituted, with Torrey, practice of using type specimens for the classification of North American flora based on natural system of classification. Chief American scientific proponent and colleague of Darwin.

BIBLIOGRAPHY

Dupree, A. Hunter. *Asa Gray. 1810–1888* (1959).

Elliott, Clark A. *Biographical Dictionary of American Science* (1979).

Watson, Sereno, and G. L. Goodale. "List of the Writings of Dr. Asa Gray, Chronologically Arranged, with an Index." *American Journal of Science* 36 (1888): 3–67, appendix.

UNPUBLISHED SOURCES

Largest collection is at Gray Herbarium, Harvard University. Additional Gray material can be found in the letters of very nearly every contemporary American biological scientist and many amateurs.

Liz Barnaby Keeney

GREEN, CHARLOTTE HILTON. Born Dunkirk, New York, 17 October 1889; died Tarboro, North Carolina, 26 March 1992. Naturalist, conservationist, nature writer, clubwoman.

FAMILY AND EDUCATION

One of seven children of William Charles Hilton, a farmer, and Mary Angeline Roscoe Hilton. Graduate of Dunkirk High School, 1908; Westfield Teachers' Training School, 1909; Chautauqua Summer School and Cornell University, 1913; B.S., North Carolina State College, 1932; additional summer study at University of North Carolina Chapel Hill, Cornell University, University of Colorado, National Autonomous University of Mexico. Married Ralph Waldo Green, 1917 (died 1946); no children.

CAREER

Taught school in one-room schoolhouse (all eight grades) in Chautauqua County, New York, 1909–1917 (?); teacher in public schools, Falls Church, Virginia, 1917–1919. Columnist for *Raleigh News and Observer*, 1932–1974.

CAREER

Green's husband's family was heavily involved in nature study movement. Following their move to North Carolina in 1920, she met various individuals involved in bird study and in 1923 began one of the earliest bird-banding stations in the state. Started writing ''Out-of-Doors-in-Carolina'' column, 1932, and in forty-two years, covered most areas of biology and conservation. Her first book, *Birds of the South* (1933, repr. 1975, 1995), was an anthology of some of her early newspaper pieces. *Trees of the South* (1939), her second book, was aimed primarily at children. Much of her other writing was devoted to conservation issues. One of her concerns was the needless destruction of birds, particularly hawks. Her pieces appeared in a variety of newspapers and periodicals, including the *Washington Post, Ladies' Home Journal, Nature*, and *Progressive Farmer*. Green and her husband, a professor of economics at North Carolina State College, actively promoted nature study throughout the state with the active assistance of many like-minded friends and associates, including Arthur A. Allen of Cornell, Roger Tory Peterson, and others. Green was a co-founder of the North Carolina Bird Club (1937). This subsequently was expanded to include South Carolina and became Carolina Bird Club, of which she was president, 1957. In 1938, she and her husband began converting large former farm they had recently purchased several miles north of downtown Raleigh into a wildlife sanctuary and arboretum. Following her husband's death, Green traveled extensively and contributed columns to the *News and Observer* in Raleigh from various parts of the world. In 1986–1987, a substantial portion of land surrounding her former home in Raleigh was acquired by the citizens of Raleigh and named Charlotte Green Park. Green was active as committee chair for both Raleigh Women's Club and North Carolina Federation of Women's Clubs. Through these and other organizations, she actively promoted nature study for both adults and youth. Green was awarded Conservation Communications Award by North Carolina and National Wildlife Federations shortly before moving to a retirement community in Tarboro, North Carolina, in 1983.

MAJOR CONTRIBUTIONS

Tireless promoter of nature study and conservation through both her individual efforts and the many organizations of which she was a member. Her newspaper columns and books did much to enhance understanding of wildlife, trees, and the values of the environment.

BIBLIOGRAPHY

Green, Charlotte Hilton. *Hardscrabble, As We Were* (1982) (autobiography).

Hoyle, Bernadette. "Charlotte Hilton Green: Nature Writer with a Pleasing Style." *The* (Raleigh, N.C.) *News and Observer*, 15 May 1955.

The (Raleigh, N.C.) News and Observer, 27 March 1992 (obit.).

Potter, Eloise F. "Introduction." *Birds of the South* by Charlotte Hilton Green (1995).

Powell, William. *North Carolina Lives: The Tar Heel Who's Who* (1962).

Wildlife in North Carolina, (October 1983): "Carolina Profile" section.

Keir B. Sterling

GRINNELL, GEORGE BIRD. Born Brooklyn, New York, 20 September 1849; died New York, 11 April 1938. Naturalist, conservationist, ethnographer, newspaper editor, author.

FAMILY AND EDUCATION

Born into upper-class family, he was son of George Blake Grinnell, head of a New York textile firm, and Helen Lansing Grinnell, and oldest of five children. Lived in several locations in early years: Brooklyn, lower Manhattan, Weehawken, New Jersey. Family moved, 1857, to upper Manhattan, near home of John James Audubon. Became student of Audubon's widow, Lucy, and friend of her grandchildren. Graduated, 1866, from Churchill Military School (Ossining, New York). B.A. (1870) and Ph.D. (1880) from Yale. Married Elizabeth Kirby Curtis Williams, 1902; no children.

POSITIONS

Volunteer assistant to paleontologist Othniel C. Marsh on his 1870 Yale expedition through American West. Assistant in osteology, Yale's Peabody Museum, 1874–1880. Naturalist with Custer's 1874 Black Hills Expedition, and with Col. William Ludlow's 1875 reconnaissance of Yellowstone National Park. Natural history editor, *Forest and Stream Weekly*, 1876–1880; editor in chief and owner, 1880–1911. Member, Harriman Alaskan Expedition, 1899. Founder, 1886, of first Audubon Society and a director of later National Audubon Society. Cofounder (with Theodore Roosevelt) of Boone and Crockett Club, 1887–1888, and one of its directors. A founder, 1895, of New York Zoological Society and one of its trustees. President Cleveland's special commissioner to treat with Blackfoot and Ft. Belknap Indians, 1895. President Roosevelt's personal emissary, 1902, to settle Indian–white controversy at Standing Rock Sioux Reservation. Member, first advisory board, Federal Migratory Bird Law. Fellow, American Ornithologists' Union, American Association for the Advancement of Science, American Society of Mammalogists, and New York Academy of Science. Trustee, American Museum of Natural History. A founder, 1911, and director of American Game Association. Chairman, Council on National Parks, Forests, and Wildlife. Succeeded, 1925, Herbert Hoover as president, National Parks Association.

CAREER

Months spent with O. C. Marsh in unmapped West, 1870, marked beginning of career as a leading interpreter and exponent of the western scene. Returned to West almost every summer for the rest of his life. On 1870 expedition met famous figures William F. ''Buffalo Bill'' Cody and leaders of Pawnee scouts, Frank and Luther North. In 1872 accompanied Pawnee on one of their last tribal buffalo hunts and, in 1873, hunted elk with Luther North. From latter trip came first publication, ''Elk Hunting in Nebraska,'' appearing under pseudonym ''Ornis'' in 2 October 1873 issue of *Forest and Stream.* In rest of decade of 1870s besides membership on government expeditions, he herded cattle and pursued big game with Cody and North on their ranch in western Nebraska in 1878, and the next year hunted in Colorado. In 1880, after receiving Ph.D. in paleontology, took over *Forest and Stream* and began editorial crusades to conserve wildlife, forests, and wilderness areas. Beginning in 1885, made regular trips to hunt in St. Mary's Lake region of northwestern Montana, where he discovered glacier named for him. In 1910 area set aside as Glacier National Park through movement he had begun in 1891. First book, *Pawnee Hero Stories and Folk Tales*, appeared in 1889, to be followed by nine others with ethnographic focus. Best known of these books are on Cheyenne, especially two-volume *Cheyenne Indians* (1923). Anthropologists Margaret Mead and Ruth Bunzel contend that no work on any other tribe ''comes closer to their everyday life than Grinnell's classic monograph on the Cheyenne.'' Besides ethnographic volumes, he edited (with Theodore Roosevelt and others) series of books containing articles by members of Boone and Crockett Club on hunting, natural history, and conservation. His *American Duck Shooting* (1901) and *American Game-Bird Shooting* (1910), written for ''the higher class of sportsman-naturalist,'' became classics. Also appealed to young readers, publishing seven books in so-called Jack series between 1899 and 1913 (*Jack, the Young Ranchman; Jack among the Indians*, and so on), which were based on own (or friends') experiences. Less popular were writings on western history, most significant of which were ''Bent's Old Fort and Its Builders,'' Kansas State Historical Society Collections (1923), and *Two Great Scouts and Their Pawnee Battalion* (1928).

Self-effacing to a fault, he preferred to work behind the scenes, allowing others to take credit for what he had initiated. As the crusading, but anonymous, editor of the leading outdoor periodical of the day, he used his weekly to protect Yellowstone National Park, to end commercial hunting and enact enforceable game laws, and to urge federal government to adopt European methods of scientific forestry. Both Yellowstone Park Protection Act, 1894, a landmark in national park legislation, and Migratory Bird Treaty with Great Britain, 1916, were result of efforts he began. Received from President Coolidge, 1925, Theodore Roosevelt Gold Medal of Honor for distinguished service in promoting outdoor life and conservation causes.

MAJOR CONTRIBUTIONS

Called "father of American conservation" in his *New York Times* obituary, he had impact on much of the early national legislation relating to wildlife, parks, and forests. Had greatest influence on development of Theodore Roosevelt's conservation philosophy before future president became well acquainted with Gifford Pinchot.

BIBLIOGRAPHY

Fisher, A. K. *Auk* (January 1939) (obit.).

New York Times, 12 April 1938 (obit.).

Reiger, J. F., ed. *The Passing of the Great West: Selected Papers of George Bird Grinnell* (1972, 1976).

Reiger, J. F. *American Sportsmen and the Origins of Conservation* (1975).

Reiger, J. F. *Arizona and the West* (Spring 1979).

UNPUBLISHED SOURCES

See letters and other manuscripts, Connecticut Audubon Society, Fairfield; various kinds of manuscripts, Southwest Museum, Highland Park, California.

John F. Reiger

GRINNELL, JOSEPH. Born Fort Sill, Oklahoma, 27 February 1877; died Berkeley, California, 29 May 1939. Zoologist, university professor.

FAMILY AND EDUCATION

Son of Fordyce and Sarah Elizabeth (Pratt) Grinnell. Father was a physician in the Indian service. The family moved to the Dakota Territory in 1880, to Pasadena, California, in 1885, to Pennsylvania in 1888, and back to Pasadena in 1891. Attended Pasadena High School. Graduated from Throop Polytechnic Institute (now California Institute of Technology) in 1897 with a B.A. Earned M.A. from Stanford University in 1901 and Ph.D. in 1913. Married Hilda Wood in 1906; three children.

POSITIONS

Assistant instructor in zoology, Throop Polytechnic Institute, 1897–1898. Assistant in embryology, Hopkins Laboratory, Stanford University, 1900; instructor in ornithology, 1901–1902. Instructor in botany and zoology, Palo Alto High School, 1901–1903. Instructor in biology, Throop Polytechnic Institute, 1903–1905; professor of biology, 1905–1908. Director, Museum of Vertebrate Zoology at the University of California, 1908–1939. Assistant professor of zoology, University of California, 1913–1917; associate professor, 1917–1920; professor, 1920–1939.

CAREER

Interest in birds and animals developed early in life, and was already collecting specimens as teenager. During years as undergraduate student, spent free time making collection of local birds. Spent the summer of 1896 in Alaska collecting ornithological information. In 1898–1899, returned there with a group prospecting for gold, but his success during that expedition was in the area of zoology. Throughout his life collected extensively in California and made several trips to museums in eastern United States. Was energetic worker, collecting over 20,000 specimens during his life, as well as producing over 3,000 pages of field notes.

Was the director of the California Museum of Vertebrate Zoology, founded by Annie M. Alexander, from 1908 until his death. It was understood that the new museum would not collect fishes, which were special preserve of David Starr Jordan at Stanford, but would confine itself to land vertebrates. In 1909, Grinnell donated 2,000 mammal specimens from his own collection to the museum, and ten years later, 8,000 birds. Under his direction the museum expanded greatly. Devised excellent curatorial system that stressed order, accuracy, and simplicity, making any specimen readily available.

Vitally interested in the various biogeographic factors that affected distribution of vertebrates. His publications "Life Zone Indicators in California" (with H. M. Hall, 1919), *Animal Life in the Yosemite* (with Tracy Storer, 1924), and "A Revised Life Zone Map of California" (1935) made clear his consistent view that the life zone concept expounded by C. Hart Merriam in 1889 had considerable validity in determining geographic ranges of species in California, despite criticisms of its use elsewhere. Grinnell was also very much concerned with relationship of geographic factors upon the evolution of geographic races.

Precise and prolific writer, he had more than 550 works published, the majority dealing with birds. Several early papers were coauthored by his mother, also a prolific writer. Some of his more important works included "An Account of the Mammals and Birds of the Lower Colorado Valley with Especial Reference to the Distributional Problems Presented" (1914); *Game Birds of California* (with H. C. Bryant and Tracy I. Storer, 1918); *Vertebrate Animals of Point Lobos Reserve*, 1934–1935 (1936); and *Fur-Bearing Mammals of California* (with J. S. Dixon and J. M. Linsdale, 1937). A number of shorter articles illustrating his views were collected in Joseph Grinnell's *Philosophy of Nature*, edited by Alden H. Miller (1941).

His writing experience and knowledge he gained as editor of *The Condor* served him well in his teaching. Frequently helped students in preparing manuscripts and offered course in scientific writing attended by graduate students and other faculty. While conducting field research, Grinnell noticed negative effect of California's growing population on natural environment and became increasingly concerned with a variety of conservation issues. Using biological data as basis for his recommendations, was influential in several areas. With Stephen Tyng Mather and C. M. Goethe, inaugurated nature guide service in national parks. Helped shape California Fish and Game Code, assisted with reformation of state statutes protecting wild game, served as consulting zoologist for state horticultural commission, and helped formulate policy concerning California state parks. At the time of his death, was actively involved in campaign opposing use of poison thallium in rodent control, the advertising of methods for killing small birds by fruit and vegetable growers, and the extension of wild animal control to public and uncultivated lands.

Active in Cooper Ornithological Society, editor, *The Condor*, 1906–1939. Librarian, California Academy of Science, 1911–1939; president, American Ornithologists' Union, 1929–1931; president, American Society of Mammalogists, 1937–1938; fellow, American Academy of Arts and Sciences; corresponding member, Academy of Natural Sciences of Philadelphia; active in some thirty additional scientific, conservation, and other organizations.

Usually limited his research to vertebrates of California and often selected a limited geographical area within the state and studied it thoroughly. In studying an area's vertebrate population, he was concerned with interrelationships and with relationship to the environment. Retiring in nature, he refused to give public lectures and would not allow his portrait to be hung in the museum. Generous with his time, resources, and influence, he often modestly concealed his part in helping students and associates. These qualities, along with his high standards, earned him the respect and friendship of colleagues and students.

MAJOR CONTRIBUTIONS

Made extensive contributions to understanding of California vertebrates, their distribution, and ecology. His emphasis on details of geographic distribution and influence of these on species formation had considerable influence on other West Coast naturalists. Named sixty-nine forms of mammals and ninety-seven forms of birds.

BIBLIOGRAPHY

Dictionary of Scientific Biography.
Fisher, Walter K. "When Joseph Grinnell and I Were Young." *Condor* 42 (1940).
Grinnell, Hilda Wood. "Joseph Grinnell: 1877–1939." *The Condor* (1940).
Hall, E. Raymond. "Joseph Grinnell (1877–1939)." *Journal of Mammalogy* (1939).
[Miller, Alden H., ed.] *Joseph Grinnell's Philosophy of Nature* (1941).
Miller, Alden H. "Joseph Grinnell." *Systematic Zoology* 13 (1964).
New York Times, 30 May 1939 (obit.).

UNPUBLISHED SOURCES

See Joseph Grinnell Papers, 1884–1938, Records of the Museum of Vertebrate Zoology, 1908–1949, Annie Montague Alexander Papers, Chester Barlow Papers, W. F. Martens Papers, Sumner Family Papers—all at Bancroft Library, University of California, Berkeley; American Ornithologists' Union, Records, 1895–1936, Smithsonian Institution Archives (correspondence, lists of birds); Joseph Grinnell letters, 1900–1936, Academy of Natural Sciences of Philadelphia collections (letters to Witmer Stone).

Cynthia D. Chambers

GUNN, DONALD. Born on a farm near Halkirk, Caithness, Scotland, September 1797; died Winnipeg, Manitoba, 30 November 1878. Fur trader, teacher, librarian, legislator, naturalist.

FAMILY AND EDUCATION

Son of William Gunn. Attended the Parish School in Halkirk, Scotland. After six years in Canada he married Margaret, daughter of James Swain and his native wife of York Factory, Hudson Bay.

POSITIONS

Beginning in 1813, Gunn worked for the Hudson's Bay Company at York Factory, Severn and Oxford House, then was promoted to assistant trader at Trout Lake, 1818–1819, Winisk Lake, 1819–1820, and Severn 1820–1821. In 1823 he settled at St. Andrew's, just north of present Winnipeg, Manitoba. After ten years as a farmer, he became the first teacher, the first librarian, and for about twenty years a magistrate and president of the Court of Petty Sessions. He was a member of the Legislative Council of Manitoba, equivalent to a senate, from 1871 until its abolition in 1876.

CAREER

Gunn responded to letters from Spencer F. Baird and began sending specimens of mammals, birds, and eggs to the Smithsonian Institution in Washington, D.C., in 1857. His first shipment of three boxes included the first examples of the American water shrew, *Sorex palustris*, and the yellow-cheeked vole, *Microtus xanthoqnathus*, that were received by the Smithsonian. In his 1858 book, Baird listed forty-eight specimens of forty-one bird species sent by Gunn. Gunn continued collecting; in 1862, aided by a fifty-dollar expense account from Baird, he took what Bendire later claimed to be the world's first set of piping plover eggs at Lake Winnipeg; this is only one of 108 sets of eggs of thirty species in the present Smithsonian catalog (a detailed list is provided in Houston and Bechard 1987).

MAJOR CONTRIBUTIONS

The first collector of natural history specimens from what is now southern Manitoba.

BIBLIOGRAPHY
Gunn, D. *History of Manitoba from the Earliest Settlement to 1835*. Ottawa: Maclean, Roger (1880).
Henderson, I. E., ''Donald Gunn on the Red River Settlement.'' *Canadian Magazine* 55 (1922).
Houston, C. S., and M. J. Bechard. ''Early Manitoba Oologists.'' *Blue Jay* 45 (1987).

C. Stuart Houston

H

HAENKE, THADDEUS PEREGRINUS XAVIERIUS (TADAEO, TEDEAS). Born Kreibitz, Bohemia (now Czech Republic), probably on 5 October 1761; died near Cochabamba, Bolivia, 4 November 1816. Botanist.

FAMILY AND EDUCATION

Son of Elias Georg Thomas and Rosalia Hanke. Choirboy and student, Prague, 1772–1773. Further preparatory study there and then student of natural and physical sciences and philosophy at University of Prague (Ph.D., 1782) until 1786. Continued as student of medicine and botany at the University of Vienna, 1786–1789. Never married, no children.

POSITIONS

Assistant to Joseph Gottfried Mikan, professor of medicine, botany and chemistry, University of Prague, for several years prior to 1786. Collected plants in vicinity of Vienna, 1786–1789. Appointed botanist to Pacific Ocean Expedition of Alessandro Malaspina, 1789, but did not join it until April, 1790. Remained with expedition until October, 1793. Botanized in Peru, Bolivia, and Brazil, 1793–1816.

CAREER

Education at St. Wenzelsseminar in Prague terminated owing to suppression of Jesuits in 1773. Completed education thanks to support of influential persons in Prague who appreciated his personal qualities and his talents as instrumentalist. Assisted Mikan at University of Prague by helping to take care of the university's botanic garden. Made extensive collections of plants in various parts of what is now Czech Republic and neighboring countries. Edited an edition of Linnaeus' *Genera Plantarum*, published in Prague in 1791. Recommended to Malaspina by Nicolaus Joseph von Jacquin, who thought highly of Haenke's botanical abilities, and by Emperor Josef II, who had also met Haenke. Haenke narrowly missed the expedition's departure from Cadiz in July 1789, took next ship to Montevideo in effort to join Malaspina but was shipwrecked near his destination. He spent some time recovering from poor health in Buenos Aires, crossed the pampas and the mountains separating Argentina and Chile (stopping here and there to collect some 1,400 plants), and caught up with Malaspina's two vessels, the *Descubierta* and the *Atrevida*, in Valparaiso in April 1790. He remained with the expedition for three years as they went up the west coast of the Americas to Alaska, crossed the Pacific, and returned. Though he worked well with Antonio Pineda, who was in charge of Malaspina's natural science work, and Luis Nee, another botanist, he was given considerable leeway to do his own collecting because of his considerable botanical expertise. Between May and August 1791, Haenke collected plants in Yakutat Bay, Alaska, and in Nootka Sound, where he made the first recorded systematic collection of Canadian flora. He focused especially on the variety of coniferous trees new to science, but found few plants that were unlike those known to him from central Europe. He also spent some time learning about local Indian culture, particularly their music. The expedition visited California from 13 to 23 September 1791, where Haenke collected various plants and was the first to collect coastal redwood seeds. He then went across the Pacific with Malaspina to the Philippines, Marianas, Australia, and New Zealand, and returned to Callao, Peru. Haenke left the ship there in October 1793, thinking to botanize his way to Buenos Aires and rejoin the expedition in the fall of 1794. He became so interested in the botany and geography of Peru, Brazil, and Bolivia, however, that he settled down in Cochabamba Province, Bolivia. There he established a botanical garden, owned a silver mine, and entered the medical profession. He remained in South America until his accidental death, probably from poisoning, in 1816. He gave some attention to other scientific studies in addition to botany, and continued to draw his salary as member of the Malaspina party. He is credited with having established the manufacture of saltpetre in Chile and having contributed to the development of the glass-making industry in Chile. Tragically, because of jealous intrigue on the part of Manuel Godoy, Spanish prime minister, most of Malaspina's hard work came to naught. He was put in prison for seven years (1796–1803); the collections, notes, and maps of his scientific party were seized; and the Spanish authorities refused to permit results of the expedition to be drafted or published. Nee and a colleague were later able to publish a few papers. Not until 1885 was an abbreviated edition of Malaspina's journal published.

Haenke's notes and specimens that were in his possession when he died were sent to Lima and evidently lost, and none sent to Europe after 1795 reached their destination. In 1821, however, seven chests belonging to Haenke, which had sat undisturbed in Cadiz and then in Hamburg, found their way to the Bohemian National Museum in Prague. They contained some 15,000 plant specimens in 4,000 species that Haenke had collected in parts of Latin America, the Pacific, California, and Vancouver Island. Nearly a dozen colleagues, principally Karel (Karl) Boriwog Presl, spent nearly fifteen years producing the *Reliquiae Haenkeanae seu Descriptiones et Icones Plantarum quas in America meridionali et boreali, in Insulis Philippinis et Marianis collegit Thaddaeus Haenke, redegit et in ordinem digressit Carolus Bor. Presl.* published in Prague between 1825 and 1835. Only a fraction of the 4,000 species received careful examination, and only six of the projected twenty parts were printed; the cost of the effort having prevented its completion. Eighteen of the forty-six genera proposed by the ten individuals who worked on the project are accepted today; the remainder were subsequently found to have been previously named. Haenke had organized his specimens by the locality where they had been collected, but did not label them individually. Errors were made (though probably not by Haenke or Presl) in the reorganization of the material, hence collection localities printed in the *Reliquiae* were not always correct. Presl also published "*Epimeliae botanicae*" (1851), which contains a reference to Haenke's collection of Redwood material, possibly near Santa Cruz, California. Haenke's other publications concerned with the Americas were few. One, an English translation based on a 1799 manuscript, was "On the Southern Affluents of the River Amazonas," published in the *Journal of the Geological Society* [of London], 1835. In 1933, some 2,400 unidentified North American specimens of Haenke's were sent to the New York Botanical Garden for identification, and approximately 1,200 duplicates remain there. Both a glacier, thirty-five miles north and east of the village of Yakutat, Alaska, and a small island, in Yakutat Bay, have been named for Haenke. A modest number of Haenke's plant specimens were purchased from an early nineteenth-century German collector and are to be found in the Missouri Botanical Garden, St. Louis.

MAJOR CONTRIBUTIONS

Haenke's collection of specimens provided the basis for the initial description of many plants and some trees from the western Pacific, Mexico, Peru, California, and Vancouver Island. He also kept notes on animals observed and minerals collected.

BIBLIOGRAPHY

Barneby, R. C. "Treasures of the Garden's Herbarium Reliquiae Haenkeanae." *The* [New York Botanical] *Garden Journal* (July–August 1963).

Dictionary of Canadian Biography.

Eastwood, Alice. "Early Botanical Explorers on the Pacific Coast and the Trees they found there." *Quarterly of the California Historical Society* 18 (1939).

Galbraith, Edith C., ed. "Malaspina's Voyage Around the World." [Translates section

of Malaspina (see below) concerning California.] *Quarterly of the California Historical Society* 3 (1924).
Jepson, Willis L. ''Early Scientific Expeditions to California-II.'' *Erythea* 7 (1899).
Malaspina, Alejandro (Alessandro). *Viaje politico-cientifico alrededor del mundo por las corbetas Descubierta y Atrevida al mando de los capitanes de navio D. Alejandro Malaspina y Don Jose de Bustamente y Guerra desde 1789 a 1794.* Publicado con una introduccion por Don Pedro de Novo y Colson teniente de navio academico correspondiente de la real de la historia (1885).
McKelvey, Susan D. *Botanical Exploration of the Trans-Mississippi West, 1790–1850* (1955; repr. 1991).
Stafleu, Frans A., and Richard S. Cowan. *Taxonomic Literature: A Selective Guide to Botanical Publications and Collections with Dates, Commentaries, and Types.* 7 vols., 3 supplements (1976–1995).
Stearn, William T. ''Introduction'' In Presl, Jarel Boriwog, ed. *Reliquiae Haenkeaniae* (repr. 1973).

UNPUBLISHED SOURCES

Haenke's papers are to be found in the Museo Naval (primarily concerned with the Malaspina Expedition), Madrid; the Archivo General de la Nacion, Buenos Aires; and the Biblioteca y Archivo Nacional de Bolivia in Sucre. Major European collections of Haenke's botanical specimens are in Prague, but some are to be found in Munich, Vienna, Geneva, and Leyden [see Barneby, Stafleu and Cowan, and Stearn, above].

Keir B. Sterling

HALL, EUGENE RAYMOND. Born Imes, Kansas, 11 May 1902; died Lawrence, Kansas, 2 April 1986. Mammalogist, conservationist.

FAMILY AND EDUCATION

Only child of Wilbur Downs Hall, a farmer, and Susan Effie Donovan Hall. Graduated from Lawrence (Kansas) High School. B.S., University of Kansas, 1924; M.A., University of California, Berkeley (UCB), 1925, Ph.D., UCB, 1928. Married Mary Frances Harkey, 1924; three sons.

POSITIONS

Curator of mammals, Museum of Vertebrate Zoology (MVZ), UCB, 1927–1944; acting director, MVZ, UCB, 1938–1944. Chair, Department of Zoology and Director, Museum of Natural History, University of Kansas, Lawrence, 1944–1967. Summerfield Distinguished Professor, 1958–1967. Professor emeritus, 1967–1986. President, American Society of Mammalogists, 1940–1941. Guggenheim Fellowship, 1942–1943; director, Kansas Biological Survey, 1946–1967; Kansas state zoologist, 1959–1967, active as director or board member of a number of other wildlife and conservation organizations.

CAREER

Spent early years in Kansas and Washington state. At UCB, studied zoology under Joseph Grinnell. Became an influential teacher and a seminal researcher in mammalogy. Active in conservation efforts, particularly in his postretirement years. Four of his doctoral students ultimately became presidents of the American Society of Mammalogists. Author of some 350 titles, most notably *Mammals of Nevada* (1946), regarded by many as the outstanding state study of the subject; also made major contributions to the mammalogy of his native state in *Handbook of Mammals of Kansas* (1955) and *Mammals of Northern Alaska* (with J. W. Bee, 1956). His *American Weasels* (1951) was an extension of studies begun as a doctoral candidate; and the *Mammals of North America* (with Keith R. Kelson, 1959, 2d ed., 1981) will probably remain the foundation work for studies in North American mammalian systematics and biogeography well into the next century. Toward the end of his life had a major revision of big brown and grizzly bears in progress. Was regarded as a leading authority on this subject. On the basis of considerable study, concluded that the number of species and subspecies of these bears worldwide was very small.

MAJOR CONTRIBUTIONS

Produced the benchmark publication in North American mammal taxonomy and distribution and did much to reduce what has been termed the ''staggering systematic complexity bequeathed by an earlier generation of systematists.'' Maintained high professional standards. From an early stage in his career, strove to cut back the role of the federal government in the control of predatory species. Was particularly critical of what he perceived as the irresponsible use of poisons in the 1930s. Devoted his postretirement years, in large part, to conservation efforts. Made financial support for his graduate students available in an era when wide-ranging federal and private programs of financial aid did not yet exist. An able, if sometimes controversial leader in his field.

BIBLIOGRAPHY

Birney, Elmer, C., and Jerry R. Choate, eds. *Seventy-Five Years of Mammalogy (1919–1994)* (1994).

Durant, S. D. ''Eugene Raymond Hall—Biography and Bibliography.'' *Miscellaneous Publications, Museum of Natural History* (1969).

Graham, Frank Jr. ''Hall's Mark of Excellence.'' *Audubon* 86. 4 (1984).

Findley, James S., and C. Jones. ''Additions to the Obituary of E. Raymond Hall.'' *Journal of Mammology* 70 (1989).

Joseph Bongiorno

HAMILTON, WILLIAM JOHN, JR. Born Corona, Queens, New York, 11 December 1902; died Ithaca, New York, 27 July 1990. Mammalogist, naturalist, teacher.

FAMILY AND EDUCATION

One of two sons of William J. Hamilton and Charlotte Richardson Hamilton. Graduate of Flushing High School. B.S., Cornell, 1926, M.S., 1928, Ph.D., 1930. Married Nellie R. Rightmyer, 12 October 1928; two daughters and a son. The son, W. J. Hamilton III, is professor of environmental studies, University of California, Davis.

POSITIONS

Held graduate assistantship in entomology, Cornell, 1926–1930. Instructor in vertebrate ecology, New York State College of Agriculture at Cornell, 1930 (subsequently joined newly established Department of Zoology), assistant professor, 1937, associate professor, 1942, professor, 1947, joined Department of Conservation, 1948, professor emeritus, 1963. With U.S. Army, 1942–1945, major, Medical Corps. Concerned with rodent and typhus control in the United States and Europe. Appointed one of several military governors of Mannheim, Germany, April 1945. President, American Society of Mammalogists, 1951–1952; president, Ecological Society of America, 1955.

CAREER

Developed early and very active interest in plants and animals as a child. Worked three summers as a teenager for Daniel C. Beard, a naturalist, artist, and cofounder of the Boy Scouts of America. A gifted teacher and researcher. Authored some 230 papers dealing with mammals, birds, reptiles, amphibians, horticulture, and conservation. Primary research interests in life histories and ecology. Published three books: *American Mammals* (1939), the first and for many years the only text book on the subject, *Conservation in the United States* (with A. F. Gustafson, H. Ries, and C. F. Guise, 1939, 2d ed., 1949), and *Mammals of Eastern United States* (with J. O. Whitaker, Jr., 1943, rev. ed., 1979, 3d ed., 1996). Among his major papers were "The Mammals of Southampton Island, Hudson Bay," with G. M. Sutton, Jr., published in the *Memoirs of the Carnegie Museum* in 1932, and "Mammalogy in North America," in a *Century of Progress in the Natural Sciences 1853–1953*, published by the California Academy of Sciences. Also completed some pioneering studies of microtine life cycles.

MAJOR CONTRIBUTIONS

Outstanding mentor, zoological researcher, and environmentalist. Author of pioneering texts in mammalogy and conservation. A leading conservationist, who stressed the vital ecological role of predators and the importance of conserving fur-bearing populations. A colorful character whose humor was legendary.

BIBLIOGRAPHY

Birney, Elmer C., and Jerry R. Choate, eds. *Seventy-Five Years of Mammalogy (1919–1994)* (1994).

Journal of Mammalogy (1990).

Stroud, Richard H., ed. *National Leaders of American Conservation* (1985).

Joseph Bongiorno

HANTZSCH, BERNHARD ADOLPH. Born Dresden, Germany, 12 January 1875; died near Nichols Bluff on the west coast of Baffin Island, about 30 May 1911. Naturalist, explorer.

FAMILY AND EDUCATION

Son of Adolph Hantzsch, a Dresden schoolmaster.

POSITIONS

Assistant teacher in the forest village of Grillenberg, then permanent teacher at Plauen, a suburb of Dresden.

CAREER

Studied waterfowl of the Danube in Slavonia, summer of 1898. During the next three summers he studied birds of prey in Southern Europe, as far east as Bulgaria. He spent the summer of 1903 studying the birds of Iceland (he listed 130 species) and the summer of 1906 in Labrador, where he recorded ninety-eight bird and fifty-three mammal taxa. He left Dresden with several tons of supplies on 4 July 1909 and sailed from Dundee, Scotland, on 29 July. On 26 September, in Cumberland Sound in southeast Baffin land, the chartered ship sprung a leak, requiring that they desert the ship. Forced to winter at Blacklead Island and short of supplies due to the shipwreck, Hantzsch nevertheless set out on 23 April 1910 with Inuit assistants. He reached the head of Nettilling Fiord on 6 June and Nettilling Lake on 21 June. He descended the Koukdjuak River between 30 August and 26 September 1910. The area of his winter camp is named Hantzsch Bay. He resumed his journey on 16 April, reaching his farthest north point on 6 May. However, soon after eating the polar bear killed on 1 May, Hantzsch began to be ill; they turned back, just short of Piling Bay, on 6 May. Hantzsch succumbed to trichinosis and died about 30 May.

MAJOR CONTRIBUTIONS

His writings and collections contributed immeasurably to our early knowledge of the birds and mammals of Labrador (1906) and of Baffin Island (1909–1911).

BIBLIOGRAPHY

Anderson, R. M. "The Work of Bernhard Hantzsch in Arctic Ornithology." *Auk* 45 (1928).

Hantzsch, B. (trans. M. B. A. Anderson). "Contributions to the Knowledge of Extreme North-Eastern Labrador." *Canadian Field Naturalist* 41,2 through 46,162 (1928 through 1932, in 26 installments). (First published in *Journal für Ornithologie* 56 [1908]).

Neatby, L. H., ed. *My Life among the Eskimos: The Baffinland Journals of Bernhard Adolph Hantzsch, 1909–1911* Saskatoon: University of Saskatchewan Press (1977).

C. Stuart Houston

HARIOT (OR HARRIOT), THOMAS. Born Oxford, England, 1560 (exact date unknown); died London, England, 2 July 1621. Explorer, astronomer, mathematician.

FAMILY AND EDUCATION

Parents are unknown. Educated at St. Mary's Hall, a branch of Oriel College. Attended 1577–1580; receiving bachelor of arts in 1580. Never married.

POSITIONS

Hariot joined Walter Raleigh's Durham house in early 1580s. He first taught Raleigh and other members navigational techniques. In 1585, was part of the Grenville Expedition to the New World. He returned to England in 1586 and stayed with Raleigh's circle for a number of years. He continued to help Raleigh with his colonization work and would continue to pursue his scientific studies while being supported by Raleigh. Eventually, after Raleigh had economic trouble, Hariot was supported by Henry Percy, ninth earl of Northumberland. This financial support allowed Hariot to devote almost all his time to his scientific and mathematical research.

CAREER

Hariot's career was filled with accomplishments in the sciences and mathematics. He was also directly involved with Raleigh's attempts to colonize North America (or the New World). In 1585, was member of the expedition, led by Richard Grenville, that was headed to Roanoke and the coastal Carolinas. Raleigh assigned Hariot with task of surveying land and studying the Indians of the area. He was assisted with the project by John White, a noted surveyor and illustrator. They made numerous trips throughout the area and took copious notes on physical features, the flora and fauna, and culture of the Indians. When they returned to England in 1586, Hariot condensed much of this material into an account entitled *A Brief and True Report of the New Found Land of Virginia* (1588). The original purpose of the work was to express the importance of colonizing the New World. Today, it is recognized as one of earliest descriptions of the New World and its inhabitants by an English settler. Hariot developed an understanding of the Indian culture and language. He was also one of first settlers to understand the devastating effect diseases were having on the Indians. After 1585 expedition, Hariot would never go to North America again but was still involved with Raleigh's colonizing efforts. He provided technical assistance with navigation as well as some administrative matters.

Hariot was also involved in mathematics and astronomy. He helped found the English school of algebra and developed greater than (>) and less than (<) signs. In field of astronomy he developed telescopes and was aware of comets and sunspots. Most of Hariot's findings were never published so he received little credit during his lifetime for his work.

MAJOR CONTRIBUTIONS

Hariot's observations of New World have allowed scholars to have an accurate portrayal of North America and its early inhabitants.

BIBLIOGRAPHY

Rukeyser, Muriel. *The Traces of Thomas Hariot* (1971).

Shirley, John W. *Thomas Harriot: A Biography* (1983).

Shirley, John W. *Thomas Harriot: Renaissance Scientist* (1974).

UNPUBLISHED SOURCES

The two largest collections of Hariot's manuscripts and unpublished scientific papers are in British Museum, London, England, and Pentworth House, Sussex, England.

Paul Cammarata

HARKIN, JAMES BERNARD. Born Vankleek Hill, Ontario, Canada, 30 January 1875; died Ottawa, Ontario, 27 January 1955. Government official, journalist.

FAMILY AND EDUCATION

Son of William and Eliza (McDonnell) Harkin. Attended public school in Vankleek Hill, Ontario, High School in Marquette, Michigan.

POSITIONS

Began journalistic career in Montreal, Quebec, probably as reporter with the *Montreal Herald*, 1892. Joined staff of *Ottawa Journal*, 1893, serving until 1901, successively as reporter, parliamentary press officer, and city editor. Appointed political secretary to Clifford Sifton, minister of the interior, Ottawa, 1901, later promoted to private secretary. Served with Sifton until 1905 and with his successor, Frank Oliver, Minister of the Interior, 1905–1911. Appointed commissioner of dominion (national) parks, 1911–1936. Also served on many boards and commissions, including Historic Sites and Monuments Board of Canada, Canadian National Parks Association, and Interdepartmental Advisory Board on Wildlife Protection.

CAREER

Following ten years' service with the Office of the Minister of the Interior in Ottawa, Harkin inaugurated the first bureau in Canada and in North America charged with administration of national parks. In 1913, Harkin advised U.S. Department of the Interior staff on creation of its own park service. When he took office, there were one park established by statute and five park reserves. There were eighteen national parks by 1936. Emphasized importance of tourism to the nation and role played by national parks in increasing tourist expenditures in Canada. Recommended establishment of national tourist bureau beginning in 1915 (this was ultimately done in 1934). Took deep interest in protection of wildlife. Canada had already experienced success in reestablishing American bison when he took office in 1911. Turned his attention to pronghorn antelope, 1914. Three antelope parks established in Alberta and Saskatchewan by 1922. Following signing of Migratory Birds Treaty with the United States in 1916 and the passing of the Canadian Migratory Birds Convention Act in 1917, Harkin established special division to administer the new legislation. Eventually, this became the Canadian Wildlife Service, which was given branch status with its own directorate advisory board on wildlife protection (1916), which functioned until 1957. On his recommendation, the minister of interior established the Historic Sites and Monuments Board of Canada. In 1920, initiated construction of the Banff-Windermere Highway, first motor road across the summit of the Canadian Rockies. Created a publicity and information division (1921), one of the first Canadian government departments to utilize motion pictures as a medium of information and education. The enactment of the National Parks Act (1930) climaxed Harkin's efforts to have park administration placed under authority of legislation concerned exclusively with parks. Concurrent legislation has four western provinces, Manitoba, Saskatchewan, Alberta, and British Columbia, in control of their own natural resources, but Harkin ensured that national parks in those provinces would remain within federal jurisdiction. Served on National Executive of Boy Scouts Association of Canada, awarded their Silver Wolf Decoration. Mount Harkin, a 9,800–foot peak in the Mitchell Mountains, Kootenay National Park, B.C., was named in his honor.

MAJOR CONTRIBUTIONS

Laid foundations for present Canadian national park system. Helped to increase their number more than threefold. Developed engineering, architectural, and information services; created and developed present system of national and historic parks and sites; extended a small migratory birds protective unit into a Canadian Wildlife Service. Played dominant role in drafting National Parks Act of 1930 and, through agreements with provinces concerned, ensured that development of natural resources within national park boundaries would be prohibited.

BIBLIOGRAPHY

Canada, Department of the Interior. *Annual Reports of the Commissioner of National Parks* (1912–1937).

Foster, Janet. *Working for Wildlife* (1978).

Harkin, J. B. *Just a Sprig of Mountain Heather* (1914).

Harkin, J. B. "Our Need for National Parks." *Canadian Alpine Journal* 9 (1918).

Hewitt, C. Gordon. *The Conservation of Wildlife of Canada* (1921).

Williams, Mabel B. *Guardians of the Wild* (1936).

Williams, Mabel B. *The Origin and Meaning of the National Parks of Canada* (1957) (extracts from the Papers of J. B. Harkin).

UNPUBLISHED SOURCES

See J. B. Harkin Papers, Public Archives of Canada, Ottawa, Ref. MG 30–Series E, vols. 1, 2.

W. Fergus Lothian

HARLAN, RICHARD. Born Philadelphia, 19 September 1796; died New Orleans, 30 September 1843. Vertebrate paleontologist, comparative anatomist, physician.

FAMILY AND EDUCATION

Eighth of ten children of Joshua Harlan, wholesale grocer and merchant, and Sarah Hinchman Harlan, both Friends. Studied medicine with Joseph Parrish in Philadelphia; served as ship's surgeon on voyage to Calcutta, India, while a student (1816–1817); earned M.D. from the University of Pennsylvania in 1818. Married Margaret Hart Simmons Howell, a widow, in 1833; four children.

POSITIONS

Practiced medicine in Philadelphia, first as demonstrator in dissecting room of Parrish's private anatomical school. Physician to Philadelphia Dispensary, 1820; professor of comparative anatomy and surgeon in Charles Willson Peale's Philadelphia Museum, 1821; physician to Philadelphia Almshouse, 1822–1838. Began practicing medicine in New Orleans, 1843; elected vice president of Louisiana State Medical Society.

CAREER

Wrote paper on animal heat in 1821 and a report of experiments on the process of absorption (with J. B. Lawrence and B. H. Coates). Was member of a commission of three sent by emergency Sanitary Board of Philadelphia to Montreal and Quebec City to study Asiatic cholera when disease threatened Philadelphia; returned to doctor its victims.

From 1832 to 1836 was corresponding secretary and curator of Geological Society of Pennsylvania, precursor of Pennsylvania Geological Survey. First publication on vertebrate paleontology was ''Observations on Fossil Elephant Teeth of North America'' (*Journal of the Academy of Natural Science of Philadelphia* [1823]) followed, in 1824, by paper on *Saurocephalus*, a fossil reptile collected by Lewis and Clark expedition. In 1824 also wrote *Observations on the Genus Salamandra*. Collected and described vertebrate fossils from the New Jersey marl pits. Most ambitious work appeared in 1825, *Fauna Americana*, the first systematic work on North American mammals; it attempted to group fossil forms and their recent representatives. It followed Desmarest's *Mammalogie* (1821–1822) closely in time and context, and reviewers questioned its originality and criticized its numerous errors. ''Genera of North American Reptilia'' appeared in 1826 and 1827 (*Journal of the Academy of Natural Science of Philadelphia*) and *American Herpetology* in 1827. Forecast discovery of prolific fossil vertebrate fauna of the West with paper read in 1832 on discovery of an *Ichthyosaurus* from Missouri (*Transactions of the American Philosophical Society* [1834]). Same year also wrote, in the *Transactions*, ''On the Structure and Teeth of the Edentata Fossils,'' early recognition of importance of odontology; also *Critical Notices of Various Organic Remains Hitherto Discovered in North America*. The latter, perhaps his most valuable work, was compiled mainly for benefit of European naturalists. In 1836 published *Medical and Physical Researches*, collection of his papers with numerous additions that covered natural history and medicine. Among his over sixty papers are works in Geological Society of London, the British Association for the Advancement of Science, the Sociéte Géologique de France. Visited Europe in 1833 and from 1838 to 1840. On second trip read paper to Geological Society of London.

Harlan accepted Cuvier's theories and methods and practiced within that framework. He did not develop his own concepts in vertebrate paleontology, as did Caspar Wistar before him and Joseph Leidy after him. He understood the concept of succession of faunas, especially of Pleistocene vertebrates. Familiar with evolutionary theories of Erasmus, Darwin, Lamarck, and Virey, he believed that evolutionary processes are continuing and considered that man may not be the ultimate perfection. However, because the fossil record was so incomplete, he could not observe the sequence of forms so equivocated on origin of species and stated that species are absolutely immutable—but if species do not change, they have repeatedly been replaced by other species. He hesitatingly proposed that species arose by spontaneous generation. His most theoretical works were published in 1835: *On the Affiliation of the Natural Sciences* and *On the Successive Formations of Organized Beings*. The first work deals with a constantly changing nature in what theologians then considered a static earth and species.

It describes the interrelationship of all creation with necessary interdependence of all branches of nature and of man.

MAJOR CONTRIBUTIONS

Harlan was the first American to devote so much effort and time to vertebrate paleontology and was its most important worker in early nineteenth century. His major contributions were in objective description, taxonomy, and nomenclature; he was first American worker in vertebrate paleontology to apply Linnaean names to fossils, the taxonomic units he recognized or described were usually valid, and the affinities usually correct in the context of incomplete taxonomic knowledge of his time. His writings were valuable in collecting and codifying much of the earlier work in American vertebrate paleontology.

BIBLIOGRAPHY

Bell, Whitfield J., Jr., and Richard Harlan. *Dictionary of Scientific Biography*. Vol. 1 (1970).

Gannal, J. N. *History of Embalming* . . . Ed. R. Harlan (1840).

Harlan, Richard. *Fauna Americana*, Philadelphia (1825).

Harlan, Richard. *Medical and Scientific Researches* (1835).

Harlan, Richard. Critical Notices of Various Organic Remains Hitherto Discovered in North America. *Transactions of the Geological Society of Pennsylvania* 1, part 1 (1834).

Hay, Oliver Perry. ''Bibliography and Catalogue of the Fossil Vertebrata of North America.'' *U.S. Geological Survey Bulletin*, no. 179 (1902).

''Richard Harlan.'' *Dictionary of American Biography*. Vol. 8 (1938).

Sharf, J. T., and Wescott Thompson. *History of Philadelphia* (1884).

Simpson, Henry. *Lives of Eminent Philadelphians Now Deceased* (1859).

Simpson, George Gaylord. ''The Beginnings of Vertebrate Paleontology in North America. *Proceedings of the American Philosophical Society* 86 (1942).

Carol Faul

HARRIS, THADDEUS WILLIAM. Born Dorchester, Massachusetts, 12 November 1795; died Cambridge, Massachusetts, 16 January 1856. Librarian, entomologist.

FAMILY AND EDUCATION

Firstborn of eight children of Thaddeus Mason Harris, minister of First Church in Dorchester (Unitarian), and Mary (Dix) Harris. From Harvard University received A.B. 1815, and M.D. 1820. Married Catherine Holbrook, 1824; twelve children.

POSITIONS

During years 1820–1831, lived in Milton, Massachusetts; practiced medicine in that town and in nearby Dorchester, associated with father-in-law, Amos

Holbrook. Librarian of Harvard College, 1831–1856; 1834 and 1837–1842 also taught Harvard natural history courses but never held professorial appointment.

CAREER

Began intensive insect studies about the time he started medical practice. Over the years, he accumulated a large insect collection, especially insects of New England. At various times in 1820s and 1830s, Harris planned several general works that never came to full fruition. Began publishing in 1823 (in the agricultural press); demonstrated from the outset a fundamental interest in, and an awareness of, the importance of studies of insect life stages and the relations of insects to the plants on which they feed. His articles in journals such as the *New England Farmer* were prepared to spread information about insects for the benefit of the public and to register tentatively his new species names. Prepared first general classified list of insects for a region of America, published in Edward Hitchcock's *Report on the Geology, Mineralogy, Botany, and Zoology of Massachusetts* (1833). In 1838, agreed to prepare section on insects for Natural History Survey of New York but apparently never contributed to the report. In late 1830s, undertook preparation of descriptive catalog of American nocturnal moths. Argued for knowledge and consideration of larvae as well as adult stages in classification of Lepidoptera, and early in private studies investigated wing neuration in this order. Published only one section of his moth study, "Descriptive Catalogue of . . . Genus Sphinx in the Cabinet of the Author" (*American Journal of Science* [July 1839]). Harris is remembered chiefly for *Treatise on Some of the Insects of New England, Which Are Injurious to Vegetation* (1842; subsequent editions 1852, 1862), originally prepared for Zoological and Botanical Survey of Massachusetts (1841). Following instructions of the governor, Harris produced a work that addressed needs of agriculture. While arranged along scientific lines by insect orders and with the inclusion of genus-species names, the *Treatise* was written in nontechnical language. Its point of departure was assumption that control of insects depended on understanding their life histories and relations to vegetation. After initial appearance of *Treatise*, duties in library, growing family, and resulting lack of leisure essentially ended his entomological ambitions. Harris produced relatively few strictly descriptive papers. In addition to dominant entomological interests, he also had secondary interest in botany and published several papers on the subject. Indirectly, Thoreau heard Agassiz say that Harris was the greatest entomologist in the world. He always had a reputation as gentleman and lover of nature, qualities he was able to project through his *Treatise* even for those who never knew him personally. One student labeled him "the Gilbert White of New England."

MAJOR CONTRIBUTIONS

Helped direct American entomological studies away from narrow taxonomic interests to consideration of life stages and botanical relations of insects. Behind his work was profound love of nature combined with feeling that nature must be kept within bounds in order that human enterprise (agriculture) might thrive.

BIBLIOGRAPHY

Dictionary of American Biography. Vol. 8 (1932).

Dow, R. P. "The Work and Times of Dr. Harris." *Bulletin of Brooklyn Entomological Society* 8 (December 1913).

Hagen, H. A. "List of Papers of Dr. T. W. Harris, Not Mentioned . . . in the 'Entomological Correspondence.' " *Proceedings of Boston Society of Natural History* 21 (1881).

Harris, E. D. "Memoir." *Proceedings of Massachusetts Historical Society* 19 (1882).

Higginson, T. W. "Memoir." In Scudder, S. H., ed. *Entomological Correspondence of Thaddeus William Harris, M.D.* (1869). (Also contains "List of the Writings of Thaddeus William Harris, M.D.").

Mallis, Arnold. *American Entomologists* (1971).

UNPUBLISHED SOURCES

Main collection of scientific correspondence, notes, and manuscripts is in Library of Boston Museum of Science. Correspondence, reports, and other records as librarian are in Harvard University Archives. Letters are in Harvard Museum of Comparative Zoology Archives. Insect collection is in Museum of Comparative Zoology.

Clark A. Elliott

HARSHBERGER, JOHN WILLIAM. Born Philadelphia, 1 January 1869; died Philadelphia, 27 April 1929. Botanist, biographer, university professor, conservationist.

FAMILY AND EDUCATION

Son of Abram Harshberger, physician, and Jane Harris Walk Harshberger. Resided all his life in Philadelphia. Interest in plants developed in boyhood; made a small herbarium at age seven. Graduated from Central High School, Philadelphia, 1888. Spent some time studying at Arnold Arboretum, Harvard University, as special student, 1890. Obtained theoretical knowledge of pedagogy as member of University Extension Seminary, 1892–1893. B.S., 1892, and Ph.D., 1893, University of Pennsylvania. Married Helen B. Cole, 1907; two children.

POSITIONS

Assistant instructor in botany, Arnold Arboretum at Harvard, 1890; instructor in botany, biology, and zoology, Veterinary Department of the University of Pennsylvania and in botany in the School of Biology, 1892–1906; assistant professor of botany, 1907–1911, professor, 1911–1929. Instructor in chemistry, physics, and astronomy, Rittenhouse Academy, Philadelphia, Pennsylvania, 1892–1895; lecturer, American Society for Extension for University Teaching, Philadelphia, 1896; instructor, Farmers' Institutions in Pennsylvania, 1904–

1906; head of nature study, Pocono Pines Assembly, Pennsylvania, summers 1903–1908; head professor in ecology, Marine Biological Laboratory, Cold Spring Harbor, Long Island, 1913–1922; director of the study of botany, Nantucket Maria Mitchell Association, 1914–1915.

CAREER

Offered his first professional position with the University of Pennsylvania upon receiving the doctorate of philosophy with the thesis "Maize; A Botanical and Economic Study," 1893. The work was important as it advanced the theory, later broadly accepted, of the origin of this crop plant from teosinte, a Mexican grass. Early career consisted mainly of observations in nature study. Visited Mexico (1896) and subsequently published several notes on the domestic and native plants of that country. During summers traveled to most areas of the United States and also botanized in the West Indies (1901), Europe (1907, 1923), Alaska (1926), South America (1927), and North Africa (1928). Research findings deposited in the herbarium of the University of Pennsylvania. Made and collected photographs, focusing on those that depicted characteristic plant formations, cataloging them in bound volumes for use by his students and peers.

Authored over 300 papers (including 62 scientific papers contributed to scientific journals, 1889–1901) on a broad range of areas connected with the study of botany in general and plant distribution specifically. Most popular contributions included *The Botanists of Philadelphia and Their Work* (1899), his monumental *A Phytogeographic Survey of North America* for the series of monographs "Die Vegetation der Erde" (1911), *The Vegetation of South Florida* (1914), *The Vegetation of the New Jersey Pine Barrens* (1916), *A Textbook of Mycology and Plant Pathology* (1917), *Colored Wall Map Vegetation of North America* (1919), and *A Textbook of Pastoral and Agricultural Botany* (1920). Served as botanical editor of *Worcester's New English Dictionary*, Funk and Wagnall's *College Dictionary*, the *Botanisches Central Blatt*, and several other dictionaries.

An enthusiastic and diverse scholar, gave more than eighty public lectures. Member of more than twenty-five scientific or conservation societies and held office in a number of them at one time or another. A prominent advocate of the conservation of natural resources in Pennsylvania, he actively served on the council of the Pennsylvania Forestry Association for many years, being among the first to discover the hazard linked to the chestnut-blight fungus.

MAJOR CONTRIBUTIONS

Participated in a wide range of activities—botanical, educational, and literary—relating to many aspects associated with the plant kingdom, most notably his innovative research in mycology and plant pathology. An indefatigable collector and inspiring teacher who, through his capacity for continued industry and devotion to conservation, instilled in many of his students a creative interest in botany and aided colleagues in future study.

BIBLIOGRAPHY

Cattell, J. M., and D. L. Brimhall. *American Men of Science*. 3d ed. (1921).
Dictionary of American Biography. Vol. 8 (1932).
Harshberger, John W. *Life and Work of John Harshberger Ph.D.* (1928).
The National Cyclopaedia of American Biography. Vol. 21 (1931).
Nichols, G. E. *Ecology* (April 1930) (obit.).
Philadelphia Enquirer, 28 April 1929 (obit.).
Stroud, Richard H. *National Leaders of American Conservationism* (1985).
Who Was Who in America. Vol. 1, 1897–1942 (1943).

UNPUBLISHED SOURCES

See correspondence, 1890–1913 (428 items), Academy of Natural Sciences of Philadelphia collections.

Lynn M. Haut

HARTWEG, (CARL) THEODOR. (At times as Karl and/or Theodore in English-language writings.) Born Karlsruhe, Baden, Germany, 18 June 1812; died Swetzingen, Baden, Germany, 3 February 1871. Horticulturalist, gardener, plant collector.

FAMILY AND EDUCATION

Son of a gardener, Andreas Hartweg, and one of a long line of gardeners. Studied horticulture and botany in Germany and France.

POSITIONS

Briefly gardener at Jardin des Plants, Paris and Kew Gardens, England, before being hired as plant collector for the Horticultural Society of London in 1836. He made two collecting trips to the Americas in 1836–1843 and 1845–1848, after which he became director (inspector) of Grand Ducal Gardens of Swetzingen, Baden, a position he held until his death.

CAREER

His first collecting trip for seeds, plants, and pressed plant specimens lasted from October 1836 to July 1843, two years and nine months of which were spent in Mexico. Landing in Veracruz, he soon headed for the higher lands and mountains, collecting plants in the states of Guanajunto, Hidalgo, Jalisco, Mexico, Michoacán, Oaxaca, Puebla, San Luis Potosi, Tamaulipas, Veracruz, and Zacatecas, as well as around Mexico City. He left Mexico on horseback, by way of Tehuantepec in Oaxaca, through Comitán in Chipas, for Guatemala. He collected for almost a year in Guatemala, making his headquarters in Guatemala City. From Sonsonate, now in the Republic of El Salvador, on the coast, he traveled along the coast to Nicaragua, where he left by ship from the port of Realejo, reaching Callao, Peru, the seaport of Lima in February 1841 and soon leaving for Guayaquil, Ecuador, where he arrived in March. After a delay because of rains, he headed for the mountains and Loja, where he made his headquarters for a time. He traveled through the mountains to Quito and spent time in that vicinity, leaving in July 1842 for Popayán Colombia. After collecting in that vicinity, he traveled to Bogotá, arriving in January 1843. In April, he left Bogotá for Honda, where he descended the Magdalena River to near the coast, traveling to catch a ship, at Cartagena, for Jamaica, where he spent a few days before leaving for England. In all he sent back to England 4,112 specimens during his travels. The second trip was to collect plants in California. He left England reaching Veracruz, Mexico, in November 1845. Crossing the country by way of Mexico City to Mazatlán in Sinola and finally Tepic in Nayarit, where he spent time until mid-March 1846 collecting plants in various localities because unrest prevented ships' taking passengers to California. He finally did obtain passage from Mazatlán in mid-May, arriving in Monterey in June, where he collected plants in that vicinity and along the coast. He made two trips to northern California, almost as far north as the 40th parallel of latitude, to what are now Marin and Sonoma counties on the first trip and on a second trip up the Sacramento Valley to the Feather River, Bear Creek, and Bear Valley. In September 1847, he initiated a southern trip up the Salinas Valley, then over the mountains and back to the coast toward San Luis Obispo and somewhat farther south near the coast. Returning to Monterey he waited until February 1848 for a ship to Mexico and continued beyond Mazatlán to Iztapa Guatemala in March, spending a time there; he then sailed to Realejo and to Lake Nicaragua and the River San Juan de Nicaragua to the east coast where he boarded a ship for England in June. Scientific reports of his plant collections were made by others but particularly by George Bentham, in *Plantae Hartwegianae* (1839–1857), where 2230 species are listed, of which more than 800 were new to science, and by John Lindley. Many of the seeds and plants were cultivated outside and in greenhouses to become important ornamental plants, particularly orchids and conifers.

MAJOR CONTRIBUTIONS

A major plant collector in Mexico, tropical America, and California, he introduced more American plants to Europe in the first half of the nineteenth century than any other collector. Many plants were named in his honor, including the genus *Hartwegia*, an orchid. His plants were frequently new to science and resulted in their being described by various botanists.

BIBLIOGRAPHY

Bulletin de la Societé Botanique de France 17 (1870) (obit.).
The Gardeners' Chronicle and Agricultural Gazette (11 March 1871) (obit.).
Hartweg, T. "Notes of a Visit to Mexico, Guatemala, and Equatorial America, during

the Years 1836 to 1843, in Search of Plants and Seeds for the Horticultural Society of London.'' *Transactions of the Horticultural Society of London* 3 (1844–1845).
Hartweg, T. ''Journal of a Mission to California in Search of Plants.'' *Journal of the Horticultural Society of London* 1 (1846); 2 (1847); 3 (1848).
Jepson, W. L. ''The Explorations of Hartweg in America.'' *Erythea* 5 (1879).
McKelvey, S. D. *Botanical Exploration of the Trans-Mississippi West 1790–1850* (1955).
McVaugh, R. *Introduction to the Facsimile Reprint of George Bentham's ''Plantae Hartwegianae''* (1970).
Reinikka, M. A. *A History of the Orchid* (1972).
UNPUBLISHED SOURCES
Original letters and journals of American travels are in Library of the Royal Horticultural Society, London.

Emanuel D. Rudolph

HARVEY, ETHEL BROWNE. Born Baltimore, 14 December 1885; died Falmouth, Massachusetts, 2 September 1965. Cell biologist, embryologist.

FAMILY AND EDUCATION

Harvey was the youngest of the five children of Bennet Bernard and Jennie R. (Nicholson) Browne. Her father was an obstetrician-gynecologist and a professor of gynecology at the Woman's Medical College of Baltimore. Harvey had two brothers and two sisters who both became physicians. Graduated from the Bryn Mawr School in 1902. Following her graduation from the Woman's College of Baltimore (later Goucher College) in 1906 with an A.B. degree, she spent a summer studying at the Marine Biological Laboratory in Woods Hole, Massachusetts. That fall she enrolled in Columbia University. Earned an A.M. in 1907 and a Ph.D. in 1913. Awarded an honorary D.Sc. by Goucher College in 1956. Elected a fellow of the American Association for the Advancement of Science, l'Institut International d'Embryologie in Utrecht, and the New York Academy of Sciences. Elected a trustee of the Marine Biological Laboratory. Married Edmund Newton Harvey, a biology professor at Princeton in 1916. They had two sons.

POSITIONS

Between 1906 and 1907 Harvey held a fellowship from Goucher College. Worked as an instructor in science and education at the Bennett School for Girls in Millbrook, New York, and received a fellowship from the Society for the Promotion of University Education for Women, 1911–1912. Worked as a laboratory assistant in biology at Princeton (1912–1913) and as a biology teacher at the Dana Hall School in Wellesley, Massachusetts (1913–1914). In 1914 she was the recipient of a one-year fellowship grant to study at the Hopkins Marine

Station at the University of California. From 1915 to 1916 Harvey worked as a laboratory assistant in histology at Cornell Medical College.

CAREER

After graduating from the Woman's College of Baltimore, Harvey spent the summer studying at the Marine Biological Laboratory before enrolling at Columbia University to study biology. While a graduate student, supported herself with fellowships. Also worked as a laboratory assistant at Princeton for one year. While at Columbia Harvey published several papers related to her doctoral research on the role of the nucleus and cytoplasm in inheritance and development. When she completed her work at Columbia, taught biology for one year at the Dana Hall School in Wellesley, Massachusetts, and then worked as a laboratory assistant in histology at Cornell Medical College.

For almost eleven years following her marriage, Harvey worked as a part-time researcher. The family spent their summers at Woods Hole, and the Marine Biological Laboratory served as a base of operations for Harvey's investigations in cytology. During the 1920s spent two years conducting research in Europe. In 1920–1921 worked at the l'Institut Oceanographique in Monaco, and in 1925–1926 was at the Stazione Zoologica in Naples. After she returned to the United States, taught biology at Washington Square College, New York University, for three years (1928–1931).

Most of Harvey's career was spent at Princeton, where she maintained an office in one of the biology laboratories. In addition, Harvey used other facilities as research bases, such as the Marine Biological Laboratory and marine stations in Bermuda, California, Japan, and North Carolina. She was awarded use of the Women's Table in Naples (1925–1926, 1931) and of the Rockefeller Institute's Jacques Loeb Memorial Table at Naples (1933–1934, 1937). Though her contributions were internationally recognized, received no major financial support except for a grant in 1937 from the American Philosophical Society.

MAJOR CONTRIBUTIONS

During the 1930s Harvey developed a technique to stimulate cell divisions of sea urchin egg fragments without nuclear material. This work generated additional studies regarding the mechanics of cell division as well as discussion of the role of cytoplasmic structures and processes.

BIBLIOGRAPHY

"The Contribution of Goucher Women to the Biological Sciences." *Goucher Quarterly* (Summer 1951).

Leaders in American Science. Vol. 1 (1953), Vol. 2 (1954), Vol. 3 (1955).

New York Times, 3 September 1965 (obit.).

Notable American Women: The Modern Period. Vol. 4 (1980).

Rossiter, Margaret. "Women Scientists in America before 1920." *American Scientists* (May–June 1974).

Domenica Barbuto

HAYDEN, FERDINAND VANDIVEER. Born Westfield, Massachusetts, 7 September 1829; died Philadelphia, 22 December 1887. Geologist, explorer, scientific administrator, teacher, physician.

FAMILY AND EDUCATION

Eldest of four children of Asa Hayden, a farmer, and Melinda Hawley Hayden. Parents divorced 1840. Lived on paternal uncle's farm near Rochester, Ohio; attended and later taught (winters) in Lorain County schools to 1845. Worked way through Oberlin College, 1845–1850, A.B. Studied medicine and natural history, privately and at Cleveland Medical College, with John S. Newberry and Jared P. Kirtland, 1851–1853. M.D., Albany (New York) Medical College, January 1854, while studying geology, paleontology, and scientific drawing with James Hall, state geologist of New York, and Fielding B. Meek, 1853–1854. Honorary LL.D., University of Rochester, 1876, University of Pennsylvania, 1886. Married Emma C. Woodruff (died 1934) of Philadelphia, 1871; no children.

POSITIONS

Field and laboratory assistant to James Hall, 1853–1854, and Chester Dewey, University of Rochester, 1854. Geologist-naturalist on U.S. Corps of Topographical Engineers' route reconnaissances by Lt. Gouverneur K. Warren (upper Missouri and Yellowstone Rivers, 1856; Niobrara River and Black Hills, 1857) and Capt. William F. Raynolds (Wind River Mountains, Big Horn and Yellowstone Rivers, 1859–1860). Acting assistant surgeon to surgeon of volunteers (mostly administrative duties in the East), 1862–1865; chief medical officer, Union Army of the Shenandoah, 1864–1865, breveted Lt. Col. for meritorious services, 1865. Adjunct professor of geology and mineralogy, University of Pennsylvania, 1865–1872. U.S. geologist, geological surveys of Nebraska, 1867, and Wyoming, 1868, General Land Office. U.S. geologist and geologist-in-charge, U.S. Geological and Geographical Survey of the Territories (Hayden Survey), Department of the Interior, 1869–1879. Geologist, U.S. Geological Survey, 1879–1886.

CAREER

Hayden's interest in natural history encouraged by George N. Allen at Oberlin. Newberry, Hall's protégé, kindled Hayden's abiding fascination with geology and recommended him to Hall to help Hayden gratify his "strong desire to labor in the field as a Naturalist . . . in that most delightful of pursuits.'' With Meek, Hall's assistant in paleontology, Hayden examined the geology of, and collected fossils in, lower Missouri Valley upstream to Fort Pierre and westward to White River badlands in 1853 to expand Hall's knowledge of the area's geology with that of areas farther east. Hayden single-handedly conducted geological and natural history reconnaissance of large parts of new Nebraska Territory, 1854–1855, mostly under auspices of American Fur Company and Upper Missouri Indian Agency. As field observer and collector, joined mentor and chief analyst Meek (1817–1876) in geologic and paleontologic reconnaissance of entire upper Missouri country, 1854–1860. Their work, based at the Smithsonian Institution after 1856, was advanced by Hayden's geological studies for topographical engineers and his and Meek's joint fieldwork in Kansas, 1858. Examined mining prospects in American South, 1865–1866. Collected vertebrate fossils in White River badlands for Academy of Natural Sciences, Philadelphia, 1866. Began initial postwar surveys for federal government by leading Meek and prewar field assistant James Stevenson (later Hayden's executive officer) in examining geology and economic resources, especially coal, in new Nebraska state, 1867. With larger staff, extended reconnaissance along Union Pacific Railroad in Wyoming, 1868. Organized, staffed, planned fieldwork, wrote, helped edit, and supervised printing of reports and lobbied Congress for increased appropriations for Hayden Survey, 1869–1879. Hayden's organization one of four federally sponsored geological, geographical, and natural resource reconnaissances conducted west of the 100th meridian before 1880. It examined and mapped in reconnaissance some 420,000 square miles in western Great Plains and Rocky Mountains from United States–Canadian boundary to New Mexico and Arizona; of this area, about 110,000 square miles were mapped more accurately by triangulation surveying after 1872. Its work included surveys of Yellowstone River headwaters and adjacent areas in 1871, 1872, and 1878 and major multipurpose mapping (at a scale of about four miles to the inch) of Colorado, 1873–1876. Member, National Academy of Sciences, 1873. Joined newly established U.S. Geological Survey (USGS), 1879, as one of five principal geologists. With Albert C. Peale (Hayden Survey, 1871–1879), formed USGS Montana Division of Geology, 1883–1886; participated in discovery of Devonian-age rocks in Montana and in mapping area around Three Forks of Missouri River.

Hayden Medal is awarded for achievement in geology by Academy of Natural Sciences, Philadelphia. Hayden's name is perpetuated in the town of Hayden, Grand County, and Mount Hayden, San Juan Mountains, Colorado; two Hayden Peaks (Uinta Mountains, Utah; Elk Mountains, Colorado); Hayden Park, Teller County, and Hayden Butte, Huerfano County, Colorado; Hayden Valley, Yellowstone National Park; and Hayden Glacier (a tributary of Malaspina Glacier), southeastern Alaska. More than forty living and fossil taxa were named for Hayden.

Published about 150 articles, reports, books, maps, catalogs, and albums (some twenty published with Meek before 1866), 1856–1888, including "Geological and Natural History Notes on Nebraska [Territory]," in G. K. Warren, "Explorations in the Dacota Country, in the Year 1855," 34th Congress, 1st Session, *Senate Executive Document*, (76) 13 (822) (1856); "Paleontology of the Upper Missouri," *Smithsonian Contributions to Knowledge*, (172) (with Meek, 1865); "The Yellowstone National Park," *American Journal of Science and Arts* (3d ser.) 3 (April 1872, with map); *The Yellowstone Park and the Mountain Regions of Portions of Idaho, Nevada, Colorado, and Utah* (1876; illustrations by Moran); *Geological and Geographical Atlas of Colorado* (1877; 2d ed., 1881); *The Great West: Its Attractions and Resources . . . A Popular Description* (1880); and "The General Geologic Map of the Area Explored and Mapped by Dr. F. V. Hayden and the Surveys under His Charge, 1869 to 1880," *Twelfth Annual Report of the United States Geological and Geographical Survey of the Territories . . . for 1878* (1883; scale 1:2,600,000).

MAJOR CONTRIBUTIONS

During 1853–1862, Hayden and Meek mapped, classified, and named Cretaceous and Tertiary formations of the upper Missouri River region; their classification of Cretaceous sedimentary rocks persists as framework of present standard geologic section in western interior. Hayden and Meek mapped and deciphered the anticlinal structure of the Black Hills; Hayden identified the post-Cretaceous interval of major mountain building in the Rocky Mountains, and he interpreted many Tertiary sedimentary rocks in its basins as lake deposits. Of the four territorial surveys, the Hayden Survey was the largest, best known, and most broadly based in natural history. In addition to reconnaissance mapping and classifying lands and resources of the western interior, the organization also provided extensive field and laboratory experience and publishing opportunities for its established and younger collaborators; eight staff members later served with the USGS. The Hayden Survey also established a significant network of publication exchanges in North America and Europe; its collections formed the nucleus of USGS library in 1882. Its work institutionalized geology as a public service in the Department of the Interior. Hayden's enthusiastic publicizing of the scenic grandeur, geologic wonders, and natural resources of the nation's western lands promoted both their use and preservation. He distributed rapidly and widely the preliminary results of his survey's work to scientific colleagues, legislators, and the public via illustrated annual reports, bulletins, maps, popular articles, and photographs. These contributions assisted settlement and development of both the agricultural and mining West.

Hayden Survey publications called for conservation of timberlands, studies of temperature and rainfall in the plains west of the 98th meridian to determine where irrigation was necessary to grow grain, and a planned system of irrigation in dry regions with governmental control of irrigation and water rights. Hayden played the leading role, 1871–1872, lobbying for congressional approval of the bill, whose language was modeled on the Yosemite Valley–Mariposa Big Tree Grove Act (1864), which established Yellowstone National Park. Hayden quickly published two articles, one illustrated by Thomas Moran, in popular and scientific journals, describing the geysers, hot springs, and other unique wonders of the lands near the Yellowstone River's headwaters explored by his survey in 1871. Hayden also distributed to Congress photographs by William Henry Jackson (q.v.) and Moran's sketches and watercolors made that summer. Hayden used his influence to aid members of Congress, the Interior Department, territorial governments, and industry in successfully promoting the reservation, withdrawal, dedication, and setting apart on 1 March 1872, of the Yellowstone lands forever as the first national "public park or pleasuring-ground for the benefit and enjoyment of the people."

BIBLIOGRAPHY

Bartlett, R. A. *Great Surveys of the American West* (1962), Pt. 1, Chs. 1–3, "Exploring with Hayden."

Bartlett, R. A. *Nature's Yellowstone* (1974).

Cassidy, G. J. "Ferdinand Vandiveer Hayden: Federal Entrepreneur of Science." Diss. University of Pennsylvania, 1991.

Chittenden, H. M. *The Yellowstone National Park*. Ed. R. A. Bartlett (1895; repr. 1964).

Cope, E. D. "F. V. Hayden, M.D., LL.D.," *The American Geologist* 1 (1888).

Goetzmann, W. H. *Exploration and Empire. The Explorer and the Scientist in the Winning of the American West* (1966), Ch. 14, "F. V. Hayden, Gilded Age Explorer."

Manning, Thomas G. *Government in Science: The US Geological Survey 1867–1894* (1967).

Nelson, C. M., M. C. Rabbitt, and F. M. Fryxell. "Ferdinand Vandiveer Hayden: The U.S. Geological Survey Years, 1879–1886." *Proceedings of the American Philosophical Society* 125 (1981).

Rabbitt, M. C. *Minerals, Lands, and Geology for the Common Defence and General Welfare. Volume 1, Before 1879* (1979); *Volume 2, 1879–1904* (1980).

Schmeckebier, L. F. "Catalogue and Index of the Publications of the Hayden, King, Powell, and Wheeler Surveys." *U.S. Geological Survey Bulletin* 222 (1904; repr., 1970).

White, C. A. "Memoir of Ferdinand Vandiveer Hayden, 1839 [*sic*]–1887." *National Academy of Sciences Biographical Memoirs* 3 (1895).

UNPUBLISHED SOURCES

Hayden's field notebooks and personal papers remain unlocated; other manuscript material and Haydeniana include Hayden and Hayden Survey (1853–1880), Microcopy No. M-623, Record Group 57 (USGS), and Record Groups 48 (Office of the Secretary of the Interior) and 49 (Bureau of Land Management; includes General Land Office records), National Archives and Records Service; Howell-Fryxell Hayden Collection, American Heritage Center, University of Wyoming; Library and Horace Albright Visitors Center, Yellowstone National Park; Joseph Leidy Papers, Academy of Natural Sciences, Philadelphia.

Clifford M. Nelson and Fritiof M. Fryxell

HEARNE, SAMUEL. Born London, England, 1745; died London, England, November 1792. Fur trader, explorer, naturalist.

FAMILY AND EDUCATION

Son of Samuel Hearne, of the London Bridge Water Works, and his wife, Diana. Schooled in Beaminster in Dorset, "without noticeable success. To the end of his life Hearne's spelling remained quaintly phonetic, his grammar erratic and his mathematics dubiously reliable." No record of marriage.

POSITIONS

Servant to Capt. Samuel Hood, Royal Navy, 1756–1763. Seaman, Hudson's Bay Company, out of Churchill, 1766–1769. Explorer for Hudson's Bay Company: following two abortive attempts, a thirty-five-day trip in November 1769, and another of eight months, twenty-two days in 1770, Hearne made his famous overland trip to the Coppermine River between 7 December 1770 and 29 June 1772. After serving briefly as mate of the *Charlotte*, Hearne was assigned to found the first inland trading post of the Hudson's Bay Company, at Cumberland House, on the Saskatchewan River, in 1774–1775. Chief at Churchill from January 1776 until he surrendered it to a French force under Jean François de Galaup, Comte de Lapérouse, on 8 August 1782. As a special dispensation from Lapérouse, Hearne was allowed to risk sailing one of the fort's little trading sloops back to England. With the British again in possession, Hearne returned to resume charge of Churchill from 1783 through 1787, then retired because of ill health.

CAREER

The first to give recognizable descriptions of the Ross goose, musk ox, and wood buffalo; to record the habits of the arctic ground squirrel and arctic hare, and to describe the nesting of the white-crowned sparrow. The first to differentiate between the Hudsonian curlew and the Eskimo curlew. Perhaps the first to appreciate that the drumming noise of the ruffed grouse was made by "clapping their wings with such a force, that at half a mile distance it resembles thunder." The summering nonbreeding Canada geese he called the "barren geese," not because they summered on the Barren lands, but because he dissected them to find "the exceeding smallness of their testicles."

MAJOR CONTRIBUTIONS

"Head and shoulders superior to every other North American naturalist who preceded Audubon" (Richard Glover).

BIBLIOGRAPHY
Glover, R., ed. *A Journey to the Northern Ocean*. Toronto: Macmillan (1958).
Hearne, S. *A Journey from Prince of Wales's Fort in Hudson's Bay to the Northern Ocean in the Years 1769—1770—1771—1772*. London: Strahan and Cadell (1795).
Houston, M., and S. Houston. "Samuel Hearne, Naturalist." *The Beaver* 67, 4 (September 1987).

C. Stuart Houston

HENSHAW, HENRY WETHERBEE. Born Cambridge, Massachusetts, 3 March 1850; died Washington, D.C., 1 August 1930. Naturalist, anthropologist, government administrator, photographer.

FAMILY AND EDUCATION

Youngest of seven children of William Henshaw, Boston book agent, and Sarah Holden Henshaw. Mother's sensitivity to nature made strong impression. Attended Cambridge High School, 1865–1869; studied briefly at Columbian College and read medicine informally with Harry Crecy Yarrow, mid-1870s. Unmarried.

POSITIONS

Bird collecting on Coast Survey schooner *Varina* in Louisiana, 1869–1870, and with Charles Johnson Maynard in Florida, 1870–1871; naturalist, U.S. Geographical Surveys West of the One Hundredth Meridian (Wheeler Survey), 1872–1879; ethnologist, Bureau of [American] Ethnology, Smithsonian Institution, Washington, D.C., 1880–1893; photographer, Hawaii, c.1895–1904; bird collecting, Hawaii, c.1900–1904; administrative assistant, U.S. Bureau of Biological Survey, Washington, D.C., 1905; assistant chief, 1905–1910; chief, 1910–1916.

CAREER

Ill health a problem throughout life and prevented advanced formal study. Developed as ornithologist through own study, fieldwork, association with naturalists in Boston area. School friend William Brewster especially influential. Henshaw claimed responsibility for meetings eventually formalized in Nuttall Ornithological Club, predecessor of American Ornithologists' Union (AOU). Served AOU on council, 1883–1894, 1911–1918; vice president, 1891–1894, 1911–1918; committee on nomenclature and checklist, 1883–1893, and committee to revise nomenclature code, 1907–1908. Appointed to Wheeler Survey on recommendation of Spencer Fullerton Baird and worked as naturalist in Colorado, Utah, Nevada, New Mexico, Arizona, and California. Contributed several new species, especially birds, to U.S. National Museum, often reported by others. Also formed private bird collection, which became first large North American collection in British Museum. Condition of collection, labeled by Robert Ridgway and Henshaw, helped demonstrate abroad soundness of rapidly developing American ornithology. Early anthropological work on Wheeler Survey in California and Southwest with Paul Schumacher and Yarrow. John Wesley Powell selected him as chief lieutenant in Bureau of [American] Ethnology. Thus took up administration, Indian work of tenth U.S. census, ethnological and linguistic fieldwork in West Coast and Plateau areas, execution of historical classification of Indian languages of United States, Alaska, Canada, and work on Indian tribal nomenclature, which he expanded into encyclopedia. Linguistic classification published in Powell, ''Indian Linguistic Families of America North of Mexico,'' *Seventh Annual Report of Bureau of American Ethnology* (1891), with map by Henshaw and his biologically derived rules of nomenclature. First comprehensive effort of its type and starting point for American historical linguistics in following decades. Encyclopedia completed under Frederick Webb Hodge and published as *Handbook of American Indians North of Mexico* (1907, 1910). (Among North Americans has been called ''museum man's bible.'') Collected for, and worked on, Smithsonian anthropological exhibits for World's Columbian Exposition, Chicago, 1893, which, in part, demonstrated effects of environment on culture.

Took up photography in ethnological work and, after retiring to Hawaii, supported self by sale of scenic and ethnological views of islands. Published *Birds of Hawaii* (1902), modest work intended to fill a void in literature. Upon return to United States, called to Biological Survey to respond to political demand for more practical work. Stressed protection of beneficial species and control of harmful ones. Had part in control of trade in plumage, negotiation of treaty to protect migratory birds with Great Britain, establishing many bird sanctuaries and big-game preserves. Prepared Department of Agriculture Farmer's Bulletin 513, ''Fifty Common Birds of Farm and Orchard'' (1913), one of the department's most successful bulletins, with 200,000 copies distributed and later republished in revision by National Geographic. After retirement, pursued study of diatoms with Albert Mann, followed by prolonged physical and mental decline.

Of a retiring, unassuming nature, Henshaw often functioned best as a second to stronger men. He apparently recognized this and declined offers to head Bureau of American Ethnology and to become president of the American Ornithologists' Union. Nevertheless, won admiration of others through gentleness, modesty, humor, extreme clarity of purpose, and dedicated, capable work and rose to leadership almost in spite of himself.

MAJOR CONTRIBUTIONS

To naturalists, was primarily a field-worker who collected several new species, labored in underworked areas, and helped form important collections. As administrator, played important role in redirecting efforts of Biological Survey to practical ends and promoting conservation. As anthropologist, salvaged from oblivion data about several moribund Indian languages, Esselen being an outstanding example. More important was association with Powell's linguistic classification, a landmark work. Generally, to Powell goes credit for idea, to Henshaw credit for its execution.

BIBLIOGRAPHY

Cameron, Jenks. *The Bureau of Biological Survey* (1929).

The Condor (1919–1920).

Heizer, Robert F. ''California Indian Linguistic Records: The Mission Indian Vocabularies of H. W. Henshaw.'' *University of California Anthropological Records* (1955).

Hodge, F. W., and C. H. Merriam. *American Anthropologist* (1931) (obit.).

Kroeber, A. L. ''Powell and Henshaw: An Episode in the History of Ethnolinguistics.'' *Anthropological Linguistics* (1960).

Palmer, T. S. *Biographies of Members of the American Ornithologists' Union* (1954) (obit.).

Sturtevant, William C. ''Authorship of the Powell Linguistic Classification.'' *International Journal of American Linguistics* (1959).

Washington Star, 2 August 1930 (obit.).

Who Was Who in America.

UNPUBLISHED SOURCES

See letters, anthropological records, photographs in Smithsonian Institution National Anthropological Archives, Washington, D.C.; photographs of Hawaii in National Geographic Society, Washington, D.C.; letters in Witmore Stone Papers in Academy of Natural Sciences, Philadelphia; letters in several collections of Smithsonian Institution Archives, Washington, D.C.; administrative materials in records of the Fish and Wildlife Service (RG 22), National Archives and Records Service, Washington, D.C.

James R. Glenn

HENTZ, NICHOLAS MARCELLUS. Born Versailles, France, 25 July 1797; died Mariana, Florida, 4 November 1856. Arachnologist, entomologist.

FAMILY AND EDUCATION

Father an advocate who had been forced for political reasons to leave Paris and live in Versailles under the assumed name of Arnould shortly before son's birth. Hentz was student of miniature painting for several years, then attended medical classes at Hospital Val de Grace, 1813–1815, as student while also working as hospital attendant. Father proscribed for political views, early 1816, and family moved to United States, finally settling in Pennsylvania. Later attended medical classes at Harvard, 1820–1821, before giving up pursuit of medical training. Married Caroline Lee Whiting, 30 September 1824; four children.

POSITIONS

Studied spiders while supporting himself as teacher of French and miniature painting, Boston, Philadelphia, and Sullivan's Island, North Carolina. Taught one year with historian George Bancroft at Round Hill School, Northampton, Massachusetts, 1824–1825. Following his marriage, taught modern languages at the University of North Carolina, 1825(?)–1830 or 1831. Subsequently lived in several other locations to establish schools for girls and other educational institutions, assisted by wife, a novelist. Moved to Covington, Kentucky, 1830 or 1831; to Florence, Alabama, 1834; to Tuscaloosa, Alabama, 1842; to Tuskeegee, Alabama, 1846; and to Columbus, Georgia, 1847. Retired from teaching in 1849 or 1850.

CAREER

Became first American arachnologist through extensive personal research and observation. Supported himself through teaching over a period of more than thirty years. Through correspondence with Thaddeus W. Harris, pioneering American entomologist and librarian at Harvard, and Thomas Say, Hentz received guidance on taxonomic questions. He, in turn, supplied Harris and Say with specimens and information. Was responsible for descriptions of several new northern species. Also painted a number of spiders for projected volume on the subject. Hentz published early papers in *Silliman's Journal* and in the *Journal of the Academy of Natural Science in Philadelphia* on beetles, and his later papers on spiders appeared in the *American Journal of Science* and the *Journal of the Boston Society of Natural History*. He had described 141 species by 1841. Spent much of his life assembling materials about spiders, published long after his death by the Boston Society of Natural History as *The Spiders of the United States, a Collection of the Arachnological Writings of Nicholas Marcellus Hentz, M.D.* (1875), edited by Edward Burgess and J. H. Emerton. Publication was delayed owing to the high cost of reproducing the twenty-one plates. Hentz retired from teaching due to the onset of a nervous disorder, which compelled him to move to his son's home in Florida.

MAJOR CONTRIBUTIONS

Pioneering American arachnologist and entomologist, whose work did much to establish the basis for the study of spiders in the United States.

BIBLIOGRAPHY

Adams, Oscar Fay. *A Dictionary of American Authors*. 5th ed., (1905).
Cobb, C. "Nicholas Marcellus Hentz." *Journal of the Elisha Mitchill Society* 47 (1932).

Cooke, John. "A Pioneering Spider Man." *Natural History* (July 1996).
Eliott, Clark M. *Biographical Dictionary of American Science: The Seventeenth through the Nineteenth Centuries* (1979).
Hart, James D. *The Oxford Companion to American Literature.* 15th ed. (1983).
Mallis, Arnold. *American Entomologists* (1971).
National Cyclopedia of American Biography. Vol. 9 (1907).
Scudder, S. H. *Entomological Correspondence of Thaddeus William Harris, M.D.* (1869).
Stewart, Wallace W. *A Dictionary of North American Authors Deceased before 1950* (1951).
Who Was Who in America. Historical Vol. (1963).

UNPUBLISHED SOURCES

Hentz's correspondence with Thaddeus W. Harris is at the Boston Museum of Science and at Harvard University. His spider portraits are at the American Museum of Natural History.

Joseph Bongiorno

HERBERT, HENRY WILLIAM (FRANK FORESTER). Born London, England, 7 April 1807; died New York City, 17 May 1858. Novelist, translator, writer on sports and nature.

FAMILY AND EDUCATION

Eldest son of Rev. William Herbert (a son of the earl of Carnarvon and member of British Parliament) and Letitia Emily Dorothea (a daughter of Joshua, fifth viscount Allen). Educated at home, then sent to Eton in 1820. Graduated with distinction from Cambridge in 1829. Emigrated to America for financial reasons in 1831. Married Sarah Baker in 1839 (deceased 1844), had one son who attended school in England and did not return to the United States; married Adela R. Budlong in 1858, who left him after a few weeks, which led to Herbert's suicide.

POSITIONS

Taught until about 1853 at Rev. R. Townsend Huddart's school in New York City. Edited *American Monthly Magazine*, 1833–1835. Did some translating of Greek classics and French novels. Spent more than twenty years writing historical romances and sketches of hunting and sporting life.

CAREER

An outstanding classical scholar, Herbert wrote a number of volumes of historical fiction, now largely forgotten, and hoped to build his reputation as an author of such works. His better novels include *The Brothers: A Tale of the Froude* (1835); *Cromwell* (1838), *Ringwood the Rover* (1843), and *The Cavaliers of England* (1852). In 1839, began a very prolific career of writing on field sports, including hunting and fishing, and used pseudonym of Frank Forester to keep his two careers separate. A number of his sketches, some of which had first appeared in the periodical press, were published in *My Shooting Box* (1843) and *The Deerstalkers: A Sporting Tale of the Southwestern Counties* (1843, with several subsequent editions). Another compilation was *The Warwick Woodlands, or Things As They Were Ten Years Ago* (1845, new edition, illustrated by Herbert, 1850). Several titles still consulted today are the two volume *Frank Forester's Field Sports of the United States and British Provinces of North America* (1849, with at least twelve later editions); and *Frank Forester's Fish and Fishing of the United States and British Provinces of North America* (1849). During the years 1850 to 1853, more than twenty of Herbert's articles concerning the hunting or fishing of various game, bird, and fish species appeared in *Graham's Magazine*. Other books published by Herbert included *American Game and its Seasons* (1853), which appeared in three other editions by 1873; *The Complete Manual for Young Sportsmen: With Directions for Handling the Gun, the Rifle, and the Rod* (1856 and at least nine subsequent editions by the mid–1870s); and *Frank Forester's Horses and Horsemanship of the United States and British Provinces of North America* (2 vols, 1857), a study still consulted by present-day afficionados of the subject. *The Silent Rifleman: A Tale of the Texas Prairie* appeared posthumously in 1870. A number of these books were illustrated with his own sketches and when discussing fishing, he depended for accuracy on specimens he had himself seen or illustrations furnished by authorities such as Louis Agassiz at Harvard. Herbert also edited sportsman's books by several other authors, notably William Post Hawes and Edward Mayhew. Herbert's collected poems were published in 1888 and a number of his other works were still appearing in reprinted editions as late as the 1920s. Herbert was a proud, ambitious, and often difficult individual with very high standards who could and often did anger those close to him. He was also a principled sportsman and a pioneering conservationist who strongly urged the wise use of fish and wildlife resources.

MAJOR CONTRIBUTIONS

A genuine lover of nature and sports, Herbert's books are still fresh and readable. As the first sportswriter in the United States, he railed against the wanton slaughter of game. Reflected in his books and essays is his alarm over the rapid depletion of wildlife resources in the Middle Atlantic states, which inspired the founding of the New York Sportsmen's Club. The club became instrumental in suing poachers and dealers for the possession of game killed out of season. So effective was the New York club in its crusade for animal life that it influenced the setting up of similar groups in Boston, Providence, Rhode Island, Toronto, Canada, and other major cities.

BIBLIOGRAPHY
Hunt, William S. *Frank Forester: A Tragedy in Exile* (1933).
Meats, Stephen. "Addendum to Van Winkle: Henry William Herbert [Frank Forester]." *Papers of the Bibliographical Society of America* 67 (1973).
Phillips, John C. *American Game Mammals and Birds: A Catalogue of Books, 1582 to 1925. Sport, Natural History, and Conservation* (1930).
Picton, Thomas. "Henry William Herbert." In David W. Judd, ed., *The Life and Writings of Frank Forester*, Vol. 1 (1882).
Van Winkle, William Mitchell. *Henry William Herbert [Frank Forester]: A Bibliography of His Writings, 1832–1858* (1936).
White, Luke. *Henry William Herbert and the American Publishing Scene, 1831–1858* (1943).

UNPUBLISHED SOURCES
Several letters are held by Yale University Library, and other materials are in Rufus Wilmot Griswold Papers, Boston Public Library, New York Historical Society, New Jersey Historical Society, and the Historical Society of Pennsylvania. Other letters and materials are widely scattered.

George A. Cevasco

HERNDON, WILLIAM LEWIS. Born Fredericksburg, Virginia, 25 October 1813; drowned in waters off Cape Hatteras, North Carolina, 12 September 1857. Naval officer, explorer.

FAMILY AND EDUCATION

One of seven children of Dabney and Elizabeth Hull Herndon. Orphaned when very young. Little information available concerning early education. Married Frances Elizabeth Hansbrough, 1836. At least one daughter, Ellen, who subsequently became the wife of Chester A. Arthur, president of the United States, 1881–1885. Was both a cousin and brother-in-law of Matthew Fontaine Maury (1806–1873), American and later Confederate naval officer and oceanographer.

POSITIONS

Entered U.S. Navy as midshipman, 1828. Promoted to passed midshipman, 1834; lieutenant, 1841; and to commander, 1855. Commander of the *Iris*, a steamer operating in the Gulf of Mexico, during the Mexican War (1847–1848). Assigned to Depot of Charts and Instruments, Washington, 1843–1846 and to Naval Observatory, Washington, 1848–1850. Sent to south Pacific on the *Vandalia*, 1850–1851; detached from that duty to lead exploring expedition to the Amazon in pursuance of scientific information required by Maury, 1851–1852. In Washington drafting reports, 1852–1853, returned to sea duty, 1854–1855. While on leave from navy, commanded Pacific Mail Company steamer *George Law* (subsequently renamed *Central America*), 1855–1857.

CAREER

Following entry into the navy, served on a succession of vessels, notably the *Guerriere* in the Pacific (1829–1832); the *Constellation* in the Mediterranean (1832–1834); and the *Independence* off the coast of South America (1834–1836). Assigned to duty under his cousin Lieutenant M. F. Maury at Depot of Charts and Instruments, with responsibility for working up astronomical calculations and performing other duties; left because of nervous breakdown, 1846; after brief period for recovery was assigned to command of USS *Isis* in waters off Mexico during Mexican War. Reassigned to Maury, whose office had now been renamed U.S. Naval Observatory in 1848, and then to the USS *Vandalia* in the south Pacific. In August 1850, while in Santiago, Chile, en route to Hawaii, he was notified by Maury of his assignment to explore the Amazon. Spent months awaiting further orders in Valparaiso, Chile, and then was directed to go to Lima, Peru, to do whatever preliminary research was possible in local monasteries and to prepare for his Amazon trip. Nine month delay in Valparaiso and Lima was occasioned by changes of federal government personnel in Washington and by Brazilian government hesitation about letting Herndon proceed. Brazilians had some concerns about Herndon's objectives. Herndon was finally joined in April 1851 by Passed Midshipman Lardner Gibbon, who had been assigned to assist him and who handed him his written orders. Herndon's official mission was to investigate the navigability of the Amazon, its geography, and its suitability for agriculture and commerce, while incidentally taking note of, or collecting, items of scientific interest. In confidential letters, Maury wrote Herndon of his thought that American slaves might in future be resettled in the Amazon and that American planters could some day be profitably engaged in agricultural development there. American interest in the area was also heightened by earlier British explorations in the region. Though the Amazon had recently been opened to steam navigation by Emperor Dom Pedro II, that privilege would not be extended to foreigners until 1867. Herndon made some confidential inquiries in furtherance of Maury's ideas, but focused on his official objectives. Maury had suggested in passing that his cousin avail himself of any opportunity to learn what he could of embryology or some other scientific field, but this Herndon had been in no position to do, either in Valparaiso or in Lima. Some scientific instructions for collecting specimens prepared by S. F. Baird of the Smithsonian probably reached Herndon in Lima, but he had no qualified biologists in his party and, lacking any scientific background, was obliged to do this work himself. The choice of route to be followed was left to Herndon's discretion. Herndon and Gibbon left Lima in May 1851, crossed the Andes and after some weeks of travel, ascended the Amazon to the river's Peruvian tributaries. In late June 1851, Herndon split his party, sending Gibbon to explore Bolivian tributaries of the Amazon and make other investigations in Peru while he continued up the river to its mouth. In addition to geographic and navigational data, Herndon collected plants, seeds, and a number of local animals along the way, some of which he skinned, others of which were taken along in cages. Unfortunately, most of the caged mammals fought or killed each other. He was left with a tortoise and only ''about a dozen mutuns, or curassows; a pair of Egyptian geese; a pair of . . . pucacunga [birds]; a pair of macaws; a pair of parrots; and a pair of large white cranes, called jaburu'' which accompanied Herndon to New York. His last caged monkey expired as his ship entered New York Harbor. Many of the specimens ultimately reached Baird in Washington and were added to the Smithsonian collections, but no money for a published catalog of the Herndon-Gibbon specimens was forthcoming. John Cassin, a Philadelphia ornithologist, was particularly impressed with Herndon's bird specimens, terming them ''uncommonly good.'' Herndon completed his Amazon trip in April, left Para for New York in May aboard the USS *Dolphin*, and arrived in New York at the beginning of July 1852. Gibbon remained in Peru and Bolivia for a somewhat longer period, investigating mining procedures and making an archaeological study of certain pre-Columbian natives before proceeding up the Amazon to Para. He also brought back natural history specimens on his return home in March 1853.

Herndon and Gibbon produced a two-volume report, *Exploration of the Valley of the Amazon: Made under the Direction of the Navy Department* (1853–1854), published as House of Representatives Executive Document no. 53, 33rd Congress, First Session. A set of maps made by Herndon and Manuel Sobreviela, and drawn by John Tyssowski and H. C. Elliott, was also published as *Maps of the Amazon, Huallaga, and Ucayali Rivers and Amazon Valley* (1852?). Gibbon's report, which took up volume two, was ''for the most part pedestrian and poorly written,'' but Herndon's account in volume one made excellent reading and had continuing value for explorers and anthropologists. Maury later based much of his own *Letters on the Amazon and Atlantic Slopes of South America* on information given to him by Herndon, which was not included in the text of Herndon's report. Herndon spent a good part of 1854 and part of 1855 on sea duty, first on the USS *San Jacinto* and then on the USS *Potomac*. In 1855 he secured a leave of absence from the navy and took command of the *George Law*, a two year old steamer (later renamed the *Central America*) owned by the Pacific Mail Company, which plied between Aspinwall (later Colon), Panama, and New York. For several years, he performed his duties without incident. In mid-September 1857, however, while enroute to New York, his vessel went down in a gale off Cape Hatteras. Something over 100 of the ship's complement, primarily women and children, were saved with the aid of several passing vessels, but Herndon and 425 others were lost. A monument to his memory stands on the grounds of the Naval Academy and a destroyer was later (1919) named in his honor.

MAJOR CONTRIBUTIONS

Herndon's report ranks as one of the better and more informative nineteenth-century American travel accounts. It was instructive and detailed, whimsical and humorous, an unusual combination in normally dry, matter-of-fact government documents. The report was also a major inspiration to Samuel L. Clemens (Mark Twain). He regarded his reading of Herndon's account in 1856 as having been a turning point in his life in that it turned him toward a career of travel writing. Herndon and Gibbon were among the first Americans to bring back valuable information and specimens concerning Amazonian fauna, flora, and anthropology.

BIBLIOGRAPHY

Basso, Hamilton, ed. "Introduction" in William Lewis Herndon, *Exploration of the Valley of the Amazon* (1952).

Bell, Whitfield, J., Jr. "The Relation of Herndon and Gibbon's Exploration of the Amazon to North American Slavery, 1850–1855." *Hispanic–American Historical Review* 19 (1939).

Dictionary of American Biography.

Dozer, D. M. "Documents: Matthew Fontaine Maury's Letter of Introduction to William Lewis Herndon." *Hispanic–American Historical Review* 28 (May 1948).

Harrison, John P. "Science and Politics: Origins and Objectives of Mid-Nineteenth Century Government Expeditions to Latin America." *Hispanic–American Historical Review* 35 (May 1955).

Kazar, John D. "The United States Navy and Scientific Exploration, 1837–1860." Unpublished doctoral dissertation, University of Massachusetts (1973).

The New York Times, 18 and 19 September 1857.

Ponko, Vincent Jr. *Ships, Seas, and Scientists: U.S. Naval Exploration and Discovery in the Nineteenth Century* (1974).

UNPUBLISHED SOURCES

There is some correspondence between Herndon and the Secretary of the Navy in Record Group 45, National Archives, and with Lt. James Gillis in Valparaiso, Chile [through whom Herndon received some of his Amazon expedition instructions] in the correspondence of the U.S. Naval Astronomical Expedition to the Southern Hemisphere, Record Group 78, National Archives. S. F. Baird's interest in the specimens brought back by this enterprise is reflected in his correspondence files at the Smithsonian Institution Archives.

Keir B. Sterling

HEWITT, CHARLES GORDON. Born near Macclesfield, England, 23 February 1885; died Ottawa Ontario, 29 February 1920. Entomologist, Canadian government official.

FAMILY AND EDUCATION

Son of Thomas Henry and Rachel Frost Hewitt. Graduate of King Edward VI Grammar School, Macclesfield. B.S., 1902, M.Sc., 1903, D.Sc., 1909, Manchester University. Married Elizabeth Borden (niece of Prime Minister Robert Borden), 1911; no children.

POSITIONS

Assistant lecturer in zoology, 1902–1904, lecturer in economic zoology, Manchester University. Dominion entomologist, Canada, 1909–1916; title changed to dominion entomologist and consulting zoologist, 1916–1920.

CAREER

During his eleven years of office in Canada, he developed the federal Entomological Service from a very small division, with one assistant and a stenographer attached to the Experimental Farms Branch, to a separate branch of the Department of Agriculture with four divisions, each headed by a highly qualified entomologist. Twelve field laboratories were established across Canada, employing sixty-three trained entomologists and ancillary staff, to investigate local insect problems. He also inaugurated an efficient quarantine system against the importation of foreign insect pests. In 1910, he was instrumental in the passage of the Destructive Insect and Pest Act by the Parliament of Canada. This act permitted the appointment of inspectors to curtail the spread of the brown-tailed moth into the Maritimes and made provision for the prohibition, fumigation, and inspection of nursery stock at designated ports of entry. In addition, his professional responsibilities embraced economic ornithology in relation to agriculture, and he had a strong interest in promoting wildlife preservation. His outstanding contributions to science are his several books: *The House Fly* (1910, 2 eds.); *House Flies and How They Spread Disease* (1912), and *Conservation of the Wild Life of Canada* (1921). His departmental publications consisted of a series of *Annual Reports* (1910–1916) and bulletins. Chief among the latter are those dealing with the honeybee and the larch sawfly. The 130 papers on various scientific subjects dealt primarily with the biology and control of pest insects, especially those related to medical entomology. A volume on the insects collected by the Canadian Arctic Expedition (1913–1918) was published in 1919, under his direction. Served as secretary of the Advisory Board on Wildlife Protection (1916–1920), an interdepartmental committee that formulated policy and advised the government on wildlife conservation matters. In 1913, initiated discussions with H. W. Henshaw, chief of the U.S. Biological Survey, concerning possibilities for Canadian–U.S. migratory bird treaty. Worked to bring this about in concert with other dominion and American authorities. Treaty was signed in 1916 and ratified in 1918, following passage of Migratory Birds Convention Act by the Canadian Parliament (1917) and the necessary enabling legislation in the U.S. Congress (1918). Hewitt received Gold Medal of the Royal Society for the Protection of Birds (London) in 1918 for his services in furthering the Migratory Bird Treaty and for his promotion of the revised Northwest Game Act (1917), which included most of the provisions he had recommended. President of the Entomological Society of Ontario (1913), the American Association of Economic Entomologists (1915), and the Ottawa Field-Naturalists' Club (1918). Active in, and officer of, several other scientific societies. His early death at age thirty-five was due to a case of influenza that developed into pleural pneumonia.

MAJOR CONTRIBUTIONS

His prestige as a scientist and administrator was widely recognized. As a resolute promoter of the well-being of wildlife, he was deeply interested in the development of game preserves, bird sanctuaries, and plans for the protection of wildlife. Greatly expanded the dominion entomological program and did much to ensure the protection of North American migratory birds.

BIBLIOGRAPHY

Criddle, Norman. *Canadian Field Naturalist* 34.9 (1920) (obit.).
Encyclopedia Canadiana.
Foster, Janet. *Working for Wildlife: The Beginning of Preservation in Canada* (1978).
Gibson, Arthur, and J. M. Swaine. *The Canadian Entomologist* 52.5 (1920) (obit.).
Howard, Leland O. *A History of Applied Entomology* (1930).
Proceedings of the Royal Society of Canada (1920) (obit.).
Riegert, P. W. *The Canadian Encyclopedia.* Vol. 2 (1985).
Riegert, P. W. *From Arsenic to DDT: A History of Entomology in Western Canada* (1980).
Spilman, T. J. *Bulletin of the Entomological Society of America* 35.3 (1989).
Wallace, W. Stewart, and W. A. McKay. *The Macmillan Dictionary of Canadian Biography* (1978).
Who's Who. London (1910).

Paul W. Riegert

HICKEY, JOSEPH JAMES. Born New York City, 16 April 1907; died Madison, Wisconsin, 31 August 1993. Ornithologist, conservationist.

FAMILY AND EDUCATION

Son of James Bernard and Sarah Theresa Mooney Hickey. B.S., New York University (NYU), 1930; additional undergraduate studies in biology, late 1930s, NYU; M.S., University of Wisconsin, 1943; Ph.D., University of Michigan, 1949. Married Margaret Brooks, 1942 (died 1976); one daughter. Married Lola Alma Gray Gordon, 1978.

POSITIONS

During the 1930s, Hickey was employed as a track coach at NYU and was later with Consolidated Edison, the New York power utility. Research assistant,

Wisconsin State Soil Conservation Commission, 1941–1943; research assistant, Toxicity Laboratory, University of Chicago, 1943–1944; assistant curator, Museum of Zoology, University of Michigan, 1944–1946; fellow, Guggenheim Memorial Foundation, 1946–1947; assistant professor, Wildlife Management, University of Wisconsin, 1947, chair of department (in succession to Aldo Leopold, 1948), later associate professor and professor to retirement in 1976; emeritus, 1977–1993. President, American Ornithologists' Union, 1972–1973; active in numerous wildlife and conservation advocacy organizations. Received numerous honors and awards from conservation groups.

CAREER

As a boy, Hickey was member of Bronx Birdwatching Club with fellow members Roger Tory Peterson and Alan Cruickshank. Following graduation from NYU, Hickey, a champion miler, was a track coach at NYU and was employed by Consolidated Edison. On recommendation of Ernst Mayr, then at American Museum of Natural History, took up formal study of ornithology. Invited by Aldo Leopold to do graduate study at Wisconsin, later invited by Leopold to join him in Wildlife Management Department at Wisconsin. Hickey's *Guide to Bird Watching* (1943, rev. 1972) served as his M.S. thesis and has since been continually in print. Completed his doctoral dissertation, ''Survival Studies of Banded Birds,'' on Guggenheim Fellowship. It was published in 1952. Shepherded Leopold's *Sand County Almanac* through to publication in 1949. Noted for his research on relationship between DDT and declines in bird populations and on effects of pesticides on wildlife generally. Committed conservationist and highly respected teacher and mentor throughout his career.

MAJOR CONTRIBUTIONS

Known for his fierce independence on issues relating to pesticides and wildlife. Organized international conference on status of peregrine falcon as outgrowth of this concern in 1965. Played major role in eventual banning of DDT. Popular and highly successful teacher at the University of Wisconsin for thirty years. His volume on bird-watching, in print for over half a century, has been an indispensable guide to the subject. Able and dedicated editor of professional journals and conference proceedings. Leading conservationist. A leader in effort to provide food and clothing to destitute European ornithologists after World War II.

BIBLIOGRAPHY

American Men and Women of Science. 15th ed. (1982).
Auk (April 1994) (obit.).
Stroud, Richard H. *National Leaders of American Conservation* (1985).
Who's Who in America (1993).

Keir B. Sterling

HIND, HENRY YOULE. Born St. Mary's Gate, Nottingham, England, 1 June 1823; died Windsor, Nova Scotia, Canada, 8 August 1908. Explorer, geologist, naturalist, educator.

FAMILY AND EDUCATION

Son of Thomas Hind, a lace manufacturer, and Sarah Youle Hind. The third child in a family of four boys and one girl. Attended the Free Grammar School of Nottingham. Entered Königliche Handelshochschule, a Leipzig commercial college, 1837–1838. Attended Caius College and later Queen's College, Cambridge, 1843–1844, but received no degree from either institution. Honorary master of arts, Trinity College, Toronto, 1853. Emigrated to Toronto, Canada, 1846. Married Katharine Cameron, 1850; seven children.

POSITIONS

Second master and lecturer of science and mathematics, Normal School, Toronto, 1847–1852; professor of chemistry and geology, Trinity College, Toronto, 1852–1864; consulting geologist, 1864–1884; secretary and managing director, Edgehill School for Girls, Windsor, Nova Scotia, 1890–1908.

CAREER

Became a member of the Royal Canadian Institute in 1851, elected a member of the council and second vice president in 1852, and served as editor of the institute's *Canadian Journal*, 1852–1855. Appointed geologist and naturalist to the Red River Expedition on the recommendation of William Logan, director of the Geological Survey of Canada, 1857. Placed in command of the Assiniboine and Saskatchewan Rivers expedition, 1858, and the Labrador expedition, 1861. Engaged by the government of New Brunswick to study the geology of the province, 1864, and by the government of Nova Scotia, 1867–1871. Collated the papers of the International Fisheries Commission, an international arbitration panel appointed to compensate Canadian fishermen for losses incurred under the Treaty of Washington, 1877.

Author of three major official government reports on the geology of the Northwest Territories and New Brunswick; also authored six books, twenty-five pamphlets, and twenty-two articles. His interests were eclectic, focusing on geology, climate, agriculture, fisheries, railway policy, and native customs. Wrote for both scientific and popular audiences. Most important works were *Narrative of the Canadian Red River Exploring Expedition of 1857 and of the Assiniboine and Saskatchewan Exploring Expedition of 1858*, 2 vols. (1860) and *Explorations in the Interior of the Labrador Peninsula, the Country of the Montagnais and Nasquapee Indians*, 2 vols. (1863).

MAJOR CONTRIBUTIONS

Hind was the scientific observer for two government-sponsored expeditions to the region of the Red, Saskatchewan, and Assiniboine Rivers in 1857 and 1858. His reports were among the first scientific works on the western frontier and did much to promote the region within British North America. Hind was also the far-seeing promoter and, unfortunately, let his enthusiasm for the economic possibilities of the frontier interfere with sound scientific investigation. He was always using hastily assembled scientific evidence to promote government policy and, in so doing, obtained prominence for himself among governments more concerned with political patronage than sound environmental management.

BIBLIOGRAPHY

Morton, W. L. *Henry Youle Hind 1823–1908* (1980).

UNPUBLISHED SOURCES

Papers of Henry Youle Hind are in Provincial Archives of Nova Scotia, Halifax. Correspondence is in records of the Provincial and Civil Secretaries' Offices: Upper Canada and Canada West, 1791–1867, Record Group 5, National Archives of Canada, Ottawa; also in papers of William Edmond Logan, McCord Museum, Montreal.

Jeffrey S. Murray

HITCHCOCK, ALBERT SPEER. Born Owosso, Michigan, 4 September 1865; died Washington, D.C., 16 December 1935. Botanist.

FAMILY AND EDUCATION

Son of Albert Hitchcock and Alice Martin Jennings, he was adopted by J. S. Hitchcock and wife. Received B.S.A. (1884) and M.S. (1886) from Iowa State Agricultural College. Granted honorary Sc.D. from Iowa State College in 1920 and Kansas State College in 1934. Married to Rania Belle, 16 March 1890; five children.

POSITIONS

Was instructor in chemistry at Iowa State College, 1886–1889; botanical assistant at Missouri Botanical Garden in St. Louis, 1889–1891; professor of botany at Kansas State Agricultural College, 1892–1901. Between 1901 and 1935, employed by U.S. Department of Agriculture, Washington, D.C., as assistant agrostologist (1901–1905); systematic agrostologist (1905–1924); senior botanist and principal botanist in charge of systematic agrostology (1924–1935).

CAREER

Was specialist on taxonomy of grasses and custodian of section of grasses in U.S. National Herbarium. Believed that taxonomist must be a field-worker.

Made expeditions to study grasses and collect specimens to Mexico and Central America (1910–1911), West Indies (1912), British Guiana (1919), China, Japan, and Philippines (1921). Made visits to all parts of United States, as well as to Labrador, South and East Africa, France, Belgium, Switzerland, Germany, Holland, and England. Author of *Textbook of Grasses* (1914), *Genera of Grasses of the United States* (1920), *Manual of Farm Grasses* (1921), *Methods of Descriptive, Systematic Botany* (1925), and *Manual of Grasses of the United States* (1935). Also wrote over 150 articles for scientific publications. As chairman of executive committee of Institute for Research in Tropical America (1920–1926), encouraged establishment of Barro Colorado Island, Panama Canal Zone, as a field station. Was member of Botanical Society of America (president, 1914), Botanical Society of Washington (president, 1916), and Biological Society of America (president, 1923).

MAJOR CONTRIBUTIONS

Became leading agrostologist in United States and his standing was acknowledged around the globe. Codified concept of a ''type specimen'' as base of a species, which was incorporated in International Rules of Botanical Nomenclature at Fifth International Botanical Congress in 1930 (thus leading to its universal acceptance). His *Manual of the Grasses of the United States*, which rested on the identity of valid species but recorded all known synonyms, contributed to the orderly handling of grass nomenclature for agronomists, foresters, and general public. *Hitchcockella*, a monotypic grass genus of Madagascar, was named in his honor.

BIBLIOGRAPHY

Dictionary of Scientific Biography. Supplement 1, Vol. 15 (1978).
National Cyclopedia of American Biography. Vol. 26 (1967).
Science 38 (6 March 1936) (obit.).
Who Was Who in America, 1897–1942. Vol. 1 (1943).

UNPUBLISHED SOURCES

Some of Hitchcock's correspondence is preserved in the special collection of the Kansas State Library, Manhattan, Kansas. Hitchcock letters are also located in the Witmer Stone Collection, Academy of Natural Sciences, Philadelphia; Marcus E. Jones Collection, Hunt Institute for Botanical Documentation, Carnegie-Mellon University, Pittsburgh; and Rolla K. Beattie, Charles V. Piper, and Harold St. John Collections, Washington State University Library, Pullman.

Richard Harmond

HOCHBAUM, HANS ALBERT. Born Greeley, Colorado, 9 February 1911; died Portage la Prairie, Manitoba, 2 March 1988. Naturalist, conservationist, artist.

FAMILY AND EDUCATION

Son of Hans Weller Hochbaum (who developed "Victory Gardens" in World War II) and Martha Schenck. Schooling in Boise, Idaho, and Washington, D.C. B.Sc. in zoology at Cornell University under Arthur A. Allen, 1933. M.Sc. in wildlife management at University of Wisconsin under Aldo Leopold, 1941. Honorary LL.D. from the University of Manitoba, 1962. Studied fine arts at Cornell and Wisconsin. Married Eleanor Joan Ward in 1939. Four children, Albert Ward Hochbaum (a Manitoba Natural Resources officer), Peter Weller Hochbaum (who wrote and illustrated a booklet about Delta Marsh), George Sutton Hochbaum (a wildlife research biologist with the Canadian Wildlife Service), and Trudi Heal. There are eleven grandchildren.

POSITIONS

Wildlife technician, U.S. National Park Service, 1934–1937; director, Delta Waterfowl Research Station, Delta, Manitoba, 1938–1970. Freelance artist and writer after 1970.

CAREER

While director at Delta, ninety-five graduate students from thirty-eight universities conducted their M.Sc. or Ph.D. thesis research at the Delta Waterfowl Research Station. Made twenty-seven field trips to the Canadian arctic. Exhibited paintings since 1933 with more than a dozen one-man shows. One of his paintings was presented to Queen Elizabeth in 1970. He gave many lectures and seminars on waterfowl, wilderness conservation, art, and writing and authored scripts for Canadian Broadcasting Corporation (CBC) television programs.

He received the Brewster Medal from the American Ornithologists' Union in 1945; the Literary Award of the Wildlife Society in 1944 and again in 1956; a John Simon Guggenheim Fellowship, 1961; the Wilderness Medal from the CBC in 1970; Manitoba Centennial Medal of Honour, 1970; Crandall Conservation Award, 1975; Canada Council Explorations Fellowship, 1975; Special Conservation Achievement Award from the National Wildlife Federation, 1986; Seton Medal, Manitoba Naturalists Society, 1986; Professional Wildlife Conservation Award, government of Manitoba, 1987; member, the Order of Canada, 1979.

MAJOR CONTRIBUTIONS

In addition to many scientific papers and popular articles, Hochbaum published three classics: *Canvasback on a Prairie Marsh* (1944, with numerous reprints); *Travels and Traditions of Waterfowl* (1956); *To Ride the Wind* (1973).

BIBLIOGRAPHY
Hochbaum, H. A. *The Canvasback on a Prairie Marsh* (1944).
Hochbaum, H. A. *Travels and Traditions of Waterfowl* (1956).
Hochbaum, H. A. *To Ride the Wind* (1973).
Houston, C. S. "In Memoriam: H. Albert Hochbaum." *Auk* 105 (1988): 769–772.
C. Stuart Houston

HOOD, ROBERT. Born Portarlington, Ireland, 1797; died twenty-eight miles northeast of Fort Enterprise, Northwest Territories, Canada, 20 October 1821. Midshipman, cartographer, artist, assistant naturalist.

FAMILY AND EDUCATION

The second son of the Reverend Richard Hood, LL.D. At fourteen he became a midshipman and continued his education on board ship. After his death, Greenstockings, maiden of the Copper Indian tribe (resident north of Great Slave Lake), bore his daughter.

POSITIONS

Hood and George Back were the two midshipmen with the first arctic exploring expedition led by John Franklin. Both were chosen because of their artistic skills.

CAREER

In the first winter at Cumberland House on the Saskatchewan River, Hood assisted John Richardson in making natural history observations. He made important contributions to anthropology, climatology, and terrestrial magnetism; he was the first to prove the action of the aurora borealis on the compass needle, demonstrating that the aurora borealis is an electrical phenomenon. Hood painted native North American Indians, mammals, and birds, including the type specimen of the North American subspecies of black-billed magpie, *Pica pica hudsonia*. He painted another four species that at the time had not yet been described to science—the black-backed woodpecker, *Picoides arcticus*, yellow-headed blackbird, *Xanthocephalus xanthocephalus*, hoary redpoll, *Carduelis hornemanni*, and the evening grosbeak, *Coccothraustes vespertinus*, but his perceptive diary and his natural history watercolors were not published until 1974.

Hood was the primary surveyor and draftsman on the Arctic expedition until his death. Whether on foot or in a canoe, Hood plotted accurately as many as thirty-three changes of course in a single day. Already failing from starvation, Hood was shot in the head by Michel, an Iroquois Indian turned cannibal. Hood's memory is perpetuated by the moss phlox, *Phlox hoodii*, and by the mighty Hood River, which plunges over Wilberforce Falls before it enters the Arctic Ocean.

MAJOR CONTRIBUTIONS

His incredibly accurate mapping, his scientific experiments, and his early paintings of birds and mammals.

BIBLIOGRAPHY

Houston, C. Stuart, ed. *To the Arctic by Canoe, 1819–1821, the Journal and Paintings of Robert Hood, Midshipman with Franklin* (1974).

C. Stuart Houston

HORNADAY, WILLIAM TEMPLE. Born Plainfield, Indiana, 1 December 1854; died Stamford, Connecticut, 6 March 1937. Conservationist, naturalist, zoological park director, taxidermist.

FAMILY AND EDUCATION

Younger of two children of William and Mary (Varner) Hornaday, grew up in large farm family with six stepbrothers and one stepsister from his parents' previous marriages. In 1858 family moved to larger farm in Iowa, where surrounding wildlife and virgin prairie made lasting impression on William. Orphaned at fifteen. Attended Oskaloosa College, Iowa, 1870–1871, and Iowa State Agricultural College, 1872–1873, but did not receive degree. Honorary Sc.D. from University of Pittsburgh, 1906, A.M. from Yale, 1917. Married Josephine Chamberlain, 1879; one daughter.

POSITIONS

Taxidermist, Iowa State Agricultural College, 1873. Assistant, Ward's Natural Science Establishment, Rochester, New York, 1873–1882; natural history collecting expeditions to Florida, Cuba, and Bahamas, 1874; South America, 1876; India, Ceylon, Malay Peninsula, and Borneo, 1876–1879. Chief taxidermist, U.S. National Museum (USNM), Smithsonian Institution, 1882–1890. Real estate executive, Buffalo, New York, 1890–1896. Director, New York Zoological Park, 1896–1926.

CAREER

Early fascination with taxidermy led Hornaday to leave college in 1873 to apprentice himself at Ward's Establishment, the foremost natural history specimen exchange in United States. Established reputation as collector by setting off on three expeditions for Henry Augustus Ward. First field trip to southeast United States in 1874 established existence of crocodile in Florida. His popular account of 1876–1879 round-the-world expedition, *Two Years in the Jungle* (1885), described how he amassed one of the largest and most varied collections ever secured by a single man. Upon return to Ward's Establishment, dissatisfaction with current taxidermy practice led him to develop more lifelike forms modeled in clay on wooden manikins and to advocate habitat exhibits of family groups. Introduced this concept to the American Association for the Advancement of Science at Saratoga Springs, New York, in 1879 with his orangutan display, "A Fight in the Treetops." Established National Society of American Taxidermists in 1880 to further these aims.

When called to Smithsonian by USNM Assistant Director G. Brown Goode, applied his taxidermy and exhibit ideas to natural history museum, notably in American bison group. An 1886 journey west to collect bison specimens revealed to him the rapid depletion of North American wildlife; thus became ardent and forceful conservationist, publishing *The Extermination of the American Bison* (1889) and numerous popular articles. Established Department of Living Animals to provide models for USNM taxidermists and as test project for national zoo in park setting where North American wildlife could be preserved. Succeeded in securing congressional approval after several years of lobbying but resigned when Smithsonian Secretary Samuel P. Langley replaced Hornaday's zoological park design with his own.

Spent the next several years in Buffalo, New York, as a real estate entrepreneur and established reputation as nature writer in many popular periodicals of the day with his innovative, casual style of adventure narrative. Also compiled his *Taxidermy and Zoological Collecting: A Complete Handbook for the Amateur Taxidermist, Collector, Osteologist, Museum-Builder, Sportsman, and Traveller* in 1891.

Accepted position in 1896 as director of a new zoo to be established by the New York Zoological Society (NYZS), organization devoted to the display and preservation of wildlife. Selected site and designed the Bronx Zoo as a nature park set in rolling hills and woods. Directed the New York Zoological Park for thirty years and established there the National Heads and Horns Museum, a trophy collection dedicated to vanishing big game of the world.

The second objective of the NYZS charter was "the preservation of our native animals," a responsibility to which Hornaday seriously committed himself. Early successes included the 1902 Alaskan Game Act and the 1911 Bayne Law, which prohibited sale of native game in New York. Massachusetts and California quickly followed suit. By 1911 his pugnacious crusade had generated sufficient outside opposition and pressure against the NYZS that a separate Permanent Wild Life Protection Fund was established by the society. After amassing sufficient endowment, Hornaday used fund as base for his wildlife protection campaigns. Hornaday directed his support to Migratory Bird Treaty Act of 1912, which protected game and non-game birds, and campaigned until its ratification in 1916. Began efforts in 1905 to create national bison preserves, successfully establishing herds in Wichita Bison Range, Montana Bison Range, Yellowstone

National Park, and other federal refuges. In 1907 took up cause of Alaskan fur seals on the Pribilof Islands, leading to ratification in 1912 of international treaty outlawing pelagic seal killing. Wrote text and campaigned for adoption of clause in 1913 tariff halting importation of wild bird plumage for commercial and millinery use. Initiated an effort in 1913 to protect birds of paradise in Dutch East Indies from plumage trade. In 1920 launched intensive drive to reduce national bag limits and open seasons for game birds. Supported legislation and brought pressure against reluctant government officials to voluntarily limit hunting. Advocated Norbeck–Andresen Sanctuary Bill of 1928, establishing migratory bird refuges.

A prolific writer, published some fifteen books as well as numerous articles on natural history and conservation, including *Free Rum on the Congo* (for the Women's Christian Temperance Union in 1887), *American Natural History* (1904), *Our Vanishing Wild Life* (1913), *Wild Life Conservation in Theory and Practice* (1914), *Tales from Nature's Wonderlands* (1924), ''My Fifty-Four Years with Animal Life'' (*The Mentor* [May 1929]), and *Thirty Years War for Wild Life* (1931).

Hornaday possessed an exceptionally strong personality, notable for deep commitment to the projects upon which he embarked. Never doubted the validity of his causes and positions or hesitated to speak out against those who opposed him. Thus gained many foes as well as many friends. Awakened popular consciousness to the destruction of wildlife and habitats. Although a good observer of nature, his attempts at scientific analysis have been criticized as amateurish, due to his lack of formal education.

MAJOR CONTRIBUTIONS

A dedicated environmentalist, Hornaday always pictured wildlife he loved in its natural settings. Through his innovative taxidermy techniques, professional society, and *Manual*, gained acceptance for family habitat groups. Pioneered zoological park designs placing animals in large, natural enclosures rather than cages. Campaigned tirelessly for protection of endangered species from hunting, as well as for an end to habitat destruction and establishment of preserves and refuges.

BIBLIOGRAPHY

Blair, W. Reid. *Journal of Mammalogy* (1937) (obit.).

Bridges, W. *Gathering of Animals: An Unconventional History of the New York Zoological Society* (1974).

Dolph, J. A. ''Bringing Wildlife to Millions: William Temple Hornaday, the Early Years: 1854–1896.'' Diss. University of Massachusetts, 1975.

Forbes, J. R. *In the Steps of the Great American Zoologist; William Temple Hornaday* (1966).

Goddard, Donald, ed. *Saving Wildlife: A Century of Conservation* (1995).

New York Times, 7 March 1937 (obit.).

Preble, E. A. *Nature Magazine* (May 1937) (obit.).

UNPUBLISHED SOURCES

See letters, diaries, manuscripts, and scrapbooks at the William Temple Hornaday Memorial Trust, Atlanta, Ga; letters, clippings, photographs, and draft manuscript of Hornaday's unpublished autobiography, "Eighty Fascinating Years," in the William Temple Hornaday Papers, Library of Congress, Manuscript Division; letters in the Henry Augustus Ward Papers, University of Rochester Library; letters and memoranda in the Smithsonian Archives.

Pamela M. Henson

HOUGH, FRANKLIN BENJAMIN. Born Martinsburg, Lewis County, New York, 22 July 1822; died 11 June 1855. Christened Benjamin Franklin but reversed name when eight years old. Physician, forester, historian.

FAMILY AND EDUCATION

Son of Horatio Gates Hough, first doctor to reside in Lewis County, New York, and Martha Pitcher Hough. Studied at Louisville Academy and Black River Institute (Watertown, New York). Admitted to advanced standing at Union College in 1840; graduated in 1843. In 1846 attended Western Reserve Medical College and received M.D. in 1848. Married in 1846 Maria S. Eggleston, of Champion, New York, who died in 1848, leaving an infant daughter. In 1849 married Mariah E. Kilham, of Turin, New York; eight children.

POSITIONS

Teacher, Academy of Champion, New York (1843–1844), then became principal of Gustavus Academy, Ohio. In 1846 left to study medicine. In 1848, established medical practice at Somerville, New York. In 1854 chosen to direct New York State census (1855). In 1861–1862, inspector for U.S. Sanitary Commission; 1862–1863, enlisted as regimental surgeon in 97th New York Volunteers. Again named to head New York State census, 1865. Supervisor of census for Washington, D.C., in 1867. Became superintendent of the U.S. census for 1870. Appointed, with George E. Emerson, to prepare memorial to Congress for enactment of laws to encourage protection of forests. Endorsed by President Ulysses S. Grant and forwarded to Congress. Was first federally appointed official forestry agent in 1876.

CAREER

Aside from vocation, had interest in history and collected and published documents on Indian wars and the American Revolution. Role as census taker alerted him to decline in nation's forests and in 1873, at meeting of American Association for the Advancement of Science, in Portland, Maine, read paper "On the Duty of Governments in the Preservation of Forests." As a result, association named Hough and an associate to prepare report to Congress. Because of his work became in 1876 forestry agent (the first) in the U.S. Department of Agriculture. Submitted first report in 1877. In 1881, received new commission with increased financing from Congress. This funding allowed travel in Europe, where he was impressed by German system of forestry and forest education. In 1882 and 1883, again submitted forestry reports, which gained wide audience among forestry experts in Europe.

Was awarded diploma of honor at international Geographical Congress few years later. When appointment as chief of the Division of Forestry went to another, Hough remained as assistant to prepare fourth volume of official forestry reports. In March 1885 drafted bill for New York State legislature to create comprehensive forestry commission for the state. In addition to role as editor of series of colonial documents, was author of another seventy-eight publications, including official government reports. Was not professionally trained forester but had genuine interest in the field. Established and published the *American Journal of Forestry* in 1862, though abandoned year later because of lack of support. Was also interested in geology. Discovered the mineral named after him, houghite. Among most significant publications were *A Catalogue of Indigenous, Naturalized and Filicord Plants of Lewis Counties, N.Y.* (1846); *A History of St. Lawrence and Franklin Counties, N.Y.* (1853); *History of Duryee's Brigade in 1862* (1864); *Washingtonia or Memorials of the Death of George Washington* (1865); *American Biographical Notes* (1875); and *Elements of Forestry* (1882). Also translated Lucien Baudens's *Guerre de Crimee* as *On Military and Camp Hospitals* (1862).

MAJOR CONTRIBUTIONS

Was among the first of American scientists to become alarmed at depletion of America's forests, and with statistical evidence accumulated from role as a local and national census taker, was able to alert Congress, president, and American people to the need for conservation of American forests. As first federal official in forestry helped prepare way for his successors.

BIBLIOGRAPHY

Fernow, B. E. *A Brief History of Forestry* (1911).

"Franklin B. Hough." *American Forests and Forest Life* (July 1922).

Harmon, F. J. "Remembering Franklin B. Hough." *American Forests* 83 (1977).

Hosmer, Ralph S. "Franklin B. Hough: Father of American Forestry." *North Country Life* 6 (1952).

Jacobsen, Edna L. "Franklin B. Hough: A Pioneer in Scientific Forestry in America." *New York History* 15 (1934).

Randall, C. E. "Hough, Man of Approved Attainments." *American Forests* 67 (1961).

UNPUBLISHED SOURCES

See Franklin Benjamin Hough Papers, 1875–1878, Burton Historical Collection, Detroit Public Library (mainly manuscripts for a proposed study of higher education as part of the Philadelphia Centennial Exposition, 1876); William Pierrepont White Papers

(1728–1939), Collection of Regional History, Cornell University Library (White, a collector, has a number of Hough letters in his collection); Hough Family Papers, Collection of Regional History, Cornell University Library (compiled by Elida C. Hough and contains letters from Franklin B. Hough).

Thomas J. Curran

HOWARD, LELAND OSSIAN. Born Rockford, Illinois, 11 June 1857; died Bronxville, New York, 1 May 1950. Entomologist.

FAMILY AND EDUCATION

Eldest of three sons of Ossian Gregory Howard, an attorney, and Lucy Dunham Thurber Howard. Family moved to Ithaca, New York, when the boy was very young. Father died when Howard was sixteen. Howard attended Ithaca Academy and another private school. B.S., Cornell, 1877, one year of premed work at Cornell (1877–1878), and a year of medical school at Columbian University in Washington, D.C., early 1880s. Did not complete work for the degree. M.S., Cornell, 1886. Six honorary degrees, including an honorary M.D., Columbian University, 1911. Married Marie Theodore Clifton, 1886; three daughters.

POSITIONS

Assistant in entomology, U.S. Bureau of Entomology, U.S. Department of Agriculture (USDA), 1878–1894; chief, Bureau of Entomology, USDA, 1894–1927; principal entomologist, Bureau of Entomology, USDA, 1927–1931. Permanent secretary, American Association for the Advancement of Science (AAAS), 1898–1919, president, AAAS, 1920.

CAREER

Interested in insects from boyhood, especially butterflies and moths. Briefly studied medicine at Cornell and again at George Washington University, but his work centered on insect study. Offered job as assistant to C. V. Riley, 1878, worked under Riley (1878–1879); J. H. Comstock (1879–1881), and again under Riley from 1881 until the latter's death in 1894. Studied role of parasitic wasps in control of gypsy and coddling moths, Mexican cotton boll weevil, European corn borer, various grain-eating insects, mosquitoes, and other insect pests. Budget of Bureau of Entomology increased 100–fold during his tenure as chief (from approximately $30,000 to over $3 million per annum). Helped bring about passage of Plant Quarantine Act of 1912. Did a considerable amount of work in seeking suitable insect parasites abroad. Author of *The Insect Book* (1901); *Mosquitoes: How They Live, How They Are Classified, and How They May Be Destroyed* (1901); *The Housefly: Disease Carrier* (1911, later translated into

Russian, Hungarian, and Spanish); a two-volume study of the Mosquitoes of the Americas for the Carnegie Institution (with H. G. Dyar and F. Knab, 1912–1917); *History of Applied Entomology—More or Less Anecdotal* (1930, later translated into French); *The Insect Menace* (1931); and was translator of E. L. Bouvier's *La Psychologie des Insectes* (1920).

MAJOR CONTRIBUTIONS

An effective administrator and popularizer of entomology and of the work of his bureau. Concerned with the biological control of insects and with the issue of insects as carriers of disease. A founder of the American Association of Economic Entomologists (president, 1894). On excellent terms with foreign entomologists, he promoted cooperative international research and control efforts. Participated in a number of international entomological, agricultural, and zoological congresses and was president of the Fourth International Entomological Congress, held at Ithaca, New York, 1928.

BIBLIOGRAPHY

Graf, John, and Dorothy W. Graf, ''Leland Ossian Howard.'' *Biographical Memoirs, National Academy of Sciences*. Vol. 33 (1959).

Howard, L. O. *History of Applied Entomology—More or Less Anecdotal*. Washington, D.C. (1930).

Howard, L. O. *Fighting the Insects: The Story of an Entomologist*. New York (1933).

Proceedings of the Entomological Society of Washington 52 (1950).

Keir B. Sterling

HUBBS, CARL LEAVITT. Born Williams, Arizona, 18 October 1894; died La Jolla, California, 30 June 1979. Biologist, educator, conservationist, writer.

FAMILY AND EDUCATION

Son of Charles Leavitt and Elizabeth (Goss) Hubbs. Father was variously farmer, merchant, real estate agent, surveyor, newspaper editor, prospector, and assayer. After divorcing her husband, mother ran private school at Redondo Beach, California, which her three children attended. Maternal grandmother, Jane Goble Goss, one of the first woman physicians, inspired young Carl to comb the area around San Diego for seashells and other specimens. At Stanford University studied zoology under Charles Henry Gilbert, outstanding protégé of David Starr Jordan, acknowledged leader of American ichthyologists. Completed A.B. at Stanford in 1916, A.M. there the following year, and Ph.D. in zoology from University of Michigan in 1927. Married Laura Cornelia Clark, 15 June 1918; two children.

POSITIONS

Launched his career as assistant curator of fishes, amphibians, and reptiles at Field Museum of Natural History in Chicago, 1917–1920. In 1920 began twenty-four-year tenure with Museum of Zoology at University of Michigan, becoming curator of fishes. First director, Institute of Fisheries Research, Michigan Department of Conservation, 1930–1935. From 1944 to 1969, was professor of biology at Scripps Institute of Oceanography at University of California at San Diego, located in La Jolla; professor emeritus, 1969–1979.

CAREER

Published 712 monographs, articles, papers, and book reviews between 1915 and 1981. During tenure at University of Michigan, concentrated on fish of Great Lakes and western intermontane basins, which resulted in many papers, practical guide entitled *Fishes of the Great Lakes Region* (1941), and monograph, "The Zoological Evidence: Correlation between Fish Distribution and Hydrographic History in the Desert Basins of Western United States" (with Robert R. Miller, 1948). After move to La Jolla, continued research on fish, but widened interests to embrace marine mammals, including the recovery of gray whales, the Guadalupe fur seal, and the northern elephant seal. Expeditions to coastal Baja California to investigate effects of temperature variations on fish resulted in papers on the history of climatic changes and the original inhabitants of the region. One of his last efforts was "List of the Fishes of California" (1979). Wrote more than one dozen articles with wife, who kept track of his notes and correspondence.

Prodigious output included articles in popular magazines and various editions of *Encyclopedia Britannica*, as well as radio talks on "The Quest for Fish Specimens" and others during the late 1920s and early 1930s. Was instrumental in protecting pupfish of Death Valley, the breeding ground of the gray whale in Scammon Bay, Baja California, and habitats of other species of mammals and fish. Believed public education was crucial to conservation.

As field representative for Department of Interior, 1939–1940, rattled cages to improve federal management of wildlife resources in Alaska. Served American Society of Ichthyologists and Herpetologists as editor of its journal, *Copeia*, 1930–1937, and president, 1934, 1946–1947. Vice president of Wildlife Society, 1939–1940. Elected to National Academy of Sciences, 1952; member of International Committee on Zoological Nomenclature, 1963–1967; elected to Linnaean Society of London, 1965. Received Henry Russell Award at University of Michigan, 1929–1930; Joseph Leidy Award of Academy of Natural Sciences (Philadelphia), 1964; Fellows Medal of California Academy of Sciences, 1966; Shinkishi Hatai Medal of Japan, 1971; and American Fisheries Society Award of Excellence, 1973. Sea World Research Institute, San Diego, dedicated in 1977 to Carl and Laura Hubbs.

MAJOR CONTRIBUTIONS

One of most prolific and influential scientists of twentieth century, was called "one of the last general naturalists and . . . the dean of American ichthyology for half a century."

BIBLIOGRAPHY

American Men and Women of Science (1979).

Dictionary of Scientific Biography. Supplement (1990).

Miller, Frances Hubbs. *The Scientific Publications of Carl Leavitt Hubbs: Bibliography and Index* (1981).

Modern Scientists and Engineers (1980).

Norris, Kenneth S. "To Carl Leavitt Hubbs, a Modern Pioneer Naturalist on the Occasion of His Eightieth Year." *Copeia* (1974).

Pister, Phil. "Carl Leavitt Hubbs." *Bulletin of the American Fisheries Society* (1979) (obit.).

Shor, Elizabeth N. "Tribute to Carl L. Hubbs." *Marine Biology* (1979) (obit.).

Shor, Elizabeth N., et al. "Carl Leavitt Hubbs." In *Biographical Memoirs, National Academy of Sciences* (1987).

Stroud, Richard H., ed. *National Leaders of American Conservation* (1985).

UNPUBLISHED SOURCES

See Carl L. Hubbs Papers, Scripps Institution of Oceanography (La Jolla), University of California, San Diego (c.120 feet). The following collections contain letters from Hubbs: James Oliver Curwood Papers, Bentley Historical Library, University of Michigan, Ann Arbor; Henry Weed Fowler Papers, Academy of Natural Sciences of Philadelphia; Jacob Ellsworth Reighard Papers, Bentley Historical Library.

Rolf Swensen

HUNTSMAN, ARCHIBALD GOWANLOCK. Born Tintern, Ontario, 23 November 1883; died Toronto, Ontario, 8 August 1973. Marine ecologist, zoologist, oceanographer, educator.

FAMILY AND EDUCATION

Son of Lution Erotas Huntsman, an accountant, and Elizabeth Gowanlock Huntsman, a schoolteacher. Grew up on family farm. Graduated with distinction from St. Catherine's Collegiate Institute in 1901. Received his bachelor of arts in 1905 at the University of Toronto, where in 1907 he also earned his bachelor of medicine. Honorary M.D., University of Toronto, 1933. Married Florence Marie Stirling, 1908; three daughters.

POSITIONS

Lecturer, 1907–1917, associate professor of zoology 1917–1927, then honorary professor of marine zoology 1927–1954, at the Department of Zoology, University of Toronto. Curator, 1911–1913, 1915–1919, and director, 1919–

1934, of St. Andrews Biological Station, New Brunswick, Canada. Director, 1924–1928, Fisheries Experimental Station, Halifax, Nova Scotia. Editor, Fisheries Research Board Publications, 1934–1949. Consulting director, Fisheries Research Board, 1934–1953.

CAREER

While still a student, spent several summers at the Canadian marine stations at Nanaimo, British Columbia, and St. Andrews, New Brunswick. Earliest researches involved systematics and distribution of ascidians of Atlantic and Pacific Coasts. Johann Hjort led the Canadian Fisheries Expedition of 1914–1915 to Gulf of St. Lawrence, and Huntsman's participation turned his interests toward fish population dynamics and physiology and distribution of marine animals. Led later local expeditions for faunistic and oceanographic surveys. During 1920s and early 1930s, led research efforts into herring fisheries of Bay of Fundy and instigated International Passamaquoddy Fisheries Commission of 1931–1932, into possible harmful effects of constructing dams across entrances of Passamaquoddy Bay to harness tidal power. Independently of Th. I. Baranov, demonstrated graphically in 1918 that fishing effort reduces mean age of fish in stock. While director of Halifax Fisheries Experimental Station, invented ''jacketed cold storage'' (1927), made other technical contributions to improve frozen seafood products. After 1930 main interest Atlantic salmon research, especially effects of environment on movement, migration, and behavior of different salmon stocks. Extensive tagging and other experiments on salmon stocks of rivers of the Maritime Provinces and pioneering investigations into saltwater acclimation of salmon. Achieved first verified return to home stream of a tagged salmon captured, retagged, and rereleased in distant waters. The salmon had traveled from Cape Breton to Newfoundland and back.

Published over 220 articles, 1912–1973, including some 62 popular articles on the Biological Board of Canada (after 1937 the Fisheries Research Board), its work, and technical guides for curing and freezing fish. Among his research papers numbered 64 articles on salmon, and 16 each on ascidians and herring. *The Passamaquoddy Bay Power Project and Its Effects on the Fisheries* (1928) sparked popular concern over the ecological consequences of the project. Also published many articles on fisheries management and research.

Recognized by many of his contemporaries and students as being ''ahead of his times'' in his research, but a weak administrator unable to delegate authority. Strongly believed that science should serve useful ends and that marine science should be applicable to the fisheries. An early critic of the ''maximum sustained yield'' concept, he believed that commercial overfishing would become unprofitable long before stocks were endangered. A leading, but unconventional, Atlantic salmon expert, he denied that salmon ''deliberately'' travel long distances in the sea before returning to their home streams. He held water movements and currents accountable for most of their distant recaptures; most, he believed, lingered in the vicinity of their home streams while in the sea. Views on importance of water movements on fish behavior then unpopular, but now gaining currency. Remembered for his love of questioning all positive statements and his predilection for philosophy and for thinking scientific problems through without experimental evidence.

MAJOR CONTRIBUTIONS

Strongly influenced the beginnings of fisheries research by the Biological Board of Canada. A great popularizer of the importance of fisheries biology to the fisheries; tried to make the science seem approachable to fishermen. Many practical contributions to fish processing, including jacketed cold storage. Emphasized importance of scientific data as a basis for fisheries legislation. Interested in ecology, the response of the entire animal to its surroundings. Taught at the University of Toronto even during periods when he had heavy duties at the marine and experimental stations; most Canadian fisheries biologists educated in the 1920s and 1930s passed, at some stage, through his hands.

BIBLIOGRAPHY

Hart, J. L. ''A. G. Huntsman—An Appreciation.'' *Journal of the Fisheries Research Board of Canada* 22 (1965).

Johnstone, Kenneth. *The Aquatic Explorers: A History of the Fisheries Research Board of Canada* (1977).

Needler, A. W. H. ''Archibald Gowanlock Huntsman, 1883–1973.'' *Proceedings of the Royal Society of Canada* 4, 13 (1975).

UNPUBLISHED SOURCES

Letters, manuscripts are in University of Toronto Archives.

Jennifer M. Hubbard

HUTCHINS, THOMAS. Born England(?); about 1742(?); died London, 7 July 1790. Surgeon, fur trader, meteorologist.

FAMILY AND EDUCATION

Not known, but perhaps a graduate of one of the London medical schools, probably in the early 1760s. In England he had a wife, Margaret.

POSITIONS

Appointed by the Hudson's Bay Company as surgeon at York Factory on 12 February 1766, at thirty-six pounds per annum, he arrived on the *King George* on 11 August 1766. After seven years as surgeon at York Factory, he took a year of furlough, then was chief at Albany Fort from 1774 to 1782. He returned to England in 1782 and on 23 July 1783 was appointed corresponding secretary of the company, a post he held until April 1790.

CAREER

Hutchins made his first careful meteorological measurements, measuring temperature and atmospheric pressure, during 1771–1772, when he was with Andrew Graham at York Factory. In 1774–1775 Hutchins added a set of observations on the dipping needle and experimented with congealing of mercury in severe cold. This led to three important publications in the *Philosophical Transactions*. For this work he was presented with the Copley Medal by the Royal Society in December 1783, only the second Hudson's Bay man to be awarded the Copley Medal in the eighteenth century. Hutchins collaborated with Andrew Graham in preparing technical descriptions of birds and mammals in their joint collection sent from York Factory in 1771. Important Hutchins observations concerning sixteen species of birds (eleven of which had not been listed by Andrew Graham, who was overall the better naturalist of the two), fourteen species of fish, and seven species of mammals were published in 1969, almost 200 years after they were written.

MAJOR CONTRIBUTIONS

Hutchins was best known for his meteorological observations. We are also grateful that he copied out the longer, more perceptive natural history writings of Andrew Graham. (For over 100 years, the writings of both Graham and Hutchins were credited to Hutchins, then in 1969 nearly full credit was mistakenly given to Graham alone.)

BIBLIOGRAPHY

Houston, C. S. *18th-Century Naturalists of Hudson Bay*. Toronto: University of Toronto Press (1994).

Hutchins, T. "An Account of the Success of Some Attempts to Freeze Quicksilver at Albany Fort, in Hudson's Bay, in the Year 1775: with Observations on the Dipping Needle." *Philosophical Transactions* 66 (1776).

Hutchins, T. "Experiments for Ascertaining the Principle of Mercurial Congelation . . ." *Philosophical Transactions* 73 (1783).

Williams, G., ed. *Andrew Graham's Observations on Hudson's Bay, 1767–91*. London: Hudson's Bay Record Society (1969).

Williams, G. "Andrew Graham and Thomas Hutchins: Collaboration and Plagiarism in 18th-Century Natural History." *The Beaver* 308, 4 (1978).

Williams, G., ed. *Andrew Graham's Observations on Hudson's Bay, 1767–1791*. London: Hudson's Bay Record Society (1969).

C. Stuart Houston

HYATT, ALPHEUS. Born Washington, D.C., 5 April 1838; died Cambridge, Massachusetts, 15 January 1902. Naturalist, paleontologist, conchologist, evolutionist, museum administrator.

FAMILY AND EDUCATION

Son of Alpheus and Harriet R. King Hyatt, members of a wealthy Baltimore family. Studied at Yale College and Lawrence Scientific School of Harvard University, A.B., 1862, under Louis Agassiz: LL.D., Brown University, 1898. Married Ardella Beeby, 1867; three children.

POSITIONS

Assistant, Museum of Comparative Zoology, 1858–1862; captain, 47th Massachusetts Volunteers, 1862–1865; honorary curator of fossil cephalopods, Museum of Comparative Zoology, Harvard, 1865–1902; curator, Essex Institute, Salem, Massachusetts, 1867; a founding curator, Peabody Academy of Science, Salem, Massachusetts (radiates, sponges, and mollusks), 1867–1870; curator, Boston Society of Natural History, 1870–1902; professor of zoology and paleontology, Massachusetts Institute of Technology (part-time), 1870–1888; professor of biology, Boston University (part-time), 1877–1900.

CAREER

Made field trip to coast of Maine to study marine invertebrates with N. S. Shaler and A. E. Verrill, 1860; made expeditions to Anticosti Island, 1861, 1885, 1886, to collect marine life and fossil cephalopods; expeditions of 1885 and 1886 made in his own vessel, the *Arethusa*, which was also used for dredging studies at his Annisquam Sea-side Laboratory. Published his first monograph on fossil cephalopods, 1866; a founder of *American Naturalist* (with Putnam, Packard, and Morse), 1867; founded and directed Teachers School of Science, Boston Society of Natural History, 1870–1900; founded and directed Annisquam Seaside Laboratory (Boston Society of Natural History and Women's Educational Association of Boston), 1880–1886, which became Marine Biological Laboratory, Woods Hole, Massachusetts, for which Hyatt was first president of Board of Trustees; elected fellow, American Academy of Arts and Sciences, 1869; elected, National Academy of Sciences, 1875; research in Europe on museum collections and fossil snails at Steinheim, 1875; founder and first president of American Society of Naturalists, 1883; palaeontologist, U.S. Geological Survey, 1889–1902; elected American Philosophical Society, 1895; corresponding member Geological Society of London, 1897; published many papers and monographs on fossil cephalopods, sponges, coelenterates, and evolutionary theory. He was member of the neo-Lamarckian school of evolution and author of Hyatt–Cope law of acceleration and retardation; he had philosophical and theoretical approach to evolution.

MAJOR CONTRIBUTIONS

Descriptions and classification of fossil cephalopods and evolutionary theory based upon them. Founder of Teacher's School of Science and Annisquam Seaside Laboratory, which developed into the Marine Biological Laboratory at Woods Hole, Mass.

BIBLIOGRAPHY

"The Annisquam Sea-side Laboratory of Alpheus Hyatt, Predecessor of the Marine Biological Laboratory at Woods Hole, 1880–1886." In *Oceanography: The Past* (1980).

Brooks, W. K. "Biographical Memoir of Alpheus Hyatt, 1838–1902." *Biographical Memoirs, National Academy of Sciences* 6 (1909).

Dexter, R. W. "Three Young Naturalists Afield: The First Expedition of Hyatt, Shaler and Verrill." *Scientific Monthly* 79 (1954).

Mayer, A. G. "Alpheus Hyatt, 1838–1902." *Popular Science Monthly* 78 (1911).

Shrock, R. R. "Alpheus Hyatt (1838–1902)." *Geology at the Massachusetts Institute of Technology, 1865–1965* (1977).

UNPUBLISHED SOURCES

Some journal notes are in Library, Peabody Museum, Salem, Massachusetts, and some correspondence is in archives of Princeton University.

Ralph W. Dexter

HYMAN, LIBBIE HENRIETTA. Born Des Moines, Iowa, 6 December 1888; died, New York City, New York, 3 August 1969. Zoologist.

FAMILY AND EDUCATION

Hyman grew up in Fort Dodge, Iowa, the only daughter of Joseph and Sabina Neuman Hyman, who were Jewish immigrants. She had three brothers. Valedictorian of her graduating class at Fort Dodge High School, 1905. Attended the University of Chicago on a scholarship. B.S., 1910, Ph.D., 1915, University of Chicago. Honorary D.Sc. from Chicago, 1941, Goucher College, 1958, and Coe College, 1959. Never married.

POSITIONS

Research assistant and laboratory instructor, University of Chicago, 1916–1931. Independent researcher, various locations including New York City, Europe, South America, Woods Hole, Massachusetts, 1931–1969. Honorary research appointment, American Museum of Natural History, 1937–1969.

CAREER

Originally interested in studying botany when she entered the University of Chicago, Hyman switched to zoology to avoid what she perceived to be an anti-

Semitic atmosphere in the Botany Department. Between 1916 and 1931 she worked as a research assistant at the University of Chicago for Charles Manning Child, under whom she had studied as a graduate student. Her work involved conducting research on physiology and morphology of lower invertebrates. Most of the articles Hyman published during this period were related to Child's research on regeneration and metabolism.

During this time Hyman also worked as a laboratory instructor for undergraduate courses in comparative vertebrate anatomy and elementary zoology. She published *A Laboratory Manual for Elementary Zoology* (1919, 2d ed. 1926) and *A Laboratory Manual for Comparative Vertebrate Anatomy* (1922), both of which were critically acclaimed and widely used. Because of her work on Child's project, Hyman acquired a reputation as a specialist in invertebrate taxonomy. Her interest in experimentation was soon eclipsed by her enjoyment of taxonomic research.

After her mother's death in 1931 Hyman resigned from the University of Chicago to pursue a career as an independent researcher. Living on the small income she earned from her books, Hyman traveled to a number of scientific research centers in Europe, including the Naples Zoological Center. In 1933 Hyman settled in New York City and began writing a reference work on invertebrates.

In 1937 she became an honorary research associate of the American Museum of Natural History. Although this was an unsalaried position, it provided Hyman with an office and laboratory space. Her work resulted in the publication of *The Invertebrates*, a comprehensive, six-volume survey published between 1940 and 1967. *The Invertebrates* was well received and praised for its scientific value. Handicapped by Parkinson's disease during the last decade of her life, Hyman was unable to complete additional projected volumes of *The Invertebrates* series.

Hyman lived and worked alone and had few close friends. Yet she corresponded with many scientists from all over the world who sent her specimens for identification. She spent several summers studying and working at the Marine Biological Laboratory in Woods Hole, Massachusetts. Her research on flatworms took her as far away as South America.

Hyman published numerous articles, including: ''Acoel and Polyclad Turbellaria from Bermuda and the Sargassum'' (December 1939); *Descriptions and Research of Fresh-Water Turbellaria from the United States* (1955); *A Few Turbellarians from Trinidad and the Canal Zone, with Corrective Remarks* (1957); ''A Further Study of Micronesian Polyclad Flatworms'' (1958); *Marine Turbellaria from the Atlantic Coast of North America* (1944); and ''A New Hawaiian Polyclad Flatworm Associated with Teredo'' (1944).

MAJOR CONTRIBUTIONS

Hyman made a large amount of information about invertebrates more accessible. She emphasized the comparative descriptive method in her research as well as close, detailed observation.

BIBLIOGRAPHY

American Women of Science (1943).
Blackwelder, Richard. *Worm Runners' Digest* (October 1970).
Encyclopedia Britannica. 15th ed. (1990).
International Dictionary of Twentieth Century Biography (1987).
Leaders in American Science. Vol. 1 (1953), Vol. 2 (1954), Vol. 4 (1961).
Lesser Known Women (1992).
Nature (24 January 1970) (obit.).
New York Times, 5 August 1969 (obit.).
Notable American Women. The Modern Period. Vol. 4 (1980).
O'Neill, Lois Decker. *The Women's Book of World Records and Achievements* (1979).
Stunkard, Horace W. "In Memoriam." In Nathan W. Riser and Patricia M. Morse, eds., *Biology of the Turbellaria* (1974).
Transactions of the American Microscopical Society (April 1970): 196 (obit.).
Washington Post, 6 August 1969 (obit.).
Who Was Who in America. Vol. 5 (1973).

Domenica Barbuto

I

ICKES, HAROLD LE CLAIR. Born Frankstown Township, Pennsylvania, 15 March 1874; died Washington, D.C., 3 February 1952. Journalist, attorney, cabinet officer, conservationist.

FAMILY AND EDUCATION

Second of seven children of Jesse Boone Williams and Martha Ann McEwen Ickes. B.A., University of Chicago, 1897; LL.B., University of Chicago, 1907. Married Anna Wilmarth Thompson, 1911 (deceased 1935); one son. Ickes also adopted her son by a previous marriage (deceased, 1936). Married Jane Dahlman, 1938; two children.

POSITIONS

Reporter for Chicago newspapers 1897–1900. Practicing attorney in Chicago, 1907–1933, and political reform activist from 1897 until his death. Chairman, Cook County Committee, Progressive Party, 1912–1914; Illinois State Committee, 1915–1916, Member Progressive National Committee and National Executive Committee, 1915–1916. YMCA service with 35th Division in France, 1918–1919. Manager of six mayoral (Chicago), U.S. senatorial and presidential campaigns, 1905–1926. Delegate to Republican, Progressive, and Democratic Nominating Conventions, 1916–1920 and 1932–1944. Led an independent Republican committee supporting Franklin D. Roosevelt in 1932. Appointed Secretary of the Interior and served from 1933 to 1946; also directed Public Works Administration (PWA), 1933–1939 and served as federal oil administrator under National Recovery Act. Chairman, National Resources Committee, 1934–1939;

chairman, National Power Policy Committee, 1934–1946; member of National Power Commission, 1938–1939. During World War II, petroleum coordinator, 1941–1946; coordinator for solid fuels, 1941–1946; coordinator for fisheries, 1942–1946; coal mines administrator, 1943–1944; chairman of American delegation to Anglo-American oil treaty negotiations, London, September, 1945. In his later years a newspaper and magazine columnist.

CAREER

Ickes' mother died when he was sixteen and he was sent to live with relatives in Chicago. Worked his way through college and law school. As Secretary of the Interior, served longer than any of his predecessors or successors. Joined with FDR in managing national parks for public use. Considered something of a ''righteous scold,'' he earned nickname of ''Honest Harold'' by restoring the good reputation of his department, but was an irascible and sometimes interfering and abusive administrator. Frequently difficult and sometimes tactless, he was outspoken in his criticism of political opponents and ''the Interests'' (big business). A dedicated conservationist, he was unsuccessful in resolving the bureaucratic dispute between his department and Henry A. Wallace's agriculture department over proper placement of conservation and non-conservation agencies. Wanted Department of the Interior transformed into Department of Conservation, but did not succeed. He did manage to persuade FDR to place some agencies from other departments under the Department of the Interior, including the Bureau of Biological Survey from Agriculture and the U.S. Fish Commission from Commerce (1940), which were combined in the Department of the Interior as the U.S. Fish and Wildlife Service. Failed in effort to have Forest Service moved, owing to opposition from some conservationists and forestry people, particularly Gifford Pinchot. Also lost newly created Soil Erosion Service to Agriculture. Oversaw radical expansion of the National Park Service. Advocated national seashore program, but it was not funded for many years. Civilian Conservation Corps, administered by the army, made many improvements in parks. Reclamation Service, a division of the Department of the Interior, built many multi-purpose dams, especially in the western United States, but some of these had adverse long-term ecological effects Ickes had not anticipated. From 1933 to 1939, Ickes demonstrated that as PWA administrator he was, in the words of William E. Leuchtenburg, ''a builder to rival Cheops.'' PWA was responsible for many public buildings, including a new Interior Department building, schools, and hospitals; major highways and tunnels; aircraft carriers and cruisers for the U.S. Navy; and planes for military and civilian use.

Ickes was a dedicated civil libertarian and an active interventionist in foreign affairs, but did not espouse the extreme social welfarism of some New Dealers. During World War II, the administration of most park, conservation, and wildlife management work was transferred out of Washington to make room for proliferating wartime agencies, an action of which Ickes disapproved. Domestic conservation programs were of secondary importance in those years. Jackson Hole, Wyoming, was the only major national monument created in that period (1943). Ickes authored *The New Democracy* (1934); *Back to Work: The Story of PWA* (1935); *America's House of Lords: An Inquiry into the Freedom of the Press* (1939); *Not Guilty: An Official Inquiry into the Charges Made by Glavis and Pinchot against Richard Ballinger, Secretary of the Interior, 1909–1910* (1940); *The Third Term Bugaboo* (1940); compiled *Freedom of the Press Today* (1941); and authored *Fightin' Oil* (1943) and *The Autobiography of a Curmudgeon* (1943). His "My Twelve Years with FDR" ran in eight parts in the *Saturday Evening Post* (June–July, 1948). His three-volume *Secret Diary of Harold L. Ickes* (1953–1954) is an often valuable insight into his personal and official lives for the years 1933–1941.

MAJOR CONTRIBUTIONS

Made Department of the Interior a major element in national conservation efforts during his tenure, though in some areas there was more style than substance. With the Collier Act of 1934, Ickes oversaw some reforms in the federal administration of Native American affairs, including repeal of the 1887 Dawes Severalty or General Allotment Act. National park lands grew from 8.2 million acres to in excess of 20 million acres between 1933 and 1941. Ickes pressed strongly for a comprehensive Wilderness Act, which was not finally enacted until the mid–1960s.

BIBLIOGRAPHY

Leuchtenburg, William E. *Franklin D. Roosevelt and the New Deal* (1963).

Nixon, Edgar B., comp. & ed. *Franklin D. Roosevelt and Conservation, 1911–1945*, 2 vols. (1957).

Riesch-Owen, A. L. *Conservation Under FDR* (1983). [Contains very useful bibliography.]

Shipley, Donald D. "A Study of the Conservation Philosophies and Contributions of Some Important American Conservation Leaders." Ph.D. Dissertation, Cornell University (1953).

Strong, Douglas H. *Dreamers and Defenders: American Conservationists* (1988).

Swain, Donald C. *Wilderness Defender: Horace M. Albright and Conservation* (1970).

Watkins, T. H. *Righteous Pilgrim: The Life and Times of Harold L. Ickes, 1874–1952* (1990). [Contains very useful bibliography]

White, Graham, and John Maze. *Harold Ickes of the New Deal: His Private Life and Public Career* (1985).

UNPUBLISHED SOURCES

Extensive personal and professional papers, Manuscript Division, Library of Congress, Washington.

Keir B. Sterling

ISHAM, JAMES. Born presumably in parish of St. Andrew's, Holborn, London, England, about 1716; died at York Factory, Hudson Bay, 13 April 1761. Fur trader.

FAMILY AND EDUCATION

Son of Whitby Isham and Ann Skrimshire. On 29 October 1748, while on furlough in London, Isham married twenty-one-year-old Catherine Mindham, who had a daughter but remained in England and predeceased Isham. Had a son by a native woman at York Factory.

POSITIONS

Employed by the Hudson's Bay Company from 11 May 1732, initially as a "writer." Arrived at York Factory, 1 August 1732. Became chief at York Factory, 4 May 1737, when only twenty-one years of age. Chief at Churchill from 16 August 1741. Again, chief at York, from summer 1746. During his second of three furloughs in England (1745–1746, 1748–1750, 1758–1759) he married Catherine Mindham and also met the naturalist George Edwards.

CAREER

Isham's first winter at Churchill, 1741–1742, was marred by the presence of the exploring expedition of Capt. Christopher Middleton, who lost thirteen of his men that winter, eleven to scurvy alone, and who in 1742 explored the west coast of the bay north to Wager Inlet and Repulse Bay. In 1746–1747 Isham hosted at York Factory the even more contentious expedition led by William Moor in the vessel *Dobbs* and Francis Smith in the vessel *California*, prior to their disappointing northward expedition of 1747.

MAJOR CONTRIBUTIONS

Isham collected more than thirty species of birds, which he brought home to London in 1745. These included officially recognized "type specimens" for thirteen species of birds (great blue heron, snow goose, surf scoter, spruce grouse, sharp-tailed grouse, whooping crane, sandhill crane, sora, marbled godwit, Hudsonian godwit, red phalarope, northern phalarope, and purple martin), which were illustrated by George Edwards in 1750 and then given binomial Latin names by Linnaeus in 1758. He also transmitted the type specimen for the North American subspecies of northern harrier, first described by Linnaeus in 1766, and has been given credit on very questionable grounds for the "Canadian" subspecies of the golden eagle. Isham's own observations, including mention of thirty-nine species of birds, but only six of the previously mentioned species, along with valuable observations on mammals and the customs and language of the the native Indians, were not formally published until 1949.

BIBLIOGRAPHY
Edwards, G. *A Natural History of Uncommon Birds*. Vol. 3. (1750).
Houston, C. S. ''Birds First Described from Hudson Bay.'' *Canadian Field-Naturalist* 97 (1983).
Rich, E. E., and A. M. Johnson. *James Isham's Observations on Hudson's Bay, 1743–49* (1949).

C. Stuart Houston

J

JACKSON, HARTLEY HARRAD THOMPSON. Born Milton, Wisconsin, 19 May 1881; died Durham, North Carolina, 20 September 1976. Mammalogist.

FAMILY AND EDUCATION

Eighth (and last) child of Harrad and Mary Thompson Jackson, English immigrants who settled in Wisconsin, and the only one of their children born in the United States. B.S., Milton College, 1904; M.S., Wisconsin, 1909; honorary M.S., Milton, 1909; Ph.D., George Washington University, 1914. Married Anna Marcia Adams, 1910 (died 1968); no children. Married Stephanie Hall, 1970.

POSITIONS

Teacher at Carthage Collegiate Institute in Missouri and at schools in Juda, Wisconsin, and Waukegan, Illinois, 1904–1908; assistant in zoology, cataloging bird collections, University of Wisconsin, 1909–1910. In charge of mammal collections, U.S. Bureau of Biological Survey (USBS), U.S. Department of Agriculture, 1910–1924; chief, Division of Biological Investigations, USBS, 1924–1926; in charge of mammal research, USBS, 1927–1936; in charge of section of Wildlife Surveys, USBS, 1936–1940; in charge of section of Biological Surveys, U.S. Fish and Wildlife Service (USF&WS), 1940–1951 (named senior biologist, 1949, and relieved of most administrative tasks); during World War II, prepared lists of rodents of East Asia, the Pacific, and Australia for the Rodent Control Subcommittee, National Research Council, and geographic information on Alaska for the Office of Strategic Services. Consulting mammalogist, Smithsonian Institution, 1951–1966.

CAREER

Developed interest in mammalogy and ornithology as a boy. Began own collection of study skins. At age fourteen, met Ludwig Kumlien, arctic explorer and Wisconsin naturalist, who gave Jackson useful guidance. At age sixteen, wrote important essay on the screech owl, later published in the *Nidologist*. His principal research interests included the taxonomy of shrews and later carnivores, biogeography, and vertebrate ecology. One of his earliest taxonomic publications was *A Review of the American Moles* (1915). His most comprehensive study of this character was *A Taxonomic Review of the American Long-Tailed Shrews (Genera Sorex and Microsorex)* (1928). A major work was *The Mammals of Wisconsin* (1961), on which he had begun research in 1912. From 1934 until 1949, Jackson's time was largely taken up with supervisory responsibilities, thus limiting research activities. From 1937 until 1946, using uniform methods of estimating the numbers of larger mammals in the United States, he published an annual compilation of "big-game surveys" with the aid of Ira Gabrielson, a colleague and later director of the USF&WS. Native Americans, trappers, wardens, forest rangers, and interested citizens were enlisted to assist with this nationwide animal census. These surveys provided a sound basis for wildlife legislation, including hunting regulations. Jackson was the major force behind the founding of the American Society of Mammalogists and served as secretary, journal editor, vice president, and president (1938–1940). Jackson worked for many years on a history of the Biological Survey, which was incomplete at the time of his death.

MAJOR CONTRIBUTIONS

Jackson's mammalian census projects, incorporating the airplane as an observation and counting tool, greatly facilitated the biological reconnaissance objectives of the U.S. Biological Survey and the later U.S. Fish and Wildlife Service. His role in the founding and early history of the American Society of Mammalogists was invaluable. His systematic revision of the shrews, with some minor changes, remains valid today. His volume on the mammals of Wisconsin was an important and useful regional account.

BIBLIOGRAPHY

American Men of Science. 10th ed. (1966).

Birney, Elmer C., and Jerry Choate. *Seventy-Five Years of Mammalogy (1919–1994)*. (1994).

Journal of Mammalogy (1977) (obit.).

Neuberger, R. L. "Great American Snout Count." *Saturday Evening Post* (22 June 1946).

UNPUBLISHED SOURCES

See H. H. T. Jackson Papers, University of Wisconsin-Whitewater Library. See also records of the American Society of Mammalogists, Smithsonian Institution Archives, Washington.

Joseph Bongiorno

JACKSON, WILLIAM HENRY. Born Keeseville, New York, 4 April 1843; died New York City, 30 June 1942. Photographer, artist, explorer.

FAMILY AND EDUCATION

Son of George Hallock Jackson, carriage maker and blacksmith, and Harriet Maria Jackson, watercolorist; oldest of eight children. Attended Fourth Ward School, Troy, New York, through eighth grade. Honorary LL.D., University of Wyoming, 1941. Married Mary (''Mollie'') Greer, 1869 (died 1872); one daughter (died 1872) and Emilie Painter, 1873 (died 1918); one son, two daughters.

POSITIONS

Freelance artist and photographic assistant, Troy, New York, and Rutland, Vermont, 1855–1862. Photographer and artist, Rutland, Vermont, 1863–1866. Operated photographic studio, Omaha, 1868–1871. Official photographer, U.S. Geological and Geographical Survey of the Territories (Hayden Survey), 1870–1879. Operated photographic studio, Denver, 1879–1897. With Detroit Publishing Co., Detroit, 1898–1924. Research secretary, Oregon Trail Memorial Association, 1929–1942. Artist with National Park Service, 1935–1937.

CAREER

Jackson termed picture making a ''kind of inheritance,'' having learned daguerreotypy from father and sketching and painting from his mother. After formal schooling ended at age fifteen, improved artistic skills by working as independent artist and assistant to professional photographers. While serving with 12th Vermont Volunteer Infantry, 1862–1863, sketched and kept diary, practices continued in later years. Opened Jackson Brothers, Photographers, Omaha, 1868, and photographed Pawnee, Winnebago, Ponca, and Omaha Indians. Photographed along the Union Pacific Railroad, 1868–1869. Systematically photographed geologic features, landscapes, mining towns, and prehistoric ruins in Rocky Mountains for Hayden Survey, 1870–1878 (F. V. Hayden, q.v.). Overcame difficulties of cumbersome wet-plate method to produce excellent photographic record of American West in stereographic pairs and single images (using 5×8–to 20×24–inch cameras); most photographs represented initial records. Organized Hayden Survey's exhibit, Centennial Exposition, 1876. Opened Jackson Photographic Co., Denver, 1879; adopted newly perfected dry-plate technology. Photographed landscapes in United States, Canada, and Mexico for major railroads, 1881–1892. Official photographer World's Columbian Exposition, 1893. Traveled around the world with World's Transportation (Railroad) Commission, 1894–1896, and recorded construction of Trans-Siberian Railroad (photographs from trip published in *Harper's*). Joined Detroit Publishing Co., 1898, which had introduced Swiss ''Photochrome Process'' for color printing to United States; Jackson, as administrator and photographer, contributed own collection and then enlarged company's holdings. Resumed work as artist, 1924; depicted historic events and places in the West, especially while research secretary, Oregon Trail Memorial Association. As artist National Park Service, 1935–1937, completed large oil murals depicting each of the four principal territorial surveys (1867–1879) in the field and many smaller paintings showing historic places and events in parks and monuments.

Jackson received the University of Colorado's gold Recognition Medal "for distinguished service to Colorado and the West" (1937). His name is perpetuated in Mount Jackson (adjacent to the Mount of the Holy Cross) and Jackson Butte (a detached remnant of the Mesa Verde), Colorado, and Jackson Canyon, near Casper, Wyoming. Published more than forty articles, catalogs, albums, and books, 1874–1940, including *Descriptive Catalogue of the Photographs of the United States Geological Survey of the Territories, for the Years 1869–1875, Inclusive. Second Edition* (1875, repr. 1978); *Descriptive Catalogue of Photographs of North American Indians* (1877); *The Pioneer Photographer: Rocky Mountain Adventures with a Camera* (with Howard R. Driggs, 1929); and *Time Exposure. The Autobiography of William Henry Jackson* (1940, repr. 1970).

MAJOR CONTRIBUTIONS

Best known and most prolific of wet-plate photographers who first recorded the scenic features of the American West. During Hayden Survey years, advanced the use of photography for scientific purposes. His photographs of the scenic and geologic wonders of the Yellowstone region in 1871 and 1872 caught the imagination of the nation. Bound into quarto volumes presented to every member of Congress, the 1871 photographs, with Hayden's report, helped win approval for the 1872 bill that established first of all national parks. Jackson's photographs of prehistoric ruins in the Southwest generated similar widespread interest and later were used in archeological fieldwork in what became Mesa Verde National Park and several national monuments (Chaco Canyon, Hovenweep, Canyon de Chelly). His photographs of Indians (about 1,000 negatives) have provided invaluable ethnological data; landscape and other "view pictures" taken between 1870 and 1924 are significant historic and scientific records. Jackson pioneered devising and testing new photographic equipment and techniques; he worked with daguerreotypes, tintypes, wet plates, dry plates, emulsion paper and film, panoramic photography, and color film. Jackson's paintings re-created frontier and Civil War scenes that contributed to the nation's growing awareness of its past and the importance of preserving historic and scenic landmarks.

BIBLIOGRAPHY

Fryxell, F. M. "William Henry Jackson, Photographer, Artist, Explorer." *American Annual of Photography* (1939).

Fryxell, F. M. "William Henry Jackson, Pioneer Teton Photographer." *Teton, the Magazine of Jackson Hole, Wyoming* 12 (1979).

Hafen, L. R., and A. W. Hafen, eds. *The Diaries of William Henry Jackson, Frontier Photographer* [1866–1867, 1873, 1874] (1959).

Hales, Peter B. *William Henry Jackson and the Transformation of the American Landscape* (1988).
Jackson, C. S. *Picture Maker of the Old West: William H. Jackson* (1947, repr. 1966).
Jackson, C. S. *Pageant of the Pioneers. The Veritable Art of William H. Jackson* (1958).
Jackson, C. S., and L. W. Marshall. *Quest of the Snowy Cross* (1952).
Naef, W. J., and J. N. Wood. *Era of Exploration. The Rise of Landscape Photography in the American West, 1860–1885* (1975).
Newhall, Beaumont, and D. E. Edkins. *William H. Jackson* (1972).
Ostroff, Eugene. *Western Views and Eastern Visions* (1981).
Taft, Robert. *Photography and the American Scene. A Social History, 1839–1889* (1938, repr. 1964).
Who's Who in America, 1942–1943. Vol. 22 (1942).

UNPUBLISHED SOURCES

See negatives (more than 85,000—not all by Jackson): Indians and Hayden Survey, National Archives and Records Service, Washington, D.C. (copy negatives, U.S. Geological Survey Library, Denver); western U.S., 1880–1920, Colorado Historical Society, Denver; eastern U.S., Library of Congress. Albumen and other Photographic Prints, Stereographs, and Albums are in Amon Carter Museum, Fort Worth; Archives, University of Wyoming; Western History Room, Denver Public Library; International Museum of Photography, George Eastman House, Rochester; Library of Congress; Colorado Historical Society; U.S. Geological Survey Library, Denver.

For manuscripts and Jacksoniana, see Archives, University of Wyoming; Coe Library, Yale University; Colorado Historical Society; Western History Room, Denver Public Library; Record Group 57, National Archives and Records Service; New York Public Library; Yellowstone National Park. For paintings, see Department of the Interior Museum; Scottsbluff National Monument, Nebraska, and other National Park Service museums; and art museums, archives, and private collections.

Fritiof M. Fryxell and Clifford M. Nelson

JAMES, EDWIN. Born Weybridge, Vermont, 27 August 1797; died Rock Springs, Iowa, 28 October 1861. Botanist, geologist, linguist.

FAMILY AND EDUCATION

Thirteenth child of Daniel and Mary James. Attended Addison Country grammar school, graduated from Middlebury College in 1816. From 1816 to 1819 studied botany and geology with John Torrey and Amos Eaton, medicine with his brother. Married Claussa Rodgers in 1827; one son.

POSITIONS

Biologist and surgeon of the Long Expedition of 1820. Assistant surgeon U.S. Army, 1823–1833. Editor *Temperance Herald and Journal*, 1833–c.1836. Subagent Potawatomie Indians, 1837–1838. Involved with Underground Railroad, 1838–1861.

CAREER

Served as surgeon, botanist, and geologist on Maj. Stephen H. Long's expedition to explore the headwaters of the Platte and Red Rivers. This expedition gave him the opportunity to study the flora and geology of the Arkansas, Red, and Canadian Rivers. With the aid of Long's notes, James published an *Account of an Expedition from Pittsburgh to the Rocky Mountains Performed in the Years 1819 and '20*, 2 vols. and Atlas (1822–1823)—the only extensive account of the expedition and the work that introduced the idea that the lands west of the Mississippi River were the Great American Desert unsuitable for settlement. While serving as an assistant surgeon in the U.S. Army, James became a student of Indian languages, prepared several Indian spelling books, and translated the New Testament into the Ojibway language.

MAJOR CONTRIBUTIONS

Climbed Pikes Peak in 1820 and became not only the first white man to ascend one of the nation's 14,000–foot peaks but also the first scientist to examine the alpine flora of the Rocky Mountains. First botanist to collect in Colorado. His specimens were studied by Torrey.

BIBLIOGRAPHY

Elliott, Clark A. *Biographical Dictionary of American Science*. Vol. 137 (1979).
Ewan, Joseph. *Rocky Mountain Naturalists* (1950).
McKelvey, Susan. *Botanical Exploration of the Trans-Mississippi West* (1955).

UNPUBLISHED SOURCES

Benson, Maxine. "Edwin James: Scientist, Linguist, Humanitarian." Ph.D. Diss., University of Colorado, Boulder (1968).

Phillip D. Thomas

JAQUES (JACQUES), FRANCIS LEE. Born Genesco, Illinois, 28 September 1887; died St. Paul, Minnesota, 24 July 1969. Artist, illustrator of wildlife and nature.

FAMILY AND EDUCATION

Son of Ephriam Parker and Emma Jane (Monninger) Jaques. A student at School of Engineering, Milwaukee, Wisconsin, 1915–1916. Married Florence Sarah Page, 12 May 1917, an author of books on nature and travel and of verse; no children. Private in U.S. Army, 1917–1919.

POSITIONS

In 1924 began work as an illustrator at American Museum of Natural History, New York City. Retired from the museum in 1942. Subsequently, illustrated works on nature written by his wife.

CAREER

At American Museum of Natural History, first assigned to paint dome of the bird hall. Painted more than half the dioramas in museum. Participated in museum expeditions to Panama, 1925; South America, 1926; Bahamas, 1926; the Arctic, 1928; England, 1932; Polynesia, 1934; South America, 1935.

Painted other exhibits for Boston Museum of Science, Peabody Museum at Yale, University of Nebraska Museum, Welder Wildlife Foundation in Texas, Minnesota Museum of Natural History.

Among books produced with wife: *Canoe Country* (1938); *The Geese Fly High* (1939); *Birds across the Sky* (1942); *Canadian Spring* (1947); *As Far as the Yukon* (1951). Other books illustrated include *South American Zoo* (1946) by Victor Von Hagen and *Mammals of North America* (1947) by Victor Cahalane.

In 1946, he and wife awarded John Burroughs Medal for best literary work on nature. Wife's book on him, *Artist of the Wilderness World* (1973), is best account of his life. Member of American Ornithologists' Union.

MAJOR CONTRIBUTIONS

Influenced by Winslow Homer, Frederic Remington, and N. C. Wyeth. Developed diorama to highest degree as an art form. Painted wilderness for both beauty and values he found there. According to Roger Tory Peterson in foreword to Don T. Luce's biography on Jaques, "His dioramas and canvases are a sensitive and joyous celebration of the wild world—a record of the way it was in his time."

BIBLIOGRAPHY

Luce, Don T., and Laura M. Andrews. *Francis Lee Jaques: Artist-Naturalist* (1982).
Nature (April 1954).
New York Times, 31 August 1934; 17 April 1935; 25 July 1969 (obit.).

UNPUBLISHED SOURCES

See Thomas Sadler Roberts Papers, 1872–1946, in University of Minnesota Library, Minneapolis. Correspondents include Francis Lee Jaques.

Carmela Tino

JEFFERSON, THOMAS. Born Shadwell, Albermarle Country, Virginia, 13 April 1743; died Monticello, Virginia, 4 July 1826. President of the United States (1801–1809), plantation and slave owner, governor of Virginia (1779–1781), drafter of the Declaration of Independence, and founder of the University of Virginia (1819).

FAMILY AND EDUCATION

Son of Peter, a Virginia surveyor, and Jane Randolph Jefferson. Traced ancestry in Virginia back to 1619, when great-grandfather was mentioned as one of delegates from Flowerdieu Hundred to first House of Burgesses. Was tutored in classics at home by the Reverend James Maury. In 1760 entered College of

William and Mary in Williamsburg, Virginia; two years later was graduated. Studied law and admitted to bar in 1767. Married Martha Skelton Wayles, 1772 (died 1782); six children, of whom only two daughters lived to adulthood.

POSITIONS

Elected to House of Burgesses (1769–1775). Delegate to the Continental Congress (1775–1776, 1783–1784); governor of Virginia (1779–1781); minister to France (1785–1789); secretary of state (1789–1793); vice president of the United States (1797–1801); president of the United States (1801–1809).

CAREER

Practiced law, 1767–1770; while member of the House of Burgesses, he published *A Summary View of the Rights of British America* (1774). When he replaced Benjamin Franklin as minister to Paris, published his *Notes on the State of Virginia* (1787). William Peden, who may well have edited the best edition of Jefferson's *Notes* (1955), observed that Jefferson reflected an understanding of his environment that indicated ecological awareness:

> From his mountain top at Monticello, overlooking the green and golden farmlands of Albermarle, he had peered in the vast laboratory of nature and had scrutinized like a lover the phenomena of the weather. He had been preoccupied with the mysteries of space and of the universe, and with that greatest of all mysteries, mankind. And for years, always the practical man rather than the dilettante, he had committed to writing any and all information which might conceivably be useful to him in "any station public or private" particularly everything pertaining to Virginia (xiii).

He studied and cataloged American plants and animals. In *Notes on Virginia*, rejected the earlier (1761) idea of Comte Georges de Buffon that plants and animals in New World were inferior to European species. Was evidently unaware that Buffon had changed his views in 1778. Jefferson, in his study of nature, was quite the outspoken American patriot. As president was able to obtain from Congress a grant for justifiably famous Lewis and Clark expedition. Proposed detailed instructions for expedition so that plant and animal life in the region would be cataloged, especially discovery of new species. In his correspondence, Jefferson was an active scientist. Suggested elaborate record keeping to demonstrate impact of clearing land of trees and the effect on climate. Was avid paleontologist; aided Charles Wilson Peale's efforts in re-creating the mammoth, a precursor of the elephant. Jefferson firmly believed in study of environment as key to understanding human nature. Also interested in archeology and anthropological study of American Indians, especially in his efforts to study Indian languages for a common origin. Took great pride in his selection as president of American Philosophical Society in 1797, a post held until 1815. Reputation as scientist was also recognized with election to the Institut National de France (1802).

MAJOR CONTRIBUTIONS

While conservation movement did not become active factor in American life until second half of the nineteenth century, Jefferson adumbrated elements that surfaced later in America's life. This feature of his life was perhaps best provided in purchase of the Natural Bridge in Virginia, which he bought with adjoining 150 acres. Maintained this natural phenomenon as public trust. In farming his own extensive acres stressed scientific management of land. Was early symbol of concern for nature in the first years of republic. Agreed with Charles Wilson Peale that man should go to nature for schooling.

BIBLIOGRAPHY

Bedini, Silvio A. *Thomas Jefferson and American Vertebrate Paleontology* (1985).

Bedini, Silvio A. *Thomas Jefferson: Statesman of Science* (1990).

Bennett, Hugh. *Thomas Jefferson, Soil Conservationist* (1944).

Boyd, Julian P., and Charles T. Cullen, eds. *The Papers of Thomas Jefferson*. 21 vols. (1950–1988). (An ongoing publication of Jefferson's papers. The first 20 volumes were edited by Boyd; vol. 21 was edited by Cullen.)

Buffon, Comte Georges de, *Histoire Naturelle, générale et particulierè,* 15 vols. (1749–1767), and *Supplément*, 7 vols. in 4 (1774–1789). [esp. *Epoques de la Nature* in vol. 5 (1778).]

Cohen, I. Bernard, ed. *Thomas Jefferson and the Sciences* (1980).

Custis, John, and Thomas Freeman. *Jefferson and Southwestern Expansion: The Freeman and Custis Accounts of the Red River Expedition of 1806*. Ed. Dan L. Flores (1984).

Jackson, Donald. *Thomas Jefferson and the Stony Mountains: Exploring the West from Monticello* (1980).

Miller, Charles A. *Jefferson and Nature* (1988).

Randall, William S. "Thomas Jefferson Takes a Vacation." *American Heritage* (June–July 1996).

Roger, Jacques. *Buffon*. Trans. S. L. Bonnefoi (1996).

Tompkins, Edmund P., and J. Lee Davis. *The Natural Bridge and Its Historical Surroundings* (1939).

UNPUBLISHED SOURCES

See Jefferson Papers, Library of Congress, Washington, D.C. (includes letters not in the published collections; also has letters not yet published in the Princeton University project edited by Boyd et al.).

Thomas J. Curran

JÉRÉMIE, CATHERINE-GERTRUDE. Born Quebec City, Quebec, 21 September 1664; buried Montreal, Quebec, 1 July 1744. Midwife, botanist.

FAMILY AND EDUCATION

Born one of fourteen children to Noël Jérémie, native language interpreter and fur trade post manager, and Jeanne Pelletier. Her brother Nicolas was distinguished for his ethnographic and wildlife studies of the Hudson Bay region published in 1720. Married Jacques Aubuchon, 28 January 1681; one daughter and Michel Lepailleur, 3 November 1688; eleven daughters, three sons.

POSITIONS

Midwife and botanist.

CAREER

After moving to Montreal in 1702, Catherine established herself as a midwife and botanist. She was notable for collecting plants to ship to the Jardin des Plantes in Paris, particularly for the notes she attached to medicinal herbs explaining their properties and uses. These shipments were among those encouraged by members of the French Académie des Sciences and the Jardin, in their attempt to collect and catalog the fauna of Canada. (Unfortunately, the Muséum National d'Histoire Naturelle in Paris has found no record of the specimens that Jérémie sent there.) Jérémie was especially interested in native medicine, perhaps inspired or facilitated by family connections, as two uncles and her brother had married native women. Her interests in this area were significant enough to be noted by the colony intendant Gilles Hocquart in his reports to France.

MAJOR CONTRIBUTIONS

Significant contribution to the pursuit of natural science in New France. One of the earliest botanists in Canada. First Canadian female naturalist of note. Demonstrated an uncommon respect for the native knowledge of plants and their medicinal properties.

BIBLIOGRAPHY

Fortin-Morisset, Catherine. "Jérémie, Catherine-Gertrude." *Dictionary of Canadian Biography*. Vol. 3 (1974).

Jetté, René. *Dictionnaire généalogique des familles du Québec* (1983).

Lynn Berry

JÉRÉMIE, NICOLAS. Born Sillery, Quebec, 14 February 1669; died Quebec City, Quebec, 10 October 1732. Fur trade post administrator, native language interpreter.

FAMILY AND EDUCATION

Born the fifth of fourteen children to Noël Jérémie, also a native language interpreter and fur trade post manager, and Jeanne Pelletier. Married Marie-Madeleine Tetaouiskoué, a Montagnais woman, 1693; annulled 1694 and Françoise Bourrot, Quebec, 1707; one daughter.

POSITIONS

Various roles at fur trade post, Fort Bourbon (present-day York Factory, Manitoba) on Hudson Bay: commercial director, 1694–1696; lieutenant commander, 1697–1707; governor, 1709–1714.

CAREER

At the age of twenty-one, Jérémie was assisting his father as fur trade clerk at trading posts in Chicoutimi, and in 1693, at Tadoussac in the Saguenay River area. He spent the next two decades in the Hudson Bay region in the service of the representative of various trading companies. At the end of this turbulent time, which included hostilities between his fort and both native groups and the English, he penned his *Relation du Détroit et de la Baie d'Hudson*, first published in 1720 at Amsterdam (English edition: *Twenty Years of York Factory. 1694–1714: Jérémie's Account of Hudson Strait and Bay*, ed. R. Douglas and J. N. Wallace. Ottawa [1926]).

The *Relation* was probably intended as a report for the French colonial administration and is significant for its coverage of the ethnology of the Amerindians and Inuit of the region, the flora, and fauna. Jérémie described the topography of the Hudson Bay territory, ice conditions, and effects of magnetic variation and documents one of the earliest attempts at agriculture in that latitude. He makes one of the first mentions of the existence of copper in the northwest territories and includes the earliest record of the musk ox in a well-written and accurate eye-witness account that reveals excellent training.

MAJOR CONTRIBUTIONS

Long-term residency in the Hudson Bay region at the turn of the seventeenth century and knowledge of native languages gave Jérémie particular insight into the natural history of the area for an early period.

BIBLIOGRAPHY

Rousseau, Jacques. "Jérémie, Nicolas." *Dictionary of Canadian Biography*. Vol. 2 (1969).

Jetté, René. *Dictionnaire généalogique des familles du Québec* (1983).

Lynn Berry

JORDAN, DAVID STARR. Born Gainesville, New York, 19 January 1851; died Stanford, California, 19 September 1931. Educator, naturalist, author.

FAMILY AND EDUCATION

Son of Hiram Jordan and Huldah Lake Hawley. Father's family originally from Devon, England. Early education was at home and in local schools. Attended Gainesville Female Seminary by special permission at fourteen. Won competitive scholarship to attend Cornell University, where received M.S. (with credit for undergraduate teaching) in 1872. Attended Indiana Medical College and received M.D., 1875; Ph.D. from Butler University, 1878. Honorary LL.D., Cornell, 1886; Johns Hopkins, 1902; Illinois College, 1903; Indiana University, 1909; University of California, 1913; and Western Reserve University, 1915.

Married Susan Bowen, 1875 (died 1885) and Jessie L. Knight, 1887. Six children.

POSITIONS

Taught at South Warsaw, Indiana, school, 1868–1869. At Cornell University was appointed instructor in botany as a junior. Professor of natural science, Lombard College, 1872–1873. Principal and science teacher, Collegiate Institute and Scientific School, Appleton, Wisconsin, 1873–1874. High school science teacher, Indianapolis, 1874–1875. Professor of natural history, Butler College, 1875–1879. Professor of zoology, 1879–1885, then president, Indiana University, 1885–1891. President Stanford University, 1891–1913, became chancellor, 1913–1916, and chancellor emeritus until his death.

CAREER

Began study of fish at urging of Louis Agassiz, with whom he studied briefly at Penikese Summer School, 1873. Collected for himself, and later for Ohio State Fish Commission and U.S. Fish Commission (cooperating assistant, 1877–1891, 1894–1909), in various parts of the United States, Mexico, Cuba, Bering Sea, Pacific islands, Japan, and Europe. Was a cooperating assistant with U.S. Fish Commission from 1877 to 1881 and again from 1894 to 1909; U.S. Commissioner in charge of fur seal and salmon investigations in the 1890s, and an international commissioner of fisheries, 1908–1910. Became leading fish taxonomist, credited with descriptions of 2,500 species and nearly half that number of genera.

First published ''Hoofrot in Sheep'' in *Prairie Farmer* (1871) while still a student at Cornell. With Balfour Van Vleck published *A Popular Key to the Birds, Reptiles, Batrachians and Fishes of the Northern United, East of the Mississippi River* (1874). Authored *Manual of Vertebrates* (thirteen editions, 1876–1929); *Synopsis of Fish of North America* (1883); *Science Sketches* (1887); *The Fishes of North and Middle America* (with Barton W. Evermann, 4 vols., 1896–1900), which was considered definitive for a number of years; *Seal and Salmon Fisheries and General Resources of Alaska* (with others, 4 vols., 1898); *Footnotes to Evolution* (1898); *Animal Life* (with Vernon L. Kellogg, 1900); *Animal Forms* (with several other authors, 1903); *Food and Game Fishes of North America* (with Barton W. Evermann, 1905); *A Guide to the Study of Fishes* (1905); *The Aquatic Resources of the Hawaiian Islands* (with others, 3 vols., 1905–1906); *The Fishes of Samoa* (with Alvin Seale, 1906); *Evolution and Animal Life* (with Vernon L Kellogg, 1907); *Fishes* (1907); *The Genera of Fishes* (4 parts, 1918–1920); *Fossil Fishes of Southern California* (9 parts, 1919–1926); and *Classification of Fishes* (with K. D. Cather, 1922). Other works dealt with education, philosophy, international arbitration, and world peace, and included *The Core and Culture of Men* (1896); *Imperial Democracy*

(1899); *The Voice of the Scholar* (1903); *Life's Enthusiasms* (1906); *Ways to Lasting Peace* (1915); two volume autobiography, *The Days of a Man* (1922); *The Higher Foolishness* (1925); and *The Trend of the American University* (1929).

Initially concerned with science, nature, and its teachings, Jordan eventually became excellent administrator. Organizational skills and ability for public speaking made him a natural leader. Was offered, but declined, Secretaryship of Smithsonian Institution, 1906. Was founding member of the Sierra Club, President, National Education Association, 1915, and advocate of international peace. Service on worldwide peace organizations included chief director, World Peace Foundation, 1910–1914; president, World's Peace Congress, 1915; vice president, American Peace Society. Additional interests included membership in the Grove Literary Society (or Struggle for Existence Club) and Simplified Spelling Board.

MAJOR CONTRIBUTIONS

Jordan was a major figure in the history of late nineteenth- and twentieth-century ichthyology. As author, his *Manual* was a standard text for half a century and considered influential in subsequent study of vertebrates in the United States. His 1925 *Plan of Education to Develop International Justice and Friendship* won him Ralph Herman Prize for a plan to create international peace. In education, his ability to utilize experiential knowledge in classroom and enthusiasm for study were his greatest assets. As an administrator, his public-speaking abilities, commanding physical presence, and intellect were respected by faculty, students, and contemporaries.

BIBLIOGRAPHY

Bailey, F. K. "Some Scientific Expeditions to the Southeastern United States Taken by David Starr Jordan." *Proceedings, Indiana Academy of Sciences* 71 (1961).

Burns, Edward M. *David Starr Jordan: Prophet of Freedom* (1953).

Dictionary of American Biography. Vol. 5 (1958).

Dictionary of Scientific Biography.

Hays, Alice N. *David Starr Jordan, A Bibliography of His Writings, 1871–1931* (1952).

Jordan, David Starr. *Days of a Man* (1922) (autobiography).

New York Times, 20 September 1931 (obit.).

San Francisco Chronicle, 21 September 1931 (obit.).

Spoehr, Luther W. "'Progress' Pilgrim: David Starr Jordan and the Circle of Reform, 1891–1931." Ph.D. dissertation, University of California (1975).

Who Was Who in America. Vol. 1 (1897–1942).

UNPUBLISHED SOURCES

At Stanford University, see: "Guide to the Microform Edition of the David Starr Jordan Papers, Stanford University Archives," 1969; David Starr Jordan Papers, President's Office Records pertaining to faculty, 1887–1918, Dora Moody Williams, Organic Evolution Class Notes, 1895–1896, David Starr Jordan's correspondence regarding purchase of Koenig Tonometer, 1901–1902, and Robert Eckles Swain, Address in Commemoration, 1932, all in Department of Special Collections, Stanford University Libraries. Elsewhere see: David Starr Jordan and Ada Morse Grose Papers at Hoover Institution Archives; Cymmodorian Society Records, c. 1898–1943, Mary E. Joralemon Papers, and Mary Downing Sheldon Barnes Papers, Bancroft Library, University of California; Stephen Alfred Forbes Correspondence relating to International Zoological Congress, 1892–1893, Archives of University of Illinois at Urbana-Champaign; Grove Literary Society Records and Simplified Spelling Board Publications, 1906–1913, Division of Rare Manuscripts Collections, Cornell University; David Starr Jordan letters to Harry P. Harrison, 1916, Special Collections Department, University of Iowa Libraries.

Geri E. Solomon

JOSSELYN, JOHN. Born, probably in Willingdale-Doe, Essex, England, 1608; died in England, probably in Willingdale-Doe, 1675 (year of death is somewhat conjectural; several sources suggest that he may have survived until 1700). Traveler, writer.

FAMILY AND EDUCATION

Son of Thomas Josselyn of Kent and Theodora Cooke Bere Josselyn: brother of Henry Josselyn, who lived in Maine and was appointed deputy governor of the province in 1645. He spelled his family name Jocelyn. No record of his early life or schooling, but evidence of training in medicine and surgery.

POSITIONS

Born to wealth, but both his father and brother suffered financial reverses, which affected his lifestyle. Evidently well-educated, with wide interests. Made two visits to New England, where he visited his brother Henry, a graduate of Cambridge University who served Fernando Gorges and other British partners in their Maine settlement enterprise in various administrative capacities. John Josselyn lived in New England for fifteen months in 1638–1639 and for eight years from 1663 to 1671. Returned to England permanently in 1671. While in New England, practiced medicine, collected plants, and assiduously gathered Indian and colonial folk lore, his detailed accounts of which often put off both lay and learned readers otherwise very interested in his scientific observations.

CAREER

Best known for his first two books, *New England's Rarities Discovered in Birds, Beasts, Fishes, Serpents, and Plants of That Country* (1672); *An Account of Two Voyages to New England* (1674). Also published *A Chronological Table of the Most Remarkable Passages from the First Discovery of the Continent of America to 1673* (1674). All three books were republished in Boston in 1865.

His pioneering accounts of New England plants were standard until the mid-1780s. His observations concerning New England birds were unscientific and incomplete, and his discussion of the mammals focused on the fur trade and the medicinal purposes to which their body parts were put. Mentioned fish and shellfish, but gave much less attention to reptiles and insects. Also discussed Indian culture, society, and government and the life of the colonists. Was fined on at least three occasions for non-attendance in church while living in New England and resented Puritan interference with his brother Henry's administration of the Maine settlements. Staunchly pro-Royalist, he attacked the New England theocracy in *Two Voyages*, which was otherwise largely promotional in nature. May have received some financial support from King Charles II after his return to England in 1671. Dedicated *Two Voyages* to the Royal Society and was hopeful of being admitted to membership, but was never selected, probably because of his penchant for mixing folklore with serious scientific observations in his books.

MAJOR CONTRIBUTIONS

His interest in botany led to fullest description of New England's flora to that date. *Rarities* focuses on natural history of upper New England, reports significantly on English and Indian folk medicine, and includes information on mineralogy and geology. Works not noted for scientific approach but considered authoritative descriptions for many decades.

BIBLIOGRAPHY

Dictionary of American Biography. Vol. 10.
Dictionary of American History. Vols. 3, 4, 6.
Elliott, Clark A., ed. *Biographical Dictionary of American Science* (1979).
Gura, P. F. "Thoreau and John Josselyn." *New England Quarterly* 48 (1975).
Lindholdt, Paul J., ed. *John Josselyn: Colonial Traveller: A Critical Edition of Two Voyages to New England* (1988).
National Cyclopedia of American Biography. Vol. 7.
Prosser, A. L. "John Josselyn." *Bulletin of the Josselyn Botanical Society of Maine* 10 (1975).
Stearns, Raymond P. *Science in the British Colonies of America* (1970).
Who Was Who in America. Historical volume (1967).

Carmela Tino

JUDAY, CHANCEY. Born Millersburg, Indiana, 5 May 1871; died Madison, Wisconsin, 29 March 1944. Limnologist.

FAMILY AND EDUCATION

Son of Baltzer and Elizabeth (Heltzel) Juday. Graduated from the University of Indiana with A.B. in 1896 and A.M. in 1897. Received an honorary LL.D.

from University of Indiana in 1933. Married Magdalen Evans, 1910; three children.

POSITIONS

High school science teacher, Evansville, Indiana, 1898–1900. Biologist for the Wisconsin Geological and Natural History Survey, 1900–1901. Professor of biology at University of Colorado, 1902–1904. Instructor in zoology, University of California, 1904–1905. Biologist for the Wisconsin Geological and Natural History Survey, 1905–1931. Lecturer in limnology at University of Wisconsin, 1908–1931, professor of limnology, 1931–1941. Director of Trout Lake Limnological Laboratory, 1925–1941. Assistant with U.S. Bureau of Fisheries, summers, 1907–1941. Research associate, University of Wisconsin, 1941–1944.

CAREER

Juday's interest in aquatic biology began at Indiana University under direction of ichthyologist Carl H. Eigenmann. After two years as high school science teacher (1898–1900), Juday accepted position as biologist with Wisconsin Geological and Natural History Survey. An attack of tuberculosis in the following year forced him to withdraw, and from 1902 to 1905, he assumed academic positions at University of Colorado and the University of California. During this period he studied the fish and fisheries of Colorado and Lake Tahoe and the smaller crustaceans of the San Diego region. In 1905, Juday resumed his position with the Wisconsin Geological and Natural History Survey, where famous collaboration with E. A. Birge began. From October 1907 to June 1908, Juday visited various limnologists and limnological laboratories in Europe. On return to the United States, he was appointed lecturer in limnology in Department of Zoology at the University of Wisconsin, where he began teaching courses in limnology and fisheries. He became professor of limnology in 1931 and directed the training of many graduate students until retirement in 1941. In February 1910, he visited Guatemala and El Salvador, making a survey of several lakes in both countries and studying their limnology. Similar research was conducted in the Finger Lakes region of New York State later in 1910. While Juday and Birge concentrated much of their early research efforts on lakes of Madison and southeastern Wisconsin, their research emphasis shifted away from Madison after 1917. From 1921 to 1924, they conducted extensive limnological research on Green Lake, and in 1925 a summer field station was established on Trout Lake. Juday served as director of the Trout Lake Limnological Laboratory during summer months, and the station attracted many biologists throughout the world.

Juday published more than 100 papers, including works on vertical migration of lake plankton, the chemistry of lake waters, hydrography and morphometry, and growth of fish and photosynthesis as indexes of the productivity of lakes.

Among the most important of these are three book-length reports, ''The Inland Lakes of Wisconsin. I. The Dissolved Gases and Their Biological Significance'' (with E. A. Birge, 1911); ''The Inland Lakes of Wisconsin. II. The Hydrography and Morphometry of Lakes'' (1914); and ''The Inland Lakes of Wisconsin. I. The Plankton. I. Its Quantity and Chemical Composition'' (with E. A. Birge, 1922).

In 1935, Juday was elected as first president of the Limnological Society of America. He served as president of American Microscopial Society in 1923 and as president of the Ecological Society of America in 1927. He was also secretary for the Wisconsin Academy of Sciences, Arts and Letters from 1922 to 1930 and acted as president from 1937 to 1939. In 1943 he was awarded the Leidy Medal of the Academy of Natural Sciences of Philadelphia, and in 1950 International Association of Limnology presented the Naumann Medal to the University of Wisconsin in recognition of his work and that of Birge.

Juday was a quiet man, modest and thoughtful, always willing to listen and give helpful advice. His outstanding reputation among students is evidenced by the number of limnologists who have been directly or indirectly influenced by his work.

MAJOR CONTRIBUTIONS

Juday's contributions to the field of limnology are numerous. His most renowned work was with E. A. Birge on mechanism of stratification and the quantitative relationships between plankton activities and dissolved gases. But perhaps more important were the extensive data and factual knowledge that he accumulated and that helped serve others in advancing the science of limnology.

BIBLIOGRAPHY

Dictionary of American Biography. Supplement 3 (1973).

Dictionary of Scientific Biography. Vol. 7 (1973).

Frey, D. G. *Limnology in North America* (1966).

National Cyclopedia of American Biography. Vol. 42 (1958).

Pearse, A. S. *Science* (23 June 1944) (obit.).

Special Publication No. 16. Limnological Society of America (1945) (obit. by Lowell E. Noland, publications of Chancey Juday by Arthur D. Hasler).

Welch, Paul S. *Ecology* (July 1944) (obit.).

Gregg A. Mitman

JUST, ERNEST EVERETT. Born Charleston, South Carolina, 14 August 1883; died Washington, D.C., 27 October 1941. Marine zoologist, educator.

FAMILY AND EDUCATION

Eldest of three children of Charles Frazier Just, a wharf builder, and Mary Matthews Just. Father died when Just was four and mother died when he was ten. Educated at Frederick Deming, Jr. Industrial School, established by mother in Maryville, South Carolina, until age twelve. Attended Classical Preparatory Department, South Carolina State College, Orangeburg, South Carolina, 1896–1899, receiving Licentiate of Instruction, 1899. Attended Kimball Union Academy, Meriden, New Hampshire, 1900–1903. A.B., Dartmouth, 1907. Began graduate studies at University of Chicago through Marine Biological Laboratory, Woods Hole, Massachusetts, and served as research assistant there, summers 1909–1914. Ph.D., University of Chicago, 1916. Honorary D.Sc., South Carolina State College, 1922. Married Ethel Highwarden, 1912 (divorced, 1939); two daughters and a son. Married Main Hedwig Schnetzler, 1939; one daughter.

POSITIONS

Instructor in English, Howard University, 1907–1909; instructor in English and biology, 1909–1910; assistant professor of biology, 1910–1911; associate professor of biology, 1911–1912; professor of biology and professor of physiology, Howard University Medical School, 1912–1920; head, Department of Zoology, Howard University, 1912–1941 (frequently on leave, 1929–1940); member of scientific staff, Marine Biological Laboratory, Woods Hole, Massachusetts, during most summers, 1917–1930; visiting professor, Kaiser Wilhelm Institut, Berlin, 1930; visiting lecturer, Oberlin, 1931.

CAREER

Just first became interested in biology during his sophomore year at Dartmouth when inspired by a course in principles of biology. Only magna cum laude graduate in his class. Began teaching at Howard in the fall of 1907 and remained on faculty for the rest of his life, though increasingly at odds with university president after 1929 and often on leave for research in Europe. Was excellent teacher; began offering graduate instruction following receipt of grant he had secured for that purpose in 1929. Was heavily dependent upon grants, 1919–1936, for his research at Howard, in Europe, and at Woods Hole Marine Biological Laboratory (MBL). Howard President M. W. Johnson placed increasing pressure on Just to teach and reduce research commitments after 1929, increasingly controlled his grant funding, and sometimes used these monies for other institutional purposes.

Just spent most summers at MBL, 1909–1930, first conducting research for his Ph.D. and as research assistant to F. R. Lillie, his mentor at Chicago. Later carried out own independent study of marine organisms, particularly their embryology. Was regarded as an authority on many marine organisms at MBL and was frequently consulted by colleagues and students. Focused on marine egg cell morphology for much of his career after 1910. Initially an experimental zoologist, specializing on fertilization in sea worms, sea urchins, and worm-like chordates. After 1928, became increasingly interested in broader biological questions, such as the role of the ectoplasm as key to behavioral questions in biology. He examined role of the nucleus in cells and possible melding of genetics and the physiology of cell development in their study. Was greatly interested in the role of biochemistry as it might apply to a theory of heredity affecting all plants and animals. More sympathetic to Kropotkin's concept of mutual aid and cooperation in biology than Darwin's survival of the fittest.

Unable because of his race to secure any prestigious post elsewhere in the United States, he became increasingly frustrated and embittered about the manner in which he had been treated by some fellow scientists in the United States. Most colleagues and prospective granting agencies felt that Just should continue to teach at Howard during the academic year and conduct research at MBL during the summers. They evidently could not appreciate, and foundations ultimately refused to support, his intense and growing eagerness to spend most of his time on research in Europe. There, he and his work were highly respected and working conditions were much more congenial for him. Between 1929 and 1940, much of his time was variously spent at Stazione Zoologica, Naples, Italy; Kaiser Wilhelm Institut, Berlin, Germany; and Station Biologique, operated by the Sorbonne at Roscoff Station, near Finistere, France. Unable to secure any grants after 1936, and increasingly unhappy with the situation at Howard, Just was virtually penniless during the last years of his life. He did have some offers of research opportunities in the United States and in France in 1937, most without salary. He stopped teaching at Howard, sought retirement with pay, which was denied, and spent 1938–1940 in Europe, conducting research under increasingly difficult conditions. Howard granted him sick leave with a very small income for 1938–1939, and he had modest support from friends. Work at Roscoff was terminated when French government closed its doors to foreigners after the German occupation began in 1940. Just was briefly detained and imprisoned by German authorities in August, 1940, and in his haste to leave for the United States he left notes on two years' research he had done on starfish and sea worms behind him in France. Obliged to return to Howard despite increasing ill health, Just taught there during academic year 1940–1941, but he could do no research after July 1941 because of terminal cancer.

During his career, Just authored some sixty papers and articles. His major work, *The Biology of the Cell Surface*, and another text, *Basic Methods for Experiments on Eggs of Marine Mammals*, were both published in 1939. He was awarded the first Spingarn Medal of the NAACP in 1915, was a vice-president of the American Society of Zoologists (1930), a fellow of the American Association for the Advancement of Science, and a member of many other scientific organizations. He was at various times on editorial boards of the journals *Physiological Zoology, Biological Bulletin, Journal of Morphology, Zeitschrift Für Zellforschung und Mikroskopische Anatomie*, and *Protoplasma*, among others.

MAJOR CONTRIBUTIONS

Achieved seminal results with research concerning the embryology of marine organisms, marine egg cell morphology, the nature and function of cell surfaces and membranes, and later the roles of genetics, the physiology of development, and biochemistry in heredity. A leading American marine biologist for a third of a century.

BIBLIOGRAPHY

American Men of Science (1927).

Boveri, Margret. *Verzweigungen: Eine Autobiographie* (1977).

Cobb, W. M. ''Ernest Everett Just, 1883–1941.'' *Journal of the National Medical Association* 49 (1957).

Dictionary of American Biography. Supplement 3.

Lillie, Frank. ''Appreciation of Just and His Career.'' *Science* 95 (1942).

Logan, Rayford. *Howard University: The First Hundred Years, 1867–1967* (1969).

Manning, Kenneth R. *Black Apollo of Science: The Life of Ernest Everett Just* (1983).

St. Laurent, Philip. ''The Negro in History: Dr. Everett E. Just.'' *Tuesday Magazine, Washington Sunday Star* (June 1970).

UNPUBLISHED SOURCES

E. E. Just Papers at Howard University and in hands of family members. See also additional listings in Manning, above.

Keir B. Sterling

K

KALM, PEHR. Born Ångermanland (Sweden), 6 March 1716; died Åbo (Turku, Finland), 16 November 1779. Naturalist, agriculturalist, topographer.

FAMILY AND EDUCATION

Son of a Lutheran priest, who died in Swedish war against Russia before Kalm was born. He grew up in northern Finland (Närpes, Österbotten), attended gymnasium at Wasa, and started his academic studies at the University of Åbo (1735). Change in his original plans to study theology influenced by Linnaeus's friend Browallius at Åbo. From 1741, attended Uppsala University, studying natural history. As protégé of Sten Carl Bielke, high judiciary official of Finland, he traveled in Russia in 1744 (St. Petersburg, Moscow). On his American journey he married in Raccoon, Pennsylvania, Anna Margareta Sandin, widow of Swedish priest. There were two sons, one of whom reached maturity.

POSITIONS

Appointed by Royal Swedish Academy to travel to South Africa in 1743, a journey he never undertook. Docent in natural history at Åbo in 1746 and professor of economy at Åbo 1747–1779. He was elected member of the Swedish Royal Academy of Sciences in 1745. The genus name *Kalmia* (mountain laurel) was given in his honor by Linnaeus.

CAREER

His travels in western Sweden in 1742 resulted in *Wästgötha och Bohusländska resa* (1745), to be compared with Linnaeus's travels in the same regions.

The great adventure of his life, which made him famous, was travel to America in 1747–1751. He visited New Sweden (Delaware), Philadelphia, Albany, Montreal, Quebec, and surroundings, getting as far north as Cap aux Oyes, and he met with American scholars, including the elder Bartram, Colden, and Franklin. The latter published his description of Niagara Falls (in the *Pennsylvania Gazette*, 1750), the first one given in English. In his journals there are accounts of Indians and Esquimaux as well of Swedish settlers. He took special interest in American flora, being educated by Linnaeus and commissioned by him to look for suitable plants to be cultivated in Sweden. Thus, there is a strong utilitarian element in Kalm's writing. The publication of Kalm's *En Resa Til Norra Amerika*, 3 vols. (1753–1761; new edition in 2 vols. based on the original manuscript, Helsinki, [1866–1870]) was interrupted by the death of the publisher. Supplement published 1929. Translated into German, Dutch, and English (the latter a new translation by Benson, 2 vols. [1937]); with a recent translation into French by Rousseau and Bethune (1977). As professor at Åbo, he industriously wrote academic papers and contributed to the *Acta* of the Royal Swedish Academy of Sciences (KVAH), publishing a considerable amount of information on American natural history and ethnography. This included articles on the rattlesnake (KVAH, 1752–1753), passenger pigeon (KVAH, 1759), and hickory (KVAH, 1778). He also worked to establish a more complete natural history for Finland, supervising dissertations concerning topography.

Trying to popularize agriculture he wrote in Swedish and Finnish, leaving Latin for more scholarly subjects. In Åbo he founded a botanical garden to cultivate American plants in Scandinavian soil. Kalm was one of the most important of Linnaeus's pupils, opening up a new continent for Linnaean science. He brought home extensive herbarias, and though he did not complete a *Flora Canadensis*, his specimens were used by Linnaeus in *Species Plantarum* (1753), where ninety of his species were cited. His journals are important sources for pioneer life as well as for eighteenth-century flora and fauna. Though written in rather dry, matter-of-fact manner, they are extremely useful to historians.

MAJOR CONTRIBUTIONS

Kalm was no man of theory, but a good observer, providing information in many fields, for example, plant geography, ethnobotany, paleontology, climatology, meteorology, and husbandry, as well as general taxonomy. His account of American natural history must be rated as one of the most important of his century.

BIBLIOGRAPHY

Dictionary of Scientific Biography (1973).

Kerkkonen, M. *Peter Kalm's North American Journey. Its ideological Background and Results* (1959).

Odhelius, J. L. *Minnestal öfwer Pehr Kalm* (1780) (obit.).

Olsson, Nils William. *Pehr Kalm and the Image of North America* (1970).

Rousseau, J., and G. Bethune. *Voyage de Pehr Kalm au Canada en 1749* (1977).
Skottsberg, C. *Pehr Kalm Levnadsteckning* (1951).
UNPUBLISHED SOURCES
See Herbarias in Linnaean Society, London, in Riksmuseet, Stockholm, Institution för systematisk botanik, Uppsala; manuscript of journey in University Library, Helsinki.
Gunnar Broberg

KALMBACH, EDWIN RICHARD. Born Grand Rapids, Michigan, 29 April 1884; died 26 August 1972. Ornithologist, mammalogist.

FAMILY AND EDUCATION

Son of Godfrey Kalmbach, a shoe merchant, and Anna Steinecke Kalmbach. Graduated from Grand Rapids High School, Grand Rapids, Michigan. Self-educated in ornithology, mammalogy, entomology, botany, ecology. Honorary D.Sc., University of Colorado, 1955. Married Kathryn Arvilla Kalmbach, 1908; three children.

POSITIONS

Assistant director of Kent Scientific Museum, 1903–1910; assistant biologist of Bureau of Biological Survey, U.S. Department of Agriculture, 1910–1924; biologist, 1924–1928; senior biologist, 1928–1932; biologist and director of Wildlife Research Laboratory, U.S. Fish and Wildlife Service, Denver, 1932–1954. Retired in 1954.

CAREER

Played an important role in establishing and encouraging the Department of the Interior program of duck stamps in the mid-1930s, and also commemorative stamp issues put out by the U.S. Post Office depicting various kinds of wildlife. Was the designer of the Ruddy Duck stamp issued by the Interior in 1941–1942. Frequently represented the concerns and priorities of the U.S. government as these related to wildlife conservation and management issues at scientific conferences. Was known among the nation's agricultural community and the mammalogists and ornithologists of his day for his work in economic mammalogy and ornithology, and also for his work as a wildlife artist.

Co-author, with W. L. McAtee and others, of *Common Birds of the Southeastern United States in Relation to Agriculture* (1916). Authored a number of valuable studies, including *The Crow in Relation to Agriculture* (1920, revised, 1939); *The Magpie in Relation to Agriculture* (1927); *The European Starling in the United States* (1928); *Homes for Birds* (with W. L. McAtee, 1930; later revised by D. D. Boone, 1979); *English Sparrow Control* (1930); *Western Duck Sickness: A Form of Botulism* (1934); *Economic Status of the English Sparrow in the United States* (1940); *The Armadillo: Its Relation to Agriculture and Game* (1943); and *Type C Botulism Among Wild Birds: A Historical Sketch* (1968). Also published *Wildlife in a Developing Hemisphere*. Was in addition an illustrator of books, including *Knowing Birds Through Stories*, by Floyd Bralliar (1922), and *Alaska Bird Trails*, by Herbert Brandt (1943). Active member of such organizations as the American Ornithologists' Union, American Society of Mammalogists, the Wildlife Society, and the Cooper and Wilson Ornithological Societies. Received a number of honors, including the Distinguished Service Award from the Department of the Interior (1955); the Wildlife Society's Aldo Leopold Award (1958), and the Izaak Walton League's Founders' Award.

MAJOR CONTRIBUTIONS

Responsible for helping to shape the U.S. government's policies in conservation and management of wildlife resources. Noted for research on economic ornithology and mammalogy.

BIBLIOGRAPHY

American Men of Science. 10th ed. (1960).

Stroud, Richard H., ed. ''Kalmbach, Edwin Richard.'' *National Leaders of American Conservation* (1985).

U.S. Bureau of Sport Fisheries and Wildlife. *The American Eagles and Their Economic Status, 1964* (1964, 1954) (Contains ''The Bald Eagle and Its Economic Status'' by R. H. Imler and E. R. Kalmbach. Cover title: Reprints of U.S. Fish and Wildlife Service Circular 30 and 27).

U.S. Fish and Wildlife Service. *Attracting Birds* (1942, 1950) (contains 3 vols. in one, including ''Homes for Birds'' by E. R. Kalmbach and W. L. McAtee.

Joan Ryan

KANE, ELISHA KENT. Born Philadelphia, 3 February 1820; died Havana, Cuba, 16 February 1857. Naval officer, physician, arctic explorer.

FAMILY AND EDUCATION

Eldest child of John Kintzing Kane (prominent attorney, attorney general of Pennsylvania, a commissioner in 1830s to settle claims with France, and U.S. district judge) and Jane Duval Leiper Kane. No information concerning early education. At University of Virginia, September 1837 (?)–November, 1839. M.D., Medical Department, University of Pennsylvania, 1842. Attended medical lectures, Jefferson Medical College, Philadelphia, fall 1843. Alleged to have married Margaret Fox, 1856.

POSITIONS

Commissioned assistant surgeon, U.S. Navy 1843. To China as unpaid physician on Caleb Cushing Expedition, 1843–1844; in brief medical practice,

Whampoa, China, last half of 1844; returned to Philadelphia via Egypt and Europe, late summer 1845. On board USS *United States* in West African waters, 1846–early 1847. To Mexico with special dispatch for Maj. Gen. Winfield Scott, late 1847, saw some brief action, returned to Philadelphia, 1848. On store ship *Supply* to Mediterranean and to Brazil, 1849. Member of first American (Grinnell) expedition searching for John Franklin, 1850–1851; commander of Second Grinnell Expedition, 1853–1855.

CAREER

Kane sustained heart damage due to rheumatic fever while at University of Virginia. Brilliant student at University of Pennsylvania. His thesis described test of new method for determining pregnancy, published at faculty's request in *Journal of the Medical Sciences* (1842). Following graduation from medical school, served as unpaid volunteer physician to Cushing Expedition to China, visited Rio (where he made some geological observations), Bombay, and Columbo en route to Macao. Secured leave of absence from Cushing Expedition, visited Philippines. Returned to Macao, resigned position with Cushing, began medical practice on hospital ship at Whampoa Reach, near Canton, with a young English physician. This venture terminated by illness. Returned home via India, Egypt (where he again became ill), and Europe. May briefly have begun medical practice in Philadelphia, 1845–1846, but joined USS *United States* on cruise off West African waters before he got practice under way, fell ill, was returned to Philadelphia early 1847. In late 1847, departed on mission to take message to Gen. Winfield Scott in Mexico, for which he had volunteered, wounded in action en route. Following some months to recover from this and from typhus, reached Scott but was ordered home because physically unfit for duty. In March 1849, sailed on naval store ship *Supply* to various Mediterranean ports and Rio. On return, volunteered to join first [Henry] Grinnell Expedition, privately financed venture searching for John Franklin, British explorer who had disappeared while searching for Northwest Passage in 1845. Departed May 1850. Kane did some botanizing and geological study, took notes on animals and plants seen. Returned to New York late 1851. Lectured widely concerning first expedition, and with aid of father (necessitated by more ill health) completed *The First U.S. Grinnell Expedition in Search of Sir John Franklin* (1853). It included some discussions of marine life, birds, geological findings, and plants collected. Excellent illustrations based on Kane's sketches. Organized and led a second expedition in privately financed vessel *Advance*, which went through Smith Sound and explored parts of what is now Kane Basin east of Ellesmere Island in hopes of finding open waters to north. A few members of expedition got as far north as Cape Constitution (80 degrees, 58 minutes north). Considerable scientific data collected, including geographic, botanical, and zoological. Crew forced to abandon icebound vessel. Kane prepared and carried out eighty-three-day, 1,200–mile trip with crew on foot and under sail, to Upernavik, Greenland, despite incredible hardships. Found near there by crew of Danish vessel, which offered to drop them off at an English port. This ship stopped briefly at Godhavn, where U.S. relief expedition found them. Published *Arctic Explorations: The Second Grinnell Expedition in Search of Sir John Franklin, 1853,'54,'55*, 2 vols. (1856). The 300 illustrations based on Kane's excellent sketches and verbal descriptions. One appendix listed some 148 species of plants found during the course of the expedition; others provided astronomical, geographical, and weather data. Presented Gold Medal of the Royal Geographical Society, 1856. Despite growing weakness from rheumatic fever, briefly went to England to support Lady Franklin's efforts to secure British government support for another effort to find possible survivors of her husband's expedition. Kane then went to Cuba in hopes of renewed health, suffered an apoplectic stroke en route, and died there.

MAJOR CONTRIBUTIONS

Explored and first mapped waterway that led to polar sea, the so-called American route to the Pole. Mapped area as accurately as possible, though it was later necessary to make corrections to some of his findings. His friendship with Eskimos left legacy of trust most helpful to later explorers. His two books, particularly his last, *Arctic Explorations, 1853,'54,'55*, a major contribution to literature of the subject. Provided useful lists of fauna and flora encountered, with some observations concerning animal life.

BIBLIOGRAPHY

Corner, George W. "Hero with a Damaged Heart: The Clinical History of Elisha Kent Kane, M.D." In *Medicine, Science, and Culture: Historical Essays in Honor of Owsei Temkin* (1968).

Corner, George W. *Doctor Kane of the Arctic Seas* (1972).

Dictionary of American Biography.

National Cyclopedia of American Biography.

Villaejo, Oscar M., ed. *Dr. Kane's Voyage to the Polar Lands* (1965).

Who Was Who in America.

UNPUBLISHED SOURCES

See Kane Family Papers, American Philosophical Society, Philadelphia; source materials and notes for an unpublished biography of Kane by M. E. Dow, Stefansson Arctic Collection, Dartmouth College; Kane journals and other papers, Historical Society of Pennsylvania; Kane journal, Stanford University Library; Polar Archives, National Archives and Records Administration; diaries and correspondence of Lady Jane Franklin, Scott Polar Research Institute, Cambridge, England.

Keir B. Sterling

KELLOGG, (ARTHUR) REMINGTON. Born Davenport, Iowa, 5 October 1892; died Washington, D.C., 8 May 1969. Zoologist.

FAMILY AND EDUCATION

Son of Rolla Remington, a printer, and Clara Louise Martin Kellogg, a school teacher. Family moved to Kansas City, Missouri, when Kellogg was six years old. He had one brother and one sister. Graduated from Westport High School, Kansas City, Missouri, in 1910. University of Kansas: A.B., 1915, M.A., 1916. University of California, Berkeley: Ph.D., 1928. Married Marguerite Evangel Henrich, 1920; no children.

POSITIONS

Taxonomic assistant, Museum of Birds and Mammals, University of Kansas, 1913–1916, teaching fellow, Department of Zoology, University of California, 1916–1919, field assistant, Bureau of Biological Survey, U.S. Department of Agriculture, summers 1915, 1916, 1917, assistant biologist, 1920–1924, associate biologist, 1924–1928, assistant curator, Division of Mammals, U.S. National Museum, 1928–1940, curator, 1941–1948, director, U.S. National Museum, 1948–1962, assistant secretary, Smithsonian Institution, 1958–1962, research associate, 1962–1969.

CAREER

After service in U.S. Army in France (1918–1919) he spent eight years in field and laboratory work in California, North Dakota, and Montana for Biological Survey, studying feeding habits of hawks and owls, taxonomy of toads and meadow mice, and dietary habits of alligators. Concurrently with this work he began studies of fossil marine mammals, which continued for the rest of his life. His early work was on pinnipeds, but his major interest was in whales. Published *Pinnipeds from Miocene and Pleistocene Deposits of California* (1922), which incorporated critical review of the literature on fossil pinnipeds of the world and is still the base on which modern research on fossil pinnipeds begins. His monograph *A Review of the Archaeoceti* (1936) is still the definitive work on that extinct suborder of whales, which includes ancestors of the living forms. He published 156 papers and two books, including 62 papers and one book on fossil whales. His other book, a major taxonomic work written with Gerrit S. Miller, Jr., is the *List of North American Recent Mammals* (1955).

In 1930 he went to Berlin as delegate to a conference of whaling experts held under the auspices of the League of Nations; this was the first in a series of meetings that led to the Washington conference of 1946, which formulated the International Convention establishing the International Whaling Commission. He was head of the U.S. delegation to the first sixteen meetings of the commission, was vice chairman of the commission from 1949 to 1951 and chairman from 1952 to 1954. Between 1931 and 1956 he contributed to many congressional hearings, protocols, and scientific reports recommending measures for preservation of whales.

He was happiest doing research; enjoyed informal scientific discussion, but seldom went to scientific meetings and gave no formal papers after his early research years. Although he spent years in administrative work, he appeared to regard it as duty rather than a pleasure. He loved to play the devil's advocate in scientific discussion; to some he appeared negative, but this at least in part stemmed from his refusal to accept a statement without proof. He was informal, sometimes even blunt, in speech, but always good-humored; unstinting in giving time and advice to fellow scientists and also to amateurs who came to the National Museum. He had vastly retentive memory that covered every phase of vertebrate zoology and was tireless and efficient in organizing large masses of data in a day when the computer was not yet available.

MAJOR CONTRIBUTIONS

Recognized for many years as dean of whale research workers. He was loath to generalize, but his description and analysis of many genera and species of fossil whales have formed the basis for greatly increased research in this field over the past two decades. He fought hard in the International Whaling Commission to obtain a rational catch limit on antarctic whaling, based upon scientific evidence. This view eventually prevailed.

BIBLIOGRAPHY

Setzer, Henry W. *Journal of Mammalogy* 58, 2 (May 1977) (obit.).

Whitmore, F. C. Jr. "Remington Kellogg, 1892–1969," in *Biographical Memoirs, National Academy of Sciences*. Vol. 46 (1975).

UNPUBLISHED SOURCES

See Files of Remington Kellogg Memorial Library of Marine Mammalogy, National Museum of Natural History, Washington, D.C. 20560; Remington Kellogg Papers, Smithsonian Institution Archives, Washington, D.C. 20560.

Frank C. Whitmore, Jr.

KENNICOTT, ROBERT. Born New Orleans, 13 November 1835; died Nulato, Alaska, 13 May 1866. Naturalist, explorer, museum director.

FAMILY AND EDUCATION

Son of John Albert Kennicott, physician/horticulturist, and Mary Ransom Kennicott. Family moved to Illinois in 1836. Four brothers, two sisters. Did not attend school regularly, no degrees. Never married; no children.

POSITIONS

Collected specimens of natural history in northern Illinois, 1852–1855; in southern Illinois, summers of 1855 and 1857; on the Red River of the North in late 1857; in the Hudson's Bay region, 1859–1862. Director of the Chicago Academy of Sciences, 1864–1866, curator, 1864. Explorer in what is now Alaska, Russian-American Telegraph Expedition, 1865–1866.

CAREER

Began study of natural history around 1852. Early education under Jared P. Kirtland and Philo R. Hoy. In 1853 established a correspondence with Spencer F. Baird of the Smithsonian Institution, who would exert a powerful influence on his career. From 1854 to 1860 Kennicott wrote a series of articles for the *Prairie Farmer*, aimed at educating farmers about local animals. In 1855 he published a list of animals from Cook County, Illinois, the first real attempt to document the fauna of Illinois. Later in 1855 Kennicott visited southern Illinois to make collections under the auspices of the Illinois Central Railroad. In 1856 he completed the first part of his work, entitled "The Quadrupeds of Illinois Injurious and Beneficial to the Farmer." Again, his emphasis continued to be the education of farmers and their children, but the accounts of each animal are still of value today. In later 1857 he traveled to the Smithsonian Institution and arranged the collection of snakes. From 1859 to 1862 Kennicott explored in the Mackenzie and Yukon Basins of the Hudson's Bay Company, his expenses paid by the Smithsonian and several Chicago groups. His journal from this period is among the finest examples of natural history writing in existence. Kennicott returned briefly to the Smithsonian in early 1863 and by 1864 was actively involved in creating a museum for the Chicago Academy of Sciences. Late in that year he accepted a position as naturalist on the Russian-American Telegraph Expedition, leaving the Chicago Academy in the capable hands of his friend William Stimpson. Kennicott departed from New York in March 1865 and traveled across Nicaragua to California and then on to St. Michael's, Alaska on the Seward Peninsula. Led a party in an exploration of the Yukon River in an attempt to chart a telegraph line to Fort Yukon. Always prone to bouts of depression, Kennicott was frustrated by myriad delays and red tape. He died presumably of heart failure.

MAJOR CONTRIBUTIONS

An inveterate explorer and acute field naturalist in both the Midwest and the Arctic. Published important contributions in herpetology, mammalogy, and ornithology. The driving force behind the creation of the Chicago Academy of Sciences and an early advocate of teaching natural science to all people. Accounts from his northern explorations played a role in convincing Congress to purchase Alaska in 1867. Also credited with naming twelve species and eighteen subspecies of North American snakes.

BIBLIOGRAPHY

Chicago Tribune, 15 November 1866 (obit.).

Committee of the Chicago Academy of Sciences. ''Biography of Robert Kennicott.'' *Transactions of the Chicago Academy of Sciences* 1, pt. 2 (1869).

James, James Alton. ''Robert Kennicott, Pioneer Illinois Natural Scientist and Arctic Explorer.'' *Transactions of the Illinois State Historical Society* (1940).

Lindsay, Debra. *Science in the Subarctic*. (1993).

Vasile, Ronald S. ''The Early Career of Robert Kennicott, Illinois' Pioneering Naturalist.'' *Illinois Historical Journal* (1994).

UNPUBLISHED SOURCES

Letters are in the Smithsonian Institution Archives; Indian vocabularies are in the National Anthropological Archives; letters are at the Grove National Historic Landmark, Glenview, Ill.; and biographical information is at the Chicago Academy of Sciences Archives.

Ronald S. Vasile

KIERAN, JOHN FRANCIS. Born New York City (Riverdale section) 2 August 1892; died 9 December 1981. Writer, radio personality, naturalist.

FAMILY AND EDUCATION

Son of James Michael and Kate (Donohue) Kieran. Father was public school principal, who later became professor and then president of Hunter College. Mother was public school teacher. Graduated from Fordham University in 1912 with B.S. Received honorary D.Sc., Clarkson, 1941, and honorary M.A., Wesleyan University, 1942. Married Alma Boldtmann, 1919 (died in 1944): three children; and Margaret Ford, 1947.

POSITIONS

Sportswriter with *New York Times*, 1915–1917, 1927–1943; between 1922 and 1926, wrote for *New York Herald/Tribune* and Hearst newspapers; columnist for *New York Sun*, 1943–1944. Panelist on ''Information, Please,'' a radio program, from 1938 to 1948.

CAREER

Family owned farm in Duchess County, and there, as he wrote in *Not under Oath*, he ''developed an ever-increasing love of nature.'' Was an avid birdwatcher who also was knowledgeable about trees and flowers. After graduating from Fordham, tried, unsuccessfully, to make living as farmer.

Before joining *Times*, worked in construction business. Began at *Times* covering golf. After army service in France in World War I, returned to *Times* and was assigned to cover major league baseball. In 1927 started ''Sports of the Times,'' first daily signed column of any kind in *New York Times*. Success on

"Information, Please" radio show, where he was panelist-authority on sports, nature, and other subjects, made him national celebrity.

Author of many articles and books. Latter included *Nature Notes* (1941); *An Introduction to Birds* (1946); *Footnotes on Nature* (1947); *An Introduction to Trees* (1954); and his magnum opus, *A Natural History Of New York City* (1959). Also an autobiography, *Not under Oath* (1964). Recipient of John Burroughs Medal, 1960.

MAJOR CONTRIBUTIONS

Amateur naturalist who was influential popular educator. *Natural History of New York City* was considered by Edwin Way Teale "in many ways the best treatment the natural history of a great city has ever received."

BIBLIOGRAPHY

"Kieran, John." *Current Biography* (1940).
New York Times, 11 December 1981 (obit.).
Who Was Who in America. Vol. 8 (1988).

Richard Harmond

KING, CLARENCE RIVERS. Born Newport, Rhode Island, 6 January 1842; died Phoenix, Arizona, 24 December 1901. Geologist, science administrator, mining consultant, alpinist, writer, art collector.

FAMILY AND EDUCATION

Only son and eldest of three children of James Rivers King and Florence (Little) King. Father a shipping merchant who died when Clarence was six and whose company (King & Co.) went bankrupt in 1857. His mother (remarried 1859) encouraged his interests in natural history and art. Attended primary and secondary schools in Pomfret and Hartford, Connecticut, at Yale (Sheffield Scientific School), 1859–1862, Ph.B., 1862, honorary LL.D., Brown, 1890. Married (?) Ada Todd of New York City (died 1930), 1888; five children.

POSITIONS

Volunteer assistant in geology, Josiah Dwight Whitney's Geological Survey of California, 1863–1865. Civilian topographer-geologist on staff of Maj. Gen. Irwin McDowell, Comm., U.S. Army's Department of California, 1865–1866. Assistant geologist, Whitney Survey, 1866. U.S. geologist and geologist-in-charge, U.S. Geological Survey of the 40th Parallel ("King Survey"), Army Corps of Engineers, 1867–1878. Director, U.S. Geological Survey (USGS), 1879–1881. Special agent, U.S. tenth census (1880), 1879–1882. Member, ex officio, [First] Public Lands Commission, 1879–1880. Manager, Prietas and Yedras Mines, and president and director, Sombrerete Mine, Mexico, 1879–1891.

Trustee, National Bank of El Paso, Texas, 1886–1893. Consultant, Anaconda Copper Co., 1900.

CAREER

Studied geology at Yale with James Dwight Dana, chemistry with George Brush. Member of first class to receive degrees from Sheffield Scientific School. In fall, 1862, King read William Brewer's letters to Brush describing Brewer's adventurous scientific work in California for J. D. Whitney's survey, heard Louis Agassiz's lectures about glaciers, and decided to become a geologist. With schoolmate James Gardiner, traveled to Virginia City and California in 1863 and joined Whitney Survey. Mapped Mt. Shasta and central Sierra Nevada areas with Brewer, Gardiner, and topographer Charles Hoffman, 1863–1864. With Gardiner, surveyed boundaries of new Yosemite Park and mapped Yosemite Valley, 1864–1865. Explored for military wagon roads in Arizona, 1865–1866. Organized, staffed, planned fieldwork, and wrote and supervised publications of the King Survey, 1867–1878, one of four federally sponsored geological, geographical, and natural-resource reconnaissances conducted west of 100th meridian before 1880. The King Survey examined and mapped, using triangulation methods throughout, some 90,000 square miles of lands flanking 40th parallel, between Sierras and Great Plains, including route of transcontinental railroad. King and his staff studied in detail principal silver and gold districts, including the Comstock, in Nevada and California, and coalfields in Wyoming. They also discovered first known active glaciers in United States on Mt. Shasta, 1870. They exposed as a fraud alleged diamond fields in northwestern Colorado, 1872, thus demonstrating quality and probity of their work; investments saved, this repaid many times over cost of the entire survey. Partner, N. R. Davis and Co., and in other cattle-ranching ventures, 1871–1882. Examined mines in Utah for British syndicate, 1873. Member, National Academy of Sciences, 1876.

With John Wesley Powell and Congressman Abram Hewitt of New York, founded U.S. Geological Survey, 1878–1879. As first director, USGS, 1879–1881, King staffed, established policies, and oriented USGS work in support of mining industry. Restricted by interpretation of USGS organic act to operations in public land states, mostly in West, King extended USGS work nationwide by organizing and personally contributing to USGS studies of mineral deposits and its gathering of statistics on U.S. mining and production of precious metals, coal, iron, and other industrial minerals in cooperation with tenth federal census of 1880, 1879–1882. Resumed full-time work as mining geologist and consultant, 1881, but kept in touch with USGS through four of the six geologists and topographers from the King Survey who had joined USGS in 1879 and through USGS geophysics-geochemistry laboratory he had established to study physical characteristics of rocks. King used results of this laboratory's experiments on cooling rates of molten igneous rock to estimate Earth's age, 1891–1893. Agreed to, but circumstances prevented, his return as director, USGS, 1892. Promoted and managed mines in United States, Mexico, and Canada for U.S. and British companies and his own interests, 1882–1901. Traveled widely, often with historian-novelist Henry Adams, in Britain, Europe, and Caribbean. Promoted conservation as "wise use" of resources in American West and social causes in United States and Cuba. His name perpetuated in Mt. Clarence King (Sierra Nevada), California; King's Peak (Uinta Mountains), Utah; King Peak (Thiel Mountains), Antarctica; King Crest (Grand Canyon rim), Arizona; and Clarence King Lake (near Mt. Shasta), California.

Published more than fifty articles, books, reports, maps and reviews, 1865–1900, including (with J. T. Gardiner) "Map of the Yosemite Valley from Surveys Made by Order of the Commissioners to manage the Yosemite Valley and Mariposa Big-Tree Grove . . . 1865" (no date, published in J. D. Whitney, *The Yosemite Book*, [1868], with twenty-four photographs by Carleton Watkins); "On the Discovery of Actual Glaciers on the Mountains of the Pacific Slope," *American Journal of Science and Arts* (3d ser.) 1 (1871, also published in *Atlantic Monthly* 27 [1871]); *Mountaineering in the Sierra Nevada* (1872; new ed., 1874, several modern reprints); *Geological and Topographical Atlas*, U.S. Geological Exploration of the Fortieth Parallel (1876; scale, 1:253, 440); "Catastrophism and Evolution," *American Naturalist* 11 (1877; published separately as *Catastrophism and the Evolution of the Environment* [1877]); "The Age of the Earth," *American Journal of Science and Arts* (3d ser) 45 (1893; reprinted in the *Annual Report of the Smithsonian Institution for 1893* [1894]).

MAJOR CONTRIBUTIONS

With William Brewer and paleontologist William Gabb (Whitney Survey), found fossil mollusks in 1863–1864 that established geological age of gold-bearing slates in the Sierra Nevadas. King's federal survey was the only one of the four territorial surveys to investigate a well-defined area for specific purposes; of the four, it did most to advance applied and basic earth science in the public service. Introduced to federal surveys mapping by baseline triangulation within distinct "quadrangles," photography as field documentation, microscopic analyses in studies of rock composition and texture, and publication of major syntheses of work as book-length "final reports." King Survey's organization, field, and laboratory methods and accomplishments were models for the Hayden, Powell, and Wheeler Surveys in the 1870s. Related the West's mining districts to its geologic and tectonic history and suggested that its mineral deposits occurred in seven longitudinal zones from California to Colorado (1870). Published in 1878 comprehensive innovative synthesis of the geology, tectonics, and geologic history of Cordilleran region flanking 40th parallel. Convinced by field evidence that major "catastrophic" geological activity of short duration produced rapid biological modifications in the form of increased speciation (his "evolution of life"). In response to critical national need in 1879 for detailed information on its mineral resources for monetary and industrial purposes, King established mission-oriented USGS program, which produced immediate practical results in support of the mining industry. Initiated pioneering studies in geophysics, geochemistry, microscopic petrography, and paleontology to support immediate solution of practical problems in mining geology and the ultimate goals of a genetic classification of ore deposits, based on their origin and relations, and a better understanding of the Earth and its history. King and Interior Secretary Carl Schurz established educational and professional requirements for USGS geologists that differ only in detail from those used today; eleven of King's principal appointees were or became members of the National Academy of Sciences. King initiated many USGS policies and philosophies still in effect today. In his day, King enjoyed no less fame as writer and raconteur than as geologist-administrator; his narrative skills in *Mountaineering in the Sierra Nevada* led one enthusiastic reviewer to consider it a pity that a man of such literary talents should waste his time on science.

BIBLIOGRAPHY

Bartlett, R. A. *Great Surveys of the American West* (1962), Part 2, Chapters 5–10, "Clarence King's Fortieth Parallel Survey."

Crosby, H. H. "So Deep a Trail: A Biography of Clarence King." Diss., Stanford University, 1953.

Emmons, Samuel F., "Clarence King," *Biographical Memoirs, National Academy of Sciences*. Vol. 6 (1909).

Goetzmann, W. H. *Exploration and Empire: The Explorer and the Scientist in the Winning of the American West* (1966), Ch. 12, "The West of Clarence King."

Manning, Thomas G. *Government in Science: The U.S. Geological Survey. 1867–1894* (1967).

Nelson, Clifford M., and Ellis Yochelson. "Organizing Federal Paleontology in the United States, 1858–1907. *Journal of the Bibliography of Natural History* 9, no. 4 (1980).

Nelson, C. M., and M. C. Rabbitt. "The Role of Clarence King in the Advancement of Geology in the Public Service, 1867–1881." In A. E. Leviton, P. U. Rodda, E. L. Yochelson, and M. L. Aldrich, eds., *Frontiers of Western Geological Exploration* (1982).

Schmeckebier, L. F. "Catalog and Index of the Publications of the Hayden, King, Powell, and Wheeler Surveys." *U.S. Geological Survey Bulletin*, No. 222 (1904, repr. 1970).

Wilkins, Thurman. *Clarence King: A Biography*. Rev. ed. (1988).

UNPUBLISHED SOURCES

See King and King Survey (1867–1881), Microcopy No. M-622, U.S. Geological Survey, Letters Sent (1879–1895), Microcopy No. M-152, and letters received (1879–1901), Microcopy No. M-590, Geologists' Field Notebooks (1867–1939), Record Group (RG)57, USGS, National Archives and Records Service (NARS); King Survey, RG 77 (Office Chief of Engineers), NARS; RG 48 (Office Secretary of the Interior), NARS; Clarence King Papers, H. E. Huntington Library, San Marino, Calif.; Clarence King Collection, Century Club, New York; Geological Exploration of the Fortieth Parallel Miscellaneous Collection, American Museum of Natural History, New York; George J. Brush, William H. Brewer, and Othniel Marsh Papers, and Class of 1862 Scrapbook, Yale University.

Clifford M. Nelson and Mary C. Rabbitt

de KIRILINE LAWRENCE, LOUISE (VENDELA AUGUSTA JANA FLACH). Born Sweden, 30 January 1894; died North Bay, Ontario, Canada, 27 April 1992. Naturalist, ornithologist, nature writer.

FAMILY AND EDUCATION

Eldest daughter of Hillevid Neergaard and Sixten Flach, a prominent family living in Sweden. Her early interest in nature was influenced by her father, a keen nature observer and conservationist. Privately educated, she later served in World War I as a nurse with the Danish Red Cross. Married Gleb de Kiriline, a White Russian officer, and later moved to Russia, where he died during the revolution. She remained there as relief worker, managing a Russian military hospital. A delegate of the Swedish Red Cross Expedition to Volga during the great famine of 1922. Returned to Sweden, then immigrated to Canada in 1927. Met and married Len Lawrence. Moved into a small log cabin at Pimisi Bay, Ontario. No children.

POSITIONS

Well known across rural northeastern Ontario as a Canadian Red Cross out-post-service nurse from 1934 to 1935 and was nurse in charge of the Dionne quintuplets, where she soon became discouraged over the publicity and resigned. Most of her naturalist/ornithological work was done in the woodlands of Pimisi Bay, near North Bay, Ontario.

CAREER

Began serious nature study in 1940. Close friend of, and correspondent with, Percy A. Taverner, Margaret Morse Nice, and Murray and Doris Speirs. Granted a government permit to band birds for research purposes. Wrote and illustrated nature stories and articles. Invited to join Canadian Authors Association in 1944. Joined American Ornithologists' Union (AOU) in 1946 and became an elected member in 1954. Rarely attended meetings, but her ''northern out-post'' hosted visits from many well-known ornithologists, conservationists, artists, and writers. Her scientific monograph ''A Comparative Life-History Study of Four Species of Woodpeckers'' was published by the AOU (1967) to favorable scientific reviews. J. David Ligon wrote, ''This work is an important contribution to the knowledge and understanding of woodpecker biology'' (*Auk* [1968]:700). A regular contributor to *Audubon Magazine* starting in 1945, she also published in *Canadian Field Naturalist*, *Wilson Bulletin*, *The Auk*, *Living Bird Quarterly*, and various other periodicals. Books include *The Loghouse Nest* (1945), *The Lovely and the Wild* (1968), *Another Winter, Another Spring: A Love Remembered* (1977), *Mar: A Glimpse into the Natural Life of a Bird* (1976), *To Whom the Wilderness Speaks* (1980). Honors included the John Burroughs Medal

(1969); Sir Charles G. D. Roberts Special Award; Francis E. Cartwright Outdoor Writing Award for *To Whom the Wilderness Speaks* (1980). She had the ability to communicate with, and inspire, a wide variety of people. Popularized ornithology and promoted conservation.

MAJOR CONTRIBUTIONS

A skilled nature observer and talented writer of scientific and popular works. Researched many woodland birds, but best known in scientific circles for her long-term, comparative life history studies of four species of woodpeckers.

BIBLIOGRAPHY

Ainley, Marianne G. "Louise de Kiriline Lawrence (1894–1992) and the World of Nature: A Tribute." *Canadian Field-Naturalist* 108 (1993).

UNPUBLISHED SOURCES

See Louise de Kiriline Lawrence Autobiographical Notes, 1975; American Ornithologists' Union Papers, Smithsonian Institution Archives, Washington, D.C.

E. Tina Crossfield

KLAUBER, LAURENCE MONROE. Born San Diego, California, 21 December 1883; died San Diego, California, 8 May 1968. Herpetologist, engineer, inventor.

FAMILY AND EDUCATION

Son of Abraham and Theresa (Epstein) Klauber. Father was born in Bohemia (then part of Austria-Hungary), and came to the United States in 1849, settling in San Diego to become a merchant. Laurence attended public schools in San Diego. A.B., Stanford, 1908; Completed two-year Westinghouse Graduate Apprenticeship course, East Pittsburgh, Pennsylvania, 1910, honorary LL.D., University of California, 1941. Married Grace Gould, 1911; two children.

POSITIONS

Associated with San Diego Gas and Electric Company from 1911 to 1953, rising from salesman to engineer, department superintendent, vice president for operations, vice president and general manager, president, 1946–1949, chairman of the board, 1949–1953, and then consultant for twelve years until 1965. Utility grew dramatically under his leadership, doubling operating revenues. Sometime lecturer in biology, Stanford University; consulting curator of reptiles, San Diego Zoological Society; consulting curator of herpetology and patron, San Diego Society of Natural History.

CAREER

Klauber's interests embraced a myriad of civic and paraprofessional concerns, from library commission to pollution control. Developed a passion for herpetology. A tireless innovator in the field, he advanced methods of taxonomy and amassed a personal collection of 36,000 preserved specimens. Chief interest was the rattlesnake, with an emphasis on venom collection and antivenin studies. Thirty-five years of research resulted in his definitive study, *Rattlesnakes: Their Habits, Life Histories and Influence on Mankind.*

Member of a large number of professional organizations, including Pacific Coast Electrical Association (president, 1923–1924) and the Pacific Coast Gas Association (president, 1927–1928). Also active in many scientific, historical, cultural, and conservation organizations representative of his wide-ranging interests. Was president, American Society of Ichthyologists and Herpetologists, 1938–1940; president, Western Society of Naturalists, 1946; president, San Diego Zoological Society, 1949–1951; and president, Society of Systematic Zoology, 1955.

MAJOR CONTRIBUTIONS

Klauber spent a lifetime gathering rattlesnakes and studying their natural history and relationship to humans. Familiar with statistical treatment of data because of his engineering background, he was the first to apply statistical methods to herpetological taxonomy and to think of species in terms of populations in nature rather than as species in museum jars. Wrote more than 100 papers on the subject. His unique two-volume compendium, *Rattlesnakes: Their Habits, Life Histories and Influence on Mankind*, remains the authoritative source for information. The work, published first in 1956, revised in 1972, then reprinted several times, was condensed from its original 1,500 pages to 400 pages in 1982.

BIBLIOGRAPHY

"Klauber, Laurence Monroe 1883–1968." *Contemporary Authors* 105 (1982) (obit.).

"Klauber, Laurence Monroe." *National Cyclopedia of American Biography*. Vol. 54 (1973).

"Klauber, Laurence Monroe." *Who Was Who in America*. Vol. 5 (1969–1973).

"Rattlesnakes: Their Habits, Life Histories and Influence on Mankind." *Choice* 19 (June 1982): 1427 (review).

George A. Cevasco

KRIEGER, LOUIS CHARLES CHRISTOPHER. Born Baltimore, 11 February 1873; died Washington, D.C., 31 July 1940. Mycologist, artist.

FAMILY AND EDUCATION

Son of Henry and Katherine Lentner Krieger. Educated in public and religious (Lutheran) schools. Studied at Maryland Institute School of Art and Design,

1886–?, Baltimore, and at Charcoal Club School of Fine Arts, Baltimore. Student at Royal Bavarian Academy of Fine Arts, Munich, 1895–1896. Married Agnes Checkley Keighler, 1904 (died 1939); one daughter.

POSITIONS

Artist assistant, Division of Microscopy, U.S. Department of Agriculture (USDA), 1891–1895. Instructor of drawing and painting, Maryland Institute School of Art and Design, 1896–1902. Mycological artist to William G. Farlow, Harvard University, 1902–1912. Artist with USDA Plant Introduction Garden, Chico, California, 1912–1918. Mycological artist for Howard A. Kelly of Baltimore, 1918–1928. Honorary curator, L.C.C. Mycological Library, University of Michigan, from 1928. Artist with Tropical Plant Research Foundation in Cuba, 1928–1929. Mycologist, New York State Museum, Albany, 1928–1929. Artist with USDA, 1929–1940.

CAREER

Early demonstrated artistic abilities. Began career with USDA by painting mushrooms found in vicinity of Washington, D.C.; also copied plates of mushrooms from various European works. Division of Microscopy eliminated in 1895; consequently Krieger took opportunity to study abroad. Taught in Baltimore while he began career as portrait painter, which he did not enjoy. While at Harvard with Farlow, produced hundreds of plates of fleshy fungi, two dozen of which appeared in *Icones Farlowianae*, edited by E. A. Burt, published in 1929. Also began indexing world literature on fleshy fungi, a project that would take up more than thirty years. During his second stint as artist with USDA, focused on cacti collected by David Griffiths. Most of the plates were still unpublished and in National Herbarium in Washington, D.C. Was employed by Howard Kelly of Baltimore, for whom he completed a large set of plates, while he continued to work on his index. Some plates appeared in *National Geographic* article ''Common Mushrooms of the United States'' (May 1920). During this period, he also published *Field Key to the Genera of the Gill Mushrooms* (1920), several articles for *Mycologia* and *Rhodora*, and *Catalogue of the Mycological Library of Howard A. Kelly* (1924). Krieger's work with Kelly ended in 1928, when Kelly's library and fungus collections were presented to the University of Michigan. Painted sugarcane diseases in Cuba for part of a year, then served as New York State mycologist for a short period, during which time he completed *A Popular Guide to the Higher Fungi (Mushrooms) of New York State*, which was not published until 1935. In 1929, returned to USDA, where he worked on several projects that were published by Division of Fruit and Vegetable Crops and Diseases. His *The Mushroom Handbook*, a somewhat modified edition of his work on the New York fungi, published in 1936. Planned and prepared illustrations for monograph on genus *Boletus*, which was not completed, though paintings were deposited in Mycological Collections at Bureau of Plant Industry. Also had under way set of paintings and atlas to illustrate inexpensive popular manual of mushrooms, which he did not live to finish. Member of half a dozen botanical and mycological societies in United States and the Deutsche Gesellschaft für Pilzkünde.

MAJOR CONTRIBUTIONS

Regarded as finest painter of North American fungi. His *Mushroom Handbook* was reprinted in 1967 and is still in use. An authority on the genus *Boletus*.

BIBLIOGRAPHY

Kelly, Howard A. Preface to *Catalogue of the Mycological Library of Howard A. Kelly*, by L. C. C. Krieger (1924).

Stevenson, John A. "Louis Charles Christopher Krieger, 1873–1940." *Mycologia* 33 (1940).

UNPUBLISHED SOURCES

Paintings are in collections of the U.S. Department of Agriculture, Washington, D.C.

Keir B. Sterling

L

LACEY, JOHN FLETCHER. Born near New Martinsville, Virginia (now West Virginia), 30 May 1841; died Oskaloosa, Iowa, 29 September 1913. Soldier, lawyer, congressman.

FAMILY AND EDUCATION

Son of John Mills and Eleanor Patten Lacey. Family moved to Oskaloosa, Iowa, when he was fourteen. He attended local schools in both Wheeling and Oskaloosa. Read law with Samuel A. Rice, then attorney general of the state, in Oskaloosa, 1861–1862, during period that he was a paroled soldier. Married Martha Newell, 1865; four children.

POSITIONS

Farmer, bricklayer, and plasterer with father in Iowa prior to the Civil War. Enlisted in 3d Regiment, Iowa Volunteer Infantry, spring of 1861, promoted to corporal, taken prisoner at Blue Mills Landing, Missouri, and paroled. Studied law during the period of his parole, then reenlisted in 33d Regiment, Iowa Volunteers, promoted sergeant major, then lieutenant; became acting adjutant and later assistant adjutant general with the grade of captain. Mustered out in September 1865 as a brevet major. Volunteered for service on Mexican border in same grade and duties under Maj. Gen. Phil Sheridan. Returned to Oskaloosa, resumed his legal studies, was admitted to the bar in 1865, and began the practice of law. Became known as an authority on railway law. Elected city solicitor, Oskaloosa, 1869; elected to the Iowa State House of Representatives, 1870; elected city councilman, 1880–1883; elected as a Republican to the U.S. House of Representatives, 1889–1891 and 1893–1907. Chairman, Committee on Public Lands, 1895–1907; also chairman, Committee on Forests. Defeated for reelection, 1907, resumed practice of law. Authored *The Third Iowa Digest* (1870) and *A Digest of Railway Decisions*, 2 vols. (1879, 1884).

CAREER

Following his military service during the Civil War, Lacey studied not only law but history and political science. Prior to the onset of his political career, he also did some traveling in the United States and abroad. Lacey is primarily known for his firm support of Native Americans, public lands, forests, and wildlife. In 1894, a bill cosponsored by Lacey and Senator Joseph M. Carey of Wyoming, calling "for the protection of the Yellowstone Park," was enacted into law. In 1897, he attempted to put through legislation that would have prohibited the interstate movement of wildlife products killed in violation of state law; it did not pass. In 1900, however, thanks to growing public sentiment for wildlife protection, the Lacey Act was passed. It not only barred interstate commerce in wildlife taken in violation of state laws but limited the importation of wildlife from other nations and explicitly banned the mongoose. The U.S. Biological Survey was for the first time given power to enforce game legislation, which it had not sought. This enactment prompted passage of similar laws by the states, particularly those that had had little such protective legislation on their books. Lacey also gave strong support to the Transfer Act for Forest Reserves of 1905, which created the U.S. Forest Service. Lacey doubted the constitutionality of legislation introduced in Congress in 1904 by Congressman George Shiras of Pennsylvania, which would have placed migratory birds under federal control, and such a bill was not signed into law until the spring of 1913. In the six months of life remaining to him, Lacey, by then back in private life, became part of an advisory board designed to assist the U.S. Biological Survey, which had power to enforce the new law, to work out problems with the new regulations.

MAJOR CONTRIBUTIONS

Lacey was known as one of the early supporters of conservation legislation, and the Lacey Act of 1900 was an important first step in curbing poaching and market hunting. It presaged a stronger role for the federal government in the protection of wildlife resources.

BIBLIOGRAPHY

Cart, Theodore W. "The Struggle for Wildlife Protection in the United States, 1870–1901: Attitudes and Events Leading to the Lacey Act." Diss., University of North Carolina, 1971.

Cart, Theodore W. "The Lacey Act: America's First Nationwide Wildlife Statute." *Forest History* (October 1973).

Dictionary of American Biography.
Gallagher, Mary Annette. "John F. Lacey: A Study in Organizational Politics." Diss., University of Arizona, 1970.
National Cyclopaedia of American Biography.
Stroud, Richard H., ed. *National Leaders of American Conservation* (1985).
Trefethen, James B. *An American Crusade for Wildlife* (1975).

Keir B. Sterling

LA GALISSONIÈRE, ROLAND-MICHEL BARRIN, MARQUIS DE. Born Rochefort, France, 10 November 1693; died Montereau, France, 26 October 1756. Naval officer, aristocratic dabbler in, and promoter of, astronomy, hydrography, and natural history.

FAMILY AND EDUCATION

Son of Roland Barrin de La Galissonière, a senior naval officer, and Catherine Bégon. Born into an influential noble family of Brittany. Studied at Collège de Beauvais, Paris, before becoming a midshipman in the navy in 1710. Married Marie-Catherine-Antoinette de Lauson in 1713; no children.

POSITIONS

Promoted to sublieutenant, 1712; lieutenant commander, 1727; captain, 1738; commissary general of artillery, 1745. Interim governor of New France, 1747–1749. Appointed commissioner in Franco-British negotiations over territorial limits in America in December 1749. Promoted to rear admiral, 1750. Placed in charge of the Dépôt des Cartes et Plans de la Marine, 1750–1753. Voted member-at-large of the Académie de Marine and the Académie Royale des Sciences in 1752. Promoted to lieutenant general of naval forces, 1755.

CAREER

Commanded French naval vessels and warships in the French navy to Canada, Cape Breton Island, the West Indies, and the Mediterranean, 1727–1746. In 1737, while commanding the *Héros*, conducted studies of a nautical instrument, probably a quadrant. Corresponded with Henri-Louis Duhamel du Monceau, member of the Académie Royale des Sciences, on matters of botany and marine biology and with other well-known scientists: Bernard de Jussieu, Pierre-Charles and Louis-Guillaume Lemonnier, Pierre Bouguer, and Jean-Étienne de Guettard. Sent exotic plants and seeds to Duhamel du Monceau, to the Jardin du Roi, and to his own garden at the château de Monnières. While interim governor of New France, promoted the gathering of botanical, zoological, and mineral specimens for study at Quebec and Paris and arranged for scientific observers to accompany expeditions to Detroit, Michilimackinac, and the Ohio Valley. Welcomed botanist Pehr Kalm, a disciple of Linnaeus, to Canada in 1749 and facilitated his investigations. As head of the Dépôt des Cartes et Plans, organized scientific expeditions to Newfoundland, Acadia, and Cape Breton (1750–1751), to the coast of Spain, and to the Cape of Good Hope. In 1754 returned to service at sea, taking part in the French expedition against Minorca in 1756, where he won a modest victory against British naval forces.

MAJOR CONTRIBUTIONS

Organized and promoted zoological, mineralogical, and, in particular, botanical investigations in Canada. Annotated a manuscript study of Canadian flora by the royal physician Jean-François Gaultier (c.1749). Cowrote with Duhamel du Monceau an opuscule, "Avis pour le transport par mer des arbres . . ." (c.1752), and provided material for latter's *Traité des arbres et arbustres . . .* (Paris [1758]). Was praised by Pehr Kalm for his courtesy, knowledge, and defense of the doctrine of strengthening France and its colonies "by adapting natural science to the service of the economy."

BIBLIOGRAPHY

Chartrand L. et al. *Histoire des sciences au Québec* (1987).

Duhamel du Monceau, H.-L. *Traité des arbres et arbustres qui se cultivent en France en pleine terre*, 2 vols. (1758).

Lamontagne, R. *La Galissonière et le Canada* (1962).

Rousseau, J. "Le Mémoire de La Galissonière aux naturalistes canadiens de 1749." In *Le Naturaliste canadien* (1966).

Rousseau, J. et al., trans. and ed. *Voyage de Pehr Kalm au Canada en 1749* (1977).

Taillemite, E. "Roland-Michel Barrin de La Galissonière." In *Dictionary of Canadian Biography*. Vol. 3 (1974).

Turgot, E.-F. *Mémoire instructif sur la manière de rassembler, de préparer, de conserver, et d'envoyer les diverses curiosités d'histoire naturelle, auquel on a joint un mémoire intitulé: Avis pour le transport par mer des arbres, des plantes vivaces, des semences, & de diverses autres curiosités d'histoire naturelle* (1758).

UNPUBLISHED SOURCES

See correspondence with Duhamel du Monceau from the archives of the château de Denainvilliers; transcriptions in the National Archives of Canada; correspondence with Michel Chartier de Lotbinière in National Archives of Canada, Papiers de Lotbinière; Gaultier's description of Canadian plants (manuscript with marginal notes by La Galissonière) in the Archives nationales du Québec; letters and documents in Muséum d'Histoire Naturelle (France); log of the *Héros*, 1737, in Archives de la Marine (France), 4JJ.

Peter L. Cook

LAING, HAMILTON MACK. Born Hensall, Ontario (Huron County), 6 February 1883; died Comox, British Columbia, 15 February 1982. Teacher, collector, naturalist, writer.

FAMILY AND EDUCATION

Son of William Oswald Laing and Rachel Melvina Mack. Attended Clearsprings School, Winnipeg Collegiate Institute, and Normal Teacher's Training School in Winnipeg, Manitoba (1900, with upgrade in 1906); studied art at the Pratt Institute, Brooklyn, New York, 1911–1915. Married Ethel Hart of Portland, Oregon, 1927 (died 1944); no children.

POSITIONS

Taught school at Glenora, Manitoba, 1901–1903; at Runnymead School near Oak Lake, 1903–1904 and 1907; at Caranton School near Boissevain, 1904–1906; principal, Oakwood High School, Oak Lake, 1907–1911; joined Royal Flying Corps and became instructor in gunnery, 1917–1918; assistant naturalist, Smithsonian Institution Expedition to Lake Athabasca, under Francis Harper, 1920; assistant naturalist with National Museum of Canada field parties: in southwest Saskatchewan, 1921; British Columbia, 1922; on the *HMCS Thiepval* expedition to Japan in 1924; Mt. Logan expedition, 1925; at Belvedere, Alberta, 1926; in southern British Columbia, 1927–1930; Vancouver Island with the Semple–Sutton expedition, 1934; northern Vancouver Island, 1935; coastal British Columbia, 1936–1940. He operated a nut farm at Comox, British Columbia, until 1949.

CAREER

Collected about 10,000 superbly prepared specimens of birds and mammals for various museums, especially National Museum of Canada and Royal British Columbia Museum. A plant, *Antennaria laingii*, and a subspecies of the Great Basin pocket mouse, *Perognathus parvus laingi*, have been named for him. He wrote a biography of Allan Brooks, noted Canadian artist.

MAJOR CONTRIBUTIONS

Notable collector of bird and mammal specimens for several Canadian museums. Artist, exponent of wilderness life, and author of many articles on Canadian birds.

BIBLIOGRAPHY

Laing, H. M. *Out with the Birds* (1915).

Laing, H. M. *Alan Brooks: Artist Naturalist* (1979).

Mackie, R. *Hamilton Mack Laing: Hunter-Naturalist* (1985).

C. Stuart Houston

LAMBE, LAWRENCE MORRIS. Born Montreal, Quebec, 27 August 1863; died Ottawa, Ontario, 12 March 1919. Paleontologist, Geological Survey of Canada.

FAMILY AND EDUCATION

Only son of William Bushby Lambe, an advocate, and Margaret Morris Lambe. Educated in private schools in Montreal and at Royal Military College, Kingston, Ontario, from which he graduated, 1883. Married Mabel Maude Schriber (Schreiber), 1902.

POSITIONS

Assistant engineer, Canadian Pacific Railway, Mountain Section, 1883–1884; joined staff of Geological Survey of Canada [GSC], 1884–1919. Scientific artist, GSC, 1885–1891; vertebrate paleontologist, GSC, 1891–1912; chief, Paleontology Division, GSC, 1912–1919.

CAREER

Though trained at the Royal Military College of Canada, Lambe suffered from impaired health, owing to which he decided to enter the service of the Geological Survey of Canada as a scientific artist. Encouraged and trained by his supervisor, the well-known zoologist and paleontologist J. F. Whiteaves, Lambe's original research concerned the recent sponges and Paleozoic corals of Canada. By the 1890s, his interests had turned to field of vertebrate paleontology. This was spurred on by several collecting expeditions for the survey's museum (now the National Museums of Canada) to the abundantly fossiliferous Cretaceous strata of the Red Deer River in Alberta and also by his association with another mentor, Henry Fairfield Osborn, then honorary vertebrate paleontologist for the GSC. Lambe studied for a time under Osborn at Columbia University in New York, where Osborn was also the principal paleontologist on the staff of the American Museum of Natural History. The majority of Lambe's nearly 100 papers were devoted to the fossil vertebrate taxa of Canada. Their scope was wide—from Pleistocene mammals to Paleozoic fishes. Toward the end of his career, he devoted his energies to describing the immense wealth of dinosaurian collections from the late Cretaceous fossil assemblages of the Red Deer River made for the museum by that family of superb collectors and preparators, C. H. Sternberg and his sons, during the golden age of Canadian vertebrate paleontology (1910–1917). Lambe was elected a fellow of the Geological Society of London, a fellow of the Royal Society of Canada and also its honorary treasurer, secretary, and later president of Section 4 (geological and biological sciences), and permanent member of the council, 1914–1919; and member-councilor of the American Society of Vertebrate Paleontologists. For many years, he held a commission as lieutenant in the Governor General's Footguards in Ottawa and was a reserve officer of the Militia of Canada as a lieutenant of engineers.

MAJOR CONTRIBUTIONS

Lambe's pioneering contributions to the study of Canadian dinosaurs and vertebrate paleontology in general have been recognized as the foundation upon which later workers have continued to build.

BIBLIOGRAPHY

The Canadian Who's Who (1910).
Kindle, E. M. *Bulletin of the Geological Survey of America* 31 (1920) (memorial).
McInnes, William. "Lawrence Morris Lambe." *Proceedings of the Royal Society of Canada* (ser. 3) 13 (1919).
Morgan, Henry J. *The Canadian Men and Women of the Time*. 2d ed. (1919).
Russell, Loris S. *Dinosaur Hunting in Western Canada* (1966).
Zaslow, M. *Reading the Rocks—The Story of the Geological Survey of Canada, 1842–1972* (1975).

Richard Day, Jeffrey S. Murray, and Keir B. Sterling

LAPHAM, INCREASE ALLEN. Born Palmyra, New York, 7 March 1811; died Oconomowoc, Wisconsin, 14 September 1875. Naturalist, organizer of science.

FAMILY AND EDUCATION

Son of Seneca Lapham, canal builder and engineer, and Rachel Allen Lapham. Family moved frequently to follow employment at canals. Seven siblings, including a brother Darius, who shared his scientific interests. Educated at home plus several months of school in Shippingsport, Kentucky, probably in 1827. Honorary LL.D., Amherst College, 1860. Married Ann M. Alcott, 1838; five children.

POSITIONS

Worker on Erie, Welland, and Miami (Ohio) Canals, 1824–1827. Worker Portland and Louisville Canal (Kentucky), 1827–1830. Worker Ohio Canal at Portsmouth, 1830–1833. Secretary, Ohio Board of Canal Commissioners, 1833–1836. Developer, surveyor, land speculator, and naturalist in Milwaukee, Wisconsin, 1836–1875. Meteorologist, U.S. Weather Bureau, 1871–1872. Wisconsin state geologist, 1873–1875.

CAREER

Spent early years along canals, beginning work on them at age thirteen. First scientific publication, 1828, on geology of Louisville and Shippingsport Canal, put him in contact with scientific community. Began serious scientific study and investigation upon moving to Milwaukee. His *A Geographical and Topograph-*

ical Description of Wisconsin with Brief Sketches . . . (1844) included first detailed map of state to be published. In 1849–1855, studied Wisconsin Native American effigy mounds, under sponsorship of American Antiquarian Society and Smithsonian Institution. In 1850s and 1860s focused on Wisconsin grasses and environmental impact of the destruction of forestland. Late 1860s and early 1870s, fought successfully for establishment of U.S. Weather Bureau, for which he then worked.

Published at least forty-five articles and nine monographs. Important are *Ancient Artificial Mounds near Madison, Wisconsin* (1859), "Antiquities of Wisconsin Surveyed and Described by I. A. Lapham on Behalf of the American Antiquarian Society," *Smithsonian Contributions to Knowledge* 7 (1855): 1–92, and *Report on the Disastrous Effects of the Destruction of Forest Fires Now Going On So Rapidly in the State of Wisconsin* (1867).

Influential organizer of science in Wisconsin and Midwest (involved in founding Wisconsin Academy of Sciences, Arts, and Letters) and facilitated exchange of specimens and information between midwestern and eastern scientists. One of last self-educated amateurs to attain stature in the American scientific community, but lack of training hindered his work with the Weather Bureau and the State Geological Survey.

MAJOR CONTRIBUTIONS

Successfully determined origin of effigy mounds. Identified impact of destruction of forestlands on erosion and water quality. One of individuals most responsible for creation of U.S. Weather Bureau. Did, or found others to do, extensive fieldwork for eastern scientists.

BIBLIOGRAPHY

Elliott, Clark, A. *Biographical Dictionary of American Science* (1979).

Hawks, Graham P. "Increase A. Lapham, Wisconsin's First Scientist." Diss., University of Wisconsin–Madison, 1960.

UNPUBLISHED SOURCES

Major collection is at the State Historical Society of Wisconsin, Madison, of thirty-one boxes and additional papers; includes correspondence and manuscripts, with finding aids. Some papers are also at the Ohio Historical Society.

Liz Barnaby Keeney

LAWRENCE, ALEXANDER GEORGE ("LAWRIE"). Born Cardiff, Wales, 24 April 1888; died Winnipeg, Manitoba, 25 August 1961. Ornithologist, naturalist, conservationist, oologist.

FAMILY AND EDUCATION

Son of Scottish parents, he spent his boyhood in Scotland and England. Received his primary and secondary schooling in Scotland and at boarding school in England. Moved to Canada in 1910. Married Selina Wallace after 1912 (died 1977); two children.

POSITIONS

Secretary, Winnipeg Department of Health, 1911–1951, secretary, Ornithology Section, Natural History Society of Manitoba (NHSM), 1920–1921, chairman, 1921–1924. General Secretary, NHSM, 1927–1929, vice president, 1924–1933, treasurer, 1933–1934, president, 1934–1936, honorary president, 1951–1954, president emeritus, 1954–1961. Also served as president of Manitoba Museum Association and Winnipeg Cine Club.

CAREER

Became interested in ornithology, botany, and geology as youth in England and pursued these interests, especially ornithology, after emigrating to Canada.

Active bander at Winnipeg with the American Banding Association before banding was coordinated by government agencies and continued bird study and photography all his life. Photography began with glass plates and continued through zoom-lens movie cameras. In 1920, was one of four people who initiated discussions leading to the formation of the Natural History Society of Manitoba and became one of its twenty-six founding members. Both he and Selina served the society throughout the rest of their lives, Lawrie receiving its bronze medal for his ornithological endeavors in 1941 and named president emeritus from 1954 until his death. Also in 1920, was active in founding the Manitoba Museum Association, through which he became an honorary curator of the museum.

Most significant contribution was his bird column, "*Chickadee Notes*," published weekly in the *Winnipeg Free Press*. Lawrence wrote 1,756 issues over a span of thirty-four years, 1921 to 1955 until temporarily incapacitated by ill health. Column served as an exchange of information for amateur and professional ornithologists all over Manitoba and adjacent parts of Saskatchewan and Ontario and remains the key source of distributional data for that area for the three decades covered. The column was widely recognized among ornithologists in North America, and Lawrence was consulted by many of the leading ornithologists of the time. "*Chickadee Notes*" also served as an important conservation voice in Manitoba and was important factor in promoting the *Blue Jay*, a Saskatchewan-based journal of nature, currently the foremost in the Canadian prairie provinces. Several ornithological institutions subscribed to the column for information on Manitoba birds, and Lawrence was among the naturalists honored with membership in the Cornell Laboratory of Ornithology at Ithaca, New York, when the laboratory was founded.

Publication outside "*Chickadee Notes*" was primarily as a contributor to cooperative projects, such as Christmas bird counts or to works by others, notably P. A. Taverner's *Birds of Canada* and A. C. Bent's life history series, the latter featuring several of Lawrence's photographs. He also compiled or cocompiled several issues of a bird checklist for Manitoba and a bird calendar and prepared bird posters for public education. "*Chickadee Notes*" was the first publication vehicle for many outstanding naturalists.

Participated widely in environmental education through lectures, broadcasts, and his columns and posters and especially enjoyed working with youngsters, for which he was awarded a silver medal by the Canadian Boy Scout organization.

Interests in other aspects of nature are best recognized in fossil cephalopod discovered by him. It was designated as the holotype in 1950 in his honor *Actinocamax manitobensis* (Whiteaves) var. *lawrencii*.

MAJOR CONTRIBUTIONS

Through his dedication and extensive communication, promoted ornithology as a science in Manitoba and surrounding areas while also heightening public awareness of natural history and conservation and bridging the gap between amateur and professional naturalists.

BIBLIOGRAPHY

Davidson, A. M., et al. *Natural History Society of Manitoba 21st Anniversary Bulletin* (1941).

"Great Naturalists of Canada. Alexander George Lawrence." *Newsletter of the Thunder Bay Field Naturalists Club* (1961).

Houston, C. S. "In Memoriam. Alexander George Lawrence, 1888–1961." *Blue Jay* 19 (1961).

Houston, C. S., and M. J. Bechard. "Later Oologists in Southern Manitoba: Forge, Norman, Lawrence and Others." *Blue Jay* 45 (1987): 155–165.

Mossop, H. "Alexander George Lawrence." *Chickadee Notes*, no. 346 (1961).

McNicholl, M. K. "The Contributions of A. G. Lawrence." *Natural History Society of Manitoba Newsletter* 17 (1968).

Shortt, A. Dedication. In A. McMaster, A. Criddle, and C. Scott. *Manitoba Naturalists Society Volume 2 (1942–1975)* (1977).

UNPUBLISHED SOURCES

Most of library is incorporated into that of Delta Waterfowl Research Station, Delta, Manitoba. Set of columns at Manitoba Museum of Man and Nature, Winnipeg.

Martin K. McNicholl

LE CONTE, JOHN EATTON, JR. Born near Shrewsbury, New Jersey, 22 February 1784; died Philadelphia, 21 November 1860. Naturalist, nature artist, military engineer.

FAMILY AND EDUCATION

Son of John Eatton and Jane (Sloane) Le Conte; brother of Louis Le Conte. Educated at Columbia College (University), New York. Married Mary Ann

Hampton Lawrence, 1821. Father of John Lawrence Le Conte; uncle of Joseph Le Conte.

POSITIONS

Captain, U.S. Topographical Engineers, 1818–1831. Vice president, Lyceum of Natural History of New York, 1852–1853. General secretary, American Association for the Advancement of Science, 1857. Vice president, Academy of Natural Sciences of Philadelphia, 1858–1860.

CAREER

Began studying natural history as boy in New York City and, in the winters, on the family plantation in Liberty County, Georgia. Assisted David Hosack in conducting the Elgin Botanic Garden. Published catalog of the plants of New York City (1811) and helped in preparation of Jacob Green's catalog of the plants of New York State (1814). Commissioned captain in Topographical Engineers, 1818 (brevet major, 1828). Explored St. John's River, Florida, 1822. Published several articles on the botanical and other specimens—several of them new to science—that he had gathered there in *Annals* of the Lyceum of Natural History of New York. The journey had shattered his health and he took leave of absence to Paris, where he shared his botanical collections with French scientists and collaborated with Jean Alphonse de Boisduval on a book on North American Lepidoptera (1829). Returned to United States and resigned his commission. A virtual invalid for the rest of his life, he wrote little until 1852, when he moved to Philadelphia. From 1852 until his death he published articles in the *Proceedings* of the Academy of Natural Sciences of Philadelphia and other journals.

A friend of John Torrey since boyhood and well known among other scientists, including Asa Gray, who described him as "a keen but leisurely observer and investigator, and still more leisurely writer. He was a man of very refined and winning manners, of scholarly habits and wide reading, of an inquiring and original turn of mind, the fruitfulness of which was subdued by chronic invalidism."

In addition to his writings (largely of a systematic nature) he produced many illustrations of plants and animals, including watercolors to illustrate his botanical articles of the 1820s and 1830s and 4,023 other watercolors, mostly of insects. His paintings of turtles earned for him the designation of "the Audubon of turtles." He gave his large herbarium and collection of mollusks to Academy of Natural Sciences of Philadelphia.

MAJOR CONTRIBUTIONS

A respected authority on the natural history of southeastern United States and the first significant American naturalist to explore Florida since Bartram.

Helped to launch and guide the career of his son, John Lawrence Le Conte, who was perhaps the most distinguished American entomologist of the nineteenth century.

BIBLIOGRAPHY

Adicks, Richard, ed. *Le Conte's Report on East Florida* (1978).

Appleton's Cyclopaedia of American Biography. Vol. 3 (1888–1889).

Baird, V. B. "The Violet Water-colors of Major John Eatton Le Conte." *American Midland Naturalist* (1938).

Gray, Asa. "John Eatton Le Conte." *Botanical Gazette* (1883).

Le Conte, Joseph, "John Eatton Le Conte, Jr." *Biographical Memoirs, National Academy of Sciences*. Vol. 3 (1895).

UNPUBLISHED SOURCES

See Le Conte Family Papers, American Philosophical Society (correspondence and entomological drawings); Linnaean Society of London, correspondence of American scientists, 1738–1872 (microfilm copies in American Philosophical Society); Lewis David von Schweinitz Papers, Academy of Natural Sciences of Philadelphia (correspondence); Scientific Papers, Academy of Natural Sciences of Philadelphia (correspondence).

Michael J. Brodhead

LE CONTE, JOSEPH. Born Liberty County, Georgia, 20 February 1823; died Yosemite Valley, California, 6 July 1901. Physician, naturalist, zoologist, author.

FAMILY AND EDUCATION

One of seven children born to Louis LeConte, plantation owner and amateur naturalist, and Ann Quarterman LeConte. Reared on 3,356–acre family estate in Liberty County, Georgia, and educated primarily at home. Entered University of Georgia in 1838 and graduated with honors in 1841. Attended lectures at College of Physicians and Surgeons in New York City in 1844–1845; received M.D. in 1845. Enrolled in Lawrence Scientific School in 1850; awarded B.S. in 1851. Married Caroline Elizabeth Nisbet, 1847; five children.

POSITIONS

Professor of science, Oglethorpe College, Milledgeville, Georgia, 1852. Professor of natural history, University of Georgia, Athens, 1853–1856. Professor of geology and chemistry, College (later University) of South Carolina, Columbia, South Carolina 1857–1869. Consulting chemist for Confederate medical manufactory, 1863–1864, and the Nitre and Mining Bureau, 1864–1865. Professor of geology and zoology, University of California, Berkeley, 1869–1901.

CAREER

During early years he collected numerous specimens (especially avifauna) in Georgia. Practiced medicine in Macon, Georgia, 1847–1850; founding member of Macon Medical Society and Georgia Medical Society. Assistant to Louis Agassiz on expedition to study marine life of Florida Keys, winter of 1851. Taught all sciences except zoology at Oglethorpe College, 1852. Held chair of natural history and supervised botanical garden at University of Georgia, 1853–1856. Delivered lectures on coral formation and origins of coal, Smithsonian Institution, 1856. Served as professor of geology and chemistry at the College of South Carolina, 1857–1869, except when college was closed because of the Civil War. He and older brother John (a physicist) were hired as original faculty members of University of California in 1869. Elected fellow of National Academy of Science, 1875; president of American Association for the Advancement of Science, 1892; president of the Geological Society of America, 1896. Held in high esteem as teacher and popular lecturer. Accepted theory of evolution about 1874 and became one of its leading proponents; articulated reconciliation of the idea with tenets of Christianity.

In addition to 190 articles on a variety of topics, including evolution, religion, education, social theory, mountain formation, and physiological optics, he published nine books. His *Elements of Geology* (1878) served as leading American introductory textbook in the field for more than three decades; his *Sight: An Exposition of the Principles of Monocular and Binocular Vision* (1881) was the first treatise on physiological optics written in the United States; and his *Evolution and Its Relation to Religious Thought* (1888) achieved notice as a highly successful defense of evolutionary theism. LeConte was an able writer and ardent camper. His *Journal of Ramblings in the High Sierra of California* (1875) as well as many articles on western cordillera of the United States, won special attention. A charter member and erstwhile officer of the Sierra Club, LeConte spoke fervently for broad preservation of California forests by government and wise use of timberlands in private enterprise. Traveled to Europe in 1892 and 1896; especially well received in Great Britain, where his works on geology and evolution were prominent. Mountains in the Sierras and Appalachians, as well as numerous other objects, were named for him.

MAJOR CONTRIBUTIONS

Leading proponent of evolutionary theory, effective popularizer of neo-Lamarckianism, and major reconciler of the idea with religious thought. One of the earliest advocates of contractional theory of mountain formation, a highly recognized textbook writer, and teacher of geology, zoology, and physiology. His facility for written expression helped immensely to popularize certain theoretical aspects of science, and his devotion to the natural beauty of the Sierras aided indirectly in the development of conservation efforts. A universalist in scope of his scientific writings, he served as a transitional figure between the age of the generalist and the era of the specialist.

BIBLIOGRAPHY

Bozeman, T. D. "Joseph LeConte: Organic Science and a 'Sociology for the South.'" *Journal of Southern History* 33 (November 1973).

Christy, S. B. ''Biographical Notice of Joseph LeConte.'' *Transactions of the American Institute of Mining Engineers* 31 (November 1902).
Fairchild, H. L. ''Memoir of Joseph LeConte.'' *Bulletin of the Geological Society of America* 26 (1915).
Hilgard, E. ''Memoir of Joseph LeConte.'' *National Academy of Sciences Biographical Memoirs* 6 (1907).
LeConte, J. *Autobiography* (1903).
Stephens, L. D. ''Joseph LeConte and the Development of the Physiology and Psychology of Vision in the United States.'' *Annals of Science* 37 (1980).
Stephens, L. D. ''Joseph LeConte on Evolution, Education, and the Structure of Knowledge.'' *Journal of the History of the Behavioral Sciences* 12 (April 1976).
Stephens, L. D. ''Joseph LeConte's Contribution to American Ornithology.'' *Georgia Journal of Science* 35 (June 1977). (corrigendum, 36 [January 1978]).
Stephens, L. D. ''Joseph LeConte's Evolutional Idealism: A Lamarckian View of Cultural History.'' *Journal of the History of Ideas* 39 (July–September 1978).
Stephens, L. D. *Joseph LeConte: Gentle Prophet of Evolution* (1982).

UNPUBLISHED SOURCES

Letters are in numerous collections, including John Lawrence LeConte Papers and Joseph Leidy Papers, Academy of Natural Sciences of Philadelphia; LeConte Family Papers, American Philosophical Society; Lester F. Ward Papers, Brown University; Alexander Agassiz Papers, Asa Gray Papers, and William James Papers, Harvard University; Daniel Gilman Papers, Johns Hopkins University; J. McKeen Cattell Papers and Lewis R. Gibbes Papers, Library of Congress; Alpheus Hyatt Correspondence, Maryland Historical Society; James Hall Papers, New York State Library; various record units, Smithsonian Institution Archives; David Starr Jordan Papers, Stanford University; Beecher Family Papers, Brush Family Papers, O.C. Marsh Papers, and Whitney Family Papers, Yale University.

Numerous other letters, personal notebooks, trustees and faculty minutes, manuscript notes, official records, and miscellany are in LeConte Family Papers and other collections, Bancroft Library, University of California, Berkeley; Minutes of the Geological Society of America; Liberty County, Ga., official records; R. Means Davis Papers, other collections, and faculty minutes, University of South Carolina; LeConte-Furman Papers, Elizabeth Furman Talley Collection, and Clifford Anderson Papers, Southern Historical Collection, University of North Carolina, Chapel Hill; Faculty minutes, trustees' minutes, and miscellany, University of Georgia; Journal of Bertha L. Chapman, Yosemite National Park; and many items in the private collections of current LeConte descendants.

Lester D. Stephens

LEIDY, JOSEPH. Born Philadelphia, Pennsylvania, 9 September 1823; died Philadelphia, 30 April 1891. Naturalist, paleontologist, physician.

FAMILY AND EDUCATION

Third of four children (2nd son) of Philip Leidy, a hatmaker, and Catherine Mellick Leidy. Mother died before he was two. Raised by maternal aunt, his stepmother. Educated at private Methodist-run school. Showed evidence of artistic ability; removed from school at sixteen by father, who wanted him to become sign painter. Stepmother encouraged Leidy in his desire to enter medical school. M.D., University of Pennsylvania, 1844. Married Anna Harden, 1864. No children, but adopted the daughter of deceased colleague.

POSITIONS

Assistant in Chemistry Laboratory, University of Pennsylvania, attempted to establish private medical practice in Philadelphia, 1844–1846; Prosector in Anatomy, University of Pennsylvania, 1845; Demonstrator in Anatomy, Franklin Medical College, 1848, 1850; Professor of Anatomy, University of Pennsylvania, 1853–1891, and Director of the Department of Biology there, 1884–1891. Army Surgeon during Civil War, Satterlee Army Hospital, Philadelphia. Professor of Natural History, Swarthmore College, 1870–1885.

CAREER

Collected minerals and plants along banks of local rivers when a child. Drew creditable drawings of shells at age ten and could assign popular and scientific names to each. His dissertation at Pennsylvania concerned ''The Comparative Anatomy of the Eye of Vertebrated Animals.'' A multifaceted scientist and an acute observer, whose first publication on fossil shells dates from 1845. In 1846, he discovered Trichinella in pork and determined that it could be controlled by adequate cooking. In 1847, completed paper ''On the Fossil Horse of America,'' which confirmed existence and extinction of native species in prehistoric times. The English geologist Charles Lyell, on the basis of Lyell's horse study, strongly recommended that Leidy focus on paleontology. Leidy's ''Ancient Fauna of Nebraska: A Description of Extinct Mammalia and Chelonia from the Mauvaises Terres of Nebraska (*Smithsonian Contribution to Knowledge*, 1854) and ''The Extinct Mammalian Fauna of Dakota and Nebraska'' (*Journal of the Academy of Natural Sciences of Philadelphia*, 1869) were considered his most outstanding works. His detailed descriptions continued to have value in the twentieth century; H. F. Osborn writing in 1935 that with the possible exception of E. D. Cope's *Tertiary Vertebrata*, Leidy had done ''the most impressive paleontological work which America has produced.'' In time, however, Leidy found his duties at the University of Pennsylvania too restrictive to allow for the caliber of paleontological work he wanted to accomplish. In addition, his growing distaste for the increasingly acrimonious paleontological feud between Cope and O. C. Marsh prompted him to drop most of his own work in the field. He turned to microscopic studies, and his ''Fresh Water Rhizopods of North America'' (*Report of the* [Hayden] *Survey of the Territories*, 1879) featured forty-eight plates of meticulous drawings that he had done himself. Authored highly praised *Elementary Treatise on Human Anatomy* (1861; rev. ed., 1889). Leidy also returned to his earlier studies in parasitology, which included his pioneering ''Flora and Fauna within Living Animals'' (*Smithsonian Contributions to*

Knowledge, 1853), with his study of human intestinal worms (1885), the first monographic account of its kind. In his "Remarks on Parasites and Scorpions" (*Transactions of the College of Physicians of Philadelphia*, 1886) he first advanced the proposition that parasites might translate from animals to man and cause disease. Leidy published some 600 titles during his career and was among the last of American zoologists to deal with much of the full range of animal life. Leidy was an assiduous collector of facts, having little time for theorizing, and wrote a friend that he was "too busy [to] make money." Leidy was elected to membership in the Academy of Natural Sciences of Philadelphia (ANSP) and the Boston Society of Natural History in his twenties, winning the latter's Walker Prize in 1880. He became president of the ANSP in 1881, serving ten years until his death, and was selected for membership in the National Academy of Sciences at its founding in 1863. He received the Lyell Medal from the Geological Society of London (1884) and the Cuvier Medal from the Institute of France (1888).

MAJOR CONTRIBUTIONS

Leidy contributed outstanding publications to the fields of human anatomy, parasitology, helminthology, microscopy, and paleontology. He was considered the leading American anatomist of his era. He did much to popularize science for the general public, and his contributions to paleontology and parasitology have had lasting value down to the present time.

BIBLIOGRAPHY

Dictionary of American Biography.
Dictionary of Scientific Biography.
Elliott, Clark A., *Biographical Dictionary of American Science: The Seventeenth Through the Nineteenth Centuries* (1979).
Leidy, Joseph Jr., "Researches in Helminthology and Parasitology by Joseph Leidy," *Smithsonian Miscellaneous Collections.* Vol. 46 (1904).
Meisel, Max, *A Bibliography of American Natural History: The Pioneer Century, 1769–1865*, vol. 1 (1924).
Osborn, Henry F., "Joseph Leidy," in *Biographical Memoirs, National Academy of Sciences.* Vol. 7 (1913).
Ruschenberger, W.S.W., "A Sketch of the Life of Joseph Leidy, M.D., LL.D.," *Proceedings of the American Philosophical Society.* Vol. 30 (1892).

UNPUBLISHED SOURCES

Leidy manuscripts are in Academy of Natural Sciences of Philadelphia. See also medical correspondence at College of Physicians, Philadelphia.

Keir B. Sterling

LEOPOLD, ALDO STARKER. Born Burlington, Iowa, 22 October 1913; died Berkeley, California, 23 August 1983. Ornithologist, naturalist, conservationist, educator.

FAMILY AND EDUCATION

Son of Aldo and Maria Alvira Estella Bergere Leopold. Father a preeminent conservationist. B.S., University of Wisconsin, 1936. Attended Yale School of Forestry, 1936–1937 (some authorities state two years), transferred to University of California, Berkeley, where he earned Ph.D. in zoology, 1944. Honorary doctorate, Occidental College, 1980. Married Elizabeth Weiskotten, 1938; two children.

POSITIONS

Junior biologist, U.S. Soil Erosion Service, 1934–1935; field biologist, Missouri Conservation Commission, 1939–1944; director of field research for Conservation Section, Pan American Union (in Mexico), 1944–1946; assistant professor of zoology, and conservationist, Museum of Vertebrate Zoology (MVZ), University of California, Berkeley (UCB), 1946–1952; associate professor, 1952–1957; professor, 1957–1967; associate director, MVZ, 1958; acting director, MVZ, 1965; transferred affiliation to Department of Forestry and Conservation, UCB, 1967; professor emeritus, 1978. Also director, Sagehen Creek Field Station, UCB, 1965–1979, and assistant to the chancellor, UCB, 1960–1963.

CAREER

Spent part of his youth in New Mexico, where his father was employed by the U.S. Forest Service, and then in southern Wisconsin, while his father was a faculty member at the University of Wisconsin–Madison. Published several books, including *Wildlife in Alaska: An Ecological Reconnaissance* (with F. F. Darling, 1953); *Wildlife of Mexico: The Game Birds and Mammals* (1959); *The Desert* (1967); *The California Quail* (1977); and *North American Game Birds and Mammals* (with R. J. Gutierrez and M. T. Bronson, 1981). Consultant to Missouri Conservation Commission and Tanzania National Parks. Presidential appointee to the U.S. Marine Mammal Commission. Chair of three committees in the 1960s and 1970s making recommendations to the secretary of interior concerning predator and rodent control, management of the national wildlife refuge system, and wildlife in national parks. Guggenheim Fellow, 1947–1948. Elected to the National Academy of Sciences, 1970, and was recipient of many other awards. President of the California Academy of Sciences, 1959–1971; of the Northern (California) Division, Cooper Ornithological Society, and of the Wildlife Society (1957–1958). Officer of a number of other wildlife and conservation organizations.

MAJOR CONTRIBUTIONS

Broad-gauged educator, ecologist, and conservationist. Active in a number of conservation and wildlife groups in the United States. Made major contributions to avian biology, particularly of game birds.

BIBLIOGRAPHY

American Men and Women of Science.

Auk (October 1984) (obit.).

Lorbiecki, Marybeth. *Aldo Leopold: A Fierce Green Fire* (1996).

McCabe, Robert A., ''Aldo Leopold,'' *Biographical Memoirs, National Academy of Sciences.* Vol. 59 (1990).

Stroud, Richard H., ed. *National Leaders of American Conservation* (1985).

Who Was Who in America.

Joseph Bongiorno

LEOPOLD, (RAND) ALDO. Born Burlington, Iowa, 11 January 1887; died Baraboo, Wisconsin, 21 April 1948. Naturalist, forester, wildlife ecologist, educator, writer.

FAMILY AND EDUCATION

Eldest of four children of Carl (furniture company executive) and Clara Starker Leopold. Graduate of Lawrenceville School, 1905. B.S., Sheffield Scientific School, Yale, 1908; master of forestry, Yale, 1909. Married Maria Alvira Estella Bergere, 1912; five children.

POSITIONS

Forest assistant, Apache National Forest, U.S. Forest Service (USFS), 1909–1911. Deputy supervisor, Carson National Forest, USFS, 1911–1912. Supervisor, Carson National Forest, USFS, 1912–1914. Forest examiner, District 3, USFS, 1915–1917. Secretary, Albuquerque Chamber of Commerce, 1918–1919. Chief of operations, District 3, USFS, 1919–1924. Assistant director, USFS Forest Products Laboratory, Madison, Wisconsin, 1924–1928. Director of game survey, Sporting Arms and Ammunition Manufacturer's Institute, 1928–1932. Consulting forester, 1932–1933. Professor of game management, University of Wisconsin, Madison, 1932–1938. Professor of wildlife management, University of Wisconsin, 1938–1948.

CAREER

Formative experiences as a young naturalist, ornithologist, and hunter along the Mississippi River bottoms near Burlington and in Les Cheneaux Islands of northern Lake Huron, where family regularly spent summers. Member of first generation of trained U.S. foresters. Was innovative force in early Forest Service, pursuing new directions in forest administration, soil conservation and range management, wilderness protection, game and wildlife protection, recreation policy, and ecological research. Spent first fifteen years of Forest Service career in national forests of American Southwest. During much of 1914–1915, forced to take eighteen months' leave due to near fatal bout of nephritis. Between 1915 and 1919, worked on grazing administration, later on recreation, publicity work, and fish and game administration. From 1915 to 1921, led successful effort to reform New Mexico's state game department as founder and secretary of New Mexico Game Protective Association. From 1919 until 1924, instigated broad innovations in forest inspection procedures, investigated ecological impacts of forest administration in southwestern watersheds, advanced cause of game protection within USFS, and advocated policy of wilderness preservation on national forestlands.

In 1921, published "The Wilderness and Its Place in Forest Recreation Policy," important early proposal to reserve public wilderness areas. As a result, Gila Wilderness Area was designated in 1924, first such in the nation. During his four years at Forest Products Laboratory, Leopold balanced official administrative duties with unofficial activities on behalf of wilderness preservation, game management, and reform of conservation administration in Wisconsin. Much of this latter activity performed through American Game Protective Association and Wisconsin Chapter of Izaak Walton League of America (IWLA); also active in national ILWA during these years. From 1928 to 1932, led Game Survey for consortium of industry representatives. Made extensive surveys of game habitat conditions in Midwest, consulted with state and federal agencies on game protection and restoration programs, promoted and supervised management-oriented research projects in cooperation with midwestern land-grant universities, and worked within conservation community to advance idea of active, scientifically informed management of game species. Summarized results of his field investigations in *Report on a Game Survey of the North-Central States* (1931), an unprecedented effort to document status and preservation of game species on regional scale. Also chaired, during these years, Game Policy Committee of American Game Conference. The committee report, "American Game Policy" (1930), was first comprehensive statement of essential aims for emerging field of wildlife management.

In all of this work, Leopold refined ideas that helped transform practice of wildlife conservation from rearguard effort concerned primarily with legal restrictions, predator "control," and game farming, to applied science that sought to understand population dynamics and ecological relationships in order to provide stable habitats for all forms of wildlife. These ideas brought together in *Game Management* (1933), first textbook on subject and standard work for several decades. Simultaneously worked to broaden philosophical dimensions of conservation and to understand ecological context of social and historical change. "The Conservation Ethic" (1933) was important early expression of his conservation philosophy. Appointed nation's first professor of game management at University of Wisconsin, 1933. Soon became influential educator and also leading figure in emergence of wildlife management as a profession. Served as member of Franklin D. Roosevelt's Committee on Wildlife Restoration (1934). Offered post as chief of U.S. Biological Survey, 1934, but declined, preferring research to policy making. With J. Norwood "Ding" Darling, also on president's committee, established Cooperative Wildlife Research Unit system, 1935. Purchased worn-out farmland along Wisconsin River north of Madison (1935) as hunting reserve and family retreat. Also used as testing ground for his theories of wildlife management, as land laboratory, where he and family restored property to ecological health, and as setting for ecological essays he began writing in the 1930s. Traveled to Germany on Carl Schurz Fellowship, 1935, to study history of forest and wildlife management in Central Europe. Visited Sierra Madre in northern Mexico, 1936 and 1937. These experiences impressed on him need for restrained and well-integrated management of natural ecosystems, as opposed to highly artificial, intensive methods of commodity production.

For the last decade of his life, stressed ecological basis of all conservation activity in his work and writing. Continued research, writing, teaching, and advising through World War II. Founding member (1936) and president (1939) of the Wildlife Society; president of Ecological Society of America (1939). Leading figure in dozens of other local, state, regional, and national professional, conservation, and scientific organizations. Adviser to United Nations International Scientific Conference on the Conservation and Utilization of Resources (1948). In addition to titles previously mentioned, published some 350 papers, reports, policy statements, reviews, popular articles, essays, editorials, and other documents. In final years, concentrated on lyrical and philosophical essays posthumously published in 1949 as *A Sand County Almanac and Sketches Here and There*. Later collections of his writings published as *Round River: From the Journals of Aldo Leopold* (1953), *Aldo Leopold's Wilderness* (1991, republished as *Aldo Leopold's Southwest*, 1993), and *The River of the Mother of God and Other Essays by Aldo Leopold* (1991).

MAJOR CONTRIBUTIONS

Endowed with a broad education, vast store of field experience, unique command of language and history, wry sense of humor, highly critical mind, and insatiable curiosity, Leopold was a gifted communicator of the scientific basis of conservation as well as the imperative need for environmental awareness, adept at modifying his style to suit his audience and able to employ these qualities in person as well as in print. Made substantial contributions to full range of conservation professions. Played critically important role in broadening narrow utilitarian foundations of the American conservation movement. Was well known only within professional community of conservationists during his lifetime but achieved broad public following with posthumous publication of *A Sand County Almanac*. "The Land Ethic," landmark essay from that book, has done much to stimulate contemporary interest in environmental ethics and philosophy.

BIBLIOGRAPHY
Bradley, H. C. "Aldo Leopold: Champion of the Wilderness." *Sierra Club Bulletin* 36 (1951).
Callicott, J. B., ed. *Companion to A Sand County Almanac* (1987).
Errington, Paul. "In Appreciation of Aldo Leopold." *Journal of Wildlife Management* 12 (1948).
Flader, Susan. *Thinking like a Mountain: Also Leopold and the Evolution of an Ecological Attitude toward Deer, Wolves and Forests* (1974).
Hott, Lawrence, and Diane Garbey, with Ken Chowder et al. *Wild by Law [Aldo Leopold, Bob Marshall, Howard Zahniser, and the Redefinition of American Progress].* Videocassette, Florentine Films, Santa Monica, Calif. Direct Cinema, 1991.
McCabe, R. *Aldo Leopold: The Professor* (1988).
McCabe, R. *Aldo Leopold: Mentor* (1988).
Meine, Curt. *Aldo Leopold: His Life and Work* (1988).
Tanner, T., ed. *Aldo Leopold: The Man and His Legacy* (1987).
UNPUBLISHED SOURCES
See Aldo Leopold Papers, University of Wisconsin–Madison Archives; other collections are in University of Wisconsin Archives; Archives of State Historical Society of Wisconsin; U.S. Forest Service records in National Archives; New Mexico State Records Center and Archives, Arizona State Historical Society; U.S. Forest Products Laboratory Library; and Yale University Archives.

Curt Meine

LESUEUR, CHARLES-ALEXANDRE. Born Le Havre, France, 1 January 1778; died Le Havre, France, 12 December 1846. Naturalist, artist.

FAMILY AND EDUCATION

Son of Jean-Baptiste-Denis Lesueur, a lieutenant general of the Admiralty duHavre, and Charlotte Geneviève Thieullent. Fourth of seven children. Previously ascribed attendance at l'École Royale Militaire de Beaumont-en-Auge, 1787–1796, seriously in question. Military service as noncommissioned officer in Garde Nationale du Havre, 1797–1799. Never married.

POSITIONS

Post with scientific expedition to explore coast of Australia, 1800–1804. Correspondant du Muséum, Muséum d'Histoire Naturelle de Paris, received coveted title 29 November 1815. Naturalist under contract to William Maclure on scientific excursion to Lesser Antilles and eastern United States, 1815–1817. Elected member Academy of Natural Sciences of Philadelphia, 31 December 1816, curator, 1817–1825. Naturalist, artist, and teacher with New Harmony, Indiana, as base, 1826–1837. First director, Muséum du Havre, 1845–1846. Also elected to membership in Société Philomatique de Paris, 1814, American Philosophical Society (Philadelphia), 1817, and Lyceum of Natural History of New York.

CAREER

In 1800, at age twenty-three, obtained post with French expedition to Australia under command of Nicholas Baudin. Advanced to position on scientific staff. With naturalist François Péron returned to France (1804) with collection of more than 100,000 zoological specimens, including some 2,500 new species. Curier credited Péron and Lesueur with discovering more new species than all the other naturalists of modern era up to their time. In collaboration with Péron and, later Louis Desalux Freycinet, prepared account of expedition, *Voyage des Découvertes aux Terres Australes*, vols. 1 and 2 (1807–1816). Atlas to vol. 1 contains 1,500 sketches by Lesueur. After death of Péron entered into two-year contract (1815–1817) with William Maclure. Departing Paris August 1815, the two naturalists embarked on scientific excursion to principal islands of the Lesser Antilles from Barbados to Santa Cruz, Maclure studying the geology, and Lesueur collecting natural history specimens, with special attention to marine life of West Indies. Produced over 100 sketches. Arriving New York 10 May 1816, the two scientists proceeded with five-month geological investigation. Maclure was anxious to revise his 1809 geological survey of the United States. Departing Philadelphia early June, they traveled through Delaware, Maryland, Pennsylvania, New York, Vermont, Massachusetts, Connecticut, and New Jersey, returning to Philadelphia 20 October.

Lesueur made sketches of the relief of country, collected natural history specimens, and filled his sketchbook with drawings of the towns through which they passed. In Philadelphia Lesueur supervised the engraving of Maclure's new geological map and geological sections for his enlarged revision of *Observations on the Geology of the United States* (1817). During nine-year period with Philadelphia as base, 1816–1825, established reputation as naturalist-engraver and professor of drawing. Spent twenty-seven days with William Maclure, Gerard Troost, and Thomas Say exploring New Jersey (spring 1817), served as cartographer for the U.S. and Canadian Boundary Commission (1819–1822), and with William Maclure and Thomas Say went on scientific excursion into northeastern Pennsylvania, northern New Jersey, and southern New York (August–September 1825). Branch of natural history that particularly attracted his attention was ichthyology. A systematic work on fishes of North America became his leading object. While curator of Academy of Natural Sciences of Philadelphia, contributed thirty-one articles to the first five volumes of the *Journal of the Academy of Natural Sciences of Philadelphia* (1817–1825). Also one article in *Nouveau Bulletin des Sciences, par la Société Philomathique* (1817), one article in *Transactions of the American Philosophical Society* (n.s.) (1818), and three articles in *Mémoires du Muséum d'Histoire Naturelle* (Paris, 1819, 1820, 1827). The most notable of his American contributions is a monographic review of the family of suckers. Joined with Welsh social reformer Robert Owen and other scientists and educators December 1825. Departing Pittsburgh, he recorded in his sketchbooks journey down Ohio River, arriving New Harmony, Indiana, January 1826. New Harmony served as base (1826–1837). Combined pursuits as field naturalist, artist, and teacher during Owen–Maclure communitarian experiment (1825–1827) and under the aegis of his benefactor Maclure until 1837.

Went on scientific excursion to observe the iron and lead mines of eastern Missouri, collect natural history specimens, and investigate archaeological prehistoric mounds in vicinity of New Harmony—the first scientific study ever made of the mounds of Indiana. Visits to New Orleans provided repeated opportunities for extending research, collecting, and sketching from the Wabash and Ohio Rivers to delta of the Mississippi River, including investigation of prehistoric sites at Bone Bank on the Wabash and Walnut Hills (Vicksburg) on the Mississippi (1828) and geological field trip deep into the mountains of Tennessee (1831). At New Harmony conducted extensive study of turtles and fish of Wabash. During five-month residency in New Harmony by Maximilian, prince of Wied Neuwied, and artist Karl Bodmer, October 1832–March 1833 (part of twenty-four-month expedition into the interior of North America), Lesueur and Say engaged in valuable scientific exchange with the visitors. On return trip, June 1834, Maximilian and Bodmer were accompanied by Lesueur from New Harmony to Vincennes, where both artists made corresponding sketches at the same time. Lesueur's close association with Say ended with Say's untimely death in 1834. Lesueur had contributed frontispiece and ten plates to Say's *American Entomology* (1824, 1825, 1828) and two plates to Say's *American Conchology* (1830–1838). Lesueur's proposed work, *American Ichthyology or Natural History of the Fishes of North America* (1827), was never completed beyond seven leaves of text and six plates, now in Muséum du Havre.

After twenty-two years in America, left New Orleans 8 June 1837, settled in Paris, and taught painting, pursued scholarly studies at Muséum d'Histoire Naturelle, and mastered technique of lithography (1837–1840). Returned to Le-Havre to investigate geological and paleontological characteristics of the cliffs at the mouth of Seine. Published *Vues et Coupes du Cap de la Hève* (1843). Appointed first director of Muséum du Havre (1845–1846) where forty of his boxes containing animal, mineral, fossil, and ethnographic specimens together with manuscripts and sketches (1,500 Australian, 1,200 American, and 400 European, West Indies) were deposited. Only the manuscripts and sketchbooks survived the destruction of the museum during World War II and are preserved in the reestablished museum today.

MAJOR CONTRIBUTIONS

His contributions in natural sciences led him to collect specimens in a wide range of fields—prehistoric artifacts, freshwater mollusks, native American vegetation and wildlife. Naturalists of Europe and America have united in giving Lesueur the highest praise for his work as painter-naturalist. Quatrefages said his watercolors were "the foremost natural history paintings of ancient or modern times." Lesueur was the first to study fishes of the Great Lakes of North America. His name will always be associated with the earliest American work on marine invertebrates and invertebrate paleontology. His ability to document faithfully what he observed has provided a remarkable collection of over 1,200 sketches of the American frontier, in many instances the earliest surviving documentation.

BIBLIOGRAPHY

Bonnemains, Jacqueline et al. "*Charles-Alexandre Lesueur* (1778–1846), peintre, voyageur, naturaliste havrais." *Bulletin Société Géologique de Normandie*. Supplement (1978).

Chinard, Gilbert. "The American Sketchbooks of Charles-Alexandre Lesueur." *Proceedings of the American Philosophical Society* 93. 2 (1949).

Hamy, E. T. *The Travels of the Naturalist Charles A. Lesueur in North America, 1815–1837*. Trans. Milton Haber; ed. by H. F. Raup (1968).

Loir, Adrien. *Charles-Alexandre Lesueur, Artiste et Savant Français en Amérique de 1816 à 1839* [*sic*] (1920).

Manneville, Philippe. "Charles-Alexandre Lesueur sa famille—son enfance Jean-Baptiste-Denis Lesueur." *Annales du Muséum du Havre*. Fascicule no. 14 (February 1979).

Maury, André. "Charles Alexandre Lesueur, voyageur et peintre-naturaliste Havrais (1778–1846)." *The French American Review* (July–September 1948).

Ord, George. "A Memoir of Charles Alexander Lesueur." *American Journal of Science and Arts* (2d ser.) 8 (1849).

Vail, R. W. G. "The American Sketchbooks of Charles Alexandre Lesueur 1816–1837." Repr. from the *Proceedings of the American Antiquarian Society*, April 1938. (1938).

UNPUBLISHED SOURCES

See Charles Alexandre Lesueur personal papers and drawings, Academy of Natural Sciences of Philadelphia; Lesueur Collection, American Antiquarian Society; Lesueur Collection, American Philosophical Society; Lesueur Collection, Historic New Harmony, Inc. Archives; Lesueur's Sketchbooks—microfilm, Illinois Historical Survey Collections; Twigg Papers—correspondence and sketches (one of Péron), Indiana Historical Society; Lesueur manuscripts—correspondence, Indiana University, Lilly Library; Center for Western Studies, Joslyn Art Museum (Omaha, Neb.) Maximilian-Bodmer Collection, Internorth Art Foundation; notebooks, sketchbooks, and unpublished articles, Muséum d'Histoire Naturelle du Havre (France); New Harmony Correspondence, Maclure–Fretageot Correspondence, d'Arusmont family papers, Twigg Scrapbook, New Harmony Workingmen's Institute; Lesueur, Dupalais sketches, Purdue University.

Ralph G. Schwarz

LEWIS, MERIWETHER. Born seven miles west of Charlottesville, Virginia, 18 August 1774; died Grinder's Tavern, on Natchez Trace, Central Tennessee, 11 October 1809. Explorer, army officer, territorial governor.

FAMILY AND EDUCATION

Son of William and Lucy Meriwether Lewis. Little is known concerning early education. Father died when Lewis was about nine years old. Mother remarried; family moved to Georgia. Lewis returned to Virginia at age thirteen, studied under local minister and private tutors until age eighteen, when stepfather (John Marks) died, and Lewis felt compelled to manage family plantation near Charlottesville, rather than go to college. Never married.

POSITIONS

Member of Virginia Militia, 1794–1795. Enlisted in U.S. Army, 1795, commissioned ensign in 2d Legion. Transferred to 1st Infantry, 1796. Promoted lieutenant, 1799. Promoted captain, 1800. Leave of absence as military aide to President Jefferson, 1801–1803. Coleader, Lewis and Clark Expedition, 1803–1806. Resigned from army, 1807, appointed governor of Louisiana, 1807–1809.

CAREER

Army officer at various military posts in Pennsylvania, Tennessee, Michigan, and in Indian Territory; served as paymaster for his regiment at one point. Invited to be military aide to Jefferson, with whom he had been friendly since at least 1792. When in 1802 Jefferson read Alexander Mackenzie's account of his 1793 expedition to the Pacific coast, he became concerned about British interests in the Pacific northwest. Following purchase of Louisiana Territory in 1803, Jefferson decided to mount a similar undertaking, and offered Lewis command of an exploratory expedition which would follow the Missouri River to its source and thence to Pacific Ocean. Lewis estimated cost at $2,500; selected ex-Lieutenant William Clark as his co-captain, though Clark was never given promised captaincy. Lacking scientific background, was directed to go to Philadelphia and secure supplies and get instruction and guidance from naturalists and other scientists there, notably B. S. Barton, C. W. Peale, Benjamin Rush, Caspar Wistar, and others. Members of expedition were enlisted and trained in St. Louis area, winter of 1803–1804. Lewis had responsibility for scientific observations, while Clark was primary cartographer. Expedition followed Missouri River to its source, which was reached in August 1805. Crossed Continental Divide and Rocky Mountains; went down Columbia River to its mouth and wintered at Fort Clatsop, which exploring party had built. Returned to St. Louis following much the same route taken out; arrived in St. Louis, 23 September 1806. Lewis kept copious notes concerning fauna and flora observed. Some 122 vertebrates new to science (undescribed as of dates found by Lewis and Clark) were discovered, mentioned, encountered, and referred to by Lewis. Some birds and mammal skins and skeletons were sent or brought back. Lewis collected many plants, some of which were lost in Columbia River, but others (117 of

178 species) now retained in collections of Academy of Natural Sciences in Philadelphia. Some bird and mammal skins and skeletal material and several live birds and mammals sent to Jefferson; one magpie and one prairie dog reached Washington alive. Most specimens relayed by Jefferson to C. W. Peale in Philadelphia. Many were ultimately prepared as museum specimens. Most passed into hands of P. T. Barnum when Peale Museum was closed and sold in 1840s; the majority of them destroyed in an 1865 fire at Barnum's American Museum, New York. A few remaining specimens, which had been at the Boston Museum and later at the Boston Society of Natural History, ended up at Harvard University in 1914. Lewis's investigations very much expanded known geographic ranges of species of birds and mammals not previously known from west of the Mississippi. Made useful observations of many vertebrates in relationship to their environment. Much taxonomic activity followed the receipt of Lewis and Clark specimens in the East, a number of works about American birds and mammals were soon published, and "with these published works American zoology, as we think of it today, had its inception" (Cutright 1969). Following resignation from army and appointment as governor, Lewis took some months on personal business, and on arranging for publication of journals in Philadelphia, arrived St. Louis in summer of 1807 and governed effectively for two years. Learning that some of his accounts with federal government were in dispute, determined to go to Washington and straighten out discrepancies. Traveling with several servants, he either killed himself or was murdered on night of 11 October 1809; but the facts are still very much in dispute. Jefferson was criticized for not having sent trained naturalist with expedition, but no academically trained zoologists or botanists existed in the United States at the time. Those who studied plants and animals had either had practical field experience or were self-taught. Frederick Pursh was the principal identifier of plant specimens (seventy-seven species). Complete journals of Lewis and Clark, edited by R. G. Thwaites, not published until 1904. A three-volume history of the expedition, drafted by Nicholas Biddle and Paul Allen from Lewis's journals and discussions with Clark, appeared in 1814. A four-volume edited version, with valuable notes concerning fauna, flora, geography, and other matters by Elliott Coues, was published in 1893. Subsequently, edited versions of other journals and field notes appeared.

MAJOR CONTRIBUTIONS

Lewis's exploration of the northern portion of the Louisiana Purchase territory, together with the exhaustive notes, observations, journals, and specimens he brought back, greatly expanded what was known about the American West and stimulated considerable zoological, botanical, and geographic research in the decades following his return. His was the first federally sponsored exploring expedition (and probably, at a total cost of $2,500, the most economical ever mounted), and it spawned a number of others, together with considerable interest in western settlement. Lewis considered his effort a failure, however, in that he did not find a feasible overland water route connecting the Missouri and the Columbia Rivers.

BIBLIOGRAPHY

Ambrose, Stephen A. *Undaunted Courage: Meriwether Lewis, Thomas Jefferson, and the Opening of the American West* (1996).

Allen, John L. *Passage Through the Garden: Lewis and Clark and the Image of the American Northwest* (1975).

[Biddle, Nicholas, and] Paul Allen. *History of the Expedition under the Command, of Captains Lewis and Clark, etc.* 2 vols. (1814).

Botkin, Daniel. *Our Natural History: The Lessons of Lewis and Clark* (1996).

Burroughs, Raymond D. *The Natural History of the Lewis and Clark Expedition* (1961).

Burroughs, Raymond D. "The Lewis and Clark Expedition's Botanical Discoveries." *Natural History* 74 (1966).

Coues, Elliott, ed. *History of the Expedition under the Command of Lewis and Clark* [by Nicholas Biddle and Paul Allen, with additional materials]. 4 vols. (1893).

Criswell, Elijah H. *Lewis and Clark: Linguistic Pioneers* (1940).

Cutright, Paul R. *Lewis and Clark: Pioneering Naturalists* (1969).

Cutright, Paul R. *A History of the Lewis and Clark Journals* (1976).

Dillon, Richard. *Meriwether Lewis: A Biography* (1965).

Jackson, Donald. *Letters of the Lewis and Clark Expedition and Related Documents, 1783–1854* (1962).

Moulton, Gary E., ed. *The Journals of the Lewis and Clark Expeditions.* 8 vols. to date (1983–).

Rudd, Velva E. "Botanical Contributions of the Lewis and Clark Expedition." *Journal of the Washington Academy of Sciences* 44 (1954).

Setzer, Henry H. "Zoological Contributions of the Lewis and Clark Expedition." *Journal of the Washington Academy of Sciences* 44 (1954).

Thwaites, R. G. *Original Journals of the Lewis and Clark Expedition.* 8 vols. (1904–1905).

UNPUBLISHED SOURCES

Original manuscript journals and other Lewis and Clark materials are at American Philosophical Society, Philadelphia. Other materials are at National Archives and Records Administration and Library of Congress, Washington, D.C., and at Missouri Historical Society.

Keir B. Sterling

LINCECUM, GIDEON. Born Hancock County, Georgia, 22 April 1793; died Washington County, Texas, 28 November 1873. Frontier naturalist-physician.

FAMILY AND EDUCATION

Eldest son of Hezekiah, a peripatetic farmer, and Sally Hickman Lincecum. Family frequently moved to various locations in Georgia, South Carolina, and Mississippi during Gideon's childhood. His playmates in youth were primarily Muskogee Indian boys. He had but five months of formal education, all in his fourteenth year. Studied medicine with a physician in Eatonton, Georgia, and also Indian herbal medicine. Married Sarah Bryan, 1814; thirteen children. Taught his surviving sons botanical medicine, which all of them practiced at various times.

POSITIONS

Assisted local land surveyor at age fourteen. Worked for short time as merchant's assistant and then as member of the Georgia militia during the War of 1812. Farmed with father for a time. Taught school for a year, briefly employed as sawyer. Became merchant and Indian trader in Columbus, Mississippi. Also held various local offices there. Demonstrated some skill in providing medical treatment from about 1818. Practiced medicine at several locations in Mississippi and Texas from 1830 to 1861 and again after the Civil War. Member of state commission empowered to organize Monroe County, Mississippi, 1821; first visited Texas for seven months in 1835 and moved there in 1848. Became landowner and, following the Civil War, sold botanical medicines to support himself. An early eugenicist, he advocated legalized sterilization in the 1850s. Moved to Tuxpan, Mexico, near Vera Cruz, in 1868, a year following the death of his wife, and lived there for five years. Returned to Texas shortly before his death.

CAREER

Largely self-taught in medicine and natural history, Lincecum practiced botanical medicine beginning in 1833 while also involved in various other means of earning a living. Spent considerable time from childhood in collecting natural history specimens. Following his move to Texas in 1848, collected plant and insect specimens, rocks, fossils, and weather data, which he sent to scientists at the Smithsonian Institution, the Academy of Natural Sciences of Philadelphia, the Jardin des Plantes in Paris, and various other individuals and institutions. Worked very much in isolation, had little personal contact with other scientists, and had no access to published works in entomology until after the Civil War. Gave some attention to the native grasses of Texas. Spent a number of years observing Texas stinging or mound-building ants. Sent several letters on this subject to Charles Darwin, who published them under the title, ''Notice on the Habits of the 'Agricultural Ant' of Texas'' in the *Journal of the Proceedings of the Linnaean Society of London* (6 [1862]:29–31). A more extended article by Lincecum, ''On the Agricultural Ant of Texas (*Myrmica molefaciens*),'' was published in the *Proceedings of the Academy of Natural Sciences of Philadelphia* in 1866 (18:323–31), though some members of the academy had serious reservations concerning his findings. The most controversial of these included claims that this species ''planted their own favorite seed-bearing grass.''

Nevertheless, he was elected a corresponding member of the academy in

1867. Many later authorities have essentially accepted his conclusions about seed planting in this species. Other critics objected to Lincecum's tendency to personalize his subjects. While his published papers dealt primarily with insects, he also gave some attention to birds, mammals, fossils, and other subjects. When he died, he left unpublished a 1,500–page manuscript dealing with the botany and materia medica of Texas and Mexico. Several other papers on the Indians of Mississippi were published posthumously by the Mississippi Historical Society three decades after his death. These included ''Chocktaw Traditions about Their Settlement in Mississippi and the Origin of Their Mounds'' (1904) and ''Biography of Apushmataha'' (1906). The society also published his ''Autobiography'' in 1904.

MAJOR CONTRIBUTIONS

Lincecum's painstaking observations of ants over many years, together with his anthropological writings, constitute his principal contributions to science.

BIBLIOGRAPHY

Bradford, A. L., and T. N. Campbell, eds. ''Journal of Linceum's Travels in Texas, 1835'' *Southwestern Historical Quarterly* 53 (1949–1950).

Burkhalter, Lois W. *Gideon Lincecum, 1793–1874: A Biography* (1965).

Geiser, Samuel W. ''Gideon Lincecum.'' In his *Naturalists of the Frontier* (1937).

Lincecum, Gideon. ''Personal Reminiscences of an Octogenarian.'' *The American Sportsman* (12 September 1874–16 January 1875).

Lincecum, Gideon. ''The Autobiography of Gideon Lincecum.'' *Publications of the Mississippi Historical Society* 8 (1904).

Lincecum, Gideon. *Adventures of a Frontier Naturalist: The Life and Times of Dr. Gideon Lincecum*, edited by Jerry Bryan Lincecum and Edward Hake Phillips (1994).

Keir B. Sterling

LINCOLN, FREDERICK CHARLES. Born Denver, 5 May 1892; died Washington, D.C., 16 September 1960. Wildlife biologist, ornithologist, government administrator.

FAMILY AND EDUCATION

Son of Fred J. and Ina Theresa (Nitschke) Lincoln. Graduated from Denver Technical High School in 1910. Intensely interested in natural history during youth. Attended University of Denver, 1912–1913. Honorary Sc.D. from University of Colorado, 1956. Married Lulu M. Lichtenheld, 1920; no children.

POSITIONS

Assistant curator of ornithology, Colorado Museum of Natural History, 1913–1914, curator of ornithology, 1914–1920. Carrier pigeon expert, U.S. Army Signal Corps, 1918–1919. Assistant biologist, U.S. Bureau of Biological Survey,

1920–1924, associate biologist, 1924–1930, biologist, 1930–1934, Senior biologist in charge of Distribution and Migration of Birds, Division of Wildlife Research, 1935–1940, senior biologist and chief, Section of Distribution and Migration of Birds, U.S. Fish and Wildlife Service, 1940–1946, assistant to the director, Bureau of Sport Fisheries and Wildlife, 1946–1960.

CAREER

Offered appointment as assistant curator upon graduation from high school because of his reputation as naturalist, based on collecting experience and extensive reading of natural history literature. Leader, Biological Survey of Colorado, 1913–1917. Ornithological expeditions to Arizona, 1916, South Carolina, 1917, 1928, Louisiana, 1919, 1929, North Dakota, 1921, 1929, Illinois, 1922, 1926, Michigan, 1923–1925, 1928, Utah, 1926, Smithsonian expedition to Haiti and Dominican Republic, 1931, 1949, Mexico, 1937, 1942, Florida, 1938, 1943–1946, and Cuba, 1948, 1949.

Organized and expanded the bird-banding program for Bureau of Biological Survey during its transfer from American Bird Banding Association in the 1920s. Most fieldwork 1920–1940 devoted to improving bird trapping, banding, and data collection techniques. Developed concept of four major migratory bird flyways in North America, which is basis for modern waterfowl management program. Secretary, Whooping Crane Advisory Committee. Considered foremost expert on migration of North American birds. Fellow, American Ornithologists' Union from 1934, served as business manager and treasurer.

Published some 300 scientific papers and popular articles in addition to *American Waterfowl* (with John C. Phillips, 1930), *Migration of Birds* (1935, 1952), and *Birds of Alaska* (with Ira N. Gabrielson, 1959). Described as quiet and studious, enjoyed writing and helping others to write. Was capable administrator of extensive bird-banding and data collection program, serving as liaison between volunteer banders and scientists. Analytical ability allowed interpretation of banding migration data and conceptualization of flyways. Aided in passage of statutes regulating hunting of waterfowl and administered flyways management program.

MAJOR CONTRIBUTIONS

Ability to manage and improve large-scale banding program, to translate massive database into migration patterns, to develop flyways concept, and to implement regulations for flyways management program.

BIBLIOGRAPHY

Aldrich, J. W. "Frederick C. Lincoln." *The Ring* (1961).
Gabrielson, Ira N. *Auk* (July 1962) (obit.).
Sunday Star, Washington, D.C., 18 September 1960 (obit.).
Who's Who in the East (1942–1943).

UNPUBLISHED SOURCES

See field notes in U.S. Fish and Wildlife Service, 1860–1961, Field Reports, Record Unit 7176, Smithsonian Archives; correspondence in Biographical File, Record Unit 7098, Smithsonian Archives; correspondence in Witmer Stone Papers, Academy of Natural Sciences of Philadelphia.

Pamela M. Henson

LLOYD, HOYES. Born Hamilton, Ontario, 30 November 1888; died Ottawa, Ontario, 21 January 1978. Conservationist, ornithologist, civil servant.

FAMILY AND EDUCATION

Moved to Toronto the year after his birth, with mother, Lizzie (Moore), and father, Henry Hoyes Lloyd. Active field observer as a youth around Toronto. Began serious collecting in 1903 after he read *Auk*. Awards for same at the Canadian National Exhibition, 1904, 1905, and 1906. Attended Harbord Collegiate Institute, Toronto, and became a chemist at the University of Toronto. Married in 1913 to Wilmot Lockwood; two daughters and one son.

POSITIONS

Forest ranger and deputy game warden, Temagami Forest Reserve, summer 1909, teaching assistant and master's candidate in chemistry, University of Toronto, 1909–1911, chemist for manufacturing firm, 1911–1912, public health chemist for the city of Toronto 1912–1918. Civil servant, 1918–1943, including secretary to the obscure but influential Federal Advisory Committee on Wildlife Preservation.

CAREER

Began work as a public health chemist, while maintaining his interest in natural history. Published his first article, "Ontario Bird Notes" in *Auk* in 1917. He eventually authored over 120 papers, dealing with policy, mammals, and birds. Hired after a competition in 1918 as the first staff member of the new Migratory Birds Unit, Department of the Interior. As such was the first professional conservationist hired by the Canadian government. From 1918 until retirement in 1943 was the civil servant responsible for administering and revising the Northwest Game Act, regulating wildlife in the North and in national parks and the Migratory Bird Convention of 1917. Educated the public, educators, sportsmen, law officers, and other government departments on federal conservation policy and regulations. Established the annual federal-provincial wildlife conferences and spoke at international wildlife conferences. Prepared the administrative and philosophic framework for what became the Canadian Wildlife Service. After retirement in 1943 was active in national and international ornithological circles. President, Ottawa Field-Naturalists' Club, 1923–1925; president, International Association of Game, Fish and Conservation Commissioners, 1929–1930; chairman, Canadian Section, International Council for Bird Preservation, 1927–1954; vice president, International Council for Bird Preservation, 1938–1950; vice president, American Ornithologists' Union, 1942, and president, 1945–1948. Fellow, New York Zoological Society. Honorary member, Ottawa Field-Naturalists' Club, Quebec Zoological Society, and Outdoor Writers Association of America. Honored with both the Leopold Award (Wildlife Society, 1956) and the Seth Gordon Award (International Association of Fish, Game and Conservation Commissioners, 1974).

MAJOR CONTRIBUTIONS

Formulated policy and promoted a vigorous, sustained national policy of bird and wildlife conservation in Canada. Shaped the administrative precursor to the Dominion Wildlife Service, later Canadian Wildlife Service. Said to have had a remarkable capacity "for unobtrusive but effective group leadership." Even in retirement he used his skills extensively to promote international cooperation for bird protection.

BIBLIOGRAPHY

Clarke, C. H. D. *Auk* (April 1979) (obit.).

Foster, Janet. *Working for Wildlife: The Beginning of Preservation in Canada* (1978).

McNicholl, Martin K. "Hoyes Lloyd: Ornithological Civil Servant." *Ornithology in Ontario* (1994).

Munro, David A., and V. M. Humphreys. *Canadian Field-Naturalist* 93. 3 (July–September 1979).

UNPUBLISHED SOURCES

The National Archives of Canada in Ottawa has extensive working files found among the Canadian Parks Service (RG84), Northern Affairs (RG85), the Canadian Wildlife Service (RG109), and the records of the Ottawa Field-Naturalists' Club. His seven decades of field books, library, and personal collections of plants, birds, and mammals are at the Royal Ontario Museum, Toronto.

Lorne Hammond

LOGAN, MARTHA DANIELL. Born St. Thomas Parish, South Carolina, 29 December 1704; died Charleston, South Carolina, 28 June 1779. Colonial horticulturist, botanist, gardener, florist.

FAMILY AND EDUCATION

Second of four children (three daughters) of Robert Daniell and his second wife, Martha Wainwright (?) Daniell; father was Lieutenant Governor of North Carolina, 1704–1705, and of South Carolina, 1715–1717. Nothing is known of her education. Married George Logan, Jr., 1719 (died 1764); eight children.

POSITIONS

Horticulturist, proprietress of nursery business. Operated a boarding school at various times from c.1742 until c.1754 and perhaps later.

CAREER

Lived on plantation at Trott's Point, ten miles from Charleston on the Wando River; there, in 1753, she was operating a nursery business, selling seeds for vegetables, flowers, and also shrubs and fruit trees. A celebrated horticulturist who, because of strained family finances, turned a pastime into a profitable business. Published ''The Gardener's Kalender'' in John Tobler's *South Carolina Almanack for 1752*, under the pseudonym of ''Lady of this Province.'' Was visited by John Bartram in 1760, and they corresponded until at least 1765. At some point, Bartram wrote Peter Collinson in London, ''The elderly widow spares no pains to oblige me,'' to which Collinson facetiously replied, ''I plainly see how thou knowest how to fascinate the longing widow.'' Logan kept in touch with Bartram, regularly sending him ''silk bags full of seeds.'' She is supposed to have managed her husband's plantation at some point, though this is not certain. She did operate a boarding school at the plantation for a time, and later one in Charlestown.

MAJOR CONTRIBUTIONS

Pioneer in colonial gardening, who provided important advice on planting and cultivating vegetable gardens. She also made valuable contributions of seeds and plants to John Bartram in Philadelphia.

BIBLIOGRAPHY

Berkeley, Edmund, and Dorothy Smith Berkeley. *Dr. Alexander Garden of Charles Town* (1969).
Darlington, William. *Memorials of John Bartram and Humphry Marshall* (1849).
Hollingsworth, Buckner. *Her Garden Was Her Delight* (1962).
Kastner, Joseph. *A Species of Eternity* (1977).
Notable American Women 1607–1950: A Biographical Dictionary (1971).

Arthur Sherman

LOGAN, WILLIAM (EDMUND). Born Montreal, Quebec, 20 April 1798; died Castle Malgwyn, Cilgerron, Wales, 22 June 1875. Geologist, geological cartographer, founder of Geological Survey of Canada.

FAMILY AND EDUCATION

Third son of nine children of Janet Edmund and William Logan, a wealthy baker and proprietor from Scotland. Primary education at Alexander Skakel's,

Montreal. High school in Edinburgh, 1814–1816. Studied mathematics and chemistry at University of Edinburgh (1816–1817), where he was exposed to the celebrated Wernerian–Huttonian debate. Never married.

POSITIONS

Bookkeeper in Uncle's countinghouse, London, England, 1817–1829. Manager, Forest Copper Works, Morriston, Wales, 1831–1838. Cofounder, curator, Department of Geology, Royal Institute, Wales, 1835–1841. Founder–director, Geological Survey of Canada, Montreal, 1842–1869. Commissioner, International Exhibitions: London, 1851, 1862; Paris, 1855. President, Canadian Institute, 1850–1852. Vice president, Natural History Society of Montreal.

CAREER

After having gained recognition as amateur geologist and stratigrapher, while managing uncle's coal interests in southern Wales, Logan assumed directorship of first geological survey for "Province of Canada," a career to which he devoted the remaining thirty-three years of his life. Early years focused on initial surveying, defining and mapping major geological formations in addition to outlining mineral potential in what is presently the southern half of Quebec and Ontario. Established first museum for geological specimens in Montreal, 1844. Organized Canadian geological displays at International Exhibitions in England (1851, 1862) and in Paris (1855), for which he issued the first geological map of Canada.

Most of his published work appears in the form of extensive government documents, entitled *Report of Progress*, 15 vols. (1844–1869). Outside of these reports, he published essentially controversial findings in *Transactions—Geological Society of London*, *Canadian Naturalist and Geologist*, and *Canadian Journal*. His most notable publication, *Geology of Canada* (1863) (coupled with *Atlas*, 1866), was intended as a compendium of current knowledge in Canadian geology and included major contributions by his associates in paleontology and mineralogy. It was widely acclaimed at the time and, 100 years later, still considered "the most outstanding single geological work in the history of Canadian geological studies" by at least one modern authority. As a stratigrapher, he excelled. Precision of coalfield maps, detailing first true scale cross-sections of Wales, *British Ordnance Survey* (1845) and Nova Scotia, *Report of Progress* (1843), was revolutionary and "greatly superior to that usual with geologists," according to De La Beche, director, British Ordnance Survey. His earlier theories relating to the in situ origin of coal, published in *Transactions–Geological Society of London* (1842), were widely accepted.

His recognition of fossil tracks, earliest known evidence of animal life at that time, in the Nova Scotian formation (1841), prompted Lyell's visit in 1842. Ten years later he documented invertebrate trails in still older Paleozoic rocks, *Quarterly of the Geological Society of London* (1851). A bitter controversy among scientists in Europe and North America was ignited, following his discovery in the Precambrian rocks of the oldest known fossil. Several years later, however, *Eozoon canadense*, as it was called by J. W. Dawson, was shown to be of inorganic origin. Logan was first to systematically differentiate the Precambrian rocks into younger "Huronian" and older "Laurentian." His terminology exists today, although with altered implications. His recognition and explanation of major break in the complicated southern Quebec formation, today called Logan's Line, established a landmark in tectonic science.

Logan was considered pragmatic, unostentatious, and "single-mindedly devoted to his work." He was teacher and friend to his self-appointed staff, most of whom had little formal training and many of whom achieved personal recognition. His notorious diplomatic skill was aptly demonstrated in his response to the Select Committee of 1854. When it had been suggested that the activities of the survey were too "scientific" and not "sufficiently practical," he answered, "Economics lead to Science, and Science to Economics." He was highly respected by scientific peers, as evidenced by his long list of honors: fellow of the Royal Society of London (1851), knighted by Queen Victoria (1856), appointed to Legion of Honor (1855), honorary D.D.L., McGill University (1856), and recipient of more than twenty medals, notably Wollaston Medal (1856), Gold Medal of Honor (1856), and Royal Gold Medal (1867). Commemorated by the Logan Gold Medal and Chair of Geology, McGill University. Logan was probably happiest sketching during his many arduous field trips. Journals and field books abound with captivating ink drawings.

MAJOR CONTRIBUTIONS

Detail and accuracy of maps, emphasis on field observation, addition of sound principles to a young discipline, and establishment of Canada's first government scientific organization.

BIBLIOGRAPHY

Canadian Journal (n.s.) 1 (1856).
Canadian Journal 3 (1855).
Dictionary of Canadian Biography. Vol 10 (1972).
Harrington, B. J. *Life of Sir William E. Logan* (1883).
Proceedings—Geological Association of Canada 16 (1965).
Proceedings—Royal Society of London 24 (1876) (obit.).
Zaslow, Morris. *Reading the Rocks* (1975).
Zeller, Suzanne. *Inventing Canada: Early Victorian Science* (1987).

UNPUBLISHED SOURCES

See Letters in Logan Papers, McGill University Archives; field notebooks, directors' letterbooks, National Archives of Canada; private journals, Baldwin Room, Metropolitan Toronto Library.

Jane Davis Nelson

LONG, STEPHEN HARRIMAN. Born Hopkinton, New Hampshire, 30 December 1784; died Alton, Illinois, 4 September 1864. Explorer, engineer, army officer.

FAMILY AND EDUCATION

Son of Moses and Lucy Harriman Long. A.B., Dartmouth, 1809. Married Martha Hodgkins, 1819; five children.

POSITIONS

Schoolteacher in Salisbury, New Hampshire, and public school principal, Germantown, Pennsylvania, 1809–1814. Commissioned 2d Lieutenant of engineers, U.S. Army, 1814. Assistant professor of mathematics at U.S. Military Academy, 1815–1816. Assigned to topographical engineers, 1816. Led expeditions to explore Fox and Wisconsin Rivers, the approximate site of present-day St. Paul, Minnesota, and Arkansas, 1817. Placed in command of expedition to Rocky Mountains, 1819–1820 with rank of brevet major. Spent several years studying railroad routes. Consulting engineer, Baltimore and Ohio Railroad (B&O), and later president, board of engineers for B&O, 1827–1830. Chief engineer, Atlantic and Great Western Railroad, 1837–1840. Superintendent of western rivers, 1853. Colonel and chief of topographical engineers, 1861–1863.

CAREER

Long's first major assignment and the one for which he is perhaps best known today was the command of the scientific expedition that traveled from Pittsburgh to the Rockies in 1819 and 1820. Secretary of War John C. Calhoun wanted the frontier explored as part of a larger plan to expand the American presence there. Expedition departed Pittsburgh, May 1819. Scientists on trip included Thomas Say, zoologist, Titian Peale, assistant naturalist, William Baldwin, botanist, and Augustus E. Jessup, geologist. There was also a painter, together with two officers and a West Point cadet on leave to assist Long. Group reached point some twenty miles above present site of Omaha on Missouri River, where they wintered, 1819–1820. Long went east to report progress to Calhoun while scientists observed Indian tribes in area. Edwin James replaced Baldwin as botanist (latter forced to resign due to ill health, died August 1819), and there were several changes among officer personnel. Long's party left for West June 1820, arrived present site of Denver early in July. Some in group climbed Pikes Peak, and all then traveled south to point near present-day Canyon City. Expedition party then divided, following different routes to present-day Fort Smith, Arkansas. Long's section, with Say and Peale, followed Canadian River to their destination, rather than Red River, which had been their objective. Group then departed Fort Smith for Philadelphia by various routes.

Numerous plant, animal and mineral specimens collected and brought back, some articles published on plants and insects collected, and geological formations observed were among principal scientific contributions. A map of region covered, published in 1821, has been described as "a landmark of American cartography [which] may have been one of the most important of all maps of the plains produced before the Civil War." Long termed much of land east of Rockies as the "Great American Desert," still a highly controversial designation. He considered it unsuited for agriculture, primarily owing to lack of water and trees. Botanist James published *Account of an Expedition from Pittsburgh to the Rocky Mountains, Performed in the Years 1819 and '20* in two volumes, with an atlas (1823). Book consisted of daily summaries of expedition findings, including information concerning Indians encountered, plants and birds seen, terrain features, and geological data. Appendixes provided lists of animals observed, vocabularies of Indian languages, astronomical and meteorological notes, and maps. Most natural history specimens were sent to Peale's Museum in Philadelphia. Long subsequently became professionally concerned with improvements in river navigation, surveying railroad routes, and the design of bridges. His military service culminated in a two-year tour as colonel and chief of topographical engineers. Long's own publications included "Voyage in a Six-Oared Skiff to the Falls of St. Anthony" (1817); *Rail Road Manual* (1829); *Narrative of the Proceedings of the Board of Engineers of the Baltimore and Ohio Rail Road Company* (with W. G. McNeill, 1830); *Description of the Jackson Bridge, Together with Directions to Builders of Wooden or Frame Bridges* (1830); and *Description of Colonel Long's Bridges, Together with a Series of Directions to Bridge Builders* (1836). Was consultant to a number of railroads when not on active duty. Named chief of topographical engineers in 1861 at age seventy-six; retired two years later.

MAJOR CONTRIBUTIONS

The geographic, geological, biological, cartographic, and ethnological information brought back from his 1819–1820 expedition was extremely important. He became an authority on railway engineering and saw service with several railroads as consulting engineer when not actively engaged in military work. He was the last chief of topographical engineers and served during first half of Civil War, after which his branch of the army was absorbed into the Corps of Engineers.

BIBLIOGRAPHY

Benson, Maxine. "Edwin James: Scientist, Linguist, Humanitarian." Diss., University of Colorado, Boulder, 1968.

Benson, Maxine, ed. *From Pittsburgh to the Rocky Mountains: Major Stephen Long's Expedition, 1819–1820*, by Edwin James (1988).

Dillon, Richard. "Stephen Long's Great American Desert." *Proceedings of the American Philosophical Society* 3 (April 1967).

Evans, Howard E. *The Natural History of the Long Expedition to the Rocky Mountains, 1819–1820* (1997).
Friis, Herrman R. "Stephen H. Long's Unpublished Map of the United States Compiled in 1820–1822 (?)." *The California Geographer* 8 (1967).
Goetzmann, William H. *Army Exploration in the American West, 1803–1863* (1959).
Goodman, G. J., and Cheryl A. Lawson. *Retracing Major Stephen H. Long's 1820 Expedition: The Itinerary and Botany* (1995).
Osterhout, George E. "Rocky Mountain Botany and the Long Expedition of 1820." *Bulletin of Torrey Botanical Club* 47 (December 1920).
Wood, Richard G. *Stephen Harriman Long, 1784–1864: Army Engineer, Explorer, Inventor* (1966).

UNPUBLISHED SOURCES

See Record Groups 77 (Records of the Office of the Chief of Engineers); 94 (Records of the Adjutant General's Office); and 107 (Records of the Office of the Secretary of War), National Archives and Records Service, Washington, D.C. Manuscript of Edwin James's diary in Rare Book and Manuscript Library, Columbia University, New York.

Keir B. Sterling

LORD, JOHN KEAST. Born Cornwall, England, 1818; died Brighton, Sussex, England, 9 December 1872. Veterinarian, naturalist, author.

FAMILY AND EDUCATION

Family unknown. Raised by an uncle in Tavistock, Devonshire. Apprenticed to a chemist in Tavistock. Entered Royal Veterinary College (1842), London. Veterinary diploma (1844). Bachelor.

POSITIONS

Veterinary surgeon, Osmani Horse Artillery, Crimean War (1854–1856). Veterinary surgeon and assistant naturalist, British North American Land Boundary Commission (1858–1862). Public lecturer, London, 1863. Contributor, *Land and Water* (1866–1872). Naturalist for viceroy of Egypt, 1868–1869. First manager, Brighton Aquarium, 1872.

CAREER

Practiced veterinary medicine at Tavistock. Left suddenly. The next decade is unclear. His colorful and perhaps questionable narratives refer variously to a shipwreck on Canada's east coast, work on an eastern arctic whaler, the Hudson's Bay Company, Bruce Mines in Ontario, travels in western Canada, Minnesota, and Arkansas. Reappeared as a veterinary surgeon during the Crimean War (1854–1856).

Appointed to the British North American Land Boundary Commission, set up to map and mark the 49th parallel between British Columbia and the United States. Assistant to botanist David Lyall, who makes little reference to him.

Returned to England with botanical and zoological collections, naming two new mammals, including the pika. In 1863 appeared on stage as a frontiersman in buckskins at the Egyptian Hall, Piccadilly, London, in a lecture series called "The Canoe, the Rifle and the Axe."

Published articles in *The Field.* Joined the staff of Frank Buckland's journal *Land and Water* under the pen name "the Wanderer" in January 1866. Published the same year *The Naturalist in Vancouver Island and B.C.*, 2 vols. (1866). Lesser works included *At Home in the Wilderness* (1867, 1876). Traveled to Egypt and Arabia collecting and writing on his experiences from 1868–1869. Returned in poor health. Appointed manager of the new Brighton Aquarium. Died four months after it opened.

MAJOR CONTRIBUTIONS

Sent several faunal collections to the British Museum. Popularized the region with his entertaining, at times pretentious, writings. His public lectures, conducted in frontier garb, are reminiscent of Grey Owl.

BIBLIOGRAPHY

Buckland, F. T. "Practical Natural History. The Late John Keast Lord." *Land and Water* (14 December 1872): 395.

Dictionary of National Biography (U.K.).

Eastman, J. W. "John Lord Keast." *Museum Notes* 2. 2 (May 1922).

Hayman, John. "The Wanderer: John Keast Lord in Colonial British Columbia, 1858–62." *Beaver* 70. 6 (December/January 1990–1991).

Johnson, Alice M. "Lord, John Keast." *Dictionary of Canadian Biography*. Vol. 10 (1972).

Lorne Hammond

LORQUIN, PIERRE JOSEPH MICHEL. Born Valenciennes, France, 2 July 1797; died Paris, France, 8 February 1873. Lawyer, entomologist, traveler.

FAMILY AND EDUCATION

Trained as a lawyer. Graduate of the University of Douay. Married, with at least one son.

POSITIONS

Began as first clerk to a notary; gradually rose to position of practicing attorney. In 1840, appointed referee in the High Tribunal, Paris. In 1848, applied for, and was given, position in French Colonial Administration in Algiers. Resigned, 1850. Evidently spent remainder of life as traveler and collector of natural history specimens.

CAREER

Lorquin is first reported to have collected natural history specimens while in Algeria between 1848 and 1850, but doubtless had begun collecting in France prior to that time. Resigned Algerian post when he heard of California gold rush, but he was principally interested in possibilities for scientific exploration in California, not gold. Remained in the state from 1850 to 1856 and returned again from 1860 to 1862, earning reputation as ''first great resident entomological collector'' there. Family joined him in California in 1852. He traveled over much of state on foot in search of insect specimens, primarily butterflies and moths, but he also collected some birds and shells. In 1852 and 1853, worked in goldfields near Sacramento and Stockton, in Carson City, Nevada, and in vicinity of Los Angeles and San Diego. Between 1854 and 1856, was active collector along the Yuba River and in the central part of the state. Met and worked with some of the members of the California Academy of Sciences, notably Hans Herman Behr (1818–1904), a German-born physician-naturalist who had himself arrived in California in 1851 and who remained there for the rest of his life. Lorquin visited the Philippines and China between 1856 and 1860. Between 1862 and 1865, he again collected and visited in China and the Philippines and continued on to various islands in the Dutch East Indies (now Indonesia). Became ill in Java, had partially recovered when he returned to France in 1865. Traveled through parts of southern France and Spain until 1870 and lived the final two and a half years of his life in Paris. Most of his specimens were combined with those of his friend and colleague the French lepidopterist J. A. Boisduval (1799–1879), who was also Lorquin's family physician, in the Charles Oberthur Collection, Rennes, France. A smaller group of specimens, principally duplicates, was given to the California Academy of Sciences in San Francisco through the good offices of Behr but were destroyed in the earthquake and fire of 1906, together with Behr's entire collection. Most of Lorquin's collections were described by his friend Boisduval in the *Annales* of the French and Belgian Entomological Societies between 1852 and 1869. Smaller amounts of Lorquin materials were described by Achille Guenee (1809–1880) and others. Sixty-two of the at least ninety-five species of butterflies and moths Lorquin collected in California were determined by Boisduval to be new to science.

MAJOR CONTRIBUTIONS

A major early collector of natural history specimens in California, particularly moths and butterflies, the majority of which were deposited in a collection in Rennes, France, following his death.

BIBLIOGRAPHY

Essig, E. O. *A History of Entomology* (1932).

Grinnell, Fordyce. ''An Early Naturalist in California.'' *Entomology News* 15 (1904).

Keir B. Sterling

LOWERY, GEORGE HINES, JR. Born Monroe, Louisiana, 20 October 1913; died Baton Rouge, Louisiana, 19 January 1978. Ornithologist, mammalogist, educator, conservationist, and museum director.

FAMILY AND EDUCATION

Son of George Hines and Pearl (Connaughton) Lowery. Attended Louisiana Polytechnic Institute, Ruston, La., 1930–1932. Received B.S. (1934) and M.S. (1936) from Louisiana State University. Studied at University of Michigan Museum of Zoology, summers 1936 and 1937. Ph.D. (1949) from the University of Kansas. Married Jean Tiebout, 1937; two daughters.

POSITIONS

Spent entire professional career at Louisiana State University. Instructor in zoology, 1936–1939; from assistant professor to professor, 1939–1955; Boyd Professor of Zoology, 1955–1978. Founder and assistant curator of the Museum of Zoology, 1936–1947; director of Museum of Natural Science, 1951–1978.

CAREER

Lowery's doctoral dissertation at the University of Kansas dealt with a new technique in nocturnal observations of migrating birds, using a new type of telescope. He later published several additional studies on this subject, one of which, ''Direct Studies of Nocturnal Bird Migration,'' completed with R. J. Newman and published in *Recent Studies in Avian Biology* (1955), won him the Brewster Medal from the American Ornithologists' Union (1956).

Associated with Louisiana State University (LSU) for over forty years. Responsible for the expansion of the LSU Museum of Zoology (which became the Museum of Natural Science), containing a fully representative collection of Louisiana birds and mammals. Prior to World War II, Lowery was extremely supportive of the relatively small number of undergraduate and graduate students in ornithology at LSU. As their numbers expanded in the 1950s and later, the geographic scope of their work was extended first to include Mexico, then Central America, and finally the northern portion of South America. By the early 1960s, Lowery's own research interests began focusing on Latin America. The LSU Museum of Natural Science became a leading center of study in neotropical ornithology, with one of the outstanding collections of bird specimens from that region. Initiated a respected new publication series, the Occasional Papers of the Museum of Zoology of Louisiana State University.

Reputation as an ornithologist and mammalist rests on such works as *Louisiana Birds* (1955, 3d ed., 1974) and *The Mammals of Louisiana and Its Adjacent Waters* (1974). Both works were honored with Louisiana Literary Awards. Among Lowery's other published studies are ''Distribution of the Flora and Fauna of Louisiana in relation to its geology and physiography'' (with W. C. Holland, W. T. Penfound, C. A. Brown, and P. Viosca, Jr., 1944); ''Birds from the State of Vera Cruz, Mexico'' (1951); *Check List of North American Birds*, 5th ed. (with Alexander Wetmore and others, 1957); and ''Family Parulidae,'' in R. L. Painter, ed., *Check List of Birds of the World* (with B. L. Monroe, 1968). He contributed chapters on woodpeckers to the National Geographic Society volume on *Song and Garden Birds of North America* (1964) and on ''The Mysteries of Migration'' to the companion volume *Water, Prey and Game Birds of North America* (1965), both edited by Alexander Wetmore.

Member of the American Ornithologists' Union (president, 1959–1962). Received the Outstanding Conservationist Award of the Year from the Outdoor Writers Association in 1965. In 1975 named Conservation Educator of the Year by the Louisiana Wildlife Federation.

MAJOR CONTRIBUTIONS

Made important contributions to the study of bird migration, developed one of the outstanding regional collections of American birds and a major center of research in Latin-American ornithology. Published authoritative accounts of Louisiana birds and mammals. Valued teacher and mentor.

BIBLIOGRAPHY

American Men and Women of Science. 13th ed. (1976).

Howell, Thomas R., and John P. O'Neill. ''In Memoriam: George H. Lowery, Jr.'' *Auk* 98 (January 1981): 159–66.

Arthur Sherman

LUTZ, FRANK EUGENE. Born Bloomsburg, Pennsylvania, 15 September 1879; died New York City, 27 November 1943. Entomologist, museum curator, educator, conservationist, writer.

FAMILY AND EDUCATION

Son of Martin Peter and Anna Amelia (Brockway) Lutz. Father successful real estate and insurance agent. As young boy in Pennsylvania, Lutz was fascinated by a caterpillar shedding its skin and developed a lifelong interest in insects. During first two years at Haverford College, planned career in insurance and specialized in mathematics, but, deciding on medical career, switched to biology, receiving A.B. in 1900. Studying on scholarship under Charles H. Davenport at University of Chicago, earned A.M. in 1902 in new field of biometrics, which emphasized statistics in biological research. The following year, became first American student of Karl Pearson, founder of biometry, at University College, London, and also briefly in Berlin. Completed Ph.D. at University of Chicago in 1907, with dissertation on variation among crickets. Married Martha Ellen Brobson, 1904; three children.

POSITIONS

First professional position was as entomologist at Biological Laboratory of Brooklyn Institute in 1902, followed next year by assistantship in Zoological Department at University of Chicago. From 1904 to 1909, conducted pioneer entomological research on evolution and genetic traits of Long Island Drosophila (fruit fly) under Davenport at Carnegie Institution's newly established Station for Experimental Evolution at Cold Spring Harbor. In 1909 joined department of invertebrate zoology at American Museum of Natural History as assistant curator, becoming curator in 1916. Five years later, was named first curator of newly created Department of Entomology, remaining in position for the rest of life; edited museum's technical papers and other publications. From 1925 to 1928, supervised Station for the Study of Insects at Tuxedo, New York. Won A. Cressy Morrison Prize of New York Academy of Sciences, 1923. Founding member and fellow, American Entomological Society, serving as president, 1927; fellow, American Association for Advancement of Science; adviser, Buffalo Society of Natural Science. Lecturer, Columbia University, 1937.

CAREER

During his long tenure at the American Museum of Natural History, he increased the museum's collections of insects and spiders from 300,000 to more than 2 million items, due in large part to his almost-annual field trips to various locations in the Americas, including British Guiana, Panama, and Cuba. The author of numerous scientific papers, as well as articles in popular magazines, believed in the "joy of research" and described travels and findings in a lighthearted, often humorous style. An innovator and educator, abjured what a museum colleague termed the usual "Noah's Ark assemblage of species" and instead created popular museum exhibits, including the first insect dioramas and "insect zoos" featuring live specimens. Unlike other curators, enjoyed addressing groups of visitors to the museum and encouraged young children to explore the world of nature. His *Field Book of Insects* (1918) was a pioneering effort written in a popular yet scientific style that went through several editions. In the 1920s, established the country's first guided nature trail in Harriman State Park, one hour north of New York City.

His backyard experiments proved that, contrary to existing scientific and popular opinion, insects are attracted to ultraviolet light, not color as are humans, and that insects can adapt to great extremes of air pressure and oxygen deprivation. Anticipating later concerns with insecticides, observed, "When man interferes with Nature's biological set-up, he is apt to make a mess of it." Acting on a bet with the director of the museum, spent three years of his spare time collecting and studying insects—which he termed "our six-footed guests"—in his own backyard in New Jersey. The resulting 1,402 varieties of insects became the basis for a popular museum exhibit, "Insects of a Suburban Yard," and a volume aptly entitled *A Lot of Insects* (1941), written in a chatty, yet descriptive and learned style. Pallister writes that Lutz had a "rich sense of humor, at times 'tart and roguish.' " A 1979 internal museum report observed that Lutz "might well be considered the [Jean Henri] Fabre of America" (Fabre [1823–1915] was a leading French entomologist).

MAJOR CONTRIBUTIONS

Pioneer in study of various insects and innovator in utilizing museum exhibits as popular educational tools. Probably the leading American entomologist of the first half of the twentieth century, who taught that insects were an integral part of the environment.

BIBLIOGRAPHY

Bacon, Annette L. "Bibliography of Frank E. Lutz." *Journal of the New York Entomological Society* (1944).

Barton, D. R. "Attorney for the Insects." *Natural History* (1941).

Dictionary of American Biography. Supplement 3, 1941–1945 (1973).

Emerson, Alfred E. "Frank Eugene Lutz." *Science* (1944) (obit.).

Gertsch, W. J. "Frank E. Lutz." *Annals of the Entomological Society of America* (1944) (obit.).

Lutz, Frank E. "Collected Works." 3 vols. New York: American Museum of Natural History Library.

Mallis, Arnold. "Frank Eugene Lutz." In *American Entomologists* (1971).

National Cyclopedia of American Biography (1958).

Pallister, John C. *In the Steps of the Great Entomologist Frank Eugene Lutz* (1966).

Schwartz, Herbert F. "Frank E. Lutz." *Entomological News* (1944) (obit.).

UNPUBLISHED SOURCES

See American Museum of Natural History, New York City, Department of Entomology, Records; approximately sixty cubic feet (contains correspondence of, and files relating to, Lutz, 1909–1980).

The following collections contain letters from Lutz: Albert Rich Brand Papers, Collection of Regional History and University Archives, Cornell University, Ithaca, N.Y.; Cornell University Press, Records, Cornell University; Charles Zeleny Papers, University of Illinois, Champaign–Urbana.

Rolf Swensen

M

MAC ARTHUR, ROBERT HELMER. Born Toronto, Ontario, 7 April 1930; died Princeton, New Jersey, 1 November 1972. Ecologist, biogeographer.

FAMILY AND EDUCATION

Youngest son of John Wood Mac Arthur (professor of genetics at University of Toronto and later at Marlboro College, Vermont) and Olive Turner Mac Arthur. Came to United States, 1947. A.B., Marlboro, 1951; M.A. (mathematics), Brown, 1953; studied under G. Evelyn Hutchinson at Yale, received Ph.D., Yale, 1957. Postgraduate study at Oxford, 1957–1958. Honorary D.H.S., Marlboro, 1972. Married Elizabeth Bayles Whittemore, 1952; four children.

POSITIONS

Service in U.S. Army, 1954–1956. Assistant professor to professor of biology, University of Pennsylvania, 1958–1965; professor of biology, 1965–1968, Henry Fairfield Osborn Professor of Biology, Princeton, 1968–1972.

CAREER

Primarily interested in birds, but preferred to push well beyond descriptive ecology, and seek mathematically based theoretical constructs. A founder of the discipline of evolutionary ecology. Three early papers demonstrated high degree of originality. In his first, ''Fluctuations of Animal Populations, and a Measure of Community Stability'' (1955), he made first formal use of information theory in examining frequency with which species occurred. In ''On the Relative Abundance of Bird Species'' (1957), he proposed what became known as the ''broken stick'' model, although this was but one of three hypotheses presented. He suggested that competing species split available territory in random manner, dependent in part on their numbers. Relative profusion could be suggested by a stick broken into varying lengths. Though this idea was controversial and later dropped by its author, he had provided new method of approaching key question in community ecology. His ''Population Ecology of Some Warblers of Northeastern Coniferous Forests'' (1958) was a unique breakdown and commentary on niche division in warblers. During the 1960s, worked on issues of species diversity with colleagues and graduate students. In ''An Equilibrium of Insular Zoology'' (1963), later elaborated in his first book, *The Theory of Island Biogeography* (with E. O. Wilson, specialist on ants, taxonomy and biogeography, 1967), MacArthur and Wilson discussed their theory of species equilibrium. Concluded that islands, both literal and figurative, tended to gain (through immigration and some little speciation) and lose (through extinction) species in approximately equal numbers over time, barring unusual natural disasters or other disruptive circumstances. Their concept of ''distance effect'' posited that the number of species an island could support varied with the extent of its remoteness. Their theory of ''area effect'' postulated that smaller islands supported fewer immigrant species relative to the extinctions they sustained.

Toward the end of his life, Mac Arthur considered himself more as a biogeographer than an ecologist. His *Geographic Ecology: Patterns in the Distribution of Species* (1972) combined the various elements of his work. Published fifty other papers, notes, and reviews, a number of them with other authors. Editor of *Monographs in Population Biology* and co-founder of the journal *Theoretical Population Biology*. Was an honorary research associate, Smithsonian Tropical Research Institute. Won Mercer Award of Ecological Society of America. Fellow, American Academy of Arts and Sciences and of National Academy of Sciences.

MAJOR CONTRIBUTIONS

Sought to make the study of biogeography more experimental and rigorous. In the words of E. O. Wilson, ''It is his distinction to have brought population and community ecology within the reach of genetics. By reformulating many of the parameters of ecology, biogeography, and genetics into a common framework of fundamental theory, Mac Arthur, more than any other person who worked during the decisive decade of the 1960s, set the stage for the unification of population biology.''

BIBLIOGRAPHY

Cody, Martin L., and Jared M. Diamond, eds. *Ecology and Evolution of Communities* (1975).

Fretwell, S. D. ''The Impact of Robert Mac Arthur on Ecology.'' *Annual Review of Ecology and Systematics* 6 (1975).

Quammen, David. *The Song of the Dodo: Island Biogeography in an Age of Extinctions* (1996).
The New York Times 2 November 1972 (obit.).
Who Was Who in America.
Wilson, Edward O. *Biophilia* (1984).
Wilson, Edward O. *Dictionary of Scientific Biography*. Supplement 2, vol. 18 (1988).
Wilson, Edward O. *Naturalist* (1994).
Wilson, Edward O., and Hutchison, E. G. ''Robert Helmes Mac Arthur.'' *Biographical Memoirs: National Academy of Sciences* 58 (1989).

Keir B. Sterling

MACFARLANE, RODERICK ROSS. Born Stornoway, Isle of Lewis, Scotland, 1 November 1833; died Winnipeg, Manitoba, 14 April 1920. Fur trader, naturalist.

FAMILY AND EDUCATION

Related on father's side to an officer of old North West Company and on mother's side to the explorer Alexander Mackenzie. He attended the parochial school and the Free Church Academy at Stornoway, then spent three years in law office of the procurator fiscal of Lewis District. Married 26 January 1870 to Ann, daughter of Chief Trader Alexander Christie, Jr.; five daughters and three sons.

POSITIONS

On 3 July 1852 sailed from Stromness for York Factory, having been hired as apprentice clerk with Hudson's Bay Company; arrived at Red River (now Winnipeg) on 14 September. After one to several years each in charge of Fort Rae, Forth Resolution, Fort Liard, and Fort Good Hope, he founded Fort Anderson on Anderson River, which he operated from 1861 to 1863 and from 1864 until it was ordered closed in 1866. Was promoted to chief trader in 1868 in charge of the MacKenzie district. Was head of the Athabasca district from 1871 until 1885, of New Caledonia district, 1887–1889, and the Cumberland district, 1889–1893. He retired in 1894 to live in Winnipeg.

CAREER

Stimulated by visit from Robert Kennicott, MacFarlane became a prolific collector of birds, bird's eggs, and mammals. Smithsonian employees, as reported by Deignan, believed that MacFarlane ''contributed more specimens in a single year than any other collector in the museum's history.'' MacFarlane's three peak years were 1863, 1864, and 1865, when he sent from Fort Anderson 1,000, 1,500 and 1,750 specimens, respectively. A single specimen might include skin, nest, and eggs. In later years contributed specimens from Fort St. James, Stuart Lake, British Columbia, and from Cumberland House on the Saskatchewan River.

MAJOR CONTRIBUTIONS

Our knowledge of wader nesting on the tundra virtually began with MacFarlane. He collected 170 sets of lesser golden plover eggs, 70 of the red-necked phalarope, 30 of the lesser yellowlegs, and 20 each of the semipalmated plover, least sandpiper, and buff-breasted sandpiper. We owe to him, apart from a single nest found by John Richardson at Point Lake in 1821, all we know about the nesting of Eskimo curlew, with data on thirty-three nests and another five with downy young.

Specimens he collected at Anderson River led Merriam to name three subspecies for him: a tundra vole, *Microtus oeconomus macfarlani* Merriam, 1900; a snowshoe hare, *Lepus americanus macfarlani* Merriam, 1900; a grizzly bear, *Ursus horribilis macfarlani* Merriam, 1918. William Brewster also honored him by naming a subspecies of what is now the western screech owl, *Otus kennicotti macfarlani*, from Walla Walla, Washington, an area not visited by MacFarlane.

BIBLIOGRAPHY

Deignan, H. C. "HBC and the Smithsonian." *Beaver* 278, 4 (1947).

Gollop, J. B. *Eskimo Curlew, a Vanishing Species?* (1986).

Lindsay, Debra. *Science in the Subarctic* (1993).

MacFarlane, R. R. "Notes on and List of Birds and Eggs Collected in Artic America, 1861–1866." *Proceedings of the U.S. National Museum* 14 (1891).

MacFarlane, R. R. "Notes on Mammals Collected and Observed in the Northern MacKenzie River District, North-west Territories of Canada, with Remarks on Explorers and Explorations of the Far North." *Proceedings of the U.S. National Museum* 28 (1905).

MacFarlane, R. R. "List of Birds and Eggs Collected in the North-west Territories of Canada, between 1880 and 1894." In C. Mair, *Through the Mackenzie Basin* (1908).

Preble, E. A. "Roderick Ross MacFarlane, 1833–1920." *Auk* 39 (1922).

Schofield, F. H. "A Brief Sketch of the Life and Services of Retired Chief Factor R. MacFarlane." (1913).

C. Stuart Houston

MACKAYE, BENTON. Born Stamford, Connecticut, 6 March 1879; died Shirley, Massachusetts, 11 December 1975. Forester, author, association executive, regional planner.

FAMILY AND EDUCATION

Son of James Morrison Steele and Mary Keith (Medbery) MacKaye. Received A.B. (1900) and A.M. (1905) in forestry from Harvard University. Married Jessie Belle Hardy in 1915.

POSITIONS

Served as research forester for the U.S. Forest Service, 1905–1917, studying forest conditions in the southern Appalachian Mountains and in the White Mountains National Forest, New Hampshire. In 1918–1919 was a policy specialist for the U.S. Department of Labor, studying, among other things, plans for the colonization of the pioneer fringe in the Great Lakes region, as well as employment and timber operations in the national forests of western Washington state, an area left bare by logging.

In 1919 established himself in private practice as a regional planner and worked for both New York State (1924) and the Massachusetts (1928). In 1932 he was involved in planning on Indian reservations for the U.S. Indian Service. From 1933 to 1936 was employed as a planner for the Tennessee Valley Authority. A consultant of the U.S. Department of Agriculture on flood-control policies from 1938 to 1941. During World War II, instrumental in formulating an Alaska–Siberia supply route. From 1942 to 1945 on the planning staff of the Rural Electrification Administration.

One of the key organizers of the Wilderness Society (1935), he was president (1945–1950) and honorary president (1950–1975). MacKaye received the Conservation Award of the Department of the Interior (1966).

CAREER

As forester and regional planner, MacKaye was concerned during his career with both urban and rural areas. In 1921 wrote an article for the *Journal of the American Institute of Architects* on the subject of the proposed Appalachian Trail. During the next ten years, worked on its development. As "father of the Appalachian Trail," was instrumental in creating this 2,000–mile footpath from Maine to Georgia. As a government planner, he spearheaded the idea of the "townless highway." His books include *The New Exploration: A Philosophy of Regional Planning* (1928) and *From Geography to Geotechnics* (1968).

MAJOR CONTRIBUTIONS

One of the founders of the Regional Planning Association of America (1923), he early on advocated public control over our geographic habitat, the preserving of cultural and recreational areas in an increasingly urbanized environment. Fostering the idea of the region as the essential component of urban development led to such foresighted planning concepts as highway placement away from human settlements.

BIBLIOGRAPHY

Encyclopedia of Urban Planning (1974).
Living Wilderness 39 (1976).

National Cyclopedia of American Biography (1982).
Stroud, Richard H., ed. *National Leaders of American Conservation* (1985).
Webster's American Biographies (1974).

Arthur Sherman

MACKENZIE, ALEXANDER. Born Stornoway, Isle of Lewis, Scotland, 1764; died Mulinearn, Scotland, 12 March 1820. Explorer, fur trader.

FAMILY AND EDUCATION

Third of four children of Kenneth and Isabella Maciver Mackenzie. Mother died at some point prior to 1774; father emigrated to New York with Alexander and two sisters in that year. Father served with loyalist unit in New York, 1776–1780. Left in custody of several aunts, Alexander was sent to Montreal, 1778, because he was son of loyalist officer which placed family under stress and this was deemed a prudent move. Mackenzie attended school in Montreal for about a year, 1778–1779, then entered fur trade. Father died on active duty in 1780, when Alexander was sixteen. In London for private study of mathematics, surveying, and to secure needed equipment, 1791–1792. Is reputed to have had a native wife by whom he had one son, who was later a clerk with the North West Company. Married Geddes Mackenzie, 1812; three children.

POSITIONS

Spent two decades in active fur trading and in management of the business in Montreal from 1779 to 1799: with firms of Finlay and Gregory, 1779–1784; Gregory, McLeod (was also partner), 1784–1787; North West Company (was also partner), 1787–1795. In 1789 and 1793, undertook two major expeditions to Arctic Sea and to Pacific coast for which he is best known; though no longer actively trading, became partner in McTavish, Frobisher & Company, 1795–1799. Returned to England and visited Canada for varying periods of time at intervals between 1799 and 1810. Knighted, 1802. Was partner in New North West Company, 1800–1804. Elected representative in Lower House of Canadian House of Assembly, 1804. Retired from business, 1811, took up residence in Scotland.

CAREER

Employed by fur trading firm of Finlay and Gregory and worked in their Montreal office, 1779–1784. Began active fur trading 1784 and made partner in same year when firm became Gregory, McLeod. Sent to what is now Saskatchewan, 1785, spent two years at Île-à-la-Crosse headquarters of English River department. Firm joined North West Company (NWC), 1787, and Mackenzie held one-twentieth share. Worked under American-born explorer–fur trader Peter Pond (1740–1807) in Athabasca region, 1787–1788, then took charge of that area for the NWC.

In 1788, Mackenzie sent his cousin Roderick, a fellow NWC employee, to establish Fort Chipewyan at western end of Lake Athabaska, in northeastern corner of what is now Alberta. From this point, on instructions from NWC, Alexander Mackenzie departed, with mixed party of French-Canadians, one German, and some Indians, to locate mouth of supposed river (now Cook's Inlet), discovered by Captain James Cook in southcentral Alaska on his last voyage in 1778. Pond had greatly underestimated distance involved and had imperfect understanding of geography of western Canadian and Alaskan terrain. Mackenzie proceeded up Slave River to Great Slave Lake, thence 1,075 miles up present-day Mackenzie River to the Arctic Sea, arriving on 12 July 1789. River portion of trip consumed but fourteen days. Return trip to Fort Chipewyan began four days later and ended 12 September. More than 3,000 miles had been covered in 102 days. In 1792, Mackenzie was given a second share in the NWC, so that he now held one-tenth of the total. He undertook a second exploratory trip (1792–1793) following the Peace River to the Parsnip River, thence via James Creek (Bad River), McGregor and Fraser Rivers to present site of Alexandria, British Columbia. On advice of local natives, Mackenzie backtracked to juncture of Fraser and West Road rivers, from whence his party continued west overland along or near West Road River, to Ulgako Creek, then to Tanya Lakes and south to and along the Bella Coola to Dean Channel, arriving at a point very near the coast on 21 July 1793. He missed by six weeks visit of Captain George Vancouver to Dean Channel, which had taken place in early June. The actions of unfriendly Nuxalk Indians toward Mackenzie may have been due to abuse received at the hands of some of Vancouver's people. After several days of tension, during which some small scale exploring was done up and down Dean Channel, Mackenzie thought it wise to retrace his steps back to Fort Chipewyan, and he brought his party safely back there in the space of one month, having traveled a total of more than 2,811 miles (sources disagree concerning this total).

Mackenzie decided to leave fur trading, which he found too confining, and left western Canada early in 1794, returned to Montreal, urging that NWC, Hudson's Bay Company (HBC), and East India Company jointly reorganize fur trade on more efficient and cooperative basis. In 1795, was offered and accepted partnership in McTavish, Frobisher Company, and held this position until its expiration late in 1799, when he went to England. It is not certain whether he left the company voluntarily or was pushed out. Published *Voyages from Montreal on the River St. Lawrence, through the continent of North America to the Frozen and Pacific Oceans in the years 1789 and 1793* in 1801, an account to which his cousin Roderick may have contributed, though William Combe has long been considered principal author. Received knighthood in February, 1802. Continued to press his idea of cooperation between trading companies. Became partner in New North West Company, 1800, which was in strong competition with NWC. The British Colonial secretary suggested that the two firms be combined as initial step toward Mackenzie's goal of wider cooperation, and Mackenzie returned to Canada in effort to accomplish this. Long-standing personal and professional antagonisms precluded Mackenzie being given a role in new joint operation when merger was achieved in 1804.

Was elected and briefly served as representative in Lower Canada House of Assembly, 1804, but was not interested in his political responsibilities and spent the remainder of his life in England and Scotland. The last of his several short visits to Canada took place in 1810. Mackenzie began buying substantial HBC stock in 1808 in effort to gain operating control and in pursuance of his plan for cooperation among trading companies. He was at first receptive to colonization scheme of Lord Selkirk, begun in the same year to buy land for extensive resettlement of poor Scottish Highlanders and Irishmen in Red River country of Canada. Mackenzie later opposed enormous scope of Selkirk's Grant (which included southern portion of present-day Manitoba, northern and eastern North Dakota, and northwestern Minnesota), because it was deemed a threat to the fur trade. Selkirk's idea was ultimately approved by HBC leadership in 1811, and Selkirk also successfully blocked Mackenzie's efforts to gain control of HBC. In that same year, British government declined to take action on his latest initiative for cooperation among large fur traders. He retired from business in 1811, when only 47 and took up residence in Scotland. His marriage to fourteen-year-old bride in 1812 produced three children prior to his death in 1820, possibly of Bright's Disease. In 1830, Mackenzie's widow received £10,000 following a lawsuit to recover the value of his interest in the old North West Company.

MAJOR CONTRIBUTIONS

Mackenzie is best known for his expeditions of 1789 and 1793, which took him through large unexplored regions of western Canada. It is perhaps significant that none of his men died or deserted him on either trip, nor was he responsible for the death of any natives. His ideas concerning a wider-ranging, more efficiently operated fur-trading operation were finally put into effect by the HBC following his death and after its amalgamation with the NWC in 1821. President Thomas Jefferson became aware of Mackenzie's 1793 trip in 1797, and his later (1802) reading of Mackenzie's *Voyages from Montreal . . . to the Frozen and Pacific Oceans* alerted the American leader to British interests in the Pacific Northwest and caused him to launch the Lewis and Clark Expedition in 1804.

BIBLIOGRAPHY

Allen, John L. *Passage Through the Garden: Lewis and Clark and the Image of the American Northwest* (1975).

Ambrose, Stephen A. *Undaunted Courage: Meriwether Lewis, Thomas Jefferson, and the Opening of the American West* (1996).

Daniels, Roy. *Alexander Mackenzie and the North West* (1969).

Fawcett, Brian. *The Secret Journal of Alexander Mackenzie* (1985).

Gough, Barry M. *The Politics of Trade, Exploration, and Territory: Alexander Mackenzie's Scheme for North Pacific Dominion* (1992).

Lamb, W. Kaye, ed. *The Journals and Letters of Sir Alexander Mackenzie* (1970).
Lamb, W. Kaye. *Dictionary of Canadian Biography.*
Mackenzie, Sir Alexander. *Voyages from Montreal on the river St. Lawrence, through the Continent of North America to the Frozen and Pacific Oceans in the years 1789 and 1793* (Radisson Society ed., 1927) (first ed. William Combe, ed., 1801).
Mirsky, Jeanette. *The Westward Crossings: Balboa, Mackenzie, Lewis and Clark* (1946).
Quaife, Milo M., ed. *Alexander Mackenzie's Voyage to the Pacific Ocean in 1793* (1931).
Sheppe, Walter, ed. *First Man West: Alexander Mackenzie's Journal of His Voyage to the Pacific Coast of Canada in 1793* (only annotated edition 1962).
Wade, Mark S. *Mackenzie of Canada* (1927).
Wagner, H. R. *Peter Pond, Fur Trader and Explorer* (1955).
Woodworth, John, and Halle Flygara. *In the Steps of Alexander Mackenzie: Trailguide* (1981).

UNPUBLISHED SOURCES

Most of Mackenzie's papers were destroyed by fire in 1833. A copy of his 1789 journal is in British Library, London. ''Sir Alexander Mackenzie, Fur Trader, Explorer and Nation Builder,'' manuscript by ''Clansman,'' (typescript, 1930), in Bancroft Library, University of California. Some faulty copies of Mackenzie's letters (no originals are known to exist), most originally written to his cousin Roderick and the majority of them transcribed by Roderick's son-in-law Louis Masson, are in National Archives Canada, Ottawa. See also bibliographies in entries by W. K. Lamb, above. Gunther Barth, ''Strategies for Finding the Northwest Passage: The Roles of Alexander MacKenzie and Meriwether Lewis,'' paper presented at American Philosophical Society, Philadelphia, 16 March 1997.

Keir B. Sterling

MACLURE, WILLIAM (JAMES). Born Ayr, Scotland, 27 October, 1763; died San Angel, near Mexico City, 23 March 1840. Geologist, reformer, patron of science and education.

FAMILY AND EDUCATION

Son of David and Ann (Kennedy) McClure. Baptized James, but later called himself William and changed the spelling of his family name. Had two brothers, Alexander and John; and three sisters, Helen, Anna, and Margaret. At some time his father removed the family to Glasgow and later to Liverpool. Received his early education at Ayr under Douglass, who was known for his classical and mathematical knowledge. It is not known if he ever attended any institution of higher learning, and the evidence of his writings suggests that he did not. Never married; no children.

POSITIONS

Elected to the American Philosophical Society in 1799 and served on the council, 1818–1829. Appointed a member of the American Spoliation Commission in France, 1803. Member of the Academy of Natural Sciences of Philadelphia in 1812 and its president, 1817–1840. First president of the American Geological Society, 1819. Founder of the Workingmen's Institute, New Harmony, Indiana, 1838. Also a member of the following organizations: Geological Society of London, Royal Geological Society (Penzance), Mineralogical Society (Jena), Philosophical Society (Moscow), Philosophical Society (Warsaw).

CAREER

Entered mercantile career at an early age and in 1782 was supposed to have made his first visit to the United States to transact business, although he may have made a prior trip in 1778. Upon his return to England he became a partner in the firm of Miller, Hart and Co. and in this capacity traveled widely throughout Europe during the next fifteen years, conducting affairs for the concern. Established residence at Philadelphia in 1796 and carried on business there and in Virginia. Became American citizen at Detroit during this same year. After having acquired a sizable fortune, he retired from business in order to pursue the study of geology and to reform education. Returned to Europe in 1799, bought home in Paris for his base of operations, and during the next fifteen years, after having mastered the geological principles and methods of Abraham G. Werner, adopted his terminology but not his entire theory, made geological surveys in most of the countries of Europe from the Mediterranean Sea to the Baltic and from the British Isles to Russia.

Appointed member of the U.S. Spoliation Commission at Paris in 1803, which was to settle the claims of American merchants and sea captains against the French government. On tour of Switzerland with Joseph C. Cabell in 1805, visited school of Johann Heinrich Pestalozzi at Yverdom and convinced of the validity of this system of education, sent Joseph Neef to Philadelphia in 1806 and in 1809 opened the first Pestalozzian school in the United States.

Returned to the United States in 1808 and during the next year and a half made (or completed) a one-man geological survey of region east of Mississippi River, traveling from Maine to Georgia, reputedly crossing the Allegheny Mountains no fewer than fifty times. The results are found in his "Observations on the Geology of the United States, Explanatory of a Geological Map," *Transactions* 6 (1809): 411–28. At this time he was responsible for bringing to the United States the scientists Gerard Troost and Silvain Godon and, later, Charles Alexandre Lesueur.

Returned to Europe in 1809 and continued to make geological tours. Apparently gave up on plans to construct a map and write a memoir on the geology of Europe but continued collecting rocks and mineral specimens for American schools, universities, and scientific institutions. Left France in the fall of 1815 and traveled with C. A. Lesueur to England and then to the West Indies. They then went to the United States in the spring of 1816 and made a tour through parts of Delaware, Maryland, Pennsylvania, New York, and a number of New England states. Lesueur conceived a plan, which he would pursue for many years with Maclure's support, of writing a monograph on the fish of North America, and he collected data for it on this tour while Maclure did research for a revised and expanded version of his "Observations on the Geology of the U.S." It was later published in *Transactions* (n.s.) 1 (1818): 1–91.

Spent the next few years at Philadelphia, where he was elected president of the Academy of Natural Sciences and held that position until his death. He donated many books and specimens to the academy and contributed an average of $1,000 a year in funds. In the winter of 1817–1818, he financed a collecting expedition to the Sea Islands along the Georgia coast and to east Florida, and his companions were Titian R. Peale, George Ord, and Thomas Say. The research of Say, an entomologist, was subsidized by Maclure over number of years, and at this time he and several others supplied funds to enable Thomas Nuttall to make a collecting expedition along the Arkansas River in the Indian Territory west of the Mississippi.

Returned to Europe in 1818, spending next two years in France and Italy. In 1820, traveled to Madrid for his health and attempted to set up agricultural schools at Alicante in southeastern Spain, but with the success of the royalist counterrevolution was forced to abandon plan in 1824.

Traveled in Ireland, Scotland, and England in 1824–1825 and while in Scotland visited model factory town and school set up by Robert Owen at New Lamark, which impressed him very much. Sold his home in Paris in 1825 and sailed to America, where his friends induced him to join forces with Owen in establishing a communitarian socialist society at New Harmony, Indiana. He was more or less responsible for the direction of education and scientific programs and to this end brought with him a number of Pestalozzian teachers, including Neef, Marie D. Fretageot, William S. Phiquepal (Phiquepal d'Arusmont) and the scientists Say, Lesueur, and Troost. Although Owen's experiment failed within a year, most of Maclure's group stayed on, and their efforts in research and publication (with Maclure's support) were to make New Harmony a sort of scientific mecca on the American frontier.

The School of Industry established by Maclure at New Harmony emphasized trades connected with publishing, and in the ensuing years the school press produced five plates of Lesueur's *American Ichthyology* (1827), a continuation of seven numbers of Say's *American Conchology* (1830–1838), and a republication of François Andre Michaux's *North American Sylva* (1841), the plates of which had been purchased by Maclure in Europe. As a result of failing health he left the United States in 1828 to spend his remaining years in Mexico, where he carried on a prolific correspondence and continued to direct his various enterprises at New Harmony and Philadelphia.

In addition to his book and map on the *Geology of the United States* (1809, 1817), published many articles on geology in the *American Journal of Science* (Silliman's journal), the *Journal of the Academy of Natural Sciences of Philadelphia*, and the press at New Harmony. Earliest known geological publication was a letter, "Sur les Volcans d'Allet en Catalone," in *Journal de Physique* 66 (1808): 219–20, which influenced Charles Lyell to visit the volcanic region of Olot in Spain. Other significant geological articles were "Observations on the

Geology of the West India Islands, from Barbados to Santa Cruz, Inclusive,'' *Journal of the Academy of Natural Sciences* 1 (1817): 134–49; and ''Essay on the Formation of Rocks, or an Inquiry into the Probable Origin of Their Present Form and Structure,'' *Journal of the Academy of Natural Sciences* 1 (1817): 261–76, 285–310, 327–345. His ideas on politics, government, economics, and the reform of education and society are to be found in *Opinions on Various Subjects, Dedicated to the Industrious Producers*, 3 vols. (1831–1838). Included in this work is a critical article on geology entitled, ''Genealogy of the Earth—Geological Observations,'' vol. 3, 175–78.

Although he followed Werner's system of geology, he was neither a ''Neptunist'' nor a ''catastrophist.'' By nature he was conservative and pragmatic geologist, deploring tendency of many of his colleagues to theorize without investigating the ''regular operations of the great laws of nature''—he believed in fieldwork. His pioneer study of the geology of the United States was primarily lithographical rather than geological, but it established his reputation and leadership in American geology. Later, become aware of importance of studies in stratigraphy and paleontology but did not utilize them to any great extent in his work. Amos Eaton, one of his successors in American geology, declared that Maclure was one of the best geologists he knew of, and although his book was a heterogeneous work, it had a better application to facts than all other American works of the period.

MAJOR CONTRIBUTIONS

Considered the ''father of American geology,'' his work remained preeminent during the first two decades of the nineteenth century. His interest in education coupled with his interest in the natural sciences caused him to become a unique one-man scientific foundation, carrying on his own work but at the same time financing the work of other scientists, various institutions, and a number of Pestalozzian teachers.

BIBLIOGRAPHY

Armytage, W. H. G. ''William Maclure, 1763–1840; A British Interpretation.'' *Indiana Magazine of History* 47 (March 1957).

Bestor, Arthur E. Jr., ed. ''Education and Reform at New Harmony: Correspondence of William Maclure and Marie Duclos Fretageot, 1820–1823.'' *Indiana Historical Society Publications* 15, 3.

Gerstner, Patsy A. ''The Academy of Natural Sciences in Philadelphia, 1812–1850.'' In A. Oleson and S. C. Brown, eds., *The Pursuit of Science in the Early American Republic* (1976).

Greene, John C., and John G. Burke. ''The Science of Minerals in the Age of Jefferson.'' *Transactions of the American Philosophical Society* 68, pt. 4 (July 1978). 1–113.

Merrill, G. P. *The First One Hundred Years of American Geology* (1924).

Moore, J. Percy. ''William Maclure—Scientist and Humanitarian.'' *Proceedings of the American Philosophical Society* 91, 3 (August 1947).

Morton, Samuel G. ''A Memoir of William Maclure.'' *The American Journal of Science and Arts* 47 (October 1844).

UNPUBLISHED SOURCES

See European journals of William Maclure (1805–1825); notes for the revised edition of the *Geology of the U.S.* (1817); Maclure–Fretageot collection of letters; and other documents and letters at the Workingmen's Institute Library, New Harmony, Indiana.

John S. Doskey

MACOUN, JOHN. Born Maralin, County Down, Northern Ireland, 17 April 1831; died Sidney, Vancouver Island, British Columbia, Canada, 18 July 1920. Botanist, naturalist, explorer.

FAMILY AND EDUCATION

Son of soldier in Princess Royal's Regiment, worked family land with two brothers. Emigrated to Northumberland County, Canada west, British North America, in 1850. Passing interest in local flora developed into full-scale study of Canadian botany with encouragement of British and American specialists. Attended Toronto Normal School in 1859. Honorary M.A. degree from Syracuse University in 1889. Married Ellen Terrill, 1862; five children.

POSITIONS

Public school teacher (Brighton, 1857–1859; Castleton, 1860; Belleville, 1860–1868). Professor of natural history, Albert College, Belleville, Ontario, 1868–1879. Subsequently known as "the Professor." Botanist, Sandford Fleming Expedition, 1872. Botanist, Selwyn Expedition, 1875. Explorer for Canadian government in Northwest Territories, 1879–1881. Dominion botanist, Geological Survey of Canada (GSC), 1882–1887. Dominion naturalist, GSC, 1887–1912. Chief, Biological Division, GSC, 1912–1917.

CAREER

Plant geographer who made five major exploratory surveys of western Canada between 1872 and 1881. Used botanical expertise to refute idea of interior desert and suggest that western conditions were ideal for large-scale agricultural settlement. Appointed dominion botanist, GSC, in 1882 in recognition of work in western Canada. Engaged in nationwide survey of range and distribution of Canadian flora. Collections formed basis of national herbarium. Promoted to dominion naturalist and assistant director, GSC, in 1887. Expanded efforts to include Canadian fauna. Named first chief of Biological Division, GSC, in 1912. Semiretired to Vancouver Island in 1913 and continued to collect until his death.

Official reports of Western Surveys contained in government publications. His ten years' work in western Canada brought together in *Manitoba and the Great North-West* (1882). Published results of his GSC fieldwork in seven-part

Catalogue of Canadian Plants (*Polypetalae*, 1883, *Gamopetale*, 1884, *Apetalae*, 1886, *Endogens*, 1888, *Acrogens*, 1890, *Musci*, 1892, *Lichens and Hepaticae*, 1902), three-part *Catalogue of Canadian Birds* (1901, 1902, 1904; revised 1909), and *The Forests of Canada and Their Distribution* (1896). *Catalogue of Canadian Mammals and Canadian Freshwater Fish* in preparation at time of his retirement. Started *Autobiography* (1922) in 1918 but died before finishing it. Completed by son, William T. Macoun.

His assessment of western lands catapulted him into national prominence but later criticized for sweeping generalizations. Regarded enumeration of Canada's flora and fauna as personal task and was willing to go to almost any length to see it carried out. Made extensive collections, but critical work performed by foreign specialists. Knew most Canadian plant forms by sight. Scorned so-called closet naturalists.

MAJOR CONTRIBUTIONS

Range of fieldwork and volume and diversity of collections. Discovered nearly 1,000 species new to science. Did much to get work of naturalists before Canadian public. Helped lay foundation for National Museum of Natural Sciences.

BIBLIOGRAPHY

Anderson, R. M. *Journal of Mammalogy* (February 1921) (obit.).

The Canadian Naturalist (September 1920) (obit.).

Macoun, J. *Autobiography of John Macoun* (1922).

Waiser, W. A. *The Field Naturalist: John Macoun, the Geological Survey and Natural Science* (1989).

Zaslow, M. *Reading the Rocks: The Story of the Geological Survey of Canada 1842–1972* (1975).

UNPUBLISHED SOURCES

See John Macoun letter books, National Museums of Canada Library; correspondence on Botany Division and Vertebrate Zoology Division, National Museum of Natural Sciences; John Macoun Papers, Public Archives of Canada.

William A. Waiser

MACOUN, JAMES MELVILLE. Born Belleville, Canada West (Ontario), November 1862; died Ottawa, Ontario, 8 January 1920. Botanist, naturalist.

FAMILY AND EDUCATION

Son of John Macoun, dominion botanist and survey naturalist, Geological Survey of Canada. Attended Belleville High School and Albert College, where his father held chair in natural history and geology. Moved with family to Ottawa in 1882. Married Mary MacLennan and Helen Scott; two children.

POSITIONS

Temporary field assistant, Geological Survey of Canada (GSC), 1882–1897; assistant naturalist, GSC (permanent), 1897–1912; acting chief, Biological Division, GSC, 1912–1918; head, Biological Division, GSC, 1918–1920.

CAREER

Joined Geological Survey of Canada as temporary employee in 1882. Spent summers in field as naturalist to various survey parties, winters in Ottawa working up natural history field collections as father's assistant. Had particular responsibility for dominion herbarium.

In 1891, appointed secretary to G. M. Dawson, one of two British investigators of Behring fur seal dispute. Divided next two years between visit to seal rookeries on Pribiloff Islands and work on British case before 1893 Paris Behring Sea Arbitration Tribunal. Appointed assistant naturalist, GSC (permanent staff) in 1898 on basis of fur seal work. Headed own biological survey parties, mostly in western Canada. Negative 1903 report on potential of Peace River country nearly cost him his position. Collected on northwest coast of Hudson Bay in 1910 and was shipwrecked. Served as expert witness to British delegation to North Pacific Fur Seal Convention in Washington, D.C., in 1911. Awarded Companion of St. Michael and St. George (CMG) for this work.

Succeeded father as acting head of biological division, GSC, in 1912. Continued to collect in British Columbia. Named chief, biological division in 1918.

Quiet, unassuming naturalist who was extremely popular with his colleagues. Strong supporter of British socialism and advocate of patronage-free civil service. Tireless field-worker but published little. Coauthored with his father *Catalogue of Canadian Birds* in 1909.

MAJOR CONTRIBUTIONS

Bridged gap between all-around generalist and biological specialist. Helped lay foundation of National Museum of Canada. Used biological expertise to help devise resource/conservation policies.

BIBLIOGRAPHY

Macoun, J. *Autobiography of John Macoun* (1922).

Waiser, W. A. *The Field Naturalist: John Macoun, The Geological Survey and Natural Science* (1989).

Waiser, W. A. ''A Bear Garden: James Melville Macoun and the 1904 Peace River Controversy.'' *Canadian Historical Review* (March 1986).

UNPUBLISHED SOURCES

See Macoun Letter books, National Museum of Canada, Ottawa.

William A. Waiser

MANNING, ERNEST CALLAWAY. Born Selwyn, Ontario, 17 April 1890; died Armstrong, Ontario, 6 February 1941. Forester, park promoter, conservationist.

FAMILY AND EDUCATION

Son of Wellington C. and Helen Brown Manning. Born on a farm, the youngest of three sons. His mother moved the boys to Toronto to assure their access to higher education. Ernest attended Jarvis Street Collegiate and University of Toronto, completing a B.Sc. of forestry, 1912. Married Loys Pettit in Calgary in 1916; three daughters and one son.

POSITIONS

Forester/timber cruiser, Canadian Pacific Railway's Department of Natural Resources, Calgary, 1912–1915. Forester, Dominion Forestry Branch, Department of Interior, Calgary, April 1915–1918. Promoted to assistant inspector of forest reserves. British Columbia Provincial Forest Service, May 1918–1941. District forester, Prince Rupert, 1920–1923. Forest Service Management Office, Victoria, 1923–1927. Assistant chief forester for B.C., Victoria, 1927–1935. Chief forester for the province, announced 27 December 1935. By 1939 duties included head of B.C. parks system. H. R. MacMillan had him appointed wartime assistant controller of timber for British Columbia (federal duties, as of June 1940).

CAREER

Undergraduate summer work with the Pennsylvania Railroad Co. in Allegheny Mountains and Dominion Forestry Branch, Western Provinces. Hired on graduation by Canadian Pacific Railway to work on B.C. forests, Crow's Nest Pass region. Dominion work included reserves in Alberta and in B.C.'s Railway Belt (federally administered provincial lands). Joined B.C. Forest Service 1918. Rose steadily through a small, underfunded department to the top post. Well informed on work by U.S. and central Canadian counterparts. Supported in provincial cabinet by Wells Gray, but forests were his only politics. M. A. Grainger called him a ''New Deal'' forester, and as chief forester he began to propound his views on forestry. In 1935 oversaw the Young Men's Forestry Training Plan, a popular relief program that put in place a network of small parks, including Englishman River, Thetis Lake, Capilano Canyon. Undertook public education through media coverage of his 1936 annual report to the legislature's Forestry Committee. Took public speaking training, advised other foresters to do the same. Proposed reforms: sustained yield planning, natural regeneration through preserving seed trees, increased fire protection budgets, regulations against slash piles, forest practices jurisdiction over private land, and park creation. Public support grew after Campbell River fire, which involved slash piles (1938). Premier T. Dufferin Pattullo accused him of "trying to increase appropriations through propaganda." Established large parks: Tweedsmuir (1937), Wells Gray (1939). In 1939 parks were brought, as he had requested, under the Forest Service. A nonskier, he promoted ski development for Vancouver's North Shore, including Grouse Mountain (1939). This early conservation movement was deflected by World War II demands for timber. Active in planning timber supply for war needs, he died in a plane crash in 1941, while returning from Ottawa to address the Canadian Association of Forest Engineers on sustained yield forestry. Ernest C. Manning Park, on the Hope–Princeton Highway, named in his honor (1941).

MAJOR CONTRIBUTIONS

Called "the father of B.C. conservation." Progressive forester who set out to change the views of government, industry, his profession, and the public. Stated foresters had an ethical responsibility to educate the public on forest policy. Promoted sustained yield forestry and multiuse forests. Developed provincial parks system.

BIBLIOGRAPHY

Akrigg, H. M. "Manning of Manning Park." *B.C. Historical News* 24, 4 (Fall 1991).

Forest Chronicle 17 (March 1941) (obit.).

Young, William A. "E. C. Manning, 1890–1941: His Views and Influences on British Columbia Forestry." B.Sc. thesis in forestry, University of British Columbia, 1982.

UNPUBLISHED SOURCES

See various records of the B.C. Forest Service, especially GR 1242 (Reports) and GR 1275 (Speeches), and Provincial Archives of British Columbia. Also useful is the B.C. Legislative Library Newspaper Index, 1900–1971, which contains some forty-seven references.

Lorne Hammond

MARIE-VICTORIN, (CONRAD KIROUAC). Born Kingsey Falls, Quebec, 3 April 1885; died Saint-Hyacinthe, Quebec, 15 July 1944. Botanist, naturalist, teacher, Christian brother.

FAMILY AND EDUCATION

Descendant of French nobility, the marquises de Keroack of Brittany. Son of an Eastern Townships merchant, Cyrille, and Philomène Luneau Kirouac. Love of flora and rural life shaped by summers in l'Ancienne Lorette and Saint-Norbert de Arthabaska. Catholic education: primary, St-Sauveur de Québec; secondary, l'Académie Commerciale, college at Mont-de-Salle, Montréal, graduated 1903; Université de Montréal, D.Sc., 1922. Joined religious order: noviate (1901), then reverend brother, Order of the Brothers of the Christian Schools.

POSITIONS

Professor, Collège de Saint-Jérôme (1903). Professor, Collège de Longueuil and at St-Léon de Westmount (1904–1920). Professor agrégé de botanique (University of Montréal, 1920–1944). Founded Institut botanique, Université de Montréal (1920). President and secretary, Société canadienne d'Histoire naturelle (SCHN). Member, Biological Board of Canada (1927–1930). Founder and secretary of the Association canadienne-français pour l'avancement des sciences (ACFAS) in 1923. General secretary, ACFAS (1928). Elected Royal Society of Canada (1923–1924), later, président, Section V (1933). Founder, director, Jardin botanique de Montréal. Canadian secretary, botany, American Society for the Advancement of Science (1938). Member, International Commission on Nomenclature (1939).

CAREER

Love of countryside combined with Catholic rural ideology (*terroir* or homeland school) to shape his popular literary sketches, which mixed history, folklore, and moral sermons: *Croquis Laurentiens* (1920) and *Récits Laurentiens* (1919), translated as *The Chopping Bee and Other Laurentian Stories* (1925). Public educational work included the *Cercles des Jeunes naturalistes* and media broadcasts, such as the "*Cité des Plantes*" series (Radio-Collège 1941, Radio-Canada 1943). Major figure in Quebec's intellectual and political culture of the 1930s.

Organized the Institut botanique at the Université de Montréal (1920). Worked with Rolland Germain, Jules Brunel, and Jacques Rousseau. Began series *Contributions de l'Institut Botanique de l'Université de Montréal.* Received doctorate and chair in botany (1922). Thesis *Les filicinées de Québec* published (1923). Undertook systematic ecologic studies of the flora of Quebec: *Études floristiques sur la région du Lac Saint-Jean* (1925), *Le dynamisme dans la flore de Québec* (1929), culminating in 1935 taxonomic masterwork, *Flore Laurentienne* (repr. 1947, revised by Ernest Rouleau and reissued in 1964). Volume received Prix Coincy de l'Académie des Sciences de Paris.

Founded the ACFAS with Léo Pariseau (1923). Internationally active in botany through late 1920s and 1930s. Canadian delegate to 1929 Capetown meeting of British Association for the Advancement of Sciences. Lectured on Gray herbarium at Harvard University, same year. Founded association in 1930 to lobby for creation of Montreal Botanical Gardens, construction begun 7 May 1935. Director of same, assisted by Jacques Rousseau. In 1939 Institut botanique moved to garden grounds. Published *Histoire de l'institut botanique de l'Université de Montréal* (1941). Began work on Caribbean botany with trip to Haiti and three-month botanizing in Cuba at the Colegio de la Salle in 1938. Published *Itinéraires botaniques dans l'Ile de Cuba* (1942) with support of Harvard University. Second botanical expedition to Cuba in 1942. Second Cuban series published (1944). Received Cuba's Order of Merit on third trip (1944). Died in automobile accident on return from collecting expedition at Black Lake, Quebec. Buried Mont-de-la-Salle, at Laval-des-rapides, Quebec. Monument unveiled by Maurice Duplessis in the Botanical Gardens (1954).

MAJOR CONTRIBUTIONS

His bioregional and phytogeographic approach to Quebec's flora prepared the way for ecologists such as Pierre Dansereau, according to Raymond Duchesne. Assembled major herbarium. As teacher trained a generation of botanists. Formalized international links for botanical research in Quebec. Established Montreal botanical gardens. Literary figure in clericonationalist movement.

BIBLIOGRAPHY

Adams, Frank Dawson. *A History of Science in Canada* (1939).

d'Aragon, Jacques. *Marie-Victorin, 1885–1985* (1985) [video].

Audet, Louis Philippe *Le frère Marie-Victorin* (1942).

Audet, Louis Philippe. "Le frère Marie-Victorin: maître littérateur." *Culture* 6, 1 (March 1945).

Brunel, Jules. "Le Frère Marie-Victorin, 1885–1944." *Revue Canadienne de biologie* 3 (November 1944).

Canadian Who's Who. Vol. 3 (1938–1939).

Chartrand, Luc, Raymond Duchesne, and Yves Gingras. *Histoire des sciences au Québec* (1987).

Gauvreau, Marcelle. *Le président de l'ACFAS* (1938).

Gingras, Yves. *Pour avancement des sciences: histoire de l'ACFAS, 1923–1993* (1994).

Marie-Victorin. *Confidence et combat: lettres (1924–1944) frère Marie-Victorin, é.c.* Ed. Gilles Beaudet (1969).

Marie-Victorin. *Pour l'amour de Québec.* Ed. Hermas Bastien (1971).

Rousseau, Jacques. "L'oeuvre du Frère Marie-Victorin." *Culture* 5, 3 (September 1944).

Rumilly, Robert. *Le frère Marie-Victorin et son temps* (1949).

UNPUBLISHED SOURCES

Unpublished works include "*Voyage à travers trois continents.*" See extensive collection of herbarium and correspondence preserved by his secretary: Marcelle Gauvreau (1907–1968), at the Université du Québec à Montréal, Montreal, Quebec.

Lorne Hammond

MARSH, GEORGE PERKINS. Born Woodstock, Vermont 15 March 1801; died Vallombrosa, Italy, 23 July 1882. Physical geographer, environmental studies, linguist.

FAMILY AND EDUCATION

Son of Charles Marsh, successful Vermont lawyer, and Susan Perkins Marsh. Poor eyesight hampered early education, but he graduated from Dartmouth with highest honors in 1820. Studied law with his father. Admitted to bar in 1825.

Married Harriet Buell, 1828; two sons. Following death of first wife, married Caroline Crone, 1839; no children.

POSITIONS

Practiced law in Burlington, Vermont, 1825–1842; served in U.S. Congress, 1843–1849; U.S. minister to Turkey, 1849–1854, 1856. Lectured at Columbia University on English Language 1860–1861; lectured at Lowell Institute. U.S. minister to Italy, 1861–1882.

CAREER

In the midst of a very busy political and diplomatic career, Marsh found time for scholarly pursuits. As a linguist he mastered twenty languages. While serving as a Whig Congressman, Marsh became a strong advocate for the establishment of the Smithsonian Institution and became one of its regents in 1847. Opponent of slavery and the Mexican War. Early recognized the danger resulting from combination of soil erosion, stream erosion, and widespread cutting of timber. Speaking in 1847, he stated that "The changes, which these changes have wrought in the physical geography of Vermont, within a single generation, are too striking to have escaped the attention of any observing person. The signs of artificial improvement are mingled with the tokens of improvident waste." Took a very definite practical approach in his conservation writings. His growing enthusiasm for conservation grew out of an early interest in geography. Was much impressed by principles of forest conservation as practiced in Europe. Saw relationships between man and his environment as the basis of geographic study. Felt that he could make a useful contribution to the early debate on conservation issues, and was uninterested in scholarly fame. While minister to Turkey, he collected specimens and antiquities for the Smithsonian. Facilitated Louis Kossuth's release from a Turkish prison and his visit to America, 1851. In 1857 prepared a detailed report on artificial propagation of fish for Vermont.

Pursued studies of history, art, and languages while minister to Italy and began work on his monumental *Man and Nature or Physical Geography as Modified by Human Action* (1864; revised ed. entitled *The Earth as Modified by Human Action* [1874]). Stated that his objective was

> to indicate the character and, approximately, the extent of changes produced by human action in the physical conditions of the globe we inhabit; to point out the dangers of imprudence and the necessity of caution in all operations which, on a large scale, interfere with the spontaneous arrangements of the organic and of the inorganic worlds; to suggest the possibility and the importance of the restoration of disturbed harmonies and the material improvement of wasted and exhausted regions; and, incidentally, to illustrate the doctrine that man is, in both kind and degree, a power of a higher order than any of

the other forms of animated life, which, like him, are nourished at the table of bounteous nature.

Final edition published as *The Earth as Modified By Human Action: A Last Revision of Man and Nature* (1885). This book has been termed ''fountainhead of the conservation movement.'' Marsh was a talented author who published in a variety of areas: *A Compendious Grammar of the Old-Northern or Icelandic Language* (1838), *The Camel, His Organization, Habits, and Uses, Considered with Reference to His Introduction into the United States* (1856), *Lectures on the English Language* (1860), and *The Origin and History of the English Language* (1862). Spent his final two decades as the first American Minister to Italy, interspersing diplomatic duties with extensive travels in Europe and lands of the eastern Mediterranean region.

MAJOR CONTRIBUTIONS

Frequent travel in America, Europe, and the Near East encouraged him to examine man's impact upon the earth and the methods by which he altered the environment through land abuse. *Man and Nature* became the first major monograph to alert society to the need to develop wise stewardship for the earth's resources. This popular work became an important source of information and inspiration for the conservation movement.

BIBLIOGRAPHY

Brooks, Paul. *Speaking for Nature* (1980).

Davis, William M., ''George Perkins Marsh,'' *Biographical Memoirs, National Academy of Sciences*, vol. 6 (1909).

Dictionary of American Biography.

Elliott, Clark A. *Biographical Dictionary of American Science* (1979).

Koopman, H. L. *Bibliography of George Perkins Marsh* (1892).

Larkin, Robert P., and Gary L. Peters. *Biographical Dictionary of Geography* (1993).

Lowenthal, David. ''George Perkins Marsh and the American Geographical Tradition.'' *Geographical Review* 43 (1953).

Lowenthal, David. *George Perkins Marsh: Versatile Vermonter* (1958).

Payne, Daniel G. *Voices in the Wilderness: American Nature Writing and Environmental Politics* (1996).

UNPUBLISHED SOURCES

Manuscript materials are in Marsh Papers at University of Vermont. See also Miner K. Kellogg Papers, Archives of American Art; Lucius E. Chittenden Papers, University of Vermont; Crane Family Papers, New York Public Library.

Phillip D. Thomas

MARSH, OTHNIEL CHARLES. Born Lockport, New York, 29 October 1831; died New Haven, Connecticut, 18 March 1899. Paleontologist.

FAMILY AND EDUCATION

Son of Caleb and Mary Gaines Peabody Marsh. Valedictorian at Phillips Academy, Andover, Massachusetts, 1856; attended Yale College, 1856–1860, and Yale's Sheffield Scientific School, 1860–1862. Studied at universities in Berlin, Breslau, and Heidelberg, 1862–1865, as well as the fossil collections of several European museums. Bachelor. Nephew of George Peabody, nineteenth-century entrepreneur and financier.

POSITIONS

First in United States to be appointed professor of paleontology, Yale, 1866–1899. Curator of Yale's geological collections, 1874–1899. Vertebrate paleontologist with U.S. Geological Survey (USGS), 1882–1899, although fieldwork terminated in 1892. Honorary curator, Section of Vertebrate Fossils, Department of Geology, U.S. National Museum (USNM), Smithsonian Institution, 1888–1899. President of National Academy of Sciences, 1883–1895.

CAREER

Became interested in trilobites and brachiopods found in construction area of Erie Canal near his home. Born into poor agricultural family, academic and European studies financed by uncle. While in Europe, convinced Peabody to donate money to Yale for natural history museum (Peabody Museum of Natural History). Later inheritance from Peabody enabled Marsh to work at Yale without accepting salary for many years.

Made first trip to fossil-rich western United States in 1868. Subsequently, conducted four summer expeditions, 1870–1873, with Yale students and alumni. Trip in 1874 consisted of trained collectors and military personnel. Collecting strategy changed after railroad reached most of the West; had local residents collect fossils and ship them to New Haven. Collected primarily in Colorado, Kansas, the Dakotas, Utah, and Wyoming.

A major discovery was extinct bird that retained reptilian teeth; bridged gap between reptiles and birds that followed reptiles in evolutionary process. Results published in *Odontornithes: A Monograph on the Extinct Toothed Birds of North America* (1880). Alexander Wetmore said this discovery from Cretaceous fields of Kansas "though made at so early a day still ranks as one of the outstanding discoveries in palaeornithology in North America." Also traced evolutionary changes from dog-sized horse to modern horse, *Equus*, of Pleistocene. Thomas Huxley said Marsh confirmed that the horse developed primarily in North America, while Charles Darwin said Marsh's fossil horse evidence was best support of evolutionary theory since Darwin's treatise in 1859. Also studied brain development in vertebrates, which showed a gradual increase in size with the "passage of geologic time."

Discovered first pterodactyl (flying dragon) in United States based on joint of wing finger but never produced a major work on this reptile.

Best known for large collection of dinosaurs. Discovery of these reptiles led Marsh away from fossil mammals on which he intended to spend most of his time. Described eighty new forms and thirty-four new genera as well as many reconstructions. Included in his collections were some of the giants of the Mesozoic and Cenozoic such as the *Brontosaurus*. One of his major works is *Dinosaurs of North America* (1896).

Involved in long-standing rivalry with Edward Drinker Cope. Vied with him in obtaining fossils, although Marsh's extensive network of collectors and personal financial circumstances enabled him to outbid Cope on most occasions. Antagonism between them detailed in series of articles in January 1890 issues of *New York Herald*. Cope requested by USGS to turn over fossils from several federal surveys to USNM, where Marsh worked. Cope accused Marsh of sending fossils from government-financed expeditions to Yale and USGS of stalling on one of Cope's papers, even though Congress authorized money for its publication. Senate Appropriations Committee later refused to fund more paleontological research under USGS. Marsh forced to resign and accept salary from Yale. Cope attempted, but failed, to deny Marsh's reelection as National Academy of Sciences president in 1889. Rivalry did not advance science but rather produced many hasty descriptions of fossils.

For distinguished research in geology and paleontology received Bigsby Medal, 1877, from the Geological Society of London, and the Cuvier Prize, 1898, from the French Academy of Sciences.

Marked by a degree of self-centeredness and fear of competition, many publications and specimens lack locality information. Many conclusions lost because he was unwilling to discuss findings with others.

MAJOR CONTRIBUTIONS

Probably did more than any other person to introduce paleontology into academic curricula. With available resources in family wealth and USGS funds, amassed fossil collections that exceeded those of other collectors and are a monument to him. One of the leading paleontologists of the nineteenth century. Contributed to the systematics of reptiles, birds, and mammals.

BIBLIOGRAPHY

Colbert, Edwin H. *Men and Dinosaurs* (1968).

Jaffe, Bernard. *Men of Science in America: The Role of Science in the Growth of Our Century* (1944).

Ostrom, John H. "Othniel Charles Marsh." *Encyclopedia of American Biography* (1974).

Romer, Alfred S. "Cope *versus* Marsh." *Systematic Zoology* 13 (1964).

Schuchert, Charles. "Othniel Charles Marsh, 1831–1899." *Biographical Memoirs National Academy of Science,* Vol. 20 (1939).

Schuchert, Charles, and Clara M. LeVene. O. C. Marsh: Pioneer in Paleontology (1940).

Shor, Elizabeth Noble. *The Fossil Feud* (1974).

UNPUBLISHED SOURCES

See Othniel Charles Marsh Papers, 1817–1940, Yale University.

William R. Massa, Jr.

MARSHALL, ROBERT. Born New York, 2 January 1901; died en route to New York from Washington, D.C., 11 November 1939. Forester, explorer.

FAMILY AND EDUCATION

Son of Louis Marshall, prominent constitutional lawyer, civil libertarian, and conservationist, and Florence Lowenstein Marshall. Attended Ethical Cultural School from third grade through high school. Spent childhood summers exploring the Adirondack Mountains of northern New York State. Graduated from New York State College of Forestry at Syracuse University in 1924. Received master of forestry degree from Harvard University in 1925; Ph.D. from Johns Hopkins Laboratory of Plant Physiology in 1930.

POSITIONS

Staff member, Northern Rocky Mountain Forest Experiment Station, Missoula, Montana, 1925–1928. Director of forestry, Bureau of Indian Affairs, 1933–1937. Chief, Division of Recreation and Lands, U.S. Forest Service, 1937–1939.

CAREER

Explored Central Brooks Range of Alaska on four different trips between 1929 and 1939. Made first map of a 12,000–square mile section of the region and named over 100 of its features. In 1933, published *Arctic Village*, a study of life and mores in and around the frontier town of Wiseman, Alaska. The book became a best-seller. Also in 1933, contributed chapters on wilderness and recreation to *A National Plan for American Forestry* and published *The People's Forests*, which analyzed destructive forestry practices by private owners and advocated public acquisition and management of American forestlands.

A tireless and very influential crusader for wilderness preservation during 1930s. Cofounded the Wilderness Society in 1935. Was responsible for getting 4.5 million acres of Indian lands set aside as "roadless" or "wild" in 1937. His pressure within U.S. Forest Service contributed directly to the expansion of the service's primitive area system during the 1930s and to the issuance of more stringent wilderness area and wild area regulations by the chief forester in 1939.

Wrote numerous articles defending the need for wilderness preservation, aesthetic considerations in public land policy development, and stricter federal control over forest resource management. Exuberant and energetic. Considered eccentric by some foresters for his liberal views, but well respected by others who shared his views toward wilderness preservation and increased federal intervention in forestry. Articulated his ideas effectively. Also became well known for his extremely long and arduous treks into America's backcountry.

MAJOR CONTRIBUTIONS

Direct influence on federal wilderness policy, role in forming Wilderness Society, articulator of the idea that wilderness is a valuable resource.

BIBLIOGRAPHY
Gilligan, James P. "The Development of Policy and Administration of Forest Service Primitive and Wilderness Areas in the United States." Diss., University of Michigan, 1953.
Glover, James M. *A Wilderness Original: The Life of Bob Marshall* (1986).
Hott, Lawrence, and Diane Garbey, with Ken Chowder et al. *Wild by Law [Aldo Leopold, Bob Marshall, Harold Zahniser, and the Redefinition of American Progress].* Videocassette, Florentine Films, Santa Monica, Calif. Direct Cinema, (1991).
The Living Wilderness, Robert Marshall Memorial Issue (July 1940).
Marshall, George. "Robert Marshall as a Writer." *The Living Wilderness* (Autumn 1951): 14–19; supplement in *The Living Wilderness* (Summer 1954): 33–35.
Nash, Roderick. *Wilderness and the American Mind.* New Haven, Conn.: Yale University Press (1967).

Jim Glover

MASSON, FRANCIS. Born Aberdeen, Scotland, August 1741; died Montreal, lower Canada, 23 December 1805. Botanist.

FAMILY AND EDUCATION

Nothing is known of Masson's background. He likely learned the basics of horticulture as an apprentice gardener. He does not seem to have married.

POSITIONS

In 1771, Masson was appointed an undergardener at the Royal Botanic Gardens, a position he held until at least 1795. From 1797 until his death, Masson worked in Canada at the request of Joseph Banks and at government expense.

CAREER

Masson impressed William Aiton, the director at Kew, who named him the garden's first official collector. He traveled with Capt. James Cook to the Cape of Good Hope in 1772. Between 1772 and 1774, Masson explored the interior of South Africa. Between that time and 1785, he made a number of collecting trips to the Azores, Canaries, Madeira, the West Indies, Portugal, Spain, Tangier, and Morocco. For ten years (1785–1795), Masson worked in South Africa, based at the Cape. He specialized in orchids and published a book on a species of *Stapelia nova*. Some of his travel accounts appeared in the *Philosophical Transactions* of the Royal Society. Masson corresponded with Linnaeus, who named a species of asphodel after him.

From 1797 to the end of his life, Masson botanized in North America. This change from semitropical to temperate botany was encouraged by Masson's friend, Joseph Banks, then president of the Royal Society, who wanted specimens from upper Canada. Masson arrived at Newark (now Niagara-on-the-Lake) in July 1798 after traveling via New York. The season's collecting concentrated upon the Niagara area and Toronto; he wintered in Montreal, from which he sent Banks seeds and specimens, including wild rice. During 1799, thanks to directors of the North West Company, Masson was able to accompany a fur-trading party up the Ottawa River and through the upper lakes to Grand Portage (now Minnesota) and back along the north shore of Lake Ontario through Kingston. By November, he was able to dispatch to Banks living plants as well as seeds.

Masson's explorations during 1800 were limited to lower Canada; he may have traveled as far as Virginia during 1801, although a return trip to the upper lake seems more likely. There is no record of his activities between 1801 and early 1805. He was still in Montreal in May 1805, when he intended to return to England. He postponed the trip because of French naval activity and botanized along the lower St. Lawrence, shipping home tree specimens that fall. Already ill during the summer, he died at the home of a friend in Montreal in December and was buried there. Canadians spoke of him as very scientific, kindly, and unassuming.

MAJOR CONTRIBUTIONS

Masson was well respected by contemporary botanists. His Canadian explorations do not appear to have been systematic, and his most important contribution to North American botany was to bring various species—such as the trillium—to the attention of British horticulturists.

BIBLIOGRAPHY

''Francis Masson.'' *Dictionary of National Biography* (U.K.).

Jarrell, Richard A. ''Francis Masson.'' *Dictionary of Canadian Biography*. Vol. 5 (1983).

UNPUBLISHED SOURCES

See Banks Papers, Brabourne Collection, State Library of New South Wales (Mitchell Library), Sydney, Australia.

Richard A. Jarrell

MATHER, STEPHEN TYNG. Born San Francisco, 4 July 1867; died Brookline, Massachusetts, 22 January 1930. Businessman; director, National Park Service.

FAMILY AND EDUCATION

Eldest of two sons of Joseph Wakeman Mather, commission merchant and later prominent borax miner, and Bertha Jemima Walker Mather. Collateral descendant of Increase and Cotton Mather, colonial Massachusetts Puritan divines. Graduate of Boy's High School of San Francisco. Bachelor of Letters, University of California, Berkeley, 1887; LL.D., George Washington University, 1921;

LL.D., University of California Berkeley, 1924. Married Jane Thacker Floy, 1893; one daughter.

POSITIONS

Reporter for *New York Sun*, 1888–1893. Clerk in Pacific Borax Company's New York Office, 1893–1894, devised "20 Mule Team Borax" brand name. National Advertising and Sales Promotion Manager, Pacific Borax, 1894–1903. Silent partner in Thomas Thorkildsen & Co., 1903–1911; President, Brighton Chemical Co. of Pennsylvania, from 1908. Vice-President, Sterling Borax, 1911–1920, President, 1920–1927, President and sole owner, 1927–1930. Assistant to the Secretary of Interior, 1915–1917. Instrumental in creation of National Park Service, 1916. First director, National Park Service, 1917–1929.

CAREER

Mather parlayed great skill in management, advertising, and public relations into early success in borax manufacturing. Spent much of the year 1903 recovering from nervous breakdown brought on by business pressures. From about 1905, established permanent residence in Chicago and was involved in civic betterment there. Staunch progressive-conservationist supporter of Theodore Roosevelt in 1912. Enjoyed a nationwide reputation as wealthy businessman when he accepted invitation to help create a national park system in 1915. Began work as $2,750 a year assistant to the Secretary of the Interior. With support of Secretary of the Interior Franklin K. Lane and his successors, Mather established foundation of present National Park System (NPS) during his twelve year tenure as director. In the early days, he sometimes spent money out of his own pocket for road improvements, supplies, and other necessities, and as a private citizen, contributed generously to conservation projects. Increased number of national parks from thirteen to twenty-one and national monuments from eighteen to thirty-three. Parks established during his tenure included Rocky Mountain, Mount McKinley, Grand Canyon, Grand Teton, and Great Smoky Mountain. Because Congress did not officially define distinguishing criteria for national parks, Mather set about getting his standards accepted. He insisted that only those areas still in an unaltered natural state, with stellar scenic importance, receive national park status. Each area so designated was forever protected from industrial and commercial development absent specific Congressional authority. Rejected candidates for National Park status that were of secondary importance, stimulated public understanding of the national park ideal, and established high standards for NPS personnel, for which he received worldwide acclaim. Enlisted the aid of first-class associates, some of whom remained with the NPS for half a century. On his retirement, Congressional resolution stated "there will never come an end to the good he has done." Congress also noted that Mather "sacrificed his money, his health, his time, his opportunities for wealth, in order that he might promote that which will mean so much to the people of this country in the future.'' Named for him were an Alaskan peak; a highway in the Cascades; a memorial arboretum at the University of California, Berkeley; a forest near Lake George, New York; and a scenic gorge on the Potomac River below the Great Falls. In addition, memorial plaques were erected at each of the then-existing national parks and monuments in the early 1930s. Active member of the Sierra Club, Save the Redwoods League, and of other conservation organizations, which helped him to achieve many of his objectives. Was obliged to retire as NPS director for reasons of health, and never fully recovered.

MAJOR CONTRIBUTIONS

Before Mather took over the direction of park service, each park worked independently and, in fact, did not possess any characteristics to distinguish them from state or municipal parks. Mather gave national parks an identity, brought them under the same umbrella, removing the designation of those that did not merit the distinction, and setting up lines of communications between those so designated. His setting and maintaining of standards for national parks ensured their continued uniqueness and protection. Nature lover and staunch defender of wildlife, he nevertheless viewed the parks in terms of their enjoyment by humans. Thus he looked upon maintaining the parks as his gift to his fellow citizens.

BIBLIOGRAPHY

Dictionary of American Biography. Vol. 7.

National Park Service. *Annual Reports*. (1916–1930).

Nature (March 1930).

New York Times 23–24 January 1930 (obit.).

Science (26 February 1932).

Shankland, Robert. *Steve Mather of the National Parks*, 3rd ed. (1970).

UNPUBLISHED SOURCES

See NPS Records, National Archives, Washington D.C., Hal E. Evarts Papers, University of Oregon Library; Louis C. Cramton Papers, University of Michigan Historical Collections.

Rodney Marve

MAXIMILIAN, ALEXANDER PHILIP. Prince of Wied-Neuwied. Born Neuwied, capital of the German principality of Wied, 23 September 1782; died Neuwied, 3 February 1867. Naturalist, ethnologist.

FAMILY AND EDUCATION

Eighth child of the reigning Friedrich Karl. Private tutors. Pursued natural history at Georgia-Augusta University in Göttingen, interrupted by Napoleonic Wars in 1806. Studied under Johann Friedrich Blumenbach (1811–1812). Never married.

POSITIONS

Major general, won the honor of Iron Cross, and entered Paris with victorious Allied army against Napoleon (1814). Naturalist, scientific expedition to South America, 1815–1817. Naturalist, head of scientific expedition into the interior of North America, 1832–1834.

CAREER

Together with two German scholars, spent two years in tropical forests of Brazil studying flora, fauna, and native races. After return to Germany published account of journey. *Reise nach Brasilien in den Jahren 1815 bis 1817* (Frankfurt, 1820–1821) and *Beiträge zur Naturalgeschichte von Brasilien* (Weimar, 1825–1833). In 1831, organized and headed his second scientific expedition to New World. Party included Karl Bodmer, Swiss artist, and David Dreidoppel, an accomplished taxidermist and hunter who had accompanied him to Brazil. The two-year scientific investigation, 1832–1834, one of the most important explorations ever made of the upper Missouri region, occurred on the eve of permanent settlement of region.

Prince Maximilian and his party arrived in Boston on 4 July 1832. They traveled to New York, then to Philadelphia, and reached Bethlehem, Pennsylvania, where they remained for six weeks. On 17 September, they proceeded to New Harmony, Indiana, where the party stayed for five months engaging in scientific studies with Thomas Say and Charles-Alexandre Lesueur, from 19 October 1832 to 16 March 1833. The party then journeyed down Ohio River and up Mississippi to St. Louis; on 10 April they traveled up the Missouri River.

On 18 June they reached Ft. Union (Williston, North Dakota), and on the 24th went on a keel boat to Ft. McKenzie on Maria's River (Great Falls, Montana) where they remained two months. On return, party spent the winter at Ft. Clark (Bismark, North Dakota) among the Mandan Indians. They returned to St. Louis in May 1834 and retraced their route to New Harmony, where they visited 9–13 June. They traveled back to New York via the Ohio River, the Ohio and Erie Canals, and the Hudson River and sailed for Europe on 16 July 1834.

Published account of journey *Reise in das Innere Nord-Amerika in den Jahren 1832 bis 1834*, 2 vols. (Coblenz, 1840, 1841), French edition, 3 vols. (Paris, 1840, 1841, 1843), English edition, 1 vol. (London, 1843). The same *Atlas* of eighty-one prints was published with all of the editions.

The prince made no further foreign explorations but lived on his ancestral estate, working with his collections and correspondence until his death at eighty-five in 1867. Published two additional reports, one on mammals of his expedition, *Verzeichniss der auf seiner Reise in Nord-Amerika beobachteten Säugethiere* (Berlin, 1862, four plates) and one on the reptiles of his expedition,

Verzeichniss der Reptilien welche auf einer Reise im Nördlichen America beobachtet wurden (Dresden, 1865, seven plates).

MAJOR CONTRIBUTIONS

The expedition up the Missouri River was timely; permanent settlement disrupted nomadic life and led to dispersal of tribes. In the case of Mandan Indians, the people most comprehensively documented by Maximilian and Bodmer, the timing was uncanny. The Mandans were almost obliterated by smallpox within few years.

While the major contribution of the work lies in its ethnological significance, it also provides highly important historical description of natural conditions west of the Mississippi and in Bethlehem, Pennsylvania, and New Harmony, Indiana, as well. Maximilian praises the work of Bodmer in the author's preface to *Travels*. Karl Bodmer "represented the Indian nations with great truth, and . . . his drawings will prove an important addition to our knowledge of this race of men."

BIBLIOGRAPHY

Davidson, Marshall B. "Carl Bodmer's Unspoiled West." *American Heritage* 14. 3 (April 1963).

Goetzmann, William H., and Joseph C. Porter. *The West as Romantic Horizon*. Omaha, Nebr.: Center for Western Studies, Joslyn Art Museum (1981).

Läng, Hans. *Indianer Waren Meine Freunde*. Bern: Hallwag Verlag (1976).

Thomas, Davis, and Karen Ronnefeldt. *People of the First Man*. New York: E. P. Dutton (1976).

Thwaites, Reuben. *Early Western Travels*. Vols. 22, 23, 24. New York: AMS Press (1966).

UNPUBLISHED SOURCES

See American Philosophical Society; APS Archives, Correspondence of American Scientists with Charles L. Bonaparte, 1824–1855—microfilm; Historic New Harmony, Inc.: Maximilian–Bodmer Collection; InterNorth Art Foundation: Center for Western Studies, Joslyn Art Museum (Omaha, Nebr.), Maximilian-Bodmer Collection, more than 400 of Bodmer's original sketches and paintings, diaries, journals, account books, correspondence.

Ralph G. Schwarz

MAXWELL, MARTHA ANN DARTT. Born Dartt's Settlement, Pennsylvania, 21 July 1831; died Brooklyn, New York, 31 May 1881. Naturalist, taxidermist.

FAMILY AND EDUCATION

Only child of Spencer and Amy Dartt. Father died when she was two, and mother remarried first husband's cousin Josiah Dartt, 1841. Received early education at home. Family moved many times during her youth; settled in Wisconsin, 1845. Attended Oberlin College, April 1851–October 1852; Lawrence University, Appleton, Wisconsin, 1853–1854; Women's Laboratory, Massachusetts Institute of Technology, fall, 1878. Married James A. Maxwell, 1854; one daughter.

POSITIONS

Taught school in Baraboo, Wisconsin, 1852–1853. Active as a social reformer there in late 1850s. Moved to Denver, 1860. Briefly co-owner of boardinghouse in Denver, 1860, and owner of another in Nevada City, Colorado, 1860–1861 (burned, late 1861); also involved in gold claims acquisition and continued social reform work. Returned to Baraboo, began work as taxidermist for Baraboo Collegiate Institute, 1862–1867. Moved to Boulder, Colorado, 1868, where she turned full-time to hunting and taxidermy. Sold some mountings, exhibited her collections at area fairs. Opened Rocky Mountain Museum, Boulder, 1874, moved it to Denver, 1875. Exhibited her work at Centennial Exhibition, Philadelphia, 1876; in Washington, D.C., 1877; again in Philadelphia, 1877–1879. Briefly managed restaurant in Philadelphia, 1879. Owned and operated bathhouse-museum, Rockaway, New York, 1880–1881.

CAREER

Interest in nature encouraged by grandmother in childhood. Struggled in early years to secure an education, raise her much older husband's half-dozen children from his first marriage, and establish her family in Wisconsin and later in Colorado. Also involved in social reform and temperance activities. Had longstanding interests in art and natural history, to which she could give little rein until her early thirties. Stimulated to begin her career as taxidermist by German claim jumper in Colorado who had prepared a collection of birds. Began this work in earnest when E. F. Hobart, teacher at Baraboo Collegiate Institute, needed an assistant. Following return to Colorado, 1868, devoted most of her time to collection of specimens, principally mammals and birds, mounting them and displaying them at local fairs. First woman to do this in United States. Conceived idea of forming her own museum, which she accomplished in 1874, but found it difficult to make it a going concern. Secured some objects for museum in California, 1873. Invited to display her mounted birds and mammals at Centennial Exposition in Philadelphia. Catalog of the forty-seven mammal and 234 bird specimens contained in her exhibit there completed by Robert Ridgway and Elliott Coues. This later incorporated as appendix to book on which she collaborated with half-sister Mary, *On the Plains, and among the Peaks; or, How Mrs. Maxwell Made Her Natural History Collection* (1878). Her work was praised by S. F. Baird, J. A. Allen, Ridgway, and Coues, but sales of book disappointing. Her mounted specimens continued on display for several years in Washington, D.C., and at permanent exposition site in Philadelphia.

Effectively separated from her husband, determined to remain in the East; built bathhouse combined with museum on Rockaway Beach, near Brooklyn, New York, but this enterprise was in debt at end of first season. Died of ovarian tumor at age forty-nine. Was first woman ornithologist to have new subspecies of bird she had identified (a burrowing owl) named for her.

MAJOR CONTRIBUTIONS

Was first woman field naturalist in United States and one of first American naturalists to focus on western mammals and birds. Expanded known ranges of several species of mammals in Colorado. Pioneering creator of habitat groups in museum exhibits. Through trial and error, worked out taxidermic concepts "similar in principle" to ones later utilized by William Hornaday, noted taxidermist at New York Zoological Society, and others. Robert Ridgway praised "her high attainments in the study of natural history," while J. A. Allen described her as "an ardent and thorough student of nature."

BIBLIOGRAPHY

Benson, Maxine. *Martha Maxwell, Rocky Mountain Naturalist* (1986).

DeLapp, Mary. "Pioneer Woman Naturalist." *The Colorado Quarterly* 13 (Summer 1964).

Ewan, Joseph, and Nesta Ewan. *Biographical Dictionary of Rocky Mountain Naturalists 1682–1932* (1981).

Henderson, Junius. "A Pioneer Venture in Habitat Grouping." *Proceedings of the American Association of Museums* 9 (1915).

Schantz, Viola S. "Mrs. M. A. Maxwell: A Pioneer Mammalogist." *Journal of Mammalogy* 24 (November 1943).

Webster, Frederick S. "The Birth of Habitat Bird Groups." *Annals of the Carnegie Museum* 30 (1945).

UNPUBLISHED SOURCES

See M. A. Maxwell Letters, Oberlin College Archives; Martha Dartt Maxwell Papers, Colorado Historical Society.

Keir B. Sterling

MCARDLE, RICHARD EDWIN. Born Lexington, Kentucky, 25 February 1899; died Washington, D.C., 4 October 1983. Forester.

FAMILY AND EDUCATION

Son of Maurice Herbert and Mildred Yarbrough Johnson McArdle. B.S., 1923, M.S., 1924, Ph.D., 1930, University of Michigan; honorary Sc.D., University of Michigan, 1953, University of Maine, 1962; honorary LL.D., Syracuse University, 1961. Married Dorothy Coppage, 1927; three children.

POSITIONS

Junior forester, U.S. Forest Service (USFS), Portland, Oregon, 1924–1927; assistant silviculturist, USFS, 1926–1930 (on leave for doctoral study, 1927–1930), associate silviculturist, USFS, 1930–1934; dean, School of Forestry, University of Idaho, 1934–1935; director, Rocky Mountain Forest and Range Experiment Station, USFS, Fort Collins, Colorado, 1935–1938; director, Appalachian Forest Experiment Station, USFS, Asheville, North Carolina 1938–1944; assistant chief for State and Private Forestry Cooperative Programs, USFS, 1944–1952; chief, USFS, 1952–1962; executive director, National Institute of Public Affairs, 1962–1964; consultant, 1965–1966; member, Board of Directors, Olinkraft, Inc., from 1967; member, Royal Commission on Forestry, Newfoundland and Labrador, 1967–1971; resources consultant, National Wildlife Federation, from 1967.

CAREER

As director of Rocky Mountain Forest Experiment Station during World War II, served as Forest Service representative for Appalachians to Committee on Post-War Planning. One of goals was to reduce unemployment of military personnel undergoing reconversion to civilian life and to limit degree of turmoil entailed in return to peacetime economy. McArdle questioned value of ''plans too vague or too general to be implemented.'' As assistant chief, had to deal with pressures from lumber industry for lessening of federal regulatory activity and felt that state forestry agencies would have to side with either the USFS or industry over issues dividing them. As chief, was affable and informal. Secured improved funding for his agency. Pressed for balance between management and research. Felt a measure of cooperation with lumbermen essential. USFS regulatory role on private lands abandoned as impractical. Oversaw study of industry concerns about proposals to expand federally held lands. Reduced abuses of grazing permits by livestock interests. Emphasized professionalization of personnel and improved grades and salaries for individuals in key jobs. Urged that more aid be given to state and private forestry. Gave more attention to recreational use of USFS lands. Was responsible for administering 153 national forests totaling some 181 million acres in forty states and for thirteen regional forest experiment stations, also 7 million acres of grassland in Great Plains. During his tenure, Multiple Use-Sustained Yield Act of 1960 codified USFS policies and traditions. Multiple use included wilderness, range, timber, watershed, wildlife and fish resources, and mineral resources. Emphasis was on equal attention to each facet of use. Smokey the Bear public education program adopted. Authored a number of technical and popular articles on forestry issues. Played important role in international forestry and was president of Fifth World Forestry Congress in Seattle (1960), which he helped organize. Awarded Order of Merit for Forestry from Mexican government, 1961, and Knight Commander Order of Merit from West German government, 1962.

MAJOR CONTRIBUTIONS

Oversaw balanced Forest Service management policy during his ten-year tenure as chief of the Forest Service.

BIBLIOGRAPHY

American Men of Science. 12th ed.

Davis, Richard C., ed. *Encyclopedia of American Forest and Conservation History* (1983).

Ogden, Gerald, comp. *The United States Forest Service: A Historical Bibliography, 1876–1972* (1976).

Steen, Harold K. *The U.S. Forest Service: A History* (1976).

Stroud, Richard H., ed. *National Leaders of American Conservation* (1985).

Who's Who in America (1974–1975).

UNPUBLISHED SOURCES

Oral history interview with McArdle was conducted by Elwood R. Maunder, 1975, Forest History Society, Durham, N.C.

Keir B. Sterling

MCATEE, WALDO LEE. Born Jalapa, Indiana, 21 January 1883; died Chapel Hill, North Carolina, 7 January 1962. Ornithologist, ecologist, conservationist.

FAMILY AND EDUCATION

Eldest child of John Henry McAtee, a carpenter, and Anna Morris McAtee. A.B., 1904, A.M., 1906, honorary D.Sc., 1961, Indiana University. Married Fannie E. Lawson, 1906; three children.

POSITIONS

Curator of birds, Indiana University Museum, 1901–1904; summer appointment as biological expert, U.S. Biological Survey (USBS), summer 1903. Assistant biologist, USBS, 1904–1916; in charge of Food Habits Research, 1916–1934 (as senior biologist, head of Division of Food Habits Research, 1921–1934); principal biologist, 1929–1936; technical adviser to the chief of the USBS and game research specialist, 1936–1940; technical adviser and senior editor of publications for U.S. Fish and Wildlife Service (USF&WS), 1940–1947; acting custodian of hemiptera, U.S. National Museum, 1920–1942; collaborator with USF&WS, 1947–1948.

CAREER

Interested in out-of-doors from childhood, when involved in walks and fishing with maternal grandfather along Mississinewa River in Marion, Indiana. Interest in birds grew from occasion when Frank M. Chapman spoke in Marion when McAtee was sixteen. McAtee's master's thesis at Indiana, "Horned Larks and Their Relation to Agriculture," was published as Bulletin 23 of USBS (1905). Involved in food habits research from onset of career with USBS and was in charge of this effort for nearly two decades. Became a leading economic ornithologist and was considered an unusually capable biologist by many of his colleagues. At least one, E. A. Preble, considered him a better-informed naturalist than C. H. Merriam, first chief of the USBS. Fieldwork for USBS involved trips to forty states, two provinces in Canada, and a dozen European nations (these last in the course of a 1927 trip). Was member of U.S. Department of Agriculture Committee on land utilization, 1922–1939. Thoroughly at home in entomology and botany, as well as ornithology. With others, he described over 460 species and twenty genera of insects. Energetic protector of predatory birds and urged protection and restoration of habitats, so that their numbers might increase. Often blunt and uncompromising in his assessments of men and situations, he was described early in his career at USBS as a man who "would rather be right than be Chief." Although he occasionally served as acting chief, he was, in fact, an unsuccessful contender for the top USBS post in 1934. Spent much of the last decade of his career as editor of technical publications for USBS. A founder of the Society of Wildlife Specialists (1936), later the Wildlife Society. Edited *Journal of Wildlife Management*, 1937–1942. Published some 1,200 technical and popular papers, notes, pamphlets, circulars, reviews, and obituaries, as well as over 6,000 abstracts for *Wildlife Review*, USBS abstracting publication begun in 1935 largely at his insistence. His "Effectiveness in Nature of the So-Called Protective Adaptations in the Animal Kingdom, Chiefly as Illustrated by the Food Habits of Nearctic Birds" (1932) was a detailed and highly controversial critique of natural selection. Contributed chapter on half century of developments in economic ornithology to *Fifty Years' Progress of American Ornithology* (1933). Also published *Wildlife Food Plants* (1939) and *A Review of the Nearctic Viburnum*. Longtime treasurer of the American Ornithologists' Union (1920–1938); president of Washington Biological Field Club, 1911–1916, active in a large number of other scientific organizations. Won several awards and was a fellow of the AAAS. Left unpublished two substantial manuscripts, a dictionary of birds' names, and "A Critique of Darwinism."

MAJOR CONTRIBUTIONS

McAtee's work on the food habits and conservation of birds was probably his most important work, but he also made valuable contributions to entomology, botany, Hoosier folklore, and other subjects.

BIBLIOGRAPHY

American Men of Science. 9th ed.
Audubon Magazine (November–December 1946).
Dunlap, Thomas R. *Saving America's Wildlife* (1988).

Kalmbach, E. R. *Auk* 80 (1963) (obit.).
Kimler, W. C. *Dictionary of Scientific Biography* 18, Supplement 2, (1990).
Stroud, Richard H., ed. *National Leaders of American Conservation* (1985).
Terres, John. *Journal of Wildlife Management* 27 (1963) (obit.).

UNPUBLISHED SOURCES

Autobiographical material is in Library of University of North Carolina, Chapel Hill. Other manuscript materials are at American Philosophical Society, Philadelphia, Fuertes and Olin Libraries, Cornell University, Ithaca, N.Y., Academy of Natural Sciences Library, Philadelphia. Substantial handwritten reminiscences about the Biological Survey, his colleagues in that organization, and other scientists whom he had known, personal correspondence, autobiographical notes, and other materials are in Manuscript Division, Library of Congress. Some ephemera are in archives of American Ornithologists' Union, Smithsonian Institution Archives.

Keir B. Sterling

MCGEE, WILLIAM JOHN. Born Farley (in Dubuque County), Iowa, 17 April 1853; died Washington, D.C., 5 September 1912. Anthropologist, geologist, hydrologist.

FAMILY AND EDUCATION

Son of James and Martha (Anderson) McGee. Father a native of Ireland, of Scottish descent (Clan Macgregor). Mother's families immigrated from Ireland and England and were pioneers in Virginia, Kentucky, Indiana, and Iowa. Self-educated except for three or four terms in pioneer public schools. LL.D., Cornell College, Iowa, 1901. Married Anita Newcomb, 1888; one daughter.

POSITIONS

Ethnologist in charge of Bureau of American Ethnology, 1893–1903. Resigned in July 1903 to become chief of the Department of Anthropology of the St. Louis Exposition. First director of the St. Louis Public Museum, 1905–1907. U.S. commander, Inland Waterways Commission, from 1907 until death. McGee was connected with all of the scientific societies of Washington, D.C., and the nation. Was lecturer, U.S. commander American International Commission of Archaeology and Ethnology from 1902. In 1904 became chairman for the International Geographic Congress and senior speaker for the Department of Anthropology, World's Congress of Arts and Sciences. Secretary for the Conference of Governors in White House in 1908. Leading founder of Columbia Historical Society and president of American Anthropological Association; acting president, American Association for the Advancement of Science, 1897–1898. Was president of the National Geographic Society, 1904–1905, establishing the *Bulletin*. Its official publication, and vice president of the Archaeological Institute of America, 1902–1905. McGee was editor of the Department of Anthropology's *International Encyclopedia* and associate editor of *National Geographic Magazine.*

CAREER

Began his career in land surveying and justice court practice between 1873 and 1875. During this period he invented, patented, and manufactured agricultural implements. Studied geology and archaeology, 1875–1877, researching Indian mounds and other relics in Iowa and Wisconsin. Made geologic and topographic survey of northeastern Iowa between 1877 and 1881 and examined and reported upon building stones and quarry industries of Iowa for the census between 1881 and 1882. In 1885 and again in 1892 became attached to U.S. Geological Survey, where he surveyed and mapped 300,000 square miles in southeastern United States. Also compiled maps of United States, including New York. In 1886 he made an on-the-ground investigation of the Charleston earthquake immediately after its occurrence. In addition to surveys between 1883 and 1893, formulated the method of correlation among geologic formations by homogeny or identity of origin; developed a genetic classification of geology; and did much to develop, and was the first to apply, the principles of geomorphy. Between 1894 and 1895 he explored Tiburon Island, home of the fierce Seri Indians and the neighboring tribes of Sonora, where he traced the early stages of agriculture, the domestication of animals, marriage conditions, and their civilization. His writings count over 200, beginning in 1878, with published papers in the *American Journal of Science* and *Proceedings of the American Association for the Advancement of Science*. Other writings include *Pleistocene History of Northeastern Iowa* (1891) and several volumes of *The History of North America*, published in the early 1900s. McGee left a will bequeathing his body and brain for anatomical and scientific study to Edward Anthony Spitzka.

MAJOR CONTRIBUTIONS

Developed method of correlating geologic formations by identity of origin; tried to classify anthropology; studied previously unknown tribes of American Indians. Did much for the Geological Society of America.

BIBLIOGRAPHY

Darton, Nelson Horatio. *Memoir of W. J. McGee* (1914).
McGee, Emma. *Life of W. J. McGee* (1915).
World Who's Who in Science. Ed. Allen G. Debus (1968).

UNPUBLISHED SOURCES

See Paul McGee Papers, 1882–1916, Library of Congress Manuscript Division, Washington, D.C.

Joan D'Andrea

MEARNS, EDGAR ALEXANDER. Born Highland Falls, New York, 11 September 1856; died Washington, D.C., 1 November 1916. Ornithologist, mammalogist, army surgeon.

FAMILY AND EDUCATION

Son of Alexander and Nancy (Carswell) Mearns. Schooling at Donald Highland Institute, Highland Falls. Medical degree from College of Physicians and Surgeons, New York City, 1881. Married Ella Wittich, 1881; one daughter and a son (latter died 1912).

POSITIONS

Commissioned assistant surgeon, U.S. Army, 1883. Retired as lieutenant colonel, 1909. Collected zoological and botanical specimens while assigned to Fort Verde, Arizona; Fort Snelling, Minnesota; the U.S.–Mexican border; Fort Myer, Virginia; Fort Clark, Texas; Fort Adams, Rhode Island; Florida; Yellowstone National Park, Wyoming; Fort Custer, Montana; Philippine Islands; also upstate New York; the Washington D.C. region; Kenya; Uganda; Ethiopia; French Somaliland; and Egypt.

CAREER

Began publishing his estimated output of 125 articles and notes in 1878. Most were on birds and mammals and appeared principally in *The Auk* and the *Proceedings of the U.S. National Museum.* His long military career took him to several posts in the West and in Philippines. His most important assignment in North America was as naturalist of Mexican–U.S. International Boundary Commission, 1892–1894, in which capacity he collected thousands of specimens of animals and plants from El Paso to the Pacific Coast, resulting in his most significant publication, *Mammals of the Mexican Boundary of the United States* (1907). A charter member of the American Ornithologists' Union.

During his tours of duty in the Philippines (1903–1904, 1905–1907), helped organize Philippine Scientific Association and made major biological and geographical reconnaissance of Mindanao. Mearns' zeal for collecting always overcame any fears he might have had for his personal safety. When President Theodore Roosevelt, an old friend, was planning his African expedition in 1909, Mearns was recommended to him "as being the best field naturalist and collector in the United States." Mearns retired from the army as a lieutenant colonel and accompanied the former president on his 1909–1910 trip to Kenya and Uganda, continuing down the White Nile to the Mediterranean coast at Port Said. He collected some 3,000 birds and a number of small mammals. Mearns undertook a second trip to east central Africa with Childs Frick in 1911–1912, bringing back some 5,200 bird specimens, many nests and eggs, and notes on his observations. By this time, Mearns was seriously ill with diabetes and he was confined to Washington save for several short trips in the area until his death. During his career, Mearns contributed some 20,000 bird, 7,000 mammal, 5,000 reptile, and 5,000 fish specimens to the U.S. National Museum, and a number of other specimens went to the American Museum of Natural History. So much of his energy was devoted to collecting that he was left with less time for writing his own scientific papers than he would have liked.

MAJOR CONTRIBUTIONS

He is remembered mostly for being an energetic and valued collector in all branches of natural history, the collections going to the U.S. National Museum and the American Museum of Natural History. At the time of his death it was estimated that one-tenth of the birds in the National Museum were collected or given by him and that "his contributions to the National Herbarium were greater than those made by any one man."

BIBLIOGRAPHY

The Evening Star, 3 November 1916 (obit.).

Hume, E. E. *Ornithologists of the United States Army Medical Corps* (1942).

Richmond, C. W. "In Memoriam: Edgar Alexander Mearns." *Auk* (January 1918).

Roosevelt, Theodore. *African Game Trails: An Account of the African Wanderings of an American Hunter-Naturalist* (1927).

UNPUBLISHED SOURCES

See Edgar A. Mearns Papers, Division of Birds, Museum of Natural History, Washington. Mearns correspondence in Manuscript Division of the Library of Congress and in Manuscript Records of the Surgeon General of the Army, National Archives. Also A. K. Fisher Papers, Library of Congress Manuscript Division (correspondence) and Naturalists' Papers, Smithsonian Institution Archives (correspondence and biographical material).

Michael J. Brodhead

MENZIES, ARCHIBALD. Born Aberfeldly, near Weem, Scotland, baptized 15 March 1754; died Notting Hill, London, England, 15 February 1842. Botanist, naval surgeon, physician.

FAMILY AND EDUCATION

Son of James and Ann Menzies. Had at least four brothers. Received early education at local parish school in Weem, and early botanical training at home. Student at Royal Botanical Garden, Edinburgh, beginning sometime prior to 1771. Student of medicine at Edinburgh University, 1771–1780. Did not receive degree there. Awarded M.D., University of Aberdeen, 1799. Married Janet (last name and date not known, but presumably either late in his naval career or soon after taking up residence in London); no children.

POSITIONS

Did work as gardener while a youngster for Menzies clan chief. Worked and studied at Royal Botanical Garden, Edinburgh, beginning in mid or late teens.

Medical student, Edinburgh, 1771–1780. Collected plants in Scottish Highlands for J. Fothergill and W. Pitcairn, London physicians, 1778. Assisted physician in Caernarvon, Wales, sometime subsequent to 1780. Assistant surgeon, Royal Navy, 1782–1786, assigned to H.M.S. *Nonsuch*, 1782–1784, to H.M.S. *Assistance*, 1784–1786; surgeon-naturalist on H.M.S. *Prince of Wales*, 1786–1789, and H.M.S. *Discovery*, 1790–1795, both on round-the-world voyages. Spent much of remainder of naval service in West Indies, 1795–1802, resigned for reasons of health. In private practice, London, 1802–1826, then retired.

CAREER

Strongly influenced by fact that four of his brothers worked as gardeners, one of whom, William, was employed at Royal Botanical Garden, Edinburgh, sometime after its establishment in 1763. Another brother, Robert, was principal gardener there in the 1790s. Menzies was encouraged to study at Edinburgh by his mentor at the botanical garden, Professor John Hope. Menzies began his twenty-year career as naval surgeon during latter stages of the American Revolution, served in combat in West Indies, 1782. Transferred to Halifax, Nova Scotia, 1784, collected specimens and seeds there, some of them for Sir Joseph Banks at Royal Botanical Garden, Kew. On his return to England in 1786, Menzie spent some time studying in Banks' library and herbarium. On Banks' recommendation, served as surgeon-naturalist on H.M.S. *Prince of Wales*, 1786–1789, making round-the-world fur-trading voyage. In course of this trip, made first visit to western coast of what is now Canada at Nootka Sound, 1787.

His second circumnavigation of the world, again at Banks' recommendation, made with Captain George Vancouver, 1790–1794. Signed on as naturalist, but from the fall of 1792 on Menzies replaced the surgeon of the *Discovery*, who had to return to England because of illness. Menzies' relations with Vancouver were often severely strained, as Vancouver was something of a martinet. Menzies made collections at various points in the Pacific and up and down the coast of both North and South America, including Nootka Sound (1792 and 1794) and what are now the states of Washington and California (1792, 1793, 1794). Menzies lived most of the time aboard Vancouver's ship, but occasionally spent some time on one of Vancouver's two other vessels. In 1792, did some botanizing (late April to late June) near Port Discovery and Admiralty Inlet, and in various parts of Puget Sound, collecting seeds of both plants and trees. Part of August was spent at Nootka, where he met Jose Mociño, the Mexican botanist, who was also collecting there. Parts of California were visited and plant collecting accomplished in late 1792 and early 1793, primarily at San Francisco and Monterey. After some months in Hawaii, Vancouver's ships again went east and anchored off northern California (now Humboldt County) in May, again went to Nootka and Puget Sound (May to October, 1793), then visited Marin County, San Francisco, and Monterey (October-December, 1793). Following another trip to Hawaii, Nootka Sound and Monterey were revisited in late 1794.

Among the species Menzies discovered were the Western Red cedar, the Douglas Fir, and the madroño. Though he had seen redwoods in 1792 and 1794, Menzies did not collect any plants of the coast redwood, *Sequoia sempervirens*, until late in 1794. Menzies' specimens are today found in a number of collections, including the Royal Botanic Gardens at Kew; the National Herbarium and British Museum, London; in Edinburgh; and at the Liverpool Botanic Garden. Menzies made very careful and accurate drawings of some of the plants he collected. In addition to his extensive botanizing, he made detailed notes of the geography of the areas he visited and included a number of faunal observations. Menzies himself published very little. Most descriptions of the plants he discovered and collected were published by others. He and Everard Home did write one article concerning a Canadian mammal, ''A description of the anatomy of the sea otter, from a dissection made November 15th, 1795,'' in the *Philosophical Transactions of the Royal Society of London* 86 (1796). Four other articles by Menzies concerning plants and animals were based on his collections and observations in the Pacific. Menzies was made a fellow of the Linnean Society of London in 1790, and he also served as its president for a time.

MAJOR CONTRIBUTIONS

Menzies is principally known today for having been the first European to make extensive collections of plants and trees and seeds from Nootka Sound, Vancouver Island, and various locations in and around Puget Sound, and smaller ones from various points in northern and central California in the early and mid-1790s.

BIBLIOGRAPHY

Desmond, Raymond, with assistance of Christine Ellwood. *Dictionary of British and Irish Botanists and Horticulturists, including Plant Collectors, Flower Painters and Garden Designers* (2d ed., 1994).

Dictionary of Canadian Biography.

Dictionary of National Biography.

Eastwood, Alice, ed. ''Archibald Menzies' Journal of the Vancouver Expedition. Extracts covering the visit to California.'' *Quarterly of the California Historical Society* 2 (1924).

Eastwood, Alice. ''Early Botanical Explorers on the Pacific Coast and the Trees they found there.'' *Quarterly of the California Historical Society* 18 (1939).

Galloway, D. J., and E. W. Groves. ''Archibald Menzies, MD, F.L.S. (1754–1842), Aspects of his life, travels, and collections.'' *Archives of Natural History* 14 (October 1987).

Jepson, Willis L. ''The Botanical Explorers of California: Archibald Menzie.'' *Madroño* 1 (1929).

McKelvey, Susan D. *Botanical Exploration of the Trans-Mississippi West, 1790–1850* (1956; repr. 1991).

Newcombe, C. F., ed. ''Menzies' Journal of Vancouver's Voyage: April to October, 1792 . . . with botanical and ethnological notes . . . and a biographical note by John Forsyth.'' *Memoir V. Archives British Columbia* (1923).

Vancouver, George. *A Voyage of Discovery to the North Pacific Ocean, and round the world: in which the coast of North-West America has been carefully examined and accurately surveyed . . . performed in the years 1790, 1791, 1792, 1793, 1794 and 1795, in the Discovery sloop of war, and armed tender Chatham*, 4 vols (1798, see also repr. ed., edited by W. K. Lamb, 4 vols., 1984).

UNPUBLISHED SOURCES

Menzies' journal is at the British Library, London. His correspondence with Sir Joseph Banks is in the Banks Correspondence, British Museum (Natural History), London, and his correspondence with Sir William J. Hooker is at Kew. Other materials are in British Columbia Provincial Archives and in the State Library of Washington, Olympia.

Keir B. Sterling

MERRIAM, CLINTON HART. Born New York City, 5 December 1855; died Berkeley, California, 19 March 1942. Mammalogist, zoologist, federal government administrator.

FAMILY AND EDUCATION

Merriam's father was a businessman, banker, and for a brief period (1871–1875), a Republican congressman from New York. He was the second of four children. Until age fourteen, he was educated by governesses. Attended Alexander Institute, White Plains, New York, 1869–1871; Dr. Pingry's School, Elizabeth, New Jersey, 1871–1872; Williston Seminary, Easthampton, Massachusetts, 1873–1874. Student at Sheffield Scientific School, Yale University, 1874–1877. Did not take a degree. Attended College of Physicians and Surgeons, New York, 1877–1879. Received M.D., 1879. Married Elizabeth Gosnell, 1886; two daughters.

POSITIONS

Naturalist, Second Division of the Hayden Survey, Northwestern Wyoming, summer, 1872; physician in an upstate New York private practice, specializing in the diseases of women, 1879–1885. Economic ornithologist, Division of Entomology, U.S. Department of Agriculture, Washington, D.C., 1885–1886; chief, Division of Economic Ornithology and Mammalogy, U.S. Department of Agriculture, Washington, D.C., 1886–1890; chief, Division of Biological Survey, U.S. Department of Agriculture, Washington, D.C., 1890–1906 (later renamed U.S. Biological Survey), 1906–1910. Private research in mammalogy and anthropology on continuing grant from Mrs. E. H. Harriman, 1910–1939.

CAREER

Introduced to S. F. Baird by father sometime in 1871 or 1872. Placed by Baird on Second Division of the Hayden Survey in the Yellowstone, summer

1872. After completing formal study at Yale and College of Physicians and Surgeons in New York, continued with collecting of birds and mammals while in private medical practice in upstate New York. Naturalist on expedition of SS *Proteus* to waters off Newfoundland, 1883. Made brief study of private and museum collections in England, the Netherlands, and Germany (his only trip abroad), spring, 1885. A charter member of the American Ornithologists' Union (AOU), 1883, chaired AOU Committee on Migration and Distribution of Birds, and with help of Baird, approached Congress for subvention to make possible analysis of data collected. Placed in charge of small government agency, 1885, for purpose of collecting data on birds and mammals helpful or inimical to the interests of farmers. With the small staff of this agency, spent much of the next twenty-five years (1885–1910) collecting and analyzing scientific data concerning biogeography, distribution, and speciation of North American vertebrates. Devised life zone theory on the basis of investigations at San Francisco Mountain, Arizona, (1889), which posited that temperature was the controlling factor in distribution of life. His concept later superseded by more sophisticated models but still considered applicable to conditions in western North America. Developed new methods of specimen collection (based in part on use of cyclone trap, new device placed on market in the 1880s) and biological reconnaissance in the field. He and his small Biological Survey staff described many hundreds of new vertebrate species, principally mammals, and the results of a number of state and regional biological surveys in the *North American Fauna* series (1889) and other Biological Survey publications. Assembled scientific team for Harriman–Alaska Expedition, 1899–1900. Under increasing pressure from Congress to undertake various regulatory responsibilities, including control of predators and other noxious species, resigned from government service, 1910. Spent remainder of productive career (to 1939) on personal research centering on mammals (principally bears) and on transcribing the vocabularies of the disappearing Indian tribes of California.

MAJOR CONTRIBUTIONS

Using his position as chief of a federal government bureau, brought major attention to the study of North American mammals and other land vertebrates and to the biogeographical setting in which they lived. Together with some of his Biological Survey associates, he published a number of biological surveys of western American states and Canadian provinces, together with major revisions of mammalian genera that were influential for many years. His thorough field methods have since set the standard for students of mammalogy. Many of his conclusions concerning speciation and the factors governing the distribution of life have undergone considerable alteration. Nevertheless, the collections he helped develop, principally at the U.S. National Museum of Natural History in Washington, D.C., have proven invaluable to later generations of mammalogists. Spent several decades painstakingly transcribing the vocabularies of disappearing Canadian Indian tribes, which remain largely unpublished. Made a study of the dynamics of fur seal populations of the Pribilof Islands beginning in 1891. This led to needed curbs on pelagic sealing and to an international agreement in 1912 that stabilized the situation there. Among his principal publications: *The Mammals of the Adirondack Region, Northeastern New York*, 2 vols. (1882–1884); ''Results of a Biological Survey of the San Francisco Mountain Region and Desert of the Little Colorado,'' *North American Fauna* 3 (1890); (with others), ''The Death Valley Expedition: A Biological Survey of Parts of California, Nevada, Arizona, and Utah (Part II only), *North American Fauna* 7 (1893); ''Laws of Temperature Control of the Geographic Distribution of Terrestrial Animals and Plants,'' *National Geographic Magazine* (December, 1894); ''Review of the Grizzly and Big Brown Bears of North America (Genus Ursus) with a description of a New Genus, Vetularctos,'' *North American Fauna* 41 (1918); ed., *An-nik-a-del: The History of the Universe as Told by the Mo-des-se Indians of California* (1928).

BIBLIOGRAPHY

Cameron, Jenks. *The Bureau of Biological Survey: Its History, Activities, and Organization*. Baltimore: Johns Hopkins University Press (1929).

Heizer, R. F., et al., eds. *Catalogue of the C. Hart Merriam Collection of Data concerning California Tribes and Other American Indians* (1989).

Kroeber, Alfred. ''C. Hart Merriam as Anthropologist.'' In C. Hart Merriam, *Studies of California Indians*. Ed. Department of Anthropology, University of California. Berkeley: University of California Press (1955).

Osgood, Wilfred H. ''Biographical Memoir of C. Hart Merriam.'' *National Academy of Sciences Biographical Memoirs* 24 (1947).

Sterling, K. B. *Last of the Naturalists: The Career of C. Hart Merriam*. Rev. ed. New York: Arno Press (1977).

Sterling, K. B. ''Builders of the Biological Survey, 1885–1930.'' *Journal of Forest History* (October 1989).

UNPUBLISHED SOURCES

See C. Hart Merriam Files of manuscripts, clippings, miscellaneous notes, and correspondence, Museum of Vertebrate Zoology, University of California, Berkeley; C. Hart Merriam Papers, Bancroft Library, University of California, Berkeley; C. Hart Merriam ''Home'' and ''Field'' journals, 1873–1938: sixteen boxes, notes concerning field observations, with some personal correspondence and papers; four boxes, Indian vocabularies and other anthropological materials; forty boxes, typewritten bibliography and complete set of publications, 1873–1934; eleven boxes, Library of Congress, Manuscript Division; C. Hart Merriam Pictorial Collections: forty-seven boxes, twenty oversize boxes, sixteen albums, one photographic print (about 15,680 photographic prints, glass negatives, drawings, paintings, and prints), Bancroft Library, University of California, Berkeley; nine boxes of glass negatives and lantern slides, principally of Native Americans and baskets, at Smithsonian Archives, Washington; collection of Indian baskets at University of California, Davis; C. Hart Merriam Letters to staff members, Museum of Comparative Zoology, Harvard University; Vernon Bailey Papers, University of Wyoming Library (Archives); A. K. Fisher and T. S. Palmer, Papers and Correspondence in

Manuscript Division, Library of Congress, Washington, D.C.; Theodore Roosevelt Papers, Library of Congress, Manuscript Division.

Keir B. Sterling

MEXIA, YNES ENRIQUETTA JULIETTA. Born Washington, D.C., 24 May 1870; died Berkeley, California, 12 July 1938. Botanical explorer.

FAMILY AND EDUCATION

Mexia was the only child of Enrique Antonio and Sarah R. (Wilmer) Mexia. Father on diplomatic mission in Washington, D.C., for the Mexican government. Her paternal grandfather, Gen. Jose Antonio Mexia, was a leader of Mexico's Federalist Party and notable participant in early history of Texas. Through her mother Mexia was related to Samuel Eccelston, the Roman Catholic archbishop of Baltimore.

Mexia spent childhood on her family's homestead near present-day Mexia in Limestone County, Texas. By the time she was sixteen, family had moved to Philadelphia, where Mexia attended private Quaker school. Later, she enrolled in Quaker-sponsored school in Ontario, Canada, as well as St. Joseph's Academy, Emmitsburg, Maryland. Attended University of California intermittently from 1921 until 1938. Did not take degree. Married Herman E. de Laue (some sources give Herman Lane), 1897 (died, 1904). Married Augustin A. de Reygados (later divorced); no children.

POSITIONS

Mexia spent her career as an independent researcher working under the auspices of several academic institutions, including Stanford University and the University of California. She also undertook an expedition on behalf of the Department of Agriculture's Bureau of Plant Industry and Exploration. Although most of her work was accomplished in Mexico and South America, Mexia also traveled as far north as Mount McKinley Park in Alaska. Mexia financed her expeditions through the sale of the specimens she collected.

CAREER

Mexia's childhood affected by parents' marital discord. She lived in Mexico City on father's hacienda from her late teens until her late thirties. Four years following death of first husband, married younger man in employ of her poultry and pet stock-raising business. Marriage not a success, so she moved to San Francisco, where she underwent medical treatment, divorced husband, and became involved in local social work efforts. Early in 1921 Mexia gained admittance as a special student to the University of California. She soon developed interest in natural sciences, particularly botany. During the next sixteen years Mexia continued her studies, frequently taking leaves of absence to pursue her botanical interests. At the time of her death she was thirty hours short of earning an A.B. In 1925 Mexia took summer course on flowering plants at the Hopkins Marine Station in Pacific Grove, California. Class was a revelation to the fifty-five-year-old student. Shortly after the course ended, Mexia joined Roxana Stinchfield Ferris on expedition to Mexico sponsored by Stanford University. In spite of the fact that she fell from a cliff, fracturing her ribs and injuring her hand, Mexia returned with samples of almost 500 species.

After this expedition Mexia established a working relationship with Nina Floy Bracelin, assistant at the University of California Herbarium. Bracelin served as Mexia's curator and agent. She handled detailed arrangements of equipping Mexia's expeditions and kept records of the sale of specimens to various institutions. Bracelin also made arrangements to have experts examine Mexia's specimens.

In 1926 Mexia set off on her own through western Mexico. When she returned to California in April 1927, she brought with her 33,000 specimens, including a new plant genus and almost fifty new species of plants. During the summer of 1928 Mexia traveled to Mount McKinley Park in Alaska. She returned with 6,100 specimens. In November of the following year Mexia set off on an expedition to Brazil and Peru, returning in March 1932 with 65,000 specimens. Mexia spent one season working for the Department of Agriculture's Bureau of Plant Industry and Exploration in Ecuador, where she searched for Cinchona, wax palm, and soil-binding herbs.

By October 1935 Mexia was back in the Andes as a member of University of California Botanical Expedition in search of a variety of items, including wild tobacco. Mexia traveled in Peru, Bolivia, Argentina, and Chile before reaching the Straits of Magellan. On this expedition, she collected approximately 15,000 specimens including plants and animals. During the winter of 1937 and the spring of 1938 Mexia embarked on her final botanical expedition. On this trip she concentrated her efforts in Guerrero and Oaxaca, Mexico. Due to a stomach ailment, Mexia was forced to return to San Francisco ahead of schedule with some 13,000 specimens. A few months later, she died of lung cancer in Berkeley, California.

Mexia related details of her expeditions in articles that appeared in a number of journals, including "Botanical Trails in Old Mexico," *Madroño* (September 1929); "Three Thousand Miles Up the Amazon," *Sierra Club Bulletin* (February 1933); "Camping on the Equator," *Sierra Club Bulletin* (February 1937); and "Ramphastidae," *The Gull* 15, 7 (1933).

MAJOR CONTRIBUTIONS

Mexia was an enthusiastic botanical explorer. She provided research institutions with extensive collections of specimens, which were supplemented with extensive, accurate annotations. Acknowledged as a careful, meticulous worker, Mexia was also considered by colleagues to be a source of invaluable practical information about primitive life in the Andes.

BIBLIOGRAPHY
Bonta, Marcia M. *Women in the Field: America's Pioneering Women Naturalists* (1991).
Goodspeed, T. Harper. *Plant Hunters in the Andes* (1941).
Madroño (October 1938) (obit.).
Notable American Women.
San Francisco News, 6 March 1937.
Science (23 December 1938) (obit.).
UNPUBLISHED SOURCES
See Letters in collections of Missouri Botanical Gardens, St. Louis; Papers in Bancroft Library and in University and Jepson Herbaria, University of California, Berkeley.

Domenica Barbuto

MICHAUX, ANDRÉ. Born Satory, France, 7 March 1746; died Madagascar, 13 November 1802. Botanist, explorer, silviculturist, farmer.

FAMILY AND EDUCATION

His father managed royal farm at Satory. Married Cécile Claye 1769; one child, François A. Michaux, also a botanist, who travelled with him. Began study of botany in order to recover from loss of wife. First studied under Le Monnier at Montreuil, then undertook more intensive study at the *Trianon* with Bernard de Jussieu, later moving home closer to the *Jardin des Plantes* in Paris.

POSITIONS

Plant collector in the Near East, the United States, Canada, Mauritus, and Madagascar, 1782–1801.

CAREER

Succeeded father as manager at Satory. Appointed secretary to French Consul at Ispahan, Persia, but soon left his post to botanize in region between Tigris and Euphrates rivers (1782–1785). On his return to France with seeds and plants, was directed by French government to go to North America, study and collect forest trees there, and determine whether any were suitable for shipbuilding and should be brought to France for transplanting. In New York, New Jersey, and the Appalachians, 1785–1787. Moved to Charleston, South Carolina, 1787, bought plantation, continued botanizing in Appalachians and in Florida. Spent several years in Bahamas and mountains of South Carolina from 1789. Went to great efforts to ship American trees to France. Obliged to return to France when government support ended with French Revolution. On return to France, discovered that some 30,000 young trees he and son François had shipped home to France were mislaid, while a like number had been given to the Austrian court by Queen Marie Antoinette. Collected in Canada, 1792. Traveled in American midwest from July 1795 to April 1796, taking message from French Minister to U.S. General George Rogers Clark and collecting plants. Expedition great botanical success, but Michaux' finances were in poor shape. Returned to France in late summer, 1796, lost some notes and plants in shipwreck. French government declined to underwrite further collections in America. Served as naturalist to Captain Nicholas Baudin during latter's expedition to Australia, but left ship in Mauritus, mid-March 1801. Began botanizing in Madagascar, fell ill and died of tropical fever there. Publications included *Histoire des Chenes de l'Amerique* (1801) and *Flora Boreali-Americana* (1803), both completed with the considerable aid of Claude Richard. He also published "Memoire sur les Dattiers" in *Journal de Physique, de Chemie et d'Histoire Naturelle* 52 (1801).

MAJOR CONTRIBUTIONS

His explorations on three continents and collections of seeds and plants. Sent 60,000 young trees back to France from North America. With instructions from Thomas Jefferson, traveled in American Midwest and far West, supported in part by American Philosophical Society. Works are considered more informative than literary.

BIBLIOGRAPHY

Dictionary of American Biography. Vol. 12.

Dictionary of American History. Vols. 2, 3, 5.

Dictionary of Scientific Biography. Vol. 9.

Elliot, Clark A., ed. *Biographical Dictionary of American Science* (1979).

New York Times Book Review (9 August 1987) (Anthony Huxley review of *André and François André Michaux* by Henry and Elizabeth J. Savage).

Savage, Henry, and Elizabeth J. Savage. *André and François André Michaux* (1986).

Who Was Who in America. Historical vol. (1967).

UNPUBLISHED SOURCES

See manuscripts at the American Philosophical Society; fifty items about Michaux Garden (Charleston) in records of the South Carolina Historical Society records, 1946–1949; records relating to restoration of this garden by Charleston Garden Club; William Thomas Arnold Papers, Ohio State University, correspondence from Michaux (one letter). Michaux's type specimens are at the Muséum d'Histoire Naturelle in Paris.

Carmela Tino

MICHAUX, FRANÇOIS ANDRÉ. Born Satory (Versailles), France, 16 August 1770; died France, 23 October 1855. Silviculturist, botanist, traveler.

FAMILY AND EDUCATION

Son of André and Cecile (Claye) Michaux. Mother died in 1771; father was an explorer and botanist. As a teenager, traveled with his father to New York,

Charleston, South Carolina, and the Keovee River, Florida; the purpose of these trips was mainly to study North American trees. Returned to France in 1790 and briefly joined the French Revolution while studying medicine with Corvisart. Married his housekeeper when an elderly man; no children.

POSITIONS

Traveled extensively, mainly in United States, and wrote about experiences, 1785–1809. Agent in the United States for the French government, 1801–1803. Administrator of a farm of the Societe Centrale de l'Agriculture, c.1820–1855. Chevalier of the Legion of Honor. Correspondent of the French Institute. Member, American Philosophical Society.

CAREER

Requested by French government to sell the tree plantations his father had established in Hackensack, New Jersey, and Charleston, South Carolina, and to determine which American species might also flourish in Europe; sent seeds and specimens to France, 1801–1803. These experiences are documented in *Travels to the Westward of the Alleghany Mountains (Voyage a l'ouest des monts Aléghanys dans les etats de l'Ohio et du Kentucky et du Tennessee, et retour a Charleston par les Hautes-Carolines)* (Paris [1804], London [1805], Weimar [1805]). Returned to France, published *Sur la Naturalisation des Arbres Forestiers de l'Amerique du Nord* (1804). En route to Charleston in 1806 was taken by British and held in Bermuda; wrote "Notice sur les Iles Bermudes, et particulierement sur l'Ile Saint-Georges." Returned to Atlantic Coast of the United States for three years of travel; was passenger on steamboat inventor Robert Fulton's preliminary Hudson River journey. Published *Histoire des arbres forestiers de l'Amerique Septentrionale* (Paris [1810–1813]); its English translation was entitled *The North American Sylva, or a Description of Forest Trees of the United States, Canada, and Nova Scotia, Considered Particularly with Respect to Their Use in the Arts and Their Introduction into Commerce* (1818–1819). Administrator of a farm of the Societe Centrale de l'Agriculture, in France, c.1820–1855. Continued to publish works on numerous subjects.

MAJOR CONTRIBUTIONS

Added considerably to the literature of North American trees begun by his father. Despite having been awarded the Legion of Honor in France, is far better known for his contributions to American silviculture. His early nineteenth-century account of his American travels was published in English, French, and German editions. Propagated American plant species in France.

BIBLIOGRAPHY
''Andre and François Andre Michaux and Their Predecessors.'' *American Philosophical Society Proceedings* (1957).
''In His Father's Footsteps.'' *Horticultural Society of New York Bulletin* (October 1961).
Humphrey, Harry B. *Makers of North American Botany* (1961).
Savage, H. *Andre and François Andre Michaux* (1986).
UNPUBLISHED SOURCES
See letters (1783–1911), American Philosophical Society Library, Philadelphia. Michaux correspondence is also found in these collections at that institution: Benjamin Smith Barton Papers (1785–1813), two collections of Caspar Wistar Pennock Papers (1758–1817, 1829–1891), Wellcome Historical Medical Society Papers (1731–1871), ''Scientists' letters'' Collection (1563–1961), and the Archives of the American Philosophical Society Library (1802–1911); letters in the papers of ''Pourret, Pierre Andre, L'Abbe'' (1746–1820), Hunt Institute for Botanical Documentation, Carnegie–Mellon University, Pittsburgh; letters, journals, notebooks in the botanical papers collection at the Academy of Natural Sciences, Philadelphia (1742–1959); letters in the Henry Jackson Papers (1800–1840), University of Georgia Libraries (Athens); letters in the Samuel Brown Papers (1817–1825), Filson Club, Louisville, Kentucky.

Jean Wassong

MILLER, ALDEN HOLMES. Born Los Angeles, 4 February 1906; died Clear Lake, California, 9 October 1965. Ornithologist, educator, administrator, editor, musician.

FAMILY AND EDUCATION

Son of Loye and Anne Holmes Miller. Father, a naturalist and paleontologist, taught at University of California at Los Angeles (UCLA) and was most influential in Miller's career choice. Earned A.B. in 1927 at UCLA, having studied zoology, chemistry, and music. In graduate school decided to become scientist and earned M.A. in 1928 and Ph.D. 1930 at the University of California at Berkeley, where he studied under Joseph Grinnell. Married Virginia Elizabeth Dove, 1 August 1928; three children.

POSITIONS

Entire professional life associated with University of California at Berkeley. Teaching fellow, 1927–1928; associate, 1930–1931; instructor, 1931–1934; assistant professor, 1934–1939; associate professor, 1939–1945; professor, 1945–1965; assistant dean, College of Letters and Sciences, 1939–1940; curator of birds, Museum of Vertebrate Zoology (MVZ), 1939–1940, and director, 1940–1965; acting chairman, Department of Paleontology, 1959–1960; curator of birds, Museum of Paleontology, 1960–1965; vice chancellor, 1961–1962.

CAREER

Strongly influenced by father, who got him interested in ornithology and introduced him to his own specialties (avian anatomy and paleornithology) and to prominent regional ornithologists. Worked closely with his mentor at Berkeley, Joseph Grinnell, whom he succeeded as MVZ Director in 1940. Fieldwork in Pacific Coast states, Mexico, Central and South America, and Australia. Awarded Brewster Medal from the American Ornithologists' Union (AOU), 1943; president of AOU, 1953–1956. Elected member of National Academy of Sciences, 1957. Guggenheim Fellow, 1958. Member California Academy of Science; American Society of Naturalists; Society for the Study of Evolution, vice president, 1956–1957; Society of Vertebrate Paleontology; American Society of Zoologists; American Society of Mammalogy; Cooper Ornithological Society, president, 1948–1951; Phi Beta Kappa and Sigma Xi. Editor, *The Condor*, 1939–1965; editorial board, *Evolution*, 1950–1953. Was Vice-President of the International Ornithological Congress, 1951–1954 and 1962–1965. Was President of the International Commission on Zoological Nomenclature from 1964 to 1965. Mentor to thirty-one Ph.D. students between 1931 and 1965, twenty-four of whom became professional ornithologists.

Published nearly 260 titles consisting primarily of results of his own research and avoided syntheses based on the work of others. Published numerous monographs and journal articles on topics including fossils, anatomy, physiology, systematics, behavior, and various areas of environmental and evolutionary biology. These included: *Passerine Remains at Rancho La Brea* (1929); *Systematic Revision and Natural History of the American Shrikes (Laniidae)* (1930); *Studies of Cenozoic Vertebrates and Stratigraphy of Western North America* (1940); *Speciation in the Avian Species Junco* (1941); *Distribution of the Birds of California* (with J. Grinnell, 1944); *Avifauna of an American Equatorial Cloud Forest* (1962); *The Lives of Desert Animals in Joshua Tree National Monument* (with R. C. Stebbins, 1964); and *The Current Status and Welfare of the California Condor* (with Ian and Eben McMillan, 1965).

MAJOR CONTRIBUTIONS

Made a number of original contributions to scientific study of ornithology. Influenced many Ph.D. students at Berkeley, who later assumed prominence in American universities. A superb and enthusiastic field man. Personally collected over 12,500 specimens, his last a song sparrow acquired in Mendocino County in 1955.

Outstanding authority on the taxonomy of birds of western North America. His works on the shrike and junco are particularly noteworthy. Ernst Mayr comments, "The most impressive aspect of his all-too-short life was the range of his activities."

BIBLIOGRAPHY

Contemporary Authors. Vol. 109.

Davis, John. "In Memoriam: Alden Holmes Miller." *The Auk* 84 (1967).

Mayr, Ernst. "Alden Holmes Miller," *Biographical Memoirs, National Academy of Sciences*. Vol. 43 (1973).

New York Times, 24 April 1957, 30.

Who Was Who in America. Vol. 4.

UNPUBLISHED SOURCES

See Alden H. Miller Papers, University of California at Berkeley; Records of the Museum of Vertebrate Zoology, 1908–1949, Bancroft Library, and files of the Museum of Vertebrate Zoology, University of California, Berkeley.

Carmela Tino

MILLER, GERRIT SMITH, JR. Born Peterboro, New York, 6 December 1869; died Washington, D.C., 24 February 1956. Mammalogist.

FAMILY AND EDUCATION

Son of Gerrit Smith Miller Sr. and Susan Dixwell Miller. The elder Miller devised the American game of football and created the first American team (1862). The younger Miller was educated in private schools and by private tutors. Studied harmony and composition in his youth and became a competent musician. Entered Harvard at age twenty-one. B.A., Harvard, 1894. Married Elizabeth E. Page, a widow with three children, 1897 (deceased 1920). Married Anne Chapin Gates, 1921. No children. Nephew by marriage and close friend of Oliver Wendell Holmes, Jr., Associate Justice of the U.S. Supreme Court.

POSITIONS

Biologist with the Bureau of Biological Survey, U.S. Department of Agriculture, 1894–1898; assistant curator of mammals, United States National Museum (U.S.N.M.), 1898–1909, Curator of mammals, U.S.N.M., 1909–1940. Occasional acting Chief Curator, U.S.N.M. associate in biology, Smithsonian Institution, 1940–1956.

CAREER

Miller's interest in animals began in boyhood on the family estate in central New York. While with the Biological Survey, completed several monographic studies for the *North American Fauna* (*N.A.F.*) series, including "The Long-Tailed Shrews of the Eastern United States," (*N.A.F.* no. 10, 1895); "The Genera and Subgenera of Voles and Lemmings," (*N.A.F.* no. 12, 1896); and "Revision of the American Bats of the Family Vespertilionidae," (*N.A.F.* no. 13, 1897). Owing to personality conflict with C. Hart Merriam, the Biological Survey chief, Miller resigned his position (1898) and became associate curator of Mammals at the U.S.N.M. His "Directions for Preparing Study Specimens of Small Mammals," originally published privately (1894), underwent six revisions as a government pamphlet, and was last revised in 1932. It was one of the earliest guides for inexperienced collectors. Miller did much to categorize North American mammals. Published *Preliminary List of the Mammals of New York* (1899); *Key to the Land Mammals of Northeastern North America* (1900), both in the context of zoogeographic zonation; and *Systematic Results of the Study of North American Land Mammals to the Close of the Year 1900* (with James A. G. Rehn, 1901). In this latter work, summarized vast increase in the number of known forms of mammals (363 to 1450) which had taken place between 1885 and 1900, owing principally to improved methods of collecting specimens. *The Families and Genera of Bats* (1907) remained, with some modifications, standard for a number of years. *Catalog of the Land Mammals of Western Europe* (1912) was based on a study of specimens in the British Museum (Natural History), the U.S.N.M., and Miller's own private collections. He began full-time work on it in 1905; was on leave in Europe doing full-time research for much of the years 1908–1911. The *List of North American Land Mammals in the United States National Museum* (1912, revisions in 1923 and 1954, the latter with Remington Kellogg) was for more than four decades an essential research tool for mammalogists. In later years, his interests in mammals widened to include such species as wolves, dolphins, bears, and primates, along with man.

In the 1920s, undertook study of West Indian and Panamanian mammals. Among other accomplishments, demonstrated that *Solenodon*, a large shrew-like mammal, had been used as food by Carib chiefs fleeing the Spaniards and hiding in caves at the beginning of the Sixteenth Century. Translated Herlauf Winge's *A Review of the Interrelationships of the Cetacea* (1921) from Danish into English. Miller's 400 titles included papers on paleontology, botany, anthropology, and music. Shy and extremely reserved, he rarely mixed with the public and once had to be escorted from a hall when he unsuccessfully attempted to deliver a scientific paper. His systematic expertise earned him the considerable respect of his scientific colleagues, though his work was little known to the general public. A charter member of the American Society of Mammalogists (1919) and an original member of its council, he was elected an honorary member in 1941. Elected a fellow of the American Association for the Advancement of Science, the Academy of Natural Sciences of Philadelphia, and the American Philosophical Society.

MAJOR CONTRIBUTIONS

One of the two or three leading American mammalogists of the twentieth century. His *Catalog of the Land Mammals of Western Europe*, his many publications on bats and other small mammals, and his *List of North American Land Mammals* were classic works to which mammalogists referred for many years.

BIBLIOGRAPHY
Birney, Elmer C., and Jerry R. Choate, eds., *Seventy-Five Years of Mammalogy (1919–1994)*, 1994.
New York Times, 26 February 1956 (obit.).
Shamel, H. H. et al. ''Gerrit Smith Miller, Jr.'' *Journal of Mammalogy* (August 1954).
Keir B. Sterling

MILLER, HARRIET MANN. (Best known by her pen name Olive Thorne Miller). Born Auburn, New York, 25 June 1831, died Los Angeles, 25 December 1918. Nature writer, author.

FAMILY AND EDUCATION

Eldest of four children of Seth Hunt and Mary Field Holbrook Mann. Father a banker, but family moved often, living for varying periods in several midwestern states. For a time, she attended a private school in Ohio. Married Watts Todd Miller, 1854; four children.

POSITIONS

Writer and author of works on natural history.

CAREER

Began writing in childhood. Her first publication for children, on the making of china, appeared in 1870. With her husband, a businessman, lived in Chicago and after about 1875, in Brooklyn, New York. Following death of her husband, spent the last fourteen years of her life in Los Angeles. Successful writer of sentimental stories for children, which appeared in books and periodicals such as *St. Nicholas* for a number of years. His first nature book, *Little Folks in Feathers and Fur, and Others in Neither* (1875), was a well-researched text covering various kinds of invertebrates and vertebrates. Began focusing on birds in 1880. Her *Bird Ways* (1885) was first title for adults, written under pen name of Olive Thorne Miller, adopted by her in 1879. Her nature books won respect of professional biologists because of high degree of accuracy in research and observation. Her other books included *Queer Pets at Marcy's* (1880); *In Nesting Time* (1888); *Our Home Pets* (1894); *A Bird-Lover in the West* (1894); *Four-Handed Folks* (c.1896); *The First Book of Birds* (1899); *The Second Book of Birds: Bird Families* (1901); *True Bird Stories from My Note-book* (1903); *With the Birds in Maine* (1904); and *The Children's Book of Birds* (1915). Was member of a number of organizations, including American Ornithologists' Union, Audubon Society of California, the Brooklyn New York Women's Club, the Meridian Club, and the Sorosis Club of New York.

MAJOR CONTRIBUTIONS

Wrote many books for children. Those about nature and wildlife reflected much care and precision. Advocated a clean environment and spoke out against the slaughter of birds to benefit the millinery trade.

BIBLIOGRAPHY

Bailey, Florence M. *Auk* (April 1919) (obit.).
Dictionary of American Biography.
Holloway, Laura C. *The Woman's Story* (1889).
Los Angeles Times, 26 December 1918 (obit.).
New York Times, 27 December 1918 (obit.).
Welker, R. H. *Notable American Women* (1971).
Who's Who in America (1918–1919).
Willard, Frances E., and Mary A. Livermore. *American Women.* Vol. 2 (1897).

UNPUBLISHED SOURCES

Papers are in Manuscript Division, Library of Congress.

Geri E. Solomon

MILLS, ENOS ABIJAH. Born near Pleasanton, Kansas, 22 April 1870; died Estes Park, Colorado, 22 September 1922. Naturalist, conservationist, writer, innkeeper.

FAMILY AND EDUCATION

Ninth of eleven children of Enos Abijah Mills, Sr., a farmer, and Ann Lamb Mills. Intermittently attended local public schools in Kansas. Student at Heald's Business College, San Francisco, 1890. Married Esther Burnell, 1918; one daughter.

POSITIONS

Did odd jobs at Elkhorn Lodge, Estes Park, Colorado, 1884; worked on cattle ranch in Eastern Colorado, 1884–1885; worked for older cousin at Lamb's Longs Peak House, Estes Park, 1885–1887; employed by Anaconda Mine, Butte, Montana, winters 1887–1902, rising from tool boy to engineer. Worked for U.S. Geological Survey group in Yellowstone, summer, 1891. Owner, Longs Peak House (Inn), 1902–1922; writer-naturalist, 1902–1922; Colorado State snow observer, 1903–1906; independent lecturer on forestry (on behalf of U.S. Forest Service), 1907–1909; led effort for establishment of Rocky Mountain National Park, 1909–1915.

CAREER

Went to Colorado at age fourteen on recommendation of mother and physician because of persistent digestive ailment. Began hiking and exploring mountains in Estes Park region, 1885. Spent seventeen years employed at Anaconda Mine on schedule that made possible his continued hiking, exploration, and reading during summer breaks. Visited Alaska, Canada, and all lower forty-eight states. Met John Muir (1889), whom Mills credited with focusing his attention in wilderness preservation. Began lecturing on forestry, wilderness, and writing on these and related subjects, 1895. Briefly visited Europe, summer, 1900. Purchased older cousin's Longs Peak House in Estes Park, 1902, renamed it Longs Peak Inn, 1904, and gradually expanded the enterprise. This became principal source of income and one of the better-known summer inns in the United States. Active as Longs Peak guide, 1902–1906.

Began serious writing for many popular periodicals and outdoor magazines, 1902. Published a number of books, including *The Story of Estes Park and a Guide Book* (1905, and a number of later editions under various titles), *Wild Life on the Rockies* (1909), *The Spell of the Rockies* (1911), *In Beaver World* (1923), *The Story of a Thousand Year Pine* (1914), *The Story of Scotch* [Mills' collie] (1916), *Your National Parks* (1917), *Being Good to Bears: And Other True Animal Stories* (1919), *The Grizzly, Our Greatest Animal* (1919), *The Adventures of a Nature Guide* (1920), *Waiting in the Wilderness* (1921), and *Watched by Wild Animals* (1922). Posthumously published titles included *Wild Animal Homesteads* (1923), *The Rocky Mountain Park: Memorial Edition* (1924), *Romance of Geology* (1926), and *Bird Memories of the Rockies* (1931). *The Grizzly* volume and the three that followed it are considered by some authorities to have been his best work. In some of his earlier books he anthropomorphized his animal subjects, but later avoided this tendency.

Undertook extensive lecture tour in eastern states, 1905. Lectured on forests at request of President Theodore Roosevelt, 1907–1909. Made great number of lectures in all parts of the country. After Roosevelt left office, fought Forest Service "wise use" policies. Pressed for creation of Rocky Mountain National Park between 1909 and 1915; some 358.5 square miles placed within park boundaries instead of the 1,000 Mills had requested. Mills and his agitation for park was opposed by many of his neighbors and ranching interests in region. Began heated battle with National Park Service over transportation concessions in park beginning 1919, which was gradually expanded into legal action by the state of Colorado, which contested federal regulatory authority over roads not formally turned over to federal jurisdiction. Injured in New York City subway accident while on business trip there early in 1922, following which he was afflicted by flu, jaw abscess, and other health problems.

MAJOR CONTRIBUTIONS

While not a scientist, and not an especially gifted thinker or writer, Mills produced clear, simple, and readable articles and books that conveyed his great enthusiasm for the out-of-doors, animals, forests, and conservation. He was a man of deep conviction and drive who successfully made both children and adult Americans conscious of their natural heritage. Began informal school of nature guiding (primarily for children) at Longs Peak Inn, one of the many means through which he hoped to publicize his views. Largely forgotten for many decades, Mills' books have begun to be reprinted in recent years. He is regarded as ''a distinctive Rocky Mountain writer, indeed one of the best . . . at a time when California and Southwest-desert writers dominated the western nature essay.'' While original, he was unable to ''explore ideas and human emotions,'' but was a ''skilled storyteller with superb descriptive abilities and a capacity for close observation'' (Drummond).

BIBLIOGRAPHY

Buchholtz, C. W. *Rocky Mountain National Park: A History* (1983).

Drummond, Alexander. *Enos Mills, Citizen of Nature* (1995) (has useful bibliography).

Hawthorne, H., and E. B. Mills. *Enos Mills of the Rockies* (primarily for younger readers, 1935).

Pickering, J. H. ''Introduction.'' In Enos Mills, *Wild Life on the Rockies* (repr. ed., 1988).

Wild, Peter. *Conservationists of Western America* (1979).

Wild, Peter. *Enos Mills* (1979).

UNPUBLISHED SOURCES

See Enos Mills Papers, Western History Department, Denver Public Library; Enos Mills Cabin Collection (papers held by Mills family).

Keir B. Sterling

MINER, (JOHN THOMAS) JACK. Born Dover Centre, Ohio, 10 April 1865; died Kingsville, Ontario, 3 November 1944. Conservationist, writer, lecturer, bander.

FAMILY AND EDUCATION

Son of John Miner, a tile manufacturer, business he took over and passed on to his sons. Mother was Anne Broadwell Miner. Family moved to Kingsville, Ontario, in 1878, when he was thirteen. Attended grade school only three months when he was twelve and could neither read nor write until he was thirty-four. An older brother, Ted, was killed in hunting accident in 1898. Married Laona Wigle, 1888. A son and daughter died at early ages; there were three more sons.

POSITIONS

Owned and operated family drain tile business most of life. Founded Jack Miner Bird Sanctuary at Kingsville in 1904, officially declared provincial crown reserve in 1917. Founded Jack Miner Migratory Bird Foundation, Inc. in 1931.

CAREER

Miner spent most of his childhood and early adult years in ''the bush'' near Kingsville, learning the ways of wildlife. He and his two brothers simultaneously helped their father in a drain tile business and sold game obtained as market hunters. Jack was widely known as proficient hunter and congenial companion and was frequently asked to guide hunting parties. The death of his brother in a hunting accident in close proximity to other family tragedies resulted in his conversion to a religious zealot about the same time that he became interested in attracting Canada geese to his property. He gave up hunting and in 1904 established sanctuary on the grounds of his tile works, but this did not succeed until 1908. The sanctuary soon became popular and was declared provincial crown reserve by the government of Ontario in 1917. In order to raise funds to help pay for food for the growing numbers of birds, he began to lecture, and his humorous and evangelical lecturing style made him very popular throughout North America and Europe, both in person and on radio. His sanctuary was, and is, visited by many people, including prominent businessmen and heads of state. He was also among first people to warn of pollution in Great Lakes.

Miner was interested in tracing the travels of his geese and ducks and began an extensive banding program in 1909, the same year that the American Bird Banding Association was founded at a meeting of the American Ornithologists' Union. He did not then enter the developing cooperative international banding program, placing verses of Scripture on his bands instead. Miner was slow to adapt his sanctuary to maximum benefit to waterfowl, having little use for "book learning" of professional biologists, but did eventually modify his property to prevent overtaming of geese, and his banding data were also eventually incorporated into international records. He never accepted predators as part of an ecological balance, regarding them as "evil vermin."

In spite of his controversial stands and practices, Miner was the most outstanding and influential promoter of both conservation and banding of his time. His sanctuary idea was not new or unique, as sometimes claimed, but it was by far the best promoted and thus resulted in similar efforts elsewhere. His efforts were recognized by honorary memberships in several hunting, business, conservation, and religious groups, and he was even dubbed the "father of conservation" by an American journalist. He was awarded the Outdoor Life Gold Medal in 1929 and the Order of the British Empire in 1943. In 1947, the government of Canada declared the week of his birthday as annual National Wildlife Week.

Miner's experiences and views were written during his lifetime with help of his sons in two books, *Jack Miner and the Birds* (1923) and *Jack Miner on Current Topics* (1929), and many of his other writings and tributes to him were gathered together in *Wild Goose Jack*, published posthumously in 1969 but prepared about thirty years earlier as *Jack Miner and His Religion and Life*.

MAJOR CONTRIBUTIONS

Enthusiastic zeal in writing, broadcasting, and especially personal appearances brought early publicity and support to banding as a research tool, especially to conservation.

BIBLIOGRAPHY
Bodsworth, F. "Billy Sunday of the Birds." *Maclean's Magazine* (1 May 1952).
Foster, J. *Working for Wildlife, the Beginning of Preservation in Canada* (1978).
Hamilton, R., ed. *Prominent Men of Canada* (1932).
Lawrence, A., ed. *Who's Who among North American Authors*. Vol. 7 (1939).
Linton, J. M., and C. W. Moore. *The Story of Wild Goose Jack. The Life and Work of Jack Miner* (1984).
Mayfield, H. "Jack Miner—and the Role of Predators in Nature." *Atlantic Naturalist* 2,4 (1968), repr. in *Ontario Nationalist*, no. 1 (1969).
McNicholl, M. K. *The Canadian Encyclopedia* (1985, 1988).
Roberts, C. G. D., and A. L. Tunnell, eds. *The Canadian Who's Who*. Vol. 2 (1937).
Wallace, W. S., and W. A. McKay, eds. *The MacMillan Dictionary of Canadian Biography* (1978).
UNPUBLISHED SOURCES
Writings by and about Jack Miner and correspondence are housed at Jack Miner Bird Sanctuary, Kingston, Ontario.

Martin K. McNicholl

MITCHELL, JOHN. Born Whitechapel Parish, Lancaster County, Virginia, 3 April 1711; died near London (?), England, 19 February 1768. Natural historian, physician, cartographer.

FAMILY AND EDUCATION

Son of Robert Mitchell, merchant, tobacco receiver for several Virginia counties, justice of the county court, merchant, and Mary Chilton Sharpe Mitchell (died when Mitchell an infant); father remarried Susannah Payne. No indication as to early schooling. Attended University of Edinburgh, perhaps as early as 1722. M.A., 1729. Studied medicine there for two years. Probably received M.D. from a European university, 1731 or 1732. Married Helen (?) at unknown date. No information as to children.

POSITIONS

Practiced medicine in Lancaster County, Virginia, 1732–1734; in Urbanna, Virginia, 1734–1746, where also operated own apothecary shop; appointed physician to the poor, Urbanna, 1735. Left Virginia for reasons of health, 1746. In London, performed various (unsalaried?) services for Board of Trade and Plantations, 1750–1768.

CAREER

Nature of his undergraduate course of study at Edinburgh unknown, save that he received thorough grounding in botany and natural philosophy (including chemistry and physics). Studied medicine at Edinburgh, but no records of his taking degree there or elsewhere have been found. Nonetheless, was fully trained physician with appropriate credentials. Practiced medicine in Virginia for some fifteen years. Developed apothecary business and botanical garden in Urbanna. Knew John Clayton; contributed to latter's *Flora Virginica* (2d ed.). Participated in Natural History Circle, sending plants and trees to J. J. Dillenius at Oxford. Mentioned in latter's *Historia Muscorum*, c.1742. Frustrated by inability to identify and classify many plants, developed own system in ''Dissertatio Brevis Botanicorum et Zoologorum . . .'' in *Acta Physico-Medica Academae Caesarae Ephimerides*, vol. 8 (1748). In this publication, also named a number of new species and genera of American flora. This was attempt to improve ''natural'' system of John Ray. Was ''first North American to write on taxonomic principles,'' also attempted to put ''taxonomy on a genetic basis.'' Also wrote famous report concerning anatomy of both male and female opossums (read to Royal Society, 1742, supplemented in 1745), of which Stearns stated that ''few, if any, zoological papers originating in the colonies . . . equaled Mitchell's observations on the opossums.''

Mitchell also completed essay concerning human pigmentation (in *Philosophical Transactions* [1744]); unpublished paper concerning Virginia pines, in which he discussed taxonomy, ecology, physiology, and medical properties of turpentine. Suffering from malaria, made brief trip to Philadelphia, 1744, where he met John Bartram, B. Franklin, and others. Drafted detailed paper on nature of yellow fever; this manuscript later used by Benjamin Rush to deal with Philadelphia epidemic in 1793; not published until 1804. Mitchell's worsening health compelled him to sell property and move to England, 1746. Ship waylaid by privateer; Mitchell lost most of his notes and herbaria and parcels for European scientists. Herbaria later returned to him in poor condition. Met many Natural History Circle colleagues. Was prominent, but unsuccessful candidate for position of deputy postmaster general in the colonies, 1751–1753, and for post of first British museum librarian, 1756. Most famous for *Map of the British and French Dominions in North America* (1755), on which he spent at least five years, based on reports and maps from various British colonies in America and extensive map files of Board of Trade. It was often copied and not infrequently plagiarized. This is considered most famous map of colonial America and was used in resolving numerous controversies and in making boundary decisions as late as the 1920s. Described as ''the Most Important Map in American History'' by Col. Lawrence Martin, chief of Division of Maps, Library of Congress (1933). From 1759 to 1762, Mitchell helped to develop Royal Botanical Gardens, Kew. Published *The Contest in America between Great Britain and France with Its Consequences and Importance, by an Impartial Hand* (1757); *The Present State of Great Britain and North America with Regard to Agriculture, Population, Trade and Manufactures, Impartially Considered* (1767, repr. in new edition edited by H. Carman, New York [1939]). In latter, urged cooperation between mother country and colonies. Elected fellow of Royal Society, 1748, to which he submitted paper on preparation and use of potash (1748).

MAJOR CONTRIBUTIONS

A colonial polymath whose contributions to botany, zoology, cartography, medicine, relations between Great Britain and its colonies, and other fields justify John Bartram's opinion of him as "an ingenious man." His famous map of 1755 has doubtless been his most lasting contribution.

BIBLIOGRAPHY

Berkeley, Edmund, and Dorothy Berkeley. *Dr. John Mitchell: The Man Who Made the Map of North America* (1974).

Carrier, Lyman. "Dr. John Mitchell, Naturalist, Cartographer, and Historian." *Annual Report American Historical Association,* pt. 1 (1918).

Dictionary of American Biography.

Horrnberger, T. "The Scientific Ideas of John Mitchell." *Huntington Library Quarterly* 10 (1947).

Stearns, R. P. *Science in the British Colonies of America* (1970).

Thatcher, Herbert. "Dr. Mitchell, M.D., F.R.S., of Virginia." *Virginia Magazine of History and Biography.* Vols. 39, 40, 41 (1931–1933).

UNPUBLISHED SOURCES

Various collections are in British Museum, Archives of Royal Botanic Gardens and of Royal Society of London, and others. See Berkeley and Berkeley (1974) for complete list.

Keir B. Sterling

MITCHELL, MARGARET HOWELL. Born Toronto, Ontario, 28 October 1901; died Victoria, British Columbia, 3 October 1988. Ornithologist.

FAMILY AND EDUCATION

Daughter of George Howell, a merchant, and his wife, née Knox, a schoolteacher from New York. Attended a small private school and Havergal College for Girls, then studied biology, geology, and paleontology at the University of Toronto. Married Osborne S. Mitchell, an English engineer, in 1927.

POSITIONS

As there were no jobs for women in science, on graduation in 1924 Mitchell took a job as a secretary of the Paleontology Department of the Royal Ontario Museum (ROM). After her marriage, she was an unpaid volunteer in the Department of Ornithology, ROM, 1927–1950.

CAREER

As an eight-year-old in May 1909, she made one of the first documented sightings of the cardinal in Toronto, confirmed by J. H. Fleming and others. Author of a six-year study of the passenger pigeon in Ontario and a monograph on the birds of Brazil. As a housewife, she studied birds in Brazil (1950–1957), England (1958–1963), Barbados (1963–c.1970), and British Columbia (1970s).

MAJOR CONTRIBUTIONS

Canada's first woman ornithologist of international repute and the first research affiliate in any natural history museum in Canada. The definitive chronicler of the demise of the passenger pigeon in Ontario. Earned elective member status in the American Ornithologists' Union in 1958.

BIBLIOGRAPHY

Ainley, M. G. ''Margaret H. Mitchell (1901–1988).'' In M. K. McNicholl and J. L. Cranmer-Byng, eds., *Ornithology in Ontario* (1994).

Mitchell, Margaret H. *The Passenger Pigeon in Ontario* (1935).

Mitchell, Margaret H. *Observations on Birds of Southeastern Brazil* (1974).

C. Stuart Houston

MITCHILL, SAMUEL LATHAM. Born North Hempstead, Long Island, New York, 20 August 1764; died Brooklyn, New York, 7 September 1831. Physician, naturalist, statesman, chemist, geologist.

FAMILY AND EDUCATION

Third son of Robert and Mary (Latham) Mitchill, both Quakers. Father was farmer as well as ''pounder'' and overseer of highways. Learned medical principles from uncle, Samuel Latham, who financed his education. Served as apprentice for Samuel Bard, New York City, 1781–1783. Received M.D. with high honors from University of Edinburgh, 1786. After touring Europe returned to New York and received honorary M.A. from Columbia University, about 1788. Set up medical practice and commenced study of law under Robert Yates, New York State chief justice. Married Catherine (Akerly) Cock, wealthy widow, 23 June 1799. Adopted two girls but had no children of their own.

POSITIONS

Member of commission to study purchase of lands from Iroquois Indian nation for New York State, 1788. Representative to New York State Assembly 1791, 1798, 1810 (single–year terms). Professor, natural history, chemistry, and agriculture, Columbia University, 1792–1807. Founder and editor, *Medical Repository*, 1797–1820. Secretary, American Philosophical Society of Philadelphia, 1797–1820. President, Society for the Promotion of Agricultural Arts and Manufacturers of New York, 1797. President, American Mineralogical Society, 1799. Member, U.S. House of Representatives, 1801–1804, 1809–1813. Member, U.S. Senate, 1804–1809. Physician, New York Hospital (dates unknown). Professor of chemistry, College of Physicians and Surgeons (later part of Columbia University), 1807–1808. Professor of natural history, College of Physicians and Surgeons, 1808–1820. Founder, New York Literary and Philosophical Society, 1814. Founder, Lyceum of Natural History of New York (later New York Academy of Sciences), 1817. Surgeon general for the New York State Militia, 1818. Professor of botany and materia medica, College of Physicians and Surgeons, 1820–1826. President, New York State Medical Society, 1821. Founder and vice president, Rutgers Medical College, New Jersey, 1826–1830.

CAREER

Began private medical practice in New York City at about age twenty-four, upon return from travels through Europe. At Columbia taught new antiphlogistic chemistry of Lavoisier. Served as mediator between the pro-Lavoisier and anti-Lavoisier forces and involved in controversy with pro-phlogistic Joseph Priestly. Published *Explanation of the Synopsis of Chemical Nomenclature and Arrangement* (1801), in attempt to defend and organize pro-Lavoisier chemical knowledge and to end controversy. Developed from the new chemistry his fallacious ''septon'' theory of disease, in which disease-carrying fluids formed from oxygen and nitrogen compounds can be prevented and eradicated with ''antiseptic'' alkalis such as lime, soda and potash. Theory served as basis for his encouragement of better sanitation and personal hygiene. With Elihu Hubbard Smith and Edward Miller, established and edited the *Medical Repository*, first medical journal in the United States and vehicle for many of his medical and natural history articles. Conducted mineralogical exploration of Hudson River basin under the auspices of the Society for the Promotion of Agricultural Arts and Manufactures of New York, 1796, and from that study published his *Sketch of the Mineralogical History of the State of New York*, a pioneer geological work (1798–1800). As New York State legislator, 1798, fought successfully to grant friends Robert Fulton and Robert Livingston exclusive rights to conduct steamboat experiments in New York waters. In Washington, D.C., as a Jeffersonian Republican, worked to improve quarantine laws, reform working conditions in the U.S. Navy, and find ways to safeguard American shipping from Tripolitan pirate raids. Offered technical assistance to secretary of the navy on gunpowder, sailors' health, and protection of naval property. During War of 1812 served on commission to oversee development of a steam-war vessel and helped develop plans for the defense of New York City against the British. As professor of natural history, collected, identified, and classified dozens of species of American flora and fauna, particularly aquatic fauna. Lectured on botany, zoology, and mineralogy. Published *Discourse . . . Embracing a Concise and Comprehensive Account of the Writings Which Illustrate the Botanical History of North and South America* (1814), which attempted to organize all knowledge of Western Hemisphere plant life, and *Fishes of New York* (1815), recognized by contemporaries as a landmark ichthyological work. Collected plant, animal, and mineral specimens throughout his career from correspondents and from personal travels through New York State and along the Atlantic seaboard from Long Island to Virginia. Published in great detail numerous articles on botany and zoology. Helped organize study of pharmacopeia in the United States, 1820. Corresponded with William Cooper, J. J. Audubon, Alexander Wilson, and C. S. Rafinesque. With colleagues J. W. Francis, David Hosack, W. J. MacNeven, and Valentine Mott, left College of Physicians and Surgeons to found the short-lived Rutgers Medical College, 1826–1830.

Published several books and at least 200 articles, 1786–1829, with most articles published in the *Medical Repository*, 1797–1821; the *American Monthly Magazine and Critical Review*, 1818–1819; and the *American Journal of Science*, 1821–1829. Publications covered wide range of topics in natural history, geology, and chemistry. Biological articles covered topics ranging from New York fish, to Galapagos tortoises, to Spanish chestnuts. Other noteworthy works include *Life, Exploits and Precepts of Tammany, the Famous Indian Chief* (1795), *A Tour through a Part of Virginia in the Summer of 1808* (1809), *Address at the Completion of the Erie Canal* (1825), and *Discourse on the Character and Services of Thomas Jefferson, More Especially as Promoter of the Natural and Physical Sciences* (1826). Edited and translated several works from European and American authors, including *Zoonomia, or Laws of Organic Life*, by Erasmus Darwin (1806) and *Elementary Introduction to the Laws of Mineralogy*, by W. Phillips (1818). Included an essay, "Observations on the Geology of North America," in American edition of George Cuvier's *Essay on the Theory of the Earth* (1818).

Wrote extensively on variety of topics, yet most articles reflect encyclopedic breadth of author's knowledge rather than originality of thought. Much of his original work superseded by more proficient scientists in later years. Widely admired in United States and Europe for ability to quickly marshal and dispense scientific information when called upon, but occasionally ridiculed for being pedantic and egotistical. Was a ready public speaker who talked on almost any subject of a scientific nature. Most efforts designed to make scientific knowledge accessible to the lay public. Affable, aristocratic, remembered more for "the goodness of his heart than the strength of his head." Influential in promoting science for its own good and for practical applications of science in everyday life.

MAJOR CONTRIBUTIONS

Encouraged widespread interest in lay and scientific circles for native American biological and geological phenomena. Aided introduction of Lavoisier's chemistry into American science. Promoted birth of disinfectants and sanitary science as means of combating disease. Organized available information on North American flora and fauna. Provided vehicles for introducing and disseminating scientific information to the public.

BIBLIOGRAPHY

Aberbach, Alan David. *In Search of an American Identity: Samuel Latham Mitchill* (1988).

Fairchild, H. L. *A History of the New York Academy of Sciences* (1887).

Francis, John Wakefield. *Reminiscences of Samuel Latham Mitchill* (1859), repr. in S. D. Gross, *Lives of Eminent American Physicians and Surgeons* (1861) and abridged in H. A. Kelly and W. L. Burrage, eds., *American Medical Biographies* (1928).

Goode, G. B. *The Beginnings of American Science: The Third Century* (1888).

Hall, Courtney Robert. ''A Chemist of a Century Ago.'' *Journal of Chemical Education* (March 1928).

Hall, Courtney Robert. ''Samuel Latham Mitchill.'' Diss., Columbia University, 1933.

Hall, Courtney Robert. ''Samuel Latham Mitchill: A Queens County Polymath.'' *New York History* (April 1933).

Hall, Courtney Robert. *A Scientist in the Early Republic: Samuel Latham Mitchill* (1934, repr. 1967).

Mitchill, Samuel L. *Some of the Memorable Events and Occurrences in the Life of Samuel L. Mitchill of New York from 1786 to 1826* (1826).

Newell, Lyman C. ''Samuel Latham Mitchill.'' *Dictionary of American Biography*. 1st ed. (1934).

Pascalis-Ouviere, Felix. *Eulogy on the Life and Character of the Hon. Samuel Latham Mitchill* (1831).

Pascalis-Ouviere, Felix. *Samuel Latham Mitchill: A Father of American Chemistry* (1922), reprinted from *Journal of Industrial and Engineering Chemistry* [June 1922]).

UNPUBLISHED SOURCES

Papers are in the Osgood Papers, Miller Papers, and Miscellaneous Manuscripts, New York Historical Society. Handwritten natural history lectures are in the Morton Pennypacker Long Island Collection, Gardiner Memorial Room, East Hampton (N.Y.) Free Library. Letters to wife while in Washington, D.C., are in Museum of the City of New York Library. Letters to Thomas Jefferson are in Thomas Jefferson Papers, Manuscript Division, Library of Congress, Washington, D.C. Notes on botany and materia medica lectures are in Thomas A. Brayton Papers, New York Public Library.

Michael J. Boersma

MOCIÑO, JOSÉ MARIANO. Born Temascaltepec, State of Mexico, Mexico, 24 September 1757; died Barcelona, Spain, 19 May 1820. Botanist, museum curator, physician.

FAMILY AND EDUCATION

From poor family, Mociño had difficulty financing his education. Graduate in humanities (*artes*), Diocesan Seminary, Mexico City, 1776. Student of theology, University of Mexico, 1776, and Diocesan Seminary, 1778, did not take degree. Medical student, University of Mexico, 1784–1787, bachelor of medicine, University of Mexico, 1787. Also studied mathematics at Royal Fine Arts

Academy, San Carlos, 1786–1787. Later studied botany at Royal Botanical Garden, Mexico City, 1789. Married Maria Rita Rivera y Melo Montaño, 1778. No information as to children. Mociño is spelling of name usually found in botanical literature. Mociño himself always signed himself Moziño. Later authors employed various spellings of the name, including Mocinno, Moçino, Moziño, and Mozinno.

POSITIONS

Professor of philosophy, Diocesan Seminary of Oaxaca, 1779–1783. Substitute professor of mathematics and astrology, University of Mexico, 1786. Appointed member of the Royal Scientific Expedition to New Spain, headed by Martín de Sessé, 1790. Member of Spanish scientific expedition to Nootka Sound, West Vancouver Island, in southwest portion of present-day British Columbia, 1791–1793. Director of Medical Police in Andalusia, Spain, 1803. Twice secretary and four times president of the Academy of Medicine of Madrid, 1803–1808. Director of Natural History Museum of Madrid and professor of zoology, 1808–1812.

CAREER

Versed in philosophy, theology, mathematics, and medicine, Mociño became the most outstanding botany student in the Botanical Garden of Mexico City. Chosen to defend the Linnaean system of plant classification before the university cloister at the public ceremony ending the course, 1789, Mociño's ability caused Martín de Sessé (died 1808) director of the Royal Scientific Expedition, to select him as the only Mexican scientist in the expedition. Immediately Mociño began a series of trips to collect plant specimens. In 1790–1791 visited western part of New Spain: Michoacán, Querétaro, Guanajuato, Colima, Guadalajara, and Tepic. His poem in Latin verse describing the volcanic eruption of Jorullo, Michoacán, was praised by Casimiro Gómez Ortega, director of Madrid's Botanical Garden, and Pablo de la Llave, noted Mexican naturalist. From San Blas in 1791 he traveled with Spanish expedition to northwest coast of North America, accompanied by Mexican artist Atanasio Echeverría. Studied culture, religion, crafts, government, and economic life of Indian whale hunters of Nootka, Canada. Learned Indian tongue and wrote dictionary. Cataloged animals and 200 species of Canadian plants according to Linnaean system. ''Notices on Nootka'' published in *Gaceta de Guatemala* (1804) and reprinted by Mexican Society of Geography and Statistics (1913). Maintained correspondence with English naturalists whose activities coincided with his own in Nootka. Alexander Von Humboldt recommended that Nootka studies be published in French. Upon returning to Mexico, continued botanical expeditions, 1793–1799: provinces of Hidalgo, Veracruz, Oaxaca, where at viceroy's request he studied eruption of San Andrés Tuxtla volcano, Tabasco, Tehuantepec, Chiapas, Guatemala, El Salvador, and Nicaragua. Mociño's travels constituted the most extensive exploration made by a naturalist in America up to that time.

From 1799 to 1803 Mociño practiced medicine in San Andres Hospital in Mexico City, conducting research on medicinal properties of plants with Luis José Montaño, who described Mociño as "the most outstanding and most agreeable genius that New Spain has produced." Used quinine to save victims of vomiting in Veracruz. Translated into Spanish and published medical works of Scottish doctor John Brown. In inaugural speech for botany course, 1801, insisted on use of Mexican medicinal plants instead of importing expensive and adulterated remedies from Europe. Traveled to Spain, 1803, when Sessé and Scientific Expedition returned. Transported immense collection of manuscripts and drawings. As director of Sanitary Police of Andalusia, combated yellow fever epidemic, 1808. Functionary of Academy of Medicine of Madrid and director of natural history museum. Taught first zoology course in Spain. During Napoleonic invasion French confirmed Mociño as museum director. When French were defeated, Mociño went to live in Montpellier, France, taking botanical manuscripts with him. Augustus De Condolle, Swiss naturalist working in Montpellier, borrowed Mociño's collection. When Mociño requested its return, 1817, De Condolle made his famous appeal to all artists of Geneva, who in a week copied at least 900 of the 1,400 drawings done by Echeverría. Thanks to this idea some drawings are known today, since De Condolle published them with Mociño's notes, explaining their origin. When Mociño, gravely ill, returned to Spain and died, the original drawings by Echeverría were lost and never recovered. Three volumes of his manuscripts on the *Flora Mexicana* are preserved in the Botanical Garden of Madrid as well as writings on Guatemala, Cuba, and Puerto Rico.

Published some scientific articles during his life in *Anales de las Ciencias Naturales*, Madrid, but his major botanical research was published over sixty years after his death. Major publications include: *Noticias de Nutka* (1804); *Noticias de Nutka. Diccionario de la lengua de los nutkenses y descripción del volcán de Tuxtla* (preliminary study by Alberto María Carreño [1913]); "Prologue" of John Brown's Epitome of *The Elements of Medicine* (Puebla [1802]); "Discurso inaugural," *Gaceta de México* (19 September, 1801); Madrid (1802); *La Naturaleza*, vol. 7 (1887); "Observaciones sobre la resina de hule" (Observations on the rubber resin) (Madrid [1804]); "La Polygala mexicana" (Madrid [1804]); "Tratado de Xiquilite y Añil de Guatemala" (Manila, Philippines [1804]); *Plantae Novae Hispniae* (México, Secretaría de Fomento [1893]); *Flora Mexicana* (with Martín de Sessé, México, Secretaría de Fomento [1895]).

First botanist to travel extensively in Canada, Mexico, and Central America, classifying specimens according to Linnaean system. Described over 2,000 specimens of Mexican plants. Investigated medicinal properties of plants included in Aztec curative practices.

MAJOR CONTRIBUTIONS

Most outstanding Mexican botanist during last years of eighteenth century and best known by the European scientific community in early decades of nineteenth century. Promoted Linnaean system in Mexico and classified over 2,000 plant species from Canada, Mexico, Central America, Cuba, and Puerto Rico.

BIBLIOGRAPHY

Arías Divito, Juan Carlos. *Las expediciones Científicas españolas durante el siglo XVIII.* Madrid (1968).

Carreño, A. M. "El Br. D. José Moziño y la Expedición Científica del siglo XVIII." In *Noticias de Nutka . . .* (1913).

Dictionary of Canadian Biography.

McVaugh, Rogers. "Mociño, Jose Mariano." *Dictionary of Scientific Biography.*

"Mociño Suárez Losada, José Mariano." *Diccionario Porrúa* (1964).

Rickett, Harold William. "The Royal Botanical Expedition to New Spain, 1788–1820." *Chronica Botánica* (1947).

Wilson, Iris H. "Scientific Aspects of Spanish Exploration in New Spain during the Late Eighteenth Century." Diss. University of Southern California, 1962.

Wilson, Iris H. "Spanish Scientists in the Pacific Northwest, 1790–1792." In J. A. Carroll, ed., *Reflections of Western Historians: Papers of the 7th Annual Conference of the Western History Association of the History of Western America, San Francisco, October 12–14, 1967* (1969).

UNPUBLISHED SOURCES

Letters and documents regarding Botanical Expedition to New Spain are in Mexican National Archives. Small number of similar items are in William C. Clements Library, University of Michigan. Manuscripts on Mexican and Guatemalan flora are in Instituto Botánico, Madrid, Spain.

Dorothy Tanck de Estrada

MOORE, RAYMOND CECIL. Born Roslyn, Washington, 20 February 1892; died Lawrence, Kansas, 16 April 1974. Geologist, invertebrate paleontologist.

FAMILY AND EDUCATION

Son of Bernard Harding Moore, Baptist minister, and Winifred Denney Moore. B.S., Denison University, 1913; Ph.D., University of Chicago, 1916. Married Georgine Watters, 1917 (divorced); one daughter. Married Lilian Botts, 1936.

POSITIONS

Assistant professor of geology, 1916–1919, professor of geology, University of Kansas, 1919–1958, department chair, 1920–1933, 1940–1941, 1952–1954. Geologist for U.S. Geological Survey Expedition surveying Grand Canyon, 1923; Kansas State Geologist, 1916–1937, State Geologist and Director, Kansas

Geological Survey, 1937–1943, 1945. On active duty with U.S. Army Corps of Engineers as assistant chief, Planning Branch, Fuels and Lubricants Division, Office of the Quartermaster General, 1943–1945. Kansas State Geologist, 1945–1954. Solen E. Summerfield Distinguished Professor, University of Kansas, 1958–1970.

CAREER

Undertook fieldwork in stratigraphy for Texas Geological Survey in early 1920s. Early became known for work on Mississippian and Pennsylvanian stratigraphy, especially in Missouri, Kansas, and other parts of the midcontinent. His work emphasized that sedimentary record was cyclic. He ultimately was able to follow some stratigraphic elements a centimeter thick for distances of several hundred kilometers. Discovered Nemaha Anticline, subterranean mountain range stretching from Oklahoma to Nebraska. Able field geologist and authority on fossil invertebrates that helped determine age of stratigraphic units he studied. Specialized in fossil sea lilies, corals, and bryozoans. His paleontological and geological papers ultimately totaled some 300. Known for his outstanding contributions as an editor. Edited *Bulletin of American Association of Petroleum Geologists*, 1920–1926; *Journal of Paleontology*, 1930–1939; *Journal of Sedimentary Petrology*, 1931–1939, and, for a number of years, the *Paleontological Contributions* of the University of Kansas. During World War II, returned to Texas, mapping strata as part of ongoing oil and gas reconnaissance that was carried on for U.S. Geological Survey. As a captain and then a major in the U.S. Army Reserves (which he had joined in 1929), served with U.S. Army in Washington, D.C. Later, in 1949, was civilian consultant to Gen. Douglas MacArthur in evaluating Japanese coal resources. From 1948, took on his most important assignment, editorship of the *Treatise on Invertebrate Paleontology*, which he initiated. Was sole editor, 1953–1964, and coeditor from 1964 until his death. Identified several hundred specialists from around the globe to complete portions of this effort, of which twenty volumes had been completed during Moore's lifetime. Often drafted sections for which contributors could not be found by himself or with a colleague. This publication, originally conceived of as a compilation, has resulted in many new and revised fossil classifications. When completed, it will also be more extensive than the thirty-six volumes originally envisioned. His other publications included *Geology of Salt Dome Fields* (1926); *Historical Geology* (1933); *Evolution and Classification of Paleozoic Crinoids* (with L. R. Laudon, 1943); *Introduction to Historical Geology* (1949, 2d ed. 1958), and *Invertebrate Fossils* (1952). Moore was cofounder of the Society of Economic Paleontologists and Mineralogists and its president in 1928. Also president of Association of State Geologists, 1941–1942; Paleontological Society, 1947, Geological Society of America, 1957–1958, and American Geological Institute, 1960. Active in a number of other professional organizations in various capacities and won a number of awards and medals for his activities.

MAJOR CONTRIBUTIONS

Outstanding paleontologist and geologist who is probably best known for his editorship of the *Treatise on Invertebrate Paleontology*. This project, of which the University of Kansas has been copublisher, was in part endowed in his will.

BIBLIOGRAPHY

Dunbar, Carl O. In Curt Teichert and E. L. Yochelson, eds., *Essays in Paleontology and Stratigraphy, R. C. Moore Commemorative Volume* (1967).

Maples, Christopher G., and Rex Buchanan. ''Raymond C. Moore (1892–1974): Memorial and Bibliography.'' *Celebration of the 100th Anniversary of the Kansas Geological Survey* (1989).

Merriam, D. F. *Annual Report, Geological Society of London, 1974* (1975).

Yochelson, E. L., *Dictionary of Scientific Biography.* Supplement 2, vol. 18 (1990).

Keir B. Sterling

MORGAN, ANN HAVEN. Born Waterford, Connecticut, 6 May 1882; died South Hadley, Massachusetts, 5 June 1966. Biologist, pond and stream ecologist, conservationist.

FAMILY AND EDUCATION

Eldest of three children of Stanley Griswold and Julia Alice Douglas Morgan. Known by the first name of Anna until age thirty. Attended Williams Memorial Institute, New London, Connecticut, Wellesley College, 1902–1904; A.B., Cornell, 1906, Ph.D., Cornell, 1912. Postgraduate studies at University of Chicago, Harvard, and Yale, summers of 1916, 1920, 1921. Unmarried.

POSITIONS

Assistant in zoology, 1906–1907; instructor in zoology, 1907–1909, Mt. Holyoke College, graduate assistant, Cornell, 1909–1912, instructor in zoology, 1912–1914, associate professor, 1914–1918, chair of the Zoology Department, 1916–1947, professor, 1918–1947, Mt. Holyoke College; on staff of Marine Biological Laboratory, Woods Hole, Massachusetts, as instructor, summers of 1918, 1919, 1921, 1923; researcher at Tropical Laboratory, Kartabo, British Guiana, summer, 1926. Taught summer sessions at Cornell and University of Chicago.

CAREER

Specialist on water biology, particularly of the northeastern United States. Described as an "exacting, but memorable" teacher. During the 1940s and 1950s, turned her attention to science curricula offered in high schools and colleges. As member of National Committee on Policies in Conservation Education, offered series of summer workshops for zoology, geography, and sociology teachers to encourage them to incorporate ecology and conservation in their courses. Authored *Field Book of Ponds and Streams* (1930), the standard guide to the subject for four decades; *Animals in Winter* (1939), which was also the title of a 1949 film she made with Encyclopedia Britannica Films, and *Kinship of Animals and Men* (1955), which was a broad-gauged text dealing with relationships within and between animal groups. Also published articles in *Annals of the Entomological Society of America, Anatomical Record*, and other journals. Described mayflies (the subject of her doctoral dissertation) as "her favorite preoccupation . . . because mayflies are fine for small boys to fish with." Spent many years researching water bugs in a pond atop Mt. Tom, near Northampton, Massachusetts. Received research grants in 1926 and in 1930 from both Sigma Xi and American Association for the Advancement of Science (AAAS), of which she was later a fellow. Strongly supported conservation efforts in the Connecticut River Valley. One of only three women selected for listing in American Men of Science in 1933.

MAJOR CONTRIBUTIONS

A notable teacher of zoology to college students and the general public. Known to her freshman students at Mt. Holyoke as "Mayfly Morgan." Specialist on aquatic insects, though she considered herself a general zoologist. Her *Field Book of Ponds and Streams* was for forty years the freshwater handbook most heavily used by both professional and amateur naturalists, together with fishermen. Always carefully set the animals she studied within context of broad ecosystem. Appointed a visiting fellow at Harvard, Yale, and Cornell Universities. Among the first biologists to make her students and readers ecologically aware.

BIBLIOGRAPHY

American Men and Women of Science. 10th ed. (1961).

Bonta, Marcia Myers. *Women in the Field: America's Pioneering Women Naturalists* (1991).

Holyoke Transcript-Telegram, 6 June 1966 (obit.).

Keene, Ann T. *Earthkeepers: Observers and Protectors of Nature* (1994).

Leaders in American Science. Vol. 1 (1953).

Notable American Women: The Modern Period (1980).

Who Was Who in America.

UNPUBLISHED SOURCES

Some of Morgan's correspondence is in Cornelia M. Clapp Papers, Mt. Holyoke College Library, South Hadley, Massachusetts.

Domenica Barbuto

MORSE, EDWARD SYLVESTER. Born Portland, Maine, 18 June 1838; died Salem, Massachusetts, 20 December 1925. Naturalist, conchologist, evolutionist, anthropologist, museum administrator.

FAMILY AND EDUCATION

A son of Jonathan Kimball and Jane Seymour Beckett Morse. Came from old New England stock. A boy naturalist who collected minerals and shells by age thirteen. At seventeen was active in Portland Society of Natural History and soon placed on Board of Managers. Was draftsman for Maine Central Railroad. Studied under Agassiz at Museum of Comparative Zoology (1859–1862); honorary Ph.D., Bowdoin College; honorary A.M., Harvard University, 1892. Married Ellen Elizabeth Owen, 1863; two children.

POSITIONS

Assistant, Museum of Comparative Zoology, 1859–1862; curator, Portland Society of Natural History, 1863–1866; curator (for mollusks), Essex Institute, Salem, Massachusetts, 1866–1867; continued as same for Peabody Academy of Science, 1867–1871; professor of comparative anatomy and zoology, Bowdoin College, 1871–1874; returned to Peabody Academy of Science, 1874; lecturer, Imperial University of Tokyo, 1877–1880; director, Peabody Academy of Science, 1880–1916; director emeritus, 1916–1925; keeper of Japanese pottery, Boston Museum of Fine Arts, 1892–1925.

CAREER

Continued to serve Essex Institute, after joining the new Peabody Academy of Science, as a curator, 1867–1925, member of council, 1888–1893, and vice president, 1894; founder (with Putnam, Packard, and Hyatt) of *American Naturalist*, 1867; elected fellow, American Academy of Arts and Sciences, 1868; instructor at Agassiz's summer school on Penikese Island, 1873; elected National Academy of Sciences, 1876. Published text on zoology, 1875, which he illustrated, as well as preparing illustrations for other authors and especially for the *American Naturalist*. Was an early environmentalist who advocated reforms through public lectures and newspaper articles. Research on mollusks and brachiopods, teaching of zoology and evolution, organization of a marine laboratory, and discovery of ancient Japanese pottery in Japan, 1877–1880 and 1881–1882. Became world's authority on Japanese ceramics and published on Japanese culture. Third-Class of the Order of the Rising Sun conferred by Emperor Mutsuhito (1899) and Second-Class Order of the Sacred Treasure conferred by Emperor Taisho for contributions to science and education in Japan (1922). President, American Association for the Advancement of Science, 1886. He was noted as a popular lecturer, especially on evolution and the environment, and for his ambidextrous drawings on the blackboard (see bibliography in Howard 1937).

MAJOR CONTRIBUTIONS

Research and publications on land snails. Determined relation of brachiopods to annelids rather than mollusks. Among first of American naturalists to support and advocate Darwinism. Books on Japanese ceramics and culture. Museum administration.

BIBLIOGRAPHY

Dexter, R. W. ''An Early Environmentalist—E. S. Morse and His One-Man Campaign to Improve the Human Environment.'' *Nature Study* 28, 1 (1974).

Dexter, R. W. ''The Impact of Evolutionary Theories on the Salem Group of Agassiz Zoologists (Morse, Hyatt, Packard, Putnam).'' *Essex Institute Historical Collections* 115, 3 (1979).

Howard, L. O. *Biographical Memoir of Edward Sylvester Morse, 1838–1925. Biographical Memoirs, National Academy of Science* 17 (1937).

Wayman, D. G. *Edward Sylvester Morse: A Biography* (1942).

UNPUBLISHED SOURCES

Correspondence is in Library, Peabody Museum, Salem, Massachusetts.

Ralph W. Dexter

MORTON, ROBIN ALLEN. Born Nelson, British Columbia, 11 November 1953; died near Robson Bight, British Columbia, 16 September 1986. Naturalist, whale researcher, film-maker.

FAMILY AND EDUCATION

Youngest of five children of Arthur and Vera (Watts) Morton. Mother staffed National Film Board office in Nelson for six years. Family moved to Victoria in 1966. Attended Oak Bay Secondary, Victoria, B.C. Married Alexandra Hubbard, 1981. One son.

POSITIONS

Film editor and then photographer, CHEK-TV, Victoria, B.C., 1974–1977. Partner in freelance underwater photography firms, Aqua Cine and Pacific Sealife Data, 1977–1986. Member of Explorer's Club.

CAREER

At age six began helping mother after school by rewinding and splicing 16 mm films. Learned to scuba dive in high school. Worked part-time for Bob Wright at commercial whale education center, Sealand (Victoria, B.C.), cleaning killer whale (*Orcinus orca*) tanks. Polished editing skills at CHEK-TV before forming Aqua Cine with Michael O'Neill, making films such as *Birth of a Salmon* (1979). Developed non-intrusive method to film orcas at Robson Bight "rubbing beach" via remote controlled underwater camera system. Met his wife while field testing it in September 1980. Influenced by John Lilly, she was in the region in preparation for graduate studies. Couple married and intended to become full-time whale researchers. Purchased 20 meter vessel *Blue Fjord* 1981. Published one article: "Encounters with the Killer Whale: Studies of *Orcinus orca* Behavior in the Wild," *Explorer's Journal*, (June 1982): 54–61. In 1982 argued for catch and release program for whales. Drew public attention to health and ethics issues involving whales at Sealand. Donated orca film footage to aid lobbying for world's first orca reserve, the Robson Bight Ecological Reserve (created June 1982, see also Michael A. Bigg). From 1984–1986, based out of Echo Bay, Gilford Island (30 km north of Robson Bight) they conducted open water work documenting orca pod histories, behavior and vocalizations. Robin died the week before a National Geographic television crew arrived to film them at work. While diving among a pod of orcas his experimental rebreather unit failed. Body recovered by wife. Memorial service held at Newcombe Auditorium, provincial museum (21 September 1986). Commemorative wall mural, Wharf Street, Victoria. Ashes scattered, Robson Bight. Alexandra continued to document language and behavior of orca pods through Raincoast Research (formerly Lore Quest), non-profit society. Posthumous documentary, *Killer Whales in the Wild* (Michael Chechik and Robin Morton, 1991) received awards: Houston International Film Festival, Earthwatch, and the Festival International du Film Maritime & d'Exploration.

MAJOR CONTRIBUTIONS

Was a young marine naturalist who combined activism, ethics, science, and film-making to increase public awareness of whales and whale research. While lacking the scholarly depth of his friend, scientist Michael Bigg, he personifies the emotive, cultural, and spiritual aspects of society's fascination with orcas. Overlooked figure in creation of world's first orca reserve.

BIBLIOGRAPHY

Morton, A. "Into the World of Orcas / Robin Morton, 1953–1986." *International Wildlife* 17 (Sept./Oct. 1987).

"Morton Had a Certain Rapport With Whales," *Times-Colonist* (Victoria), 19 September 1986.

"Noted B.C. Filmer of Killer Whales Drowns," *Sun* (Vancouver), 18 September 1986.

Obee, B. and G. Ellis. *Guardians of the Whales: The Quest to Study Whales in the Wild* (1992).

UNPUBLISHED SOURCES

Source material limited to film footage. Contact: Alexandra Morton, General Delivery, Simoom Sound, B.C. VOP 1SO, Canada.

Lorne Hammond

MOSELEY, EDWIN LINCOLN. Born Union City, Michigan, 29 March 1865; died Bowling Green, Ohio, 6 June 1948. Educator, scientist, naturalist.

FAMILY AND EDUCATION

Born to William Augustus and Sophia (Bingham) Moseley; father a merchant, maternal grandfather a missionary. Youngest of nine children, at the third-grade level became interested in natural history through efforts of a teacher who took pupils on field trips and identified plants. Primary and secondary education at Union City schools where he studied the sciences, the classics, and philosophy and graduated (1880). Too young to enter the university, he spent a postgraduate year in high school and after four years of study was awarded the M.A. from the University of Michigan (1885), the youngest member of his class. Honorary Doctor of Humane Letters from Bowling Green State University (1943), the first faculty member to be so honored. Never married.

POSITIONS

Science teacher, Grand Rapids High School, Michigan (1885–1887), and Sandusky High School, Ohio (1889–1914); member of the J. B. Steere Zoological Expedition to the Philippine Islands (1887–1888); became one-man science department and member of the original faculty of Bowling Green State Normal College (now Bowling Green State University), Ohio (1914–1936); professor emeritus of biology and curator of the university museum (1936–1948).

CAREER

Traveled widely in the United States and in foreign countries. Cataloged the vascular plants of northern Ohio's most diverse floristic areas, the island and Sandusky Bay region of Lake Erie and the Oak Openings west of Toledo. A prodigious collector of all kinds of scientific and historic materials, he developed a museum in Sandusky of over 17,000 specimens which was considered the best high school museum in the country; later transferred to Bowling Green State University. A pioneer in outdoor science education, he guided his students to make careful, original observations and to use independent thinking. Taught all sciences, including astronomy, biology, chemistry, geography, geology, hygiene, physics, philosophy and even some courses in English, Latin, and geometry.

Conducted the first satisfactory experiments to demonstrate that trembles in animals result from their grazing on a common woodland plant, white snakeroot, and that the poison enters the milk, causing the disease known as milk sickness in human beings. Stimulated much interest locally and nationally by making long-range weather forecasts, based on detailed studies of tree-ring widths and lake-level records; formulated a theory that rainfall in certain areas repeats itself in cycles of 90.4 years or four times the period of the magnetic sun-spot cycle. Described the formation of Sandusky Bay and Cedar Point and mapped the preglacial river valleys in the region; this information was useful for practical engineering developments in the city of Sandusky and its environs. Of frail physique throughout life, he lived somewhat as a recluse and was abstemious to the comforts of life, yet unostentatiously philanthropic in certain ways, financially helped students who showed promise of achievement; in accordance with this philosophy willed his entire estate, valued in 1974 at over $120,000, to Bowling Green State University to aid worthy students.

Author of three general science textbooks and over 100 articles in professional and popular journals, making original contributions to botany, geology, meteorology, medical science, science education, and zoology, including ''Climatic Influence of Lake Erie on Vegetation'' (1897), ''Sandusky Flora'' (1899), ''Original Work in High Schools'' (1903), ''Formation of Sandusky Bay and Cedar Point'' (1905, reprinted 1973), ''The Cause of Trembles and Milk Sickness'' (1909), *Trees, Stars, and Birds: A Book of Outdoor Science* (1919, repr. 1925, 1927; rev. ed. 1935), ''A Plea for More Outdoor Science Teaching'' (1924), ''Some Suggestions for Outdoor Science Teaching'' (1925), *Our Wild Animals* (1927), ''Flora of the Oak Openings West of Toledo'' (1928), *Other Worlds* (1933), ''Blue Heron Colonies in Northern Ohio'' (1936), ''What May be Learned from Stumps?'' (1938), ''Long Time Forecasts of Ohio River Floods'' (1939), ''The Ninety-Year Precipitation Cycle'' (1940), ''Milk Sickness Caused by White Snakeroot'' (1941), ''Sun-spots and Tree Rings'' (1941), ''Solar Influence on Variation in Rainfall in the Interior of the United States'' (1942), ''Precipitation Prospects, 1943–47, for Ohio and Near-by States'' (1944), ''Recurrence of Floods and Droughts after Intervals of about 90.4 Years'' (1944), and ''Variations in the Bird Population of the North-central States due to Climatic and Other Changes'' (1947).

MAJOR CONTRIBUTIONS

A truly devoted teacher who received overwhelming loyalty from his students, he pioneered in the teaching of natural science by the experimental method in the field. As stated by F. J. Prout, his lifelong friend and former president of Bowling Green State University, ''his greatness as a teacher lay in his emphasis on always searching for the reason of things.'' In his time, he was one of Ohio's greatest all-around naturalists.

BIBLIOGRAPHY
''Edwin Lincoln Moseley.'' *National Cyclopaedia of American Biography* 41 (1956).
Mayfield, H. F. ''Edwin Lincoln Moseley, Naturalist and Teacher, 1865–1948.'' *Northwest Ohio Quarterly* 56 (1984).
Niederhofer, R. ''Edwin Lincoln Moseley: An Internationally Known Naturalist.'' *Bartonia* 54 (1988).
Niederhofer, R. E., and R. L. Stuckey. ''Edwin Lincoln Moseley (1865–1948): Naturalist, Scientist, Educator.'' (In Press).
Otis, C. H. ''Edwin Lincoln Moseley.'' *Ohio Journal of Science* 49 (1949).
Prout, F. J. ''The Inquiring Mind of Mr. Moseley.'' *Bowling Green State University Magazine* (November 1959).
Stuckey, R. L. ''Edwin Lincoln Moseley's Contributions to Science.'' *Ohio Journal of Science* 83, 2 (1983) (abstract).
True, J. ''Edwin Lincoln Moseley: The Biography of an Educator.'' *Nature Magazine* 38 (1945).
Tudury, M. ''Long-range Weather Man.'' *Country Gentleman* 103 (November 1943).
Van Tassel, C. S. ed. ''Edwin Lincoln Moseley.'' In *The Ohio Blue Book, or Who's Who in the Buckeye State: A Cyclopaedia of Biography of Men and Women of Ohio* (1917).
Williams, H. B. ''Dr. Williams lauds Moseley.'' *Bee Gee News* 21 (31 March 1937).
Wolfe, D. ''The Legend and Legacy of a Farsighted Professor.'' *Toledo Blade Sunday Magazine* (9 April 1972).

UNPUBLISHED SOURCES

No major collection of correspondence or manuscript material known to exist; library of Bowling Green State University and public libraries of Sandusky and Toledo maintain files of notes and newspaper clippings; the C. E. Frohman Collection at the Rutherford B. Hayes Memorial Library, Fremont, Ohio, has an index of Moseley's articles written in the Sandusky newspapers.

Ronald L. Stuckey

MOUSLEY, (WILLIAM) HENRY. Born Taunton, England, 17 February 1865; died Montreal, Quebec, 22 September 1949. Ornithologist, botanist, entomologist, nature photographer.

FAMILY AND EDUCATION

Eldest of three sons of a prosperous railway contractor. Educated Clifton College, Bristol, 1876–1882. Self-taught naturalist. Married Alice Maud Mary Lake in 1885. Of their twelve children three died in infancy.

POSITIONS

Agent and civil engineer for his father's company in Yorkshire, England, 1882–1909. After losing their business, the family moved to Hatley, Quebec, in 1910. Freelance naturalist, 1910–1924. Assistant librarian, Emma Shearer Wood

Library of Ornithology, McGill University, 1926–1938. Honorary consultant of ornithological collection, McGill University from 1938 to 1949.

CAREER

Early interest in ornithology and natural history while student and later engineer in various parts of England. After settling in Hatley, Quebec, in 1910, he carried out extensive local fieldwork, discovering large number of orchids and ferns, in addition to nesting birds and some butterflies, not known to have occurred in the region. As associate (1915) and member (1926) of the American Ornithologists' Union (AOU) he traveled to many meetings in New England, New York, Pittsburgh, Philadelphia, Quebec, and Ontario. After moving to Montreal in 1924, took up nature photography. He first exhibited his bird pictures at the AOU's first Canadian meeting in Ottawa in 1926. His large collection of mounted photographs, chiefly of Canadian birds (sixty-six species), orchids (fifty-eight species), and ferns (six species) was donated to the Provincial Museum of Quebec in 1948. His herbarium of pressed plants is at Montreal Botanical Gardens, and his large collection of birds' nests and eggs is in part in National Museum of Canada, Ottawa, and in part in Redpath Museum, McGill University.

Published 131 articles on birds, plants, and butterflies in the *Canadian Field Naturalist*, the *Auk*, the *Wilson Bulletin*, *Bird-Lore*, *Oologist*, *Torreya*, *American Fern Journal*, *Canadian Entomologist*, and *Orchid Review*. His collection of articles from the various journals was published in one volume in 1929. While annotating Casey A. Wood's *Catalogue of Vertebrate Zoology*, discovered interesting material that led to the research and eventual publication of his "Historical Review of the Woodcock," published as special number of the *Canadian Field Naturalist* in 1935. Among ornithologists he is best known for his studies of the "home life" of birds. In his two famous papers, "The Singing Tree, or How near the Nest Do the Male Birds Sing," *Auk* (1919) and "Which Sex Selects the Nesting Locality," *Auk* (1921), he independently evolved same theory as Eliot Howard in his *Territory of Bird Life*. According to one of his American peers, Mousley's studies "dove-tailed in beautifully at a most fortunate time with the growing tendency to break away from the purely descriptive presentation of ethological data." Mousley believed in publishing his findings and was greatly encouraged in his research by various ornithologists and botanists, including Charles W. Townsend, Percy A. Taverner, and Brother Marie-Victorin.

MAJOR CONTRIBUTIONS

Thoroughness of field observations, detailed studies of natural history of the Hatley region of southern Quebec. Theoretical contributions to the "territoriality" of birds. While he was general naturalist in the nineteenth-century British tradition and expert on orchids and ferns, his theoretical contributions place him in the mainstream of twentieth-century ornithological research.

BIBLIOGRAPHY

Ainley, M. G. "Henry Mousley and the Ornithology of Hatley and Montreal." *Tchébec* (1981).

Terrill, L. McI. "William Henry Mousley." *Province of Quebec Society for the Protection of Birds (PQSPB) Annual Report* (1949) and *The Canadian Field Naturalist* (1950).

UNPUBLISHED SOURCES

See letters in the Lewis McIver Terrill Correspondence, National Museum of Natural Sciences, Ottawa; letters in the Percy A. Taverner Correspondence, National Museum, Ottawa; letters in the Casey A. Wood Correspondence, McGill Library Archives and Blacker Wood Library Archives, McGill University, Montreal.

Marianne Gosztonyi Ainley

MUIR, JOHN. Born Dunbar, Scotland, 21 April 1838; died Los Angeles, 24 December 1914. Conservationist, geologist, nature writer.

FAMILY AND EDUCATION

Third child in family of eight and eldest born to Daniel Muir, former British army sergeant-major and Scottish grain merchant, and Anne Gilrye Muir. Educated in local schools until age eleven. Family emigrated to the United States in 1849 and settled near Portage, Wisconsin. Naturalized, 1903. Attended five semesters at University of Wisconsin, 1861–1863; did not take degree. Honorary M.A., Harvard, 1896; honorary LL.D., University of Wisconsin, 1897, University of California, 1913; honorary Litt.D., Yale, 1913. Married Louise Strentzel, 1880 (died 1905); two daughters.

POSITIONS

Held various odd jobs, for example, mechanic, sawyer at a lumberyard, sheepherder, wilderness guide, as a means of providing income, which were often interrupted by his travels. Rented and later purchased father-in-law's California fruit farm from 1880. Achieved considerable success as fruit farmer. This made possible his later explorations, travels, writing, and involvement in major conservation battles.

CAREER

Worked on family farm as teenager. Father, a harsh, bigoted, and overly strict Campbellite, disapproved of secular learning and books. His early schooling in Scotland, the reading and study done on his own, excellent memory, and comparative maturity (he was nearly twenty-three when he entered the University of Wisconsin) got him through necessary preparatory training and enabled him to take regular university classes. Continued to demonstrate great skill at inventing, begun in his teens. Guided in his reading by Jean Carr, faculty wife and first in a series of several surrogate mothers in his life. A firm pacifist, Muir left Wisconsin when it appeared that he might be drafted for service in Civil War. Traveled and botanized in Canada until war's end, worked in sawmill near Georgian Bay, went to Indianapolis, where he worked as mechanic at carriage factory and botanized before leaving as result of accident and eye injury. Made famous 1,000–mile walk to the Gulf, fall of 1867. West on to Cuba and New York, taking ship for California, arriving March 1868. In years following, alternately worked in sawmills (until 1871) and as wilderness guide and explored up and down California. Spent some time studying glacial erosion in Yosemite. Began publishing articles on nature in late 1871 (his first, in *New York Daily Tribune*, was on "Yosemite Glaciers"). Continued to write newspaper and magazine pieces concerning conservation issues, including forest management, development of national parks, and wilderness preservation. Published seven-part "Studies in the Sierra" in *Overland Monthly* (1874–1875). This later brought out in book form by Sierra Club in 1911. Muir's basic conclusions concerning glaciation essentially confirmed by later study of François E. Matthes between 1913 and 1930. Muir traveled through southern sequoia region, 1875. Began career as public speaker on forestry, geology, and conservation issues, 1876. Assisted with mapping of 39th parallel across Nevada, 1878. Made first trip to Alaska, 1879, and is often credited with discovery of Muir Glacier, but this had previously been found by other explorers. Following marriage in 1880, engaged in fruit farming. Traveled on U.S. Revenue cutter *Thomas Corwin* to Aleutians, through Bering Strait and northwesterly along Siberian coast, thence east toward Point Barrow in search for vessel *Jeannette*, which was not located (it had sunk in polar seas further up north Siberian coast). Spent much of 1880s as fruit farmer on Strentzel Ranch, but this responsibility passed to others in early 1890s.

Instrumental in pressing for establishment of Yosemite National Park, 1890, creation of Sierra Club (1892), and expansion of Sierra Forest Reserve (1893). Traveled to Europe, 1893. Adviser to National Forestry Commission, 1896. His differences with Gifford Pinchot and others led to split in developing conservation movement between preservationists, such as Muir, and Pinchot's more utilitarian "greatest good for the greatest number." Participated in Harriman–Alaska Expedition, 1901. Guided President Theodore Roosevelt through Yosemite, 1903; made around-the-world trip, including England, France, Germany, Finland, Russia, Manchuria, Japan, Malay Peninsula, India, Egypt, Australia, New Zealand, Philippines, and the China Coast, 1903–1904. Instrumental in pressing President Roosevelt to make Petrified Forest of Arizona and Grand Canyon national monuments in 1906 and 1908 (both subsequently would become national parks). Led long but ultimately unsuccessful struggle to defend Hetch-Hetchy Valley against flooding to supply water for San Francisco, 1906–1913. Took trip to Amazon Valley and to Africa, 1911–1912.

Published a number of books (many of them containing articles that had previously appeared in newspapers and magazines), including *Picturesque Cal-*

ifornia, 2 vols. (1888); *The Mountains of California* (1894); *Our National Parks* (1901); *Stickeen* (1909); *My First Summer in the Sierra* (1911); *Edward Henry Harriman* (1911); *The Yosemite* (1912); *The Story of My Boyhood and Youth* (1913). Posthumously published books include *Travels in Alaska* (1915); *A Thousand-Mile Walk to the Gulf* (1916); *The Cruise of the Corwin* (1917); *Steep Trails, Utah, Nevada, Washington, Oregon, the Grand Cañon* (1918). Later compilations included *John of the Mountains: The Unpublished Journals of John Muir*, edited by L. M. Wolfe (1938); *Studies in the Sierra* (1950); and at least eight anthologies, which have appeared since 1954. Muir was a fellow of the American Association for the Advancement of Science and a member of the American Academy of Arts and Letters.

MAJOR CONTRIBUTIONS

Muir was a seminal figure in the development of American conservation thought. His writings and his public speaking helped to generate a strong public interest in conservation issues. He was a pioneer in both glaciology and forestry. He stressed the beauty of wilderness areas and discussed the need to preserve them. He influenced the conservation thinking of a number of individuals in his lifetime and later, including Theodore Roosevelt, who gave Muir credit for helping him to develop his own views. Muir was a founder of the Sierra Club (and president, 1892–1914) and helped promote creation of a number of national monuments, reserves, and parks, including Yosemite, where he helped persuade California legislature to convey state park to surrounding national park, Sequoia, Mount Shasta, Petrified Forest (Arizona), Grand Canyon, and others.

BIBLIOGRAPHY

Badé, William Frederic. *The Life and Letters of John Muir* (1924).
Clarke, James M. *The Life and Adventures of John Muir* (1979).
Cohen, Michael P. *The Pathless Way: John Muir and the American Wilderness* (1984).
Cohen, Michael P. *The History of the Sierra Club* (1988).
Fox, Stephen. *John Muir and His Legacy: The American Conservation Movement* (1981).
Fryxell, Fritiof. *The Incomparable Valley: A Geologic Interpretation of the Yosemite* (1950).
Goetzmann, William H., and Kay Sloan. *Looking Far North: The Harriman Expedition to Alaska* (1982).
Jones, Holway. *John Muir and the Sierra Club: The Battle for Yosemite* (1965).
Kimes, William F., and Maymie B. Kimes. *John Muir: A Reading Bibliography*. 2d ed. (1986).
Limbaugh, Ronald H. *John Muir's "Stickeen" and the Lessons of Nature* (1996).
Lynch, Ann T. "Bibliography of Works by and about John Muir, 1869–1978." *Bulletin of Bibliography* 36 (April–June 1979).
Matthes, François. *Geologic History of the Yosemite Valley* (1930).
Payne, Daniel G. *Voices in the Wilderness: American Nature Writing and Environmental Politics* (1996).
Turner, Frederick. *Rediscovering America: John Muir in His Time and Ours* (1985).
Wolfe, Linnie Marsh. *Son of the Wilderness: The Life of John Muir* (1945).
Wilkins, Thurman. *John Muir: Apostle of Nature* (1995).

UNPUBLISHED SOURCES

See John Muir Papers, Bancroft Library, University of California–Berkeley (includes papers, correspondence, and notes); John Muir Papers, Holt–Atherton Pacific Center for Western Studies, University of the Pacific, Stockton, Calif. (the most extensive collection of Muir materials, including notes by Badé and Wolff, his first two biographers).

Paul Cammarata and Keir B. Sterling

MURIE, OLAUS JOHAN. Born Moorhead, Minnesota, 1 March 1889; died Moose, Wyoming, 21 October 1963. Biologist.

FAMILY AND EDUCATION

Son of Joachim D. and Marie Frimanslund Murie. Student at Fargo College (North Dakota), transferred to, and received A.B. from, Pacific University, 1912; M.S., University of Michigan, 1927; honorary D.Sc., Pacific, 1949. Married Margaret E. Thomas, 1924; three children.

POSITIONS

Conservation officer with Oregon State Game Commission, 1912–1914; field naturalist and curator of mammals, Carnegie Museum, Pittsburgh, 1914–1917. Served with U.S. Army Balloon Service, 1917–1919. Field biologist for U.S. Biological Survey (USBS) in Alaska, 1920–1926; field naturalist, USBS, 1927–1946.

CAREER

Brought up in rural environment, so had enormous appreciation for nature and wilderness preservation from youth. Explored Hudson Bay region (1914–1915) and later Labrador Peninsula (1917) while with Carnegie Museum, studying birds and mammals in natural habitat. Conducted somewhat similar research in British Columbia and in far North of Canada and Alaska with Biological Survey, 1920–1927. Some fieldwork in Alaska done with younger brother Adolph (1899–1974). Also while in Alaska, acted as fur warden. In 1927, undertook major research effort centering on North American elk in Wyoming, where populations were declining, and there he made his home for remainder of his life. A ''tolerated maverick'' within Biological Survey, he disagreed with its policy of eradicating predators in late 1920s and early 1930s. In 1945, however, Fish and Wildlife Service declined to publish major study he had completed on coyotes because it reflected his continuing criticism of this policy. Conducted biological survey of Aleutian Islands (two expeditions, 1946–1947). Visited national forests, Indian reservations, and various other public lands throughout western states in interests of making land-management recommendations. At invitation of New Zealand government, led scientific expedition there

(1948–1949) to propose solutions to problem of introduced elk in that country. Leader of expedition to Brooks Range in Alaska, 1956, which was cosponsored by Wilderness Society, Conservation Foundation, and New York Zoological Society; resulted in establishment of Arctic Wildlife Range, 1960. Discussions with like-minded conservationists, including Aldo Leopold, Robert Marshall, and Benton MacKaye, led to creation of Wilderness Society in 1935, a major activity of Murie's for thirty years. Elected to Council of Society, 1937, on staff as part-time director following retirement from USBS and served as president, 1950–1957. Also president of the Wildlife Society, national director of the Izaak Walton League of America. Author of seven major publications, including *Alaska-Yukon Caribou* (North American Fauna [NAF] No. 54, 1935); *Food Habits of the Coyote in Jackson Hole, Wyoming* (1935); *The Elk of North America* (1951); *Field Guide to Animal Tracks* (1954); *Fauna of the Aleutian Islands and Alaska Peninsula* (NAF No. 61, 1959); *Jackson Hole with a Naturalist* (1963); *Wapiti Wilderness* (with Margaret Murie, 1966). Also numerous popular and technical articles. Won a number of awards from wilderness and conservation organizations. One of five trustees selected by Robert Marshall to administer Robert Marshall Wilderness Fund to promote preservation of wilderness. In presenting him Audubon Medal, its highest award (1959), Audubon Society described Murie as "personification of the spirit of wilderness." Did not live to see federal Wilderness Act passed in 1964, but its enactment was in part fulfillment of his work and convictions.

MAJOR CONTRIBUTIONS

Authority on wildlife and wilderness preservation. Was convinced that success of wilderness movement over time was dependent on extent to which it received support of general public. Worked to expand base and enhance effectiveness of citizen advocacy organizations.

BIBLIOGRAPHY

Davis, Richard C., ed. *Encyclopedia of American Forest and Conservation History* (1983).

Dunlap, Thomas R. *Saving America's Wildlife* (1988).

Fox, Stephen. *John Muir and His Legacy: The American Conservation Movement* (1981).

Glover, James M. *A Wilderness Original: The Life of Bob Marshall* (1986).

Living Wilderness (Summer–Fall, 1963).

Murie, Margaret E. *Two in the Far North* (1962, rev. ed., 1978).

Sherwood, Morgan. *Big Game in Alaska: A History of Wildlife and People* (1981).

Stroud, Richard H., ed. *National Leaders of American Conservation* (1984).

Who Was Who in America.

UNPUBLISHED SOURCES

See Murie Papers and Wilderness Society Archives, Conservation Center, Denver Public Library; Records of the U.S. Fish and Wildlife Service (which includes older U.S. Biological Survey materials), Record Group 22, National Archives, Washington, D.C.

Margaret E. Murie and Keir B. Sterling

MURPHY, ROBERT CUSHMAN. Born Brooklyn, New York, 29 April 1887; died Stony Brook, Long Island, 19 March 1973. Zoologist, ornithologist, oceanographer, geographer, writer.

FAMILY AND EDUCATION

Son of Thomas D. Murphy, lawyer and educator, who commuted weekends to Mount Sinai, Long Island, after family moved to country in 1894. Mother was Augusta Cushman Murphy. Fourth-generation Long Islander, oldest of eleven children, two of whom died in infancy. Graduated from Port Jefferson High School in 1906; Ph.B., Brown University, 1911; M.A., Columbia, 1918; honorary Sc.D., San Marcos University, Lima, Peru, 1925, and Brown, 1941. Phi Beta Kappa, Sigma Xi. Married Grace Emeline Barstow, 1912 (died 1975); three children.

POSITIONS

Brooklyn Museum: curator of mammals and birds, 1911–1917; curator of the Department of Natural Science, 1917–1920. American Museum of Natural History (AMNH): associate curator of birds, 1921–1926; associate director, 1924–1936; curator of oceanic birds, 1927–1942; chairman of the Department of Birds, 1942–1954; Lamont Curator, 1948–1955; Lamont Curator Emeritus and research associate, 1955–1973.

CAREER

Assistant to Frank M. Chapman at AMNH for year before college. Signed as "assistant navigator" (actually naturalist) on brig *Daisy*, on last of whaling voyages under sail, 1912–1913, for research east coast of South America and south Georgia. Brought back specimens new to United States of seabirds, seals, porpoises, plants, also superb photographic record. Charted Bay of Isles and other parts of south Georgia, named many geographic features. *Logbook for Grace*, written many years later from his journal, was called *success d'estime*. Led Lower California Expedition for Brooklyn Museum, 1915. Research in Peru (1919–1920, 1924–1925, 1953–1954) led to improved guano production. Other Western Hemisphere research: Ecuador (1925), Pacific Coast of Colombia (1937, 1941), Pearl Islands (1945), Bermuda (1951, 1972), Venezuelan Islands (1952). His work with pelagic birds, previously considered mysterious vagrants, placed, in *Oceanic Birds of South America*, the families and orders of seafowl "in their natural relations to their environment and opened up an entirely new dimension of life on the planet." "These two fat volumes are living proof that the most exhaustive scientific studies can also make splendid reading." Twice in New Zealand and islands to the south (1947, 1949), he and Mrs. Murphy excavated sixty-four skeletons of the extinct moa (eight were brought back to

United States) and collected specimens for dioramas at AMNH. In Bermuda (1951) rediscovered the cahow, a petrel considered extinct since 1625. Consulted with U.S. government during World War I (camouflage) and World War II; in 1960 returned to Antarctic with Operation Deep Freeze. Delegate to many international scientific conferences, including as U.S. delegate to 8th, 9th, 12th Pacific Science Congress. President Cold Spring Harbor Biological Laboratory, National Audubon Society (also honorary president), American Ornithologists' Union. Vice president New York Academy of Science. Member Antarctic Programs Committee (1963–1973); Advisory Committee Fire Island National Seashore, he and Mrs. Murphy leading campaign for its establishment; Executive Council Long Island University at Brookhaven (1964–1973). Consultant on conservation problems to governments of New Zealand, Peru, Chile, Venezuela, and Bharatpur. Recipient of many awards: Congressional Medal for Antarctic Service; American Geographical Society Cullum Medal; American Ornithologists' Union Brewster Medal; John B. Burroughs Memorial Association Medal; Garden Club of America Hutchinson Medal; Geographical Association of Lima Raimondi Medal; National Academy of Science Elliott Medal; Explorers Club Medal; Long Island Press Distinguished Service Award; included with Mrs. Murphy in Long Island Hall of Fame. Member or fellow of thirty-six scientific organizations in nine countries. Published some 600 articles and twelve books, including *A Report on the South Georgia Expedition* (1914), *Bird Islands of Peru* (1925), *Problems of Polar Research* (with others, 1928), *Oceanic Birds of South America*, 2 vols. (1936), *Logbook for Grace* (1947, 1965), *Land Birds of America* (with Dean Amadon, 1953), *John James Audubon 1785–1851*, an evaluation (1956), text for *Audubon* (1957), *Fish-Shape Paumonok* (1964), *Rare and Exotic Birds* (1964), *A Dead Whale or a Stove Boat* (1967), *Larousse Encyclopedia of Animal Life* (foreword, 1967).

Under his leadership, AMNH Department of Birds became one of the world's foremost, with over a million specimens, including Lord Rothchild's collection at Tring, which he cataloged and packed. Manager of Whitney South Sea Expedition for ten years, he created Whitney Memorial Hall of Ocean Life, AMNH. First to speak out against use of DDT. One of world's foremost authorities on marine birds, "among the earliest and most vocal advocates of conservation, talking about ecology long before it became fashionable." Maintenance of a viable environment was one of his major concerns. Lover of poetry, music, language, gave Brown valedictory on value of Latin and Greek in a liberal education. Named for him were a spider, two mountains, an antarctic inlet, a louse, a plant, a fish, several birds, and a school, the last "[T]he most meaningful honor of my life."

MAJOR CONTRIBUTIONS

Pioneered in study of ecology and biogeographic zonation of oceanic birds. "By following the seabirds to their haunts, he has deepened our knowledge of the world and its waters'' (Cullum Medal inscription). As one of the literary naturalists, with broad scientific knowledge in many interrelated fields, with outstanding facility in writing and speaking, helped spark the environmental movement.

BIBLIOGRAPHY

Amadon, Dean. *Auk* (January 1974) (obit.).
Brooks, Paul. *Speaking for Nature* (1980).
New York Times, 21 March 1973 (obit.).
Time, 2 April 1973 (obit.).
Washington Post, 21 March 1973 (obit.).
World Who's Who in Science. 1st ed. (1968).

UNPUBLISHED SOURCES

See diaries, journals, and observations of a naturalist, generously illustrated with newspaper clippings, photographs, letters, sketches, maps, and charts; thirty-five volumes; other letters, Library of the American Philosophical Association, Philadelphia; *A List of Tring Ornithological Collection*, pts. 1–7, 1932; letters, AMNH.

Alison Murphy Conner

N

NELSON, EDWARD WILLIAM. Born Manchester, New Hampshire (in a district called Amoskeag), 8 May 1855; died Washington, D.C., 19 May 1934. Naturalist, government administrator, meteorological observer, rancher.

FAMILY AND EDUCATION

Eldest of two sons of William Nelson, a butcher who died in the Civil War. Mother, Nancy Martha Wells Nelson, was Civil War nurse and, later, dress-maker in Chicago. Part of childhood spent with grandparents in Franklin County, New York. Joined mother in Chicago in 1868. Attended public schools in New York State and Chicago. Graduated from Cook County Normal School, 1875; briefly attended Northwestern University, 1875, and Johns Hopkins University, 1876–1877; honorary M.A. from Yale, 1920 and honorary Sc.D. from George Washington University, 1920. Unmarried.

POSITIONS

Teacher, Dalton, Illinois, 1875–1876; private (weather observer), U.S. Signal Corps, St. Michael, Alaska, 1877–1881; contributor, Smithsonian Institution, 1881–1882; rancher, county clerk, Apache County, Arizona, 1882–1890; special field agent, U.S. Department of Agriculture, Division of Ornithology and Mammalogy (redesignated Bureau of Biological Survey, 1905), 1890–1906; chief field naturalist, 1907–1912; head, Division of Biological Investigations, 1913–1914; assistant chief, 1914–1916; chief, 1916–1927; principal biologist, 1927–1929; research associate, Smithsonian Institution, 1930–1934.

CAREER

Introduced to ornithology by childhood friends in Chicago and became weekend collector. First extended field experience with Samuel Garman and Edward Drinker Cope in Nevada, Utah, and California, 1872. Began correspondence with Henry Wetherbee Henshaw, who urged him to go to Washington, D.C., to meet Spencer Fullerton Baird. Baird recommended him for position as weather observer at St. Michael, Alaska, and to study zoology, ethnology, and geography and collect for Smithsonian. Explored relatively unknown areas of western and northwestern Alaska and joined U.S. Revenue cutter *Corwin* on voyages to Siberia, Diomede Islands, St. Lawrence Island, and Alaska coast from mouth of Yukon to Barrow. Published *Report upon Natural History Collection Made in Alaska between the Years 1877 and 1881* (1887) and *The Eskimo about Bering Strait* (1899). The latter, in spite of notable lacunae, is still one of richest sources of data on people of western Alaskan coast. Also collected data on northern Athapascan Indians. In United States, developed tuberculosis and taken by family to Arizona to recuperate. Resumed biological work as member of Clinton Hart Merriam's Death Valley Expedition and then explored California and all states of Mexico as government naturalist, often accompanied by Edward Alphonso Goldman, whom Nelson trained. In later years, concerned with administration, rising to chief of Biological Survey. Became concerned with conservation. Recognized effects of drainage of watered areas on wildlife and urged restoration. Worked for Migratory Bird Treaty with Great Britain, law to implement treaty, and Alaska Game Law. Instrumental in efforts to improve Alaskan reindeer herds and use of banding for ornithological studies. President, American Ornithologists' Union, 1908–1911; Biological Society of Washington, 1912–1913; American Society of Mammalogists, 1920–1923.

Published over 200 articles and books, ranging in subject from meteorological observations to descriptions of several new species and subspecies and advocacy of conservation, including *Lower California and Its Natural Resources* (1921); *The Rabbits of North America* (1909), believed by mammalogists to be his finest contribution; and *Wild Animals of North America* (1919). Upon reading the latter, Theodore Roosevelt declared that Nelson was one of America's "keenest naturalists."

Nelson has been described as a "determined, intense worker, with great restless energy," so completely devoted to work that neither illness nor other physical discomfort could deter him for long. His intensity was coupled with a direct, sometimes abrupt, manner that alienated some, especially politicians. Others saw him, however, as basically kind, helpful to young scientists, and possessing many sterling characteristics.

MAJOR CONTRIBUTIONS

Collected and reported many new forms, especially from areas that had been relatively unexplored and could be explored only with great hardship. As administrator, promoted conservation through legislation, new practices, and popularization of effort. Some believe the conservation work to be his finest contribution. His anthropological work, although confined to Alaska, is noteworthy. He made one of the largest and best-documented collections of Eskimo artifacts and provided invaluable information on nineteenth-century Eskimo culture.

BIBLIOGRAPHY

Cameron, Jenks. *The Bureau of Biological Survey* (1929).

Collins, Henry B. ''The Man Who Buys Good-for-Nothing Things.'' In William W. Fitzhugh and Susan A. Kaplan, *Inua: Spirit World of the Bering Sea Eskimo* (1982).

Jackson, Donald D. ''A Stout Ship's Heartbreaking Ordeal by Ice.'' *Smithsonian* (March, 1997).

Lantis, M. ''Edward William Nelson,'' *Anthropological Papers of the University of Alaska.* (1954).

Nelson, E. W. ''Narrative.'' In *Report upon Natural History Collection Made in Alaska between the Years 1877 and 1881* (1887).

Nelson, E. W., and Goldman, E. A. Chapter on Mexican biogeography in V. E. Shelford, ed., *Naturalist's Guide to the Americas.* (1926).

Oehser, Paul. ''Nelson and Goldman.'' *Cosmos Club Bulletin* (1951).

Palmer, T. S. *Biographies of the Members of the American Ornithologists' Union* (1954, obit. by Witmer Stone).

Sterling, Keir B. *Last of the Naturalists: The Career of C. Hart Merriam* (rev. ed., 1977).

Vanstone, James W. ''Introduction.'' In *E. W. Nelson's Notes on the Indians of the Yukon and Innoko Rivers* (1978).

Who Was Who in America.

UNPUBLISHED SOURCES

Anthropological material, photographs, letters, field notes, journals, and other papers are in several collections of the Smithsonian Institution National Anthropological Archives, Washington, D.C. Letters are in Smithsonian Archives, Washington, D.C. Some letters in A. K. Fisher Papers, Manuscript Division, Library of Congress. Administrative materials are in records of the Fish and Wildlife Service (RG 22), National Archives and Records Service, Washington, D.C. See also records of the United States Fish and Wildlife Service, National Archives and Records Administration, Washington.

James R. Glenn

NEWBERRY, JOHN STRONG. Born Windsor, Connecticut, 22 December 1822; died New Haven, Connecticut, 7 December 1892. Geologist, paleontologist, naturalist, educator, scientific society executive.

FAMILY AND EDUCATION

Son of Henry and Elizabeth Strong Newberry. Spent most of childhood in northeastern Ohio, where father was developer of coal mines. Graduated from Western Reserve College in 1846; M.D. from Cleveland Medical School in 1848. Further medical and botanical study in Paris, 1849–1850, with Adolphe Brongniart, among others. LL.D. from Western Reserve in 1867. Married Sarah Gaylord, 1848; seven children.

POSITIONS

Naturalist and surgeon on three western exploring expeditions, 1855–1859. Professor of chemistry and natural history, Columbian College, Washington, D.C., 1856–1857. Administrator, western district, U.S. Sanitary Commission, 1861–1865. Professor of geology and paleontology, Columbia University School of Mines, 1866 to death.

CAREER

Newberry's interest in geology and natural history dates from childhood, when he collected plant fossils from Ohio coal strata. Participation as naturalist on expeditions of Lt. Williamson (San Francisco Bay to Columbia River, 1855–1856), Lt. Ives (up Colorado River, 1857–1858), and Capt. Macomb (northern New Mexico and Arizona, 1859) established Newberry's reputation as a geologist, the report of the Ives expedition (published 1861) being the most important in this regard. After the Civil War Newberry settled in New York, where he became a leader of the local scientific community. He was professor of geology at Columbia University for twenty-six years. Most of Newberry's more than 200 papers date from the postwar period. He published on manifold aspects of geology—coal and gas geology, deep-sea dredgings, theories of climatic change, ocean currents, glacial stratigraphy, the antiquity of man—as well as botanical and zoological topics. His most thorough studies were of fossil plants and fishes. After 1880 Newberry was increasingly active as mining consultant.

Newberry was an " 'all-round' man of science." While his eclecticism may have hindered his development as a specialist, a broad perspective was also strong point of Newberry's work. Among paleontologists Newberry was said to be atypically concerned with fossils as remains of once-living organisms, rather than merely as taxonomic raw material or stratigraphic indicators. This approach is exemplified in his 1873 paper on "Cycles of Deposition" (*Proceedings of the A.A.A.S.*, pt. ii, 22 [1873]: 185–96), in which biological and physical considerations were united to explain sequences of sedimentary rocks as results of rising and falling sea level. Newberry may be regarded as a pioneer paleoecologist.

President, American Association for the Advancement of Science, 1867. President, Lyceum of Natural History of the City of New York (became New York Academy of Sciences in 1876), 1868–1892. President, Torrey Botanical Club, 1880–1890. President, International Geological Congress, 1891. Was charter member of both the National Academy of Sciences in 1863 and the Geological Society of America in 1888.

MAJOR CONTRIBUTIONS

Newberry's chief contributions are as a scientific explorer of the American West, as a paleontologist, and through his activities as teacher, lecturer, administrator of scientific societies, and assembler of museum collections, a booster of natural history. Newberry was among a small number of individuals whose experience in western exploration spanned both the Pacific Railway Surveys and others led by the Topographical Engineers in the 1850s and those led by John Wesley Powell, Hayden, King, and Wheeler in the late 1860s and 1870s.

BIBLIOGRAPHY

Aldrich, Michele L. ''Newberry, John Strong.'' *Dictionary of Scientific Biography*. Vol. 10.

Fairchild, H. L. ''Memoir of Prof. John Strong Newberry.'' *Transactions, New York Academy of Sciences* 12 (1892–1893).

''John Strong Newberry.'' *Bulletin of the Torrey Botanical Club* 20 (1893).

Meisel, Max. *A Bibliography of American Natural History: The Pioneer Century, 1769–1865*. Vols. 2–3 (1924–1929).

Merrill, G. P. *Contributions to the History of American Geology* (1906).

Merrill, G. P. *The First One Hundred Years of American Geology* (1924).

Schmeckebier, Lawrence. *Catalog and Index of the Hayden, King, Powell, and Wheeler Surveys*. Bulletin of the U.S. Geological Survey. Vol. 222.

Sloan, Douglas. ''Science in New York City, 1867–1907.'' *Isis* 71 (1980).

Stevenson, J. J. ''John Strong Newberry.'' *American Geologist* 12 (1893).

White, Charles A. ''John Strong Newberry.'' *Biographical Memoirs, National Academy of Sciences*. Vol. 6 (1909).

UNPUBLISHED SOURCES

Notes taken by Newberry while in Paris (1849–1850) are at New York Botanical Garden.

Christopher Hamlin

NEWCOMBE, CHARLES FREDERICK. Born Newcastle on Tyne, England, 14 September 1851; died Victoria, British Columbia, 19 October 1924. Physician, anthropologist, naturalist, museum collector.

FAMILY AND EDUCATION

Eighth of fourteen children born into the Victorian middle-class family of Eliza and William Lister Newcombe. Studied medicine, University of Aberdeen; M.B. (1873), M.D. (1878), postgraduate work in Germany. Interned, psychiatric medicine, West Riding Pauper Insane Asylum, Wakefield, Yorkshire. Married Marion Arnold (1879); three daughters and two sons. Marion died of childbirth complications (1891).

POSITIONS

Investigator, Inquiry into Conditions at the Provincial Asylum, New Westminster. Appointed superintendent, immediately revoked for political reasons. His relationship to the Provincial Museum was limited to informal understandings and commissions to collect. Founding member of the Victoria Natural History Society and president, 1900. Chairman, Biological Board of Canada Commission on the Sea-Lion Question in British Columbia, 1915–1916.

CAREER

First post as M.D. at Rain Hill Mental Hospital, Liverpool. Left to join a general practice at Windermere, in the Lake District, where he learned to sail and began collecting botanical specimens. Traveled to the United States to study American medicine, reached San Francisco in 1882. Second trip to Pacific Northwest in 1883. Returned for his family and set up a general practice, Hood River, Oregon, 1884–1888(?). Continued natural history interests, collecting wildflowers on Mount Hood, and began studying native artifacts. Moved to Victoria by 1889. Traveled after his wife's death: Ottawa to meet with G. M. Dawson; New York to visit museums; London to study geology at the University of London and natural history at the British Museum.

On return to Victoria pursued botany and marine zoology in the mid 1890s, organizing ocean-dredging expeditions. In 1895–1897, with W. Francis Kermode, provincial museum curator, sailed around the Queen Charlotte Islands in a series of natural history collecting expeditions. Also collected native artifacts, including totem poles, assisted by two native friends, Henry Moody and Charles James Nowell of Alert Bay. Sent Kew Gardens a botanical collection relating to native plant use. On at least one occasion robbed graves. Railed against the looting of coastal culture by foreign anthropologists (Boaz, Kroeber, Sapir, and others) and museums. Yet, assisted the same in building collections in Berlin, at the American Museum of Natural History, the Smithsonian, Kew Gardens and organized the Ethnological Hall of the Northwest Coast, Field Museum, Chicago (1905). His drive to preserve "vanishing culture" created a holistic sense of regional identity, placing it on a world stage.

Never a prolific writer. Produced a museum guide and papers on local marine shells and fossils, a work on Capt. Vancouver, *First Circumnavigation of Vancouver Island* (1914), and *Botanical and Ethnological Appendix to Menzies' Journal of Vancouver's Voyage, April to October, 1792* (1923). In 1915–1916 conducted field studies of sea lions and public hearings into the new federal bounties on them, designed to aid the fishing industry.

Died of pneumonia following another nautical expedition. The journal *Nature* regretted "the closing of so rich a storehouse of knowledge" and praised his "many-sided, humorous, and charming personality." His son William, ally of painter Emily Carr, continued his father's work.

MAJOR CONTRIBUTIONS

Primarily famous (or infamous) for his role in ethnographic collecting. Established the province's first extensive botanical, paleontological, and marine collections. Chaired the region's first wildlife hearings, which were controversial. Promoted an inclusive environmental awareness, and the public programs in the auditorium of the Royal British Columbia Provincial Museum, named in his honor, continue that philosophy.

BIBLIOGRAPHY

B.C. Sessional Papers. "In Memoriam." *Provincial Museum of Natural History and Anthropology Report for the Year 1924.*

Cole, Douglas L. *Captured Heritage* (1985).

Corley-Smith, Peter. *White Bears and Other Curiosities: The First 100 Years of the Royal British Columbia Museum* (1989).

Low, Jean. "Dr. Charles Frederick Newcombe." *Beaver* 312 (Spring 1982): 32–39.

Nature (20 December 1924) (obit.).

UNPUBLISHED SOURCES

See MG 1077, Newcombe Family (7.5 m.); Natural History Society of British Columbia; Provincial Museum—Correspondence Inward; Provincial Secretary; B.C. Archives and Records Service, Victoria, B.C.

Lorne Hammond

NICE, MARGARET MORSE. Born Amherst, Massachusetts, 6 December 1883; died Chicago, Illinois, 26 June 1974. Ornithologist, zoologist.

FAMILY AND EDUCATION

Daughter of Anson Daniel Morse and Margaret Ely Morse. Father was a professor of history at Amherst College. She attended Amherst High School. Received a B.A. from Mt. Holyoke College in 1906, was a fellow in biology at Clark University from 1907 to 1909, and received an M.A. in psychology from Clark University in 1915. Honorary D.Sci. from Holyoke College in 1955, honorary D.Sci. from Elmira College in 1962. Married Leonard Blaine Nice, 1909; five daughters, one of whom died in childhood.

POSITIONS

Leading student of bird behavior. President of the Wilson Ornithological Club, 1937–1939.

CAREER

Originally trained in psychology, she applied her research in child rearing to behavioral studies of birds. She worked at home while raising her children. Member of the Oklahoma Academy of Science, Cooper Ornithological Society, Audubon Society, Wilderness Society, National Parks Association, and the Linnaean Society of New York. She was a fellow of the American Ornithologists' Union and received its Brewster Medal in 1942. Her first publication appeared in 1910, her final one in 1965. Author of *Birds of Oklahoma* (1924). Especially noteworthy were her studies of song sparrows, reflected in her *Population Study of Song Sparrow* (1937); the semi-popular *Watcher at the Nest* (1939); and *Behavior of the Song Sparrow* (1943). An important later work was *Development of Behavior in Precocial Birds* (1962). Shorter papers dealt with a variety of subjects, including ethology, ecology, territorial behavior, sexual behavior, incubation of young, and conservation. Some were published in foreign languages. She was associate editor of the *Wilson Bulletin* from 1939 to 1949 and associate editor of *Bird Banding* from 1935 to 1942 and 1946 to 1974. Nice was the first woman president of one of the principal American ornithological organizations, The Wilson Ornithological Society, in 1938–1939. In her later years, she became outspoken about conservation issues and condemned the indiscriminate use of pesticides and inappropriate uses of wildlife refuges. Nice researched in seven languages. Ernst Mayr said of her that she ''almost single-handedly initiated a new era in American ornithology.''

MAJOR CONTRIBUTIONS

Specialized in ecology and behavior of the song sparrow. Developed the concept of territoriality to explain birds' nesting behavior. Was considered one of the world's foremost ornithological behaviorists. Nice's accomplishments were impressive when it is considered that she spent much of her adult life as wife to a medical school professor and mother to their five children. Though her husband and family were very supportive of her work, she received very little outside financial support or secretarial aid, and never held any academic or other institutional appointments.

BIBLIOGRAPHY

Mayr, Ernst. ''Foreword and Epilogue.'' In Stresemann, Erwin, *Ornithology: From Aristotle to the Present* (1975).

Nice, M. M. *Research is a Passion with Me* (1979).

Notable American Women: The Modern Period: A Biographical Dictionary (1980).

Parkes, Kenneth. *Wilson Bulletin* (September 1974) (obit.).

Rossiter, Margaret W. *Women Scientists in America: Struggles and Strategies to 1940* (1982).

Terres, John K. *Discovery: Great Moments in the Lives of Outstanding Naturalists* (1961).

Trautman, Milton B. *Auk* (July 1977) (obit.).

Who Was Who in America. Vol. 6 (1966–1974).

UNPUBLISHED SOURCES

Principal collection of Nice's personal and professional papers at Cornell University Library. Small collection of correspondence concerning editorial matters in files of Wilson Ornithological Society. ''A Partial Bibliography of Margaret Morse Nice'' by Milton B. Trautman, available at Museum of Zoology, Ohio State University, Columbus.

Roberta Pessah

NICHOLSON, HENRY ALLEYNE. Born Penrith, Cumberland, England, 11 September 1844; died Aberdeen, Scotland, 19 January 1899. Paleontologist, professor of natural history.

FAMILY AND EDUCATION

Son of John Nicholson, distinguished biblical scholar, and Annie Elizabeth Waring. Childhood spent in Cumberland and Westmoreland. Early education at

Appleby Grammar School. Received Ph.D. in zoology from University of Göttingen in 1862. Studied medicine and natural science at University of Edinburgh, received B.Sc. in 1866, D.Sc. in 1867, and M.D. in 1869.

POSITIONS

Lecturer on natural history, Edinburgh School of Medicine, 1869–1871. Professor of natural history, University of Toronto, 1871–1874. Professor of biology, Durham College of Physical Science, 1874–1875. Professor of natural history, St. Andrew's University, 1875–1882. Regius Professor of Natural History, Aberdeen University, 1882–1899.

CAREER

At Toronto, wrote primarily on paleontology. At request of Ontario government, investigated bottom fauna of Lake Ontario and Silurian and Devonian fossils of the province. While in North America collected fossil corals and monticuliporoids, which provided many years' study after return to Britain. Left Toronto to take position at Royal College of Science, Dublin, but was then offered position at Durham. Was instrumental in creating school of zoology at St. Andrew's. Delivered in London annual course of Swiney lectures in geology, 1878–1882, 1890–1894. Paleontological research in England noted particularly for work on graptolites, corals, monticuliporoids, and stromatoporoids and the geological succession of the Paleozoic rocks of the lake district.

Wrote more than 150 papers and monographs, of which the most important include *A Monograph of the British Graptolidæ* (1872), *A Monograph of the British Stromatoporoids* (1886), and *The Phylogeny of the Graptolites* (1899). Published twenty-six papers, mostly of paleontology, while at Toronto, most appearing in *Journal of the Canadian Institute*. Best known for widely used textbooks of zoology and paleontology, including *A Manual of Zoology for the Use of Students* (seven ed. between 1870 and 1887), *A Manual of Palæontology for the Use of Students* (eds. in 1872, 1879, 1889), described as "the most complete general work on Invertebrate Palæontology in the English language," *Introduction to the Study of Biology* (1872), *The Ancient Life-History of the Earth* (1877), and *Synopsis of the Classification of the Animal Kingdom* (1882). Contemporaries noted that "he seemed to radiate energy" and had a keen sense of humor.

MAJOR CONTRIBUTIONS

Made important contributions through his paleontological research, but noted particularly for his zoological and paleontological textbooks.

BIBLIOGRAPHY

Craigie, E. Horne. *A History of the Department of Zoology of the University of Toronto* (1967).

Dictionary of National Biography. Supplement (1921–1922).
Ellis, W. Hodgson. *The University of Toronto Monthly* (October 1902) (obit.).
Macallum, A. B. Appendix B, "Publications by Members of Staff." In *The University of Toronto and Its Colleges, 1827–1906* (1906) (includes Nicholson's papers published while at University of Toronto).

Stephen Bocking

NIGRELLI, ROSS FRANCO. Born Pittston, Pennsylvania, 12 December 1903; died Southampton, Long Island, 4 October 1989. Marine biologist.

FAMILY AND EDUCATION

Son of Castrenza and Emanuela Dobrin Nigrelli. B.S., Pennsylvania State University, 1927; M.S., 1929, Ph.D., 1936, New York University. Married Margaret, 1927; one daughter.

POSITIONS

Field assistant, U.S. Department of Agriculture, 1927; pathologist, New York Aquarium, 1934–1939; director, Osborn Laboratories of Marine Sciences, 1964–1973, senior scientist, Osborn Marine Laboratories, 1973–1989; instructor, evening and summer sessions, City College of New York, 1939–1943; visiting instructor, 1943–1946, visiting assistant professor, 1946–1948, visiting associate professor, 1948–1949, adjunct associate professor, 1949–1958, adjunct professor, from 1958; consultant U.S. Fish and Wildlife Service, 1943–1948; U.S. Food and Drug Administration, 1945; director, Laboratory of Marine Biochemistry and Ecology, New York Aquarium, 1957–1964; director, New York Aquarium, 1966–1970.

CAREER

Held fellowship at New York University (1927–1931) and New York Zoological Society fellowship at New York Aquarium, 1931–1932 and was longtime pathologist at the New York Zoological Society from 1934. An authority on the diseases of fish and other aquatic life. He was one of the first to discover virus-induced tumors in fish. Was also expert on factors affecting health of marine fauna, including pollution, changes in salinity, and alterations in water temperature. Among the first to study poisons discharged by marine organisms; investigated blooms of plankton, such as so-called red tides that decimated fish along Florida's Gulf Coast in the 1940s and along shores of Long Island during the 1950s. Experimented with utilization of marine species as possible sources of drugs and as factors in medical research. In one case, discovered that sea cucumbers exude substance that can be fatal to fish in minute quantities, while this same excretion slows growth of tumors in mice. Determined that chemical substances secreted by sea sponges have antibacterial properties and that blood poisoning in man can be detected by using blood of horseshoe crabs. Former president, New York Academy of Sciences. Member of Committee on Animals from Nature, Institute of Animal Resources, Commission on Oceanography, Office of Science and Technology, 1958–1959, and member Subcommittee on Marine Biology, Committee on Oceanography, Office of Science and Technology, from 1965; member of advisory panel, Sea Grant Project, Office of Sea Grant Programs, from 1969; Mayor's Oceanography Advisory Committee, from 1969; member, Advisory Council, Hudson Marine Institute, from 1970; director and treasurer, Executive Committee, New York Institute of Ocean Resources, from 1970; member, President's Council, New York Ocean Science Laboratory, from 1971; active member of other scientific groups. Published several books and several hundred articles on fish diseases.

MAJOR CONTRIBUTIONS

Made notable discoveries in disease among marine organisms and factors influencing their health.

BIBLIOGRAPHY

American Men of Science. 12th ed.

Goddard, Donald, ed. *Saving Wildlife: A Century of Conservation* (1995).

New York Times, 13 October 1989 (obit.).

Keir B. Sterling

NUTTALL, THOMAS. Born Long Preston, Yorkshire, England, 5 January 1786; died Nutgrove, near Liverpool, England, 10 September 1859. Naturalist.

FAMILY AND EDUCATION

Son of James and Mary Hardacre Nuttall. Born to parents who were not prosperous, father's occupation unknown. Youth spent at Long Preston. Apprenticeship in printing business with his uncle at Liverpool (1800–1807), alone in London (1808). Self-educated; became interested in natural history through field experiences, gardening, visits to museums, and attendance at public lectures on botany. Honorary M.A., Harvard College (1826). Zealous enthusiasm and strong curiosity were exclusively devoted to his scientific pursuits. Although often described as shy, unsocial, and eccentric, he readily established acquaintanceships and had numerous lifelong friends. Never married, no children.

POSITIONS

Headquartered at Philadelphia (1808, 1815, 1817–1818, 1820–1822); took botanical expeditions to Delaware (1809), Niagara Falls (1809), the Great Lakes region, St. Louis, up Missouri River to Fort Mandon, New Orleans (1810–1811),

Savannah (1815–1816), Ohio River, Lexington, Charleston (1816–1817), Arkansas (1818–1820). Curator of the Botanic Garden and instructor in natural history and botany at Harvard College, Cambridge (1823–1834); while there took expeditions to northern New England and New York (1824), New Hampshire (1828, 1931), southeastern United States (1829–1830, 1832), Azores (1832). England (1812–1815, 1823–1824, 1832–1833). Transcontinental expedition across the Rockies, Oregon, Hawaiian Islands, California (1834–1836). Boston, Cambridge, New York, Philadelphia (1836–1841), with one trip to the southeast (1838–1839). Nutgrove, England (1842–1859); while there made one visit to America (1847–1848).

CAREER

Hazardous scientific exploration far beyond the frontiers of his day; perhaps the most widely traveled naturalist, who saw more of North America in its primeval condition than any other naturalist. Not only a collector but a thorough scholar who named and described numerous genera and hundreds of species of plants. Publication of his *Genera*, the first botanical taxonomic work of original and broad scope published in America, brought him immediate recognition on both sides of the Atlantic as an established, respected botanist. At Harvard College, conscientiously improved the botanical garden, was well received as a teacher, and published extensive descriptions of birds arranged in systematic order. Pursued horticultural interests in later years. Extensive collections of vascular plants deposited in the herbaria of the Academy of Natural Sciences of Philadelphia and the British Museum, London.

Publications include *The Genera of North American Plants and a Catalogue of the Species, to the Year 1817*, 2 vols. (1818, repr. with Introduction by J. Ewan, 1971); *A Journal of Travels into the Arkansas Territory during the Year 1819* (1821, repr. 1905, 1966, 1980); *Observations on the Geological Structure of the Valley of the Mississippi* (1821); *Introduction to Systematic and Physiological Botany* (1827, 2d ed., 1830); *A Manual of the Ornithology of the United States and Canada, I. Land Birds* (1832, 2d ed. with additions, 1840), *II. Water Birds* (1834), (rev. eds. of both vols., 1891, 1896, 1903); *A Catalogue of a Collection of Plants Made Chiefly in the Valleys of the Rocky Mountains or Northern Andes Towards the Sources of the Columbia River by Mr. Nathaniel B. Wyeth* (1834); *Collections Towards a Flora of the Territory of Arkansas* (1837 [1835–1836]); *The North American Sylva*... vol. 1, pt. 1 (1842), pt. 2 (1843); vol. 2 (1846); vol. 3 (1849) (3–vol. eds. repr. 1852, 1853, 1855; 2–vol. eds. [three vols. in two] repr. 1857, 1859, 1865); *Descriptions of Plants Collected by William Gambel, M.D. in the Rocky Mountains and Upper California* (1848). Early in career, elected to membership in the Linnaean Society of London (1813), the Academy of Natural Sciences of Philadelphia (1817), and the American Philosophical Society (1817). Commemorated in the genus *Nuttalia*, a member of the Loasa family, and in the epithets of numerous species, for example, *Cirsium nuttallii*, *Elodea nuttallii*, *Viola nuttallii*.

MAJOR CONTRIBUTIONS

Made important contributions to the fields of botany, zoology, ornithology, geology, mineralogy, ecology, and horticulture through his pioneering scientific explorations in North America that have never been equaled in extent or productivity. His *Genera*, wrote J. Torrey, "contributed more than any other work to advance the accurate knowledge of plants in this country," a statement echoed by later botanists E. L. Greene and M. L. Fernald. "His observations uncommonly excellent . . . the fruits of real personal acquaintance with the plants in nature," stated Schweinitz. On birds, Audubon noted, "none can describe the songs of our different species like Nuttall."

BIBLIOGRAPHY

Beidleman, R. G. "Some Biographical Sidelights on Thomas Nuttall, 1786–1859," *Proceedings of the American Philosophical Society* 104 (1960).

[Durand, E.]. "Biographical notice of the late Thomas Nuttall," *Proceedings of the American Philosophical Society* 7 (1860).

Graustein, J. E. "Nuttall's Travels into the Old Northwest, An Unpublished 1810 Diary," *Chron. Bot.* 14 (1951).

Graustein, J. E. *Thomas Nuttall, Naturalist: Explorations in America 1808–1841* (1967).

MacPhail, Ian. *Thomas Nuttall*. Lisle, Illinois: The Sterling Morton Library Bibliographies in Botany and Horticulture, The Morton Arboretum (1983).

Pennell, F. W. "Travels and Scientific Collections of Thomas Nuttall." *Bartonia* 18 (1936).

Smith, Jr., C. E., and J. W. Thieret. "Thomas Nuttall (1786–1859): An Evaluation and Bibliography." *Leaflets of Western Botany* 9 (1959).

Stuckey, R. L. "Biography of Thomas Nuttall; A Review with Bibliography." *Rhodora* 70 (1968).

UNPUBLISHED SOURCES

Widely scattered, consult Graustein (1967) for details.

Ronald L. Stuckey

O

OLMSTED, FREDERICK LAW. Born Hartford, Connecticut, 26 April 1822; died Waverley, Massachusetts, August 1903. Landscape architect, environmentalist, author, abolitionist, U.S. Sanitary Commission executive secretary.

FAMILY AND EDUCATION

Son of John and Charlotte Law Hull Olmsted. Father was partner in successful dry-goods store. Had one brother and six half-brothers and half-sisters. Olmsted boarded with succession of minister-schoolmasters, attended day schools, and studied for one semester at Yale. Honorary degrees from Harvard (M.A., 1864; LL.D., 1893); LL.D., Yale, 1893. Married Mary Cleveland Olmsted, his brother's widow, 1859; three stepchildren and four children.

POSITIONS

Dry-goods clerk, New York City, 1840–1841; sailor, 1843–1844; scientific farmer in Syracuse and on Long Island, New York, 1846–1855. Began career as author and journalist, 1850; assistant editor and publisher, *Putnam's Monthly Magazine*, as partner in firm of Dix and Edwards, New York City, 1855–1857; began career as landscape architect with Calvert Vaux, 1857. Superintendent of Central Park, New York City, 1857–1861; continued with Vaux as landscape architect and superintendent for the park after Civil War and until 1872. Appointed (with Vaux) landscape architect and designer to Commissioners North of 155th Street, New York City, 1860. Resident secretary, U.S. Sanitary Commission, Washington, D.C., 1861–1863. Superintendent, Mariposa Company, Bear Valley, California, 1863–1865. Resumed professional office career as landscape architect with Vaux (to 1872), in New York City until 1881, and then in Brookline, Massachusetts, 1865–1895.

CAREER

Olmsted endured brief career in dry-goods business in New York City, sailed to Canton, and attempted scientific farming before he came into his own as landscape architect and author. His English travels resulted in *Walks and Talks of an American Farmer in England* (1852), and he recounted his journeys through the American pre-Civil War South (1853–1855) in series of letters to *New York Daily Times*, later reprinted in three volumes as *The Cotton Kingdom* (1861), considered strong antislavery statement. A publishing venture as associate editor and part owner of *Putnam's Literary Magazine* (1855–1857) was a financial failure, but he was able to gain a significant reputation as a writer with a progressive social conscience. In 1857, due in part to his knowledge of British landscape gardening and his background in scientific farming, he became superintendent of Central Park in New York and, with Calvert Vaux, won competition for landscape plan for new park in 1858. Olmsted spent much of the next three years implementing his and Vaux's plans for the park. In 1861, Olmsted left park work to lead U.S. Sanitary Commission in effort to alleviate suffering of ill and wounded northern soldiers. Resigned this post in 1863 to become superintendent of Mariposa Company, mining operation with headquarters near Yosemite Valley, California. At this point, he became involved in pressing for national reservation for Mariposa Big Trees and Yosemite Valley as state parklands; resulting legislation, signed into law in June 1864, eventually provided basis for U.S. national park system. As member of California's state Yosemite Commission (1864–1867), he produced report for preserving park, but report was suppressed, apparently due to rivalries for state funds. Mariposa Company's already tenuous financial situation became desperate in 1865, at which point Olmsted resigned and returned to New York. There he reestablished partnership with Vaux, and the two resumed working together as landscape architects to Central Park and began work on Brooklyn's Prospect Park, among other commissions. Their work expanded to include city, campus, residential, resort, and park planning.

Olmsted continued interest in journalism, now with *The Nation*. In 1869, he and Vaux completed plan for commuter "garden subdivision" of Riverside, Illinois, a model for many later U.S. subdivisions. In 1871, Olmsted Vaux and Company finished plans for Chicago's Jackson and Washington Parks, linked by a midway. Partnership with Vaux ended in 1872, and Olmsted himself continued as landscape architect, at first in New York City (working with Boston park system among other projects) and later (in 1880s) in Brookline, Massachusetts. Between 1873 and 1879, he worked on development of grounds of the Capitol at Washington, D.C. Other commissions included Mont Royal Park, Montreal (1874) and scenic preservation of lands around Niagara Falls. He eventually brought his nephew and stepson John Charles and finally his son Frederick Law Olmsted, Jr., into his firm. During the late 1880s, Olmsted developed campus plan for Stanford University. In 1892, Olmsted and Harry Codman reworked Olmsted and Vaux's earlier design for Chicago's Jackson Park as site of World's Columbia Exposition, for which Olmsted's firm was chosen as landscape architects. From 1888 to 1895, Olmsted worked on landscape plan for ''Biltmore,'' estate of G. W. Vanderbilt in Asheville, North Carolina, where Olmsted convinced Vanderbilt to hire Gifford Pinchot as a ''scientific forester,'' thus aiding in establishment of this profession in the United States. Due to ill health, Olmsted forced to retire in 1895. The firm he began continued under the leadership of his children until the 1950s, eventually producing over 10,000 landscape plans. Olmsted's Brookline office now preserved by National Park Service as Frederick Law Olmsted National Historic Site. Olmsted publications included article on cultivation of pears (*The Horticulturist* [January 1852]) and *Walks and Talks of an American Farmer in England* (1852); series of letters to *New York Daily Times* on slave states signed ''Yeoman'' (1853–1854); *A Journey in the Seaboard Slave States with remarks on Their Economy* (1856); *A Journey through Texas, or, a Saddle-Trip on the Southwestern Frontier* (1857); *A Journey through the Back Country* (1860) (these last three titles republished as *The Cotton Kingdom*, 1861); various articles in *The Nation*; and ''The Spoils of the Park'' (1882).

MAJOR CONTRIBUTIONS

Olmsted was widely considered the ''father of landscape architecture in the United States,'' though that title probably should be shared with Andrew Jackson Downing (1815–1852). However, Olmsted's wide-ranging interests and general organizational ability allowed him to make substantial contributions in other areas as well, notably abolition (with his antislavery writings); public health (with the U.S. Sanitary Commission); scientific forestry (through his sponsorship of Gifford Pinchot); and the U.S. national park system (through his work with the protection of Yosemite and the Mariposa Big Trees). To do this, Olmsted often worked to the point of nervous and physical exhaustion but was nonetheless noted for his stability and endurance. His sense of the basic worthiness of human beings—his idealism—continued to give him inner strength for his work, while he himself often felt unworthy of accolades received for his contributions.

BIBLIOGRAPHY

Beveridge, Charles E., and Paul Rocheleau. *Frederick Law Olmsted: Designing the American Landscape* (1995).

Fein, Albert, ed. *Landscape into Cityscape: Frederick Law Olmsted's Plans for a Greater New York City* (1981).

Lee, Hall. *Olmsted's America: An Unprecedented Man and His Vision of Civilization* (1995).

McLaughlin, Charles Capen, ed. *The Papers of Frederick Law Olmsted* (1977–).

Olmsted, Frederick Law, Jr., and Theodora Kimball, eds. *Forty Years of Landscape Architecture: Being the Professional Papers of Frederick Law Olmsted, Sr.* (1928).

Roper, Laura Wood. *FLO: A Biography of Frederick Law Olmsted* (1973).

Stevenson, Elizabeth. *Park Maker: A Life of Frederick Law Olmsted* (1977).

Todd, John Emerson. *Frederick Law Olmsted* (1982).

UNPUBLISHED SOURCES

Many of the papers of Frederick Law Olmsted, Sr., are held in the Manuscript Division, Library of Congress. Many of his and his firm's drawings are held by the Frederick Law Olmsted National Historic Site, Brookline, Mass.

Noel Dorsey Vernon

OLSON, SIGURD FERDINAND. Born Chicago, 4 April 1899; died Ely, Minnesota, 13 January 1982. Nature writer, conservationist, biologist, zoologist, educator.

FAMILY AND EDUCATION

Son of Lawrence J. and Ida May (Cedarholm) Olson. Father held succession of Baptist pastorates in northern Wisconsin; his library instilled a ''craving for the frontier—the wilderness'' in young Sigurd. Attended Northland College, 1916–1918. B.S., University of Wisconsin, 1920; postgraduate work, 1922–1923; M.S. (biology), University of Illinois, 1931. Master's thesis a pioneer study of wolves. Honorary degrees included D.Sc., University of Wisconsin, 1972; L.H.D., University of Minnesota, 1979. Married Elizabeth Dorothy Uhrenholdt, 1921; two children.

POSITIONS

County agent and agricultural instructor, Nashwauk, Minnesota, 1920–1921. Head of biology department, 1922–1935, and dean, 1935–1947, Ely (Minnesota) Junior College, where he held many classes outdoors and offered controversial field trips. Served in biology department at U.S. Army University, Shrivenham, England, 1945. Worked for Information and Employment Division of U.S. Army, European Theater of Operations, Germany, Italy, France, and Austria, 1945–1946. Ecological consultant to Izaak Walton League of America, 1947–1982.

CAREER

Although spent much of his life as educator, Olson's first love was wilderness, particularly the Quetico-Superior area of lakes and forests northwest of Lake Superior. Early in his career, spent summers as outfitter, which gave first-hand experience with the north country and put him in contact with other wilderness devotees. Beginning in 1920s, wrote dozens of articles on ecology, sport fishing, and hunting for such publications as *Wilderness*, *Ecology*, *Field and Stream*, and *Sports Afield*. After resigning position at Ely Junior College in 1947, became regular contributor to Izaak Walton League's magazine *The Living Wilderness* and worked to preserve part of the vast Quetico-Superior area on both sides of the Canadian border.

Through his books Olson had his biggest impact. Among his early works were *The Singing Wilderness* (1956), *Listening Point* (1958), and *The Lonely Land* (1961), which describe the "song" of the woods, exploits of French voyagers, and his own cabin. *Runes of the North* (1963) sketches trips to Saskatchewan, Hudson Bay, Northwest Territories, Yukon Territory, and Alaska. *The Hidden Forest* (1969) is organized around the four seasons, while *Open Horizons* (1969) is largely autobiographical. *Reflections from the North Country* (1976) and *Of Time and Place* (1982) contain distillations of his philosophy of life and the wilderness. Olson's works are a lyric exploration of man's primordial relationship with the sights, sounds, feel, touch, and exquisite fragrance of the wilderness. One writer terms him a "child of nature" who had the "soul of a Wordsworth mixed with the rough-and-ready streak of the American frontier."

Member of President's Quetico-Superior Committee, 1947; member of National Advisory Board of Parks, Monuments, and Historic Sites, 1950; president of National Parks Association, 1953–1958; member Wilderness Society Council, 1956, vice president, 1963–1968, president, 1968–1971; member of secretary of interior's Advisory Committee on Conservation, 1960–1966; consultant to director of National Park Service, 1962. Awards and honors included election to Izaak Walton League Hall of Fame plus Founder Award, 1963; Medal of John Burroughs Memorial Association, 1974; and Robert Marshall Award of Wilderness Society, 1981. Northland College opened Sigurd Olson Environmental Institute in 1972, dedicating the building to him in 1981.

MAJOR CONTRIBUTIONS

Of Thoreauian bent, he was perhaps the preeminent American nature writer of the mid-twentieth century and a strong force in wilderness preservation. A pioneer ecologist, one of the first to realize that wolves and other animals are part of a natural ecosystem.

BIBLIOGRAPHY

American Men and Women of Science (1972).

Audubon (1982) (obit.).

Contemporary Authors. Vol. 105 (1982) (obit.).

Graham, Frank, Jr. " 'Leave It to the Bourgeois': Sigurd Olson and His Wilderness Quest." *Audubon* (1980).

New York Times, 15 January 1982 (obit.).

Olson, Sigurd F. *Collected Works of Sigurd F. Olson: The Early Writings, 1921–1934*. Ed. Mike Link. Intro. Robert K. Olson (1988).

Olson, Sigurd F. *Collected Works of Sigurd F. Olson: The College Years: 1935–1944.* Ed. Mike Link. Intro. Jim Klobuchar (1990).
Olson, Sigurd F. *Songs of the North.* Ed. Howard Frank Mosher (1987).
Searle, R. Newell. *Saving Quetico-Superior: A Land Apart* (1977).
Stroud, Richard H., ed. *National Leaders of American Conservation* (1985).
Vickery, Jim dale [*sic*]. *Wilderness Visionaries* (1986).

UNPUBLISHED SOURCES

See Sigurd F. Olson Papers, Minnesota Historical Society (MHS), St. Paul (100 linear feet; includes correspondence, literary manuscripts, book drafts, reports, surveys, maps, and printed material covering all facets of his career). The following collections contain letters from Olson: Alden J. Erskine Papers, Iowa State University, Ames; Friends of the Wilderness, Records, MHS; Quetico-Superior Council, Records, MHS.

Rolf Swensen

ORD, GEORGE. Born Philadelphia, 4 March 1781; died Philadelphia, 24 January 1866. Naturalist, philologist.

FAMILY AND EDUCATION

Ord was the son of George and Rebecca (Lindemeyer) Ord. When Ord was seventeen years old, his father, who had been a sea captain, established himself as a ship chandler and rope maker in Philadelphia. The younger Ord joined the family business in 1800 and continued to run it after his father's death six years later. He eventually retired, probably around 1829, and for almost forty years enjoyed the life of a gentleman of leisure. There is no reliable record of Ord's early formal education. However, he possessed a wide knowledge of literature and science. His contributions were recognized by several scientific societies in Philadelphia. Ord married in 1815. He and his wife had two children, a son, Joseph Benjamin Ord, who became a portrait painter, and a daughter, who died in infancy.

POSITIONS

Ord was a businessman, not a professional scientific researcher. His interests in philology and ornithology are probably best described as lifelong avocations, but this does not imply that his work could be described as exhibiting less than professional qualities.

CAREER

Ord went to work with his father in the family's chandlery when he was nineteen years old. After his father's death six years later he took over the responsibility of running the operation full-time. Ord seems to have been fairly successful. He retired at age forty-eight (1829), apparently financially independent, and for the next thirty-seven years pursued his personal interests of ornithology and philology.

By the time he reached his mid-twenties, Ord had established a close friendship with Alexander Wilson, an ornithologist who was fifteen years his senior. At about the time that the two men became acquainted, Wilson had begun working on his multivolume study *American Ornithology; or, the Natural History of the Birds of the United States* (1808–1814). Ord often went along with Wilson on his expeditions near Philadelphia, and his contributions appear frequently in the pages of Wilson's *Ornithology*.

After Wilson's death in 1813 Ord set about completing the final two volumes of the *Ornithology*. This involved editing volume 8 and assuming responsibility for supplying the text to accompany Wilson's drawings for volume 9. Ord also included a biography of Wilson in the ninth volume. In 1824, ten years after the publication of the last volume of the *Ornithology*, Ord brought out a new, enlarged edition. However, in deference to Wilson, he went to great lengths to conceal the substantial contribution that he made to enlarging his late friend's life's work. At about the same time that the new edition of the *Ornithology* was published, John James Audubon's illustrations of American birds appeared in print. Ord, zealous in his allegiance to his late friend, began a campaign to discredit Audubon and preserve what he believed was Wilson's claim to preeminence.

Ord undertook only one lengthy scientific expedition. This was a trip to Georgia and Florida along with Titian Peale, William Maclure, and Thomas Say in 1818. In later years Ord published memoirs of both Say and Charles A. Lesueur, another close associate. In addition to these personal tributes, Ord contributed articles to scientific journals and wrote a description of North American zoology that was included in the second American edition of William Guthrie's *New Geographical and Commercial Grammar* (1815, repr. 1894). This account included the first comprehensive and systematic listing and discussion of North American mammal species ever published by an American. In 1828, Ord published a book-length biography of his late friend Wilson. Later, Ord's interest in philology inspired him to revise both Samuel Johnson's and Noah Webster's dictionaries.

Ord was a member of the Linnaean Society of London and the American Philosophical Society. He also served as president of the Philadelphia Academy of Sciences (1851–1858). Having outlived his many friends, Ord died a recluse in Philadelphia in 1866. Two noteworthy bequests in his will were the donation of his library to the College of Physicians in Philadelphia and a donation of $16,000 to the Pennsylvania Hospital.

MAJOR CONTRIBUTIONS

Perhaps Ord's major contribution was the completion of Wilson's *Ornithology*. His wide-ranging scientific knowledge was recognized by a number of scientific societies in Philadelphia, and his contributions to philology were recognized in the positive reaction to his new edition of Johnson's dictionary.

BIBLIOGRAPHY

Dictionary of American Biography.

Elliott, Clark A. *Biographical Dictionary of American Science, the Seventeenth through the Nineteenth Centuries* (1979).

National Cyclopedia of American Biography.

Rhoads, Samuel N. *A Reprint of the North American Zoology by George Ord . . . to which is added an Appendix on the More Important Scientific and Historic Questions Involved* (1894).

Rhoads, Samuel N. "George Ord." *Cassinia: Proceedings of the Delaware Valley Ornithological Club* 12 (1908).

UNPUBLISHED SOURCES

The following materials are located in the collections of the American Philosophical Society Library, Philadelphia: Ord's letters to his lifelong friend, Charles Waterton, are located in the George Ord Letters Collection; selected correspondence is located in the John Torrey Collection of Autograph Letters of Naturalists 1744–1894; materials related to Ord's memoir of Thomas Say are included in the Thomas Say Papers; additional correspondence is to be found in the William Darlington Letters Collection.

Correspondence between Ord and Titian Peale is included in the Peale Family Papers at the Historical Society of Philadelphia. Information about Ord's bequests to the Library of the College of Physicians may be found in the records of the Library Committee of the College of Physicians in Philadelphia.

Domenica Barbuto

OSBORN, HENRY FAIRFIELD. Born Fairfield, Connecticut, 8 August 1857; died Garrison, New York, 6 November 1935. Vertebrate paleontologist, museum administrator.

FAMILY AND EDUCATION

Second of four children of William Henry Osborn, president of Illinois Central Railroad, and Virginia Reed Sturges, chief founder of Virginia Day Nursery. Spent early life in the vicinity of New York City. Father built "Castle Rock" on a hilltop at Garrison above the Hudson River; it became Osborn's favorite residence. Osborn attended Columbia Grammar School and M. W. Lyon's Collegiate Institute in New York before entering College of New Jersey (now Princeton University) in 1873, received A.B. 1877. From spring 1879 to autumn 1880, Osborn studied at Cambridge University before returning to Princeton, where he was appointed to special biological fellowship; received Sc.D. in 1881. Married Lucretia Perry in 1881 (died 1930); five children, one of whom died in infancy.

POSITIONS

Instructor, assistant professor of natural sciences and professor of comparative anatomy at Princeton University (1881–1890); Da Costa Professor of Zoology and dean of the Faculty of Pure Science at Columbia University (1891–1910); president, New York Academy of Sciences (1898–1899); founder, curator, and honorary curator of the Department of Vertebrate Paleontology, American Museum of Natural History (AMNH), New York (1891–1908); president, New York Zoological Society (1909–1924); president, American Association for the Advancement of Science (1928); president of AMNH (1908–1933).

CAREER

As boy Osborn showed no inclination toward science fields. Transition probably initiated at Princeton University by James McCosh and later boosted by Arnold Guyot, professor of geology. McCosh believed in evolution as God's means of creation and possibly influenced Osborn's rejection of idea of fortuity or chance; this and his family's religious background made it impossible for him to accept natural selection in its Darwinian form. These views were later manifested in Osborn's classifications, which were highly polyphyletic. First paleontological expedition was in 1877 to the Bridger Eocene Basin of Wyoming. Together with William B. Scott and Francis Speir, Jr., they collected many fossils, especially mammals. In the following year (1878), same three students conducted second expedition to Wyoming. These trips captivated Osborn and Scott and were beginning of lifelong friendship and devotion to vertebrate paleontology.

While in Cambridge, England (1879–1880), Osborn studied comparative anatomy under Thomas H. Huxley and embryology under Francis M. Balfour. Also met Charles Darwin. In 1891 founded the Department of Biology at Columbia University and Department of Vertebrate Paleontology (DVP) at the AMNH. Osborn was also dean of graduate faculty at Columbia University and for many years was Da Costa Professor of Zoology, training numerous students, many of whom became distinguished zoologists and paleontologists (1891–1910). During this time he served as head of DVP, AMNH, building a collection of worldwide importance.

Osborn was primarily research scientist in spite of his involvement with several concurrent careers. Continuously studied fossil vertebrates and published 940 articles, books, and monographs (totaling over 12,000 printed pages), many of which were completed with the help of assistants and colleagues. His books and monographs have stimulated popular interest and curiosity in anthropology and paleontology; they are praised as true accomplishments, some were labeled as "classics" and served as models for subsequent works; recognized by scholars and laymen alike. Some of these books have been translated into French, German, Russian, and Japanese. Two of the titles in his bibliography (the monographs on titanotheres and proboscideans) stand out above the rest and will continue to be "consulted as long as Paleontology remains a living science."

A sample of Osborn's astonishing output includes the following best-known books and monographs (some titles are shortened): *On the Structure and Clas-*

sification of the Mesozoic Mammalia (1888), *From the Greeks to Darwin* (1894), *Evolution of Mammalian Molar Teeth to and from the Triangular Type* (1907), *The Age of Mammals in Europe, Asia and North America* (1910), *Men of the Old Stone Age* (1915), *The Origin and Evolution of Life on the Theory of Action, Reaction and Interaction* (1917), *Equidae of the Oligocene, Miocene, and Pliocene of North America* (1918), *The Earth Speaks to Bryan* (1925), *Evolution and Religion in Education* (1926), *Creative Education in School, College, University, and Museum* (1927), *Man Rises to Parnassus* (1927), *Impressions of Great Naturalists* (1928), *The Titanotheres of Ancient Wyoming, Dakota, and Nebraska* (1929), *Cope: Master Naturalist* (1931), *Proboscidea. Vol. I: Moeritherioidea, Deinotherioidea, Mastodontoidea* (1936), *Proboscidea. Vol. II: Stegodontoidea, Elephantoidea* (1942, published posthumously). These selected works plus those that were never completed justified George G. Simpson's statement that Osborn "planned as if he were to live forever and he laid out more work for himself than could have been completed in ten lifetimes."

Osborn's paleontological writing displayed a vast panorama of the evolution and dispersal during the Age of Reptiles and the succeeding Age of Mammals. His outstanding principles of evolution (or laws, as he called them) include the following: (1) the law of continental and local adaptive radiation; (2) the law of homoplasy or parallel but independent evolution in related lines of descent; (3) the law of tetraplasy, whereby evolution results not from the operation of single causes but as the result of forces from four principal directions (external environment, internal environment, heredity, selection); (4) the law of allometry, or adaptive modification of dimensions of the skull, feet, or other parts, arising independently in different lines of descent; (5) the law of rectigradation or aristogenesis, that is, gradual appearance during long ages of new structural units of adaptive value, predetermined in germ plasm and in their initial stages independent of natural selection; and (6) the law of polyphyly, that is, the normal occurrence of many related lines of descent.

Writings included both scientific and popular subjects. He was much respected in the scientific community, and the many species, subspecies, and genera named after him serve as silent testimony. Osborn himself named many new species, genera, subfamilies, families, superfamilies, and suborders (many of his species were also subsequently designated as synonyms), many of which are proboscidean taxa.

Contributions to science were also recognized by the many memberships in learned societies, honoraria, fellowships, and medals he received in United States and in many countries in Europe (especially Great Britain), USSR, China, Argentina, Mexico, and Cuba. Some medals he received include the Darwin Medal of the Royal Society of London (1918), the Albert Gaudry Prize (1918), Commandeur de l'Ordere la Couronne, decoration conferred by Albert, king of the Belgians (1919), Medal of Pasteur Institute of Paris (1921), and Wollaston Medal of Geological Society of London (1926).

On 8 April 1942 a bust of Osborn was unveiled to commemorate his contribution to the AMNH. The bust is exhibited at the Hall of North American Mammals, AMNH.

MAJOR CONTRIBUTIONS

Greatest single achievement is probably the development of the Department of Vertebrate Paleontology at the American Museum of Natural History, New York; with help of exceptionally able assistants and staff during forty-five years of his dynamic, indefatigable contribution, built one of the finest collections in the world. He envisioned the museum not only as a research collection institution but mostly as educational display cases for laymen. Wrote much to build the bridge between religion and science; popularized paleontology and made ''dinosaur'' a household word. Much interested in larger forms of life, he was a theorist who proposed explanations for many aspects of evolution, most of which were based on fossil evidence. He was master of synthesis, as illustrated by his monumental monographs.

BIBLIOGRAPHY

Colbert, Edwin H. ''Osborn, Henry Fairfield.'' *Dictionary of Scientific Biography* (1974).

Colbert, Edwin H. ''Henry Fairfield Osborn.'' *Science* 82 (1935).

Colbert, Edwin H. ''Bibliographical Memoir of Henry Fairfield Osborn 1857–1935.'' *Biographical Memoirs, National Academy of Sciences* 19 (1938).

Colbert, Edwin H. ''Henry Fairfield Osborn and the American Museum of Natural History.'' *Nature* 150 (1942).

Colbert, Edwin H. ''Henry Fairfield Osborn and the Proboscidea.'' in *The Proboscidea: Evolution and Paleoecology of Elephants and Their Relatives.* Ed. J. Shoshani and P. Tassy (1996), pp. xxiii–xxvii.

Fuller, Robert N. ''Henry Fairfield Osborn, the Man and His Books: A Bibliographical Sketch and a Survey of His Published Work'' (n.d.).

Goddard, Donald, ed. *Saving Wildlife: A Century of Conservation* (1995).

Gould, Stephen J. ''An Essay on a Pig Roast.'' *Natural History* 1, 89 (1989).

Milligan, Florence. ''Henry Fairfield Osborn, Man of Parnassus.'' *Bios* (1936).

Osborn, Henry F. *Fifty-two Years of Research, Observation and Publication 1877–1929* (1938).

Rainger, Ronald. *An Agenda for Antiquity: Henry Fairfield Osborn and Vertebrate Paleontology at the American Museum of Natural History, 1890–1935* (1991).

Shoshani, J., and Tassy, P., eds. The Proboscidea: Evolution and Paleoecology of Elephants and Their Relatives (1996).

Simpson, George G. ''Osborn, Henry Fairfield.'' *Dictionary of American Biography.* Vol. 21, Supplement 1 (to 31 December, 1935) (1944).

Woodward, A. Smith. ''Henry Fairfield Osborn, 1857–1935.'' *Obituary Notices of Fellows of the Royal Society* 2 (1936).

UNPUBLISHED SOURCES

Collected literature and correspondence to and from H. F. Osborn (in fourteen filing drawers) and accompanying Index at the Archives/Manuscript Collection are at the Library of the American Museum of Natural History (AMNH), New York.

Note: Many thanks to Mary Genett and Pamela Haas of the AMNH Library for allowing access to the Reference and Archive collections and to Malcolm C. McKenna for the unlimited access to the Osborn Library.

Jeheskel (Hezy) Shoshani

OSBORN, (HENRY) FAIRFIELD, JR. Born Princeton, New Jersey, 15 January 1887; died New York City, 16 September 1969. Naturalist, conservationist, zoo administrator.

FAMILY AND EDUCATION

Son of Henry Fairfield Osborn, biology professor and president of the American Museum of Natural History, and Lucretia Perry. Graduate of the Groton School, 1905, A.B., Princeton, 1909. Graduate study at Cambridge University, England, 1909–1910. D.Sc., New York University, 1955, Princeton, 1957, University of Buffalo, 1962; LL.D., Kenyon, 1959, Hofstra, 1966. Married Marjorie Mary Lamond, 1914; three daughters.

POSITIONS

Worked in railway freight yards in San Francisco and on railroad crew in Nevada as a young man. Briefly treasurer, Union Oil Company before World War I; treasurer of a label-making firm, 1914–1917. Served as captain, U.S. Army during World War I. Partner, Redmond and Company, investment bankers, 1919–1935. With banking firm of Maynard, Oakland, and Lawrence during part of 1935. Retired from business, 1935. Treasurer and member of the Executive Board, New York Zoological Society, 1923–1935, secretary and board member, 1935–1940, president and board member, 1940–1969. Member, Conservation Advisory Committee, U.S. Department of the Interior, 1950–1957, also the Planning Committee, Economic and Social Council of the United Nations.

CAREER

Associated with the New York Zoological Society (NYZS) from 1923 until his death. Placed animals in zoo settings approximating their natural habitat and was responsible for other innovations. Edited *The Pacific World* (1944), a guide to the people, geography, animal and plant life of the region for servicemen and others. Published *Our Plundered Planet* (1948), *The Limits of the Earth* (1953), and *Our Crowded Planet* (1962). With other colleagues, organized the Conservation Foundation as part of the NYZS and served as its president, 1948–1962, and as chairman of the board, 1962–1969. Beginning in the 1950s, published articles on ecological issues for various periodicals. Produced a number of studies and films concerning endangered species, flood control, and water resources. Worked with Laurance Rockefeller, a vice president of the NYZS, to establish Jackson Hole Wildlife Park near Moran, Wyoming. Member of the executive committee of the Advisory Board, American Committee for International Wildlife Protection, and was active in a number of other conservation organizations.

MAJOR CONTRIBUTIONS

Helped develop New York Zoological Park into one of the world's leading zoos. Instrumental in creating Jackson Hole Wildlife Park. Played valuable role in a number of conservation organizations and initiatives.

BIBLIOGRAPHY

Bridges, William. *A Gathering of Animals: An Unconventional History of the New York Zoological Society* (1974).

Dictionary of American Biography. Supplement 8 (1988).

Goddard, Donald, ed. *Saving Wildlife: A Century of Conservation* (1995).

The New Yorker, 9 March 1957 (interview).

New York Times, 17 September 1969 (obit.).

Rockefeller, Laurance. "My Most Unforgettable Character." *Reader's Digest* (October 1972).

Keir B. Sterling

P

PACKARD, ALPHEUS SPRING, JR. Born Brunswick, Maine, 19 February 1839; died Providence, Rhode Island, 14 February 1905. Naturalist, entomologist, evolutionist, University professor.

FAMILY AND EDUCATION

Son of Alpheus Spring Packard, Bowdoin College professor, and grandson of Bowdoin College president and Frances Elizabeth Appleton. Raised by his aunt Sara S. Packard, following death of his mother few months after his birth. A boy naturalist who collected minerals and shells and issued his own publication at an early age. As a college student went on Williams College Expedition to Labrador (summer, 1860). Graduated from Bowdoin College 1861 and studied under Agassiz at Museum of Comparative Zoology, Harvard University, 1861–1864. Also studied medicine, Maine Medical School (Bowdoin), receiving M.D., 1864. Married Elizabeth Derby Walcott, 1867; four children.

POSITIONS

Assistant, Maine Geological Survey (under Edward Hitchcock), 1861–1862; assistant, Museum of Comparative Zoology (under Louis Agassiz), 1861–1864; assistant surgeon, 1st Maine Veteran Volunteers, 1864–1865; acting custodian and librarian, Boston Society of Natural History, 1865–1866; curator, Essex Institute, Salem, Massachusetts, 1866–1867; curator of anthropods, Peabody Academy of Science, Salem, Massachusetts, 1867–1878; director, Peabody Academy of Science, 1877–1878; professor of zoology and geology, Brown University, 1878–1905.

CAREER

Returned to Labrador (Williams College Expedition), summer 1864; a founder (with Putnam, Morse, and Hyatt) and principal editor, *American Naturalist*, 1867–1886. Collected marine invertebrates on Florida reefs and at Beaufort, North Carolina, winter 1869–1870. Lecturer on economic entomology, Maine College of Agriculture, 1870, and Massachusetts Agricultural College, 1870–1878; lecturer on entomology, Bowdoin College, 1871–1874; state entomologist for Massachusetts, 1871–1873. Published *Annual Record of American Entomology*, 1871–1873. Elected to National Academy of Sciences, 1872. Taught (insects and crustaceans), Agassiz's summer school on Penikese Island (Anderson School of Natural History), 1873, 1874; assistant, Kentucky Geological Survey, summer, 1874. In charge, entomological work of U.S. Geological and Geographical Survey (Hayden Survey), 1875–1877. Organized and directed Summer School of Biology, Peabody Academy of Science, 1876–1881; member and secretary, U.S. Entomological Commission, 1877–1882; a founder of neo-Lamarckian school of evolution (with Hyatt and Cope); biographer of Lamarck (1901). Published nearly 600 works, especially on economic entomology, lepidoptera, coleoptera, crustacea, glaciers, and cave life; see bibliography in Cockerell (1920). Described over fifty genera and about 580 species of invertebrates. He had a very quiet and retiring personality, but sociable and helpful to his students.

MAJOR CONTRIBUTIONS

Published five textbooks on entomology, four textbooks on zoology, six monographs on lepidoptera. He trained many of the leading entomologists of the succeeding generation. Leader among the neo-Lamarckian evolutionists. Author of *Lamarck, The Founder of Evolution, His Life and Work* (1901).

BIBLIOGRAPHY

Cockerell, T. D. A. "Biographical Memoir of Alpheus Spring Packard, 1839–1905," *Biographical Memoirs, National Academy of Science* 9 (1920).

Cockerell, T. D. A. "Alpheus Spring Packard." *Bios* 14 (1943).

Dexter, R. W. "The Development of A. S. Packard, Jr. as a Naturalist and an Entomologist." *Bulletin of the Brooklyn Entomological Society* 52 (1957).

Dexter, R. W. "Contributions of Dr. A. S. Packard, Jr. to Entomology." *Bios* 52, 1 (1981).

Kingsley, J. S. "Alpheus Spring Packard." *Science* 21 (1905).

Mead, A. D. "Alpheus Spring Packard." *Popular Science Monthly* 67 (1905).

UNPUBLISHED SOURCES

Copy of early diary is in possession of family.

Ralph W. Dexter

PAGE, THOMAS JEFFERSON. Born Shelby, Virginia, 4 January 1808; died Rome, Italy, 26 October 1899. Naval officer, explorer.

FAMILY AND EDUCATION

Eighth son of Mann and Elizabeth Nelson Page. Paternal grandfather was congressman and three-term Virginia governor. Maternal grandfather signed the Declaration of Independence and was Revolutionary War officer. Father died when Page was young. Thomas' mother, in recognition of her father's services to the nation, was offered places at military academy for two of her sons. Thomas preferred the navy and became a midshipman. Details of his earlier education not known. Married Banjamina Price, 1838; seven children.

POSITIONS

Appointed midshipman, U.S. Navy, on school ship at Norfolk, Virginia, 1827. Passed midshipman, 1833. Promoted lieutenant, 1837 (some sources say 1839); commander, 1855. Posted to U.S.S. *Erie* for several years. Assigned to Coast Survey, 1833–1842. Assigned to U.S.S. *Columbus*, 1842–1844. Assigned to U.S. Naval Observatory, Washington, 1844–1848. Commanded U.S.S. *Dolphin* on round-the-world cruise, 1848–1851. Commanded steamer *Water Witch* on expedition to Paraguay, 1853–1856. Again explored in Paraguay, 1859–1860. Resigned commission and joined Confederate navy, 1861. In Europe, 1863–1864. Commanded C.S.S. *Stonewall* (ex-*Sphinx*), December 1864–April 1865. Cattle rancher in Argentina, 1865–1880. Thereafter resident of Florence, Italy.

CAREER

Won plaudits of his superiors when, during cruise of flagship U.S.S. *Erie*, he took command when all officers were sick with yellow fever, and, though he was the most junior midshipman, brought vessel safely into port with aid of another fellow midshipman. While in command of the U.S.S. *Dolphin*, with which he circumnavigated the globe, he recognized need for thorough survey of Chinese waters in support of American commerce and whaling industry. The navy agreed, and the Ringgold expedition was organized, but a more senior officer was placed in charge, and Page declined post of second-in-command. He was instead given charge of expedition to conduct "survey and explor[ation of] the river La Plata and its tributaries," which departed Norfolk in February, 1853. This mission had several purposes. One was to support efforts of the United States and Paraguay Navigation Company, an American firm, to promote trade in Paraguay. A second was to secure better information about that nation's geography and natural resources. The company's managing director, Edward A. Hopkins, who knew the region well, was also American vice-consul in Asuncion. Hopkins presented a speech to American Geographical and Statistical Society in 1852, calling attention to Paraguay, and that organization in turn strongly urged the U.S. Navy to send an exploring expedition to the area. Between his arrival in Buenos Aires in May 1853, and his departure more than two years later, Page deftly managed to accomplish his mission despite a number of diplomatic complications, which involved Hopkins, several American diplomats in the region, the navy, the strong men of Argentina and Paraguay, and the sometimes conflicting objectives of both of those nations and of Brazil. Between May 1853 and February 1856 Page traveled 3,600 miles by water and 4,400 miles on land in parts of Argentina, Brazil, and Paraguay. At one point in 1854 a quarrel developed between dictator Carlos Antonio Lopez of Paraguay and Hopkins. Page sided with Hopkins, the *Water Witch* was denied use of Paraguayan waters, and his vessel was fired on by a Paraguayan fort while going up the River Paraná. Page demanded support from the commander of the U.S. Navy Brazilian squadron but without success. When Page returned to the United States in 1856, he demanded action, and newly inaugurated President James Buchanan sent a nineteen vessel fleet with Commodore William B. Shubrick in command and Page as his fleet commander to demand satisfaction. In these circumstances, Lopez was very quickly persuaded to sign a new treaty with the United States, and Page was released from sea duty to continue his Paraguayan explorations from March 1859 to October 1860.

Prior to his first trip, Page had been given instructions on collection of natural history specimens by S. F. Baird of the Smithsonian and by Asa Gray of Harvard. Edward Palmer, a young horticulturist, accompanied Page and collected seeds, plants, and roots. Various rivers were mapped for future navigational purposes. Crew members were detailed by Page to collect animals, plants, and geological specimens. Some live animals were sent to the Navy Department in 1854, including a tiger, nutria, capybara, coati, kinkajou, and deer. Page was careful to note where each had been captured, an essential point sometimes missed by collectors on other early expeditions. Because the government did not then have a zoological park, those animals that survived the trip were sent to the Government Asylum for the Insane in Washington, there to amuse the patients. Later shipments included eighteen varieties of beans and peas, Yerba Mate seeds, preserved fish and reptiles, and insects and butterflies, most of which were sent to Baird.

No government funding was available for publication of Page's report of his first expedition, which was issued commercially as *La Plata, The Argentine Confederation, and Paraguay, Being a Narrative of the Exploration of the River La Plata and Adjacent Countries, during the Years 1853,'54, and '56, under the orders of the United States Government* (1859). John Cassin, the Philadelphia ornithologist, and Charles Girard, a noted herpetologist, provided appendices concerning the birds, reptiles, and fish collected. The report went through several editions and also appeared in a Spanish translation. Baird warmly acknowledged excellence and variety of the specimens brought back by Page, but no public or private monies were ever made available for the comprehensive scientific account of the plants and animals collected, as Page had wished. His second report, covering the work done in 1859 and 1860, was never published owing to the onset of the Civil War.

Page entered Confederate service in 1861, was in command of coastal defenses for a time, and then went to England to take command of an ironclad vessel under construction there (1863). When this was rendered impossible due to American diplomatic pressure, he went to Florence, and later to Copenhagen, where in December 1864, he took charge of the *Stonewall*, a vessel originally built in France for the Confederacy, then sold to Denmark. Following the War of 1864 with Prussia, the Danes sold the vessel to the Confederates. Page and his crew saw no action during the less than three months he was at sea, and on hearing of the Civil War's end, surrendered the ship to the Spanish authorities in Cuba late in March 1865. Page held the rank of Commodore in the Confederate naval service, and was also an artillery colonel in the Confederate army. He spent most of the years 1865–1880 in Argentina as a rancher in collaboration with former President Justo Urquiza (to 1870) and then on his own, but was in England for a time supervising construction of two ironclads and two gunboats for the Argentine navy. After 1880, he lived in Florence and then in Rome.

MAJOR CONTRIBUTIONS

Of the fifteen pre-Civil War exploring expeditions mounted by the U.S. Navy, the La Plata River surveys led by Page yielded one of the most outstanding natural history collections. It was the leading collection of South American natural history made by a U.S.-led group to that time. The charts produced of the rivers Page and his men traversed were very well done and were of continuing value to travelers and to commercial interests in and around Paraguay for a number of years.

BIBLIOGRAPHY

American Biographies (1940).

Dictionary of American Biography.

Kazar, John D. "The United States Navy and Scientific Exploration, 1837–1860." Diss., University of Massachusetts (1974).

National Cyclopedia of American Biography. Vol. 10.

Page, Thomas J. "Autobiographical sketch." *Proceedings of the United States Naval Institute* 44 (October, 1923).

Ponko, Vincent Jr. *Ships, Seas, and Scientists: U.S. Naval Exploration and Discovery in the Nineteenth Century* (1974).

The Richmond Times, 29 October 1899 [account of Page's Confederate naval service], repr. in R. A. Brock, ed. *Southern Historical Society Papers* 27 (1899).

UNPUBLISHED SOURCES

Letters of the Exploration and Survey of the River Plata, Record Group 45; Records of the Hydrographic Office, Record Group 37; manuscript maps in Cartographic and Audiovisual Records Division; and Muster Roll of the *Water Witch* in Old Military Records Division, National Archives, Washington. U.S. Exploration and Government Reports (bound letters), and in Baird correspondence, Smithsonian Institution Archives.

Keir B. Sterling

PALLISER, JOHN. Born Dublin, Ireland, 29 January 1817; died Comragh House, Kilmacthomas, County Waterford, Ireland, 18 August 1887. Big-game hunter, explorer.

FAMILY AND EDUCATION

Eldest son of Anne Gledstanes and Col. Wray Palliser. Educated abroad. Attended Trinity College, Dublin, 1834–1838, without completion. Bachelor.

POSITIONS

In Ireland: high sheriff (1844), deputy lieutenant and justice of the peace. Commission, captain, Waterford Artillery Militia (20 September 1839–14 July 1864). Minimal regular military service. Fellow, Royal Geographical Society (24 November 1856). Commander, British North American Expedition, 1857–1860.

CAREER

His taste for adventure and big-game hunting was inspired by friend William Fairholme's 1840 buffalo-hunting expedition to Missouri. Visited North America (1847–1848) to hunt big game and observe Indians and fur traders. Returned to London via New Orleans and Panama (1849). Wrote popular travelogue, *Solitary Rambles and Adventures of a Hunter in the Prairies* (1853). Elected to Royal Geographical Society, he proposed a return expedition to explore the plains and locate passes in the Rocky Mountains. Aware of the American search for a railway route to the Pacific, the British society proposed larger scientific expedition. Expedition planners consulted Edward Sabine, William Jackson Hooker, John Richardson, and George Simpson of Hudson's Bay Company. Palliser took with him James Hector, geologist and naturalist; Eugène Bourgeau, botanist; Lt. Thomas Blakiston, magnetic observer and ornithologist; and John W. Sullivan, secretary and astronomer. John Ball of the Colonial Office, who provided funding, instructed Palliser to make an impartial assessment of the region—gathering information on soils, climate, flora and fauna, farming potential, general geography, and possible transportation routes.

Palliser left for New York 16 May 1857, traveling to Sault Ste. Marie, picking up voyagers and reaching Lower Fort Garry in the Red River Colony (Manitoba). Began work at the southern boundary before moving west into Saskatchewan to winter at Carlton House on the North Saskatchewan River. Ranged extensively over modern Saskatchewan and Alberta, between the 49th and 54th parallels, spending next winter at Edmonton. Quarrels broke out among the members. Hector Blakiston traveled into the Rockies through Vermilion Pass and returned through Kicking Horse Pass. In 1859 the expedition attempted several routes to reach Pacific: Hector tried Howse Pass unsuccessfully; Palliser and Sullivan followed the North Kootenay Pass down to Bonner's Ferry Idaho, reaching Fort Colville. Sullivan then went east, Palliser went west, where he met Lt. Henry Spencer Palmer, American Boundary Survey party, near Midway, B.C. Palliser, Sullivan, and Hector reunited at Fort Colville, descended the Columbia, and returned via Victoria, San Francisco, and Panama. Palliser stopped in Montreal to brief Simpson, reaching Liverpool in June 1860. Expedition gave briefings to various scientific societies and published reports in 1859, 1860, and 1863, culminating in the first comprehensive map of the region, published 1865. Palliser received Royal Geographical Society's Patrons Gold Medal and a CMG (1877).

Expedition reports gave first comprehensive survey of geology, drawing attention to rich soil on the banks of the Saskatchewan River, as did the Canadian expedition of Henry Youle Hind (1858). Palliser identified the drought region now called Palliser's Triangle. Argued railway construction faced higher costs than the American route, due to difficult terrain between the Great Lakes and Red River and the Rocky Mountain passes. Identified the problem of native and Métis dependence on the vanishing buffalo. Although he passed among the plains tribes without conflict, he foresaw future conflict with settlers. The reports maintained their value through the century. Information on Palliser's later life is fragmentary. May have run Union blockade to visit Charleston in 1862. In 1869 undertook walrus-hunting expedition to Novaya Zemlya and the Kara Sea. He died at his estate, impoverished.

MAJOR CONTRIBUTIONS

Provided exhaustive scientific reports on the Canadian plains. The reports identified the long-term problems and issues involved in the political, social, and economic processes that dispossessed some peoples, established others, and transformed the great grasslands into a region of monoculture. His reports were invaluable to policy planners and railway engineers. Left a remarkable portrait of a landscape.

BIBLIOGRAPHY

Haig, Bruce. *James Hector, Explorer* (1983).

Lytton, Edward Bulwer. *Progress of the British North American Exploring Expedition under the Command of Captain John Palliser, F.R.G.S.* (1859?).

Palliser, John. *Solitary Rambles and Adventures of a Hunter in the Prairies* (1853).

Palliser, John. *Papers Relative to the Exploration by Captain Palliser of That Portion of British North America Which Lies between the Northern Branch of the River Saskatchewan and the Frontier of the United States; and between the Red River and Rocky Mountains* (1859, 1969).

Palliser, John. *Further Papers Relative to . . . River and Rocky Mountains; and thence to the Pacific Ocean* (1860).

Palliser, John. *The Journals, Detailed Reports and Observations Relative to the Exploration . . . during the Years 1857, 1858, 1859, and 1860* (1863).

Palliser, John. *Index and Maps . . . Laid before Parliament on the 19th May 1863* (1865).

Spry, Irene M. "John Palliser." *Dictionary of Canadian Biography.*
Spry, Irene M. *The Palliser Expedition: An Account of John Palliser's British North American Expedition, 1857–1860* (1963).
Spry, Irene M. "The Pallisers' Voyage to the Kara Sea." *Musk-Ox* 26 (1980).
Spry, Irene M. *The Papers of the Palliser Expedition, 1857–1860* (1968).

Lorne Hammond

PALMER, E(PHRAIM) LAURENCE. Born McGraw, New York, 8 July 1888; died Ithaca, New York, 18 December 1970. Naturalist, educator.

FAMILY AND EDUCATION

Son of Ephraim Clark and Laura Lincoln Darrow Palmer. Graduate of Cortland (New York) State Teacher's College, 1908; B.A., Cornell, 1911, M.A., 1913, Ph.D., 1917 (some sources indicate his Ph.D. was from Columbia University). Married Katherine Evangeline Hilton Van Winkle (herself an able paleontologist), 1921, (born 1895, with a Ph.D. from Cornell, 1925); two sons.

POSITIONS

Assistant in botany, Cornell, 1910–1913; instructor in botany and elementary agriculture, 1913, assistant professor of botany, 1914, professor of natural science and university extension instructor, 1915–1919, Iowa State Teachers College; Goldwin Smith Fellow in Botany, Cornell University, 1916–1917, enlisted in U.S. Naval Reserve, 1918; assistant professor of rural education, 1919–1922, professor of rural education, Cornell, 1922–1952, professor emeritus of nature and conservation education, Cornell, 1952–1970. Editor, Cornell University Rural School Leaflets, 1919–1952; director of nature education, *Nature Magazine*, 1925–1959; contributing editor, *Nature Magazine*, 1960–1962. On summer teaching staffs of several universities, including University of California, Los Angeles, 1922, 1924; Iowa Agricultural College, 1923, 1933; Utah Agricultural College, 1925–1927; University of Hawaii, 1931; University of Washington, 1937. Extension lecturer at University of California, Berkeley, 1926–1927 and at University of Minnesota, 1959. Director of conservation education, National Wildlife Federation, 1950–1957. President, American Nature Study Society, 1936–1937, and National Association of Biology Teachers, 1947. Fulbright Fellow in New Zealand, 1949.

CAREER

At Cornell on state scholarship during years 1908–1912 and was Richard Robin Assistant in Botany there, 1910–1913. Deeply involved in nature and conservation education throughout his career. Many of the graduate students he trained at Cornell (fifty doctoral and seventy-five master's candidates) made up a substantial percentage of the next generation of conservation educators. Productive and creative author of articles for a wide range of periodicals and journals. Published 700 articles and papers, which appeared in *Nature and Science Education Review, Nature Study Review, Childhood Education*, and the *Biology Teacher*. Contributed to *This Is Nature* (1959) and *Suburban Conservation* (1961). His books included *Camp Fire Nature Guide* (1925), *Field Book of Nature Study* (1925, rev. ed. 1928), *Nature Magazine's Guide to Science Teaching* (1936), *Aids to Knowing Natural History*, 3 vols.: *Birds* (1944), *Mammals* (1945), *Fossils* (1965). His *Field Book of Natural History* (1949, rev. ed. 1975) is regarded as a classic. He also authored *Guide to Conservation* (1954) and *Fieldbook of the Mammals* (1957). Was also a systematic botanist and ichthyologist with an interest in limnology. President, National Council of Supervisors of Nature Study and Gardening (later Council for Elementary Science International), 1927; Department of Science Instruction of National Education Association, 1929; American Nature Study Society, 1936–1937; National Council of Nature Study Supervisors, 1936–1937. Chair, Wildlife Committee, National Research Council, 1943–1945, director, National Audubon Society, 1943–1949. Made important contributions to the educational programs of such organizations as the Boy Scouts of America, Ecological Society of America, International Union for the Conservation of Nature and Natural Resources, and National Committee on Policies in Conservation Education. Received numerous awards for his work from these and other organizations. Owner-manager of Slingerland-Comstock Publishing Company, Ithaca, New York, for a time in the 1940s.

MAJOR CONTRIBUTIONS

An outstanding nature and conservation educator for nearly half a century. Placed strong emphasis on ecology in classroom. Pioneer of modern teaching methods at elementary level. Prolific writer. Active in both higher education and a variety of conservation and education advocacy organizations.

BIBLIOGRAPHY

American Biology Teacher (March 1971).
American Men of Science. 8th ed. (1949).
National Cyclopedia of American Biography. Vol. 56 (1976).
Science Teacher (March 1971).
Stroud, Richard H., ed. *National Leaders of American Conservation* (1985).
Who Was Who in America.

UNPUBLISHED SOURCES

Letters are in William Irving Myers Collection (1923–1959) and in Records of American Nature Study Society (1908–1956), Cornell University Libraries, Collection of Regional History and University Archives, Ithaca, N.Y.

Transcript of oral history interview is in New York State College of Agriculture, Agricultural Leaders Project: oral history interviews, 1962–1969, Cornell University Libraries, Regional History and University Archives.

Jean Wassong and Keir B. Sterling

PALMER, THEODORE SHERMAN. Born Oakland, California, 26 January 1868, died Washington, D.C., 24 July 1955. Ornithologist, naturalist.

FAMILY AND EDUCATION

Son of Henry Austin Palmer and Jane Olivia Day Palmer. A.B., University of California, Berkeley, 1888, M.D., Georgetown University, 1895 (never practiced medicine). Married Bertha M. Ellis, 21 November 1911; no children.

POSITIONS

First assistant ornithologist, U.S. Biological Survey (USBS), 1890–1896; in charge, Death Valley Expedition, USBS, 1891; assistant chief, USBS, 1896–1902, 1910–1914; assistant in charge of game preservation, USBS, 1902–1910, 1914–1916; expert in game conservation, USBS, 1916–1924; biologist, USBS, 1924–1928; senior biologist, USBS, 1928–1933; associate in zoology, U.S. National Museum, 1933–1955.

CAREER

Father intended Palmer to follow him in operating family bank, which Palmer did briefly. Joined the two-man original scientific staff of the USBS (C. H. Merriam and A. K. Fisher) in 1890. Author of *The Jack Rabbits of the United States* (1896, rev. 1897), *Index Generum Mammalium*, 1904. Palmer soon turned his attention to the subject of legislation affecting wildlife. Published *Hunting Licenses: Their History, Objects, and Limitations* (1904); *Chronology and Index of the More Important Events in American Game Protection, 1776–1911* (1912); *Game as a National Resource* (1922). Involved in conservation efforts and often pressed for conservation legislation in the District of Columbia, many of the states, and at the national level, 1900–1924. Author of preliminary draft of the Migratory Bird Treaty of 1916, affecting birds migrating between Canada and the United States. For many years an officer of the American Ornithologists' Union, he also served on a number of AOU committees. Long involved in indexing the AOU's journal, *Auk*. He also authored hundreds of biographical sketches of deceased members for that publication and published *Biographies of Members of the American Ornithologists' Union* (1954), a compilation of most obituaries completed to that time. This earned him the nickname of "Tombstone Palmer," but he had performed a vital service for the union.

Vice president, National Association of Audubon Societies, 1905–1936; vice president, American Society of Mammalogists, 1928–1934; secretary, American Ornithologists' Union (AOU), 1917–1937; treasurer, AOU, 1920–1938; active in some twenty-nine North American and foreign scientific or conservation organizations and served as officer of a number of these.

MAJOR CONTRIBUTIONS

Effective at compiling and summarizing information essential for early twentieth-century federal and state conservation activities. Persistent in pressing for enactment of necessary legislative enactments in support of the conservation effort.

BIBLIOGRAPHY
McAtee, W. L. "In Memoriam: Theodore Sherman Palmer." *Auk* (1956).
Sterling, K. B. *Last of the Naturalists: The Career of C. Hart Merriam*. Rev. ed. (1977).
UNPUBLISHED SOURCES
See C. H. Merriam, A. K. Fisher, and T. S. Palmer Papers, Manuscript Division, Library of Congress.

Keir B. Sterling

PARKER, ELIZABETH. Born Colchester County, Nova Scotia, 19 December 1856; died Winnipeg, Manitoba, 26 October 1944. Journalist, writer, founder of Alpine Club of Canada, social reformer, conservationist.

FAMILY AND EDUCATION

Daughter of Mary Tupper and George Fulton. Educated at home, at public school, and at normal school in Truro, N.S., where she graduated with a first-class teaching certificate. Married Henry J. Parker, c.1874; two sons and one daughter. Moved to Winnipeg in 1892.

POSITIONS

Schoolteacher, Nova Scotia, c.1873. Journalist, *Manitoba Free Press*, Winnipeg, 1904–1940. Secretary, Alpine Club of Canada, Winnipeg, 1906–1910.

CAREER

Taught school for a year prior to marriage and raising a family in Halifax, where she attended lectures at Dalhousie University. Participated in various Winnipeg literary and philanthropic groups, first secretary of the Travelers' Aid Society, instrumental in organizing YWCAs, member of University Women's Club, founder of Winnipeg Women's Canadian Club (1907). Reviewed a poetry reading in her first published article (1904), leading to a regular column lasting thirty-six years in the *Manitoba Free Press*. Columns titled "Literary Causerie" and "A Reader's Notes." First traveled to the Canadian Rockies in late 1880s and in 1904 sojourned to Banff and wrote about travel. Founded the Alpine Club of Canada (ACC) with Arthur O. Wheeler, Winnipeg, 1906. Served as the first ACC secretary, housed club headquarters at home, daughter Jean ACC librarian and avid climber. Valuable club organizer who attended early ACC camps and remained a lifetime ACC supporter.

Columns in the *Manitoba Free Press* (later, the *Winnipeg Free Press*) included ''The Alpine Club'' (31 March 1906), ''A Backward Look at a Midsummer Holiday'' (30 September 1905), ''A Holiday Tour in the West'' (23 September 1905), ''Another Rocky Mountain Book'' (16 December 1905), ''The Canadian Rockies: A Joy to Mountaineers'' (23 September 1905), ''The Mountaineering Club'' (17 February 1906), and ''A Reader's Reminiscent Note'' (9 November 1922). Contributed secretary's reports, articles, book reviews, and obituaries to the *Canadian Alpine Journal*: ''Alpine Club of Canada'' (1907), ''Tyndall's Alpine Books'' (1909), ''Upper Columbia'' (1911), ''Mountaineering Classics'' (1921–1922), ''Some Memories of the Mountains'' (1928), ''The Approach to Organization'' (1938), ''Early Explorers of the West'' (1944–1945 to 1950). Coauthored *The Selkirk Mountains: A Guide for Mountain Climbers and Pilgrims* (with Arthur Wheeler, 1912). Also contributed to *Scribner's* (1914), *The University Magazine* (1908), *Dalhousie Review*. Reviewed every issue of the *Canadian Alpine Journal* until her later years.

A literary romantic, nature lover, social reformer, and Canadian nationalist with a wry sense of humor and abiding love for the mountains. Highly esteemed among early ACC members, but since then somewhat overshadowed by Arthur Wheeler. A persuasive writer and skilled organizer with a knack for publicity who worked to build lasting institutions. Across the prairies, even her newspaper columns took on this aura. Believed strongly that Canadians should awaken to their ''mountain heritage'' and seek enlightenment in nature. Advocated walking and preserving mountains as a retreat from mercenary urban life.

MAJOR CONTRIBUTIONS

Through her columns, she spread the ''gospel'' of the mountains to Canadians. As an architect of the ACC, she built principles of sexual egalitarianism, nationalism, and wilderness preservation into a long-standing national organization.

BIBLIOGRAPHY

''Elizabeth Parker.'' *Canadian Alpine Journal* (1944–1945).

''Elizabeth Parker Hut.'' *Canadian Alpine Journal* (1931).

''Mrs. Elizabeth Parker Dies.'' *Winnipeg Free Press*, 27 October 1944.

Reichwein, P. A. ''Guardians of the Rockies.'' *The Beaver* (August/September 1994).

Wheeler, A. O. ''Origin and Founding of the Alpine Club of Canada, 1906.'' *Canadian Alpine Journal* (1938).

UNPUBLISHED SOURCES

See Whyte Museum of the Canadian Rockies Archive, Banff, Alberta; letters, photos, Alpine Club of Canada Collection.

PearlAnn Reichwein

PEALE, CHARLES WILLSON. Born Queen Annes County, Maryland, 15 April 1741; died Philadelphia, 22 February 1827. Artist, naturalist, museum keeper, inventor.

FAMILY AND EDUCATION

Son of Charles and Margaret Triggs Matthews Peale. Father emigrated from England and became a schoolmaster in Maryland. Family moved to Chestertown when Peale was a young child. Father died in 1750, leaving a widow in straitened circumstances with five children, of whom Peale was the oldest. Family moved to Annapolis. Peale's formal schooling ended at age thirteen, when he was apprenticed to a saddler. Began work as saddler, 1762. Taught himself painting. With financial help from a group of wealthy Marylanders, he was able to study with painter Benjamin West in London, 1765–1767. Married Rachel Brewer, 1762, Elizabeth DePeyster, 1791, and Hannah Moore, 1805. Sixteen children by first and second wives, of whom eleven lived past childhood.

POSITIONS

Portrait painter in Maryland. Moved to Philadelphia, 1776. Fought in Revolutionary War as member of Philadelphia militia. Member of the Pennsylvania legislature, 1779–1780. Opened a portrait gallery in an addition built to his home, 1782. Opened an exhibition of "moving pictures," 1785. Opened a museum of natural history in his home, 1786. Made several long trips to Maryland to paint portraits and collect for museum, 1788–1791. After 1791, supported himself and large family primarily from the proceeds of the museum rather than by his painting. Museum moved to rooms at the American Philosophical Society, 1794. Museum moved to the State House (now Independence Hall) in space provided gratis by the city, 1802. Retired to farm in Germantown, 1810, leaving management of museum to his son, Rubens. Returned to resume active management of the museum, 1821–1827.

CAREER

Idea of forming a museum said to date from 1783. Museum officially opened in 1786, and donations solicited. Museum soon came to be known as the Philadelphia Museum or, more popularly, as Peale's Museum. Museum exemplified Enlightenment and republican ideals and was intended to be a school of nature for the new nation. At first arranged animals in an artistic display and, after 1788, more formally, according to the Linnaean system. Experimented with preservation techniques and was a pioneer in use of arsenic as a preservative. Exhibited animals in lifelike poses with painted backdrops showing typical habitat. Made collecting trips locally and to Delaware, Maryland, New Jersey, and New York. Sons collected for him on their travels. Set up correspondence and exchange with other naturalists in America and abroad. In 1801, excavated the skeletons of three mastodons in Orange and Ulster Counties, New York. Set up one skeleton for exhibit at his museum and formed a composite of the other two skeletons that his sons exhibited in various American cities and in England. Peale's mastodons were the first complete skeletons in existence, and drawings of them were incorporated into the works of such major naturalists as George Cuvier. Skeletons did much to convince the public of the reality of extinction.

Peale's Museum housed the first large and scientifically arranged collection of animals in America. Early naturalists considered it an important resource and aided it with their donations. Museum received the specimens from the Lewis and Clark Expedition and later the specimens from the Long Expedition. Peale's collections of birds and mammals were especially noteworthy. Alexander Wilson (q.v.) incorporated the birds of the museum into his *American Ornithology*. Benjamin Smith Barton (q.v.) taught his zoology classes for the University of Pennsylvania at the museum. Peale himself emphasized the educational function of the museum more than its value in advancing science. Influenced by Enlightenment thought, Peale envisioned the museum as a ''Great School of Nature'' and a means of revealing the Creator's handiwork. Peale gave two series of public lectures on natural history illustrated by specimens in 1799 and 1800. After his retirement in 1810, the museum became increasingly a place of entertainment, rather than a scientific institution. In 1821, the museum was incorporated, and Peale resumed the directorship. For a short period after 1821, John Godman, Richard Harlan, Thomas Say, and Gerard Troost were appointed to give lectures. After Peale's death, his museum was moved from Independence Hall to privately owned quarters in the Arcade, and in 1848 it was disbanded and sold at auction.

Known as an excellent ''practical naturalist,'' but while familiar with standard works of the day, such as Buffon, he was not considered a scholarly naturalist. Discovered several undescribed species but did not have the patience or ability to comb through the literature and write formal descriptions. Wrote no scholarly papers on natural history. Mammals in the museum were described by Palisot de Beauvois, an exiled French naturalist whom Peale supported financially, in the catalog by C. W. Peale and A. M. F. J. Palisot de Beauvois, *Scientific and Descriptive Catalogue of Peale's Museum* (1796) (also an edition in French). The opening lectures of Peale's two courses were published as C. W. Peale, *Introduction to a Course of Lectures on Natural History Delivered in the University of Pennsylvania, Nov. 16, 1799* (1799), and C. W. Peale, *Discourse Introductory to a Course of Lectures on the Science of Nature; with Original Music, Composed For and Sung on the Occasion. Delivered in the Hall of the University of Pennsylvania, Nov. 8, 1800* (1800).

MAJOR CONTRIBUTIONS

Founder and proprietor of the first scientifically organized museum of natural history in America. Museum served as an important resource to early naturalists and as a major public attraction. Pioneer in museum preservation and exhibition techniques. Excavated and displayed the first complete mastodon skeleton in the world. Played a key role in early popular education in science.

BIBLIOGRAPHY

Appel, T. A. "Science, Popular Culture and Profit: Peale's Philadelphia Museum." *Journal of the Society for the Bibliography of Natural History* (now *Archives of Natural History*) 9 (1980).

Miller, L. B. ed. *The Collected Papers of Charles Willson Peale and His Family* (1980) (microfilm edition of all known Peale manuscripts).

Miller, L. B., and D. C. Ward, eds. *New Perspectives on Charles Willson Peale: A 250th Anniversary Celebration* (1991).

The Selected Papers of Charles Willson Peale and His Family. Vol. 1, 1735–1791, ed. L. B. Miller, S. Hart, and T. A. Appel (1983); vol. 2, 1791–1810 (1988); vol. 3, 1810–1820 (1991) (vols. 2, 3 ed. L. B. Miller, S. Hart, and D. C. Ward); vol. 4, 1821–1827 (1996).

Sellers, C. C. *Charles Willson Peale* (1969).

Sellers, C. C. *Mr. Peale's Museum* (1980).

UNPUBLISHED SOURCES

Manuscripts and correspondence are published in a comprehensive microfilm edition (see earlier). Main collection of original documents is in the American Philosophical Society.

Toby A. Appel

PEALE, TITIAN RAMSAY. Born Philadelphia, 17 November 1799; died 13 March 1885. Artist, explorer, naturalist.

FAMILY AND EDUCATION

Son of Charles Willson Peale, artist and founder of Philadelphia Museum, and Elizabeth De Peyster Peale. Attended lectures at University of Pennsylvania and developed skill in preservation of specimens for Philadelphia Museum and in making drawings of subjects for museum's records. Married in 1822, Eliza Cecilia La Forgue, by whom had six children, and Lucinda Mac Mullen in 1848.

POSITIONS

In 1819, appointed assistant naturalist and painter with U.S. Expedition under Maj. Stephen H. Long; ordered to explore the country between Mississippi River and Rocky Mountains. In 1821 became assistant manager and in 1833 manager of Philadelphia Museum. Between 1838 and 1842 was member of civil staff of U.S. Exploring Expedition to South Seas, under Charles Wilkes, Jr. From 1849 to 1872, was examiner in U.S. Printing Office in Washington, D.C.

CAREER

Worked briefly in a spinning machine factory, before joining Philadelphia Museum. In 1818, was member of expedition to coast of Georgia and eastern

Florida with William Maclure, Thomas Say, and George Ord, to study fauna and secure specimens for Academy of Natural Science in Philadelphia. With U.S. Army Expedition under Maj. Long explored, collected animal specimens, and made sketches used in illustrating papers by members of Long's party. In 1824 was sent to Florida by Charles Lucien Bonaparte to collect specimens and make drawings for his *American Ornithology*, 4 vols. (1825–1833), of which colored plates in volumes 1 and 4 were Peale's work. Also drew plates for Thomas Say's *American Entomology*, 3 vols. (1824–1828). Visited Columbia in 1832 to collect specimens and in 1833 published a *Prospectus* for his projected *Lepidoptera Americana*, a massive project on which he labored many years, but never completed. Through his position with the Wilkes expedition, Peale made it possible for the Academy of Natural Sciences of Philadelphia to acquire its notable collection of Polynesian ethnica. Also authored articles in various scientific publications. As zoologist to the expedition, Peale discovered a number of new species of birds and mammals (although the precise number is still in dispute), together with a number of insects. Many of his specimens and his personal research library were lost on the USS *Peacock*, which foundered off the coast of Oregon on the return trip from the Pacific. Peale was obliged to work on the volume on mammals and birds in Washington, rather than Philadelphia, then the center of American scientific research, because the Senate Committee on the Library of Congress wanted no expedition specimens lost. Lacking the books he needed (only one library in the nation—the Academy of Natural Sciences of Philadelphia—had an adequate selection), Peale labored under very real difficulties. Because he had been primarily an artist and field naturalist, some of Wilkes' advisors felt Peale was not up to the scientific requirements of his task. Relations with Wilkes, acting for the Library Committee, grew steadily worse. Wilkes first suppressed Peale's Introduction to the book *Mammalia and Ornithology*, in which Peale had tried to explain some of the difficulties he had labored under, and then the book itself. A pitifully small number of the prefaceless books were distributed, but most were destroyed and others burned in a warehouse fire, making the volume one of the rarest American natural history titles. The book was rewritten by John Cassin, then a leading American ornithologist, and reissued in 1858 as *Mammalogy and Ornithology*. The majority of plates (32 of 53) in the Cassin book were originally done by Peale for his own suppressed volume. Most modern critics feel that Peale's book was quite acceptable. Its shortcomings were ''obviously those of haste rather than of knowledge, of inexperience in editing and proofreading.'' Peale not only lacked access to needed books, but to specimens he might have compared with what he had brought back from the Pacific. The deaths of his wife, eldest daughter, and infant son between 1844 and 1847 also sorely afflicted Peale. Cassin, a difficult person and something of a politician, was the curator and later corresponding secretary and vice-president of the Academy of Natural Sciences, and he regarded the bird books and specimens there as his personal fief. Peale spent the rest of his working life as an examiner for the U.S. Patent Office. He became one of the first amateur photographers in the country, and was a founder of the Philosophical Society of Washington, D.C., in 1871.

MAJOR CONTRIBUTIONS

Explorer who helped to open up the West. Careful observer and collector, his drawings and paintings of animal life, known for their accuracy and beauty, illustrated various publications, including *American Ornithology* and *American Entomology*. Also produced drawings for a number of plates in the narrative volumes (vols. 1–5) of the *Wilkes Expedition Reports*, his own suppressed volume on mammals and birds of the expedition, and Cassin's replacement text.

BIBLIOGRAPHY

Dictionary of American Biography. Vol. 7 (1927–1936).

Dictionary of Scientific Biography. Vol. 10 (1974).

Evans, Howard E. *The Natural History of the Long Expedition to the Rocky Mountains, 1819–1820* (1997).

Haskell, Daniel C. *The United States Exploring Expedition, 1938–1842 and its Publications, 1844–1874: A Bibliography* (1942).

Miller, L. B. ed. *The Collected Papers of Charles Willson Peale and His Family* (1980) (microfilm edition of all known Peale manuscripts).

National Cyclopedia of American Biography. Vol. 21 (1967).

Poesch, Jessie. *Titian Ramsay Peale, 1799–1885, and his Journals of the Wilkes Expedition* (1961).

Ponko, Vincent Jr. *Ships, Seas, and Scientists: U.S. Naval Exploration and Discovery in the Nineteenth Century* (1974).

Porter, Charlotte M. "Bibliography and Natural History: New Sources for the Contributions of the American Naturalist, Titian Ramsay Peale." In Alwyne Wheeler ed., *Contributions to the History of North American Natural History* (1983).

The Selected Papers of Charles Willson Peale and His Family. Vol. 1, 1735–1791, ed. L. B. Miller, S. Hart, and T. A. Appel (1983); vol. 2, 1791–1810 (1988); vol. 3, 1810–1820 (1991) (vols. 2, 3 ed. L. B. Miller, S. Hart, and D. C. Ward); vol. 4, 1821–1827 (1996).

Stanton, William. *The Great United States Exploring Expedition of 1838–1842* (1966).

Sterling, Keir B. "Introduction." In Titian R. Peale, *Mammalia and Ornithology* (repr. ed. 1978).

Viola, Herman, and Carolyn Margolis, eds. *Magnificent Voyagers: The U.S. Exploring Expedition, 1838–1842* (1985).

Who Was Who in America, 1607–1896 (1963).

UNPUBLISHED SOURCES

Peale's papers can be found in the collections of the American Philosophical Society (Philadelphia), Library of Congress, Historical Society of Pennsylvania, and Academy of Natural Sciences of Philadelphia. Peale's unpublished manuscript on Lepidoptera, with three volumes of drawings and paintings, is at the American Museum of Natural History.

Richard Harmond

PEARSE, THEED. Born Bedford, England, 26 October 1871; died Comox, British Columbia, 25 May 1971. Lawyer, ornithologist.

FAMILY AND EDUCATION

He was educated in Bedford and then articled with his father, who was practicing law. He married Elizabeth Margaret Llewelyn in 1919.

POSITIONS

Barrister and solicitor in Vancouver and then in Courtenay, British Columbia, where he served one term as mayor.

CAREER

In 1906 he sold out his law practice in Bedford, moved to an ill-fated venture in Virginia, and then to a fruit farm in Nova Scotia. After moving to British Columbia in 1909, he practiced law in the Courtenay-Comox area until he retired in 1941 to devote his full time to ornithology. He moved from Courtenay to Comox in 1945. He was an active bander of birds, especially gulls.

MAJOR CONTRIBUTIONS

"He published 103 papers in *Murrelet*, *Condor*, and *Auk*. He was the first of the non-collecting but meticulous recorders of birds in British Columbia" (Campbell et al. 1990). His book on the history of birds in the North Pacific, representing fourteen years of research, appeared when he was ninety-seven years old.

BIBLIOGRAPHY

Pearse, T. *Birds of the Early Explorers in the Northern Pacific* (1968).

C. Stuart Houston

PEARSON, T(HOMAS) GILBERT. Born Tuscola, Illinois, 10 November 1873; died New York, 3 September 1943. Biologist, ornithologist, wildlife conservationist, author.

FAMILY AND EDUCATION

Son of nomadic farm family, one among five children of Thomas Barnard and Mary (Eliott) Pearson. Family moved to Indiana and in 1882 to Archer, Florida, to join other Quakers in citrus farming. Graduated from Guilford College in 1897 with B.S. and achieved another B.S. from the University of North Carolina in 1899. Received honorary LL.D. in 1924 from the University of North Carolina and the medal of the Societe Nationale d'Acclimation of France in 1937. Married Elsie Weatherly in Greensboro, North Carolina, 1902; three children.

POSITIONS

Professor of biology at Guilford College, 1899–1901; and from 1901 to 1904, taught biology and geology at the State Normal and Industrial College for Women, Greensboro, North Carolina. Formed North Carolina Audubon Society in 1902. Acted as secretary and then president of National Association of Audubon Societies from 1905 until retirement in 1934. Founder and president of the International Committee for Bird Preservation from 1922 to 1938 and chairman of the U.S. section until his death.

CAREER

Studied for six years at Guilford College, continuing his work in natural history, where he developed a talent for public speaking that was to assist him in career devoted to supporting legislation for the protection and preservation of birds and wildlife. Established refuges and maintained favorable environments for birds and other wildlife in the United States during forty years of work with the Audubon movement. Lobbied for protective laws supporting the Audubon Plumage Bill. Campaigned for international control by visiting President Diaz of Mexico in 1909. Lobbied for the Federal Migratory Bird Law of 1913. Negotiated with Canada, resulting in the important Migratory Bird Treaty of 1916. Visited Europe in 1922 and at a London meeting founded International Committee for Bird Preservation. Lectured in Europe; North, Central, and South America; and West Indies.

Published autobiography, *Adventures in Bird Protection* (1937); edited *Portraits and Habits of Our Birds* (1920); senior editor of *Birds of America*, 3 vols. (1917); coauthor of *The Birds of North Carolina* (1919); and coeditor of the *Book of Birds* (1937). Books for children include *Stories of Bird Life* (1901), *The Bird Study Book* (1917), and *Tales from Birdland* (1918). Contributed notes, articles, reports, editorials to *Bird-Lore*, 1905–1940, *Audubon Magazine*, 1941–1942, and *National Geographic Magazine*, 1933–1939.

Devoted his life work to public activities, educational, organizational, and political, with a moral background founded in Quaker and southern upbringing. Exuded the self-confidence of a natural leader accompanied by sensitivity of human frailty and with an understanding of the limitations of human endeavor. As young boy learned to shoot birds and collect eggs, which, ironically, led to the dedication of his professional career to the protection and preservation of birds and wildlife.

MAJOR CONTRIBUTIONS

Developer of the National Association of Audubon Societies into the largest organization in the world interested in the protection of wildlife. Extended its influence throughout the world by traveling and lecturing in Europe and by organizing (1922) an international committee for bird preservation. Official collaborator of the National Park and Fish and Wildlife Services. Organized and presided over world conference for bird protection at Geneva, Switzerland, in 1928 and international bird protection conference in Amsterdam in 1930 and was U.S. delegate to 9th International Ornithological Congress in Rouen, France, in 1938. Chaired both the U.S. and Pan-American sections of International Committee for Bird Preservation in 1938 and founded and chaired the National Conference on Wild Life Legislation. Served on President Hoover's Yellowstone Park Boundary Commission and was national director of the Izaak Walton League of America.

BIBLIOGRAPHY

Dictionary of American Biography. Supplement 3 (1973).
National Cyclopaedia of American Biography. Vol. 33 (1947).
New York Times, 5 September 1943 (obit.).
Orr, Oliver H, Jr. *Saving American Birds* (1992).
Who Was Who in America. Vol. 2 (1950).

Linda L. Reesman

PERLEY, MOSES HENRY. Born Maugerville, New Brunswick, 31 December 1804; died on HMS *Desperate* off Labrador, 17 August 1862. Naturalist, lawyer, author, officeholder.

FAMILY AND EDUCATION

Son of Moses and Mary Perley. Father died before his birth so was raised by his well-to-do mother in Saint John, New Brunswick. Educated in New Brunswick in public schools. Studied law and was called to the bar in 1830. Adopted two orphaned English immigrants, 1847.

POSITIONS

Businessman in lumber industry, 1835, in several failed and then later successful ventures. New Brunswick commissioner of Indian affairs, 1841–1848. New Brunswick agent of immigration, 1843–1858. British agent of emigration to New Brunswick, 1847–1858. New Brunswick commissioner of fisheries, 1855–1862. Owner of the *Colonial Empire*, Saint John.

CAREER

As a boy, interest in nature was sparked by hunting, fishing, and trading with Native Americans. Unfortunate accidental killing of a Native American during target shooting induced a lifelong interest in the welfare of Native Americans. As commissioner of Indian affairs, he often lived among the Micmac and Malecite tribes. Made a chief of both tribes for attempts to halt government abuses of treaty lands. Constant travels around the reserves made him best-informed man about provincial rivers, natural resources, and fisheries. Became an avid promoter of developing these resources, tied in with his work to settle New

Brunswick with British immigrants. Often rendered personal financial assistance to poor and ailing immigrants. Began on his own initiative a study of Gulf of St. Lawrence fisheries in 1843. In 1846 New Brunswick government requested him to study New Brunswick mineral resources, soil fertilities, and settlement possibilities for a railway survey. In 1849 he reported on New Brunswick fisheries of the Gulf of St. Lawrence. He also wrote a preliminary report of Bay of Fundy fisheries and in 1852 published a catalog of Bay of Fundy fisheries, received favorably by British and New Brunswick governments. From 1849 to 1850 Perley covered 900 miles, over 500 by canoe, to study inland and sea fisheries. Drew up regulations for fishery wardens and recommended fishery societies modeled on agricultural societies. In 1849 and 1852 he drafted legislation for regulating sea and river fisheries. Compiled fisheries statistics between 1852 and 1854 for Reciprocity Treaty negations between United States and British North America. Appointed fishery commissioner in New Brunswick in 1855 to enforce Reciprocity Treaty. As fishery commissioner, Perley visited Prince Edward Island, Newfoundland, and Labrador. He died after an illness during a fisheries survey off Labrador aboard the HMS *Desperate*. Buried at Forteau, Labrador.

Published prolifically on many subjects, including fisheries, forestry, railways, industry and natural resources, and Indian settlements. Fisheries publications included *Report on the Fisheries of the Gulf of Saint Lawrence* (Fredericton [1849]); *Report on the Sea and River Fisheries of New Brunswick, within the Gulf of Saint Lawrence and the Bay of Chaleur* (Fredericton [1850]); *Report on the Fisheries of the Bay of Fundy* (Fredericton [1851]); *Reports on the Sea and River Fisheries of New Brunswick* (Fredericton [1852]); and *Descriptive Catalogue (in part) of the Fishes of New Brunswick and Nova Scotia* (Fredericton [1852]).

Concerns included neglect of sea fisheries and rapid decay of river fisheries, threatened by overfishing, lack of regulation, and sawdust and other pollution from uncontrolled logging and lumber industries. Issued dire warnings about the future of Canadian fisheries if pollution and fishing regulations not enacted and enforced. On his recommendations fisheries societies created by government funding to increase fishermen's knowledge of conservation, better fishing methods and processing techniques. Called for outlawing fish manures, both to conserve stocks and prevent soil deterioration. Called for government fisheries inspection and quality control to increase revenues and trade levels and better conserve the resources through their more effective use.

MAJOR CONTRIBUTIONS

Perley made the first in-depth study of the fishes and fisheries of the Canadian Maritime Provinces by a Canadian and recommended regulation of all aspects of the industry, for conservation and more efficient utilization of the fishes. His work was later built upon by the Biological (Fisheries Research) Board of Canada (1898–1973).

BIBLIOGRAPHY
Allardyce, Gilbert. "The Vexed Question of Sawdust River Pollution in Nineteenth-century New Brunswick." *Dalhousie Review* 52 (Summer 1972).
Johnston, Kenneth. *The Aquatic Explorers: A History of the Fisheries Research Board of Canada* (1977).
Spray, W. A. "Perley, Moses Henry." *Dictionary of Canadian Biography*, Vol. 9 (1976).
UNPUBLISHED SOURCES
Huntsman, and Margaret Rigby, A. G., "Materials Relating to the History of the Fisheries Research Board of Canada (Formerly the Biological Board of Canada) for the Period 1898–1924." Fisheries Research Board of Canada, Manuscript Report No. 660 (1958).

Jennifer M. Hubbard

PETERSON, RANDOLPH LEE. Born Roanoke, Texas, 16 February 1920; died Rockwood, Ontario, 29 October 1989. Mammalogist.

FAMILY AND EDUCATION

One of five children of Omas and Margaret Francisco Peterson. B.S., Texas A&M, 1941, Ph.D., University of Toronto, 1950. Married Elizabeth Fairchild Taylor, 1942; one daughter.

POSITIONS

Assistant curator, Texas Cooperative Wildlife Research Collection, 1937–1941. With U.S. Air Force, 1942–1945; acting curator, Mammal Division, Royal Ontario Museum (ROM), 1946–1950; curator, 1950–1985. Special lecturer, 1949–1962; associate professor, 1962–1968; professor, Department of Zoology, University of Toronto, 1968–1985. Curator emeritus, ROM, and professor emeritus, University of Toronto, 1985–1989.

CAREER

Was raised on family farm, which provided introduction to mammals, plants, and ecology. Began career of museum curating at Texas A&M University, simultaneously assisting W. B. Davis with similar work at Texas A&M Museum. During World War II, was pilot and instructor; later, operations officer. Offered position at ROM at annual meeting of American Society of Mammalogists in Toronto in spring of 1946 and remained there for rest of his career. Survey of mammals of Ontario, begun at ROM in the 1920s, continued and was expanded to include Quebec and the Maritimes. His research on moose culminated in *North American Moose* (1955), a revision of his doctoral dissertation, since reprinted several times. Published *Mammals of Eastern Canada* (1966). Specialist on bats and developed collection at ROM into one of finest anywhere, with many varieties and a total of some 35,000 specimens. Over one-third of his published articles dealt with bats, but he also contributed chapters to several books and authored scientific articles on a variety of other mammals. He described five new species of bats and was responsible for several taxonomic revisions. Led expeditions to various parts of Canada and the western United States, Jamaica, Trinidad, Guyana, Belize, Kenya, Madagascar, Cameroon, and Zimbabwe. Sponsored eight doctoral and eight master's students at University of Toronto. Long an active member and officer of the American Society of Mammalogists (president, 1966–1968), and honorary member (1986). Also founder of Future Mammalogists Fund (1986), designed to assist students of mammalogy and facilitate their professional preparation. Chairman of the board, Metropolitan Toronto Zoological Society, 1977–1979. Member Board of Management, Metropolitan Toronto Zoo, 1978–1982. Was councillor, Society of Systematic Zoology, 1966–1969, and active in a number of other professional organizations in Canada, the United States, Australia, and South Africa. With his wife, founded Boreal Biological Laboratories, Ltd. in the 1950s, a biological supply firm (later sold, 1974, though Peterson remained a director until his death). An active promoter of bat conservation and public teaching. Provided many interviews to Canadian press, especially about bats. Responsible for concept and design of a bat cave at ROM, which became most popular gallery there, particularly for children.

MAJOR CONTRIBUTIONS

Responsible for substantial research and published literature concerning bats, moose, and other mammals.

BIBLIOGRAPHY

Birney, Elmer C., and Jerry R. Choate, eds. *Seventy-Five Years of Mammalogy, 1919–1994* (1994).

Eger, Judith L., and Lorelie Mitchell. *Journal of Mammalogy* 71.4 (1990) (obit.).

Time, 12 August 1996.

Keir B. Sterling

PETERSON, ROGER TORY. Born Jamestown, New York, 28 August 1908; died Old Lyme, Connecticut, 28 July 1996. Ornithologist, artist.

FAMILY AND EDUCATION

Son of Charles Gustav Peterson, craftsman for Art Metal Construction Company, and Henrietta Bader Peterson. Educated in local public schools. Attended Art Student's League classes, New York City, 1927–1928; National Academy of Design, New York City, 1929–1931. Honorary D.Sc. degrees from Franklin and Marshall, 1952, Ohio State, 1962, Fairfield University, 1967, Allegheny College, 1967, Wesleyan University, 1970, Colby College, 1974, Gustavus Adolphus College, 1978, Connecticut College, 1985, University of Connecticut, 1986, Memorial University of Newfoundland, 1987, Bates College, 1991; Hum.D., Hamilton College, 1976; D.H.L., Amherst, 1977, Skidmore College, 1981, Yale, 1986, Southern Connecticut State University, 1991; D.F.A., University of Hartford, 1981, State University of New York, 1986, Middlebury, 1986, Long Island University, 1987, MacMurray College, 1989. Married Mildred Warner Washington, 1936 (divorced); Barbara Coulter, 1943 (divorced), two children; Virginia Westervelt, 1976, two stepchildren.

POSITIONS HELD

Decorative artist for a furniture company, 1926; instructor in science and art, Rivers School, Brookline, Massachusetts, 1931–1934; bird painter and illustrator, 1934–1996; in charge, editorial activities, National Audubon Societies and Art Editor, *Audubon Magazine*, 1934–1943; enlisted man with U.S. Army Engineers and U.S. Army Air Corps, 1943–1945; Editor, Houghton Mifflin Field Guide Series, 1946–1996; Audubon Lecturer from 1946; Art Director, National Wildlife Federation, 1946–1975; Fellow, Davenport College, Yale University, 1966–1996; First Distinguished Scholar in Residence, Fallingwater–Western Pennsylvania Conservancy, 1968; Founder, Roger Tory Peterson Institute for Study of Natural History, 1986; frequent lecturer and contributor to journals and books on natural history subjects.

CAREER

Interest in birds began at age eleven with formation by elementary school teacher of a Junior Audubon Club. Made frequent field trips near his home. Won art contest run by Buffalo Times, 1922. Briefly an artist who decorated furniture after graduation from high school, then studied art in New York for four years. Studies terminated because of the Depression; taught art and science at private school in Massachusetts for three years, 1931–1934. While teaching, continued personal study of bird identification, influenced in part by work earlier done by Ernest Thompson Seton. Active in Bronx County Bird Club from c.1928 until it was dissolved following World War II. Developed his system of identification in several magazine articles, was then encouraged to complete field guide incorporating it by William Vogt, later to be a colleague at National Audubon Societies. After four publishers rejected it fearing low sales, *A Field Guide to the Birds* was published (1934) becoming instant and continuing success. Revised editions appeared 1939, 1947, and 1980. Peterson was working on fifth edition at his death. Appointed Education Director by National Audubon Societies, 1934, he revised Junior Audubon leaflets and was art editor of *Audubon Magazine*. Authored *Field Guide to Western Birds* (1941; revised edition, 1961; 3rd edition, 1990). During World War II, worked for U.S. Army at Fort Belvoir, Virginia, on several projects, including Army camouflage program, the writing of manuals on defusing land mines, and on plane spotting techniques based on his bird identification system, employing size, shape, color pattern, and characteristic markings. Later studied effects of DDT for the Army Air Force in Orlando, Florida, subsequently joining other environmentalists in firm opposition to the indiscriminate use of this pesticide. Resumed nature writing career after the war, authoring *Birds Over America* (1948), *How To Know the*

Birds (1949; 2nd ed., 1957), *Wildlife in Color* (1951), and a *Field Guide to the Birds of Britain and Europe* (with G. Mountfort and P.A.D. Hollom, 1954; revised three times; 5th ed., 1993). Completed *Wild America* (with James Fisher, 1955) based on extensive travels through much of North America; *The Bird Watcher's Anthology* (1957); *A Field Guide to the Birds of Texas* (1960; revised 1963); *The Birds* (1963; revised 1968); and the *World of Birds* (with J. Fisher, 1964). *A Field Guide to the Wildflowers* (Northeastern and North Central North America), completed with Margaret McKenny (1968) took two decades to complete. He wrote a merit badge pamphlet on *Bird Study* for the Boy Scouts of America, *A Guide to Mexican Birds* (with E. L. Chalif, 1973), *Birds of America* (1978), *Penguins* (1979), and *The Field Guide Art of Roger Tory Peterson* (1992). Peterson also produced several field guides to the songs of eastern (1959; 2nd ed., 1983) and western birds (1962; new ed., 1975). He was also instrumental in producing a *Videoguide to the Birds of North America* (1985–1988, Michael Godfrey, writer and narrator). Edited some fifty other field guides in the Houghton Mifflin series, including one written by his son Lee. Many of these went through more than one edition. Also edited and contributed text or illustrations to a number of other works on birds and wildlife. Maintained home and studio in Old Lyme, Connecticut from 1953 until the end of his life, but travelled extensively, eventually visiting virtually every part of the world. After 1971, turned increasingly to full-time painting. Was a notable field observer and wildlife photographer. Completed four major film projects. During his lifetime, Peterson personally observed approximately half of the 9,000 known world species of birds. Peterson found his knowledge of bird habits socially useful when in 1971 he finally learned how to dance. He was advised to "think like a bird dance," proved adept at improvisation, and rarely encountered any difficulties thereafter. Peterson was a delegate to six international ornithological congresses between 1950 and 1970. He was a member of some fifty scientific and conservation organizations in the United States and abroad, and held office in a number of them. He was first vice-president, American Ornithologists' Union (1962–1963); president, Wilson Ornithological Club (1964–1965), vice-president of the Society of Wildlife Artists; president, American Nature Study Society (1952–1953); member of the board of directors of the World Wildlife Fund (1962–1976); president of the Intrepids Club (1974–1979); honorary vice-president of the New Jersey, Massachusetts, District of Columbia, and Rhode Island Audubon Societies; honorary president, International Council of Bird Preservation (1986); fellow, American Association for the Advancement of Science, and honorary member of the British Ornithologist's Union and the Spanish Ornithological Society. His many honors and awards included the William Brewster Medal from the American Ornithologists' Union (1944); the John Burroughs Medal (1950); the Geoffrey St. Hilaire Gold Medal of the Societé National de Acclimation de France (1955); the Arthur A. Allen Award, Cornell Laboratory of Ornithology (1967); the Audubon Medal of the National Audubon Society (1971); the Gold Medal of the World Wildlife Fund (1972); the Explorer's Club

Medal (1974); the Linnaean Gold Medal of the Swedish Academy of Science (1976); Officer, Order of the Golden Ark (Netherlands), 1978; the Medal of Freedom from President Jimmy Carter (1980); the Smithsonian Medal (1984); and the Silver Buffalo Award (Boy Scouts of America) (1986).

MAJOR CONTRIBUTIONS

Peterson revolutionized bird watching and wildlife observation techniques and subsequently extended his methods to virtually all areas of natural science in North America and abroad. His many field guides became standard references for, and influences on, millions of professionals and amateurs alike, for whom it was second nature to say, "I'll just get my Peterson." The leading bird watcher of his generation. He insisted that birding be an activity for anyone interested, and helped make acceptable to amateurs and professionals alike the point that a bird need not be shot in order to be positively identified. Stimulating and highly effective writer and lecturer. Translated into several languages, his books heightened public awareness of nature. He was one of the principal exemplars of environmental thinking in the United States and abroad.

BIBLIOGRAPHY

Note: this sketch was written within days of Roger Tory Peterson's death, just as this volume went to press, hence no notice could be taken of the many memorials and appreciations subsequently published.

Contemporary Authors, New Revision Series, Vol. I (1981).

Cowan, Mary S., "Golden Anniversary for Birder's Bible," *Christian Science Monitor*, 6 July 1984.

Devlin, John C., and Grace Naismith, *The World of Roger Tory Peterson* (1977).

Lehner, Urban C., "Roger Tory Peterson, After 40–Year Reign, Remains Birder's Guru," *Wall Street Journal*, 20 December 1974.

New York Times, 30 July 1996 (obit.).

Washington Post, 30 July 1996 (obit), and appreciation (commentary), 31 July 1996.

Who's Who in America, 1995–1996.

Zinsser, William, *Roger Tory Peterson: The Art and Photography of the World's Foremost Birder* (1994).

Zusi, Richard L., *Roger Tory Peterson at the Smithsonian* (1984).

UNPUBLISHED SOURCES

See Peterson correspondence in Arthur A. Allen Papers, Cornell University Archives; Houghton Mifflin Publishing Company Correspondence, Houghton Library, Harvard University; Louis J. Halle Papers, Alderman Library, University of Virginia, Charlottesville.

Keir B. Sterling

PHELPS, ALMIRA HART LINCOLN. Born Berlin, Connecticut, 15 July 1793; died Baltimore, 15 July 1884. Educator, popularizer of science.

FAMILY AND EDUCATION

Daughter of Captain Samuel Phelps, a farmer, and Lydia Hingdale Hart Phelps, youngest of seventeen children. Early education at home and at Berlin,

Connecticut District school under the tutelage of her sister, Emma Hart Willard. Lived with Willard, 1810–1812, in Middlebury, Vermont, to study. In 1812 attended a cousin's female academy at Pittsfield, Massachusetts. Studied natural science, especially botany and chemistry, under Amos Eaton in Troy, New York, 1823–1831. Married Simeon Lincoln, 1817 (died 1823); three children; and John Phelps, 1831 (died 1849); two children.

POSITIONS

Teacher, district school near Hartford, Connecticut, 1809. Teacher, New Britain, Connecticut, district winter school, c.1812 and 1823. Teacher, Berlin Academy, 1812–1813. Teacher and head, small boarding school for young ladies at her family home in Berlin. Head, Academy, Sandy Hill, New York, 1816. Homemaker, 1817–1823. Teacher, Troy Female Seminary (now Emma Willard School), Troy, New York, 1823–1831, vice principal, 1824–1831, acting principal, 1830. Homemaker, wrote *Lectures to Young Ladies* (1833), and taught at home. Principal, female seminary, West Chester, Pennsylvania, 1838–1839. Superintendent, Rahway (New Jersey) Female Institute, 1839–1841. Principal, Patapsco Female Institute, Ellicott's Mills (now Ellicott City), Maryland, 1841–1856. Retired, wrote, active in antisuffrage movement and religious causes, 1856–1884.

CAREER

Under influence of parents and sister, Emma Hart Willard, began career as a teacher while still a student. At Troy, following first husband's death, began study of science under Amos Eaton and immediately became a champion of introducing science to academy, seminary, and primary school curricula. Began writing texts for use in common schools and at home. Most popular *Familiar Lectures on Botany* (1st ed. 1829; seventeen editions in all), which sold 275,000 copies by 1872, and *Botany for Beginners*. Books gave many professional botanists their first exposure to science, despite complaints of poor science and extreme nationalism by professionals, including Asa Gray. After second marriage and second round of child rearing, Phelps was encouraged to reenter teaching by husband, who moved law practice several times and eventually abandoned it to accommodate her career. Her Patapsco Female Institute, modeled after Troy, became the most respected female academy in the South, offering science, music, art, modern and ancient languages, as well as basic academic courses, in attempt to produce "good women rather than fine ladies." Elected American Association for the Advancement of Science, 1859 (second female member). Spent retirement writing, studying, and working against suffrage.

Published twenty books and uncounted articles. In addition to those already mentioned are *Dictionary of Chemistry* (trans. from French [1829]), *The Female*

Student or Fireside Friend (1833), *Caroline Westerly, or the Young Traveller* (1833), *Geology for Beginners* (1834), *Chemistry for Beginners* (1835), *Lectures on Natural Philosophy* (1836), *Natural Philosophy for Beginners* (1836), *Lectures on Chemistry* (1837), *Ida Norman* (1848), and *Hours with My Pupils* (1858). Her articles appeared in *Godey's Lady's Book, National Quarterly Magazine*, *Church Review*, the *Philadelphia Home Weekly*, and others.

Her ideas about education were spread through grammar, academy, and home study curricula, designed especially for women. Reputed to be strong-willed and commanding, Phelps was credited with ''[a] soaring mind, a benevolent disposition, an investigating judgment, and devout piety, patience the most untiring.''

MAJOR CONTRIBUTIONS

Prime moving figure in popularization of science, especially botany, in nineteenth-century America. *Familiar Lectures on Botany* the most widely used text for fifty years. Patapsco Female Institute provided a quality female school in the Southeast, comparable to the best northeastern schools.

BIBLIOGRAPHY

Bolzau, Emma L. *Almira Hart Lincoln Phelps* (1936).

Elliott, Clark A. *Biographical Dictionary of American Science* (1979).

Rudolph, Frederick. *Notable American Women*. Vol. 3 (1971).

Scott, Anne Firor. ''Almira Lincoln Phelps: The Self-Made Woman in the Nineteenth Century.'' *Maryland Historical Magazine* 75 (1980).

UNPUBLISHED SOURCES

Little manuscript material has survived. Phelps material can be found in the correspondence of other female educators and in the Maryland Diocesan Archives, on deposit in the Maryland Historical Society.

Liz Barnaby Keeney

PICKERING, CHARLES. Born Susquehanna County, Pennsylvania, 10 November 1805; died Boston, 17 March 1878. Physician, naturalist.

FAMILY AND EDUCATION

Son of Timothy Pickering, Jr., and the former Lurena Cole. Received A.B. from Harvard University in 1823 and studied at Boston Medical College, graduating with M.D. in 1826. In 1851, married Sarah Stoddard Hammond. No children.

POSITIONS

Physician in Philadelphia, 1826–1837; librarian, 1828–1833, and curator, 1833–1837, at Academy of Natural Sciences in Philadelphia. Chief zoologist of

U.S. Exploring Expedition, which sailed to the south seas in 1838 under Lt. Charles Wilkes. Following travels abroad in mid-1840s spent most of his life in Boston practicing medicine and writing.

CAREER

Developed interest in natural history as a child in rural Massachusetts. Secured medical degree so as to be able to make a living while pursuing his scientific interests. Practiced medicine in Philadelphia for about eleven years after receiving medical degree. Spent much of the time not devoted to his practice in research at Academy of Natural Sciences in Philadelphia, and by his mid-thirties had a reputation for wide-ranging knowledge of plant and animal science. Contributed valuable papers to transactions of Academy of Natural Sciences and American Philosophical Society. Pickering resigned from the American Philosophical Society in 1837 on grounds that its membership, primarily old-fashioned naturalists and laymen, did not sufficiently support its scientific objectives.

Pickering spent much time during a four-year exploring expedition on plant identification and sought evidence supporting his views on the geographic distribution of plants and animals and the races of man. He was appointed superintendent of the collections by the Congressional Library Committee, 1842, but soon resigned to focus on his own research. Wilkes then took over aggressive oversight of the project. Some of the plants collected reached Washington by a most circuitous route. Pickering wrote a colleague of one extreme case about a group of plants supposedly shipped from Hawaii direct to the United States. The vessel instead went to Chile, then to China and Europe, where the vessel was sold. Under new ownership, the ship then went to the West Indies, where the American authorities were requested to send a vessel to fetch the plants.

After a trip to south seas under Wilkes, Pickering visited Malta, Egypt, Zanzibar, and Bombay in the mid-1840s. Pickering's *Races of Man and Their Geographical Distribution* (1848), published as volume 9 in the expedition's reports, concerned itself with question of mankind's single or multiple origin, and enjoyed some vogue before Charles Darwin's later writings rendered Pickering's work out-of-date. He originally was to have written the volume on fish (later expanded to fish, reptiles, and insects) collected by the expedition, but fish were turned over to Louis Agassiz (whose two volume work was never published) and reptiles to Charles Girard. Insects were never covered at all, in large part because so many of the collections had been lost with the sinking of the expedition vessel *Peacock* off the coast of Oregon. Pickering's *Geographic Distribution of Animals and Plants* (volume 15 of the reports, published in two parts, 1854 and 1876), was a disappointment. No official issue of the book was ever printed and the two portions that did appear were published commercially. In them he focused on mankind's effect on plant and animal distribution. Little evidence from work done during the expedition appeared in the first part, and the second essentially contained edited excerpts from his expedition journal. It was supplemented by *Plants in the Wild State* (1876). Some of Pickering's sketches may be found in volume 5 of Captain Wilkes' *Narrative* of the exploring expedition, and he was responsible for the four maps published with his *Geographic Distribution of Plants and Animals*. His magnum opus, *Chronological History of Plants: Man's Record of His Own Existence Illustrated Through their Names, Uses and Companionship* (1879), ran to more than 1,200 pages, and still has much to interest scholars today. Pickering spent his final decades in Boston, again with a medical practice.

MAJOR CONTRIBUTIONS

Pickering spent his final sixteen years meticulously researching his massive *Chronological History of Plants*, left in manuscript at his death in 1878 and published by his widow the next year. Together with the plants and other items he collected during the expedition, most of which were reported on by Asa Gray and others, these constitute his greatest contribution to American science. His anthropological observations are principally of interest because he was one of the first American scientists to address the biological basis for variations among mankind.

BIBLIOGRAPHY

Bartlett, Harley H. ''Reports of the Wilkes Expeditions, and the Work of the Specialists in Science.'' *Proceedings of the American Philosophical Society* 82 (1940).

[Dana, James D.] ''United States Exploring Expedition.'' *American Journal of Science* 44 (1893).

Dictionary of American Biography. Vol. 7 (1928–1937).

Haskell, Daniel C. *United States Exploring Expedition, 1838–1842, and its Publications, 1844–1874: A Bibliography* (1942).

National Cyclopedia of American Biography. Vol. 13 (1967).

Stanton, William. *The Great United States Exploring Expedition of 1838–1842* (1975).

Tyler, David B. *The Wilkes Expeditions: The First United States Exploring Expedition (1838–1842)* (1968).

Viola, Herman J., and Carolin Margolis, eds. *Magnificent Voyagers: The U.S. Exploring Expedition, 1838–1842* (1985).

UNPUBLISHED SOURCES

Pickering letters can be found in the William D. Brackenridge Collection at the Smithsonian Institution Archives (Washington, D.C.). See also Pickering Papers, Massachusetts Historical Society. Pickering journal is at Academy of Natural Science of Philadelphia. Some letters in Torrey Papers. Also consult records of the U.S. Exploring Expedition (1838–1842) (Microcopy 75, Rolls 1–25), National Archives, Washington, D.C. See also his manuscript journal of plants collected during the exploring expeditions and a few pieces of correspondence at the Gray Herbarium, Harvard University; some letters at the Maryland Historical Society; and a few letters and an invoice of specimens at the National Museum of Natural History, Washington, D.C.

Richard Harmond

PIKE, ZEBULON MONTGOMERY. Born Lamberton, New Jersey, 5 January 1779; died York (now Toronto), Canada, 27 April 1813. Army officer, explorer.

FAMILY AND EDUCATION

Son of Zebulon Pike, army officer, and Isabella Brown Pike. Attended country schools in New Jersey and Pennsylvania. Married Clarissa Brown, 1801. Had several children; only one daughter survived infancy.

POSITIONS

Cadet in father's military unit as teenager, commissioned first lieutenant, 1799. Spent several years in routine frontier assignments, 1799–1805. Selected to lead exploring expedition to locate source of Mississippi River, 1805–1806. Selected to lead exploring party to the West, 1806–1807. Promoted captain, 1806, major, 1808, colonel, 1812, brigadier general, 1813. Killed while leading attack on York, April 1813.

CAREER

Pike is probably best known for his 1806–1807 expedition, which was intended to supplement findings of Lewis and Clark's 1805–1806 trip to the Northwest. Specifically, he was to explore and map Arkansas and Red Rivers, while avoiding any confrontation with Spanish colonial authorities. Left St. Louis in July 1806, reached Colorado in November. Discovered, but did not succeed in reaching, summit of Pikes Peak. Promoted captain while en route west (August 1806). Reached present site of Pueblo, Colorado, and located source of the Arkansas, but apprehended by Spanish authorities while seeking source of Red River to the south. Taken to Santa Fe and subsequently to Chihuahua, where he was obliged to surrender his papers. Did encounter and describe some mammals, including unfamiliar deer and several species of birds new to science. Several grizzly bear cubs sent east to Thomas Jefferson, who forwarded them to C. W. Peale in Philadelphia. There they were chained to tree in Independence Hall Park but subsequently (as adults) frightened Peale family and passing citizens and became mounted specimens in Peale's Museum. Engraving of these bears later appeared in John Godman's *American Natural History* (1826–1828). One bird identified by Pike as being ''of the carnivorous species''; later identified as Carolina parakeet by Elliott Coues, but this designation has since been in question. Pike's papers were kept by Mexican authorities for a century; not returned to the United States until after their discovery in Mexican archives by an American historian early in this century. Despite loss of papers, published *An Account of Expeditions to the Sources of the Mississippi and through the Western Parts of Louisiana* (1810), which was later published in England and in several other European editions. Pike remarked that the Great Plains ''[m]ay become in time equally celebrated as the sandy desarts [*sic*] of Africa,'' anticipating Stephen Long's designation of them as the ''Great American Desert.'' Pike's expedition had been ordered by Brig. Gen. James Wilkinson, later implicated in alleged conspiracy of Aaron Burr's to create independent empire in Southwest from Mexican and U.S. territory. Some therefore suspected Pike of being involved in that plot. He insisted on his noninvolvement and continued military career, though doubts persisted in some quarters. ''Nothing that Pike ever tried to do was easy, and most of his luck was bad'' (Jackson). Killed in explosion of British powder magazine in attack upon York during War of 1812, aged thirty-four. The ornithologist Alexander Wilson applied to Jefferson for permission to accompany the 1806 expedition. Jefferson later insisted that he never received the letter, and no copy has been found in his papers. Plans for Pike's expedition were supposedly known to only a handful of men before it began, because government did not want it known to Spanish authorities. Some modern writers believe Wilson's application may have been deliberately kept from the president.

MAJOR CONTRIBUTIONS

Best known for western explorations, mapping of areas traversed, and for discovery of Pikes Peak, originally variously termed by Pike ''Blue Peak'' and ''Grand Peak.'' His expedition was, with Lewis and Clark's and Long's, source of stimulation for further exploration in the American West.

BIBLIOGRAPHY

Cantwell, Robert. *Alexander Wilson, Naturalist and Pioneer* (1961).

Carter, H. L. *Zebulon Montgomery Pike: Pathfinder and Patriot* (1956).

Coues, Elliott, ed. *Expeditions of Zebulon Montgomery Pike*. 3 vols. (1895).

Ewan, Joseph, and Nesta Dunn Ewan. *Biographical Dictionary of Rocky Mountain Naturalists* (1981).

Goetzmann, William H. *Army Exploration in the American West, 1803–1863* (1959).

Hunter, Clark. *The Life and Letters of Alexander Wilson* (1983).

Jackson, Donald L., ed. *The Journals of Zebulon Montgomery Pike, with Letters and Related Documents*. 2 vols. (1966).

Jackson, Donald L. *Thomas Jefferson and the Stony Mountains: Exploring the West from Monticello* (1981).

UNPUBLISHED SOURCES

Pike Papers are in National Archives, Washington, D.C.; American Philosophical Society and Historical Society of Pennsylvania, Philadelphia.

Keir B. Sterling

PINCHOT, GIFFORD. Born Simsbury, Connecticut, 11 August 1865; died New York City, 4 October 1946. Conservationist, forester.

FAMILY AND EDUCATION

Eldest son of wealthy wallpaper executive, James Wallace Pinchot, and Mary Jane (Eno) Pinchot. Gifford spent much of youth at Grey Towers, the family's

French chateau-styled country home located near Milford, Pennsylvania. Received private tutoring in French and English and attended Phillips Exeter Academy and other private schools in Paris and New York. Graduated from Yale University in 1889 with B.A. Attended classes at French National Forestry School at Nancy for year but did not receive a degree. Honorary Sc.D. from Michigan Agricultural College, 1907; honorary LL. D. from Yale, 1925. Married Cornelia Elizabeth Bryce, 1914; one child.

POSITIONS

Consulting forester for Phelps, Dodge and Company 1891. Inspected timber stands in Alabama, Arizona, Arkansas, California, Oregon, Pennsylvania, Washington, and other states. Managed forests of Biltmore, North Carolina estate of George W. Vanderbilt, and developed timber management plans or forest surveys for other private tracts in Adirondacks and New Jersey, 1892–1898. Appointed chief, Division of Forestry of U.S. Department of Agriculture, 1898; presided over the elevation of the division to the U.S. Forest Service (1905) and remained as head of Forest Service until his dismissal by President Taft in 1910. Served as president of the Society of American Foresters, 1900–1908, 1910–1911, and the National Conservation Association, 1910–1923. Professor of forestry at Yale, 1903–1936, professor emeritus, 1936–1946. Governor of Pennsylvania, 1923–1927, 1931–1935.

CAREER

Obtained post of resident forester at Biltmore Estate through Frederick Law Olmsted. Developed Biltmore along lines suggested by the German forester Dietrich Brandis, an advocate of demonstration forestry. The Biltmore forest display at the World's Columbian Exposition in Chicago and Pinchot's *Biltmore Forest* (1893) publicized this first application of scientific forestry in United States and did much to convince American lumbermen that this new forestry was cost-effective. Appointed to National Forest Commission of the National Academy of Sciences in 1896. The commission's study led to passage of Forest Management Act of 1897, enabling legislation for the rational commercial exploitation of public timber reserves.

Began government service in 1898 as chief of the tiny Division of Forestry. Built an efficient, decentralized agency responsive to regional needs yet able to regulate use of natural resources. With backing of President Theodore Roosevelt and lumber interests he obtained the 1905 transfer of forest reserves from General Land Office of the Interior Department to his own division, which became the Forest Service. Strongly influenced conservation views of Theodore Roosevelt. Favored use of Hetch-Hetchy Valley as reservoir by city of San Francisco, 1906. Helped organize 1908 White House Conference on the Conservation of Natural Resources. Clashes over conservation policy, hydroelectric power regulation, and Alaskan coal lands with Secretary of Interior Richard A. Ballinger caused his dismissal in 1910.

Published numerous articles and books, including *Biltmore Forest* (1893), *The White Pine* (with Henry S. Graves, 1896), *Timber Trees and Forests of North Carolina* (with W. W. Ashe, 1897), *The Adirondack Spruce* (1898), *A Study of Forest Fires and Wood Production in Southern New Jersey* (1899), *A Primer of Forestry* (1899–1905), *The Fight for Conservation* (1910), *The Training of a Forester* (1914), *The Power Monopoly, Its Make-up and Its Menace* (1928), *Just Fishing Talk* (1936), *Breaking New Ground* (1947).

Retained interest in conservation and forestry after leaving the Forest Service. Lobbied successfully for passage of the Weeks Act in 1911 and Waterpower Act of 1920. These laws allowed for expansion of forest reserves by purchase and began federal regulation of the power industry. Unsuccessful bids for U.S. Senate (1920, 1926, 1932), presidential aspirations, and two terms as governor of Pennsylvania claimed much of his energy in the 1920s and 1930s. His terms as governor marked by vigorous enforcement of Prohibition, increased state regulation of electrical utilities, and construction of thousands of miles of rural roads.

MAJOR CONTRIBUTIONS

As chief forester, decided issues based on ''greatest good'' for ''the greatest number.'' Catalyst to the conservation movement and father of scientific forestry in America. His concern for organizational efficiency and insistence on competency tests for employees became models for the civil service system. Pinchot's successful campaign to transfer forest reserves from Interior Department to Department of Agriculture was a watershed in the history of natural resource management. This shift signaled acceptance of his idea that American forests should be managed and harvested like other agricultural crops. Committed to both the regulated full use of forest resources and the principle of sustained yield, Pinchot's utilitarian philosophy appeased lumber interests but angered preservationists like John Muir.

BIBLIOGRAPHY

Fausold, M. L. *Gifford Pinchot: Bull Moose Progressive* (1961).

McGeary, M. N. *Gifford Pinchot: Forester-Politician* (1960).

New York Times, 6 October 1946 (obit.).

Penick, J., Jr. *Progressive Politics and Conservation: The Ballinger-Pinchot Affair* (1968).

Pinchot, G. *Breaking New Ground* (1947).

Pinkett, H. T. *Gifford Pinchot: Private and Public Forester* (1970).

Reiger, John, ''*Gifford Pinchot With Rod and Reel''/Trading Places: From Historian to Environmental Activist: Two Essays on Conservation History* (1994).

Steen, H. K. *The U.S. Forest Service: A History* (1976).

Yale University. *Obituary Record of Graduates* (1946–1947).

UNPUBLISHED SOURCES

Gifford Pinchot Papers are in the Library of Congress, Washington, D.C. Records of the Forest Service are in the National Archives, Washington, D.C.

Michael A. Osborne

PITTMAN, HAROLD HERBERT. Born London, England, June 1889; died Regina, Saskatchewan, 29 August 1972. Naturalist, photographer, writer.

FAMILY AND EDUCATION

The son of John and Clara (née Wakelin) Pittman. Educated at King's College School on the Strand. He married Elizabeth O'Higgins in 1914; three children.

POSITIONS

Visited western Canada in 1905, 1907, and 1912 and emigrated permanently in 1913. Briefly a farmer and railway worker, then a notary public in the village of Wauchope, Saskatchewan, after 1923. He listed himself in a directory as a journalist.

CAREER

He sold over 100 nature articles, illustrated with his photographs, to *Nature Magazine*, *Outdoor Canada*, *Canadian Geographical Journal*, *Country Life*, *Bird-Lore*, and *Audubon Magazine* and contributed scientific articles to *Condor*, *Canadian Field-Naturalist*, *Journal of Mammalogy*, and *Blue Jay*. His avowed aim was "to popularize the study of natural history."

MAJOR CONTRIBUTIONS

Compiled breeding records for seventy-four species of birds on the prairie lands around Wauchope, Saskatchewan, and documented early records for the ferruginous hawk, greater prairie chicken, and sandhill crane. Some of his bird's eggs have been donated to the University of Saskatchewan. He had exhibitions of bird photography at Winnipeg in 1913 (Winnipeg Camera Club), in St. Louis, Missouri (Bird Art and Photography Exhibition), in March 1942, and in New York City (1947). His photographs of sharp-tailed grouse in winter and downy killdeers were published in Bent's *Life Histories*.

BIBLIOGRAPHY

Houston, C. S. "Harold Herbert Pittman, 1889–1972: A Memorial Tribute." *Blue Jay* 30 (1972).

Pittman, H. H. "Some Canadian Grouse." *Bird-Lore* 18 (1916): 1–6.

Pittman, H. H. "Sandhill Cranes in Retrospect." *Nature Magazine* 49 (1956) : 237–39.

C. Stuart Houston

POPE, CLIFFORD HILLHOUSE. Born Washington, Georgia, 11 April 1899; died 3 June 1974. Herpetologist, writer.

FAMILY AND EDUCATION

Son of Mark Cooper and Harriet (Hull) Pope. Attended University of Georgia for two years. Received B.S., University of Virginia, 1921. Married Sarah Haydock Davis, 1928; three sons.

POSITIONS

Herpetologist, Chinese Division, Central Asiatic Expeditions, American Museum of Natural History, New York, 1921–1926; assistant curator, Department of Amphibians and Reptiles, American Museum of Natural History, 1928–1935; curator, Division of Amphibians and Reptiles, Field Museum of Natural History, Chicago, 1941–1953.

CAREER

Key member of herpetological staff at American Museum of Natural History and at Field Museum in Chicago. Led three major expeditions to Mexico and various parts of the United States. Freelance writer on popular natural history subjects. Books include *The Reptiles of China: Turtles, Crocodilians, Snakes, Lizards* (1935), *Snakes Alive and How They Live* (1937), *Turtles of the United States and Canada* (1939), *China's Animal Frontier* (1940), *Amphibians and Reptiles of the Chicago Area* (1944), *The Reptile World: A Natural History of the Snakes, Lizards, Turtles and Crocodilians* (1955), *Reptiles Round the World: A Simplified Natural History of the Snakes, Lizards, Turtles and Crocodilians* (1957), *The Giant Snakes: The Natural History of the Boa Constrictors, The Anaconda, and the Largest Pythons* (1961). Contributed to: *Encyclopedia Britannica*, *World Book Encyclopedia*, and other reference works, *The Boy Scout Book of True Adventure* (1927), Roy Chapman Andrews's *The New Conquest of Central Asia* (1932), *Island Life* (1948), *The Care and Breeding of Laboratory Animals* (1950). Honorary member, Boy Scouts of America; fellow of the New York Zoological Society; president and journal editor, American Society of Ichthyologists and Herpetologists. President, Kennicott Club. Contributed many articles to magazines and professional journals, including *Natural History*.

MAJOR CONTRIBUTIONS

Prolific writer and popularizer of information concerning reptiles and amphibians, natural history, and exploration.

BIBLIOGRAPHY
American Men and Women of Science (1973).
Author's and Writer's Who's Who (1971).
Contemporary Authors. Vols. 1–4 (1967).
McSpadden, J. W. *To the Ends of the World and Back* (1931).
Ward, Martha E., et al. *Authors of Books for Young People* (3d ed., 1990).
Who Was Who in America.

Jean Wassong

PORSILD, (ALF) ERLING. Born Copenhagen, Denmark, 17 January 1901; died Vienna, Austria, 13 November 1977. Botanist, taxonomist, phytogeographer.

FAMILY AND EDUCATION

Son of Morten Pedersen Porsild, director of the Danish Biological Station on Disko Island, Greenland. Received part of his early education in Copenhagen and spent much of youth on Disko collecting and studying plants with brother, Robert Thorbirn. In 1955 University of Copenhagen granted him Ph.D. based mainly on submission and defense of a thesis, "The Vascular Plants of the Western Canadian Archipelago." Awarded M.B.E., 1946, for his wartime service to Canada; elected fellow of the Royal Society of Canada, 1946, the Arctic Institute of North America, 1946, the Swedish Phytogeographical Society, Uppsala, 1958; granted honorary D.Sc. from Acadia University, Wolfville, Nova Scotia, 1967, and from University of Waterloo, Waterloo, Ontario, 1973; Massey Medal (Royal Canadian Geographical Society) for arctic researches, 1966; George Lawson Medal (Canadian Botanical Association) for contributions to botany, 1971.

Married Gertrude Jorgensen, 1922, Asta Kofoed-Hansen, 1929, Elizabeth Williams, 1948, and Margrit Stoeffel, 1958. One daughter by his first wife; one adopted daughter.

POSITIONS

Assistant botanist, Danish Biological Station, Godhavn, Disko Island, Greenland, 1922–1925. Botanist with Canadian government, 1926–1935; acting chief botanist, National Museum of Canada, 1936–1946; chief botanist, National Herbarium of Canada, 1946–1967; curator emeritus, Botany Division, National Museum of Natural Sciences, Ottawa, Canada, 1967–1977.

CAREER

Because of his knowledge of arctic flora and his fluency in the Eskimo language gained in Greenland, he was selected to work for Canadian government on series of reindeer-grazing surveys in western arctic North America, 1926–1931; led several expeditions to Alaska, 1926–1927, 1929, and to the Mackenzie District of arctic Canada, 1927–1928, to James Bay, 1929, and to Keewatin and Ungava, 1930; visited Lapland, 1931, to hire local reindeer herders to teach their methods to Canadian Eskimo; supervised driving of reindeer purchased by Canadian government from Alaska and delivered from Kotzebue Sound to the Mackenzie River Delta, where he established and directed Dominion Government Reindeer Experiment Station, Killigarjuit, (Northwest Territories), 1931–1935; on Bartlett Expedition to Greenland and in Labrador, 1937; vice-consul, 1940, and then consul, 1941–1943, representing Canada in Greenland; collected plants in Alaska and Yukon, 1944; Canadian delegate to 220th anniversary celebration of U.S.S.R. Academy of Sciences, Moscow and Leningrad, and collected plants in Russia and Siberia, 1945; fieldwork in Banff National Park and other parts of Rocky Mountains in Alberta, 1945–1946, 1951, 1955–1960, in Mackenzie River region, 1947, on Banks and Victoria Islands, NWT, 1949; attended the Seventh International Botanical Congress, Stockholm (vice president), 1950; collected on Axel Heiberg Island, 1953; Eighth International Botanical Congress, Paris, 1954, and studied at herbaria in London, Paris, Stockholm, and Oslo; fieldwork in Hudson Bay Lowlands, 1956–1957; leader of field excursions to Rocky Mountains and Canadian Arctic for Ninth International Botanical Congress, Montreal, 1959; collected also in western United States, 1948, 1961, and in Mexico, 1963.

Published over 100 articles and six books, the latter on flora of southern and central Yukon, Canadian arctic archipelago, and wildflowers of the Rocky Mountains (the latter based on series of watercolor paintings by Dagny Tande Ltd of Oslo, Norway). Senior author (with W. J. Cody) of major work (published posthumously, 1980) on *Vascular Plants of Continental Northwest Territories, Canada* (for list of publications see Soper and Cody, 1978). In addition to the floristic and phytogeographical approaches used in many of his publications, he did some monographic studies, notably in the genus *Antennaria* (family *Compositae*) and the *Stellaria longipes* complex (family *Caryophyllaceae*).

According to Raup, "Erling Porsild was well known and highly respected among biologists in both America and Europe. His research was meticulous, and rested solidly upon a clear understanding of the materials he worked with and of the theoretical concepts within which he operated. He made major contributions to knowledge not only in the taxonomy of boreal American plants but also in their geographic and circumpolar relationships." As noted by Russell, "Erling Porsild was a Danish-Canadian botanist who achieved international recognition as an authority on Arctic and alpine floras and on the northern regions of Canada." He was a quiet and reserved man with an innate sense of humor. He had many friends in Canada and abroad. He hated the bureaucratic constraints of administrative procedures and was concerned solely with getting on with the job.

Porsild recognized the relationships between the Arctic and alpine floras of Canada and also focused attention on the circumpolar element in the flora of arctic North America and on its significance phytogeographically. His sound taxonomic work on vascular plants of Canadian Arctic laid foundations of our knowledge of the flora of that region. His *Illustrated Flora of the Canadian Arctic Archipelago* and the posthumous *Vascular Plants of Continental Northwest Territories* contain keys, descriptions, illustrations, and distribution maps and are the best manuals available for identification of the vascular flora of arctic North America.

BIBLIOGRAPHY

Raup, H. M. *Arctic* 31, 1 (1978): 67–68 (obit.).

Russell, L. S. *Proceedings of the Royal Society of Canada.* (4th ser.) 18 (1980), 111–14.

Soper, James H., and William J. Cody. *The Canadian Field-Naturalist* 92, 3 (1978), 298–304.

UNPUBLISHED SOURCES

Professional correspondence is on file at the National Museums of Canada, Ottawa, K1A OM8, Canada.

James H. Soper

PORSILD, ROBERT (THORBJORN). Born Denmark, 28 December 1898; died Whitehorse, Yukon, 30 December 1977. Botanist.

FAMILY AND EDUCATION

Son of botanist Morten Pedersen Porsild, director of the Biological Institute of Greenland. One of four children; childhood spent at Godhavn, Greenland. Brother of (Alf) Erling Porsild. Moved back to Denmark in 1914 to finish high school and start studies at the University of Copenhagen, cut short by illness, did not receive a degree. Married Elly Rothe-Hansen from Denmark in 1930; five children, one died in infancy.

POSITIONS

Contract with Canadian government, Department of the Interior, to explore western Arctic for suitability as reindeer-grazing ground and subsequently prepared area for arrival of herd near Mackenzie delta, 1926–1933, gold miner and trapper at Sixtymile, Yukon; position with Yukon Consolidated Gold Corporation, Dawson, Yukon; construction worker on Snag Airport (Yukon); boat builder; carpenter for the Canol Project; built Johnson's Crossing Lodge on the Alaska Highway (Yukon) in 1947 and operated it for almost twenty years; retired 1965; contract with National Museums of Canada to collect, catalog, and preserve wildflowers in central Yukon, 1966–1968.

CAREER

Robert and his brother Alfred (Alf) Erling Porsild were offered a three-year contract with the Canadian government in 1926 to explore the Mackenzie Delta and to determine its suitability as a reindeer-grazing ground. They spent two years traveling from western Alaska to the Mackenzie Delta, some 1,500 miles by dog team, to make arrangements for the planned trek of 3,442 reindeer from near Nome, Alaska, to the Mackenzie Delta, Northwest Territories. The reindeer were to provide relief for the starving Inuvialuit of the area. During their journey they collected some 16,00 specimens of plants. Report and recommendations submitted to government in 1928. Subsequent reindeer drive under leadership of Andrew Bahr took over five years instead of the projected twenty-three months. The Porsild brothers prepared for the arrival of the herd on the Canadian side of the border; built huge corrals and housing for the herders; herd of 2,370 heads was delivered 6 March, 1935. Alfred Erling had recruited Laplanders in Norway to instruct Canadian natives in the care of the reindeer. In 1933, Robert left the employ of the government and, after a short stay in Vancouver, spent the rest of his life in the Yukon. After his retirement in 1965, Robert and his wife, Elly, were contracted by National Museums of Canada to collect, catalog, and preserve wildflowers in the central Yukon. They spent three summers collecting 464 species, some extremely rare, and found the Yukon to be the home of twelve kinds of orchids. The couple found fifty plant species previously unknown to exist in the Yukon. Several entirely new discoveries were named by the Porsilds. Alfred Erling, who, in 1937, was appointed dominion botanist with the government in Ottawa (successor of Malte), wrote an account of Robert's and Elly's collections of 1966–68 (see bibliography). Prior to his death in 1977, ''Bob'' Porsild was nominated as a member of the Order of Canada.

MAJOR CONTRIBUTIONS

Robert Porsild was instrumental in the success of the reindeer drive of 1929–1935. A scientist and a passionate botanist, he added significantly to knowledge of the flora of the Yukon, continuing fieldwork long after his contract with the National Museums ended. He never missed an opportunity to share his knowledge of botany and his love of the North.

BIBLIOGRAPHY

Burns, Mary. ''A Visit with the Plant Man.'' *Yukon News* (17 April 1977).
North, Dick. *Arctic Exodus, The Last Great Trail Drive* (1991).
Porsild, Alfred Erling. *Materials for a Flora of Central Yukon Territory* (1975).
Whyard, Flo. ''Porsild: A Giant of a Man.'' *Whitehorse Star* (3 January 1978).

UNPUBLISHED SOURCES

See Lyn Harrington, '' 'Botany Bob' Porsild'' (n.d.), Manuscript 82/137, Yukon Archives.

Felicitas Tangermann, with Elly Porsild and Ellen Davignon

POTTER, LAURENCE BEDFORD. Born Monmouth, England, 4 November 1883; died Eastend, Saskatchewan, 5 November 1943. Rancher, naturalist.

FAMILY AND EDUCATION

Son of the Reverend Peter Potter and Georgiana Potter of St. Thomas vicarage, Monmouth, the ninth child in a family of twelve. Educated at King Edward School, Bromsgrove, Worcestershire. Unmarried.

POSITIONS

On 15 June 1901 he took over the ranch, established in 1897 by his eldest brother, Ernest S. Potter. Twelve years later the railway came through to Eastend, five miles to the east of his ranch.

CAREER

When he arrived in Canada in 1901, "there was nobody either in this province or anywhere else who could tell me much about the bird life here." He began keeping bird records in 1906 and published the first of twenty-seven notes and articles (chiefly in *Condor* and *Canadian Field-Naturalist*) in 1922.

MAJOR CONTRIBUTIONS

He was the first naturalist to reside in southwestern Saskatchewan, adding many "firsts" to the provincial bird list and documenting the changes in birdlife that occurred with settlement. He acted as host and guide to professional naturalists, including H. Hedley Mitchell, Fred Bradshaw, Fred G. Bard, P. A. Taverner, and J. Dewey Soper.

BIBLIOGRAPHY

Houston, C. S., and M. I. Houston. "Four Rancher-Naturalists of the Cypress Hills, Saskatchewan." *Blue Jay* 37 (1979).

Potter, L. B. "Bird-Life Changes in Twenty-Five Years in southwestern Saskatchewan." *Canadian Field-Naturalist* 44 (1930).

Potter, L. B. "Bird Notes from South-Western Saskatchewan." *Canadian Field-Naturalist* 57 (1943).

Soper, J. D. "Laurence Bedford Potter, 1883–1943." *Canadian Field-Naturalist* 58 (1944).

C. Stuart Houston

POWELL, JOHN WESLEY. Born Mount Morris, New York, 24 March 1834; died Haven, Massachusetts, 23 September 1902. Explorer, geologist, anthropologist, government science administrator, philosopher.

FAMILY AND EDUCATION

Son of Joseph Powell, Methodist exhorter, and Mary Dean Powell. As a youth introduced to natural science and specimen collecting by family friend. Largely self-educated, he qualified as country school teacher and amateur scientist specializing in collection of mollusks in association with Illinois Natural History Society. Attended Illinois College, Wheaton (Illinois), Oberlin; no degrees. Honorary Ph.D., Heidelberg, 1886, Honorary LL.D., Columbian College, 1882; Harvard, 1886. Married Emma Dean, 1861; one child.

POSITIONS

Served with 2nd Illinois Artillery during Civil War. Professor, Illinois Wesleyan College, 1865. Lecturer and museum curator, Illinois Normal University. Director, Geographical and Geological Survey of Rocky Mountain Region, 1870–1879. Member, U.S. Public Lands Commission, 1879. Director, U.S. Geological Survey (USGS), 1881–1894. Director, U.S. Bureau of Ethnology, Smithsonian Institution, 1879–1902.

CAREER

Explored 900 miles of Grand Canyon, May–August, 1869. Founder of American School of Geology, his own contributions as scientist based on two treatises: his geological description of the Colorado Plateau appearing in *Exploration of the Colorado* (1875) and *Geology of the Uinta Mountains* (1876). He introduced new branch of geology, geomorphology, noting stream flow in genesis of topographical features. During his directorship of USGS, government-sponsored geology and paleontology became primary research sciences in the nation. Powell's greatest achievement was in organizing his bureau to be the foremost promoter of basic government research.

His vision of the role for public science, his recruitment of leading scientists, his enhanced congressional budgetary and statutory support brought widespread initial public acceptance for its scientific informational role. Topographical mapping project covered 20 percent of the nation's land surface. His geological surveys, built on cooperation with states, encouraged mining and his publishing program enhanced the dissemination of specialized knowledge. The culmination of USGS scientific research occurred with the irrigation survey, 1888–1890, as Powell used science to reform land settlement laws by implementing the program outlined in his *Report on the Lands of the Arid Region of the United States* (1878). Outraged by prolonged withdrawal of public lands, a western congressional bloc terminated the reservoir survey and reduced the agency appropriations to the point where the director felt compelled to resign in 1894. Under his leadership the Bureau of Ethnology became the center of organized anthropological research in America. Eminent scientists were hired to direct compilation of American Indian bibliographic sources, tribal names, and vocabularies. Artifacts were accumulated, and field studies added data that made possible a systematic classification of Indian language groups and cultures. Publications of the bureau reflected growing professional quality of anthropology inspired by the director.

MAJOR CONTRIBUTIONS

Explorer and pioneer scientist of the Colorado plateau region. Organizer of government science in the USGS and Bureau of Ethnology. Leadership crucial in scope and productivity of agencies' research, which advanced disciplines of geology, paleontology, and anthropology to professional status. Advocate of government reforms of land settlement process in arid West; his hydrographic survey formed the nucleus of the future Reclamation Service.

BIBLIOGRAPHY

Aton, James M. *John Wesley Powell* (1994).

Cooley, John, ed., *The Great Unknown: The Journals of the Historic First Expedition Down the Colorado River* (1988).

Darrah, W. C. *Powell of the Colorado* (1951).

Davis, W. M. "John Wesley Powell." *Biographical Memoirs of the National Academy of Sciences* (1913).

Dictionary of American Biography (1935).

Dupree, A. H. *Science in the Federal Government* (1957).

Fowler, D. D. and C. S. Fowler. "John Wesley Powell, Anthropologist." *Utah Historical Quarterly* (Spring 1969).

Gaines, Ann. *John Wesley Powell and the Great Surveys of the American West* (1991).

Hinsley, C. M. *Savages and Scientists* (1981).

James, Preston E. "John Wesley Powell." *Geographer's Bibliographical Studies* 3 (1979).

Manning, T. C. *Government in Science* (1967).

Nelson, Clifford M., and Ellis Yochelson. "Organizing Federal Paleontology in the United States, 1858–1907." *Journal of the Bibliography of Natural History* 9, no. 4 (1980).

Rabbitt, Mary C. *John Wesley Powell's Exploration of the Colorado River* (1981).

Rabbitt, Mary C. "John Wesley Powell: Pioneer Statesman of Federal Science." *Geological Survey Professional Paper* (1969).

Reisner, Mare. *Cadillac Desert: The American West and its Disappearing Water* (1986).

Stegner, W. *Beyond the Hundredth Meridian* (1954).

Terrell, John U. *The Man Who Rediscovered America: A Biography of John Wesley Powell* (1969).

Worster, Donald. "The Legacy of John Wesley Powell." In Donald Worster, *An Unsettled Country: Changing Landscapes of the American West* (1994).

Zernel, John J. "John Wesley Powell: Science and Reform in a Positive Context." Ph.D. Dissertation, University of California, Irvine (1983).

UNPUBLISHED SOURCES

See Fritiof Fryxell, "Memorandum: Photographs and Document Collections in Washington [D.C.] relating to the Hayden, King, Powell and Wheeler Territorial Surveys, 1869–1879," United States Geological Survey, 1935 [another copy in Bancroft Library, University of California, Berkeley]. Letters received by John W. Powell, 1869–1879 are in National Archives. Some papers are at Library of Congress. See also Stephen Powers Letters and Walter C. Powell Journal, both at University of California, Berkeley.

Lawrence B. Lee

PREBLE, EDWARD ALEXANDER. Born Somerville, Massachusetts, 11 June 1871; died Washington, D.C., 4 October 1957. Naturalist, conservationist, government biologist, editor.

FAMILY AND EDUCATION

Oldest of three sons of Edward Perkins and Marcia (Alexander) Preble. Family moved to Wilmington, Massachusetts, the year after his birth. Summers spent at family farm in Ossipee, New Hampshire. Graduated from Woburn High School in 1889. Did not attend college. Married Eva A. Lynham, 1896 (died 1952) and her cousin, Minnie R. Setz, 1952 (died 1954). Three daughters by first marriage; two survived infancy.

POSITIONS

Field naturalist, U.S. Bureau of Biological Survey, 1892–1902, assistant biologist, 1902–1924, biologist, 1924–1928, senior biologist, 1928–1935. Consulting naturalist, *Nature Magazine*, 1925–1934, associate editor, 1934–1957.

CAREER

Spent much of youth roaming and hunting in surrounding woodlands, thus established friendships with other New England naturalists. Appointed field naturalist in newly created survey under Clinton Hart Merriam at the suggestion of Frank Harris Hitchcock. Began fieldwork in Texas with Vernon Orlando Bailey and conducted life zone samplings in Georgia, Maryland, Oregon, Washington, and Utah. Noted as a rugged outdoorsman, his 1900 expedition to Hudson Bay region began a series of faunal studies of northwest North America. Explored Athabaska-Mackenzie region, 1901, 1903, 1904, 1907. His report is considered by colleagues the best of the North American fauna series. Explored parts of Alaska, Alberta, British Columbia, Montana, and North Dakota, 1910. After a 1911 journey to Jackson Hole, Wyoming, issued report on status of the elk with recommendations for its management. Accompanied Wilfred H. Osgood and George H. Parker in 1914 to Pribilof Islands to study fur seals, and their report provided the framework for herd management. Published a faunal study of the Pribilofs as well. Last major field trip, 1934, was investigation of status of waterbirds of Athabaska and Peace River Deltas with Luther J. Goldman and led him to recommend series of closed seasons for hunting.

Became deeply concerned with conservation of natural resources during fieldwork years. Appointed consulting naturalist to *Nature Magazine* shortly after it commenced publication; edited, reviewed, and wrote articles for the American Nature Association until his death. Served as Chairman, Editorial Committee, *Journal of Mammalogy*, 1930–1935. Fellow, American Ornithologists' Union from 1935 and member of its Bird Protection Committee. Served on Committee on Wildlife and Committee on Preservation of Natural Conditions, National Research Council. For many years, judged humane trap designs for the American Humane Association. Inherited and purchased land in Ossipee, New Hampshire, and established Ossipee Wildlife Sanctuary there.

Published some 239 items, 1893–1956, including *A Biological Investigation of the Hudson Bay Region*, North American Fauna (NAF) 22 (1902), *A Biological Investigation of the Athabaska-Mackenzie Region*, 27 NAF (1908), *Report on Condition of Elk in Jackson Hole, Wyoming, in 1911* (1911), *The Fur Seals and Other Life of the Pribilof Islands, Alaska, in 1914* (with W. H. Osgood and G. H. Parker, 1915), *A Biological Survey of the Pribilof Islands*, NAF 46 (with Waldo L. McAtee, 1923), "The Lover of Nature" (*Nature Magazine*, 1925), and "Report of the Committee [of the American Society of Mammalogists] on Conservation of Land Mammals" (with T. S. Palmer and Lee R. Dice, *Journal of Mammalogy*, 1928). His prose poem, "The Lover of Nature," was widely quoted by conservationists.

Scientific publications produced primarily in early years, with later career devoted to popular writing on conservation. Although not noted as a systematist, described as new two subgenera, seven species, and twelve subspecies, most of which have stood the test of time. Island in Great Slave Lake, bay in Great Bear Lake, one species of fish, two species and five subspecies of mammals named for him. Lack of formal education compensated for by comprehensive self-education. Known for his meticulous checking of details, knowledge of sources and bibliography, and excellent prose style; reviewed and edited publications for many colleagues at the survey and *Nature*. A skilled writer, as well as ardent lover of nature and conservationist, wrote some 200 articles on wildlife and the environment for *Nature* and other popular publications.

MAJOR CONTRIBUTIONS

Produced major faunal studies of northwest North America. Exceptional skills as outdoorsman and observer of nature, combined with comprehensive scientific knowledge, love of nature, and felicitous prose style produced an articulate and effective conservationist. Noted for championing conservation of woodcock, elk, and fur seals.

BIBLIOGRAPHY

Birney, Elmer C., and Jerry R. Choate, eds. *Seventy-Five Years of Mammalogy, 1919–1994* (1994).

The Evening Star, Washington, D.C., 5 October 1957 (obit.).

McAtee, W. L. Obituary of E. A. Preble. *Auk* 79 (October 1962).

McAtee, W. L., and Francis Harper. "Published Writings of Edward Alexander Preble (1871–1957)." *Miscellaneous Publication 40* (August 1965).

Walter, E. P., and P. F. Hannah. "A Modern Thoreau, Edward A. Preble, Friend of Wild Life." *Nature Magazine* (August 1932).

Westwood, R. H. "Edward A. Preble—An Appreciation." *Nature Magazine* (December 1957).

Westwood, R. W. *This Is Nature. Thirty Years of the Best from Nature Magazine* (1959).
Who Was Who. Vol. 3.

UNPUBLISHED SOURCES

Correspondence, field notes, manuscripts, and photographs are in the Edward Alexander Preble Papers, Record Unit 7252, Smithsonian Archives. Field notes are in U.S. Fish and Wildlife Service, 1860–1961, Field Reports, Record Unit 7176, Smithsonian Archives.

Pamela M. Henson

PRIESTLY, ISABEL M. Born Isabel Mary Adnams, at Speenhamland, near Newbury, Berkshire, England, 25 July 1893; died Yorkton, Saskatchewan, 23 April 1946. Housewife, naturalist.

FAMILY AND EDUCATION

Daughter of Frank H. and Isabel Mary Dreweatt Adnams. Attended school at Newbury and then studied botany in England and Germany, the latter interrupted by the outbreak of war. Giving up her plans for further botanical studies, she instead married a Canadian soldier, Robert J. Priestly, on New Year's Eve of 1918 and accompanied him to Canada; one daughter, two sons.

POSITIONS

Founder and president of the Yorkton Natural History Society, 1942–1946. Editor of the *Blue Jay*, 1942–1946.

CAREER

After living in Calgary and Victoria, where she was active in the Natural History Society of British Columbia, working for the preservation of the lily, *Erythronium oreganum*, the Priestlys moved in 1929 to Winnipeg. Here she joined the botanical section of the Manitoba Natural History Society, where she served as chairman (1932–1934) and as secretary (1934–1935). On the society outings she felt the ornithology section people were having at least an equal amount of fun, so she undertook to learn her Canadian birds. She contributed nature articles to the *Winnipeg Free Press* and *Winnipeg Tribune*. Soon after her husband was transferred from Winnipeg to Yorkton, she undertook to coauthor a weekly nature column in the *Yorkton Enterprise*. She founded an informal group of nature enthusiasts and in July 1942 compiled ''A Preliminary List of the Birds of the Yorkton District, Saskatchewan.'' This created so much interest and correspondence that within two months the Yorkton Natural History Society was formed, and its quarterly bulletin, the *Blue Jay*, began. Her fresh, newsy style made it readable and interesting; her ability to collect and organize scientific facts made it of scientific interest and importance; her enthusiasm made it a medium for the exchange of observations, which proved stimulating to amateur and professional alike. Soon after Priestly's death, the Saskatchewan Natural History Society was formed to carry on publication of this respected and widely circulated regional journal.

MAJOR CONTRIBUTIONS

Ability to interest the general public in natural history, launching a successful society and journal.

BIBLIOGRAPHY

Houston, C. S. ''How the Blue Jay Got Its Name.'' *Blue Jay* 35 (1977).
Houston, C. S., ed. *Blue Jay* 50 (1992).

C. Stuart Houston

PRINCE, EDWARD ERNEST. Born Leeds, England, 23 May 1858; died Ottawa, Ontario, 10 October 1936. Marine biologist, ichthyologist, fisheries administrator, educator.

FAMILY AND EDUCATION

Son of George Augustus Prince, Leeds tradesman, and Harriette Rothery Prince. Graduated from University of St. Andrews, Scotland, 1884. Received additional education at Edinburgh and Cambridge Universities. Honorary LL.D., University of Aberdeen, 1911. Married Bessie Morton Jack, 1894; two daughters and one son.

POSITIONS

Senior assistant in zoology, Edinburgh University, 1884. Naturalist, Marine Laboratory, St. Andrews, Scotland, 1885–1892. Professor of Natural History, St. Mungo's College, Glasgow, 1890–1893. Dominion commissioner of fisheries, Canada, 1893–1924. Founder and chairman of the Biological Board of Canada, 1898–1921. Chairman of fifteen fisheries commissions in Canada, 1898–1924.

CAREER

While naturalist at St. Andrews Marine Station, Scotland, his mentor was William Carmichael M'Intosh, then regarded as ''the most prominent fishery scientist'' in the world. Together they issued voluminous joint writings on the eggs and life history of fishes, regarded at the time of Prince's death as ''still the most important works on that technical subject.'' M'Intosh recommended Prince for the position of dominion commissioner of fisheries, in Canada, a post he occupied for thirty-one years. His monographs on the Atlantic and Pacific fisheries gave the Canadian government a store of information, and he served as chairman of fifteen fishery commissions. He represented Canada on the International Fish Commission under the Fishery Treaty of 1908 and made a survey with U.S. ichthyologist David Starr Jordan of all waters lying on the Canadian–U.S. boundary from the Atlantic to the Pacific. He was the chief adviser to successive ministers of marine and fisheries on fishery legislation. He was founder–chairman of the Board of Management of the Marine Biological Stations (Canada), which became the Biological Board of Canada in 1912 and which inaugurated marine science and fisheries biology in Canada. In 1914–1915 he investigated the fisheries of New Zealand at the request of its government; he authored the first fisheries report ever issued by the government of New Zealand.

Published more than 170 reports and papers, 1885–1929. He excelled at writing accounts of the Canadian fisheries, among which are "Canada's Fishery Resources," from the *Proceedings of the Fisheries Congress at Rome* (1911), and *The Fisheries of Canada* (1925). With W. C. M'Intosh, he wrote "On the Development and Life-Histories of the Teleostean Food- and Other Fishes," *Transactions of the Royal Society of Edinburgh* 35 (1889).

A recognized fisheries expert, he was also an exceptional popularizer, writing nearly forty popular articles, mainly for the *Ottawa Naturalist*, and was much in demand as a lecturer from coast to coast in Canada. He was an energetic and dedicated scientist who did great work in promoting accurate knowledge of fish life as a basis for good fisheries laws. He was immensely popular and was known in both Britain and Canada as "the genial Professor Prince."

MAJOR CONTRIBUTIONS

Built up the role of scientific research in Canadian fisheries management. Established the precursors of the Fisheries Research Board of Canada. Great popularizer of the cause of fisheries research and world emissary on importance of Canadian fisheries.

BIBLIOGRAPHY

Gunther, A. E. *William Carmichael M'Intosh, FRS* (1977).

Huntsman, A. G. "Edward Ernest Prince 1858–1936." *Canadian Field Naturalist* 59 (1945).

Johnstone, Kenneth. *The Aquatic Explorers: A History of the Fisheries Research Board of Canada* (1977).

Morgan, Henry James. *Canadian Men and Women of the Times* (1912).

Scott, Duncan Campbell. "Edward Ernest Prince." *Proceedings of the Royal Society of Canada* 31 (1937).

UNPUBLISHED SOURCES

Letters are in the National Archives of Canada.

Jennifer M. Hubbard

PROVANCHER, LÉON. Born Bécancour, Lower Canada, 10 March 1820; died Cap-Rouge, Quebec, 23 March 1892. Roman Catholic priest, naturalist.

FAMILY AND EDUCATION

Son of Joseph-Étienne and Geneviève Hébert Provancher. Attended Séminaire de Nicolet, 1834–1844. Completed two-year program in philosophy there, 1840; undertook study of theology. Ordained 1844. Honorary D.Sc. from Université Laval.

POSITIONS

Successively held six curacies, 1844–1847. Sent to Gross Île, 1847–1848. Parish priest, St. Victor, Beauce Region, 1848–1852. Curé at Saint Jean Baptiste, L'Isle Verte, 1852–1854; parish priest, Saint Joachim, 1854–1862, and Notre-Dame de Portneuf, 1862–1869. Resigned from parish work, 1869, spent most of the rest of his life at Cap-Rouge, near Quebec.

CAREER

Interest in nature began in boyhood, when he was shown fossil shellfish uncovered by workmen excavating a well. Won scholarship, which enabled him to attend Séminaire de Nicolet. There he often won horticulture prizes. While curé at Grosse Île (1847), provided important support to Irish immigrants suffering from typhus epidemic. Published *Essai sur les insectes et les maladies qui affectent le blé* (1857) and *Traité élémentaire de botanique* (1858), which enjoyed wide use as a textbook in schools until the beginning of the 1870s. Began study of insects with William Couper and was in touch with Spencer F. Baird, assistant secretary of the Smithsonian Institution in Washington, seeking American texts that could be of help to him in his insect research. First important published work in botany was *Flore canadienne* (1862), which was the first comprehensive attempt to deal with the flora of Canada. In this text, he made extensive use of American materials in such a way as to cause Gray to take Provancher to task and ignore what he had done. Gray's lead was followed by others, with the result that *Flore canadienne* was soon no longer in regular use. It remained the standard reference for French Canadians, however, until the beginning of the 1930s. His *Le verger canadien* (1862) suggested best ways to grow fruit, vegetables, and flowers in Quebec. *Le Naturalist canadien* was a French-language natural science journal launched by Provancher in 1868 and edited by him until 1891, when volume 20 appeared. By the time declining strength compelled him to turn over the reins to another editor, some 8,000 pages of text had appeared under his aegis. This journal has continued to be published to the present. By 1869, Provancher had grown disenchanted with parish work, as his increasingly short-tempered impatience with parishioners demonstrated. He moved first to Saint-Roch and in 1872 to Cap-Rouge. There he edited the *Gazette des familles acadiennes et canadiennes* for about a year and a half (1875–1876) and undertook (1888) publication of *La Semaine reli-*

gieuse de Québec, mostly for the benefit of the clergy. The rest of his life, however, was largely devoted to the natural sciences, in particular, entomology. His early papers in the field of entomology were published in his *Petite faune entomologique du Canada*, a summary of what was then known concerning every species of insect in Canada. This was issued in three volumes, comprising some 2,506 pages, between 1877 and 1890. A taxing assignment, it was made more difficult because Provancher often lacked access to libraries, other necessary source materials, including insect collections, and fellow entomologists. Nonetheless, his work demonstrates that he took great care to be precise and accurate, since more recent specialists have upheld the validity of his conclusions. E. O. Essig has stated that Provancher's errors were "small enough, compared to the difficulties he encountered."

Among the Hymenoptera (bees, wasps, ants, and their allies), he was responsible for identifying and naming some 923 new species, and many of his descriptions were of higher caliber than contemporary work that was being done in the United States. Though sometimes criticized for his conclusions, he was a flexible person, able to change his mind when presented with better information. He was an opponent of Charles Darwin's ideas concerning evolution and the origin of species, although, unlike many of Darwin's critics at the time, he had taken the trouble to read Darwin's books on these subjects. Concerned that science did not enjoy sufficient respect and attention among educational authorities in Canada outside the universities, he made it his business to familiarize the public with the problem in an effort to get it addressed. He also argued in favor of some practical solutions to this issue, including establishment of an agricultural museum, a journal of agriculture, a botanical garden in Quebec, and a natural history museum at the Université Laval. In addition to botany and entomology, he had considerable interest in conchology. His collections of plants, mollusks ("tens of thousands of specimens"), and about 30,000 insects are at the Université Laval in the city of Quebec.

MAJOR CONTRIBUTIONS

Provancher's writings, publications, and collections are of enormous value to researchers interested in both the extent of his accomplishments and the nature of Canadian biological research in the last century. His work with insects has had lasting value, and most of his taxonomic conclusions have stood the test of time. His varied collections have been well cared for and are probably more complete than those of any other nineteenth-century naturalist.

BIBLIOGRAPHY

Béique, René. "L'oeuvre et l'héritage de l'abbé Léon Provancher." *Le Naturaliste canadien* 95 (1968).

Bernard, Jean-Guy. "Les 'collections Provancher.' " *L'Echo du collège dé Lévis* 48, 3 (January–February 1969).

Essig, E. O. *A History of Entomology* (1931).

Mallis, Arnold. *American Entomologists* (1971).
Perrin, Jean-Marie. *Dictionary of Canadian Biography*. Vol. 11 (1990).
Rousseau, Jacques, and Bernard Boivin. ''La contribution à la science de la *Flore canadienne* de Provancher.'' *Le naturaliste Canadien* 95 (1968).
UNPUBLISHED SOURCES
Manuscripts and correspondence for years 1847–1892 are in the archives of the Séminaire de Chicoutimi, Quebec. A 500–volume library, comprising the books owned by Provancher, has ''been reconstituted'' at the Université Laval, Quebec, Canada. See also J. B. Lévéille, ''Bio-bibliographie de M. l'abbé Léon Provancher, docteur ès sciences,'' thèse de bibliothéconomie, Université de Montréal (1949).

Keir B. Sterling

PURSH, FREDERICK TRAUGOTT. Born Grossenhain, Saxony, Germany, 4 February 1774; died Montreal, Canada, 11 July 1820. Botanist.

FAMILY AND EDUCATION

Details of his life are meager. Attended public school at Grossenhain, studied botany and horticulture under Johann Heinrich Seidel of the Royal Botanic Gardens at Dresden. Details of possible marriage to a barmaid are scarce.

POSITIONS

On the staff of the Royal Botanic Gardens at Dresden prior to 1799. Traveled to America and found employment in a Baltimore botanic garden in 1799. From 1803 to 1805 supervised the gardens of William Hamilton's elaborate estate, the Woodlands, near Philadelphia and met many botanists while in this position, including Benjamin Smith Barton. In 1806 Pursh became a plant collector for Barton, obtaining specimens west and south of Philadelphia to the mountains of Maryland and border of North Carolina. In 1807 collected in Pennsylvania and north to Niagara Falls and west through the Green Mountains of Vermont. Worked briefly with Barton on the flora collected by Lewis and Clark in the Pacific Northwest. In 1809 entered the employment of David Hossack, who was establishing the Elgin Botanic Gardens. After traveling in the West Indies, 1810–1811, and after a brief return to the United States, Pursh sailed for England. Employed by Aylmer Bourke Lambert in Britain. By 1816, he had settled in Montreal, Canada, and was beginning work on Canadian flora.

CAREER

Pursh was an able plant collector for Benjamin Smith Barton, 1806–1809. With Barton's aid, he became the first botanist to study the plants of the Pacific Northwest collected by Lewis and Clark. This project was abandoned when Barton became involved in other tasks. While in England from 1811 to 1814, he completed his *Flora America Septentrionalis* (1814) by using British materials.

MAJOR CONTRIBUTIONS

Pursh's *Flora America Septentrionalis* not only provided a description of the flora of the Pacific Northwest but also was the first attempt to publish a complete flora for North America.

BIBLIOGRAPHY

Elliot, Clark A. *Biographical Dictionary of American Science* (1979).

Ewan, Joseph. ''Frederick Pursh, 1774–1820, and His Botanical Associates.'' *Proceedings of the American Philosophical Society* 96 (1952).

Ewan, Joseph. ''Frederick Pursh.'' *Dictionary of Scientific Biography*. Vol. 11 (1975).

Phillip D. Thomas

PUTNAM, FREDERIC WARD. Born Salem, Massachusetts, 16 April 1839; died Cambridge, Massachusetts, 14 August 1915. Naturalist, ichthyologist, archeologist, museum founder and administrator, Harvard professor.

FAMILY AND EDUCATION

Son of Ebenezer and Elizabeth Appleton Putnam. From old New England stock: tutored by his father and developed interest in nature as a boy roaming seashores, fields, and woods and in his father's greenhouse. Influenced by Henry Wheatland at Essex Institute. Published first work at age sixteen on fishes of Salem Harbor. Entered Lawrence Scientific School, Harvard University, to study under Louis Agassiz, 1856–1864. B.S. (1862). Married Adelaide M. Edmands, 1864 (died 1879); three children. Married Esther Orne Clarke, 1882 (died 1922); no children.

POSITIONS

Assistant, Museum of Comparative Zoology (in charge of fishes), 1856–1864; curator of vertebrates, Essex Institute, Salem, Massachusetts, 1856–1867; curator of ichthyology, Boston Society of Natural History, 1859–1868; superintendent and director, Essex Institute, 1864–1870; superintendent, Museum of East India Marine Society, Salem, Massachusetts, 1867–1869; director and curator of vertebrates, Peabody Academy of Science, Salem, Massachusetts, 1867–1874; assistant in ichthyology, Museum of Comparative Zoology, 1876–1878; curator, Peabody Museum of American Archaeology and Ethnology, Harvard University, 1875–1909; honorary curator, 1909–1913; honorary director, 1913–1915.

CAREER

A founding curator and first director, Peabody Academy of Science, Salem, Massachusetts, 1867–1874; a founder and publisher, *American Naturalist*, 1967; founder and publisher, *The Naturalist's Directory*, 1865; founder, Naturalists' Agency, 1867; instructor, Anderson School of Natural History (Agassiz's summer school of biology on Penikese Island), 1873–1874; permanent secretary, American Association for the Advancement of Science, 1873–1898; assistant, Kentucky Geological Survey (Summer 1874); assistant, U.S. Geological and Geographical Survey (Wheeler Survey), 1876–1879; member, Massachusetts State Commission of Fish and Game, 1882–1889; appointed Peabody Professor of American Archaeology and Ethnology, Harvard University, 1885 (but not confirmed until 1887); established Division of American Archaeology and Ethnology, Harvard, 1890 (changed to Division of Anthropology, 1903); elected, National Academy of Sciences, 1885; president, Boston Society of Natural History, 1887. Tireless worker, especially in the organization and administration of museums: Peabody Academy of Science; Peabody Museum of American Archaeology and Ethnology; Field Museum (from the World's Columbian Exposition in Chicago); American Museum of Natural History; Museum and Department of Anthropology, University of California; and for twenty-five years the operation of the American Association for the Advancement of Science. He published over 400 works on natural history, archeology, and especially administrative reports for the various museums with which he was connected and the proceedings of the American Association for the Advancement of Science (AAAS) (1873–1898) (see bibliography in Mead [1909] and Tozzer [1936]).

MAJOR CONTRIBUTIONS

He founded or developed museums and departments of natural history and anthropology (Peabody Academy of Science, Peabody Museum of American Archaeology and Ethnology, Department of Anthropology at the World's Columbian Exposition, American Museum of Natural History, and University of California at San Francisco). He also edited publications of those institutions while he was affiliated, as well as those of the AAAS for twenty-five years. He was often referred to by his contemporaries as the "father of American archeology."

BIBLIOGRAPHY

Boas, Franz. "Frederic Ward Putnam." *Science* 42 (1915).

Dexter, R. W. "Frederic Ward Putnam and the Development of Museums of Natural History and Anthropology in the United States." *Curator* 9, 2 (1966).

Kroeber, A. L. "Frederic Ward Putnam." *American Anthropologist* 17, 4 (1915).

Mead, F. H. *The Putnam Anniversary Volume* (1909).

Tozzer, A. M. "Biographical Memoir of Frederic Ward Putnam, 1839–1915," *Biographical Memoirs, National Academy of Science* 16 (1936).

UNPUBLISHED SOURCES

Records and correspondence are deposited in Archives of Harvard University, Peabody Museum of Archaeology and Ethnology, Smithsonian Institution, and Lowie Museum of Anthropology, University of California.

Ralph W. Dexter

R

RAE, JOHN. Born near Stromness, Orkney Islands, 30 September 1813; died London, England, 22 July 1893. Surgeon, fur trader, explorer, naturalist.

FAMILY AND EDUCATION

Son of John Rae, Sr., who was factor to William Honeyman, Lord Armadale, and Margaret Glen Rae. After four winter sessions at the Edinburgh Medical School, Rae qualified as licentiate of the Royal College of Surgeons of Edinburgh on 18 April 1833. Honorary M.D. from McGill University, 6 May 1853. Honorary LL.D. from Edinburgh University, 1866. Elected Fellow of the Royal Society, 1880. Married Kate Thompson (1839–1919) in January 1860, in Toronto, Ontario; no children.

POSITIONS

Rae lived twenty-two years in what is now Canada. Surgeon on *Prince of Wales*, 1834–1835; clerk and surgeon, Hudson's Bay Company; at Moose Factory, 1835–1844, York Factory, 1845–1846, Fort Simpson, 1849–1850. Private surgeon, Hamilton, Ontario, 1857–1860. Ship's surgeon and surveyor, Atlantic Telegraph Survey, 1860. Guide for Chaplin–Johnstone expedition, western Canada, 1861. Explorer for Canadian Telegraph Survey, 1864.

CAREER

His first arctic expedition, 1846–1847, crossed Rae Isthmus to reach Lord Mayor's Bay, mapping the shore of Simpson Peninsula and then the west coast of Melville Peninsula for a total of 1,050 kilometers of new coastline mapped. In 1848, he was second in command of the first search expedition for the missing expedition led by John Franklin. John Richardson and Rae searched the arctic coastline between the mouths of the Mackenzie and Coppermine Rivers. In 1851 Rae explored 1,010 kilometers of Victoria Island coastline. In 1853, Rae explored the Quoich River for 335 kilometers, mapped 430 new kilometers of the west side of Boothia Peninsula.

Published one book, *Narrative of an Expedition to the Shores of the Arctic Sea*, in 1846 and 1847 (1850), twenty papers, and forty-five letters in *Nature*. His *Correspondence with the Hudson's Bay Company on Arctic Exploration, 1844–1855*, edited by E. E. Rich, was not published until 1953. Natural history observations were incidental to exploration, but he published observations of birds and mammals and submitted plant collections to Joseph Hooker. Rae also cured his men of scurvy the first winter by feeding them cranberries gathered from beneath the snow and sprouts of the vetch or wild pea.

MAJOR CONTRIBUTIONS

Rae left his mark as the most cost-efficient northern explorer ever, learning native methods of traveling light and "living off the land." He found the first evidence of the fate of the missing Third Franklin Expedition; for this he received a £10,000 reward.

BIBLIOGRAPHY

Houston, C. S. "John Rae, 1813–1893." *Arctic* 40 (1987).

Houston, C. S. "Dr. John Rae: The Most Efficient Arctic Explorer." *Annals of the Royal College of Physicians and Surgeons of Canada* 20 (1987).

Hooker, J. D. "On Some Collections of Arctic Plants." *Journal of the Proceedings of the Linnaean Society* 1 (1857).

Rae, J. "Notes on Some of the Birds and Mammals of the Hudson's Bay Company's Territory, and of the Arctic Coast of North America." *Journal of the Linnaean Society, Zoology* 20 (1890).

Rae, J. *Narrative of an Expedition to the Shores of the Arctic Sea in 1846 and 1847.* (1850).

Rae, J. "Journey from Great Bear Lake to Wollaston Land." *Journal of the Royal Geographical Society* 21 (1851).

Rich, E. E., ed. *John Rae's Correspondence with the Hudson's Bay Company on Arctic Exploration, 1844–1855.* (1953).

Richards, R. L. *Dr. John Rae.* (1984).

C. Stuart Houston

RAFINESQUE (-SCHMALTZ), CONSTANTINE SAMUEL. Born Galata, a European suburb of Constantinople, 22 October 1783; died Philadelphia, 18 September 1840. Naturalist, botanist, ethnologist, economist, poet.

FAMILY AND EDUCATION

Son of François Georges Anne Rafinesque, a French merchant from Marseilles, who married Magdeleine Schmaltz, a Constantinople-born Greek national of German descent. Family returned to Marseilles when Rafinesque was a baby; had younger brother and younger sister. Father died of yellow fever in Philadelphia in 1793 while on a voyage to China; during his absence mother, to escape the dangers of the Reign of Terror, removed her family to Livorno, where they remained until 1796. In Italy the boy was taught by private tutors and, though the mother remarried, the reduced circumstances of family prevented his attending college in Switzerland as had been planned. Joining his paternal grandmother in Marseilles in 1797, continued his education by omnivorous reading. With his stepfather, also a merchant, returned to Livorno for two years, where his reading took special slant toward botany and the other natural sciences. Departed for United States in 1802 with brother Anthony. Two honorary degrees were conferred upon him; a Ph.D. from the University of Bonn in 1820 and an M.A. from Transylvania University in 1822. Also accompanied by Anthony, left for Sicily late in 1804, where remained until 1815. During this period added hyphenated mother's name Schmaltz to avoid being taken for a Frenchman. A Protestant, he could not formally marry Josephine Vaccaro, with whom he lived from 1809 to 1815 in Catholic Sicily; two children.

POSITIONS

Professor of botany and natural history, Transylvania University, 1819–1826. Tried unsuccessfully to join any federal expedition that would enable him to explore the trans-Mississippi west; also failed to obtain academic appointments at the Universities of Palermo, Pennsylvania, Virginia, and North Carolina. Botanized in New Jersey, Pennsylvania, and as far south as Virginia, 1802–1804; throughout Sicily, 1805–1815; Ohio Valley as far west as the Mississippi, 1817–1825; after 1826, East Coast between Massachusetts and Virginia. Estimated his excursions, mostly on foot, covered 25,000 miles. Because of exchanges with other botanists, his writings touch on most floristic areas of the world.

CAREER

In Sicily, was secretary to Abraham Gibbs, American consul, who was also an exporter. Relinquished this post in 1808 to Anthony and set up his own successful business, which included exporting medicinal plants and distilling brandy. Returning to United States in 1815, was shipwrecked off Long Island, losing his collections. Briefly a tutor for daughters of Robert L. Livingston at Clermont. Learning of his financial reverses, Josephine married and later refused to send their daughter to join him. Through the assistance of John D. Clifford of Lexington, for whose Philadelphia firm Rafinesque had worked during his first visit to America, obtained only real academic appointment at Transylvania University, where he also taught modern languages. One of the earliest such professorships in United States and certainly the earliest in the West, the position paid no salary but permitted Rafinesque to sell lecture tickets, as was common in medical schools at the time. After dispute with Transylvania's president, returned to Philadelphia, which was his base for remainder of his life. Naturalized in 1832. Lectured occasionally at Franklin Institute, bought and sold natural history specimens, marketed a tuberculosis nostrum called Pulmel, publicized his patented ''Divitial Institution''—a plan to make the value of both goods and labor serve as currency, on the principle of which he founded Six Per Cent Savings Bank—and tried to promote a western land settlement scheme.

Voluminous publishing began in Sicily, where Rafinesque issued two volumes of the periodical *Specchio delle scienze* (1814) and such studies in biosystematics as *Précis des découvertes et travaux somiologiques* (1814), *Principes fondamentaux de somiologie* (1814), and *Analyse de la nature* (1815). During these years also published descriptive works such as *Caratteri di alcuni nuovi generi e nuove specie* (1809), as well as many articles both in Europe and in United States. After his return to United States, most of his scientific publications were descriptive, with emphasis on botany, though he also contributed to ichthyology, ornithology, mammalogy, herpetology, malacology, and entomology. Major books were *Florula Ludoviciana* (1817), *Ichthyologia Ohiensis* (1820), *Monograph of the Fluviatile Bivalve Shells of the River Ohio* (1820, 1832), *Medical Flora* (Vol. 1, 1828; Vol. 2, 1830), *New Flora and Botany of North America* (1836–1838), *Flora Telluriana* (1837–1838), *Sylva Telluriana* (1838), and *Autikon Botanikon* (1840). In addition, launched three periodicals, *Western Minerva* (first issue suppressed, 1821), *Atlantic Journal* (1832–1833), and *The Good Book* (1840), for which he was the principal contributor. Published his autobiography, *A Life of Travels* (1836); a book-length philosophical poem, *The World or Instability* (1836); two volumes of pre-Columbian history, *The American Nations* (1836); and countless pamphlets, essays, reviews, poems, and broadsides. Though often said to have published nearly 1,000 titles, he himself claimed authorship of only ''220 works, pamphlets, essays and tracts.''

His undisciplined self-education, unshakable high opinion of himself, irascible temperament, and humorless intolerance led to many woes while he lived. These traits, coupled with an innocent credulity, willingness to print at his own expense what no one would publish for him, and the barely adequate style of his scientific writing, caused Rafinesque's contributions to be discounted by his contemporaries, ignored as often as possible by his successors. Yet, in botany alone, he deserves credit for at least 740 generic names, 2,560 binomials, which were legitimately published according to Merrill but are not listed in the standard indexes.

MAJOR CONTRIBUTIONS

In all, originated about 2,700 new generic names, about 6,700 new binomials for plants, and had great talent in devising appropriate names. Had advanced ideas about biosystematics; was one of the earliest botanists in America to advocate the French natural system of classification. Though marred by descriptions of nonexistent species, his book on the Ohio fishes is fundamental to freshwater ichthyology. A pioneer ethnographer, he made an early effort to order the multifarious Indian languages by family groups and recorded hundreds of sites of prehistoric earthworks since obliterated. Rejecting Mosaic history, he theorized on peopling of the New World by migrations from the Old. A protoevolutionist, he understood taxa as convenient fictions, believing that varieties become species which, in turn, become genera; but did not "anticipate" Darwin, as is sometimes claimed. Because his conception of the time frame of evolution was too narrow, he had no explanation for the mechanics of variation, nor did he argue his theory at length. Rather, was an evolutionist like Darwin's own grandfather and, like him, expressed his views in long poem.

BIBLIOGRAPHY

Boewe, C. *Fitzpatrick's Rafinesque: A Sketch of His Life with Bibliography* (1982).
Call, R. E. *Life and Writings of Rafinesque* (1895).
Elliott, Clark, ed. *Biographical Dictionary of American Science* (1979).
Dictionary of American Biography.
Dictionary of Scientific Biography.
Merrill, E. D. *Index Rafinesquianus* (1949).
Pennell, F. W. "The Life and Work of Rafinesque." *Rafinesque Memorial Papers, Transylvania College Bulletin* 15 (1942).
Sterling, Keir, ed. *Rafinesque: Bibliography and Lives* (1978).

UNPUBLISHED SOURCES

The largest cache, consisting of essays, journals, and letters—mostly unpublished—is in Philadelphia, at the American Philosophical Society and the Academy of Natural Sciences. The next largest, consisting of collections made by the bibliographer Fitzpatrick, is at the University of Kansas. The New York Botanical Garden has a notable group of letters written to John Torrey. Other letters are widely scattered, including in Europe, where the Linnaean Society of London has the greatest number; it also has several unpublished botanical essays. The Institut de France has his unpublished, book-length submission for the 1835 Prix Volney.

Charles Boewe

RAINE, WALTER. Born Leeds, England, 15 September 1861; died Toronto, Ontario, 26 July 1934. Lithographic engraver, oologist.

FAMILY AND EDUCATION

Eldest son of T. Raine of Leeds, England. Educated in Leeds. Married 1885 to Florence Selina Grimwade; five daughters, two sons.

POSITIONS

In England, he was a lithographic engraver and won the Queen Victoria Medal for ornamental designing, before emigrating to Canada about 1886. In addition to lithographic engravings, Raine worked with watercolors, drew blueprints, and made numerous bird's-eye-view drawings.

CAREER

Raine's obsessive lifelong hobby, avocation, and after-hours business was oology. He bought and sold bird's eggs from his home, one of the largest oology businesses in North America at the turn of the century, when egg collecting was more popular than stamp collecting. He made several important collecting trips to western Canada in 1891, 1893, and 1894; the 1891 trip was published in book form, and a portion of the 1893 trip was published in *Nidiologist*. Many of Raine's records were quoted by John Macoun in his *Catalogue of Canadian Birds* (1900–1904). Nine of Raine's beautiful black-and-white photographs of bird's nests were published by Chester A. Reed in *North American Birds Eggs* in 1904. Business sometimes triumphed over science when Raine trusted the identifications of his enthusiastic but often inexperienced collectors; not all Raine records can be accepted at face value.

MAJOR CONTRIBUTIONS

The first four authentic nest records of the solitary sandpiper involved sets taken by Raine's collector, Evan Thompson of Red Lodge, Alberta, in 1903 and 1904.

BIBLIOGRAPHY

Barnes, R. M. "Walter Raine." *Oologist* 27 (1910).
Houston, C. S. "An Assessment of Walter Raine and His Saskatchewan Records." *Blue Jay* 39 (1981).
Macoun, J. *Catalogue of Canadian Birds*. Ottawa: Queen's Printer (1900–1904).
Raine, W. *Bird Nesting in North West Canada*. Toronto: Hunter Rose (1892).
Raine, W. "Discovery of the Eggs of Solitary Sandpiper." *Oologist* 21 (1904).
Reed, C. *North American Birds Eggs*. New York: Doubleday, Page (1904).

C. Stuart Houston

RAND, AUSTIN LOOMER. Born Kennville, Nova Scotia, 16 December 1905; died Lake Placid, Florida, 6 January 1982. Zoologist, explorer, writer.

FAMILY AND EDUCATION

Son of Stanley Bayard Rand and Carrie Forsythe Rand. Received B.Sc. from Acadia University in 1927 and Ph.D. from Cornell University in 1932. Honorary

D.Sci. from Acadia University in 1961. Married Rheua Medden in 1931; two sons.

POSITIONS

Assistant ornithologist, Cornell University 1927–1929, 1931–1932; assistant, Franco-American Zoological Mission to Madagascar, 1929–1931; American Museum of Natural History Expedition to New Guinea, 1932–1934; on staff, American Museum of Natural History, 1934–1935; research associate, Department of Ornithology, American Museum of Natural History, 1935–1942; assisted with establishment of Archbold Biological Station, Florida, 1941–1942; assistant zoologist, National Museum of Canada (NMC), 1942–1947, associate zoologist and acting chief of biology, NMC, 1946–1947; curator of birds, Field Museum of Natural History, 1947–1954, chief curator of zoology, 1955–1970; research associate, Archbold Biological Station, 1971–1982.

CAREER

Began career in 1929, when, at suggestion of his mentor A. A. Allen of Cornell, he was asked to replace C.G. Herrold of Winnipeg as collector of birds on expedition to Madagascar. Rand selected to draft account of ornithological collections, which became basis of his Cornell Ph.D. As coleader and ornithologist, participated in three expeditions to New Guinea, 1933–1934, 1936–1937, 1938–1939, and published accounts of ornithological collections and observations. During his tenure at National Museum of Canada, spent roughly half his time on mammals and the other half on birds. Spent some time in fieldwork in northwestern Canada. During tenure as curator and chief curator at Field Museum, ornithological specimens in collections increased by 132,000. Completed over 100 papers on wide range of ornithological subject matter based on work done at Field Museum and also completed a number of books for popular consumption, including *Stray Feathers from a Bird Man's Desk* (1955) and *Ornithology: An Introduction* (1967). Also authored *Distribution and Habits of Madagascar Birds* (with Richard Archbold, 1936); *New Guinea Expedition* (1939); *Development and Enemy Recognition of the Curve-Billed Thrasher* (1941); *The Southern Half of the Alaska Highway and Its Mammals* (1944); *Mammals of Yukon, Canada* (1945); *Mammals of Eastern Rockies and Western Plains of Canada* (1947); *A Midwestern Almanac* (1961); and *Handbook of New Guinea Birds* (with E. T. Gilliard, 1967). Was editor of *Canadian Field-Naturalist*, 1942–1947. Contributed to periodicals such as *National Geographic*. President, American Ornithologists' Union, 1962–1964.

MAJOR CONTRIBUTIONS

Enjoyed a ''consuming interest in all aspects of birds.'' Author of authoritative technical and popular studies of birds and mammals. Was vitally interested in sharing research results with interested public.

BIBLIOGRAPHY
New York Times, 8 November 1982 (obit.).
Traylor, M., Dean Amadon, and W. E. Godfrey. *Auk* (July 1984) (obit.).
Who Was Who in America. Vol. 8 (1985).
World Who's Who in Science (1968).

Roberta Pessah

RANDALL, THOMAS EDMUND. Born Rodmersham Green, Kent, England, on 21 June 1886; died in Calgary, Alberta, 20 December 1984. Farmer, oölogist, naturalist.

FAMILY AND EDUCATION

Educated in England. Emigrated to Saskatchewan, Canada, in 1912. Married Ruth Ross in 1926; three children.

POSITIONS

Farmer near Drinkwater, Saskatchewan, 1912–1914. In Canadian army, 1914–1919. Farmer near Castor, Alberta, 1919–c.1927. Househusband for teacher wife, 1928–1935. Collector for National Museum of Canada at Churchill, Manitoba, summer 1936. Park warden, Elk Island National Park, 1937–1939. Bander for Ducks Unlimited, 1940–1945. Employee of Eastern Irrigation District, Brooks, Alberta, 1946–1951, when he retired.

CAREER

Collected many sets of bird's eggs of all species, including over fifty nests of the Gray jay, but especially of marsh birds and waders, which are now in museums, including the Provincial Museum of Alberta and the National Museum of Canada. Contributed life histories on Dowitcher (he had found forty-three nests of the Short-billed Dowitcher), Wilson's Phalarope (based on over 1,000 nests found by himself), Lesser Yellowlegs, Greater Yellowlegs, Bonaparte's Gull, and Sora to Bannerman's *Birds of the British Isles* (vols. 9–12). Wrote regional bird lists for Brooks, Castor, Elk Island, Alberta, and Kazan Lake, Saskatchewan.

MAJOR CONTRIBUTIONS

Whether as farmer, park warden, local public official, or collector for the National Museum of Canada, published valuable additions to the literature of ornithology in western Canada and the British Isles. Notable oölogist.

BIBLIOGRAPHY
Bannerman, D. A., ed. *The Birds of the British Isles*. Vols. 9–12. London: Oliver and Boyd (1961–1963).
Houston, C. S., M. J. Bechard, and P. R. Stepney. "Thomas Edmund Randall, Nest-Finder Supreme." *Blue Jay* 42 (1984).
Randall, T. E. "Birds of the Kazan Lake Region, Saskatchewan." *Blue Jay* 20 (1962).
Randall, T. E. "The Birds of Elk Island National Park, Alberta, Canada." *Canadian Wildlife Service Management Bulletin* (ser. 2, no. 3) (1952).

C. Stuart Houston

RATHBUN, MARY JANE. Born Buffalo, New York, 11 June 1860; died Washington, D.C., 4 April 1943. Marine zoologist, carcinologist.

FAMILY AND EDUCATION

Youngest of five children (second daughter) of Charles Howland and Jane Furey Rathbun. Her elder brother, Richard, also a zoologist, rose to become assistant secretary of the Smithsonian and director of the U.S. National Museum. Completed her formal high school education at Central School, Buffalo, 1878. Honorary M.A., University of Pittsburgh, 1916, Ph.D., George Washington University (for her monograph, *The Grapsoid Crabs of America*). Never married.

POSITIONS

Volunteer assistant to brother, Woods Hole, Massachusetts, summers 1881–1884. Employed by U.S. Fish Commission at U.S. National Museum (USNM), 1884–1886, to catalog and take care of collections. Clerk-copyist, Department of Marine Invertebrates, 1886–1898, second assistant curator, 1898–1907, assistant curator in charge of department, 1907–1914, USNM; resigned position to devote to full-time research as honorary associate in zoology, Smithsonian Institution, 1914–1939.

CAREER

Began working at Woods Hole after observing elder brother Richard, summer, 1881. Employed by S. F. Baird, 1884. From 1886, had effective charge of the department, though Richard held title of curator until 1897. Curated collections, maintained card catalog of departmental library. Began self-education in zoology; her first publication appeared in 1891. These totaled 158, and the majority were concerned with classification and description of specimens of recent and fossil shrimps and crabs. Her major publications included *Les Crabes d'Eau Douce* (1904–1906); *The Grapsoid Crabs of America* (1918); *The Spider Crabs of America* (1925); *The Cancroid Crabs of America* (1930); and *The Oxystomatous and Allied Crabs of America* (1937). Her paleontological studies included *The Fossil Stalk-Eyed Crustacea of the Pacific Coast of North America* (1926) and *The Fossil Crustacea of the Atlantic and Gulf Coastal Plain* (1935).

MAJOR CONTRIBUTIONS

Her work in nomenclature brought high degree of order to classification of crabs and shrimps, both recent and fossil. Her meticulous descriptions, notes, and records, both published and unpublished, continued to be used by students and colleagues for many years after her death.

BIBLIOGRAPHY

Schmitt, Waldo. *Notable American Women* (1971).
Science (14 May 1943) (obit.).
Washington Evening Star, 6 April 1943 (obit.).

Keir B. Sterling

RATHBUN, RICHARD. Born Buffalo, New York, 25 January 1852; died Washington, D.C., 16 July 1918. Geologist, invertebrate zoologist, museum administrator.

FAMILY AND EDUCATION

Son of Charles Howland Rathbun, a quarry owner and Jane Furey Rathbun. Mary Jane Rathbun, one of five brothers and sisters, was a renowned carcinologist. Attended Buffalo public schools until the age of fifteen, when he went to work for his father's quarry business. Attended Cornell University, 1871–1873, studying geology and paleontology, but did not receive a degree. Honorary M.S. from Indiana University, 1883; honorary Sc.D. from Bowdoin College, 1894. Married Lena Augusta Hume, 1880; one son.

POSITIONS

Assistant in zoology, Boston Society of Natural History, 1873–1875; geologist, Geological Commission of Brazil, 1875–1878; scientific assistant, U.S. Fish Commission, 1879–1896; honorary curator, Division of Marine Invertebrates, U.S. National Museum (USNM), 1880–1896; assistant secretary in charge of office and exchanges, Smithsonian Institution, 1896–1898; assistant secretary in charge of the USNM, Smithsonian Institution, 1898–1918.

CAREER

Developed an early interest in paleontology, studying fossils found in quarries around Buffalo, New York. His fossil collections became foundation for the paleontology section of the museum of the Buffalo Society of Natural Sciences. Entered Cornell University in 1871, on recommendation of Charles F. Hartt, to study geology and paleontology. Was assigned to work up the Devonian and Cretaceous fossils collected by Hartt in Brazil. Resulted in his first publication, ''On the Devonian Brachiopoda of Erere, Province of Para, Brazil'' (1874). From 1873 to 1875 he studied paleontology at the Museum of Comparative Zoology and acted as assistant in zoology at Boston Society of Natural History. During summer months of those years he served as volunteer with the U.S. Fish Commission on the invitation of Spencer F. Baird. From 1875 to 1878, he conducted fieldwork in Brazil as geologist to Geological Commission of Brazil. Investigations focused on the geology and coral reefs of the Bay of Bahia region and the mineral resources of Sao Paulo province. Accepted salaried post with the U.S. Fish Commission in 1880. Examined American fisheries for crustacea and other invertebrates for the fisheries investigations of the tenth census, 1880–1881. Results published in *Fisheries and Fishing Industries of the United States* (1884) and according to one authority ''form one of the most important of all contributions to marine economic zoology.'' Prepared and described collections of crustacea, worms, echinoderms, and sponges for the Great International Fisheries Exhibition in London, 1884. Assisted Secretary of State John W. Foster in gathering information for the U.S. case at the Paris Fur Seal Tribunal, 1891. Appointed by President Harrison to be American representative on the Joint Commission with Great Britain to examine fisheries in the boundary waters between the United States and Canada, 1892. Served as chief adviser to the International Commission to the Fur Seal Islands, 1896. In addition to his duties with the Fish Commission, he was in charge of curating the national collections of marine invertebrates, 1880–1896.

Published numerous papers on fossil and recent invertebrates, American fisheries, and economic aspects of zoology between 1874 and 1896. After joining Smithsonian Institution, scientific investigations became subordinate to administrative duties. At Smithsonian he was instrumental in the planning and construction of the Natural History Building of the U.S. National Museum, completed in 1911, and the development of the Institution's National Gallery of Art (now the National Museum of American Art). His major publications after 1896 concentrated on these efforts and included *The United States National Museum. An Account of the Buildings Occupied by the National Collections* (1903), *Descriptive Account of the Natural History Buildings of the United States National Museum* (1913), and *The Gallery of Art, Department of Fine Arts of the National Museum* (1916).

MAJOR CONTRIBUTIONS

Analysis of economic aspects of marine zoology, especially his investigations of American marine invertebrate fisheries and his work with international commissions on fisheries and fur seals. Built up national collections of marine invertebrates. Was chief administrative officer of the U.S. National Museum during a period of intense growth and diversification of viewpoint.

BIBLIOGRAPHY
Benjamin, Marcus. *Science* (6 September 1918) (obit.); repr. in *Annual Report of the Smithsonian Institution* (1919).
Bowdoin College Bulletin (January 1920) (obit.).
Coe, Wesley R. ''Richard Rathbun and His Contributions to Zoology.'' *The American Journal of Science* (December 1918)
Dictionary of American Biography. Vol. 8 (1964).
UNPUBLISHED SOURCES
See Smithsonian Institution Archives, Record Unit 7078, Richard Rathbun Papers, 1870–1918 (includes correspondence, manuscripts, notebooks, research notes, sketches, and photographs); Smithsonian Institution Archives, Record Unit 54, assistant secretary in charge of the U.S. National Museum, 1897–1918 (contains Rathbun's correspondence as assistant secretary); Smithsonian Institution Archives, Record Unit 192, U.S. National Museum, Permanent Administrative Files, 1902–1977 (includes Rathbun's correspondence as assistant secretary).

William Cox

RAVENEL, HENRY WILLIAM. Born Woodville Plantation, near Charleston, South Carolina, 19 May 1814; died Aiken, South Carolina, 17 July 1887. Botanist, agriculturist.

FAMILY AND EDUCATION

Son of Henry Ravenel, a physician, and Catherine Stevens Ravenel. His mother died when he was two, and a stepmother died soon thereafter. His father married for a third time in 1821, but young Henry was raised by his grandmother until her death in 1826. He then rejoined his father's household. Attended Pineville Academy, Pineville, South Carolina, 1820–1828. Privately tutored for college, Columbia, South Carolina, 1829. Entered South Carolina College (later University of South Carolina) with sophomore standing, 1829. B.A., 1832. Honorary LL.D., University of North Carolina, 1886. Married Elizabeth Gaillard Snowdon, 1835 (died 1855); five daughters, one son. Married Mary Huger Dawson, 1858; five daughters.

POSITIONS

Plantation owner and agriculturist, 1832–1853; farmer and pomologist, 1852–1873; botanical scholar and collector, 1842–1887; agricultural editor, *Charleston* [S.C.] *News and Courier*, 1882–1887.

CAREER

Sought to enter medical profession but was dissuaded by father, who argued that it might be too demanding a profession for his ''delicate'' health, though there is no evidence that young Ravenel had any physical disabilities. Given a plantation by father where he raised rice, some cotton, and other crops. Developed interests in botany and soil improvement, early 1840s. Became aware of the nature of ecological succession, was interested in questions of plant distribution. Was not primarily concerned with the utility of botany, though he was interested in promoting agriculture. Gave some attention to the efficacy of marl and other fertilizers. Began study of fungi, though first publication (1849) was list of seed-bearing plants not previously known in South Carolina. Moved to Aiken, South Carolina, for reasons of health, 1853. There, became interested in trees of his region, and in the mid-1850s, fruit-tree classification. From this point until well into the Civil War years, fruit production brought him a considerable income.

Was elected corresponding member of Academy of Natural Sciences of Philadelphia and of American Association for the Advancement of Science (1849). Occasionally presented papers before various scientific organizations and published a number of papers on botanical and other subjects. Corresponded and exchanged plant specimens with considerable number of botanists in other parts of the United States and Western Europe, to include Asa Gray, John Bachman, W. S. Sullivant, John Torrey, Miles J. Berkeley, and others. While he was regularly in touch with other botanists and naturalists in his own state and region, he was not greatly involved in the scientific organizations of Charleston because of its distance from his home, his dislike of cities, and his relatively narrow field of specialization. Produced five fascicles (with index) of *Fungi Caroliniani Exsiccati* between 1852 and 1860, consisting of volumes of dried specimens. Two earlier sets of exsiccati, covering mosses and lichens, had been assembled or were in production by others in the United States, but Ravenel's was the first to feature fungi. During Civil War he was effectively cut off from northern and European colleagues. Too frail physically to take active part in the conflict, he did make heavy investment in Confederate bonds. This and the emancipation of his slaves left him in reduced circumstances for the rest of his life. Compelled to give up fruit production because he no longer had the personnel required to handle this labor-intensive work. Ravenel sold plant specimens and seeds together with some of his books, launched a nursery operation, and did some writing to make ends meet. Essentially became a collector, giving up personal research in part for reasons of health. Declined two possible offers of professorships (at Washington College in Virginia and at the nascent University of California) because he doubted his capacity to do academic work, preferred to remain where he was, and had concerns about his growing deafness. Investigated Texas Fever in cattle for Federal Bureau of Agriculture, 1869; demonstrated that it was not caused by a fungus. Had over 11,100 species of fungi in personal collection by 1881. Regarded as leading southern botanist from the early 1870s. Contributed to several parts of Baron Felix von Theuman's *Myotheca Universalis* (1876–1884) and *Herbarium Mycologicum Oeconomicum*. Co-author (collector) with M. C. Cooke of *Fungi Americani Exsiccati* (8 parts, 1878–1882). Contributed botanical articles to *Land We Love*, edited by General D. H. Hill

(1867–1869), and was agricultural editor for the *Charleston* [S.C.] *News and Courier* (1882–1887).

MAJOR CONTRIBUTIONS

A leading American authority on mycology for nearly half a century. Also made some contributions to improvements in agriculture. Completed most of his scholarly work during the twenty years preceding the Civil War, in large part because he had the energy, resources, and leisure to do so. Open-minded concerning evolutionary theory, though he finally concluded its proponents gave it more weight than was justified. Suspended most of his botanical activity during the war, but gave some attention to practical matters, such as encouragement of mushroom consumption. Because of declining health and the need to produce income, spent less time with scholarly aspects of botany after 1865. Focused instead on collecting, sales of plants and seeds, and on writing. Most of Ravenel's collections of fungi were sold to British Museum and his other plant materials to Converse College following his death.

BIBLIOGRAPHY

Childs, Arney Robinson, ed. *The Private Journal of Henry William Ravenel, 1859–1887* (1947).

Dictionary of American Biography.

Haygood, T. M. *Henry William Ravenel, 1814–1887: South Carolina Scientist in the Civil War Era* (1987) (includes bibliography of Ravenel's publications).

Rogers, D. P. *A Brief History of Mycology in North America* (1981).

Stevens, Neil E. "The Mycological Work of Henry W. Ravenel." *ISIS* 18 (1932).

Who Was Who in America.

UNPUBLISHED SOURCES

Henry Ravenel papers in collections of Charleston Museum; Clemson University; Gray Herbarium, Harvard University; South Carolina Historical Society; and South Caroliniana Library, University of South Carolina. See also bibliography in Haygood.

Keir B. Sterling

RAWSON, DONALD STRATHEARN. Born near Uxbridge, Ontario, 19 May 1905; died Saskatoon, Saskatchewan, 16 February 1961. Limnologist.

FAMILY AND EDUCATION

Son of Reuben Richardson and Mary (McFarlane) Rawson. Early education at Claremont and at Harbord Collegiate Institute, Toronto. Studied at University of Toronto, received B.A. in 1926, M.A. in 1927, Ph.D. in 1929. Married Hildred Patton, 1932; three children.

POSITIONS

Professor of Biology, University of Saskatchewan, 1928–1961. Head of Department of Biology, 1949–1961. Fellow, Royal Society of Canada, 1944–1961.

Member, Royal Commission on the Fisheries of Saskatchewan, 1946–1947. President, Limnological Society of America, 1947. President, Canadian Committee for Freshwater Fisheries Research, 1951. Member, Fisheries Research Board of Canada, 1959–1961.

CAREER

First research, on Lake Simcoe, done while a student at the University of Toronto's Ontario Fisheries Research Laboratory. As the first professional limnologist in western Canada, conducted pioneering studies of many lakes in that region. From 1928 to 1935 studied the physics, chemistry, biology, and fisheries of lakes in Prince Albert National Park, north of Saskatoon, and in British Columbia. Between 1935 and 1941 researched lakes in the Canadian Rockies and in Riding Mountain National Park in Manitoba. From 1942 to 1947 studied lakes in northern Canada, including Lake Athabaska and Great Slave Lake. After 1947, devoted most attention to Lac la Ronge, Amisk Lake, and other lakes in Saskatchewan.

Published widely in the fisheries and limnological literature. Doctoral dissertation (1929) on Lake Simcoe was viewed by many as model of limnological research. Views on lake productivity presented in ''Some Physical and Chemical Factors in the Metabolism of Lakes,'' in *Problems of Lake Biology* (1939). Reported on results of Great Slave Lake survey in ''Estimating the Fish Production of Great Slave Lake,'' *Transactions of the American Fisheries Society* (1947). An example of conservation work was Rawson et al., *The Big River Survey: A Comprehensive Study of Natural Resources as an Aid to Improved Utilization* (1943). Interest in holistic comparison of lakes indicated in ''A Limnological Comparison of Twelve Large Lakes in Northern Saskatchewan,'' *Limnology and Oceanography* (1960).

Most distinctive feature of his research was its significance to both limnology and fisheries management. Limnological contributions included greater understanding of factors influencing lake productivity (a chief concern of this discipline during Rawson's career). He identified four types of factors: human, edaphic conditions, shape of basin, and climate, and he gathered data from dozens of lakes to estimate their relative significance. Studies of lake productivity also provided a basis for assessing fisheries potential. For example, his study of Great Slave Lake (sponsored by the Fisheries Research Board of Canada) helped justify establishment of a commercial fishery. Other researchers, building on his ideas, eventually developed the ''morphoedaphic index,'' widely used to estimate potential fish yield. A dedicated teacher, he was perhaps most influential through his many students, who populated numerous universities in Canada. Also noted as a highly able administrator, he built a strong university biology department and established and directed a provincial fisheries research program at his university, under the jurisdiction of the Saskatchewan Department of Natural Resources.

MAJOR CONTRIBUTIONS

Important studies of limnology and fisheries biology in western Canada; development of the University of Saskatchewan as an important regional center for this research; contributed to establishment of a scientific basis for fisheries management in this region.

BIBLIOGRAPHY

Bocking, S. A. "Fisheries and Fundamental Science: Donald Rawson's Studies of Lake Productivity." *Scientia Canadensis* (1990).

Journal of the Fisheries Research Board of Canada (1961).

Northcote, T. G., and P. A. Larkin. "Western Canada." In *Limnology in North America*. Ed. D. G. Frey (1963).

Transactions of the Royal Society of Canada (1961) (obit.).

UNPUBLISHED SOURCES

Extensive research and administrative records are in Donald S. Rawson fonds (papers), MG 8, University of Saskatchewan Archives, Saskatoon.

Stephen Bocking

RICHARDSON, JOHN. Born Nith Place, Dumfries, Scotland, 5 November 1787; died (Lancrigg), Grasmere, Lake District, England, 5 June 1865. Explorer, naturalist, physician.

FAMILY AND EDUCATION

Eldest son of twelve children of Gabriel Richardson, a brewer-provost-magistrate, and Anne Mundell. Attended Dumfries Grammar School until apprenticed to his surgeon uncle just before thirteenth birthday. Attended Medical School at University of Edinburgh, 1801–1804, 1806–1807, returning in 1814 to complete M.D., graduating in 1816, with a thesis on yellow fever. Honorary LL.D., University of Dublin, 1857. Married Mary Stiven, 1818 (died 1831), Mary Booth, 1833 (died 1845), Mary Fletcher, 1847. Six children with Mary Booth.

POSITIONS

House surgeon, Dumfries and Galloway Royal Infirmary, 1804–1806. Assistant surgeon/surgeon, British navy, 1807–1815. Second to Franklin on two Arctic Exploring Expeditions, 1819–1822, 1825–1827. Surgeon, Chatham Division, Marines, 1824–1828. Chief medical officer, Melville Hospital, Chatham, 1828–1838. Physician, Royal Naval Hospital, Haslar, 1838–1855. Inspector of hospitals for the Admiralty, 1840–1855. Leader, expedition to search for John Franklin, 1848–1849.

CAREER

Encouraged to scholarship by his parents and their close friend, poet Robert Burns, could read at four. Just prior to thirteenth birthday was apprenticed to uncle, a surgeon, and at fourteen entered medical school in 1801, studying Greek, botany, and geology in addition to medical subjects. At seventeen, became house surgeon in Dumfries in 1804 but returned to medical school in 1806, receiving his license to practice in 1807. After graduating, was admitted to Royal College of Surgeons and began long career with Royal Navy as assistant surgeon on the boat *La Nymphe*, followed by five other ships during the Napoleonic Wars, 1806–1814. During War of 1812–1814, was also a surgeon with Royal Marines in North America, where he found war distasteful. Returned to England on half-pay, completing his M.D. thesis at Edinburgh on yellow fever.

Explored vast areas of present-day Manitoba, Saskatchewan, Alberta, and especially Northwest Territories, first as second to John Franklin on two Arctic Land Exploring Expeditions and later, when sixty, to search for Franklin. During expeditions, prepared maps of major part of northern Canada and made detailed observations on geology, botany, zoology, customs of natives, and meteorology and collected large numbers of plants, insects (especially beetles), birds, mammals, and other animals, many of which were housed in a museum at Haslar. Although he suffered a mild heart attack during the search for Franklin, had great stamina, walking hundreds of miles overland and setting a record for rapidity of canoe travel.

Remained in the employ of the navy until retirement in 1855, serving as surgeon of the Chatham division of Marines between the two Franklin expeditions, then rising through various positions until becoming senior physician at Royal Naval Hospital at Haslar, near Portsmouth, and also inspector of hospitals for the Admiralty. Was influential in converting the treatment of mentally ill from confinement to humane ward care, in improving and promoting nursing as a career, and in pioneering the use of general anesthesia. Consulted Florence Nightingale on nursing.

Writings were prolific, based partly on own expeditions and partly in preparing zoological appendixes, especially of fish, to reports of expeditions by other explorers elsewhere. Charles Darwin consulted him on details of arctic fauna, and Thomas Huxley was Richardson's student. Described large numbers of plants, birds, mammals, and especially fish, of which he named over 200 species and forty-three still-extant genera. These activities kept him busy long after retirement, when he also contributed to the *Dictionary of Philology*, precursor to *Oxford English Dictionary*, compiling an index of words used by Burns, with their cognates in Norse, Icelandic, and Gaelic.

Bibliographies of his books and papers in scientific journals appear in *Journal of the Society for the Bibliography of Natural History* 5 (1969) and 6 (1972). Subjects include fossil mammals, soils, aurora borealis, birds, and especially fish. Best known are the four volumes of *Fauna Boreali—Americana*, mammals and fish on his own, birds with W. Swainson, and insects with W. Kirby (1829–1838), and contributed most of specimen material for accompanying flora by Hooker (1840). Other notable titles include *Report on North American Zoology* (1836), *An Arctic Searching Expedition*, 2 vols. (1851), and *The Polar Regions* (1851). Zoological appendixes to reports of other explorers include expeditions of Franklin, Parry, and Ross.

Many faunal and floral species and races bear his name, including an owl, vole and ground squirrel, as do a river, bay, lake, and mountain range, all in northern Canada. Knighted by Queen Victoria, 1846, made Companion of the Bath, 1850, awarded medal of the Royal Society, 1856, and received LL.D., University of Dublin, 1857.

MAJOR CONTRIBUTIONS

In medicine, left legacy of improved treatment of mentally ill, improved nursing, and use of general anesthetics. Not a theorist, contributions to zoology, botany, and geology lie in detailed descriptions and naming of many forms and their habitats, with early information on their ranges. Leading ichthyologist of his time.

BIBLIOGRAPHY

Gill, T. "Life and Ornithological Labors of Sir John Richardson." *Osprey* (n.s.) 24 (1931).

Houston, C. S. "John Richardson (1787–1865)." *Arctic* 36 (1983).

Houston, C. S., ed. *Arctic Ordeal, The Journal of John Richardson, Surgeon-Naturalist with Franklin 1820–1822*. (1984).

Houston, C. S. "John Richardson—First Naturalist in the Northwest." *Beaver* 315, 2 (1984).

Houston, C. S. "Dr. John Richardson: Arctic Doctor." *Canadian Family Physician* 34 (1988).

Johnson, R. E. *Sir John Richardson, Arctic Explorer, Natural Historian, Naval Surgeon, Naturalist with Franklin* (1976).

Kupsch, W. O. *The Canadian Encyclopedia* (1985, 1988).

McIlraith, J. *Life of Sir John Richardson* (1868).

Stephen, L., ed. *The Dictionary of National Biography*. Vol. 16 (1917, repr. 1949–1950).

Stewart, D. A. "Sir John Richardson. Surgeon, Physician, Sailor, Explorer, Naturalist, Scholar." *British Medical Journal* (1931). *Canadian Medical Association Journal* (n.s.) 24(1931).

Swinton, W. E. "Physicians as Explorers. Sir John Richardson: Immense Journeys in Rupert's Land." *Canadian Medical Association Journal* 117 (1977).

Wallace, W. S., and W. A. McKay, eds. *The MacMillan Dictionary of Canadian Biography* (1978).

UNPUBLISHED SOURCES

Letters in Burgh Museum, Dumfries, Scotland, and Richardson-Voss Collection of John Voss, Chestham Bois, Amersham, England. Record of service from April 1807 to June 1855 in Public Records Office (England) Adm 8/196.

Martin K. McNicholl

RIDGWAY, ROBERT. Born Mount Carmel, Illinois, 2 July 1850; died Olney, Illinois, 25 July 1929. Ornithologist, museum curator, scientific illustrator and artist, botanist, horticulturist.

FAMILY AND EDUCATION

Son of David Ridgway, a druggist, and Henrietta James Reed Ridgway. Oldest of ten children; educated in local school and by parents, who encouraged his interests in natural history and hunting. Honorary M.Sc., Indiana University, 1884. Married Julia Evelyn Perkins (died 1927) of Washington, D.C., 1875; one son.

POSITIONS

Zoologist, U.S. Geological Exploration of the 40th Parallel (King Survey), 1867–1876; from assistant in ornithology to curator of birds, U.S. National Museum (USNM), 1869–1929 (museum-salaried from 1874, nonresident after 1915).

CAREER

Showed precocious talent for collecting, drawing, and painting (mixed own watercolors) birds and other animals near home in Wabash County, Illinois. To identify local birds scientifically, first corresponded with, and then collected for, Spencer Baird, assistant secretary, Smithsonian Institution, in charge of U.S. National Museum, who encouraged Ridgway's fledgling efforts in ornithology, 1864–1866. Baird, who advised King Survey on natural history operations, secured Ridgway's position as zoologist on this geological and mineral resources reconnaissance, 1867. After brief training at Smithsonian, Ridgway collected and explored for King Survey, often with staff botanist Sereno Watson and staff photographer Timothy O'Sullivan, in California, Nevada, Idaho, Utah, and Wyoming, 1867–1869. Described his specimens from West and learned taxidermy, curation, and scientific techniques as assistant to Baird, who placed Ridgway in charge of National Museum's bird collections, 1869–1874. Under contract to Boston publisher, wrote descriptions, and, with Henry Elliott, drew illustrations for Baird, Thomas Brewer, and Ridgway's *A History of North American Birds. Land Birds* (1874) and *Water Birds of North America* (1884). Classified American falcons, 1874–1876. Ornithologist, 1874–1879; curator, Department of Ornithology, 1880; and curator, Division of Birds, 1881–1929, U.S. National Museum. Founding member, American Ornithologists' Union (AOU), 1883 (vice president, 1883–1891; president, 1898–1900), and, with Elliott Coues and others, Member AOU Committee on Check-List of North American Birds (report published 1886, revised 1895, 1910). Described Edward Palmer's collections from Guadalupe Island, 1876; Leonhard Stejneger's (Ridgway's assistant curator, 1884–1889) birds from Kamchatka and Komandor Islands, 1883; *Al-*

batross collections from Cozumel, 1885; Galápagos and South American birds, 1888; and Indian Ocean collections, 1896. Member, Permanent Committee, First International Ornithological Congress, Vienna, 1885. Assisted Smithsonian Secretary Samuel Langley's experiments on powered flight by providing data on wing characteristics of soaring birds, 1890s. At Smithsonian's request, began extensive systematic catalog of North American avifauna in National Museum's collections, 1894. To aid classification, continued fieldwork from Virginia to California, Florida in 1895–1897, Alaska (with Louis Agassiz Fuertes on Harriman Expedition) in 1899, and Costa Rica (with José Zeledon) in 1904–1905 and 1908. As part of efforts in wildlife preservation, purchased (1906) and maintained until 1929 property outside Olney, Illinois, as Bird Haven, an arboretum and bird sanctuary. Walker Grand Prize, Boston Society of Natural History, 1913. Continued investigations of trees of Illinois and Indiana and horticultural interests, 1915–1929. Daniel Giraud Elliot Medal, National Academy of Sciences, 1920. William Brewster Medal, American Ornithologists' Union, 1921. Member, National Academy of Sciences, 1926.

Some thirty-five neontologic taxa named in his honor. Published more than 500 books, articles, catalogs, checklists, and other works, 1869–1929, including "Notes on the Vegetation of the Lower Wabash Valley," *American Naturalist* 6 (1872) and 7 (1873); "The Birds of Colorado," *Bulletin of the Essex Institute* 5 (1873); *Ornithology*, Report of the U.S. Geological Exploration of the Fortieth Parallel, vol. 4, pt. 3 (Professional Paper No. 18 of the Engineer Department, U.S. Army) (1877); "Song Birds of the West," *Harper's New Monthly Magazine* 56 (1878); "Nomenclature of North American Birds Chiefly Contained in the United States National Museum," *USNM Bulletin* 21 (1881); "Notes on the Native Trees of the Lower Wabash Valley and White River Valley," *Proceedings of the United States National Museum* 5 (1882); *A Nomenclature of Colors for Naturalists and Compendium of Useful Information for Ornithologists* (1886); *A Manual of North American Birds* (1887; 2d ed., 1896, repr. 1900); *The Ornithology of Illinois Part I. Descriptive Catalogue* (1889); "The Humming Birds," *Report of the United States National Museum for 1890* (1892); "Additional Notes on the Trees of the Lower Wabash Valley," *Proceedings of the United States National Museum* 17 (1895); "Birds of the Galápagos Archipelago," *Proceedings of the United States National Museum* 19 (1897); "The Birds of North and Middle America," *United States National Museum Bulletin*, 50, pts. 1 (1901) to 8 (1929) [manuscripts and notes for additional volumes completed by Herbert Friedmann, USNM curator of birds, 1929–1957, pts. 9 (1941) to 11 (1950), pts. 12, 13 unfinished]; *Birds of the World: A Popular Account* (by Frank H. Knowlton, ed. Ridgway) (1909); *Color Standards and Color Nomenclature* (1912).

MAJOR CONTRIBUTIONS

Provided detailed description of birds of eastern Sierra Nevada, Great Basin, and west slope of Rocky Mountains, expanding knowledge gained by Pacific

Railroad Surveys in 1850s. Made fundamental contributions to American systematic ornithology, insisting on accurate morphological descriptions as basis for work and applying results in studies of zoogeography and variation. After deaths of Baird (1887) and Coues (1899), foremost descriptive ornithologist in North America—knowledge based on fieldwork from Alaska to Costa Rica and studies of worldwide collections at Smithsonian. As ultimate product of Bairdian tradition in ornithology, contributed major systematic-distributional catalog of birds of North and Central America. A taxonomic discriminator as a systematist; long favored but ultimately warned against excessive use of ''subspecies'' category in classification. Effective popularizer of bird studies in United States through publications in magazines and standard reference manuals of birds of North America and the world. Talented scientific illustrator of birds, emphasizing accuracy in morphologic detail over compositional aspects. Helped to train younger brother, John Livesy Ridgway, as illustrator and colorist in Division of Birds, U.S. National Museum and, with William Henry Holmes, in U.S. Geological Survey (USGS) (1880s); John Ridgway later USGS's chief illustrator and author of *Scientific Illustration* (1938). Completed scientific analysis and description of colors; used color wheel, Maxwell discs, and sets of two to three colors, each hand-mixed in definite percentages, to produce uniform series of standard named colors—from 186 (watercolors) in 1886 to 1,115 (pigments) in 1912. Color system widely used in ornithology, by florists, and by paint and chemical industries, before supplemented by newer systems in the 1930s. His color names cross-referenced by USGS in standard series of colors for its geologic map units. Most of Bird Haven property purchased by Olney, Illinois, in 1970 and a part restored, as Richland County's Bicentennial Project in 1976, as the Robert Ridgway Memorial Arboretum and Bird Sanctuary.

BIBLIOGRAPHY

Bullard, A. K. *Olney* (Illinois) *Daily Mail*, 12 January 1976.

Deiss, W. A. ''Spencer F. Baird and His Collectors.'' *Journal of the Society for the Bibliography of Natural History* 9 (1980).

Harris, Harry. ''Robert Ridgway.'' *The Condor* 30 (1928, extensive bibliography includes Ridgway's taxa; published separately 1928).

New York Times, 26 March 1929 (obit.).

Oberholser, H. C. ''Robert Ridgway: A Memorial Appreciation.'' *Auk* 50 (1933).

''Robert Ridgway.'' In C. G. Abbot, ''Report of the Secretary,'' *Annual Report of the Board of Regents of the Smithsonian Institution . . . for the Year Ending June 30, 1929* (1930).

''Robert Ridgway, Ornithologist.'' *Audubon Bulletin* (Winter 1917–1918).

Wetmore, Alexander. ''Biographical Memoir of Robert Ridgway.'' *National Academy of Sciences Biographical Memoirs* 15 (1932).

UNPUBLISHED SOURCES

See Robert Ridgway Papers (RU 7167), Division of Birds (RU 105, 7215, 1060102, 1060104), U.S. National Museum Bulletin 50 Collection (RU 7133), and Spencer F. Baird Papers (RU 7002), Smithsonian Institution Archives, Washington, D.C.; King Survey (1867–1881), Microcopy No. M-622, Record Group 57 (U.S. Geological Survey), National Archives and Records Service, Washington, D.C.; Robert Ridgway Letters, Collection No. 681, Library, Academy of Natural Sciences of Philadelphia; George N. Lawrence Collection, Library, American Museum of Natural History, New York; Robert Ridgway Papers, Blacker-Wood Library, McGill University, Montreal.

Clifford M. Nelson

RILEY, CHARLES VALENTINE. Born Chelsea, London, England, 18 September 1843; died Washington, D.C., 14 September 1895. Entomologist, journalist.

FAMILY AND EDUCATION

Son of Charles and Mary Valentine Cannon Riley. Father was clergyman in Church of England who died early in his son's life; mother remarried. Attended school in Dieppe, France, and in Bonn, Germany. Developed interest in natural history, drawing, and painting and created prizewinning sketches of the insects he studied. Left school at age seventeen because of financial problems. Married Emilie Conzelman, 1878. Honorary A.M., Kansas State Agricultural College. Honorary Ph.D., Missouri State University and Washington University, St. Louis.

POSITIONS

Editor, artist, and reporter, *Prairie Farmer*, Chicago, 1864. Member of Union Army, 134th Illinois Volunteers regiment, 1864–1865. State entomologist of Missouri, 1868–1877. Chief, U.S. Entomological Commission, 1877–1879. Entomologist to U.S. Department of Agriculture, 1878–1879. Chief, Federal Entomological Service, 1879–1894. Founder, *Insect Life*, 1889–1894. Cofounder and editor of *American Entomologist*, 1868–1880. Honorary member, Entomological Society of London. Honorary fellow, Royal Agricultural Society of Great Britain. Founder and president, Entomological Society of Washington. President, Academy of Science of St. Louis. Lecturer, Cornell University.

CAREER

Traveled to the United States in late teens (c.1860). Settled near Chicago at a farm, where he studied the effects of insects on crops; reported his findings to agricultural journal *Prairie Farmer*. Joined *Prairie Farmer*'s staff as editor, artist, and reporter of entomology, 1864. Joined Union Army near the end of the Civil War; his regiment (134th Illinois Volunteers) served until November 1865. In France studied phylloxera, a lice harmful to grapevines; proposed solution of grafting them with phylloxera-free American vines. Began working with state entomologist B. D. Walsh. Appointed Missouri's first state entomologist, spring 1868; this post established his reputation. During his stay he published nine annual reports entitled, *On the Noxious, Beneficial and Other Insects of the State of Missouri*. Studied the ''grasshopper plague'' in the West, 1873–1877, and brought it to Congress's attention. A bill was passed forming the U.S. Entomological Commission, of which Riley was appointed chief, March 1877. First report of commission, Spring 1878. Appointed entomologist to U.S. Department of Agriculture, 1878; resigned in 1879. Reinstated in 1881 as chief of Federal Entomological Service until second resignation June 1894. Published the journal *Insect Life*, 1889–1894. Established the journal *American Entomologist* with B. D. Walsh, 1868; became its editor after Walsh's death and published three volumes in the years 1868–1880. Honored by the French government for his study of the grapevine pest *Phylloxera*.

Published 1,657 works, 479 in coauthorship with B. D. Walsh, 364 papers coauthored with L. O. Howard. The Missouri annual reports were later published in book form. Authored the books *Potato Pests: The Colorado Potato-Beetle and Other Insect Foes in North America: With Suggestions for Their Destruction* (1876) and the *Locust Plague* (1877). Notable papers include ''The Caprification of the Fig,'' ''The Yucca Moth and Yucca Pollination,'' and ''Some Interrelations of Plants and Insects.'' Darwin praised Riley's work, especially his observational ability, 1871.

MAJOR CONTRIBUTIONS

Published work was distinguished by its attention to detail, strength of knowledge, insight, observation, and quality and accuracy of the drawings. Significant study of *Phylloxera*, American grasshopper plague, yucca moths and plants, hypermetamorphoses of blister beetles. His efforts moved Congress to form U.S. Entomological Commission. Status of Federal Entomological Service grew under his leadership; it later became a division of the Department of Agriculture. Responsible for advances in development of insect control methods, such as arsenic-based insecticides and use of kerosene emulsions. Considered ''the foremost entomologist of the day'' by his peers (from *Entomologists Monthly*, London).

BIBLIOGRAPHY

Dictionary of American Biography. Vol. 15.

Mallis, A. *American Entomologists* (1971).

UNPUBLISHED SOURCES

See correspondence, notes, drawings, data, scrapbooks (1866–1895), Smithsonian Institution Archives, Washington, D.C., letters in LeConte family papers (1830–1897), American Philosophical Society Library, Philadelphia; letters in Charles Edwin Bessey Collection (1870–1915), University of Nebraska Archives, Lincoln; letters in John Henry Comstock Collection (1885–1931), Cornell University Libraries, Regional History and University Archives, Ithaca, N.Y.; Letters in the Stephen Alfred Forbes Collection (1892–1893), University of Illinois at Urbana–Champaign, Archives.

Jean Wassong

ROBERTSON, A(BSALOM) WILLIS. Born Martinsburg, West Virginia, 27 May 1887; died Lexington, Virginia, 1 November 1971. Senator, lawyer, congressman, bank consultant.

FAMILY AND EDUCATION

Son of Franklin Pierce and Josephine Ragland (Willis) Robertson, distinguished Virginia family. Moved to Lynchburg, Virginia, with his parents in 1891. Attended public schools of Lynchburg and Rocky Mount, Virginia. Graduated from the University of Richmond, Virginia, in 1907 and from its law department in 1908. Admitted to the bar in 1908 and began to practice in Buena Vista, Rockbridge County, Virginia. Moved to Lexington, Rockbridge County, Virginia, in 1919. Honorary LL.D., University of Richmond and College of William and Mary. Married Gladys Churchill, 1920; two children.

POSITIONS

Democratic member of the Virginia Senate, 1916–1922. Commonwealth attorney for Rockbridge County, 1922–1928. Chairman of the state commission of game and inland fisheries, 1926–1932. Elected to U.S. House of Representatives, 1932. Elected to U.S. Senate, 1946, reelected 1948, 1954, 1960. Chairman of the Senate Banking and Currency Committee in 1958. Consultant to the International Bank for Reconstruction and Development, 1966–1968.

CAREER

Served in Virginia Senate from 1916 to 1922 and as member of the House of Representatives from 1933 until his election to the U.S. Senate in 1946. Developed the reputation as champion of conservationist causes during his tenure in the House. Sponsored an influential and policy-setting House resolution in 1934 providing for the establishment of a Select Committee on Conservation of Wild Life Resources and helped lead the fight for the resultant Wild Life Conservation Act of 1937. Was named nation's leading conservationist in 1946 by *Field and Stream* magazine. Was assigned to the powerful Ways and Means Committee, where he spent the next ten years in the House focused on the problems of taxation. Voted against the Roosevelt administration on central elements of the New Deal and on questions of defense and foreign affairs. Worked for the preservation of states' rights and individual constitutional freedoms. Opposed social and labor legislation supporting the Taft–Hartley Act in 1947. During coal strike of 1950, introduced a bill to subject unions to civil and criminal action under the antitrust laws if they threatened the nation's economy, health, or safety. Strongly supported the North Atlantic Treaty Organization (NATO), endorsing Truman's recommendation for universal military training and the draft. Backed foreign aid programs. Opposed making Marshall Plan aid contingent on the foreign trade policies of prospective recipient nations, calling it "dollar diplomacy." Supporter of Eisenhower and continued to maintain expertise on foreign trade, banking and currency, tariffs and taxation. Obtained chairmanship of the Banking and Currency Committee in 1959, criticizing and opposing programs of the Kennedy New Frontier and the Johnson Great Society. Became a consultant to International Bank for Reconstruction and Development.

Respected reputation in the Senate as an expert on foreign trade, tariffs, banking, currency, and taxation, subjects many senators seemed reluctant to pursue. While he supported his political party in presidential elections, he was outspoken in maintaining strong opinions on economic policies; regarded as an economic conservative. At the University of Richmond, he was a football star and hammer thrower; later he became an avid huntsman and fisherman.

MAJOR CONTRIBUTIONS

Championed conservationist causes during a ten-year tenure in the House of Representatives, establishing a Select Committee on Conservation of Wild Life Resources in 1934. Promoted the Wild Life Conservation Act of 1937.

BIBLIOGRAPHY

Biographical Directory of the U.S. Congress 1774–1989 (1989).
New York Times, 2 November 1971 (obit.).
Political Profiles (1978).
Who Was Who in America. Vol. 5 (1973).
Who's Who in American Politics (1967).

Linda L. Reesman

ROBINSON, HARRY. Born Oyama, British Columbia, 8 October 1900; died Hedley, British Columbia, 25 January 1990. Naturalist, storyteller, rancher.

FAMILY AND EDUCATION

Born on a potato farm, where his mother Arcell Newhmkin (and grandparents, Louise and Joseph Newhmkin of the Lower Similkameen Indian Band) had stopped to earn some money digging and packing. Father, Jimmy Robinson, born at Ashnola of an Okanagan mother and a Scottish father. Parents separated before Harry's birth. Spent childhood with mother and grandparents on a small ranch at Chopaka, in Similkameen Valley of south central British Columbia. Learned to ride and tend horses, bring in the cattle, fencing, haying, and participated in rodeos and stampedes. Attended school for five months at age thirteen. In 1924 married Matilda Johnny of Chopaka. No children. Widowed 1971.

POSITIONS

Ranch hand, 1913–1924. Rancher, 1924–1971. Storyteller, 1971–1990.

CAREER

While mother worked, Harry spent long hours with his partially blind grandmother Louise. Raised in Brewster, Washington, she moved to Chopaka after her marriage in 1852. During her many hours with her young grandson, Louise told stories, which would later become central to her grandson's life. Harry's storytelling circle consisted of others such as Mary Narcisse (reputed to be 116 years old when she died in 1944) and John Ashnola, with whom Harry lived for almost a year when he was fourteen. "Every night when I come in from working, he always tell me stories until late." Ashnola died in the 1918 flu epidemic, age ninety-eight. Other names include Alex Skeuce, old Pierre, and old Christine.

Ranching was the main occupation of the inhabitants of the Similkameen Valley of Harry's youth. He began attending school at Cawston at the age of thirteen, but found the twelve mile return trip taxing. After five months he quit and began to work for local ranchers. With his marriage to Matilda Johnny, Harry got his own ranch. Through buying, selling, and trading cows and horses and through various seasonal jobs, Harry and Matilda prospered. They gradually acquired four large ranches between Chopaka and Ashnola, with sixty horses and one hundred fifty head of cattle. Harry sustained a serious hip injury in 1956 that never healed properly. With no children to help them, they began gradually to scale down their ranch. By 1971, they were down to fifty head of cattle. Two years after Matilda's death in March 1971, Harry sold everything and rented a small bungalow near Hedley on the property of longtime friends, Slim and Carrie Allison.

With more time on his hands, he began to reflect on the hours and hours of stories he had heard as a child. Year after year he began telling these and other stories until by the mid-1970s, he became known as one of the most prolific storytellers in the southern interior. People visited him regularly to listen to his stories that covered a wide range of topics and periods. Anthropologist Wendy Wickwire met him in 1977 and until his death in 1990, visited him regularly and recorded on audiotape over 100 of his stories. In 1989, she compiled and edited a general cross-section of these stories and in 1992, published a second volume of his stories.

MAJOR CONTRIBUTIONS

As a keeper of oral traditions Harry was a living repository of knowledge, traditions, and natural history of his region. His stories explained the origins of the Okanagan people and their habitat. Others were about the early "animal-people" and their ways. His favorite stories, about "Indian doctors" and their abilities to impart wide range of cures, told how they acquired their power from rivers, birds, deer, and various other natural beings. Harry kept extensive records of his community's history of births, deaths, and family genealogies dating back three or four generations. He knew the Similkameen Valley intimately, recording the details surrounding floods and rock slides; the Okanagan names for every creek, hill, and valley; and he could also identify and explain the intricate details of all of its local flora and fauna.

BIBLIOGRAPHY

Bouchard, R. and D. Kennedy, *Indian History of the Lower Similkameen River-Palmer Lake Area: Report Prepared for the United States Army Corps of Engineers, Seattle District* (1984).

Robinson, H. and W. Wickwire. *Write It on Your Heart: The Epic World of an Okanagan Storyteller* (1990).

Robinson, H. and W. Wickwire. *Nature Power: In the Spirit of an Okanagan Storyteller* (1992).

Turner, N. J., D. Kennedy and R. Bouchard, *Ethnobotany of the Okanagan-Coville Indians of British Columbia and Washington* (1980).

UNPUBLISHED SOURCES

Primarily consists of recordings and transcripts of interviews and stories. Contact Wendy Wickwire, Department of History, P.O. Box 3045, University of Victoria, B.C., V8W 3P4, Canada.

Wendy Wickwire

ROOSEVELT, FRANKLIN DELANO. Born Hyde Park, New York, 30 January 1882; died Warm Springs, Georgia, 12 April 1945. Thirty-second President of the United States, conservationist.

FAMILY AND EDUCATION

Only son of James and Sara Delano Roosevelt. Tutored at home until age fourteen. Graduate of Groton School, 1900, B.A., Harvard, 1904. Attended Columbia University Law School, 1904–1907, did not graduate. Honorary LL.D., Pennsylvania Military College, 1920, Dartmouth, 1929, Harvard, 1929, Fordham, 1929, Oglethorpe University, 1932, Catholic University of America, 1933, Rutgers, 1933, Washington College, 1933, Yale, 1934, William and Mary College, 1934, American University, 1934, University of Notre Dame, 1935, University of Southern California, 1935, Rollins College, 1936, University of Georgia, 1938, University of North Carolina, 1938, Queens University (Canada), 1938, University of Puerto Rico, 1939, University of Pennsylvania, 1940, London University, 1941, (conferred 1943), McGill University, 1944, University of Louvain (Belgium), 1945; D.H.L., Hobart, 1929; D.C.L., Syracuse, 1930, Oxford, 1941; doctor of jurisprudence, Temple, 1936; honorary doctorate, University of Rio de Janeiro (Brazil), 1936, University of Uruguay, 1936, University of Buenos Aires (Argentina), 1936, University de Leon (Nicaragua), 1943; University of Brussels (Belgium), 1944. Married Anna Eleanor Roosevelt, 1905; six children (one died in infancy).

POSITIONS

Admitted to New York bar, 1907. Practiced with Carter, Ledyard, and Milburn, New York City, 1907–1910; member firm of Roosevelt and O'Connor, New York, 1924–1933. Elected to New York State Senate (Democrat), 1910, relected 1912, resigned March 1913; assistant secretary of the navy, 1913–1920. Democratic nominee for vice president, 1920; elected governor of New York, 1928, 1930, elected president of the United States, 1932, 1936, 1940, 1944.

CAREER

As a child, FDR watched and collected birds, became convinced of the importance of forests and forestry, and developed lifelong interest in the subject. As state senator, chaired Committee on Forest, Fish and Game. Attempted without success to push through legislation that would have limited tree cutting on privately owned land. As governor, he had unemployed men plant trees under aegis of the State Temporary Emergency Relief Administration. Also concerned with state water resources. Taking advantage of a nationwide urge to conserve and reform during his first two terms, FDR sought to tie resource development and management to his economic reform program. Turf battles between competing New Deal agencies got in the way of coordinated efforts, and FDR would not approve proposal of Interior Secretary Harold Ickes that his department be renamed the Department of Conservation with expanded responsibilities. There were successes, however. FDR initiated the Civilian Conservation Corps, which did put several million young men to work in national parks, forests, and refuges, together with other projects. The Tennessee Valley Authority effectively harnessed water resources in the south-central United States. The Army Corps of Engineers and Bureau of Reclamation embarked on tens of dam projects that limited potential flood damage, provided cheaper public power, and, in some cases, made possible recreational opportunities for the public. Unfortunately, some unanticipated long-term ecological damage resulted. Soil conservation was a major priority of the Roosevelt administration, both before and after the dust bowl crisis of the late 1930s. Shelter belts of trees in the upper Midwest (another proposal of FDR's) and efforts to reclaim exhausted farmland through reforms in agricultural methods enjoyed some success. Later generations deplored the New Deal emphasis upon a managed environment and demanded adherence to what conservationist Aldo Leopold termed a "land ethic." FDR was author of five books: *Whither Bound* (1926); *The Happy Warrior: Alfred E. Smith* (1928); *Government—Not Politics* (1932); *Looking Forward* (1933); *On Our Way* (1934).

MAJOR CONTRIBUTIONS

As governor of New York and particularly as president of the United States, FDR oversaw an essentially pragmatic approach to the preservation and effective management of the nation's natural resources. Often conservation objectives and initiatives had to take a backseat to political and economic considerations, such as efforts to relieve unemployment. On balance, however, the first eight years of FDR's presidency saw the most concerted effort of any American chief executive before or since to emphasize the wise use of natural resources to benefit the greatest number of citizens. FDR's appointment of, and strong support for, Harold L. Ickes, his only interior secretary, together with other dedicated conservationists in the administration, ensured that many valuable programs were vigorously maintained and pursued. Endorsed earlier conservation ethic of Gifford Pinchot—the greatest good for the greatest number.

BIBLIOGRAPHY

Burns, James M. *Roosevelt: The Lion and the Fox* (1956).

The Columbia River Basin Project. Washington, D.C. (1949).

Leuchtenburg, William E. *Franklin D. Roosevelt and the New Deal, 1932–1940* (1963).

Nixon, Edgard B., ed. *Franklin Roosevelt and Conservation, 1911–1945*. 2 vols. (1957).

Riesch-Owen, Anna L. *Conservation under FDR* (1983).

Salmond, John. *The Civilian Conservation Corps, 1933–1942: A New Deal Case Study* (1967).

Smith, Frank E. *The Politics of Conservation* (1966).

Swain, Donald C. "The Bureau of Reclamation and the New Deal, 1933–1940." *Pacific Northwest Quarterly* (July 1970).

Trani, Eugene. "Conflict or Compromise: Harold L. Ickes and Franklin D. Roosevelt." *North Dakota Quarterly* (Winter 1968).

Watkins, T. L. *Righteous Pilgrim: The Life and Times of Harold L. Ickes, 1874–1952* (1990).

Wessel, Thomas R. "Roosevelt and the Great Plains Shelterbelt." *Great Plains Journal* (Spring 1969).

Worster, Donald L. *Dust Bowl: The Southern Plains in the 1930s* (1979).

UNPUBLISHED SOURCES

See FDR Papers, Roosevelt Library, Hyde Park, N.Y.; Harold Ickes Papers, Library of Congress; Interior Department Records, National Archives and Records Administration (NARA), Washington, D.C.; Agriculture Department Records, NARA.

Keir B. Sterling

ROOSEVELT, ROBERT BARNWELL. Born New York City, 7 August 1829; died Sayville, Long Island, 14 June 1906. Author, lawyer, conservationist, politician.

FAMILY AND EDUCATION

Son of Cornelius Van Schaack and Margaret Barnhill Roosevelt. Educated at home by private tutors. Married Elizabeth Thorn Ellis, 26 November 1850 (died 1887); one son and two daughters; and Marion O'Shea (Fortescue), 14 August 1888; two sons and one daughter.

POSITIONS

Practiced law from 1850 to about 1872. Had one term in Congress (1871–1873). Served as a New York City alderman (1882) and as U.S. minister to the Netherlands (1888–1889). From 1868 to 1888 was member of New York State Fisheries Commission (serving 1879–1888 as president).

CAREER

As young man Roosevelt evinced a love of fishing and hunting, and his early books—especially *Game Fish of the Northern States of America . . .* (1861), *Superior Fishing . . .* (1865), and *Game Birds of the Coasts of the Northern States of America* (1866)—reveal how much he was absorbed in their activities. His love of sport also drew him into early conservation movement, as he saw and deplored damage being done to wildlife by overfishing and overhunting, as well as by industrial development. He became active in, and between 1877 and 1906 served as president of, the New York Association for the Protection of Game, a pioneer conservation organization in New York. Moreover, Roosevelt also believed that pisciculture was an ''absolute necessity'' for ''the preservation of the fish of the county from total destruction,'' sponsored various projects in the field of fish culture (such as the establishment of fish hatcheries) during his years on the New York State Fisheries Commission. His later writings included *Fish Hatching and Fish Culture* (coauthor, 1879), two books of satire, *Five Acres Too Much* (1869) and *Progressive Petticoats*, and a novel, *Love and Luck* (1886).

MAJOR CONTRIBUTIONS

Helped to introduce pisciculture in New York State. Won public support for conservation through his books and other writings on fishing and hunting and through his activities on the Fisheries Commission. The New York Association for the Protection of Game, which he headed for many years, lobbied regularly in New York State legislature for laws to conserve wildlife.

BIBLIOGRAPHY

Dictionary of American Biography. Vol. 8 (1935).

Harmond, Richard. ''Robert Barnwell Roosevelt and the Early Conservation Movement.'' *Theodore Roosevelt Association Journal* 14 (Summer 1988).

New York Times, 15 June 1906 (obit.).

Richard Harmond

ROOSEVELT, THEODORE. Born New York City, 27 October 1858; died Oyster Bay, New York, 6 January 1919. Twenty-sixth president of the United States, writer, naturalist, historian, rancher, soldier, explorer.

FAMILY AND EDUCATION

Son of Theodore Roosevelt, Sr., and Martha Bulloch. A.B., Harvard, 1880. Honorary LL.D.'s from Northwestern, 1893, Columbia, 1899, Hope College, 1901, Yale, 1901, Harvard, 1902, University of Chicago, 1903, University of California, 1903, University of Pennsylvania, 1905, Clark, 1905, George Washington University, 1909, Cambridge University, 1910. Honorary D.C.L., Oxford, 1910; Honorary Ph.D., University of Berlin, 1910. Married Alice Hathaway Lee, 1880. She died 1884 after giving birth to their daughter. Married Edith Kermit Carow, 1886; five children.

POSITIONS

Elected to New York State Assembly 1881, 1882, 1883; U.S. Civil Service Commission, 1889–1895; president of Board of Police Commissioners, New York City, 1895–1897; assistant secretary of the navy, 1897–1898; lieutenant colonel, then colonel, First U.S. Volunteer Cavalry Regiment, 1898; governor of New York State, 1898–1900; vice president of United States, 1901; twenty-sixth president of the United States, 1901–1909.

CAREER

Theodore Roosevelt (TR) may be described as an American "Renaissance man." He followed many and varied interests in living what he called the "strenuous life." As a boy, he became a taxidermist and avid collector of birds and other specimens. His father was a founder of the American Museum of Natural History, beginning TR's lifelong association with that institution. He entered Harvard with the intention of becoming a professional field naturalist and published two papers on birds while in college, but soon after graduation he entered politics. In the 1880s he donated his boyhood collection to the Smithsonian Institution (622 bird skins) and the American Museum of Natural History (about 125 bird skins). As a rancher in the Dakota Territory in the 1880s and through frequent hunting trips over the years, Roosevelt became a noted authority on North American mammals, about which he wrote extensively. In 1888, with other sportsmen, he founded the Boone and Crockett Club (B & C) serving as first president, 1888–1894. The club promoted "manly sport with the rifle" and also the preservation and study of wildlife. With TR at the helm, the B & C was instrumental in the passage of legislation protecting Yellowstone and establishing the forest reserve system, as well as in founding the New York Zoological Society. Roosevelt and George Bird Grinnell edited and contributed to three books published by the B & C: *American Big-Game* (1893), *Hunting in Many Lands* (1895), and *Trail and Campfire* (1897). As president, TR carried out extensive and multifaceted environmental program. Established first federal bird reservations (fifty-one), reclamation-irrigation projects (twenty-four) under the Newlands Act of 1902), and game preserves (four, including the National Bison Range in Montana). Greatly expanded forest reserves, adding over 150 million acres and establishing 150 national forests. He withdrew many mineral, oil, phosphate, coal, and waterpower site lands from private exploitation. Set up five national parks and in 1905 founded the present U.S. Forest Service. Under the Antiquities or National Monuments Act (1906), Roosevelt established the first eighteen national monuments, which included Arizona's Petrified Forest, the Grand Canyon, and Mount Olympus. In all, Roosevelt was responsible for placing under public protection approximately 230 million acres of the United States. TR appointed Public Lands Commission in 1903, which led to new policies on use of the open range and federal lands; the Inland Waterways Commission in 1907, which studied river systems, waterpower, flood control, and reclamation; and the Country Life Commission in 1908 to study rural conditions. TR called first national conference of governors, which met May 1908 to consider the problems of conservation. This led to the creation of thirty-eight state conservation commissions, the annual governors' conferences, and the appointment (June 1908) of the National Conservation Commission to prepare first inventory of U.S. natural resources. Next came the North American Conservation Congress, which met in February 1909. In 1909–1910 Roosevelt led an expedition to Africa, sponsored by Smithsonian Institution, collecting some 5,013 mammals, 4,453 birds, 2,322 reptiles and amphibians as well as thousands of fish, insects, plants, and other specimens. Many of the species and subspecies were new to science. In 1914 Roosevelt and Candido Rondon led expedition, sponsored by the American Museum of National History and the Brazilian government, to explore the River of Doubt, subsequently renamed by Brazil the Rio Roosevelt. TR was a friend of, influenced, and was influenced by, most of the noted naturalists and conservationists of his time, including C. Hart Merriam, Henry Fairfield Osborn, John Burroughs, Gifford Pinchot, Frank M. Chapman, John Muir, and William Beebe.

Roosevelt's books on outdoor life and nature, in addition to the Boone and Crockett books, are *Hunting Trips of a Ranchman* (1885), *Ranch Life and the Hunting Trail* (1888), *The Wilderness Hunter* (1893), *Some American Game* (with T. S. Van Dyke, D. G. Elliot, and A. J. Stone, 1897) *The Deer Family* (1902), *Outdoor Pastimes of an American Hunter* (1905), *Good Hunting* (1907), *African Game Trails* (1910), *Through the Brazilian Wilderness* (1914), with Edmund Heller, *Life-Histories of African Game Animals*, 2 vols. (1914), and *A Book-Lover's Holidays in the Open* (1916). TR wrote numerous introductions to books, articles, book reviews, and papers on outdoor life, nature, and conservation, some collected in *Papers on Natural History*, in vol. 6 of the *Memorial Edition* and vol. 5 of the *National Edition* of *The Works of Theodore Roosevelt* (1923–1926).

MAJOR CONTRIBUTIONS

Theodore Roosevelt saw government as the ''steward of the people'' with responsibility to protect, husband, and develop rationally the natural resources of the nation. He was both a ''preservationist'' and a ''use-conservationist,'' and his policies as president come under both categories. His vision was broad and not confined to any one issue or aspect of environmentalism. ''The conservation of natural resources is the fundamental problem. Unless we solve that problem it will avail us little to solve all others,'' TR said in 1907. ''The 'greatest good for the greatest number' applies to the number within the womb of time, compared to which those now alive form but an insignificant fraction. Our duty to the whole, including the unborn generations, bids us restrain an unprincipled present-day minority from wasting the heritage of these unborn generations,'' TR wrote in 1916. Charles Van Hise, conservationist, geologist, president of the University of Wisconsin, wrote in 1911 that what Theodore Roosevelt had done as president to advance conservation and ''to bring it into the foreground of the consciousness of the people will place him not only as one of the greatest statesmen of this nation but one of the greatest statesmen of any nation of any time.''

BIBLIOGRAPHY

Cutright, Paul Russell. *Theodore Roosevelt: The Making of a Conservationist.* Urbana: University of Illinois Press (1985).

Cutright, Paul Russell. *Theodore Roosevelt the Naturalist.* New York: Harcourt Brace, (1947).

Gable, John A. ''President Theodore Roosevelt's Record on Conservation.'' *Theodore Roosevelt Association Journal* 10. 3 (Fall 1984):2–11.

Payne, Daniel G. *Voices in the Wilderness: American Nature Writing and Environmental Politics* (1996).

Pinchot, Gifford. *Breaking New Ground.* New York: Harcourt, Brace (1947).

UNPUBLISHED SOURCES

The Theodore Roosevelt Papers are at the Library of Congress, and many other important papers and materials are part of the Theodore Roosevelt Collection at the Houghton Library, Harvard University.

John Allen Gable

ROSS, BERNARD ROGAN. Born Londonderry, Northern Ireland, 25 September 1827; died Toronto, Ontario, 21 June 1874. Fur trader, naturalist.

FAMILY AND EDUCATION

Son of James Ross and Elizabeth Rogan. Educated at Foyle College, Londonderry. In 1860 he formally married Christina or Christiana Ross, daughter of Chief Trader Donald Ross (no relation) of the Hudson's Bay Company (HBC), Norway House. They had three children. Ross studied surveying under John Rae of the HBC and Capt. W. J. S. Pullen of the Royal Navy, probably in 1849–1850.

POSITIONS

Entered the service of the Hudson's Bay Company at the suggestion of George Simpson, who had met him at the Londonderry home of his uncle, Frank Rogan. Taught school at Cornwall during his first winter in Canada at age sixteen. Became an apprentice clerk at Norway House in 1843, at Fort Frances in 1845–1846, and at York Factory in 1846–1847. From 1847 through 1862 he was employed in the Mackenzie River district, first as clerk in charge of Frances Lake, 1847–1848, then at Fort Simpson 1848–1850 and 1851–1852. He was at Fort Norman, 1852–1854, and then was promoted to chief trader at Fort aux Liards, 1855–1856, Fort Resolution, 1856–1858, and Fort Simpson, 1858–1862. After a two-year furlough to Dublin, he returned as chief trader at Mingan in the Montreal Department, 1864–1865, Ruperts House, 1865–1869, and Fort Alexander on Lake Winnipeg, 1871–1872. He retired in 1872.

CAREER

Joseph Henry of the Smithsonian Institution wrote to enlist Ross's help in obtaining a mountain goat specimen in 1858, and by the next year Ross was corresponding regularly with Spencer Fullerton Baird, Henry's assistant secretary. Ross submitted at least 1,200 specimens to the Smithsonian, but also sent 124 specimens to Andrew Murray at Edinburgh and 476 to John Edward Gray, Keeper of Zoology, at the British Museum (Natural History). Each specimen was accompanied by Ross's determination of the scientific name of the bird (usually correct) and extensive notes as to such things as the habitat, range, and nesting of the bird.

Ross's interest was further stimulated by the visit of the naturalist Robert Kennicott, who stayed with the Rosses during the winter of 1859–1860 and left a butterfly net and pins with Christina Ross. Christina Ross sent large numbers of butterfly specimens to W. H. Edwards, then of Newburgh, New York. A specimen Christina collected near present Fort Smith on the Alberta–Northwest Territories boundary in 1862 was named a new species, the Christina sulphur, *Colias christina* (now considered a subspecies, *Colias alexandra christina*), by Edwards in her honor.

In 1861, John Cassin, in assigning a Latin name to the small goose that had first been described by Samuel Hearne, gave it the name of *Anser Rossii* (now *Chen rossii*), based on a specimen collected by Ross at Fort Resolution, Great Slave Lake.

Ross became a corresponding member of the Natural History Society of Montreal and of the Academy of Natural Sciences of Philadelphia (1861), of the New York Historical Society (1863), a foundation fellow of the Anthropological Society (1863), and a fellow of the Royal Geographical Society (1864).

MAJOR CONTRIBUTIONS

Ross and Roderick Ross MacFarlane were the two Hudson's Bay Company factors who contributed the greatest number of specimens of birds and mammals from a part of the world known previously only from the collections made in the 1820s by John Richardson. Ross published four scientific articles, the most important the manuscript he transmitted to *Natural History Review* through John Richardson.

BIBLIOGRAPHY

Bowsfield, H. "Bernard Rogan Ross." *Dictionary of Canadian Biography*. Vol. 10, 1871–1880. Ed. Marc Le Terreur. Toronto: University of Toronto Press (1972).

Lindsay, D. "The Hudson's Bay Company—Smithsonian Connection and Fur Trade Intellectual Life: Bernard Rogan Ross, a Case Study." *Le Castor Fait Tout, Selected Papers of the Fifth North American Fur Trade Conference*. Montreal: Lake St. Louis Historical Society (1987).

Ross, B. R. "List of the Mammals, Birds, and Eggs, Observed in the McKenzie's River District, with Notices." *Natural History Review* (2d ser.) 2 (1862), and *Canadian Naturalist and Geologist* 7 (1862).

C. Stuart Houston

ROUSSEAU, JACQUES. Born Saint-Lambert Chambly, Quebec, 5 October 1905; died Lac Ouareau, Quebec, 4 August 1970. Botanist, ethnobiologist, ethnohistorian, scholar.

FAMILY AND EDUCATION

The third of fourteen children of Gabrielle (Fafard) Rousseau and Lacasse Rousseau, a pioneer electrical engineer. Educated at Saint-Lambert, Montreal, Sainte-Thérèse, and then La Pocatière. Health problems led to college studies with private tutors. Classical B.A. (Université de Montréal, 1926); License ès Sciences (Université de Montréal, 1928); D.Sc. (Ph.D.), Université de Montréal, 1934; Honorary D.Sc., Laval (1966). Married Madeleine Aquin, 1934; three children.

POSITIONS

Demonstrator, Botany Lab, Université de Montréal, 1926–1928; lecturer, head lecturer and head of practical work, Institute of Botany, Université de Montréal, 1928–1935; professor of natural science, Collège Sainte-Marie, 1929–1931; professor, Institut de Botanie, Université de Montréal, 1935–1944; founder and secretary-general of L'Association canadienne-française pour l'avancement des sciences, 1930–1947; fellow, Royal Society of Canada, 1942; advising botanist, United Nations, 1946; botanist to Canadian law courts; assistant director, Botanical Gardens, Montreal, 1938–1944; director, 1944–1957; director, Canadian Museum of Human History, 1957–1959; governor, Arctic Institute of North America, 1954–1957; executive, 1957; director, National Museum of Man, 1957–1959; governor, Arctic Institute of North America, 1954–1957; executive, Marie-Victorin Foundation; member and sometimes officer in over seventy scientific organizations. Numerous honors, including Order of Canada, 1969.

CAREER

Captivated by the 1923 lectures of Marie-Victorin at the Université de Montréal Botanical Institute. Studied full-time, 1926–1928. Doctoral work included studies and teaching at Cornell University (1931, 1933), University of New Mexico (summer, 1932), University of Vermont, and Harvard. Thesis on the genus *Astragalus* in Quebec. Appointed *professeur agrégé* (Université de Montréal, 1935–1944). Wrote several chapters for Marie-Victorin's *Flore laurentienne.* Began public science education in the 1930s through lectures, radio, and organizational work. Assistant director, Montreal Botanical Gardens (1938–1944), director (1944–1956). Taught at Laval (1949, 1952, 1957); Port-au-Prince, Haiti (1944); Institut Français de l'Amérique Latine, Mexico City (1945); McGill Geography Summer School (1952, 1953); and the Sorbonne, Paris (1950). Director of the Human History Branch, National Museum of Man, Ottawa (November 1956–1959). Professeur invité at the Sorbonne and the Ecole Pratique des Hautes Etudes (1959–1962). Returned to Université Laval, Quebec City, where he worked until his death as professor of ethnobiology at the newly established Centre d'Études nordiques.

Prior to 1945 his fieldwork focused on the lower St. Lawrence. Made several extended expeditions on foot across the Gaspé Peninsula and Anticosti. The summer of 1944 began his long association with Lake Mistassini. After 1945 northern Quebec became the focus of his fieldwork. In 1947 canoed down the George River to Ungava Bay and down lesser-known rivers, such as the Kogaluk and Arnaud (1948). First heart attack in the field at the Otish Mountains (1949), returning two years later to work in Ungava Bay and at Chubb Crater. Fieldwork described 130 botanical species. Eight were named in his honor. Last fieldwork was with archeologists at Payne Lake (1965). A prolific author, whose eighty-five-page résumé lists some 550 scholarly publications and over 100 conference papers. He died while completing a coedition with Guy Béthune of Pehr Kalm's 1749 *Voyage*, finished by Pierre Morissette (1977). He took quietly passionate public stands against war, for ''Amérindiens,'' for students, and placed the case for the natural world before urban peoples. Humanist, scientist, and teacher.

MAJOR CONTRIBUTIONS

Quebec's most influential twentieth-century scientist. Promoted interdisciplinary scholarship through example. His merging of ethnology and botany has yet to be fully acknowledged. He oversaw both the professionalization and internationalization of science in Quebec. He focused Quebec's vision northward, just prior to large-scale hydroelectric development. A major cultural figure whose loss was deeply felt.

BIBLIOGRAPHY

Audet, Louis-Philippe. ''Jacques Rousseau.'' *Les cahiers des dix* 35 (1970).

Canadian Who's Who. Vol. 11 (1967–1969).

Caron, Fabien. ''Jacques Rousseau (1905–1970).'' *Arctic* 24 (June 1971).

Chartrand, Luc, Raymond Duchesne, and Yves Gingras. *Histoire des sciences au Québec* (1987).

Gingras, Yves. *Pour avancement des sciences: histoire de l'ACFAS, 1923–1993* (1994).

Hamelin, Louis-Edmond. ''Notice nécrologique, Jacques Rousseau (1905–1970).'' *Cahiers de géographie de Québec* 14. 32 (September 1970).

Laverdière, Camille. ''Jacques Rousseau (1906–1970) N'est Plus . . .'' *Revue de géographie de Montréal* 25. 1 (1971).

Pomerleau, René. ''Jacques Rousseau, 1905–1970.'' *Naturaliste canadien* 98, 3 (May–June 1971).

Raymond, Marcel. ''Jacques Rousseau (1905–1970).'' *Proceedings and Transactions of the Royal Society of Canada* (4th ser.) 9 (1971).

Tremblay, Marc-Adélard, and Josée Thivierge. ''La nature et la portée de l'oeuvre amérindienne de Jacques Rousseau.'' *Anthropologie et Sociétés* 10, 2 (1986).

UNPUBLISHED SOURCES

Field journals and academic papers, some 3.3 meters wide on a shelf, are at the Archives de l'université Laval, Sainte-Foy, Quebec. Early correspondence is in the Marcelle Gauvreau (secretary to Marie-Victorin) Collection, Université du Québec à Montréal. Additional material is at the Archives nationales du Québec, Sainte-Foy, Quebec, the National Archives of Canada, Ottawa, and in the collections of the numerous institutions and organizations in which he served.

Lorne Hammond

ROWAN, WILLIAM ROBERT. Born Basel, Switzerland, 29 July 1891; died Edmonton, Alberta, Canada, 30 June 1957. Biologist, ornithologist, conservationist, wildlife artist, popularizer of science.

FAMILY AND EDUCATION

Son of Gerdine Atalia (Jacobsen) Rowan and William Robert Rowan. Family lived in Switzerland until 1898, then France, and moved to England in 1901. Attended Bedford public school, 1901–1908. M.Sc., 1919, D.Sc., 1929 University College London. Married Reta Bush in 1919; six children.

POSITIONS

Lecturer in zoology at University of Manitoba in 1919. Founded Department of Zoology at University of Alberta, which he headed from 1920 until 1956.

Associate professor, 1921. Elected president, University Science Association, 1928. Appointed full professor in 1931.

CAREER

Worked on a cattle ranch in Alberta, 1908–1911. Enrolled in honors zoology at University College, London, England, 1911–1917. Served in the London Scottish Regiment for one year. Obtained a B.Sc. in 1917. Member of the British Ornithologists' Union. Taught biology at two separate English secondary schools. Returned to Manitoba in 1919 as lecturer of zoology at University of Manitoba.

Charter member of Natural History Society of Manitoba. Appointed by H. M. Tory to chair the new Zoology Department at University of Alberta. Close friends with Percy A. Taverner, J. B. Collip, Francis Harper, Ray Salt, and Julian Huxley. Member of Alberta Fish and Game Association. Began historic experiments on bird migration in September 1924. His two papers, ''Relation of Light to Bird Migration and Developmental Changes,'' *Nature* (1925) and ''On Photoperiodism, Reproductive Periodicity, and the Annual Migrations of Birds and Certain Fishes,'' *Proceedings of the Boston Society of Natural History* (1926), met with favorable reviews among international scientific community. Elected member of American Ornithologists' Union.

Thesis, ''Experiments on Bird Migration I: Manipulation of the Reproductive Cycle: Seasonal Histological Changes in the Gonads,'' published in *Proceedings of the Boston Society of Natural History* (1929). Attended Canadian Matamek Conference on conservation in 1931. Received the Royal Society of Canada's Flavell Medal in 1946. Publications include *The Riddle of Migration* (1931), a book that appealed to scientific as well as popular readers, and articles in *British Birds*, *Nature*, *Canadian Field-Naturalist*, *Science*, *Journal of Wildlife Management*, and *Journal of Mammalogy*. Mentored many students who later became biologists and conservationists. Enlisted the help of farmers, trappers, and settlers in collection of field data. Supported other naturalists, such as Elsie Cassels, Frank Farley, Tom Randall, and A. D. Henderson. A Renaissance man known for his kindness, excellent teaching, creative research, diversity of interests and talents, ability to inspire and challenge others, and love of nature and outdoors.

MAJOR CONTRIBUTIONS

Conducted fieldwork in many habitats and studied a diverse selection of wildlife. A pioneer in research on bird migration. Studied cyclic population fluctuations in birds and mammals. Ardent conservationist who strove to protect the threatened wood buffalo herds in the 1920s. Used national radio broadcasting as a means to popularize science and conservation. Instrumental in changing ornithology into a complete science composed of fieldwork, laboratory work, and experimentation.

BIBLIOGRAPHY
Ainley, Marianne G. *Restless Energy: A Biography of William Rowan 1891–1957* (1993).
Ainley, Marianne G. "William Rowan: Canada's First Avian Biologist." *Picoides* 1 (1987).
Ainley, Marianne G. "Rowan vs Tory—Conflicting Views of Scientific Research in Canada, 1920–1935." *Scientia Canadensis* 12, 1 (1988).
Hohn, Otto. "Professor William Rowan: Ornithologist and Artist." *Alberta Naturalist* 14, 4 (1984).
McNicholl, Martin K. "Rowan, William." *The Canadian Encyclopedia* (1988).
Salt, Ray W. "In Memoriam: William Rowan." *Auk* 75, 4 (1958).
Salter, F. M. "William Rowan, 1891–1957." *Proceedings and Transactions of the Royal Society of Canada* (3d ser.) 52 (1958).

E. Tina Crossfield

RUDOLPH, EMANUEL DAVID. Born Brooklyn, New York, 9 September 1927; died Columbus, Ohio, 22 June 1992. Polar lichenologist, historian of botany and biology, educator.

FAMILY AND EDUCATION

Youngest of three children of Jack, inventory manager in cheese-making company, and Sarah (Wolfe) Rudolph. Graduate of Erasmus Hall High School, Brooklyn, 1945. B.A., New York University, 1950; Ph.D., Washington University of St. Louis, 1955. Married Ann Herrick Waterman, 1962 (died); no children.

POSITIONS

Instructor to assistant professor of biology, Wellesley College, 1955–1961; assistant professor to professor of botany, 1961–1989, and professor emeritus, 1990–1992, Ohio State University. Research associate, Institute of Polar Studies, 1961–1992, director of the institute, 1969–1973, director of the Environmental Biology Program, 1972–1978, chairman of the Department of Botany, 1978–1987, Ohio State University.

CAREER

Influenced in high school by Phyllis Busch, noted authority on teaching basic science. At Wellesley, taught courses in plant sciences. Held National Science Foundation Postdoctoral Fellowship at University of Wisconsin, 1959, where he studied sclerotium formation in the mold *Aspergillus* and examined arctic lichens in the family *Teloschistaceae*. At Ohio State, taught courses in introductory biology, botany, and history of biology, the latter course for twenty-two years. Participated in five research expeditions to Antarctica during austral summers. Field seasons of 1961–1962, 1962–1963, and 1963–1964 spent at Hallett Station, Ross Sea Sector. There he mapped cover vegetation, studied seed germination of flowering plants, and gathered microclimatic data on growth rates, succession, moisture content, and substrate types of lichens. Participated in Byrd Land Botanical Survey in 1967–1968 field season, collecting cryptogamic plants on a seven-mile coastal strip, previously unexplored for plants. Fifth Antarctic trip, 1968–1969, at McMurdo Station, focused on ecology and distribution of lichens, algae, and fungi. Continued fieldwork on lichens in arctic coastal tudra of Prudhoe Bay and Point Barrow, Alaska. In 1972 and 1975, with two graduate students, studied productivity of lichens to determine effects on vegetation that could result from construction of crude oil pipeline from Point Barrow to Valdez, Alaska. From 1972 to 1981 investigated stress conditions on lichens in Ohio strip mines. Supervised fourteen graduate students (four Ph.D., ten M.S.) in botany and environmental biology. With his wife, amassed personal scientific and technical library of 53,000 volumes, a major resource in his own research, which was willed to various library units at Ohio State on his death. Recipient of U.S. Antarctic Service Medal, 1969. Rudolph Glacier in Victoria Land, Antarctica, commemorates him. A lichen species, *Catillaria rudolphia* Dodge (1955), and a lichen genus, *Edrudia* (1980), were named in his honor. Promoted the cause of libraries in Ohio by his service on committees and active participation in their development.

MAJOR CONTRIBUTIONS

The first botanist to conduct diverse experiments on the total biology of lichens in both polar regions, he also produced the first detailed map of Hallett Station (1963). His original studies in the history of science defined the role of nineteenth-century women in the development of popular and professional botany and elucidated the influence of children's books in the popularization of botany.

BIBLIOGRAPHY

Kiss, John S. *Bulletin of the Torrey Botanical Club* 120 (1993) (obit.).

Stuckey, Ronald L., and Barney L. Lipscomb. *Sida* 15, (1992) (obit.).

Stuckey, Ronald L. *The Ohio Journal of Science* 93 (1993) (obit.).

Stuckey, Ronald L. "Emanuel David Rudolph (1927–1992): Polar Lichenologist and Historian of Botany." *The Bryologist* 97 (1994).

Stuckey, Ronald L. "Emanuel David Rudolph (1927–1992): Book Collector and Library Friend." *Michigan Botanist* 34 (1995).

White, Sidney E. "In Memoriam." *Arctic and Alpine Research* 25 (1993).

UNPUBLISHED SOURCES

Principal collections of papers and other materials are at the Herbarium and in the Rare Books and Manuscripts Library, Ohio State University.

William R. Burk

RUTHVEN, ALEXANDER G(RANT). Born Hull, Iowa, 1 April 1882; died Ann Arbor, Michigan, 18 January 1971. University president, administrator, zoologist.

FAMILY AND EDUCATION

Son of John, a native of Scotland, and Katherine Rombough Ruthven, daughter of pioneer family. Received B.S. from Morningside College in 1903 and Ph.D. from University of Michigan in 1906. Married Florence Hagle, September 1907; three children.

POSITIONS

Instructor in zoology and curator of University of Michigan's Museum of Zoology, 1906–1911; professor and director of the museum in 1911 and eleven years later director of all the university's museums, 1922–1936; dean and chairman of Department of Zoology and director of zoology laboratory, 1927–1929; administrator, 1928–1929, and then president of University of Michigan, 1929–1951, emeritus after 1951. Chief field naturalist, Michigan Geological and Biological Survey, 1908–1912. Directed various scientific expeditions in North, South and Central America during the 1920s and 1930s.

CAREER

Spent most of career associated with the University of Michigan. Began as instructor in zoology and then curator of the university's Museum of Zoology. Became assistant professor and director of the museum in 1911 and later director of all university's museums.

Led eighteen zoological expeditions sponsored by the university both in the United States and in South and Central America. Published hundreds of scholarly studies dealing with the ichthyosaurus, a prehistoric marine reptile with a fishlike body and dolphinlike head. Before becoming an administrator, he won reputation in fields of ichthyology and herpetology. Was president of the University of Michigan from 1929 until his retirement in 1951. Had also been chief naturalist of the Michigan Geological and Biological Survey since 1908.

Published *Herpetology of Michigan* (1912; 2d ed. 1928); *Laboratory Directions in the Principles of Animal Biology* (1919); *Principles of Animal Biology* (1924); the annual reports of the museum of zoology from 1908 to 1928; and about 120 scientific papers and bulletins on his special scientific studies.

Handled the intricate problems of the university with a skill that drew admiration of all acquainted with difficulty of his position. Practiced thrift as president of Michigan, abolishing post of dean of administration, claiming it was waste of money. An ardent prohibitionist, he helped authorities dry up the Michigan campus and yet maintained good student relations. Declared in 1939 that football should be abolished unless it was given back to the students.

MAJOR CONTRIBUTIONS

Published hundreds of scholarly studies dealing with the ichthyosaurus. Led eighteen zoological expeditions, resulting in his best-known findings from an expedition to the Sierra Nevada de Santa Marta, in Colombia, where he assembled a collection of mollusks, spiders, fish, birds, insects, and mammals. Made profound study of animal life in Michigan, particularly of amphibians and reptiles. As curator and director of the university museum, developed zoological collection inadequately housed and overcrowded into a well-arranged and scientifically adequate university museum, unique as working laboratory for the study of the natural sciences.

BIBLIOGRAPHY

National Cyclopaedia of American Biography (1930).

New York Times, 20 January 1971 (obit.).

Ruthven, A. G. *Naturalist in Two Worlds: Random Recollections of a University President* (1963).

Who Was Who in America. Vols. 5, 7 (1973, 1981).

Van de Water, Peter E. *Alexander Grant Ruthven of Michigan, Biography of a University President* (1977).

Linda L. Reesman

S

SARGENT, CHARLES SPRAGUE. Born Boston, 24 April 1841; died Jamaica Plains, Massachusetts, 22 March 1927. Botanist, arborculturist.

FAMILY AND EDUCATION

Son of Ignatius Sargent, prominent Boston merchant engaged in trade with East Indies, and Henrietta Gray Sarge. Graduated from Harvard University in 1862. LL.D. from Harvard, 1901. Married Mary Allen Robeson, 1873; two sons and three daughters.

POSITIONS

Director, Harvard Botanic Garden, 1873–1879; director, Arnold Arboretum, 1872–1927; professor of horticulture, Harvard University, 1872–1873; professor of arborculture, Harvard University, 1879–1927.

CAREER

From 1863 to 1865 Sargent served in the Union forces during the Civil War and obtained the rank of major. His interest in botany was stimulated by association with Asa Gray, and by 1873 he had begun a professional relationship with Harvard University that would last for the balance of his life. As head of the newly established Arnold Arboretum, he built its library and herbarium into the nation's major center for the study of woody plants. In 1880 he became a special agent for the tenth census charged with preparing a survey of the forest resources of the United States. After coordinating the endeavors of numerous field botanists, he published in 1884 *Report on the Forests of North America* (vol. 9 *10th Census*). From 1882 to 1883 served as member of Northern Pacific Transcontinental Survey and sought support for establishment of western parks and reserves. From 1881 to 1885 he collected specimens of American trees for a major display at the American Museum of Natural History. Edited *Garden and Forest*, 1887–1897.

The author of numerous articles, Sargent established his reputation with the editing and publishing of the fourteen-volume *Silva of North America* (1891–1902), which replaced Michaux's *American Sylva*. He also authored *A Catalogue of the Forest Trees of North America* (1880), *Report on the Forests of North America* (1884), *The Woods of the United States* (1885), *Forest Flora of Japan* (1894), *A Guide to the Arnold Arboretum* (1911) and edited E. H. Wilson's monumental study of the trees of China, *Plantae Wilsonianae*, 3 vols. (1913–1917).

MAJOR CONTRIBUTIONS

Led to greater professionalism in the study and practice of arborculture. Established the Arnold Arboretum as the major center for the study of trees and shrubs in the United States. Coordinated the activities of numerous field botanists and plant collectors and alerted the nation to the need for more prudent forest management policies.

BIBLIOGRAPHY

Boston Transcript, 23 March 1927 (obit.).

Dictionary of American Biography (1935).

Sargent, Charles Sprague. "The First Fifty Years of the Arnold Arboretum." *Journal of the Arnold Arboretum* 3 (1922).

Sutton, S. B. *Charles Sprague Sargent and the Arnold Arboretum* (1970).

Trelease, William. "A Biographical Memoir of Charles Sprague Sargent." *National Academy of Sciences Biographical Memoirs* 12 (1929).

UNPUBLISHED SOURCES

Letters and other manuscript materials are at Arnold Arboretum, Harvard University.

Phillip D. Thomas

SARRAZIN, MICHEL. Born Nuits-sous-Beaune, Burgundy, France, 5 September 1659; died Quebec City, Quebec, 8 September 1734. Physician, surgeon, botanist, naturalist.

FAMILY AND EDUCATION

Son of a bailiff, Claude Sarrazin, and Madeleine de Bonnefoy. He received his first medical diploma from the École de Médicine in Paris in 1685. Received his doctorate in medicine from the University of Rheims in 1697. In 1712, at age fifty-three, he married Marie-Anne Hazeur, with whom he had seven children. As the result of financial reverses, he died in abject poverty, survived by his wife, two sons, and two daughters.

POSITIONS

Made surgeon major of the colonial regular troops in Quebec, 1686, and promoted physician in chief of the king to the hospitals of New France, 1689; made the king's doctor of botany in New France, 1699. Member of the French Royal Academy of Science, 1669–1734. Appointed to the Quebec Superior Council, 1707.

CAREER

Sarrazin is known as the father of French Canadian science. He traveled to Canada in 1685. Served as surgeon major to the French troops in Quebec, 1691–1692. Dissatisfied with his medical education, he returned to Paris in 1694 to pursue doctoral studies in medicine. During study for his doctorate in Paris, he developed a deep and lasting interest in medicinal plants, which he grew, gathered, and regularly sent to the Jardin Royale in Paris. He sent samples of novelties—blueberries, the pitcher plant, and maple sugar—and often included native medical lore in his observations. Sarrazin also dispatched specimens of rocks and minerals to Paris but played only a secondary role in mineralogy. He contributed to the *Transactions of the Academy of Science*, giving original dissertations on the general and minute anatomy and physiology of animals indigenous to Canada, such as the beaver and muskrat.

MAJOR CONTRIBUTIONS

He contributed greatly to the early development of medicine in Canada, serving with distinction as physician to the king. He was central in the development and understanding of Canadian botany and natural history. His physiological reports and observations on the flora and fauna of Canada show the observation and approach of modern biology. Sarrazin was likely the first to grow winter wheat successfully on Canadian soil and is commonly given credit for the industrialization of maple sugar.

BIBLIOGRAPHY

Abbott, Maude E. "An Early Canadian Biologist—Michel Sarrazin (1659–1735): His Life and Times." *Canadian Medical Association Journal* 19 (1928).

Leblond, Sylvio. "Le Testament de Michel Sarrazin." *La Vie Medicale au Canada français* 3 (May 1974).

Leblond, Sylvio. "Michel Sarrazin: Un Document Inédit." *Laval Médical* (March 1961).

Rousseau, Jaques. "Michel Sarrazin." *Dictionary of Canadian Biography*. Vol. 2 (1969).

Vallée, Arthur. *Un biologiste Canadien, Michel Sarrazin (1659–1735). Sa vie, ses travaux, et son temps* (1927).

Sasha Mullally

SAUER, CARL ORTWIN. Born Warrenton, Missouri, 24 December 1889; died Berkeley, California, 18 July 1975. Geographer.

FAMILY AND EDUCATION

Son of William Albert Sauer, a college professor, and Rosetta J. (Vosholl) Sauer. Preparatory study in Germany, 1898–1908. Received A.B. from Central Wesleyan College (now defunct) in 1908. Did graduate work at Northwestern, then transferred to University of Chicago, from which he received Ph.D. in 1915. Honorary degrees included LL.D. from Syracuse University (1958) and University of California at Berkeley (1960). Also awarded Ruris Uteriusque Doctorem from University of Glasgow (1965). Married Loren Schowengerdt, 1913; one son, one daughter.

POSITIONS

Map editor at Rand McNally Company, Chicago, 1912–1913. Instructor at State Normal School, Salem, Massachusetts, 1913–1914; instructor of geography (1915), assistant professor (1918), associate professor (1920), professor (1922) at University of Michigan; and professor of geography and founding chairman of department at University of California, 1923–1957. Also assistant geologist, Illinois Geological Survey, 1910–1912; agent in agricultural economics, U.S. Department of Agriculture, 1919–1920. A founder of Michigan Land Economic Survey, 1922. Consultant to Social Conservation Service; member, Committee on Latin-American Studies of the American Council of Learned Societies (1940). Also, member of advisory board of Guggenheim Memorial Foundation and cochairman of Wenner-Gren Princeton Conference (1955).

CAREER

Did most important work at Berkeley, where he was founder of the distinguished "Berkeley School of Geography" exported to various other American campuses by the fifty Ph.D.s awarded during his tenure, including the thirty-seven who completed doctoral dissertations under his personal supervision. This distinct and well-recognized group of geographers was one of the most influential within the profession. Undertook field work in various parts of Mexico during most years from 1925 to 1945. Maintained that geography was study of human history at least as much as of physical terrain and that it was also strongly related to biology. Written documents, archeological evidence, plant data, soil profiles, the wisdom of the small farmer—all provided him with clues for unraveling problems of human global environment. Was leading and widely published authority on desert studies, tropical areas, and human geography of Indian populations, agriculture, and native crops of the New World.

Writings were numerous and often distinguished. Wrote twenty-one books and prepared over ninety papers and articles. These included "*Morphology of Landscape*" (1925), which was turning point in development of American geography. In it, Sauer urged his colleagues to "think of man and the land in terms of cause and result," rather than focusing exclusively on "environmental determinism." Other important articles included "Theme of Plant and Animal Destruction in Economic History" (1938), "Environment and Culture during the Last Glaciation," (1948), "Cultivated Plants of South and Central America" (1950), "Age and Area of American Cultivated Plants" (1959), "Concerning Primeval Habit and Habitat" (1964), and "Plants, Animals, and Man" (1970). His books included *Man in Nature: America Before the Days of the White Man: A First Book of Geography* (1939); *Agricultural Origins and Dispersals* (1952), later revised as *Agricultural Origins and Dispersals: The Domestication of Animals and Foodstuffs* (1969) and as *Seeds, Spades, Hearths, and Herds: The Domestication of Foodstuffs* (1972); *Land and Life: A Selection from the Writings of Carl Ortwin Sauer*, J. Leighly, ed. (1963); *The Early Spanish Main* (1966); *Sixteenth Century North America The Land and the People as seen by the Europeans* (1976); *Seventeenth Century North America*, 1980; *Selected Essays, 1963–1975*, Bob Callahan, ed. (1981).

Was President, Association of American Geographers, 1940, and its Honorary President from 1955. Winner of Vega Medal in 1957, regarded as world's most coveted honor in sciences of earth and sea. Special award from Association of American Geographers "for a lifetime of exceptional achievement as a scholar in geography," 1974. Awarded Victoria Medal by Royal Geographic Society, 1975.

MAJOR CONTRIBUTIONS

Considered one of most influential geographers of his generation. Perhaps no other American geographer influenced so wide a spectrum of scholars. Was first-rate teacher who "tried to encourage students to keep on thinking." For fifty years was advocate of environmental protection. Report he wrote in 1934 on land use, for U.S. Science Advisory Board, was instrumental in establishing U.S. Soil Conservation Service.

BIBLIOGRAPHY

Geographic Journal 141 (November 1975).

Kenzer, Martin S. *Carl O. Sauer: A Tribute* (1987).

Larkin, Robert P., and Gary L. Peters. *Biographical Dictionary of Geography* (1993).

Leighly, John. "Carl Ortwin Sauer." *Annals of the Association of American Geographers* 66 (1976).

New York Times, 21 July 1975 (obit.).

Parsons, James J. *Fifty Years of Berkeley Geography, 1923–1973* (1973).

Parsons, James J. "Carl Ortwin Sauer." *Geographical Review* 66 (1976).

Who Was Who in America. Vol. 6 (1976).

UNPUBLISHED SOURCES
Carl Ortwin Sauer Papers are at University of California, Berkeley. There are also some letters there in the Isabel Kelly Correspondence. There are Sauer letters in the Edward Louis Papers at the University of Washington Library in Seattle. Other correspondence can be found in the Marston Bates Papers at Bentley Historical Library, University of Michigan, Ann Arbor.

Richard Harmond

SAUNDERS, WILLIAM. Born Devonshire, England, 16 June 1836; died London, Ontario, 13 September 1914. Agricultural scientist, entomologist, pharmacist.

FAMILY AND EDUCATION

Moved with parents to Canada at age twelve. Little formal education. Honorary LL.D., Queen's University, 1896, University of Toronto, 1904. Married Sara Agne Robinson. Had at least five sons, several of whom became active horticulturalists and naturalists.

POSITIONS

Druggist in London, Ontario, and from 1868 a scientific farmer. Director, Experimental Farms Branch, Dominion Department of Agriculture, 1886–1911.

CAREER

Secured some training in chemistry following his arrival in Canada in 1848. Purchased farm to carry out scientific experiments. Collected plants and insects in vicinity of his London, Ontario, home and became student of botany and entomology. This led him into manufacture of plant extracts on both wholesale and retail basis. Became professor of materia medica at University of Western Ontario in London. Served as founding president, Ontario College of Pharmacy. Cofounder with C. J. S. Bethune of Entomological Society of Canada, 1863, and of *Canadian Entomologist*, 1868. Served as editor of this journal from 1873 to 1886. Published *Insects Injurious to Fruits*, 1883 (2d ed. 1892). Frequently went out on collecting trips and had particular interest in Lepidoptera. Raised caterpillars and carefully described results. His appointment as director of Experimental Farms followed trip to United States to visit a number of experiment stations, which was undertaken at request of Canadian government. Considered an authority on fruit insects, forestry, insectivorous birds, and beekeeping. Developed a number of new fruit strains, including raspberries, gooseberries, and grapes. His initial researches later led to development of Marquis wheat by his son, Charles Edward Saunders. Well regarded as teacher of pharmacy. Studied sprays and other insecticides for control of harmful insect species. Was selected charter member of Royal Society of Canada (1882), elected its president, 1906. Published various scientific articles and reports.

MAJOR CONTRIBUTIONS

Authority on Canadian insects, particularly Lepidoptera, and plants, especially fruit varieties. Noted pharmacist and developed several varieties of cereal plants and other grains.

BIBLIOGRAPHY

Bethune, C. J. S. "Dr. William Saunders, CMG." *Canadian Entomologist* 46 (1914).
Goding, F. W. "A Pen Sketch of Prof. William Saunders, FRSC, FLS." In *Annual Report Entomological Society of Ontario* 25 (1894).
Judd, W. W. *Early Naturalists and Natural History Societies of London, Ontario* (1979).
Mallis, Arnold. *American Entomologists* (1971).
Osborn, H. *Fragments of Entomological History* (1937).
Pomeroy, Elsie. *William Saunders and His Five Sons* (1956).

Keir B. Sterling

SAUNDERS, WILLIAM EDWIN. Born London, Ontario, 16 August 1861; died London, Ontario, 28 June 1943. Naturalist, horticulturalist, ornithologist, pharmacist, conservationist.

FAMILY AND EDUCATION

Son of Sara Agnes and William Saunders. Father was a native of Crediton, Devonshire, England, who immigrated to Canada in 1848 and settled at London, Ontario, to become an accomplished pharmaceutical manufacturer, founder of the Entomological Society of Canada, 1863, and first director of the Dominion Experimental Farm, 1886–1911. "W. E." was eldest of five sons, all of whom excelled in science or music. His brother Charles Edward Saunders was knighted for his development of Marqúis wheat. Educated at London Central School and Collegiate Institute and Trinity College, Port Hope, Ontario. Graduated from Philadelphia College of Pharmacy in 1883. Honorary LL.D. from University of Western Ontario, 1936. Married Emma Lee, 1885, and had one daughter.

POSITIONS

One of seven active members elected to the American Ornithologists' Union at the founding meeting in 1883. Founding president of the Ornithological Section of the Entomological Society of Ontario, 1890–1894 (now the McIlwraith Field Naturalists, Canada's oldest existing natural history club). Secretary-treasurer of the Entomological Society of Ontario, 1887–1906. Professor of practical chemistry at the London (University of Western Ontario) Medical School, 1884–1889. Charter member, 1919, and a director, 1931–1943, of the American Society of Mammalogists. First Canadian president of the Wilson Ornithological

Club, 1912–1913. Founding president of the Federation of Ontario Naturalists, 1931–1943.

CAREER

Developed a lasting interest in birds and nature in general, as a child. Began keeping detailed species lists and notes in 1873. Graduated from Pharmacy College in 1883 and began a career in pharmaceutical supply that he continued until retiring in 1929. As a dedicated amateur naturalist, he made his most significant contributions. Taught himself taxidermy and amassed a considerable collection of ornithological specimens. His first oological specimens date from 1873, and the earliest bird skins from 1879. He built cases and displays for these specimens to illustrate his many presentations. His first publication appeared in the *Canadian Entomologist* in 1879. Published ''Birds of Western Ontario'' with J. A. Morden in 1882 and established the biological importance of Point Pelee for its bird migration and Carolinian flora. Presented a paper to the Ontario Agriculture Commission in 1881, outlining the importance of insectivorous birds to biological control. Catalyst and founding secretary of the Great Lakes Ornithological Club in 1905. Following his retirement he became much more active in promoting the conservation of wildlife and habitat. Instrumental in gaining legislated protection for eagles and osprey. Published a nature column in the *London Free Press*, ''Nature Week by Week,'' 1929–1943. In addition to these 700 articles, he published nearly 300 papers. Also an active member of the London and Middlesex Historical Society and the London Horticultural Society and secretary of the American Peony Society.

MAJOR CONTRIBUTIONS

One of Canada's leading early conservationists. Adviser to the Canadian Commission on Conservation. A strong advocate for the protection of birds. One of the first to promote the use of nest boxes and a leader in the campaign to prevent the wanton destruction of raptors. Recognized as the ''dean of Ontario ornithologists.'' A careful ornithological observer responsible for eleven first records for Ontario, five of which were first records for Canada. A dynamic and active speaker across the province of Ontario. To his credit, he spoke to a wide range of groups, especially horticultural and agricultural, on the value of conservation, not just naturalist groups.

BIBLIOGRAPHY

Dale, E. M. S. ''William Edwin Saunders 1861–1943.'' *The Canadian Field-Naturalist* 57, 6 (1943): 99–100.

Judd, W. W. *Early Naturalists and Natural History Societies of London, Ontario* (1979).

Pomeroy, E. M. *William Saunders and His Five Sons* (1956).

Rutter, R. J., ed. *W. E. Saunders: Naturalist* (1949).

Wake, D. G. "William Edwin Saunders (1861–1943)." In M. K. McNicholl and J. L. Cranmer-Byng, eds., *Ornithology in Ontario* (1994).

UNPUBLISHED SOURCES

Papers, letters, field books, and various manuscripts are held by the Royal Ontario Museum, Archives, Special Collections (SC47). London Public Library, Archives of the McIlwraith Field Naturalists, also has reminiscences, photographs, and transcripts (Accessions 506, 507).

Michael S. Quinn

SAY, THOMAS. Born Philadelphia, 27 June 1787; died New Harmony, Indiana, 10 October 1834. Naturalist.

FAMILY AND EDUCATION

Son of Benjamin Say, successful physician, select councilman of Philadelphia, state senator, and member of Congress, and Ann Bonsall, granddaughter of John Bartram, the early botanist; married 1776. Second of four children. (Father remarried 1795; three children by second union.) In 1799 at age twelve enrolled Westtown (Friends) Boarding School. Probably remained until age fifteen. Interest in natural history encouraged by William Bartram of Kingsessing, his great-uncle. Independent studies at Academy of Natural Sciences of Philadelphia, organized 1812. Active duty as private with First City Troop (Philadelphia), September–December 1814. Married Lucy Way Sistare, of New York City, at New Harmony, Indiana, 4 January 1827; no children.

POSITIONS

Cofounder, Academy of Natural Sciences of Philadelphia (1812), curator, 1812–1826. Professor of zoology, Philadelphia Museum, 1821–1825. Curator, American Philosophical Society, 1821–1827. Professor of natural history, including geology, University of Pennsylvania, 1822–1828. First president, Maclurian Lyceum of the Arts and Sciences (Philadelphia), 1826. Zoologist and antiquary, Maj. Long's Expedition to Rocky Mountains, 1819–1820, and Maj. Long's Expedition to the Source of St. Peter's River, 1823. Naturalist, teacher, and administrator, New Harmony, Indiana, 1826–1834. Also correspondent of the Société Philomatique de Paris, 1824, and foreign member of the Linnaean Society of London.

CAREER

Worked as apothecary, first assisting his father, later in brief partnership with John Speakman in business as Speakman and Say. Cofounder of the Academy of Natural Sciences of Philadelphia (1812), curator of the museum (1817–1826), and on Committee of Publication of *Journal of the Academy of Natural Sciences of Philadelphia* (1817–1826), contributing forty-one articles on entomology, conchology, paleontology, mammalia, reptilia, and crustacea (1817–1842). Contributed articles on conchology to first American editions of *Nicholson's British Encyclopedia* (Philadelphia, 1816, 1818, 1819), first conchological work by an American published in America. On twenty-seven-day scientific excursion to New Jersey with William Maclure, Charles-Alexandre Lesueur, and Gerard Troost, spring 1817. Began *American Entomology*, six plates with their letterpress (1817), incorporated into first of three volumes of *American Entomology, or Descriptions of the Insects of North America Illustrated by Coloured Figures from Original Drawings from Nature* (Philadelphia, 1824, 1825, and 1828). Numerous descriptive papers on entomology also appeared in *Transactions of the American Philosophical Society*, the *Western Quarterly Reporter, Contributions of the Maclurian Lyceum to the Arts and Sciences, Annals of the Lyceum of Natural History of New York*, and the *Boston Journal of Natural Science*. On scientific expedition to sea islands off the coast of Georgia and penetrating into eastern Florida with William Maclure, George Ord, and Titian R. Peale, December 1817–May 1818. Appointed zoologist on Maj. Stephen H. Long's western expedition to the Rocky Mountains, April 1819–November 1820. Printed account from notes of Maj. Long and Say, *Account of an Expedition from Pittsburgh to the Rocky Mountains Performed in the Years 1819, 1820* by Edwin James (1823). Say also furnished catalog of animals at Engineer Cantonment and other places, "Vocabularies of Indian Languages," "Indian Language of Signs," and "Indian Speeches." The plants collected by James and Say were sent to John Torrey to classify, whose third paper on the Long expedition plants abandoned traditional Linnaean method, adopting the more complicated, more accurate natural system, the first time an American had made use of it with American plants. Lectured at Philadelphia Museum on zoology (1821–1825), served as curator of the American Philosophical Society (1821–1827), and appointed professor of natural history at the University of Pennsylvania (1822–1828). Appointed zoologist and antiquary on Maj. Stephen H. Long's expedition to the source of St. Peter's River, April 1823–October 1823. Printed account from notes of Maj. Long, Say, W. H. Keating, and J. C. Calhoun, *Narrative of an Expedition to the Source of St. Peter's River, Lake Winnepeek, Lake of the Woods Performed in the Year 1823* by William Keating (1824). Say provided everything relating to the zoology and botany of the country, extensive material relating to Indians, and an extensive appendix on the natural history. The plants Say collected were examined and described by Lewis David von Schweinitz. Say corrected and arranged for publication all of Charles Lucien Bonaparte's papers for *Journal of the Academy of Natural Sciences* and Bonaparte's first volume of *Natural History of Birds Inhabiting the United States* (continuation of Alexander Wilson's *American Ornithology*). On twenty-three-day scientific excursion to northeastern Pennsylvania, northern New Jersey, and southern New York with Maclure and Lesueur, 23 August–14 September 1825. With Maclure and Lesueur joined Welsh social reformer Robert Owen and other scientists and educators at Pittsburgh, arriving at New Harmony, Indiana, late January 1826.

New Harmony served as base until his untimely death. On scientific excursion to Ohio and Kentucky with Maclure, June–October 1826. *American Entomology* 3 (1828) published in Philadelphia from Say's prior work during this period. Final trip with Maclure, December 1827–April 1828, to Mexico. In New Harmony, assumed full charge of Maclure's press in June 1830 and served as editor. Frequent contributor to the *New Harmony Disseminator* (1828–1834). Eleven monographs were printed on press (in part, takeouts from *Disseminator*), six on entomology, five on conchology. On this press he produced his most important work, *American Conchology, or Descriptions of the Shells of North America, Illustrated by Coloured Figures from Original Drawings Executed from Nature* (1830–[1838]): six parts were issued before his death in 1834, a seventh posthumously in Philadelphia (1838). It is now definitely established that Say made one return visit to New York and Philadelphia between July and November 1833.

MAJOR CONTRIBUTIONS

Say has been called the "father of American conchology," the "father of American entomology," and the "father of American descriptive entomology." According to John G. Morris, Say described 1,150 species of Coleoptera, 225 species of Diptera, 100 species of Hemiptera, and 100 species belonging to other groups. Say was the leader in descriptive studies during early part of the century. His article in *Nicholson's Encyclopedia* was the foundation of American conchology. Although Say was not the first American to publish on American insects, he is regarded as the first efficient and extensive describer of North American insects, especially Coleoptera. He belongs to the descriptive period of American natural history.

BIBLIOGRAPHY

Bestor, Arthur E., Jr., ed. "Education and Reform at New Harmony." *Indiana Historical Society Publications* 15, 3 (1973).

Coates, Benjamin H. *A Biographical Sketch of the Late Thomas Say, Esq.* (1835).

Evans, Howard E. *The Natural History of the Long Expedition to the Rocky Mountains, 1819–1820* (1997).

Evermann, Barton Warren. "A Century of Zoology in Indiana, 1816–1916." *Proceedings of the Indiana Academy of Science* (1916).

Kane, Lucille M., et al., eds. *The Northern Expeditions of Stephen H. Long: The Journals of 1817 and 1823 and Related Documents* (1978).

Kastner, Joseph. *A Species of Eternity* (1977).

Reingold, Nathan, ed. *Science in Nineteenth-Century America, a Documentary History* (1979).

Sellers, Charles Coleman. *Mr. Peale's Museum: Charles Willson Peale and the First Popular Museum of Natural Science and Art* (1980).

Stroud, Patricia T. *Thomas Say: New World Naturalist* (1992).

Weiss, Harry B., and Grace M. Ziegler. *Thomas Say, Early American Naturalist* (1931).

UNPUBLISHED SOURCES

See Academy of Natural Sciences of Philadelphia: Isaac Lea Papers—conchological annotations; Thomas Say Papers—scientific correspondence and notes; notes on scientific expeditions; American Philosophical Society: Thomas Say Correspondence, 1820–1883—correspondence, published and unpublished drawings for *Conchology*; Haverford College: Robert B. Haines III Collection (Reuben Haines)—correspondence; Historic New Harmony, Inc. Archives: Thomas Say Collection; InterNorth Art Foundation: Center for Western Studies, Joslyn Art Museum (Omaha, Neb.), Maximilian-Bodmer Collection; Museum of Science, Boston: Thaddeus William Harris manuscripts; New Harmony Workingmen's Institute: New Harmony Correspondence, Maclure-Fretageot Correspondence; University of Michigan—Michigan Historical Collections, Alexander Winchell Papers.

Ralph G. Schwarz

SCHMIDT, KARL PATTERSON. Born Lake Forest, Illinois, 19 June 1890; died Chicago, 26 September 1957. Herpetologist, zoogeographer, taxonomist.

FAMILY AND EDUCATION

Oldest son of George Washington Schmidt, professor of German at Lake Forest College. Awarded Haven Medal as scholastic leader of class upon graduation from Academy of Lake Forest College (1906). After first year at college, father moved family to Wisconsin to start a family farm. Took correspondence classes through University of Chicago. B.A., 1916, Cornell University. Honorary D.Sc., 1952, Earlham College. Married Margaret Wightman, 1919; two children.

POSITIONS

Farmer, Clark County, Wisconsin, 1907–1913. Teaching assistant, Cornell University, 1913–1916. Research assistant, American Museum of Natural History, 1916–1922. Served in U.S. Army, Camp Grant, Illinois, 1918. Curator, Division of Amphibians and Reptiles, Field Museum of Natural History (Chicago), 1922–1940, chief curator of zoology, 1941–1955, curator emeritus of zoology, 1955–1957.

CAREER

When James G. Needham, Schmidt's former biology teacher at Lake Forest, moved to Cornell, he encouraged his former student to return to his studies from family farm. At Cornell, lived in Needham's house and assisted in some of his classes. Originally planning to be a geologist, changed focus to herpetology during a four-month job (1915) with Pardee Oil Company near swamps of Louisiana.

Offered assistantship in Department of Herpetology at American Museum at recommendation of Mary C. Dickerson, working on collection from Congo expedition, about which he wrote two papers in the museum's *Bulletin* (vols. 39, 49). At Field Museum, built one of the foremost herpetological departments in world. Led or participated in many scientific expeditions, including Santo Domingo (1916), Puerto Rico (1919), Honduras (1923), Brazil (1926), and the Pacific Islands (1928–1929).

Wrote over 200 articles and books, for both specialists and general public, including seventy-five articles on new species of reptiles and amphibians. Author of *Homes and Habits of Wild Animals* (1934), *Our Friendly Animals and Whence They Came* (1938), *Field Book of Snakes of the United States and Canada* (with D. Dwight Davis, 1941), *Principles of Animal Ecology* (with W. C. Allee and Alfred E. Emerson, 1949), *A Check List of North American Amphibians and Reptiles* (6th ed., 1953), and *Living Reptiles of the World* (with Robert F. Ingar, 1957). Translated several important German works into English, including Richard Hesse's *Tiergeographie auf oekologischer Grundlage*, which he completely revised with W. C. Allee as *Ecological Animal Geography* (1937; 2d ed. 1951). From 1941 to 1955, edited the zoological publications of the Field Museum. Contributed articles to *Chicago Naturalist*, *Texas Geographic Magazine*, *Copeia*, and *Scientific Monthly* and served as an editor for *Biological Abstracts*, *Copeia*, *American Midland Naturalist*, and *Encyclopedia Britannica*.

Although he made no important discoveries and promulgated no important new theories, he was universally held in high esteem by fellow biologists.

MAJOR CONTRIBUTIONS

Work on coral snakes and crocodiles. Regional studies and geographical distribution of reptiles and amphibians. Turned herpetology from a hobby into a branch of biology. Influenced numerous young biologists as editor, teacher, author, and curator.

BIBLIOGRAPHY

Davis, D. Dwight. *Copeia* (9 October 1959) (obit.).
Dictionary of Scientific Biography. Vol. 18, Supplement 2 (1990).
New York Times, 27 September 1957 (obit.).
Who Was Who in America. Vol. 3 (1960).
Wright, A. G. *In the Steps of the Great American Herpetologist Karl Patterson Schmidt* (1967).

John A. Drobnicki

SCHMITT, WALDO LA SALLE. Born Washington, D.C., 25 June 1887; died Takoma Park, Maryland, 5 August 1977. Zoologist, carcinologist.

FAMILY AND EDUCATION

Second of five children of Ewald, an engineer, and Fanny Mathilde Hesselbach Schmitt. Graduate of Central High School, Washington, D.C., 1907. B.S., George Washington University, 1913; M.A., University of California, Berkeley,

1916; Ph.D., George Washington University, 1922. Honorary D.Sc., University of Southern California, 1948. Married Alvina Stumm, 1914; two children.

POSITIONS

Aide in economic botany, U.S. Department of Agriculture (USDA), 1907–1910; scientific assistant, 1910–1913, and naturalist, 1913–1914, U.S. Bureau of Fisheries; assistant curator, 1915–1920, associate curator, 1920, curator, 1920–1943, Division of Marine Invertebrates, U.S. National Museum (USNM); Instructor in zoology, George Washington University, 1923–1924; head curator of biology, 1943–1947, head curator of zoology, 1947–1957, U.S. National Museum; research associate, Smithsonian Institution, 1957–1977.

CAREER

Had considerable interest in nature from childhood. Became progressively deaf from age eleven because of scarlet fever. Took position with USDA with the thought of becoming a forester. At the suggestion of Paul Bartsch, his mentor at George Washington, applied for opening at USNM to assist Mary Jane Rathbun, assistant curator in charge of marine invertebrates in her work with decapod Crustacea (shrimps, crabs, and their relatives, living and fossil). The two worked well together, and Rathbun resigned her post in 1914 to make possible the hiring of an assistant, becoming an honorary associate in zoology. This in essence enabled USNM to employ Schmitt as assistant curator and later as curator, though Rathbun "was effectively in charge" until her death in 1943. The collecting work he did on expeditions made from 1924 until 1940 primarily done to enable Rathbun to accomplish her scientific objectives. One of the primary aims of his career was to collect and properly curate specimens on which later workers based their monographic studies, though he published some taxonomic descriptions. Schmitt was a highly respected systematic zoologist whose understanding of the Crustacea, an extremely large group of organisms, was wide-ranging. His research work was much more limited following his appointment as head USNM curator in 1943. He was a cofounder of the Society of Systematic Zoology, 1946, provisional chairman, 1947, and first president, 1948, and was instrumental in founding of the journal *Systematic Zoology* in 1952. Schmitt participated in numerous research and collecting efforts and expeditions from the mid-1920s on. These included work at the Tortugas Marine Laboratory in 1924–1925 and 1930–1932 and collecting along the east and west coasts of Latin America as Walter Rathbone Travelling Scholar, Smithsonian Institution, 1925–1927. Member and, in some cases, leader of Hancock Pacific Expeditions, 1933–1935; Smithsonian-Hartford Expedition to the West Indies, 1937; naturalist on President F. D. Roosevelt's cruise to the Galapagos, 1938; Hancock Atlantic Expedition, 1939; Alaskan King Crab investigations for U.S. Fish and Wildlife Service in Alaska, 1940; U.S. Navy-sponsored trip to Galapagos to arrange for naval field station, 1941; U.S. Army-sponsored trip to Galapagos to arrange for army field station, 1942; goodwill mission to Latin America for U.S. State Department, 1943; Smithsonian-Bredin Congo Expedition, 1955; Smithsonian-Bredin Caribbean Expedition, 1956, 1958, 1959; Smithsonian-Bredin Society Island Expedition, 1957; Smithsonian-Bredin Expedition to Yucatan, 1960; Palmer Peninsula Survey, Antarctica, 1962–1963. Published *Marine Decapod Crustacea of California* (1921), "Crustaceans: Shelled Invertebrates of the Past and Present" (1931), issued in revised and extended form as *Crustaceans* (1965, British ed. 1973), and many technical papers dealing with marine invertebrates. Made a number of technical and popular presentations concerning the subject. Three genera (one of Crustacea, one of sponges, and one of mollusks), named for him, together with at least six marine vertebrates and invertebrates. Active member of a number of professional organizations.

MAJOR CONTRIBUTIONS

The many thousands of specimens Schmitt collected made the USNM a major repository for marine invertebrates. He was known for the thoroughness of his fieldwork and for encouraging other taxonomists to prepare descriptive monographs based on the USNM and other resources. His understanding of the Crustacea was unusually broad, though there were many groups on which he did no work. Helped to develop other important collections not part of the USNM.

BIBLIOGRAPHY

Blackwelder, Richard E. "Twenty-Five Years of Taxonomy." *Systematic Zoology* 26 (1977).

Blackwelder, Richard E. *The Zest for Life, or Waldo Had a Pretty Good Run: The Life of Waldo La Salle Schmitt* (1979).

Chace, F. A., Jr. *Crustaceana* 1 (1978) (obit.).

Cox, William E. *Guide to the Papers of Waldo LaSalle Schmitt, Smithsonian Archives* (1983).

Washington Star, 6 August 1977 (obit.).

Who Was Who in America.

Keir B. Sterling

SCHREIBER, RALPH W. Born Wooster, Ohio, 6 July 1942; died Los Angeles, 29 March 1988. Ornithologist.

FAMILY AND EDUCATION

One of four sons of William I. (professor of German and department chair, College of Wooster), and Clare Adel Schreiber. B.A., College of Wooster, 1964; M.S., University of Maine at Orono, 1966; Ph.D., University of South Florida, Tampa, 1974. Married Elizabeth Anne Ferguson, 1972.

POSITIONS

Research biologist, Pacific Ocean Biological Survey Program, Smithsonian Institution, 1966–1969; cofounder Biological Research Associates, Inc., Tampa,

Florida, 1969–1976; Cofounder Seabird Research Associates, Inc., Tampa, Florida, 1972–1976; adjunct researcher, University of South Florida, 1974–1976; member, Brown Pelican Recovery Team, U.S. Department of Interior, 1975–1976; curator of ornithology, Natural History Museum, Los Angeles County (California), 1976–1984; head of the Section of Birds and Mammals, Natural History Museum, Los Angeles County (California), 1984–1988. Adjunct associate professor, Department of Zoology, University of Southern California.

CAREER

Involved in Central Pacific faunal research for more than twenty years. At Smithsonian Institution, carried out self-designed projects concerned with breeding biology of seabirds and marine mammals of the Central Pacific. Developed particular interest in colonial seabirds. Noted that DDT affected brown pelicans and subsequently fought to ban that chemical. In early 1980s, noticed profound effect of El Niño on seabird breeding on Christmas Island in the South Pacific. He determined that their populations are largely influenced by random catastrophic climatic events, not competition for food or nesting sites. Other projects in Pacific region included study of albatross breeding and survival on Midway Island. In Florida, his Biological Research Associates was an environmental consulting firm. Had great research interest in biology and conservation of waterbirds, notably pelicans, which was enhanced as cofounder (with his wife) of Seabird Research, Inc., a nonprofit research and educational organization. Charter member and vice president (1975–1977), Florida Ornithological Society. During his twelve years in California, active member of International Council for Bird Preservation (chairman, Seabird Specialist Group, 1984–1987). President, Cooper Ornithological Society, 1983–1985. Active in various capacities in other organizations, including American Ornithologists' Union, Pacific Seabird Group, and Hubbs Seaworld Research Institute. Had more than 100 publications to his credit, among them his dissertation on brown pelicans. Some were coauthored with others, including his wife, the majority of them dealing with ornithological subjects. His excellent photographs appeared in a number of periodicals, including *National Geographic*. The Ralph W. Schreiber Hall of Birds at Los Angeles County Natural History Museum named in his memory.

MAJOR CONTRIBUTIONS

Contributed much to understandings of the breeding, biology, and means of conserving tropical seabirds of the Pacific and various parts of the United States.

BIBLIOGRAPHY

American Men and Women of Science. 13th ed. (1976).

Wolfenden, Glen E. *Auk* (January 1989) (obit.).

Keir B. Sterling

SCHWEINITZ, LEWIS DAVID VON. Born Bethlehem, Pennsylvania, 13 February 1780; died Bethlehem, Pennsylvania, 8 February 1834. Moravian clergyman, botanist, mycologist.

FAMILY AND EDUCATION

Son of Baron Hans Christian Alexander von Schweinitz and Anna Dorothea Elizabeth de Watteville von Schweinitz. Mother was a granddaughter of Nicholas Lewis Count Zinzendorf, founder of the Moravian Church, Unitas Fratrum, in North America; father was in charge of the church's American fiscal and secular affairs. Between age seven and eighteen, educated under Samuel Kramsch at the Moravian School in Nazareth, Pennsylvania, where he became interested in botany. Moved with parents to Germany and entered the Moravian Theological Seminary at Niesky and Silesia (1798); continued formal education under Johannes Baptista von Albertini, the distinguished theologian, philosopher, and botanist, with whom he studied fungi; graduated (1801). Married Louisa Amelia le Doux (1812) of French ancestry whose parents resided in Stettin; four sons. Described as a retiring, but most genial, peaceful, charitable, and lovable man; communicated with ease in social situations.

POSITIONS

Teacher and preacher in the Moravian schools in Germany; Moravian Academy at Niesky (1801–1807), at Gnadenberg (1807–1808), and at Gnadau (1808–1812). Administrator of Church Estates in North Carolina at Salem (1812–1821). Proprietor of the Church Estates in the North and senior pastor at Bethlehem (1821–1834).

CAREER

Although a devoted pastor and dedicated administrator of the Moravian Church Estates in North America, his avocation was botany, especially the study of fungi. Named and described as new to science about 1,400 species of fungi in four major publications and documented by an extensive collection of specimens now at the Academy of Natural Sciences, Philadelphia. Made significant contributions to knowledge of vascular plants, being among the first in North America to publish monographic treatments of genera, including *Viola* and *Carex*, and first to prepare analytical tables or nearly dichotomous keys to vascular plants, which accompanied these monographs. Named and described fifty-five species of vascular plants as new to science. Prepared local floras of the communities where he lived. Extensively studied North American vascular plants, leaving a 357–page unpublished manuscript, *Synopsis Plantarum Americanum*, containing the names, descriptions, and habitat notes on known flowering plants and ferns of the continent. Studied the invasion and migration of foreign vascular plants into North America and wrote a second paper on the subject for the continent. Traveled widely in connection with his church work but always studied botany while in route, to Herrnhut, Saxony, Germany (1817–1819; 1825), to Gnadenheutten, Ohio (1823), to Erie, Pennsylvania (1827), to Goshen (now Hope), Indiana (1831), and northeast of Albany, New York (1832). Extensive and important herbarium of some 23,000 species of vascular plants (including numerous type specimens) acquired by his own collecting, through exchange with some 108 contributors, and by purchase of the William Baldwin herbarium, was presented to the Academy of Natural Sciences, Philadelphia.

Wrote *Conspectus Fungorum in Lusatiae Superioris Agro Niskiensi Crescentium* (with J. B. von Albertini, 1805); *Specimen Florae Americae Septentrionalis Cryptogamicae*; . . . [hepatic mosses] (1821); *Attempt of a Monography of the Linnean Genus Viola, . . .* (1822); *Synopsis Fungorum Carolinae Superioris . . .* (1822); *An Analytical Table to Facilitate the Determination of the Hitherto Observed North American Species of the Genus Carex* (1824); *List of the Rarer Plants Found near Easton, Pennsylvania* (1824); *Description of a Number of New American Species of Sphaeriae* (1825); *A Monograph of the North American Species of Carex* (with John Torrey, 1825); *Synopsis Fungorum in America Boreali Media Degentium . . .* (1832); *Remarks on the Plants of Europe which have become Naturalized in a More or Less Degree, in the United States* (1832); *The Journey of Lewis David von Schweinitz to Goshen, Bartholomew County, in 1831* (1927).

Ordained a deacon (1808), presbyter (1818), and *Senior Civilis* (1825), the last person to whom the latter office was bestowed. Elected memberships in the Academy of Natural Sciences of Philadelphia, American Philosophical Society, Linnaean Society of Paris, and Society of Natural Sciences of Leipzig. Awarded honorary doctor of philosophy degree from University of Keil, Germany (1817), believed to be first American to hold a Ph.D. from this university. Commemorated in the genus *Schweinitzia*, a member of the Heath family, with rare species in the mountains of the Carolinas and in Florida.

MAJOR CONTRIBUTIONS

As a mycologist, he was America's first authority in the field, being referred to as the "Father of American Mycology"; his publications were the most extensive and important in the field at that time. They have been described by W. Johnson as works that "indicate, not only great industry and perseverance in the collection of facts, but a judicious *method* in the prosecution of his labours." E. Fries, noted mycologist, wrote that the *Conspectus Fungorum* "taught me knowledge of more things than any other."

BIBLIOGRAPHY

Barnhart, J. H. "The Botanical Correspondents of Schweinitz." *Bartonia* 16 (1935).

Bynum, F. A. "Lewis David von Schweinitz: Father of American Mycology." *The Three Forks of Muddy Creek* [Winston-Salem, North Carolina] 2 (1975).

Johnson, W. R. *A Memoir of the Late Lewis David von Schweinitz, P.D. with a Sketch of His Scientific Labours* (1835).
Pennell, F. W. "The Botanist Schweinitz and His Herbarium." *Bartonia* 16 (1935).
Rogers, D. P. "L.D. de Schweinitz and Early American Mycology." *Mycologia* 69 (1977).
Shear, C. L., and N. E. Stevens. "Studies of the Schweinitz Collections of Fungi." *Mycologia* 9 (1917).
Shear, C. L., and N. E. Stevens, eds. "The Correspondence of Schweinitz and Torrey." *Memoirs of Torrey Botanical Club* 16, 3 (1921).
Stuckey, R. L. "Type Specimens of Flowering Plants from Eastern North America in the Herbarium of Lewis David von Schweinitz." *Proceedings of the Academy of Natural Science of Philadelphia* 131 (1979).

UNPUBLISHED SOURCES

Most extensive collection of manuscripts is in the archives of the Academy of Natural Sciences, Philadelphia, including biographical data and personalia, correspondence, drawings, floras, herbarium catalogues and indexes, and synopses of various groups of plants as described by V. T. Phillips and M. E. Phillips, *Guide to the Manuscript Collections in the Academy of Natural Sciences of Philadelphia*, Special Publ. No. 5 (1963); F. W. Pennell, *Bartonia* vols. 13, 19 (1932, 1938). Manuscript materials are also in the libraries of the University of Michigan, University of North Carolina, and the New York Botanical Garden.

Ronald L. Stuckey

SCOTT, WILLIAM BERRYMAN. Born Cincinnati, Ohio, 12 February 1858; died Princeton, New Jersey, 29 March 1947. Paleontologist, geologist.

FAMILY AND EDUCATION

Youngest of three surviving sons of William McKendree Scott, Presbyterian minister, and Mary Elizabeth Hodge Scott. An older brother, Hugh Lenox Scott (1853–1934), was superintendent of West Point and Chief of Staff of the Army (1914–1917). Family moved to Chicago and then to Princeton, New Jersey before father's death when William was three. Educated by his mother until age nine, then attended several private schools in Princeton and Philadelphia before being placed in the hands of private tutors. Attended College of New Jersey (now Princeton), received A.B., 1877. Honorary LL.D., University of Pennsylvania, 1906; Honorary Sc.D., Harvard, 1909, Oxford, 1912, and Princeton (on the occasion of his retirement from its faculty), 1930. Married Alice Adeline Post, 1883. Seven children.

POSITIONS

Instructor in geology, 1881–1884, Blair Professor of Geology, College of New Jersey (later Princeton University), 1884–1930; Chair, Department of Geology, Princeton 1904 (some sources suggest 1909) to 1930. Blair Professor Emeritus, 1930–1947.

CAREER

Made ten collecting trips for fossils to the western United States between his graduation from college in 1877 and 1893. Most of this field work was done in Wyoming, South Dakota, Montana, and Oregon. His earliest publication based on this work was ''Paleontological Report of the Princeton Scientific Expedition of 1877,'' with H. F. Osborn and F. Speir, Jr., *Contributions from the Museum of Geology and Archaeology, Princeton College* (1878). Later studies included *On the Osteology of Mesohippus and Leptomeryx, with observations of the modes and factors of evolution in the Mammalia* (1891), and *The Osteology and Relations of Protoceras* (1895). Scott continued to work with materials gleaned from these early expeditions for much of his life. After 1893, however, Scott was primarily a laboratory scientist, and did little field work. He did much to uncover, describe, and classify vertebrate fossils, with particular attention to mammals of the Tertiary period. He sided with E. D. Cope in the latter's paleontological quarrels with O. C. Marsh of Yale, and thus his membership in the National Academy of Sciences was blocked until seven years following Marsh's death in 1899. Scott spent his entire academic career at Princeton, though he traveled extensively and did much museum research in Europe and attended several international geological and zoological congresses there. He also made one research trip to Latin America in 1901 and another to South Africa in 1905. Published *An Introduction to Geology*, a well-known text in 1897 (new editions in 1907 and 1932), and *Physiography* (1922). His *A History of Land Mammals in the Western Hemisphere* (1913; revised edition, 1937) covered both North and South America and is still useful. In his *The Theory of Evolution* (1917), Scott posited that the underlying causes of evolution had not yet been uncovered. Between 1905 and 1932, he edited and partially wrote the eight volume *Reports of the Princeton University Expeditions to Patagonia*, which had taken place between 1896 and 1899. In the early 1930s, he undertook a five part study of the Oligocene mammals of the White River ''Badlands'' country of South Dakota, Colorado, and Nebraska with the assistance of G. L. Jepsen and Albert E. Wood, two former students. In doing so, he reexamined specimens he had collected beginning in the early 1880s, and studied paleontological materials at various museums and other institutions in various parts of the United States. This was published as *The Mammalian Fauna of the White River Oligocene* (five parts, 1936–1941). He then began a study of Eocene Uinta vertebrates, and in 1945, published *The Mammalian Fauna of the Duchesne River Oligocene*. Scott's bibliography totaled some 177 titles. His many awards included the E. K. Kane Medal of the Geographic Society of Pennsylvania (1905); the Wollaston Medal of the Geological Society of London (1910); and the F. V. Hayden Medal of the Academy of Natural Sciences of Philadelphia (1926). Elected to the National Academy of Sciences in 1906, he was subsequently awarded the Academy's Mary Clark Thompson (1931) and Daniel Giraud Elliott (1940) Gold Medals. He also received the Walker Grand Prize from the Boston Society of

Natural History (1934) and the Penrose Medal of the Geological Society of America (1936). He was president of the Paleontological Society of America (1911), was elected a vice president (1903) and later president of the American Philosophical Society (1918–1925), and was president of the Geological Society of America (1925). Elected Honorary member, American Society of Mammalogists, 1936.

MAJOR CONTRIBUTIONS

Made important studies of western American paleontology and geology over a period of nearly sixty-five years, and was one of the leading American paleontologists of his day. An outstanding teacher of paleontology and geology at Princeton University for half a century. His *History of the Land Mammals of the Western Hemisphere, Reports of the Princeton University Expeditions to Patagonia*, and *Mammalian Fauna of the White River Oligocene* were among his most important publications.

BIBLIOGRAPHY

Birney, Elmer C., and Jerry R. Choate, eds., *Seventy-Five Years of Mammalogy (1919–1994)* (1994).

Dictionary of American Biography.

Dictionary of Scientific Biography.

Jepson, Glenn C., Memorial in *Proceedings of the Geological Society of America* (1948).

New York Times, 30 March 1937 (obit.).

Scott, William B., *Memoirs of a Paleontologist* (1939).

Simpson, George G., ''William Berryman Scott,'' *Biographical Memoirs, National Academy of Science*, 25 (1948).

UNPUBLISHED SOURCES

See Scott correspondence in the Oliver Perry Hay Papers, University of Florida.

Keir B. Sterling

SCUDDER, SAMUEL HUBBARD. Born Boston, 13 April 1837; died Cambridge, Massachusetts, 17 May 1911. Entomologist.

FAMILY AND EDUCATION

Son of Charles and Sarah Lathrop (Coit) Scudder. Father was well-to-do hardware and commission merchant. Received A.B. in 1857 and A.M. in 1860 from Williams College; B.S. in 1862 from Lawrence Scientific School, Harvard University, where he studied under Louis Agassiz. Married Ethelinda Jane Blatchford, 1867 (died 1872); only son predeceased his father.

POSITIONS

Assistant to Louis Agassiz, 1862–1864; custodian, Boston Society of Natural History, 1864–1870; assistant librarian of Harvard University, 1879–1882; paleontologist in U.S. Geological Survey, 1886–1892. Vice president (1874–1880) and president (1880–1887) of Boston Society of Natural History.

CAREER

Considered ''perhaps the greatest American entomologist of his time,'' was chiefly interested in diurnal Lepidoptera, the Orthoptera, and fossil insects. Bibliography comprises 791 scientific titles. Papers were mainly of descriptive character, though he wrote many popular articles. His monumental treatise on butterflies, *The Butterflies of the Eastern United States and Canada, with Special Reference to New England*, 3 vols. (1888–1889), was the product of thirty years of research in life histories and affinities of these insects. Other publications include *The Tertiary Insects of North America* (1891); *Revision of the American Fossil Cockroaches* (1895); *Excursions into the World of Butterflies* (1895); *Everyday Butterflies* (1899). Received Walker Prize of Boston Society of Natural History in 1898 for his ''Contributions to Entomology Recent and Fossil.''

During his six years as paleontologist with the U.S. Geological Survey, named and described more than 1,100 species of fossil insects. His work on fossil species, notably the beetles and their allies, was done with greater exactitude than the available data probably warranted. His pioneering correlation of fossil and modern species has not yet undergone the modern revision it requires. Largely inactive in his last decade because of illness.

Member of American Academy of Arts and Sciences, National Academy of Sciences, American Association for the Advancement of Science, American Society of Naturalists, and Appalachian Mountain Club.

MAJOR CONTRIBUTIONS

Work on fossil insects was ''profound, very extensive and of a pioneer character.'' In course of his career, named and described 1,884 species of fossil insects, Orthoptera, and butterflies. Was early leader in movement for more precise classification.

BIBLIOGRAPHY

Dictionary of American Biography. Vol. 8 (1927–1936).

Dictionary of Scientific Biography.

Mayor, A. G. ''Samuel Hubbard Scudder,'' *Scientific Memoirs, Biographical Memoirs, National Academy of Sciences*. Vol. 16 (1936).

National Cyclopedia of American Biography. Vol. 24 (1967).

New York Times, 18 May 1911 (obit.).

''Samuel Hubbard Scudder.'' *Science* 34 (15 September 1911).

Richard Harmond

SEARS, PAUL BIGELOW. Born Bucyrus, Ohio, 17 December 1891; died Taos, New Mexico, 30 April 1990. Botanist, ecologist, conservationist.

FAMILY AND EDUCATION

Son of Rufus Victor Sears, an attorney, and Sallie Jane (Harris) Sears. Graduated from Bucyrus High School, 1908. B.S. in zoology, 1913, and B.A. in economics, 1914, Ohio Wesleyan University; M.A. in botany, University of Nebraska, 1915; Ph.D. in botany, University of Chicago, 1922. Honorary D.Sc., Ohio Wesleyan, 1937, Oberlin College, 1958, Bowling Green State University, 1968; Litt.D., Marietta College, 1951; LL.D., University of Arkansas, 1957, University of Nebraska, 1957, and Wayne State University, 1959. Married Marjorie Lee McCutcheon, 1917 (died 1982); three children. Married Marguerite Saxer.

POSITIONS

Instructor in botany, The Ohio State University, 1915–1919 (military service, 1917–1919); assistant professor, 1919–1925, and associate professor, 1925–1927, of botany, University of Nebraska, Lincoln; professor of botany and department chairperson, University of Oklahoma, 1927–1938, also botanist for the State Biological Survey of Oklahoma; research associate, Teacher's College, Columbia University, 1936–1938. Professor of botany and department chairperson, Oberlin College, 1938–1950; professor and chairperson of Conservation Program at Yale University, 1950–1960, also chairperson of Department of Botany and the Yale Nature Preserve, 1953–1955, professor emeritus, Yale University, 1960. Member, National Science Board, 1958–1964; Guggenheim Fellowship, 1958. Several visiting professorships during the 1960s, including two years at the University of Louisville.

CAREER

Active as teacher of botany, ecology, and conservation. For many years he conducted research in various areas of botany, biogeography, post-glacial vegetation, and inferred climate in glaciated and nonglaciated areas of eastern Canada and the United States. From his initial work on the natural vegetation of Ohio and his experience with the great dust storms of the 1930s, his professional awareness of conservation was strengthened as it related to issues of human ecology. Established first graduate program in conservation of natural resources at Yale, which did not, unfortunately, survive after his retirement.

Author of some fifty papers in scientific journals and more than 100 articles in general periodicals concerning ecology and conservation. His major selected journal publications are *Vegetation Mapping* (1921), *Variations in Cytology and Gross Morphology in Taraxacum* (1922), *The Natural Vegetation of Ohio*

(1925–1926), *Common Fossil Pollen in the Erie Basin* (1930), *A Record of Postglacial Climate in Northern Ohio* (1930), *History of Conservation in Ohio* (1942), *Xenothermic Theory* (1942), *Conservation in Theory and Practice* (1950), *Palynology in North America* (1951), *Ohio's Conservation Record 1908–1958* (1958), and [*History of*] *Plant Ecology* [*in the United States*] (1969). His published books are *Deserts on the March* (1935, several times republished), which has been described as "his most successful book on conservation awareness"; *This is Our World* (1937); *Life and Environment* (1939); *Who Are These Americans?* (1939); *This Useful World* (with others, 1941); *Charles Darwin: The Naturalist as a Cultural Force* (1950); *The Ecology of Man* (1957); *Where There Is Life* (1962, revised and expanded as *The Living Landscape* in 1966); *The Biology of the Living Landscape* (1964); *Lands Beyond the Forest* (1969); and *Wild Wealth* (with others, 1971).

Was president of the Ecological Society of America, 1948; American Association for the Advancement of Science, 1856; American Society of Naturalists, 1959; honorary president and board chairman, National Audubon Society, 1956–1959. Numerous honors and awards from professional societies and nature and conservation advocacy groups.

MAJOR CONTRIBUTIONS

Sears was the first to make use of witness tree information recorded by federal land surveyors, that when plotted on a map, determined the distribution of various tree species from which the original natural vegetation could be inferred. He was the first to create a set of symbols and methods used in making maps of natural vegetation, and the first to publish a map of the virgin forest of Ohio. He made the first contributions to the literature in the United States on the technique of methodology, identification, and interpretation in pollen analysis. Sears was also the first American botanist to outline in detail the probable succession in postglacial forest vegetation (with the inferred accompanying changes in climatic conditions for north-central Ohio) since the retreat of the Wisconsinian glacier, and he later did this for all of North America. In addition, he first described in depth the concept of xerothermic theory. Sears had the remarkable ability to explain complex environmental problems clearly and simply to colleagues, students, and citizens. He devoted his life to furthering man's understanding of the delicately balanced ecosystems on which mankind's very survival exists. The keenness of his mind, the warmth of his personality, the quality of his writing, and his capacity to relate scientific problems to human affairs earned him the distinction of an exemplary individual in American science. For his many accomplishments, he was the recipient of many honors and awards.

BIBLIOGRAPHY

Dexter, R. W. "Conservation and The Ohio Academy of Science—An Historical Review." *Ohio Journal of Science* 62 (1962).

Forristal, L. J. "Paul Bigelow Sears." *The World and I* (February 1988).
Moore, R. H. "The Paul Sears I Know." *Transactions of the Nebraska Academy of Science* 13 (1985).
National Cyclopedia of American Biography (1964).
"Paul B. Sears, Eminent Ecologist—1965," *Bulletin of the Ecological Society of America* 46, 4 (1965).
Sinnott, E. W. "Paul B. Sears." *Science* 121 (1955).
Stuckey, R. L. "Paul Bigelow Sears (1891–1990): Eminent Scholar, Ecologist, and Conservationist." *Ohio Journal of Science* 90 (1990).
Stuckey, R. L. *Tribute to an Eminent Ohioan: Paul Bigelow Sears (1891–1990): Contributions to Vegetation Mapping* (1991).
UNPUBLISHED SOURCES
Archives at Oberlin College and Yale University.

Ronald L. Stuckey and Keir B. Sterling

SESSIONS, KATHERINE (KATE) OLIVIA. Born San Francisco, 8 November 1857; died La Jolla, California, 24 March 1940. Horticulturist, nurserywoman.

FAMILY AND EDUCATION

Daughter of Josiah and Harriet (Parker) Sessions, both natives of Connecticut. From an upper-middle-class Unitarian family, she became familiar with flowers at an early age. Attended Brooklyn Grammar School in Oakland and in 1876 graduated from Oakland High School. Entered the University of California at Berkeley in 1877 and received the Ph.B. degree in 1881, having majored in chemistry. Never married.

POSITIONS

Worked as a substitute teacher in Oakland until 1883, when she moved to San Diego to teach algebra and geometry in the newly built Russ School (later the San Diego High School). Also taught for a while in San Gabriel and resigned her position in 1885. Between 1915 and 1918 served as supervisor of agriculture for the San Diego grammar schools.

Was one of the founding (1909) leaders of the San Diego Floral Association and served as an officer or board member for over twenty years. Contributed to the association's journal, *California Garden*, over 250 articles in which she discussed new plant introductions and promoted horticulture. Occasionally, she wrote for local newspapers. As a participant in flower shows and exhibits, she displayed unusual plants and foliage. Was very much interested in the local landscape and would draw plans and elevations for gardens working with architect Irving Gill on the landscaping of many homes. Was the founder of Arbor Day in San Diego and was honored by several tree plantings during her lifetime. In 1935, in San Diego, a "K. O. Sessions Day" was celebrated at the California–Pacific International Exposition. In 1939 she held classes for adults in gardening at University of California Extension Division. Also, in 1939 she received the Meyer Medal, given by American Genetic Association for notable service in the field of foreign plant introduction, in recognition of her ''outstanding contributions to the horticulture of her native state.'' In 1956 and 1957, in Pacific Beach, an elementary school and a memorial were named in her honor.

CAREER

Though she loved teaching, real interests were in horticulture, and in 1885 she became a partner in a well-established nursery in Coronado, with an office and shop in San Diego. Her interest in plants had begun on a trip to Hawaii in 1876, and during this trip her interest in the poinsettia was aroused. Was among the first nurserywoman to cultivate the plant for wide distribution. In 1892 she leased from San Diego thirty acres of land in the city park to use as nursery, provided that she plant 100 trees there each year and donate about 300 more to the city. This area eventually became the San Diego Balboa Park and introduced many exotic greens to the city. For this she became known as the ''mother of Balboa Park.'' In 1903 her nursery moved to Mission Hills section of the city. Later, in 1927, it was moved to Pacific Beach. In 1902 she brought back seeds from Mexico and pictures of the palm described as *Erythea brandegeei*, which was introduced into the nursery trade by Sessions. Also introduced the *Fremontia mexicana*, a large flowering shrub, and sent it to Theodore Payne, a nurseryman, who distributed it in 1916. Always looking for better selection of cultivated material, she introduced into horticulture a form of *Romneya coulteri*, the Matiliji poppy, and *Ceanothus cyaneus*, a native shrub. These plants are now widely cultivated in California. Through her endeavors plants such as the queen palm (*Arecastrum romanzoffianum*, known in the area as *Cocos plumosa*), silk oak (*Grenvillea robusta*), flame eucalyptus (*Eucalyptus ficifolia*), Chinese twisted juniper (*Juniperus chinensis torulosa*), bunyabunya tree (*Araucaria bidwillii*), camphor tree (*Cinnamomum camphora*), silver tree (*Leucadendron argenteum*), cork oak (*Quercus suber*), many acacias, bougainvilleas, hibiscus, pride of Madeira (*Echium fastuosum*), vines, aloes, mesembryanthemums, and other succulents were popularized and introduced into the San Diego area. During 1926–1927 traveled through Europe and Hawaii, which widened her perspective on experimentation and introduction of new plants. She obtained rare plants and seeds from several botanists, including Francesco Franceschi and his nursery in Santa Barbara and La Mortola Gardens in northern Italy. Sessions always kept her hand in the nursery business and continued to operate a small one until her death.

MAJOR CONTRIBUTIONS

Major contribution is the introduction of new varieties of plants into the San Diego region and into the horticultural business. She fostered colorful and drought-resistant varieties of plants, and her exotic landscaping beauty is still evident throughout the San Diego area.

BIBLIOGRAPHY

California Garden. "Kate Sessions Issue." San Diego: San Diego Floral Association (1953).

MacPhail, Elizabeth C. *Kate Sessions, Pioneer Horticulturist* (1976).

Notable American Women 1607–1950. Ed. Edward T. James (1971).

UNPUBLISHED SOURCES

See Kate Sessions Papers, San Diego Historical Society, Sierra Museum and Library (Calif.): mostly correspondence relating to botanical studies; diaries, travel notes, and pencil sketches.

Joan D'Andrea

SETON, ERNEST (EVAN) THOMPSON. Born South Shields, Durham, England, 14 August 1860; died, Santa Fe, New Mexico, 23 October 1946. Naturalist, artist, author, explorer, youth worker.

FAMILY AND EDUCATION

Eighth of ten sons of Joseph Logan Thompson, ship broker, insurance agent, later a shipowner, and Alice Snowden (a daughter died at age six). Family moved to Lindsay, Ontario, in 1866, and to Toronto, 1870. Educated in Toronto public schools, the Toronto Collegiate Institute, and the Ontario School of Art (Gold Medalist, 1879). Moved to London, England, 1879, admitted to Royal Academy (RA) School, 1880, declined seven-year RA scholarship, returned to Canada, 1881. Studied at Art Student's League, New York, 1884, Academie Julian, Paris France, intermittently, 1891–1896. Married Grace Gallatin, 1896 (divorced, 1935), one daughter; married Julia M. Buttree, 1935; one adopted daughter. (Although his given name was Ernest Evan Thompson, his work until 1901 bore various signatures: Ernest E. Thompson, E. E. Thompson, Ernest E. Seton-Thompson, Ernest E. T. Seton, Ernest Seton-Thompson. "Seton" was adopted to reflect claim to ancient family name. He legally took name of Ernest Thompson Seton in 1901 and kept it until his death.)

POSITIONS

Worked as freelance artist, 1880–1896, preparing illustrations for publishers in London, lithographers in New York, the U.S. Biological Survey, the Smithsonian Institution, and the American Museum of Natural History in New York. Also freelance author of animal and Indian stories and popular science articles, primarily for children, from 1886 to 1946. Periodicals included *Scribner's, Century, St. Nicholas, Ladies' Home Journal, Boy's Life, American Boy, Recreation, Country Life in America, Field and Stream,* and *Forest and Stream.* Subsidized his own scientific research, 1884–1946, publishing field notes and articles in *Auk, Journal of Mammalogy, Proceedings of the U.S. National Museum, Pro-*

ceedings and Transactions of the Canadian Institute, Annual Reports of the New York Zoological Society, Annual Reports of the Smithsonian Institution, and *Transactions of the Manitoba Historical and Scientific Society*. Official naturalist, government of Manitoba, 1893–1946; founder and chief, Woodcraft League of America, 1902–1934; president, Campfire Club of America, 1910; executive chairman, Organizing Committee, Boy Scouts of America (BSA), 1910; chief scout, Boy Scouts of America, 1910–1915; editor, Totem Board, 1917–1934; founder and president, Seton Institute/College of Indian Wisdom, 1932–1946.

CAREER

Trained in the best academies of Toronto, London, New York, and Paris, 1877–1896, exhibited his animal and bird paintings in London, Montreal, Toronto, Chicago, and Paris. Homesteaded in Manitoba and Saskatchewan, 1882–1886. Thenceforth, though he lived and traveled extensively in Canada, Britain, and Europe, resided primarily in New York, New Jersey, Connecticut, and, after 1930, the year in which he became a U.S. citizen, New Mexico. Abandoned professional artistic career in his thirties to learn language and methods of biology. Guided by Elliott Coues, Spencer F. Baird, J. A. Allen, C. Hart Merriam, Frank Chapman, William Hornaday, Theodore Roosevelt, and many others, used money generated from scientific illustrations as vehicle to subsidize field travel and writing. His *Studies in the Art Anatomy of Animals* (1896), product of a painstaking regimen of detailed dissections done in Paris, was applauded by artists and scientists alike. His career as father and most respected raconteur of realistic animal story was launched with *Wild Animals I Have Known* (1898), still in print after nearly a century. This book followed by others, all built on animal "biographies" drawn from field observations entered in his thirty-six volumes of handwritten journals. After being lumped with Nature Fakers by John Burroughs (1903), Seton responded with *Life Histories of Northern Animals*, 2 vols. (1909), detailing mammals of Manitoba. Still considered a biology classic. It won Seton the Gold Medal of the Camp Fire Club of America (1909) and contributed to his winning Silver Medal of the French Societe d'Acclimatation (1918). This success led him to expand scope of the work. Published *Lives of Game Animals*, 4 vols. (1925–1928), which won Daniel Giraud Elliott (National Institute of Science) Medal (1927) and, most important, the John Burroughs Medal (1928). *The Arctic Prairies* (1911) describes scientific expedition to Northwest Territories done with E. A. Preble, during which Seton mapped Aylmer and Clinton-Colden Lakes for Royal Geographical Society. His *Two Little Savages* (1902–1903), a combination of fiction, autobiography, natural history, and ethnography, provided pedagogical framework for youth organization he called Woodcraft Indians, precursor of BSA. After leaving BSA in 1915, continued his fieldwork and scientific and popular writing, but became increasingly interested in cultures of North American First Nations. As practical outlets for this interest, established Woodcraft League of America and Seton

Institute. Published 400 articles and short stories and over forty books, in hundreds of editions, many translated into several languages, including *The Birds of Manitoba* (1891); *Trail of the Sandhill Stag* (1899); *Biography of a Grizzly* (1900); *Lives of the Hunted* (1901); *Monarch, the Big Bear of Tallac* (1904); *Animal Heroes* (1905); *Woodmyth and Fable* (1905); *The Natural History of the Ten Commandments* (1907); *Biography of a Silver Fox* (1909); *The Forester's Manual* (1910); *Rolf in the Woods* (1911); *The Book of Woodcraft and Indian Lore* (1912); *Wild Animals at Home* (1913); *Wild Animal Ways* (1916); *The Preacher of Cedar Mountain* (1917); *Sign Talk* (1918); *Woodland Tales* (1921); *Animals Worth Knowing* (1934); *The Gospel of the Red Man* (with Julia M. Seton, 1936); *Great Historic Animals* (1937); *Animal Tracks and Hunter Signs* (1958).

MAJOR CONTRIBUTIONS

Made extensive and detailed field observations, especially useful in study of morphology, physiology, distribution, and behavior. A generalist, able to analyze and communicate complex biological information whether in sophisticated scientific monographs or in fiction for children. A radical who believed, with Peter Kropotkin, that ethology and ecology identified essential natural laws by which humans should govern themselves politically; with this end in mind launched Boy Scout movement, using North American Indian as model. That the Indian model was dropped by American organization in favor of British military structure devised by Baden-Powell was source of great disillusionment for Seton. An extraordinary bird and mammal painter whose scientific illustrations supported scholarship of many senior mammalogists and ornithologists.

BIBLIOGRAPHY

Andersen, H. Allen. *The Chief: Ernest Thompson Seton and the Changing West* (1986).
Current Biography (1943).
Dictionary of American Biography. Supplement 4.
Keller, Betty. *Black Wolf: The Life of Ernest Thompson Seton* (1984).
McMullen, Lorraine. *Ernest Thompson Seton and His Works* (1989).
New York Times, 24 October 1946 (obit.).
Redekop, Magdalene. *Ernest Thompson Seton* (1979).
Sampson, John G. *The Worlds of Ernest Thompson Seton* (1976).
Seton, Ernest Thompson. *Trail of an Artist-Naturalist* (1940).
Seton, Julia M. *By a Thousand Fires* (1967).
Wadland, John Henry. *Ernest Thompson Seton, Man in Nature and the Progressive Era, 1880–1915* (1978).

UNPUBLISHED SOURCES

See journals (1879–1946), American Museum of Natural History, New York; papers (manuscripts, correspondence, illustrations, photographs), National Archives of Canada, Ottawa.

John Henry Wadland

SHALER, NATHANIEL (SOUTHGATE). Born Newport, Kentucky, 20 February 1841; died Cambridge, Massachusetts, 10 April 1906. Geologist, geographer, conservationist, paleontologist, educator.

FAMILY AND EDUCATION

Son of Nathaniel Burger Shaler, medical doctor, and Ann Hinde Southgate Shaler. Eldest of family died in infancy. Shaler and three younger children survived to maturity. Due to ill health, was educated informally by a Swiss tutor, Johannes Escher, before entering Harvard's sophomore class of 1859. Soon abandoned humanities and enrolled as student under Louis Agassiz at Harvard's Lawrence Scientific School, where he specialized in geology. In 1862 received S.B. *summa cum laude*. In 1903, honorary LL.D. from Harvard for his work as "naturalist and humanist." Married Sophia Penn Page, 1862; two daughters.

POSITIONS

Assistant in paleontology, Museum of Comparative Zoology, Harvard University, 1864–1868; lecturer in paleontology and animal life, 1868–1869; professor of paleontology, 1869–1888; professor of geology, 1888–1906; director, Kentucky Geological Survey, 1873–1880; director, Atlantic Coast Division, U.S. Geological Survey (USGS), 1884–1900; dean, Lawrence Scientific School, 1891–1904; president, Geological Society of America, 1895. At various times commissioner of agriculture for Massachusetts; member of Topographical Survey Commission; member of Gypsy Moth Commission; member of Massachusetts Highway Commission; vice president, Immigration Restriction League of Boston; vice president, Massachusetts Society for the Promotion of Good Citizenship.

CAREER

One of the first generation of Americans to accept the theory of evolution, although his early work on the Brachiopoda phylum was an attack on natural selection. Continuing influence of Agassiz, of whom he was student, colleague, and successor, is reflected in his adoption of the neo-Lamarckian version of evolution associated with his lifelong friend Alpheus Hyatt. After graduation returned to home state of Kentucky, where he obtained commission as captain of the Fifth Kentucky Battery, but his participation in the Civil War was short-lived due to illness. Returning to Cambridge in 1864, assisted Agassiz at the Museum of Comparative Zoology (MCZ), mainly classifying accumulated store of fossils. During Agassiz's expedition to Brazil in 1865, Shaler took charge of geology and zoology teaching at the Lawrence Scientific School. 1866, due to ill health, extended travels in Europe, where he worked on Alpine glaciers and visited many museums. On his return, participated in fossil excavations at Big

Bone Lick, Kentucky in 1868; this research led to publications on historical range of bison in Southeast. As professor at Harvard, further developed his educational theories; undergraduate summer field excursions provided stimulus for the famous Anderson School of Natural History at Penikese—a precursor of the summer school. Returned to Europe, 1872, and met such outstanding British naturalists as Darwin, Lyell, Huxley, Tyndall, and Galton.

While in England, appointed director of Kentucky Geological Survey, and under his administration the first triangulation survey of state was completed; publications on aspects of the state's environment from geology to forestry; an inventory of its natural resources for 1876 Philadelphia Centennial Exposition. In 1874, publication of study of ''Recent Changes of Level on the Coast of Maine'' (*Memoirs of the Boston Society of Natural History* 2), which constituted one of the earliest systematic formulations of the principle of glacial isostasy. Through his 1875 summer camp at Cumberland Gap, W. M. Davis began his long distinguished career. After a third visit to Europe in 1881, he worked during the 1880s and 1890s for the USGS and published, *inter alia*, reports on seacoast swamps of the eastern United States (6th report); geology of Martha's Vineyard (7th report); Nantucket (Bulletin No. 53); the island of Mount Desert, Maine (8th report); Cape Ann (9th report); freshwater morasses of the Dismal Swamp region (10th report); the origin and nature of soils (12th report), described as ''a landmark in the history of soil concepts''; the geology of the common roads (15th report); and survey of peat deposits (16th report). Further, research along New England-Acadian coast led to early contributions to the understanding of isostatic and eustatic changes of sea level and to theories of shoreline development.

Aside from technical reports, Shaler published many popular pieces in such journals as the *Atlantic Monthly*, *North American Review*, *Scribner's Magazine*, and *Chautauquan*, and, while many dealt with popularizations of physiography, he also addressed numerous social issues from race to education. In 1891 his celebrated *Nature and Man in America* first appeared—a work reissued in new editions until well into the twentieth century. As well as outlining the continent's geological structure, it dealt with such geographical themes as prairie homesteading, folk tillage practices, the ''Great American Desert,'' the use of anthropometric and actuarial data to measure population quality, and aesthetic responses to landscape. The following year he delivered series of lectures at Andover Theological School, the published version of which, entitled *The Interpretation of Nature* (1893), revealed a teleological view of man's place in nature. His 1905 *Man and the Earth* represents the culmination of his thinking on resources, going beyond Marsh's contribution by focusing on mineral exhaustion, land reclamation, need for alternative energy sources and by reiterating his earlier warnings about soil erosion and deforestation.

Overall, Shaler published over 300 articles and books. In addition to those specified he authored *Kentucky: A Pioneer Commonwealth* (1884); *The Story of Our Continent* (1892); *Sea and Land* (1894); *Domesticated Animals* (1895); *American Highways* (1896); *Outlines of the Earth's History* (1898); a trilogy on social theory: *The Individual. A Study of Life and Death* (1900), *The Citizen. A Study of the Individual and the Government* (1904), *The Neighbor. The Natural History of Human Contacts* (1904); a five-part dramatic romance, *Elizabeth of England* (1903); volume of poems from the Civil War, *From Old Fields* (1906); and edited the three-volume *United States of America* (1894).

Shaler had a warm and engaging personality—around him was spun a whole web of anecdote and myth in Harvard at the turn of the century. One of the university's most popular teachers, annual enrollments for his famous course Geology 4 reaching 500. The broad, prolific, and lucid nature of his writings fostered his image as a purveyor of science to the nation, especially in providing a ''scientific'' perspective on current sociopolitical questions.

MAJOR CONTRIBUTIONS

Foundational statement on principle of glacial isostasy and on origin and nature of soils; basic research on Atlantic coastlands; synthetic approach to natural resources, balancing conservationist and preservationist perspectives; early regional geography of United States; important contribution to study of American frontier, distinguishing between forest and prairie homesteading; originator and developer of the Harvard summer school; founding father of American historical geography.

BIBLIOGRAPHY

Berg, W. ''Nathaniel Southgate Shaler: A Critical Study of an Earth Scientist.'' Diss., University of Washington, 1957.

Berg, W. ''Shaler, Nathaniel Southgate.'' In C. C. Gillispie, ed., *Dictionary of Scientific Biography* (1975).

Davis, W. M. ''Professor Shaler and the Lawrence Scientific School.'' *Harvard Engineering Journal* 5 (1906).

Haller, J. S. Jr. ''Nathaniel Southgate Shaler: A Portrait of Nineteenth-Century Academic Thinking on Race.'' *Essex Institute Historical Collections* 107 (1971).

Koelsch, W. A. ''Nathaniel Southgate Shaler.'' In T. W. Freeman and P. Pinchemel, eds., *Geographers. Biobibliographical Studies* (1979).

Livingstone, D. N. ''Nature and Man in America: Nathaniel Southgate Shaler and the Conservation of Natural Resources.'' *Transactions of the Institute of British Geographers* (n.s.) 5 (1980).

Livingstone, D. N. ''Environment and Inheritance: Nathaniel Southgate Shaler and the American Frontier.'' In B. Blouet ed., *The Origins of Academic Geography in the United States* (1981).

Livingstone, David N. *Nathaniel Southgate Shaler and the Culture of American Science* (1987).

Love, J. L. *The Lawrence Scientific School of Harvard University 1847–1906* (1944).

M[errill], G. P., and D[obson], E. R., ''Shaler, Nathaniel Southgate.'' *Dictionary of American Biography*, Vol. 17 (1935).

UNPUBLISHED SOURCES

Shaler's papers are widely scattered, but the following contain relevant materials: Nathaniel S. Shaler Papers at the Harvard University Archives; Official Correspondence of the Lawrence Scientific School, Harvard University Archives; N. S. Shaler File, Museum of Comparative Zoology; N. S. Shaler File, Kentucky Historical Society.

David N. Livingstone

SHELDON, CHARLES. Born Rutland, Vermont, 17 October 1867; died Kedgemakooge, Nova Scotia, 21 September 1928. Conservationist, naturalist.

FAMILY AND EDUCATION

Son of John A. and Caroline A. Eastman Sheldon. Graduate of Phillips Academy, Andover, Massachusetts, 1887. A.B. (civil engineering), Yale, 1890. Married Louise Walker Gulliver, 1909; no children.

POSITIONS

Assistant superintendent, Toledo Division, Lake Shore and Michigan Shore Railway, 1893; general manager, Consolidated Car Heating Co., Albany, New York, 1894–1898; general manager, Chihuahua and Pacific Railway, Mexico, 1898–1903. Retired from business at age thirty-six. Privately engaged in exploration and conservation activities, 1903–1928.

CAREER

Earned enough from his business pursuits to be able to devote the rest of his life to exploration, scientific conservation, and big-game hunting. While in Mexico from 1898, became interested in desert bighorn sheep and developed considerable expertise on the subject. Following his retirement from business in 1903, worked closely with staff of the U.S. Biological Survey and with the Boone and Crockett Club (of which he became a member in 1903) in developing a personal program of natural history fieldwork. Collected zoological specimens for the U.S. National Museum of Natural History (USNM), ultimately contributing 554 scientific specimens (including 120 big-game mammals). First visited Mt. McKinley in 1906, spent a year camping in vicinity. Pressed to have it made a national park for ten years (accomplished in 1917). Author of *The Wilderness of the Upper Yukon: A Hunter's Exploration for Wild Sheep in Sub-Arctic Mountains* (1911), *The Wilderness of the North Pacific Coast Islands: A Hunter's Experience While looking for Wapiti, Bears and Caribou on the Larger Coast Islands of British Columbia* (1912), and *The Wilderness of Denali* (1930). Edited *Hunting and Conservation* (a publication of the Boone and Crockett Club, with George B. Grinnell, 1925).

MAJOR CONTRIBUTIONS

Contributed to scientific knowledge of big-game and other mammal species; made major efforts to understand and conserve big-game animals; instrumental in establishment of Mt. McKinley area as a national park. Added much to the literature of exploration in sub-arctic North America.

BIBLIOGRAPHY
Dexter, F. B. *Biographical Sketches of the Graduates of Yale College, 1885–1912.*
Journal of Mammalogy (November 1928) (obit.).
Story, Norah. *The Oxford Companion to Canadian History and Literature* (1967).
Stroud, Richard H., ed. *National Leaders of American Conservation* (1985).
Trefethen, James B. *Crusade for Wild Life* (1961).
Trefethen, James B. *An American Crusade for Wildlife* (1975).

Joseph Bongiorno

SHELFORD, (ERNEST) VICTOR. Born Chemung, New York, 22 September 1877; died Urbana, Illinois, 27 December 1968. Animal and community ecologist, founder of Ecologist's Union (Nature Conservancy).

FAMILY AND EDUCATION

Son of Alexander Hamilton Shelford and Sarah Ellen Rumsey Shelford. Attended West Virginia University, 1899–1901. Student at University of Chicago, 1901–1907, where he received B.S. in 1903 and Ph.D. in 1907. Married Mary Mabel Brown, 1907 (died 17 August 1940); two children.

POSITIONS

Teacher in public schools, Chemung County, New York, 1895–1897. Associate and instructor in zoology at University of Chicago, 1903–1914. Assistant and associate professor of zoology at University of Illinois, 1914–1927, professor, 1927–1946. Biologist in charge of the research laboratories of the Illinois Natural History Survey, 1914–1229. In charge of Marine ecology, Puget Sound Biological Station, alternate summers, 1914–1930. First president of Ecological Society of America, 1915. Chairman of a Committee on the Preservation of Natural Conditions of the Ecological Society, 1917–1938. Chairman of National Research Council Committee on Grasslands, 1932–1939. Founder of Grassland Research Foundation, 1939, president, 1958, chairman of scientific advisory board, 1959–1968. Founder of Ecologist's Union (now Nature Conservancy), 1946.

CAREER

Shelford's early work in ecology at University of Chicago was greatly influenced by Henry C. Cowles. His doctoral thesis on "Tiger Beetles of Sand Dunes," completed in 1907, described the parallels of beetle populations and vegetational succession, an avid interest of Cowles. This work led to series of five publications on "Ecological Succession" in the *Biological Bulletin* in 1911 and 1912. In 1913, Shelford published one of the monumental works in ecology, *Animal Communities in Temperate America*. He helped organize the Ecological

Society of America and became its first president in 1915. Was also concerned with experimental studies in both the laboratory and field, and in 1929 published *Laboratory and Field Ecology*, which served as a methods book for animal ecology. Started the "century-cycle" project in 1933 in the William Trelease Woods at the University of Illinois, designed to study correlations of vertebrate and invertebrate populations with environmental factors. Summary of the first fifteen years appeared in two publications in *Ecological Monographs* for 1951.

Shelford's major interest and reputation were in the field of community ecology. His work included descriptions of both aquatic and terrestrial communities. The summers served as director of marine ecology at Puget Sound (1914–1930) led to a publication in *Ecological Monographs* in 1935. Further publications in this area include a description of stream communities in 1929, one of bottom communities in western Lake Erie in 1942, analysis of tundra communities in 1935 and of Mississippi floodplains in 1954, all published in *Ecology*. He collaborated with F. Clements, and in 1939 *Bio-Ecology* was published, which was an attempt to incorporate animal, plant, and aquatic ecology within the community ideal. His trips throughout North America eventually resulted in his last major work, *The Ecology of North America*, published in 1963. Shelford also emphasized the need for preservation of whole communities and in 1946 founded the Ecologist's Union, which is now known as the Nature Conservancy.

Was recognized as a major figure in American ecology. He was an outstanding teacher with a large number of graduate students but was not considered a polished lecturer. His ideas, enthusiasm, and leadership qualities proved invaluable in helping establish ecology as a recognized scientific discipline.

MAJOR CONTRIBUTIONS

Considered by many to be the "father of animal ecology" in America. Contributed to the areas of physiological and population ecology, but his most significant work was in the field of community ecology. Did much to ensure the preservation of natural communities through the founding of the Ecologist's Union.

BIBLIOGRAPHY

Croker, Robert A. *Pioneer Ecologist; The Life and Work of Victor Ernest Shelford, 1877–1968* (1991).

Kendeigh, S. Charles "Victor Ernest Shelford, Eminent Ecologist—1968." *Bulletin of the Ecological Society of America* 49 (September 1968).

McIntosh, R. P. "Ecology since 1900." In *Issues and Ideas in America* (1976).

Who Was Who in America. Vol. 5 (1973).

Gregg A. Mitman

SHORT, CHARLES WILKINS. Born Greenfield near Versailles, Kentucky, 6 October 1794; died Louisville, Kentucky, 7 March 1863. Physician, botanist, educator.

FAMILY AND EDUCATION

Son of Peyton and Mary Symmes Short. Father was a well-to-do Virginia planter who migrated to Kentucky, mother was eldest daughter of John Cleves Symmes, Revolutionary War colonel, later congressman and judge, who was also a colonizer of vast areas of ''military land'' between the Miami Rivers in Ohio. Early education was at the Joshua Fry School near Danville, Kentucky. At age thirteen, entered Transylvania University and graduated with honors (1811). Studied medicine with his uncle, Frederick Ridgely (1811–1813), and took medical degree at the University of Pennsylvania (1815), where he was much influenced by Caspar Wistar, Nathaniel Chapman, Abbe Jose Francisco Correa de Serra, and Benjamin Smith Barton, the latter being especially important for developing his interest in botany. Married Mary Henry Churchill (1815) of Lamington, New Jersey; had ten children, of whom six (one son, five daughters) lived to maturity. A quiet, calm, modest, and dignified gentleman, he shunned society, gave generously to his family and friends, and was of untarnished moral character.

POSITIONS

Accepted chair of materia medica and medical botany at Transylvania University (1815), but soon thereafter declined and declined a second time (1818). Practiced medicine in Lexington (1816), then in Hopkinsville in western Kentucky (1817–1825). Occupied chair of materia medica and medical botany in the Medical Department of Transylvania University (1825–1837), served as dean (1828–1837). Louisville Medical Institute (1839–1849) served as dean (1840–1849). Professor emeritus of materia medica and medical botany; retired to Hayfield, a beautiful, 230–acre country estate five miles south of Louisville (1849–1863).

CAREER

Published a total of thirty papers appearing between 1828 and 1845 on the vascular-plant flora of Kentucky, botanical nomenclature, botanical history, bibliography, biography, and public lectures to medical students. With John Esten Cooke, founded in 1828 and edited jointly for the next four years *The Transylvania Journal of Medicine and the Associate Sciences*. Named and described six taxa of vascular plants as new to science and had many taxa described from his specimens. Herbarium of about 17,000 species, the largest west of the Allegheny Mountains, was assembled through his own collecting, principally in Kentucky in the 1830s and by contributions from distinguished correspondents in America and Europe and presented to the Academy of Natural Sciences, Philadelphia; between 1833–1838 distributed some 28,000 specimens of Kentucky plants to his correspondents. During retirement, curtailed his writing, shrank from public life, worked in his garden and herbarium, financed botanical expeditions, and entertained friends and relatives.

Publications include *Prodromus Florulae Lexingtoniensis, . . .* (1828); *Florula Lexingtoniensis, . . .* (1828, 1829); *Instructions for the Gathering and Preservation of Plants for Herbaria; . . .* (1833); *A Catalogue of the Native Phaenogamous Plants, and Ferns of Kentucky* (with Robert Peter and Henry A. Griswold, 1833); *Supplementary Catalogues* (1834, 1835, 1837, 1840); *Remarks on the Nomenclature of Botany* (1835); *A Sketch of the Progress of Botany in Western America* (1836); and *Observations on the Botany of Illinois, More Especially in Reference to the Autumnal Flora of the Prairies* (1845).

Elected to membership in various domestic and foreign scientific societies, including the Western Museum Society of Cincinnati, Academy of Natural Sciences of Philadelphia, American Philosophical Society, American Academy of Arts and Sciences. Commemorated in the genus *Shortia*, a member of the Diapensia family with one species, *S. galacifolia*, endemic to the mountains of North Carolina; and in the epithets of several species; for example, *Aster shortii, Astragalis shortianus, Carex shortiana, Solidago shortii.*

MAJOR CONTRIBUTIONS

Best known for his pioneering floristic studies of Kentucky and for his well-prepared prolific plant collections which he liberally distributed worldwide to botanists. As noted by Asa Gray, foremost contemporary authority on American botany, Short "was a very industrious botanist, and an effectual promoter of our science in this country . . . the first in this country to prepare on an ample scale dried specimens of uniform and superlative excellence and beauty." Quality of his specimens set example for future botanists. As a lecturer his style was described as chaste, concise, and classical, and his manner always grave and dignified. Medical students considered him "universally esteemed as a dignified and amiable gentleman, and as a teacher at once sound, practical, able, and instructive!"

BIBLIOGRAPHY

Ahlquist, I. F. "Ohio Valley Culture as Reflected in the Short Family, 1700–1860." Ph.D. Diss. University of Illinois, Urbana (1947).

Coker, W. C. "Letters from the Collection of Dr. Charles Wilkins Short." *Journal of the Elisha Mitchell Science Society* 57 (1941).

Davies, P. A. "Charles Wilkins Short, 1794–1863: Botanist and Physician." *Filson Club Historical Quarterly* 19 (1945).

Gray, A. "Dr. Charles Wilkins Short." *American Journal of Science* 86 (1863) (obit.).

Gross, S. D. "Obituary of Charles Wilkins Short, M.D." *Proceeding of the American Philosophical Society* 19 (1865).

Skaggs, D. S. "Charles Wilkins Short: Kentucky Botanist and Physician 1794–1863." M.A. Thesis, Dept. of History, University of Louisville, Kentucky (1982).

Stuckey, R. L., ed. *Scientific Publications of Charles Wilkins Short . . . with Introduction* (1978).

Titley, J. "Dr. Charles Wilkins Short and the Medical Journals: 1820–1831." *Stechert-Hafner Book News* 19, 3 (1964).

UNPUBLISHED SOURCES

Correspondence with botanists, friends, and family members is widely disbursed but housed principally in the libraries of the Filson Club, American Philosophical Society, Academy of Natural Sciences, Library of Congress, University of North Carolina, and Cincinnati Historical Society. Described by Ahlquist (1947) and Davies (1945).

Ronald L. Stuckey

SHUFELDT, ROBERT WILSON. Born New York City, 1 December 1850; died Washington, D.C., 21 January 1934. Ornithologist, physician, army officer.

FAMILY AND EDUCATION

Son of Rear Admiral Robert Wilson Shufeldt, U.S. Navy, an amateur ornithologist, and Sarah H. Abercrombie Shufeldt. Family moved to Stamford, Connecticut, in late 1850s. During the Civil War accompanied family to Cuba, where Rear Admiral Shufeldt served as consul general. Had two brothers and one sister, who died at birth. Attended public schools in Stamford. Attended Cornell University, 1870–1873. Received M.D. from Columbian College (now George Washington University) in 1876. Married Catherine Babcock, 1876 (died 1892), Florence Audubon, 1895 (divorced 1896?), and Alfhild Dagny Lowum, 1898. Two sons.

POSITIONS

First lieutenant, Medical Department, U.S. Army, 1876–1881, captain, 1881–1904, major, 1904–1934. Retired, medical disability, 1891. Returned to active duty, 1918–1919, Medical Corps U.S. Army, in charge of war collections of Army Medical Museum.

CAREER

As a youth in Connecticut and Cuba developed an interest in ornithology and entomology and began assembling a specimen collection. Learned to prepare bird skins in Stamford under the tutelage of James Jenkins, who had collected in South America and elsewhere in the tropics. Through his father's influence, was invited by Albert S. Bickmore of the American Museum of Natural History to join his polar expedition as naturalist but was unable to join the expedition because of a delay in the telegram notifying him of his appointment. Published his first scientific paper, the "Osteology of the Burrowing Owl" (1881), while on active duty in the Department of the Platte. During the early 1880s served in various capacities at the Army Medical Museum and, informally, at the Smithsonian Institution. Many of his early papers were published by the U.S. Geological Survey through the good offices of Elliott Coues. An accomplished scientific illustrator, furnished many figures for Coues's *Key to North American Birds*, 2d ed. (1884). In late 1880s served at Fort Wingate, New Mexico, before being retired from the military on medical disability. Returned to active duty in 1917–1918 as a curator at the Army Medical Museum.

Major work was in the osteology, morphology, and paleontology of birds. Dealt thoroughly with the osteology of most orders of birds, and many of his accounts remain of value. Was a pioneer in North American paleornithology and named over sixty new fossil birds. Was an early exponent of bird photography and called attention to the value of photography in ornithological studies.

By 1924 had published over 1,500 books and articles and probably amassed some 1,800 by the time of his death. Works were illustrated by some 20,000 figures, most of which were done by Shufeldt. Most were on biological topics, but other topics included law, travel, taxidermy, fine arts, anatomy, museology, botany, entomology, and anthropology. Major publications include *Contributions to the Anatomy of Birds* (1882), *Myology of the Raven* (1890), and *Osteology of Birds* (1909).

MAJOR CONTRIBUTIONS

Pioneering work in osteology and morphology of birds, paleornithology, and natural history photography. *Myology of the Raven* (1890), first complete account of the musculature of a single species of bird, remains a classic.

BIBLIOGRAPHY

Hume, Edgar Erskine. *Ornithologists of the United States Army Medical Corps: Thirty Six Biographies* (1942).

Lambrecht, Kahman. "In Memoriam: Robert Wilson Shufeldt, 1850–1934." *Auk* (October 1935).

The National Cyclopaedia of American Biography. Vol. 6 (1896).

Palmer, T. S., et al. *Biographies of Members of the American Ornithologists' Union* (1954).

Shufeldt, R. W. "Complete List of My Published Writings, with Brief Biographical Notes." *Medical Review of Reviews* (January–September 1920).

Shufeldt, R. W. "Life History of an American Naturalist." *Medical Life* (February, March, April, May, August 1924).

UNPUBLISHED SOURCES

Correspondence with Witmer Stone is at Academy of Natural Sciences of Philadelphia; additional correspondence is in Smithsonian Archives.

William A. Deiss

SIMPSON, GEORGE GAYLORD. Born Chicago, Illinois, 2 June 1902; died Tucson, Arizona, 6 October 1984. Paleontologist, naturalist, educator.

FAMILY AND EDUCATION

Only son and youngest of three children of Joseph Alexander Simpson, lawyer and later a western land developer and railroad claims adjuster who became involved in mining, and Helen Julia Kinney Simpson. Family moved in his infancy to Denver, Colorado, where Simpson graduated from East Denver High School, 1918. Attended University of Colorado, 1918–1919, 1920–1922; Yale University, 1922–1923. Ph.B., Yale, 1923, Marsh Fellow, 1924–1926, Ph.D., 1926. Honorary Sc.D., Yale, 1946, Princeton, 1947, Durham (England), 1951, Oxford, 1951, University of New Mexico, 1954, University of Chicago, 1959, Cambridge, 1965, York University, 1966, Kenyon, 1968, University of Colorado, 1968. Honorary LL.D., Glasgow, 1951; Doctor *honoris causa*, Paris, 1965, Universidad de la Plata, Argentina, 1977. Married Lydia Pedroja, 1923 (divorced, 1938); four daughters. Married Anne Roe, 1938.

POSITIONS

Field assistant under W. D. Matthew on American Museum of Natural History fossil collecting expedition in Texas and New Mexico, New York, 1924. National Research Council and International Education Board Fellow, 1926–1927; researched paleontological materials at British Museum, London, and other collections in England, France, Germany, and Switzerland. Assistant curator of vertebrate paleontology, 1927; associate curator of vertebrate paleontology, 1928–1942; curator of fossil mammals, American Museum of Natural History, 1942–1959. On leave with U.S. Army, 1942–1944, as captain and later major. Chairman, Department of Geology and Paleontology, American Museum of Natural History, New York City, 1944–1958. Professor of vertebrate paleontology, Columbia University, 1945–1959. Agassiz Professor of Vertebrate Paleontology, Museum of Comparative Zoology, Harvard, 1959–1970. Professor of geosciences, University of Arizona, 1967–1982. Professor emeritus, 1982–1984. President and trustee, Simroe Foundation, 1968–1984.

CAREER

Grew interested in nature as result of hiking and camping in Colorado mountains with father. Completed high school at sixteen, despite having had to remain out of school for one year owing to illness. Entered college, briefly dropped out and worked owing to financial difficulties. Transferred to Yale in his senior year on recommendation of a geology professor who had done the same. Dissertation later published as *American Mesozoic Mammals* (1929). Studied Mesozoic fossils at British Museum and elsewhere in Europe on National Research Foundation fellowship, 1926–1927, and by the age of twenty-six was considered an authority on the subject. Published *A Catalogue of the Mesozoic Mammalia in the Geological Department of the British Museum* (1928). During thirty-two year career at American Museum of Natural History, Simpson participated in or led a number of expeditions to Patagonia (1930–1931, 1933–1934); Venezuela (1938–1939); and Brazil (1954, 1956). In addition, there were numerous expeditions to various parts of the western (1929, 1932, 1935, 1936, 1946–1950, and 1952–1954) and southeastern United States (1929–1930). Simpson was professor of vertebrate paleontology at Columbia through cooperative program between Columbia and the American Museum, 1945 to 1959, then moved on to Harvard as Agassiz Professor of Vertebrate Paleontology from 1959 to 1970. Was offered but declined directorship of Agassiz Museum at Harvard. Moved to Arizona for reasons of health in 1967, and became professor of geoscience at University of Arizona, 1970 to 1982, then emeritus professor. His Patagonian journals provided the foundation for his book *Attending Marvels: A Patagonian Journal* (1934). His two-part ''The Beginning of the Age of Mammals in South America'' was published by the American Museum in 1948 and 1967. A projected third part was never published. Permission to do field work in Mongolia, 1934, denied by Soviet Union, although Simpson later (1977) cordially received in Moscow. Badly injured by a falling tree during the 1956 Brazilian trip, Simpson underwent a dozen operations, and was lame in one leg for the remainder of his life. This effectively ended his career of strenuous field work. Simpson uncovered a fossil leaf-eating mammal which was named *Meniscotherium* in Northern New Mexico (1947) and in Colorado (1953), excavated eight skulls of Eohipus, the so-called ''dawn horse.'' Simpson took a number of visits to Europe and at least four to Africa (1961, 1967, 1970, 1972). In 1961, while in Kenya, Simpson was with anthropologist Louis B. Leakey when his host uncovered part of a skull of a fourteen-million-year-old *Ramapithecus*, thought to be an ancestor of humankind. Simpson's three trips to Antarctica (1970, 1971, and 1972) centered on the study of penguins, fossil forms of which he had originally examined in Patagonia. Other trips in the 1970s took him to Central America, the Galapagos, Arctic Canada, Greenland, and the Pacific. Simpson was among an important group of zoologists, including Julian Huxley and Ernst Mayr, among others, who strongly believed that evolution took place within animal populations, rather than with individual organisms. He also felt that quantitative analyses were essential to an understanding of the evolutionary process. In the 1930s, he opposed the theory of continental drift but later ''fully accept[ed]'' plate tectonics, arguing, however, that continental drift had an uneven effect in different parts of the world during the early Age of Mammals (the Cenozoic).

Simpson was a highly prolific author during his career. With his second wife Anne Roe, a clinical psychologist, he published *Quantitative Zoology: Mumerical Concepts and Methods in the Study of Recent and Fossil Animals* (1939, rev. ed., 1960). One early seminal work was *Tempo and Mode in Evolution* (1944), revised and expanded as *Major Features of Evolution* (1953). In this significant study, he discussed the rate at which evolutionary changes had progressed and suggested ways of synthesizing paleontology and genetics. *Principles of Classification and a Classification of Mammals* (1945) was for many years the standard text on the subject. Other titles included *The Meaning of Evolution* [the Terry Lectures] (1949, rev. ed., 1967); *Horses: The Story of the*

Horse Family in the Modern World and Through Sixty Million Years of History (1951); *Evolution and Geography: An Essay on Historical Biogeography, with Special Reference to Mammals: Condon Lectures* (1953); *Life of the Past: An Introduction to Paleontology* (1953); *Life* (with other authors, 1957, rev. ed., 1965); *Behavior and Evolution* (with Anne Roe, 1958); *Quantitative Zoology* (1960); *Principles of Animal Classification* (1961); *This View of Life: The World of An Evolutionist* (1964); *The Geography of Evolution: Collected Essays* (1965); *Biology and Man* (1969); *Penguins: Past and Present, Here and There* (1976); *Concession to the Improbable: An Unconventional Autobiography* (1978); *Why and How: Some Problems and Methods in Historical Biology* (1980); *Splendid Isolation: The Curious History of South American Mammals*, (1980); *Fossils and the History of Life*, (1983); and *Discoverers of the Lost World: An Account of Some of Those Who Brought Back to Life South American Mammals Long Buried in the Abyss of Time* (1984).

Simpson was a co-founder of the Society of Vertebrate Paleontology (1941) and its president (1942), and also helped found the Society for the Study of Evolution (1946), of which he was the first president. He was also president of the Society of Systematic Zoology (1962) and the American Society of Zoologists (1964). He held memberships in, and was a fellow of, a number of other professional organizations, including the American Philosophical Society and American Academy of Arts and Sciences. Simpson won many awards and medals from organizations in North and South America and in Europe, including the Gaudry Medal of the Geological Society of France (1947); Darwin-Wallace Medal (1958), and Gold Medal (1962) of the Linnaean Society of London; the Darwin Medal of the Royal Society; and the National Medal of Science (1966). He was elected to the National Academy of Sciences in 1941, and was later named a fellow of the NAS.

MAJOR CONTRIBUTIONS

Simpson was one of a handful of leading American vertebrate paleontologists active during the late nineteenth and twentieth centuries. He was unique in publishing pioneering studies which demonstrated the compatibility of genetics and paleontological data. His work in the biogeography of North and South America, in evolutionary theory, and in taxonomic work greatly advanced all three fields. In his final decades, new ideas concerning numerical taxonomy and Hennigian, or phylogenetic systematics (cladistics), prompted some to think that his conclusions concerning evolutionary systematics were outmoded. Simpson regarded the principles of Hennigian taxonomy as wrong and vacuous (*Concession to the Improbable*, p. 271).

BIBLIOGRAPHY

American Men and Women of Science. 15th ed.

Birney, Elmer C., and Jerry R. Choate, eds. *Seventy-Five Years of Mammalogy (1919–1994)* (1995).

Gould, Stephen J. "G. G. Simpson, Paleontology, and the Modern Synthesis." In E. Mayr and W. B. Provine, eds. *The Evolutionary Synthesis: Perspectives on the Unification of Biology* (1980).
Laporte, Leo F., ed. *Simple Curiosity: Letters from George Gaylord Simpson to his Family, 1921–1970* (1987).
Mayr, Ernst. "G. G. Simpson." In E. Mayr and W. B. Provine, eds. *The Evolutionary Synthesis: Perspectives on the Unification of Biology* (1980).
New York Times, 8 October 1984 (obit.).
Olson, Everett C. "George Gaylord Simpson." In *National Academy of Sciences Biographical Memoirs* 60 (1991).
Schaeffer, Bobb, and Malcolm McKenna. [Memorial]. *News Bulletin*, Society of Vertebrate Paleontology, no. 1933 (1985).
Who Was Who in America.

UNPUBLISHED SOURCES

Some collected papers, 1926–1968, are in the Science Library at the University of California, Berkeley. Additional papers, including correspondence with his daughter Martha Lee Simpson Eastlake, the manuscript of the book *Meaning of Evolution* (1949), and other writings are at the American Philosophical Society Library, Philadelphia. Some correspondence is in the Julian Huxley Papers, Rice University, Houston, Texas.

Keir B. Sterling

SMITH, JEDEDIAH STRONG. Born Jericho, New York, 24 June 1798 (some sources give 6 January 1799); died Cimarron River, Kansas, 27 May 1831. Fur trapper and trader, explorer.

FAMILY AND EDUCATION

Son of Jedediah Smith, a merchant. Family moved to Western Reserve in Ohio, c.1815. Strongly held Methodist beliefs. Curiosity about West stimulated by reading about Lewis and Clark. Left home 1821. Never married.

POSITIONS

Fur trapper, trader, explorer upper Missouri River, northern Rocky Mountains, Great Basin, Sierra Mountains, Oregon, 1822–1830.

CAREER

Hired by William H. Ashley in 1822 as a trapper. Trapped on upper Missouri and Yellowstone Rivers, 1822–1823. After wintering on Wind River, crossed the South Pass to trap on Green, Bear, and Snake Rivers and visited Hudson's Bay Company post on Clark's Fork of Columbia River, 1824–1825. Fur-trading partnership with Ashley, 1825–1826, and David Jackson and William Sublette, 1826–1830. Explored in Salt Lake area and from Salt Lake to southern and central California, 1826–1827. Made first west-to-east crossing of Sierras and Great Basin from California to Salt Lake, 1827. Returned to California by 1826 route. Trapped Sacramento Valley, crossed coastal range to Pacific via Virgin and Klamath Rivers. Went up coast of Oregon and on to Hudson's Bay Company post, Fort Vancouver, 1827–1828. Returned to North Rockies and trapped, 1829–1830. Ended partnership with Jackson and Sublette and returned to St. Louis, 1830. With intent of publication prepared a map of the West based on knowledge from his own travel and exploration and that of fellow trappers. On trading trip to Santa Fe killed by Comanches, 1831.

Only two of his writings published during his life. An 1827 letter to William Clark summarizing his travels from the Salt Lake to California and return across Great Basin published in slightly condensed form in (St. Louis) *Missouri Republican*, 11 October 1827. Widely reprinted in other newspapers. Translated and published in France in *Nouvelles Annales Des Voyages* (1828). Report (with Jackson and Sublette) to Secretary of War, dated 29 October, 1830, published in U.S. Senate (22d Cong., 2d Sess.), Senate Doc. 39 (Serial 203), 21–23. Untimely death ended plans to publish map and journal. However, A. H. Brue published in France a map of North America that incorporated information from letter to Clark, 1833. Albert Gallatin and David H. Burr used the 1830 manuscript map in preparing maps they published, respectively, in 1836 and 1839.

Strongly held religious beliefs, personal bravery, and great physical endurance carried him through several major Indian attacks, a mauling by a grizzly, starvation and thirst, desert heat and mountain cold. Sought financial fortune through trapping, and fame through exploration.

MAJOR CONTRIBUTIONS

Did much to clarify knowledge of geography of the West from Rocky Mountains to the Pacific Ocean.

BIBLIOGRAPHY

Dale, H. C. *The Ashley-Smith Explorations and the Discovery of a Central Route to the Pacific* (1941, rev. ed.).

Dictionary of American Biography.

Morgan, D. L. *Jedediah Smith and the Opening of the West* (1953).

Morgan, D. L., and C. I. Wheat. *Jedediah Smith and His Maps of the American West* (1954).

Sullivan, M. S. *Jedediah Smith, Trader and Trailbreaker* (1936).

Sullivan, M. S. *The Travels of Jedediah Smith, a Documentary Outline, including the Journal of the Great American Pathfinder* (1934).

UNPUBLISHED SOURCES

See Jedediah Smith Letters, Kansas State Historical Society; reports from Firm of Smith, Jackson, and Sublette in William Clark Papers, Kansas State Historical Society; letter to John H. Eaton, secretary of war, in National Archives.

Richard R. Wescott

SMITH, JOHN. Born Willoughby, England, 9 January 1580; died London, England, 21 June 1631. Explorer, adventurer, writer.

FAMILY AND EDUCATION

Son of George, yeoman farmer, and Alice Smith. Received elementary education in village schools.

POSITIONS

Apprenticed to merchant in King's Lynn, England, 1595. Involved in various European wars and adventures in various capacities, 1597–1606. Councillor, London Company; in Jamestown settlement, 1607–1609; president of the council, 1608–1609; in charge of land defenses of colony, 1609; in England, 1609–1614; captain of three-ship expedition to Plymouth, exploring and mapping coast, 1614; returned to England and began second expedition but became French captive, 1614–1615, returned to England, and remained there until his death.

CAREER

Smith's is one of the more fantastic lives connected to North American colonial history. His father died when he was seventeen, leaving him some property. He soon left the service of the man to whom he had been apprenticed and spent much of the years 1597–1604 as a soldier in Europe. There continues to be much disagreement as to how much of what he supposedly experienced there actually took place and how much of it was the fruit of his very active imagination. His connection with American history began in 1606, when, at the age of twenty-six, he left England with 144 colonists embarked for Virginia. Arriving in the spring of 1607, Smith became one of a seven-man council, though he spent much of his time exploring the surrounding region, learning something of the Indian language, and attempting to keep the colonists supplied with food. During one of his visits with the Powhatans, he allegedly was condemned to death and supposedly rescued by Pocahontas, the eleven-year-old daughter of the tribe's chief. Scholars have for 375 years debated whether Smith was, in fact, slated for execution or undergoing some form of tribal initiation rite. In any event, he was allowed to depart and returned to the settlement early in 1608. Smith was a highly controversial figure in the colony, in part because of his habit of bluntly criticizing individuals and policies with which he disagreed. Conflicting directions from London Company directors in London and opposition from those who opposed him on personal or policy grounds took up considerable time during the settlement's critical early period. Smith spent much of the year 1608 in trading with the Indians and in several expeditions up the Chesapeake. During the summer of 1608, he reached the head of the bay near the present site of Havre de Grace, Maryland, and brought back useful geographical information. Smith found the time to compile and forward to London a "Mappe of the Bay and Rivers, with an Annexed Relation of the Countries and Nations That Inhabit Them" late in 1608. He was in effective charge of the colony from late 1608 until the late summer of 1609. Relieved of his presidency of the colony, Smith was briefly given charge of land defense before again being discharged from all responsibilities owing to fresh instructions from the London directors. The accidental explosion of a gunpowder sack left him badly injured, and some of his enemies tried to dispose of him during his convalescence, so he departed Jamestown for England in October 1609 and did not return. During much of the next several years, he drafted *A Map of Virginia, with a Description of the Country*, an expanded version of the 1609 ''Mappe of the Bay'' (published 1612), an excellent geographic and ethnological account with a fine map, which also served to ''sell'' the idea of colonization to many people. He described animals and plants but tended to follow what Thomas Hariot had to say on these subjects, while adding some of his own observations. His physical description of the opossum, however, was evocative and quite original. Twice in 1614, he led expeditions to the New England colonies. During the first, he made detailed maps and charts of the coast. His second ended in disaster when he was captured by the French. For much of the rest of his life following his return to England in 1615, Smith churned out a succession of books. *Description of New England* appeared in 1616, *New England's Trials*, 1620 and 1622, and *The Generall Historie of Virginia, New England, and the Summer Isles*, 1624. This last was for the most part an anthology of much that Smith had previously published. The two printers who did the first and last parts of the job clearly were not in good communication, since pages 97 through 104 do not exist in any extant copy of the book. Though dismissed by many who had experienced Smith's biting criticism, it established his standing as an author. Three other volumes followed during 1624–1627, all providing useful guidance for the practical soldier or sailor. These included *An Accidence or the Path-way to Experience* (1626) and *The Seaman's Grammar* (expanded version of *An Accidence*) (1627). *The True Travels, Adventures, and Observations of Captain John Smith in Europe, Asia, Africa, and America, from Anno Domini 1593 to 1629*, an interesting mixture of verifiable fact and much that was not, appeared in 1630. His final work, *Advertisements for the Unexperienced Planters of New England, or Any Where*, was published in 1631. Smith sometimes provided multiple versions of some of his life experiences in his various publications, which has led to confusion and many lingering doubts concerning some of the events he describes.

MAIN CONTRIBUTIONS

Smith was an indefatigable promoter of colonization in the early seventeenth century and a knowledgeable historian (at a time when history as a discrete field of inquiry was in its infancy), geographer, geologist, cartographer, and ethnologist. His writings show some interest in the relationship of Native Americans to their environment. He was the first to make anything approaching a comprehensive exploration of Chesapeake Bay and among the first to provide detailed descriptions of the coastal areas of New England. He also was the first to give the region known as New England its name (1616). A practical man, his observations on plants and animals tended more toward discussions of the uses to which they could be put.

BIBLIOGRAPHY

Arber, Edward, ed. *Travels and Worlds of Captain John Smith* (1884, 1910).
Barbour, Philip L. *The Three Worlds of Captain John Smith* (1964).
Barbour, Philip L., ed. *The Complete Works of Captain John Smith, 1580–1631*. 3 vols. (1986).
Davis, Richard B. *Intellectual Life in the Colonial South, 1585–1763*. 3 vols. (1978).
Dictionary of American Biography.
Emerson, Everett H. *Captain John Smith* (1993).
Hayes, Kevin J. *Captain John Smith: A Reference Guide* (1991).
Lemay, J. A. L. *The American Dream of Captain John Smith* (1991).
Ristow, Walter W. *Captain John Smith's Map of Virginia* (1957).
Stearn, Raymond P. *Science in the British Colonies of America* (1970).
Vaughan, Alden T. *American Genesis: Captain John Smith and the Founding of Virginia* (1975).

Keir B. Sterling

SOPER, J. DEWEY. Born near Guelph, Ontario, 5 May 1893; died Edmonton, Alberta, 2 November 1982. Ornithologist, mammalogist, artist.

FAMILY AND EDUCATION

Son of Joseph Soper and Ilonia Elizabeth McLaughlin. Studied zoology at the University of Alberta, graduated 1923. Awarded LL.D., honoris causa, University of Alberta, 1960. Presented with the Douglas Pimlott Award from the Canadian Nature Federation, 1980. Married Carolyn Freeman, 1927; two children.

POSITIONS

Naturalist with the Canadian Government Arctic Expedition, 1923. Collected and explored in southern Baffin Island, 1924–1925, 1926, 1928–1929, 1930–1931. Boundary wildlife survey, along 49th parallel of Canadian prairie provinces, 1927. Wildlife officer studying bison in Wood Buffalo National Park, 1932–1934. Chief migratory bird officer for the Prairie Provinces, 1934–1948. Chief migratory bird officer for Yukon and the Northwest Territories, 1948–1952. Honorary research associate, University of Alberta Museum of Zoology, 1961–1970. Dewey Soper Bird Sanctuary on Baffin Island was named for him, in addition to Soper River and Soper Highlands, in 1957.

CAREER

In four arduous years in the Arctic, in which he covered 30,000 miles by water and by land, much of it by dog team, he collected birds and mammals and finally, in June 1929, was the first to discover the breeding grounds of the blue goose, then considered a separate species. He made surveys of many parts of the Canadian prairies, Yukon, and Northwest Territories.

MAJOR CONTRIBUTIONS

Finder of the breeding grounds of the blue goose. Made extensive surveys of birds and mammals within Canada's western national parks. Published 120 articles, two books (one on mammals and one on Baffin Island), and ten monographs of the birds and mammals within the following national parks: Riding Mountain, Prince Albert, Elk Island, Waterton Lakes, Wood Buffalo, Banff, Jasper. Promoted the need for preservation of wetlands habitat and upland nesting cover on the prairies. Collected more than 10,000 scientific specimens of mammals and birds. His collection of over 1,000 arctic photographs has been donated to the Boreal Institute for Northern Studies, University of Alberta.

BIBLIOGRAPHY

Ealey, D. M., G. W. Scotter, W. A. Fuller, and L. N. Carbyn. "D. Dewey Soper, a Life 1893–1982." *Alberta Naturalist* 14 (1984).

Soper, J. D. *The Blue Goose*. Ottawa: Department of Interior (1930).

Soper, J. D. *Mammals of Alberta* (1964).

Soper, J. D. *Canadian Arctic Recollections: Baffin Island, 1923–1931*. Ed. S. Milligan. University of Saskatchewan (1981).

Soper, R., and T. Beck. "Joseph Dewey Soper, 1893–1982." *Arctic* 36 (1983).

C. Stuart Houston

SPREADBOROUGH, WILLIAM. Born Farnham, England, 12 November 1856; died Esquimalt, British Columbia, 30 March 1931. Natural history collector, naturalist.

FAMILY AND EDUCATION

Emigrated to Bracebridge, Canada West (Ontario), at age five with family. Raised in backwoods along south branch of Muskoka River. Educated at local public school. Married Jessie Allen and later Jessie Dumbreck; no children.

POSITIONS

Field assistant, temporary staff, Geological Survey of Canada, 1889–1919.

CAREER

Taken on as summer field assistant in 1889 by John and James Macoun of Natural History Division, Geological Survey of Canada. Attached to various survey parties in summer, returned to odd jobs in Bracebridge in winter. Responsible for zoological specimens, mostly birds. Often sent ahead to field during spring migration. Collected largely in western Canada. Accidentally shot man while collecting at Indian Head in 1892. Moved to Victoria in 1901 and began to collect year-round. Provided much of field data for *Catalogue of Canadian Birds* (1900, 1904). Refused to take part in any more fieldwork following death of Macouns in 1920. Committed suicide the day he retired from his job with the Esquimalt municipality.

Efficient, though relatively unknown, field-worker who was at home in the wilds. Skilled with the skinning knife. Able to make the most unpromising camp comfortable. Remarkable recall of specimens and location. Devoted to his work and the Macouns.

MAJOR CONTRIBUTIONS

Bird specimens and field notes marked beginnings of systematic study of Canadian ornithology at Geological Survey of Canada.

BIBLIOGRAPHY

Macoun, J. *Autobiography of John Macoun* (1922).

Taverner, P. A. "William Spreadborough—Collector." *The Canadian Field-Naturalist* (March 1933).

Waiser, W. A. *The Field Naturalist: John Macoun, the Geological Survey and Natural Science* (1989).

UNPUBLISHED SOURCES

See National Museums of National Sciences historical correspondence, National Archives of Canada, Ottawa.

William A. Waiser

STANSBURY, HOWARD. Born New York City, 8 February 1806; died Madison, Wisconsin, 17 April 1863. Army officer, explorer.

FAMILY AND EDUCATION

Son of Arthur Joseph and Susanna Brown Stansbury. Received education as civil engineer, but no information as to where this was done. Married Helen Moody, 1827; at least two children.

POSITIONS

As civilian engineer in 1828–1832, had charge of succession of surveys examining possibility of connecting Lakes Erie and Michigan and Wabash River with canals. Surveyed route of Mad River and Lake Erie Railroad and mouths of several rivers and in charge of several public works projects in Indiana, 1832–1835. Had charge of surveys for improving harbor of Richmond, Virginia, 1836, and rail route from Milwaukee to the Mississippi River, 1838. Appointed lieutenant, Corps of Topographical Engineers, U.S. Army, 1838, promoted captain,

1840. Involved in surveys of Great Lakes and harbor of Portsmouth, New Hampshire, 1840–1848. Selected by Col. J. J. Abert, chief, Corps of Topographical Engineers, to survey and explore Great Lake and surrounding area, 1849–1850. Made further surveys in Great Lakes region and constructed military railroads in Minnesota, 1851–1861. Army mustering officer in Columbus, Ohio, 1861; promoted major. Retired, September 1861, but subsequently reentered army as mustering and disbursing officer for Wisconsin and served 1862(?)–1863; died while on active duty.

CAREER

Stansbury is perhaps best known for the reconnaissance of the Great Salt Lake made by him and his eighteen-man party in 1849–1850. He was to have joined a regiment of mounted rifles on Oregon Trail to Fort Hall, in what is now southeastern Idaho, but missed connection with armed party at Fort Leavenworth, Kansas, and proceeded on his own. Reached Fort Bridger, now in extreme southwestern corner of Wyoming, from which he identified new route to Salt Lake, reached it, and then went completely around it. Remained in area longer than planned and returned to Fort Leavenworth following previously discovered route, which he later recommended for use by Union Pacific Railroad in November 1850. Stansbury followed up on explorations previously made by Capt. Benjamin Bonneville in 1831 and by Capt. John Fremont during 1843–1844. His was the first detailed study of the area, however. His interestingly written *Exploration and Survey of the Valley of the Great Salt Lake of Utah, including a Reconnaissance of a New Route through the Rocky Mountains*, originally published as a Senate document in 1852, proved to be a great success both in the United States and in England and was several times republished commercially in the years that followed. Stansbury's was the first detailed account, not only of the geography and resources of the area but of Mormon Society. This last was made easier because Stansbury took pains to assure Mormon leader Brigham Young that he had not come to the area to spy, and, in fact, Stansbury's report revealed that he had some respect for what the Mormons had accomplished. The rest of Stansbury's career was spent out of the public spotlight. He performed a variety of surveying and engineering assignments, mainly in the Great Lakes region, and then was involved in recruiting and disbursing duties in Wisconsin, with an office in the city of Madison.

MAJOR CONTRIBUTIONS

Stansbury's report is probably his most significant achievement. In his text, he included discussions of what were then unusual fauna, including rattlesnakes, prairie dogs, burrowing owls, and vertebrate fossil animals. Several appendixes, by contemporary authorities such as S. F. Baird, Charles Girard, Titian Peale, John Torrey, and James Hall, listed the mammals, birds, reptiles, insects, plants, and fossils collected in the course of the expedition. Four reptiles had previously been undescribed. Stansbury's geographical and geological comments were also of considerable interest. His discussion of the Mormons was doubtless the one element in his book that elicited the greatest public interest five years before the federal government felt it necessary to suppress the so-called Mormon insurrection in the late 1850s.

BIBLIOGRAPHY

Dictionary of American Biography.

Ewan, Joseph, and Nesta Dunn Ewan. *Biographical Dictionary of Rocky Mountain Naturalists* (1981).

Fowler, Don D. "Introduction" to reprint edition of *Exploration of the Valley of the Great Salt Lake*, by Howard Stansbury (1988).

Goetzmann, William H. *Army Exploration in the American West, 1803–1863* (1959).

Who Was Who in America.

Keir B. Sterling

STEFANSSON, VILHJALMUR. Born Arnes, Manitoba, 3 November 1879; died Hanover, New Hampshire, 26 August 1962. Arctic explorer, ethnologist, writer.

FAMILY AND EDUCATION

Son of Ingiborg (Johannesdottir) and Johann Stefansson, Icelandic immigrants who settled near Gimli, Manitoba, in 1875. Family moved to North Dakota the year after his birth. Father died 1892. Little formal schooling, attended preparatory school and the State University of North Dakota (1898–1902) until expelled; State University of Iowa (A.B., 1903); scholarship, Harvard University Divinity School (1903–1904); Harvard University Graduate School (1904–1906, A.M. 1923). Numerous honorary degrees, medalist of all major geographical societies. Fathered one child, Alec, with Fanny Pannigabluk, in 1910. Married Evelyn Baird, 1941; no children.

POSITIONS

Ethnologist, Anglo-American Polar Expedition (1906–1907); Stefansson-Anderson Expedition (1908–1912); commander, Canadian Arctic Expedition (1913–1918); commissioner, Royal Commission to Investigate the Possibilities of the Reindeer and the Musk-Ox Industries in the Arctic and Sub-Arctic Regions of Canada (1919–1920); president, Stefansson Arctic Exploration and Development Company, Ltd.; president, Explorers Club, 1919, 1932, 1937; adviser, Pan American Airways, 1932–1945; adviser, U.S. Office of the Coordinator of Information, 1941–1942; editor, *Encyclopedia Arctica* for the U.S. Office of Naval Research, 1946–1949; consultant, Polar Studies Program, Dartmouth College, 1952–1962.

CAREER

Visited Iceland (1904), archeological expedition there with Peabody Museum of Harvard University (1905). Boaz-influenced ethnologist to Anglo-American Polar Expedition (Harvard/University of Toronto 1906–1907). Lived with the Inuit (Eskimos) of the Mackenzie Delta. Second expedition with University of Iowa classmate Rudolph M. Anderson (American Museum of Natural History/ Canadian Geological Survey 1908–1912). First contact with "Blonde Eskimos," the Copper Inuit. Gained media and public attention by claiming they were related to lost Scandinavian colony. Published *My Life with the Eskimos* (1913). Commanded Canadian Arctic Expedition (1913–1918); inauspicious start—loss of vessel *Karluk* trapped in the ice, eleven crew died. Clashed with Anderson and others in dispute over instructions and authority. Ordered dangerous drift survey among ice floes; discovered last unknown large islands in the Arctic. Expedition ended with accusations of mutiny and antagonism. Promoted economic development as member of Royal Commission on reindeer-and musk ox-ranching. Advised Hudson's Bay Company in disastrous Lapland-style reindeer project (1921–1925). In 1920 drew Canada's attention to the fragile nature of claims to arctic islands, such as Ellesmere. Offered to lead expedition to strengthen same but was discredited and distrusted in Ottawa. Canadian claims enforced without him. Attempted to wrest Soviet claims to Wrangel Island by forming a company and sending a small expedition in 1921. The 1923 news of their death by starvation discredited Stefansson in Canada.

Thereafter worked from New York City's Greenwich Village, as an arctic adviser, public lecturer, and popular writer. After Wrangel, other explorers attacked his book *The Friendly Arctic* (1921) as dangerously romantic. However, his visionary writings, beginning with *Northward Course of Empire* (1922), emphasized the importance of the Arctic in the geopolitics of submarines and Great Circuit Route air travel. Spent the 1930s and 1940s as an adviser to Pan American Airways and the U.S. military on survival and travel in the Arctic. From 1945 to 1949 worked on the *Encyclopedia Arctica*, a U.S. Navy-funded research group that sought Soviet cooperation on arctic information. Cold war sensitivity in 1948 led the navy to cancel the project. Stefansson became a victim of the "Red Scare," having written positively on the Soviet Arctic, as in *New Compass of the World* (1949); for his comments on Eskimos as "perfect communists"; his friendship with Owen Lattimore, declared a Soviet spy master by Joseph McCarthy; and accusations against his wife (unfounded). This ended official government support. Sold his extensive arctic research library to Dartmouth College (1952). Consultant to its Polar Studies Program until his stroke.

MAJOR CONTRIBUTIONS

An egotistical arctic visionary who captured the public's imagination. His insistence that adaptation to local ecology, indigenous diet, and traditional knowledge would assure survival set him apart from other explorers. His exploration and advice secured Canada's northern boundaries after World War I. First to grasp the strategic implications of the Arctic, his views, expressed to governments that came to mistrust him, directly shaped state policies for the Arctic.

BIBLIOGRAPHY

Allen, Phillip E. *One Came Late* (1992).

Diubaldo, Richard J. *Stefansson and the Canadian Arctic* (1978).

Dictionary Catalog of the Stefansson Collection on the Polar Regions in the Dartmouth College Library. 8 vols. (1967).

Hunt, William R. *STEF: A Biography of Vilhjalmur Stefansson, Canadian Arctic Explorer* (1986).

Le Bourdais, D. M. *Stefansson, Ambassador of the North* (1962).

Lloyd, Trevor. ''Vilhjalmur Stefansson.'' *Polar Notes*, no. 4 (November 1962).

Mattila, Robert W. *Chronological Bibliography of the Published Works of Vilhjalmur Stefansson, 1879–1962* (1978).

Stefansson, Georgina. ''My Grandfather, Dr. Vilhjalmur Stefansson.'' *North* 8 (July–August 1961).

Stefansson, Vilhjalmur. *Discovery: The Autobiography of Vilhjalmur Stefansson* (1964).

Stefansson, Vilhjalmur. *Stefansson–Anderson Arctic Expedition* (1919, 1978).

Who's Who in Canada. Vol. 50 (1962–1963).

UNPUBLISHED SOURCES

Diubaldo and Hunt have extensive references. The Stefansson Collection is at the Baker Library, Dartmouth College, Hanover, New Hampshire. National Archives of Canada, Ottawa, has collections such as the Canadian Arctic Expedition, government files, and key material in the Rudolph M. Anderson and J. B. Harkin Collections.

Lorne Hammond

STEGNER, WALLACE EARLE. Born Lake Mills, Iowa, 18 February 1909; died Santa Fe, New Mexico, 13 April 1993. Novelist, teacher, conservationist.

FAMILY AND EDUCATION

Son of George and Hilda (Paulson) Stegner. Family moved several times during 1909–1921 among North Dakota, Washington, Montana, Wyoming, Saskatchewan and settled in Salt Lake City, Utah in 1921. Graduated from high school and entered University of Utah in 1927; received B.A. in 1930. Received M.A. in English literature from the University of Iowa in 1932 and Ph.D. in English in 1935. Married Mary Stuart Page in 1934 and had son, Stuart Page, in 1935.

POSITIONS

Taught one semester at Augustana College in Rock Island, Illinois, 1934. Teacher at the University of Utah, 1934–1937, and at University of Wisconsin, 1937–1939. From 1939 to 1945 was Briggs-Copeland instructor of Writing at

Harvard University. In 1945 became Professor of English at Stanford University. From 1946 to 1971 was Director of Stanford's Creative Writing Program.

West Coast editor for Houghton Mifflin, 1945–1953. In 1961, became special assistant to Secretary of the Interior Stewart Udall. From 1962 to 1966 served as member of the National Parks Advisory Board. Editor in chief of magazine *American West*, 1966–1968.

CAREER

A prominent American writer, often referred to as ''the dean of western writers,'' who wrote more than two dozen works of fiction, history, biography, essays, and numerous articles. Most of his writings, both fiction and nonfiction, dealt with the West. Was dedicated to environmental causes, which he supported through his writings and speeches. As a result, he contributed to changed perception of the West and raised nation's consciousness about the environment.

His novel *Angle of Repose* (1971) brought him the Pulitzer Prize for fiction in 1972; and another, *The Spectator Bird* (1976), won National Book Award for Fiction in 1977. Earlier, his first novel, *Remembering Laughter* (1937), won Little Brown Prize and brought Stegner to national attention.

During his Harvard years wrote *On a Darkling Plain* (1940); *Fire and Ice* (1941); *Mormon Country* (1942); *The Big Rock Candy Mountain* (1943), which was heavily influenced by his family experiences; and *One Nation* (1945), dealing with racial minorities, for which he received the Houghton Mifflin Life-in-America Award and shared the Anisfield-Wolfe Award.

At Stanford was teacher and founder of Creative Writing Program, which produced such writers as Larry McMurtry, Robert Stone, Wendell Berry, Tom McGuane, and others. In his own words, ''I had a sense . . . that I was seeing American literature before it was in print.''

During the next decades he continued addressing old and new themes, such as the complicated aspects of human nature, past and present, death, old and young, human isolation and inequality, skimming of the land, and reverence for nature.

Among titles he published were *The Preacher and the Slave* (1950); *The Woman on the Wall* (1950); *Beyond the Hundredth Meridian* (1954); *This Is Dinosaur* (1955); *A Shooting Star* (1961); *Wolf Willow* (1962); *All the Little Live Things* (1967); *The Sound of the Mountain Water* (1969); *The Uneasy Chair* (1974); and *Crossing to Safety* (1987).

In 1981, together with son Page, published *American Places*, essays on the wilderness areas in United States, and in 1992 published *Where the Bluebird Sings to the Lemonade Springs: Living and Writing in the West*, a collection of autobiographical essays.

Received numerous awards, grants, and fellowships, including O. Henry Memorial Short Story Award in 1942, 1950, and 1954; Rockefeller Fellowship to conduct seminars for writers in Far East, 1950–1951; Guggenheim Fellowships, 1950, 1959; Wenner-Gren Foundation Grant, 1953; Fellow at the Center for

Advanced Studies in the Behavioral Sciences, 1955–1956; Writer in Residence, American Academy in Rome, 1960; received Senior Fellowship from National Endowment for the Humanities, 1972; Robert Kirsh Award for Life Achievement, 1980; Life-Time Achievement Award from the PEN USA Center West, 1990; and Senior Fellowship from National Endowment for the Arts, 1990.

Member of National Institute of Arts and Letters, American Academy of Arts and Sciences, Board of Directors of the Sierra Club, and Governing Council of Wilderness Society.

The editor of *Wilderness* magazine called him ''the modern equivalent of Thoreau.'' Supported legislation to prohibit construction of dams in the national parks. Joined a campaign of support for ''Wilderness Act,'' which was eventually signed into law by President Johnson in 1964. Except for trips to Far East and Europe, he did not travel abroad extensively.

MAJOR CONTRIBUTIONS

Was a major contributor to the western literature of United States and Canada. Brought attention to fragility of the environment and need for its protection. Identified strong connection between the land and people. Became one of the important champions of the conservation movement in United States, providing its inspiration, encouragement, and conviction.

BIBLIOGRAPHY

Arthur, Anthony. *Critical Essays on Wallace Stegner* (1982).

Barnes, Bart. ''Wallace Stegner Dies; Prize-Winning Author.'' *The Washington Post*, 15 April 1993 (obit.).

Benson, Jackson. *Wallace Stegner: His Life and Work* (1996).

Colberg, Nancy. *Wallace Stegner: A Descriptive Bibliography* (1990).

Current Biography (1977).

Honan, William H. ''Wallace Stegner Is Dead at 84; Pulitzer Prize-Winning Author.'' *The New York Times*, 15 April 1993 (obit.).

Stegner, Page, and Mary Stegner, eds. *The Geography of Hope: A Tribute to Wallace Stegner* (1996).

Watkins, T. H. ''Wallace Stegner 1909–1993.'' *Wilderness* 56. 201 (Summer 1993) (obit.).

Willrich, Patricia Rowe. ''A Perspective on Wallace Stegner.'' *Virginia Quarterly Review* 67. 2 (Spring 1991).

UNPUBLISHED SOURCES

Wallace Earl Stegner Correspondence, literary manuscripts, drafts, printer's copies, and galley proofs relating to Stegner's works, including *Beyond the Glass Mountain, Beyond the Hundredth Meridian, The Big Rock Candy Mountain, Mormon Country, The Preacher and the Slave, Remembering Laughter*, and *The Woman on the Wall*; Special Collections Department, University of Iowa Libraries, Iowa City, Iowa; Wallace Earle Stegner Manuscript of *The Gathering of Zion* and Wallace Earle Stegner Papers, 1868–1879 (typescripts and photocopies relating to John Wesley Powell and the Colorado River), University of Utah, Marriot Libraries, Manuscript Division, Salt Lake City; Wallace Earle Stegner Creative Writing Program (correspondence and manuscripts, 1949–

1992); Wallace Earle Stegner Papers, 1961–1965 (includes manuscripts, typescripts, correspondence, photographs, and documents relating to *A Shooting Star* and *The Gathering of Zion*); Stegner Conservation Collection, 1956–1974 (includes printed materials; reports, articles, magazines, and so on relating to conservation), Department of Special Collections, Stanford University Libraries, Stanford, Calif. Bernard Augustine De Voto Papers, 1885–1974 (includes correspondence with Wallace Stegner), Department of Special Collections, Stanford University Libraries, Stanford, Calif.

Joseph Henry Jackson Papers, c.1931–1955 (includes correspondence with Stegner), Bancroft Library, University of California, Berkeley.

Andrew Sankowski

STEJNEGER, LEONHARD (HESS). Born Bergen, Norway, 30 October 1851; died Washington, D.C., 28 February 1943. Zoologist, especially in ornithology and herpetology, museum curator, systematist.

FAMILY AND EDUCATION

Son of Bergen businessman Peter Stamer and Ingeborg Catharina Hess Stejneger. Attended Smith Theological School, 1859–1860, and Latin School, 1860–1869, in Bergen; University of Kristiania in 1870 as a candidate in art; candidate in philosophy, 1872; later studied medicine and received law degree in 1875. Received honorary Ph.D. from Kristiania in 1930. Spent winters of 1869 to 1871 in Meran due to mother's delicate health. Emigrated to United States in 1881 and naturalized on 4 February 1887. Married Anna Normann, 1876 (divorced in ?), and Helene Maria Reiners, 1892. Adopted Inga in Kristiania in 1907.

POSITIONS

Observer, U.S. Signal Service and Smithsonian representative at Kamchatka and Commander Islands, 1882–1883; assistant curator of birds, U.S. National Museum (USNM), Smithsonian, 1884–1889; curator of reptiles, USNM, 1889–1943; head curator of biology, USNM, 1911–1943. Member of U.S. Fur-Seal Commission, 1896–1897; International Committee on Zoological Nomenclature, 1898–1943 and its Permanent Committee. Received Knight First Class of the Royal Norwegian Order of St. Olaf, 1906, and Commander of the Order, 1939. Chaired Smithsonian committee to consider manuscripts for publications.

CAREER

Began publishing scientific works at age sixteen with notes on the ornithology of Meran and described first species in 1878. Left Europe because of meager returns received by naturalists. Continued to study avifauna, with particular interest in aquatic birds, after immigration to United States. Worked with Robert Ridgway and Albert Kendrick Fisher at USNM on birds. Initial exploration work took him to Commander Islands in 1882, where he established first weather observation station for Signal Service on Bering Island and studied fur seal rookeries. Interest in Georg Wilhelm Steller resulted from this trip with an exhaustive biography, *Georg Wilhelm Steller, the Pioneer of Alaskan Natural History* (1936). Returned in 1883 to inspect weather stations and studied fur seals on Copper Island. Collected bones of the seacow, skeleton of extinct Pallas's cormorant, and birds, with last described in *Results of Ornithological Explorations in the Commander Islands and in Kamtschatka* (1885). Also made substantial contributions to the *Natural History of Birds* in John Sterling Kingsley's *Standard Natural History* (1885). Interest in fur seals continued, even though scientific research and administrative responsibilities were directed in other areas. Served as attaché, 1895, to Bering and Copper Islands for Fish Commission, with *The Russian Fur Seal Islands* published that year; member of International Fur Seal Commission, 1896, to study rookeries of Pribilof and Kurile Islands, Robben Island, and Japan; and a follow-up trip in 1897 to Bering, Copper, and Commander Islands and Japan. For U.S. Department of Commerce, 1922, examined northern fur seal rookeries to note conditions developed since last international agreement in 1911.

Also studied extensively birds of Japan and Hawaii until named acting curator of the Department of Reptiles and Batrachians, USNM, in 1889, when a former curator resigned. This position resulted in a new focus of scientific research for him, for which he is best known. According to one colleague, investigations in herpetology resulted in his becoming "one of the foremost authorities in the systematics of this branch of science." Major works in this area include *Herpetology of Porto Rico* (1904), his widely cited *Herpetology of Japan and Adjacent Territories* (1907), and *Check-List of North American Amphibians and Reptiles* (with Thomas Barbour, 1917, 1923, 1933, 1939, 1943), which provides a "standard for forms, names and distribution" for these animals. Published over 100 works on reptiles and amphibians. Fifteen reptiles and amphibians named in his honor.

Knowledge of several languages made him well-suited representative to International Zoological Congresses. Attended the 4th, Cambridge, England, 1898; 5th, Berlin, 1901; 6th, Bern, 1904; 7th, Boston, 1907; 9th, Monaco, 1913; 10th, Budapest, 1927; 11th, Padua, 1930; and 12th, Lisbon, 1935. Also attended International Fisheries Exposition, Bergen, 1898; Fourth International Ornithological Congress, London, 1905; and International Entomological Congress, Madrid, 1935. While in Europe, studied correlation between European and North American life zones.

Was exempted from compulsory federal retirement in 1932 by executive order because of his outstanding contributions to his field.

MAJOR CONTRIBUTIONS

Publications marked by quality research span seventy-two years; most of which still hold up today for research value and nomenclature. Laid foundation for later studies in herpetology. Played prominent role in standardizing zoological nomenclature because of breadth of scientific knowledge and literature coupled with ability as linguist.

BIBLIOGRAPHY

Fisher, Albert Kendrick. ''Leonhard Stejneger.'' *Copeia* 3 (30 October 1931).

Schmitt, Waldo L. ''Leonhard Stejneger.'' *Systematic Zoology* 13 (1964).

Wetmore, Alexander. *Biographical Memoir of Leonhard Hess Stejneger, 1851–1943.* Washington, D.C.: National Academy of Sciences (1946).

UNPUBLISHED SOURCES

See Division of Reptiles and Amphibians, 1873–1968, Records, in Smithsonian Archives, Record Unit 161. In addition to containing official administrative records of the division, a successor of the Department of Reptiles and Batrachians, records contain material pertaining to the International Zoological Congresses Stejneger attended as a representative of the USNM.

Also see Leonhard Stejneger Papers, 1753, 1867–1943, in Smithsonian Archives, Record Unit 7074. Papers consist of manuscripts on herpetology, ornithology, fur seals, life zones studies, and Stellar; diaries; notebooks and account books from field trips and attendance at zoological congresses; photographs of fur seals and inhabitants of North Pacific–Bering Sea area; scrapbooks; correspondence; personal material; and a copy of Stellar's *Beschreibung von sonderbaren Meerthieren* (1753).

William R. Massa, Jr.

STERNBERG, CHARLES HAZELIUS. Born Middleburg, New York, 15 June 1850; died Toronto, Canada, 20 July 1943. Naturalist, paleontologist.

FAMILY AND EDUCATION

Son of Levi Sternberg, minister and principal of a seminary, and Margaret Levering Miller Sternberg. Educated at Hartwick Seminary, New York, Iowa Lutheran College, and one year at Kansas State Agricultural College as student of Benjamin Mudge (1875–1876); did not take a degree. Honorary A.M., Midland College, Atchison, Kansas, 1911. Married Anna Musgrove Reynolds, 1880; three sons (all of whom followed their father into paleontological work) and one daughter (died).

POSITIONS

Collected fossils in West for E. D. Cope, 1876–1879, 1894, 1896, 1897; for Alexander Agassiz, 1881, 1882; for O. C. Marsh, 1884. Collected fossils for several Canadian and British museums under contract with Geological Survey of Canada, 1912–1916. Later collected fossils for the University of California. Conducted own Paleontological Laboratory, 1917–1939. Collected on contractual basis for museums in the United States and Europe over a period of more than forty-five years.

CAREER

Sternberg early began developing an interest in nature while a child in upstate New York. He began collecting fossils when he moved with his family to Kansas in 1867, starting with fossil plants (which he sent to Smithsonian Institution) and soon moving on to fossil vertebrates. He began his professional career in 1876, when he appealed to E. D. Cope for funds to secure a wagon, horses, and equipment for collecting in western Kansas. Cope responded favorably and later employed Sternberg on a number of occasions. This helped Sternberg get over his failure to have Mudge, his mentor at Kansas State College, sign him up with Cope's great rival O. C. Marsh that same year. Marsh, however, later made use of Sternberg's services. Sternberg early developed standard technique of burlap strips with rice paste (later flour paste or plaster of paris) for encasing and preserving fossil specimens, used well into the twentieth century. Much of Sternberg's time was devoted to the services of Cope, the younger Agassiz, and Marsh until the early 1890s, when he began supplying specimens to museums here and abroad as independent contractor. His association with the Geological Survey of Canada ended in 1916, owing to a disagreement with Lawrence Lambe concerning plans for fieldwork. One famous shipment of fossils destined for the British Museum became a war casualty when the vessel *Mount Temple* was torpedoed in 1917. Sternberg's three sons, George, Charles, and Levi, followed their father into paleontological work, George at Fort Hays State College in Kansas, Charles with the Canadian Paleontological Survey, and Levi as paleontologist with the Royal Ontario Museum in Toronto. Two of the sons, George and Levi, discovered a complete duck-billed dinosaur (*Anatosaurus*) in southern Wyoming in 1908, while their father and other brother were off getting supplies for several days. This specimen ended up in the American Museum of Natural History in New York. The elder Sternberg told his son, "George, this is a finer fossil than I have ever found. And the only thing which would have given me greater pleasure would have been to have discovered it myself" (G. F. Sternberg 1930). A later specimen of the same species ended up in a museum in Frankfurt, Germany. The elder Sternberg retired following the death of his wife in 1939.

MAJOR CONTRIBUTIONS

A redoubtable fossil collector and father of three sons who continued the family tradition well into the mid-twentieth century. The discovery by two of Sternberg's sons, George and Levi, of several mummified *Anatosaurus* (duck-billed dinosaurs) in Wyoming in 1908, complete with impressions of the skin, was a dramatic major find; the family contributed many other successes to the field of vertebrate paleontology over a period of more than seventy-five years. Specimens found by the Sternbergs can be found today in a number of museums in the United States, Canada, and Europe.

BIBLIOGRAPHY
Colbert, Edwin H. *Men and Dinosaurs: The Search in Field and Laboratory* (1968).
Ewan, Joseph, and Nesta Dunn Ewan. *Biographical Dictionary of Rocky Mountain Naturalists* (1981).
Howard, Robert W. *The Dawnseekers* (1975).
Lanham, Url. *The Bone Hunters* (1973).
Raff, Rudolf A. "Foreword" to reprint edition of Charles H. Sternberg, *The Life of a Fossil Hunter*, 1990 (1909).
Sternberg, Charles H. *Hunting Dinosaurs on Red Deer River, Alberta, Canada* (1917).
Sternberg, C. M., and R. S. Lull. *American Journal of Science* 241 (1943).
Sternberg, George F. "Thrills in Fossil Hunting." *Aerend* 1 (1930)
Who Was Who in America.
UNPUBLISHED SOURCES
See Katherine Rogers, unpublished biography of Sternberg family (information through Indiana University Press).

Keir B. Sterling

STIMPSON, WILLIAM. Born Roxbury, Massachusetts, 14 February 1832; died Ilchester, Maryland, 26 May, 1872. Naturalist, explorer, museum director.

FAMILY AND EDUCATION

Son of Herbert Hathorne Stimpson, a merchant, and Mary Ann Devereau Brewer. At least two brothers and one sister. Graduated from Cambridge High School, 1848, then attended Cambridge Latin for one year. Received honorary Doctor of Medicine from Columbian University (now George Washington University), 1860. Married Annie Gordon, 1864; two sons and a daughter.

POSITIONS

Naturalist on the U.S. North Pacific Exploring Expedition, 1853–1856. From late 1856 through 1864 affiliated with the Smithsonian Institution. January 1865 took over as temporary director of the Chicago Academy of Sciences. On the death of Robert Kennicott in 1866 Stimpson was appointed secretary, director, and trustee in January 1867, and served until his death in 1872.

CAREER

Began formal study of natural history as a special student of Louis Agassiz in October 1850. Stimpson evinced a particular interest in dredging for marine invertebrates and published a number of papers on the subject. In late 1852 he received an appointment as naturalist on the U.S. North Pacific Exploring Expedition, which departed from Virginia in June 1853. Over the next three years Stimpson made significant collections of marine invertebrates in southern Africa, Australia, China, Japan, off the Siberian coast and in northern California. In late

1856 he traveled to Washington D.C. to work up the collections of the expedition at the Smithsonian Institution. The coming of the Civil War disrupted plans to publish the results of the expedition, which remains little known. During his time in Washington Stimpson became one of the central figures in several social gatherings of naturalists, including the Megatherium Club and the slightly more formal Potomac Side Naturalist's Club. During the war years Stimpson studied freshwater snails, publishing several articles on the subject. In 1862 he traveled to Europe where he visited the great museums and dredged in the north Atlantic. In late 1864, at the behest of his friend Robert Kennicott, Stimpson agreed to come to Chicago temporarily to oversee the newly reorganized Chicago Academy of Sciences. On Kennicott's death in 1866 Stimpson's position became permanent. Aided by Spencer F. Baird and Joseph Henry of the Smithsonian, Stimpson soon amassed a large collection at the Academy, one especially rich in marine invertebrates. Other institutions, including the Smithsonian and the Museum of Comparative Zoology, loaned valuable specimens to the Academy. Continuing his own investigations, in 1870 Stimpson became among the first to dredge and describe animals from the Great Lakes. He was elected to the National Academy of Sciences in 1868 at the young age of thirty-six. Tragedy struck in the form of the great Chicago fire of October of 1871, which destroyed all of the collections as well as Stimpson's many unpublished manuscripts. In late 1871 he traveled to Florida in an effort to rebuild the collections, but his already poor health soon failed him, and he succumbed to tuberculosis seven months after the fire.

Among his notable publications are "Synopsis of the Marine Invertebrata of Grand Manan; or the Region About the Mouth of the Bay of Fundy, New Brunswick," *Smithsonian Contributions to Knowledge* 6, 5 (1854); "Prodromus Descriptionis Animalium Evertebratorum in Expeditione ad Oceanum Pacificum Septentrionalem missa, C. Ringgold et Johanne Rodgers, ducibus, observatorum et descriptorum," *Proceedings of the Academy of Natural Sciences of Philadelphia* (eight parts, 1857–1860), "The Crustacea and Echinodermata of the Pacific Shores of North America," *Journal of the Boston Society of Natural History* 6 (1857); "Researches upon the Hydrobiinae and Allied Forms," *Smithsonian Miscellaneous Collections* 7, 4 (1866). Stimpson's work in Florida was published as "Preliminary Report on the Crustacea Dedged in the Staits of Florida, edited by I. F. de Pourtalès," *Bulletin of the Museum of Comparative Zoology* 2 (1871). Many of Stimpson's findings concerning the crustacea found in the course of the Ringgold Expedition and thought to have been lost in the Chicago fire were later found and edited by Mary Jane Rathbun as "Report on the Crustacea (Brachyura and Anomura) Collected by the North Pacific Exploring Expedition, 1853–1856," *Smithsonian Miscellaneous Collections* 49 (1907).

MAJOR CONTRIBUTIONS

By the age of twenty he established himself as one of the foremost dredgers in the world. On the North Pacific Exploring Expedition Stimpson was one of the first naturalists to dredge in and around Japan. As an editor for the *American Journal of Science* from 1858 to 1863, Stimpson championed the work of American naturalists, while at the same time showing little patience for works that did not meet professional standards. He helped build the museum of the Chicago Academy of Sciences into one of the largest and most important collections of natural history objects in the United States, only to see it all destroyed. He published over fifty scientific papers, and described species in ten different phyla. An excellent taxonomist, particularly in the field of marine crustacea. Stimpson described nearly 1,000 new invertebrates, and most of the names are still in use today.

BIBLIOGRAPHY

Chicago Tribune, 12 June 1872 (obit.).

Dall, William H. "Some American Conchologists." *Proceedings of the Biological Society of Washington* 4 (1888).

Deiss, William A., and Raymond B. Manning. "The Fate of the Invertebrate Collections of the North Pacific Exploring Expedition, 1853–1856." In A. C. Whaler and J. H. Price, eds., *History in the Service of Systematics* (1981).

Dictionary of American Biography

Mayer, A. G. "William Stimpson." *Biographical Memoirs of the National Academy of Sciences* 8 (1918).

Who Was Who in America.

UNPUBLISHED SOURCES

The most significant collections of Stimpson letters are at the Smithsonian Institution archives. Stimpson's journal from the North Pacific Exploring Expedition is also at the Smithsonian Institution archives. Other important repositories of Stimpson letters are the Academy of Natural Sciences of Philadelphia and the Chicago Academy of Sciences.

Ronald S. Vasile

STÖHLER, GEORG WILHELM. (later changed to Steller.) Born Windesheim, then a Free Imperial City in Franconia, now Bavaria, 10 March 1709, died Tyumin, Siberia, 12 November 1746. Naturalist, explorer, first European to visit Alaska.

FAMILY AND EDUCATION

One of ten children of Johannes Jacob Stöhler (later Stöller), Cantor at Windsheim Gymnasium, and his second wife, Susanna Louysa Baumann Stöhler. Attended Windsheim Gymnasium, 1714–1729. Attended University of Wittenberg as theology student on scholarship, 1729–1731. Attended University of Halle as student of medicine, anatomy, botany, and natural history, 1731–1734. Did not secure medical degree, but passed special examination establishing eligibility for a professorship, 1734. Married Brigitta Elena Messerschmidt, widow of older colleague, in St. Petersburg, 1737. No children. Changed spelling of name to Steller after arrival in Russia, 1734, to accord with Russian pronunciation of it.

POSITIONS

Teacher at Francke's Waisenhaus (Latin school and orphanage), Halle, and private tutor to students in that city, 1731–1732. Privatdozent in Botany, Francke's Waisenhaus, 1732–1734. Stubenpraeceptor (room-inspector), University of Halle, c.1731–1734. Secured appointment as Russian Army physician, Danzig; served on Russian Army transport ship en route Danzig to St. Petersburg, 1734. Assisted Johann Amman, botanist at Academy of Sciences, St. Petersburg. Physician to, and member of household of, Archbishop Theophan Prokopovitch, St. Petersburg (also assisted in Archepiscopal herbarium and botanized in region), 1734–1737. Adjunct in natural history, Academy of Sciences, St. Petersburg, and member of Vitus Bering's Second Kamchatka Expedition, 1737–1746.

CAREER

Steller was first European to set foot in what is now Alaska, and first naturalist who collected specimens of plants and animals there before the existence of Alaska was generally known. Prepared for professorial career, but employment prospects were uncertain. Impatient and seeking adventure, he interrupted his studies and went to St. Petersburg. There he was befriended by Archbishop of Novgorod, by whom he was employed as physician, botanist, and herbarium worker. In 1737, sought and secured place as naturalist with Vitus Bering's second Kamchatka Expedition, took place of Academician Johan Georg Gmelin (1709–1755), who did not go to Kamchatka or Alaska. Bering's second expedition had been under way since 1733. Its purpose was to prove Bering's contention that Asia and America were not joined between Siberia and Alaska and to complete other exploratory and scientific objectives. Steller departed St. Petersburg January, 1738, made extensive plant collections in and around Lake Baikal, and did not reach Bering in Kamchatka until August, 1740. Spent winter there and accompanied Bering in the vessel *St. Peter* on expedition that reached Alaska coast near Cape St. Elias, July, 1746. Steller landed on Kayak Island for some hours on 20 July; discovered several previously unknown species, including raspberry plant and bird later named Steller Jay. During ensuing six weeks, observed other various animals, including some new species, while visiting Shumagin Islands and vicinity. Among those seen were "sea apes" (Northern Fur Seal), sea lions, porpoises, sea otters, and previously unknown fish (sculpins). Many of crew of the *St. Peter* ill with scurvy, and with season advancing, Bering directed return to Kamchatka. Bering's ship badly damaged and crew overwintered on Bering Island, in the Komandorskiyes, just east of Kamchatka, where Bering himself and thirty-one others died. Steller first and only naturalist to observe and study Steller's Sea Cow (exterminated 1768) and Spectacled Cormorant (exterminated 1850). He prepared specimens of these and other animal and plant species, but could not take them with him when surviving crew members returned to Kamchatka in August 1742 because new vessel built from remains of the *St. Peter* was too small. He is said to have begun heavy drinking at this time, affecting his health. Steller explored Kamchatka and Kuril Islands, making collections of plants and seeds, 1742–1744. Began return trip to St. Petersburg with specimens and journals in August, 1744, judging that he would soon be recalled (letter of recall had in fact been issued at end of January, 1744, but had not reached him).

Spent almost a year studying natural history of Yakutsk region, 1744–1745. Gmelin's four volume *Flora Sibirica* (1747–1769) later incorporated many of Steller's findings. Reached Solikamsk by April, 1746, where Steller planted some of the seeds carefully brought back from Kamchatka and eastern Siberia and botanized for three months. Here, an Imperial courier intercepted him and told him he must return to Irkutsk to face old (and ill-founded) legal charges centering on his allegedly freeing certain natives of Kamchatka imprisoned on suspicion of rebellious activity. This matter had already been investigated and resolved, though Imperial authorities had not been informed. Once they received updated information, new instructions for Steller's release were sent, which reached him in town of Tara. Steller then began retracing his steps to Tobolsk, where he came down with fever. He pressed on to Tyumin, reaching it at the point of death 12 November 1746. Expired same day and was buried there. Some of his reports and specimens ultimately reached Academy of Sciences in St. Petersburg, though others were lost.

At his death, Steller was well-known among the scientific fraternity as an outstanding botanical collector and one who well understood the importance of environmental and biogeographical factors in botany and zoology. Many of Steller's writings remained in manuscript. Most of what was published appeared after his death. *De bestiis marinus* [*The Beasts of the Sea*] appeared in 1751, German and English translations followed. Steller's *Beschreibung von dem Lande Kamtschatka* (1774) is an illustrated account of the people and customs there. His *Journal of Bering's Voyage of Discovery from Kamtchatka to the Coast of America, in 1741* appeared in several languages in various editions. First complete English edition not published until 1925. Steller often had considerable difficulty working with others, but his intelligence, initiative, and prodigious capacity for hard work made him superbly qualified to be an explorer–naturalist.

MAJOR CONTRIBUTIONS

Steller was the first trained naturalist to set foot on Alaska soil. He observed, collected, and took detailed notes on plants and animals observed, including a number of new species. His are the only available descriptions of some animals, notably the Steller Sea Cow and Spectacled Cormorant, because these species were soon extirpated. Steller's *Beschreibung von dem Lande Kamtschatka*, his *Journal of Bering's Voyage*, and his *De bestiis mar-*

inus are valuable accounts of his work in the North Pacific and in Siberia. Many of his collections either had to be left behind on Bering Island or were lost in transit from Siberia to St. Petersburg. Some later writers used his unpublished descriptions without proper attribution, though Linnaeus, J. G. Gmelin and P. S. Pallas made a point of seeing to it that Steller received appropriate credit for what he had accomplished.

BIBLIOGRAPHY

Dictionary of Scientific Biography.

Ford, Corey. *Where the Sea Breaks Its Back: The Epic Story of a Pioneer Naturalist and the Discovery of Alaska* (1966).

Frost, O. W., and K. E. Willmore. *Description of Unpublished Steller Papers in the Smithsonian Institution and the Library of Congress* (1983).

Frost, O. W., ed. *Steller's Journal: The Manuscript Text* (with transliteration by M. A. Engel and K. E. Willmore, 1984).

Frost, O. W., ed. *Journal of a Voyage with Bering, 1741–1742 /Georg Wilhelm Steller* (translated by M. A. Engel and O. W. Frost, 1988).

Golder, F. A. *Bering's Voyages: An Account of the Efforts of the Russians to determine the Relation of Asia and America. Log Books and Official Reports of the First and Second Expeditions, 1725–1730 and 1733–1742, with Steller's Journal of the Sea Voyage from Kamchatka and America and Return on the Second Expedition, 1741–1742* (translated and in part annotated by Leonhard Stejneger, 2 vols., 1922–1925) (has useful bibliographies).

Stejneger, Leonhard. *George Wilhelm Steller* (1936) (has useful bibliographies).

Waxell, Sven. *The American Expedition* (edited and translated by M. A. Michael, 1952).

UNPUBLISHED SOURCES

See Frost, 1983, and bibliographies in Golder and Stejneger, above. Records relating to Russians in Alaska, 1732–1796, 2 microfilm reels, Bancroft Library, University of California, Berkeley.

Keir B. Sterling

STORER, DAVID HUMPHREYS. Born Portland, Maine, 26 March 1804; died Boston, 10 September 1891. Physician, naturalist, ichthyologist.

FAMILY AND EDUCATION

Son of Woodbury Storer, shipowner, merchant, and chief justice of common pleas, and Margaret Boyd Storer. Graduated Bowdoin College, 1822. Studied medicine with John Collins Warren and graduated Harvard Medical School, 1825. Married Abigail Jane Brewer, sister of Thomas Mayo Brewer, 1829; five children.

POSITIONS

Distinguished career as a physician, specializing in obstetrics. One of the founders of Tremont Street Medical School and professor of obstetrics, 1837–1854. Professor of obstetrics and medical jurisprudence, Harvard Medical

School, 1854–1868, and dean, 1855–1864. Member of staff, Massachusetts General Hospital, 1849–1858. Member of staff, Lying-In Hospital, 1854–1868.

CAREER

Interest in natural history began at an early age. Original member of Boston Society of Natural History, 1830, and served as recording secretary, 1830–1836, curator of ichthyology and herpetology, 1838–1843, and second vice president, 1843–1860. Fellow of the American Academy of Arts and Sciences, 1837. At first, his chief interest was in conchology. Translated part of French naturalist L. C. Kiener's *General Species and Iconography of Recent Shells* (1837). Selected to survey the reptiles and fishes of Massachusetts for state Zoological and Biological Survey, 1837, and thereafter specialized in ichthyology. His survey report, *Report on the Fishes and Reptiles of Massachusetts* (1839), continued the American tradition of describing animals by states. It was considered the best American work in ichthyology to date. Described a number of new species of fishes, especially saltwater fishes. Devoted his leisure in the next twenty-five years to perfecting the ichthyological part of his report. Revisions appeared in parts, 1853–1867, in *Memoirs of the American Academy of Arts and Sciences* and were published separately as *A History of the Fishes of Massachusetts* (1867). This work, describing all the fishes in the state with careful consideration of synonymy and attention to economic uses, was said to be "a land-mark in the ichthyological literature of this country" and "a storehouse of facts." Gathered information directly from fishermen and fish markets. Work was a forerunner of reports of government fishery commissions. His other major work on ichthyology, *A Synopsis of the Fishes of North America* (1846), was a dictionary of all known species with tables summarizing geographical distribution. Was regarded of lesser value because it was more of a compilation than an original work. Remembered as a genial, conscientious, but impulsive man, "an enthusiast in every part of his being."

MAJOR CONTRIBUTIONS

One of the foremost American ichthyologists of the early nineteenth century and author of a landmark regional survey of fishes. Concern with economic uses of fish was in advance of all previous works on fishes.

BIBLIOGRAPHY

Bowdoin College Library Bulletin, no. 2 (August 1892).

Gifford, G. E., Jr. "The Ichthyologist Dean." *Harvard Medical Alumni Bulletin* (Fall 1964).

Proceedings of the Boston Society of Natural History 25 (1892) (obit. by J. C. White and article on his work by S. Garman).

Scudder, S. H. *Proceedings of the American Academy of Arts and Sciences* 27 (1893) (obit.).

Viets, H. R. *Dictionary of American Biography*.

UNPUBLISHED SOURCES

Collection of letters received is in Boston Museum of Science; some correspondence is in Countway Library, Harvard Medical School.

Toby A. Appel

STREET, MAURICE GEORGE. Born Whiteparish, Wiltshire, England, 28 September 1910; died Nipawin, Saskatchewan, 27 October 1966. Farmer, truck driver, naturalist.

FAMILY AND EDUCATION

Son of Henry James Street and Annie Flora Stride, who emigrated to a farm near Moose Range post office, Saskatchewan, in November 1913. Attended rural schools near Aylsham, Codette, and Tisdale, attaining grade 8 in five years. Married to Rose Short; two daughters.

POSITIONS

Farmed south of Nipawin, 1927–1938, then worked for the Imperial Oil bulk dealership at Nipawin, delivering oil products to farmers.

CAREER

Self-taught, Maurice learned the songs and habits of local birds through careful observation. His method was to sit still on a stump or log, all the while watching with his keen eyes and listening with his gifted ears. He kept records of bird migration and bird nests from 1922, using a series of five-cent notebooks, one of which was always in his pocket. Alone in an area of thirty-mile radius, he recorded over 3,000 nests of 133 species and found another 10 species with flightless young. He banded over 13,000 birds of 101 species, including over 4,000 redpolls and over 2,275 Tennessee warblers. For several years he banded more Tennessee warblers than all other banders on the continent together.

MAJOR CONTRIBUTIONS

His one-man record of breeding birds in a local area is one of the most complete for any region in North America. He found Saskatchewan's first nest of the whippoorwill, on 27 June 1956. He was coauthor of one of Canada's most important regional bird lists, recording the changes of bird life over 200 years on the Saskatchewan River—from presettlement observations of Hearne, Richardson, and Blakiston to the late 1950s.

BIBLIOGRAPHY

Houston, C. S. "Maurice G. Street, 1910–1966." *Blue Jay* 24 (1966).

Houston, C. S., and M. G. Street. *Birds of the Saskatchewan River, Carlton to Cumberland* (1959).

Mary I. Houston

STRONG, HARRIET WILLIAMS RUSSELL. Born Buffalo, New York, 23 July 1844; died near Whittier, California, 16 September 1926. Inventor, engineer, horticulturist, civic leader, music composer.

FAMILY AND EDUCATION

Fourth daughter of Henry Pierrepont Russell and Mary Guest (Musier) Russell. Father had a wide variety of jobs, including being postmaster of Buffalo and state adjutant general in Nevada. Family (of seven children) moved west to improve the state of Mary Russell's health. They settled first in Wisconsin, then California, and finally, in the 1860s, Nevada. Chronic problems with her spine limited Harriet's childhood activities, but during periods of confinement she was tutored privately. She attended Miss Mary Atkins's Young Ladies' Seminary, Benicia, California, 1858–1860. Married Charles Lyman Strong, prominent Nevadan, 1863 (committed suicide, 1883); four daughters.

POSITIONS

Owner and operator of Ranchito del Fuerte, San Gabriel Valley, California, 1883–1929. Established and owned Paso de Bartolo Water Company, San Gabriel Valley, California, 1897–1901. Composer, civic and cultural leader.

CAREER

Following husband's death, dealt with eight years of litigation over his land holdings. Was obliged to become a vigorous, versatile, and dominant individual, in contrast to her earlier semi-invalid state. Managed husband's estate, Ranchito del Fuerte, or "the Strong Ranch," in San Gabriel Valley, California. Pioneered in walnut production there. Grew award-winning walnut and orange trees, pomegranates, and pampas grass. Nicknamed the "Walnut Queen" and "Pampas Lady" in recognition of her success. Became interested in questions of marketing, irrigation, and flood control. Supported water conservation by means of water stored in dams at origins of streams. Secured patent for dam design that provided irrigation and flood control, 1887. Secured a second patent for hydraulic mining procedure involving the impounding of water and debris in 1893. Her efforts brought about election as first female member of Los Angeles Chamber of Commerce, and her pampas grass and water storage concept won her nationwide attention and two World's Columbia Exposition Medals in Chicago (1893). Encouraged business education for women; named president of Business League of America, a feminist organization. Concerned about cultural opportunities for women, was founder of Ebell Club, Los Angeles, in 1894. Active supporter and vice president, Los Angeles Symphony Association. Had a number of artesian wells drilled near her property and established Paso de Bartolo Water Company with two of her daughters, 1897, but sold it in 1901. Grew increasingly concerned about water supply and flood control in and about Los Angeles in her later years, was also member of Executive Board of Inland Waterways Association in San Francisco. Sought development of Colorado River, proposed dams and water impoundment in Grand Canyon before Congress, 1918. Wrote and published a number of songs. Organized successful effort to preserve as national monument old home that had belonged to former Spanish governor Pio Pico near Whittier, California.

MAJOR CONTRIBUTIONS

Encouraged and proposed innovative designs for irrigation, flood control, and waterpower procedures. First to suggest flood prevention by means of source conservation. Won several patents for her designs. Successful agriculturist, accomplished businesswoman, and cultural leader. Pioneered in business education for women.

BIBLIOGRAPHY

Leitch, F. G. *Los Angeles Examiner*, 1 April 1923.
Los Angeles Times, 17 September 1926 (obit.).
McGroarty, John S., ed. *History of Los Angeles County*. Vol. 3 (1923).
National Cyclopedia of American Biography.
Notable American Women.
Southern California Business (November 1926).
Who's Who in the Pacific Southwest (1913).

UNPUBLISHED SOURCES

Substantial collection of papers is held by family.

Jean Wassong

SULLIVANT, WILLIAM STARLING. Born Franklinton (now part of Columbus), Ohio, 15 January 1803; died Columbus, Ohio, 30 April 1873. Businessman, self-taught botanist.

FAMILY AND EDUCATION

Son of Lucas Sullivant, a surveyor, businessman, and landowner, and Sarah (Starling) Sullivant. Attended private academy of Samuel Wilson in Jessamine County, Kentucky, one year at Ohio University, Athens, and earned A.B. from Yale College (1823). Received honorary L.L.D. from Kenyon College (1863) and elected to the National Academy of Sciences (1872). Married Jane Marshall, 1824 (died 1825), Eliza Griscom Wheeler, 1834 (died 1850), and Caroline Eudora Sutton, 1851 (died 1891). Had thirteen children from these three marriages.

POSITIONS

At father's death in 1823 inherited extensive lands and business, which he expanded into mills, stone quarries, banking interests, stage coach, canals, and other companies while engaged in civil engineering and surveying. First became interested in botany about 1834, partly influenced by his second wife and brother, Joseph.

CAREER

First studied flowering plants, particularly grasses and sedges, and published *A Catalogue of Plants . . . in the Vicinity of Columbus, Ohio* (1840); traveled extensively in Ohio acquiring a large herbarium of flowering plants emphasizing grasses and sedges, suspected hybridization in *Carex*, and named and described flowering plants new to science. Gained much encouragement through correspondence with Asa Gray and John Torrey and turned to the study of mosses and liverworts. His first major works were an exsiccatae set of mounted and labeled moss specimens, the *Musci Alleghanienses* (1845, 1846), and his treatment of mosses and liverworts in the first and second editions of Asa Gray's *Manuel of Botany of the Northern United States* (1848, 1856), the illustrations in the latter believed to have been drawn by his second wife, Eliza Griscom Wheeler. His greatest work of unrivaled execution, the *Icones Muscorum* (1864, facsimile, 1969), contains 129 copper plate illustrations of most of those mosses peculiar to eastern North America and a posthumous *Supplemental* volume (1874, facsimile 1969) by Leo Lesquereux and Asa Gray. His *Manual of Mosses of North America* (1884) was completed by Lesquereux and Thomas P. James. The Sullivant Moss Chapter was named in his honor (1899), then changed to the Sullivant Moss Society (1908), and is now known as the American Bryological and Lichenological Society. His name is commemorated in the flowering-plant genus *Sullivantia, Carex* x *sullivantii,* in which the *x* denotes a species of hybrid origin, *Lonicera sullivantii, Rudbeckia sullivantii*; selected mosses are *Fontinalis sullivantii, Hookeria sullivantii, Hypnum sullivantii, Neckera sullivantii.*

MAJOR CONTRIBUTIONS

Recognized worldwide as the first and foremost authority on the bryology of the United States during the latter half of the nineteenth century; named and described 270 species of bryophytes. His extensive moss herbarium of approximately 18,000 species is housed at the Farlow Reference Library and Herbarium of Cryptogamic Botany, Harvard University. Asa Gray (1873) wrote that Sullivant was "the most accomplished bryologist which this country has produced."

BIBLIOGRAPHY

Gray, A. "William Starling Sullivant." *American Journal of Science and Arts* 106

(1873); *Proceeding of the American Academy of Arts and Sciences* 9 (1873); *Biographical Memoirs of the National Academy of Science*. 1 (1877).
Rodgers III, A. D. "*Noble Fellow*:" *William Starling Sullivant* (1940, facsimile, 1968).
Rodgers III, A. D. "Forward." In *Icones Muscorum* (facsimile ed., 1969).
Sayre, G. "Cryptogamae Exsiccatae, An Annotated Bibliography . . . Bryophyta IV." *Memoirs of the New York Botanical Garden* 19, 2 (1971).
Sayre, G. "Publication Dates of W. S. Sullivant's New Bryophytes." *The Bryologist* 84 (1981).
Sayre, G. *Index to the Moss herbarium of William Starling Sullivant (1803–1873)* [*at the Farlow Herbarium of Harvard University*] (May 1984).
Stuckey, R. L. comp. *First Botanists of Columbus: A Sullivant Family Scrapbook Compiled for Field Trip to Green Lawn Cemetery, 38th Annual AIBS Meeting, Columbus, Ohio, 9 August 1987* (1987).
Stuckey, R. L. "William S. Sullivant and His Central Ohio Botanical Associates." *Bartonia* 54 (1988).
Stuckey, R. L., and M. L. Roberts. "Frontier Botanist: William Starling Sullivant's Flowering Plant Botany of Ohio (1830–1850)." *Sida, Botanical Miscellany* 6 (1991).
Sullivant, J. *William S. Sullivant*. In *Genealogy and Family Memorial* (1874).
UNPUBLISHED SOURCES
Correspondence in various libraries: to Asa Gray, Gray Herbarium of Harvard University; to John Torrey, The New York Botanical Garden; to George Engelmann, The Missouri Botanical Garden; to Charles Wilkins Short, The Filson Club, Louisville, Kentucky; to Increase A. Lapham, Historical Society of Wisconsin, Madison.

Ronald L. Stuckey

SWALLOW, ELLEN HENRIETTA (ELLEN RICHARDS). Born Dunstable, Massachusetts, 3 December 1842; died Jamaica Plains, Massachusetts, 30 March 1911. Chemist, educator.

FAMILY AND EDUCATION

Only child of Peter and Fanny Gould (Taylor). Attended Westford Academy. Vassar College (A.B. 1870, M.A. 1873). Special student (chemistry) Massachusetts Institute of Technology (S.B. 1873). Honorary D.S., Smith College (1910). Married Robert Hallowell Richards, 1875. No children.

POSITIONS

At M.I.T.: Instructor, Womens' Laboratory (1876–1883), Instructor in Sanitary Chemistry (1884–1911). Member, Women's Education Association and Association of Collegiate Alumnae (later American Association of University Women). Elected, American Institute of Mining Engineers (1879). Director, New England Kitchen, 1890. President, American Home Economic Association, 1908–1910. Consultant to private industry, state and federal agencies.

CAREER

Blocked by gender from an M.I.T. doctorate, she applied chemistry to sanitation, nutrition, and environmental issues. Her pioneer work in environmental assessment mapping chlorine and water quality in Massachusetts created the world's first water purity tables. Assisted in establishing limnology and oceanography programs at Woods Hole Marine Research Laboratory (1888). Taught in first U.S. sanitary engineering program (1890) and trained first generation of specialists. Author of numerous textbooks and publications on policy and research. Active public speaker on women, science, and environmental issues.

MAJOR CONTRIBUTIONS

Pioneer in environmental public policy, water quality standards, and public health. Promoted home economics. M.I.T's first woman student, graduate, and faculty member. Actively promoted access for women to science education, teaching and research. Her 1893 definition of Oekology as ecology within a social policy framework sought to bridge the growing gap between natural sciences and the gendered experience of individuals, households, and urban centers.

BIBLIOGRAPHY

Clarke, R. *Ellen Swallow: The Woman Who Founded Ecology* (1973).

Craven, Hamilton. ''Establishing the Science of Nutrition at the USDA: Ellen Swallow Richards and Her Allies,'' *Agricultural History* 64, 2 (1990).

Dictionary of American Biography.

Hunt, Caroline L. *The Life of Ellen H. Richards* (1912).

Hynes, H. Patricia. ''Ellen Swallow, Lois Gibbs and Rachel Carson: Catalysts of the American Environmental Movement,'' *Women's Studies International Forum*, 8, 4 (1985).

National Cyclopedia (1897).

Notable American Women, 1607–1950.

UNPUBLISHED SOURCE MATERIALS

Letters and manuscripts in the Sophia Smith Collection, Smith Library; Vassar College Library and Archives, Edward Atkinson Papers; Massachusetts Historical Society; and University of Chicago.

Lorne Hammond

T

TAVERNER, PERCY (ALGERNON). Born Guelph, Ontario, 10 June 1875; died Ottawa, Ontario, 9 May 1947. Ornithologist, conservationist, museum curator.

FAMILY AND EDUCATION

Son of Edwin Fowler, a high school principal, and Ida Van Cortland Buckley Fowler. After his parents' divorce his mother married Albert Taverner, who unofficially adopted Percy. Had unsettled childhood; mother and stepfather traveled as actors. Schooling in Port Huron and Ann Arbor, Michigan. After high school tried business, taxidermy and, around turn of the century, took correspondence course in architecture. Self-taught naturalist and ornithologist. Married Martha H. Wiest, 1930; no children.

POSITIONS

He practiced architecture in Chicago, 1902–1907, and Detroit, 1907–1911. Assistant naturalist and curator, National Museum of Canada, 1911–1914, zoologist 1914–1919, ornithologist, 1919–1936, chief, Division of Ornithology, 1936–1942. Retired as honorary curator of ornithology, 1942.

CAREER

Interested in birds and flowers at an early age. Learned taxidermy while attending school in Ann Arbor and sold early collection of study skins to University of Michigan. Greatly inspired by Canadian ornithologists J. H. Fleming and W. E. Saunders and founded the Great Lakes Ornithological Club with them in 1905. He was one of the pioneers of bird banding in North America. After moving to Ottawa as curator, he organized extensive fieldwork for faunistic studies. Collected in Ontario, Quebec, and New Brunswick, 1913–1915, in the Western Provinces of Canada, 1915–1920, Alberta and British Columbia, 1925. Traveled to Anticosti Island and Labrador in 1928, eastern arctic islands and Greenland 1929, Churchill, 1930, and western Manitoba, 1937. Through his personal collecting efforts and with cooperation from other divisions of the museum, Taverner increased the bird collection from fewer than 5,000 to over 30,000. The great amount of organization and compilation involved in running, almost single-handedly, the ornithology division of the museum resulted, as a well-known American ornithologist said, in Taverner's succeeding "with a minimum amount of help doing for Canadian birds . . . what the Biological Survey with the participation of numerous persons . . . had not been able to complete for birds north of Mexico since 1885." His early scientific interest focused on faunal lists. Well aware of the unknown nature of much of Canada's avifauna, he encouraged many ornithologists, vocational and avocational, to carry out independent projects and to contribute information to the museum. He was interested in subspecies versus species question and carried out taxonomic research on the red-tailed hawk, Canada goose, and great horned owl.

Wrote several books on Canadian ornithology, notably *Birds of Eastern Canada* (1919), which was intended as a popular book on Canadian birds and became highly successful. *Birds of Western Canada* (1926) and *Birds of Canada* (1934) were equally well regarded. Two pocket field guides, *Canadian Water Birds* and *Canadian Land Birds*, were published in 1939. Wrote nearly 300 papers, most of which were published in the *Canadian Field Naturalist*. His interest in the economic status of birds and in conservation is evident in his special reports and in several chapters in his books on the birds of Canada. An accomplished bird illustrator, while still in Michigan, Taverner contributed many of the plates and line drawings to W. B. Barrows, *Michigan Bird Life* (1912). Throughout his long career as ornithologist for the National Museum of Canada, Taverner, in addition to his scientific work, popularized the study of ornithology and promoted in official and private capacity the conservation of birds. He was ornithological editor of the *Canadian Field Naturalist*, 1912–1942, and largely thanks to him the journal became a national, rather than a local, publication. He was active in American Ornithologists' Union, was a member of the Canada Advisory Board of Wildlife Protection and a Fellow of the Royal Society of Canada. By all accounts he was a charming, cultured, versatile person, with a good sense of humor and a "passion for music and literature." In spite of his busy professional life he remained active as an artist, gardener, bookbinder and had many friends.

MAJOR CONTRIBUTIONS

Pioneer of bird banding in North America, museum builder, conservationist, popularizer of ornithology in Canada.

BIBLIOGRAPHY

Canadian Who's Who. (1938–1939).

Cranmer-Byng, John L. ''A Life with Birds: Percy A. Taverner, Canadian Ornithologist, 1875–1947.'' *Canadian Field-Naturalist*, 110, 1 (1996).

McAtee, W. L. ''Percy Algernon Taverner, 1875–1947.'' *Auk* 65 (1948) (obit.).

Saturday Night, 1 February 1936, 4 November 1944.

Wilson Bulletin 59 (1947) (obit.).

UNPUBLISHED SOURCES

See P. A. Taverner Correspondence, National Museum of Natural Sciences, Ottawa; J. H. Fleming Correspondence, Royal Ontario Museum Archives, Toronto.

Marianne Gosztonyi Ainley

TEALE, EDWIN WAY. Born Joliet, Illinois, 2 June 1899; died Norwich, Connecticut, 18 October 1980. Naturalist, photographer, illustrator, author.

FAMILY AND EDUCATION

Son of Oliver Cromwell and Clara Louise Way Teale. Had a happy boyhood on his grandfather's farm in Indiana, where he developed an early interest in prairie insects and photography. In twelfth year changed his name from Edwin Alfred to Edwin Way as more distinguished for the photographer/writer he determined to become. Graduate of Joliet Township High School; student at University of Illinois; A.B., Earlham College, 1922, L.H.D., 1957; A.M., Columbia University, 1926; honorary L.H.D., Indiana, 1970; Sc.D., University of New Haven, 1978. Married Nellie Imogene Donivan, 1923; one son, killed in action during World War II.

POSITIONS

Instructor in public speaking, Friends University, Wichita, Kansas, 1922–1924; editorial assistant to Frank Crane, religious journalist, 1925–1927; staff feature writer, *Popular Science Monthly*, 1928–1941; freelance writer from 1941.

CAREER

Wrote thirty books, most on how to experience and appreciate nature. His first book, *Book of Gliders* (1930), was revised in 1939. His *Grassroots Jungles* (1937) set pattern for works that followed: combination of observations of nature with philosophical musings. In 1943 became full-time popularizing naturalist and completed an autobiography, *Dune Boy*: *The Early Years of a Naturalist*.

Among his more successful books that followed are *The Lost Woods* (1945), *Days without Time* (1948), *North with the Spring* (1951), *Autumn across America* (1956), *Journey into Summer* (1960). Also edited such volumes as Hudson's *Green Mansions* (1944), Thoreau's *Walden* (1946), *The Insect World of J. Henri Fabre* (1949), and *The Wilderness of John Muir* (1955).

In 1966 awarded a Pulitzer Prize for his *Wandering through Winter* (1965). Completed his last book, *A Walk through the Year*, in 1978, a distinguished addition to American nature writing that echoed views of Thoreau and Muir and the value they placed on the meaning and beauty of the natural world.

A number of his books were also published in English, Swedish, Spanish, French, Finnish, Arabian, and Braille editions. President, Nature Study Society, 1947; president, Thoreau Society, 1958; member of a number of scientific societies in the United States and abroad. Recipient of a number of honors and awards, including John Burroughs Medal, 1943, Eva L. Gordon Award, American Nature Study Association, 1965; Ecology Award of the Massachusetts Horticultural Society, 1975; Conservation Medal of the New England Wildflower Association, 1975. Fellow of the American Association for the Advancement of Science.

MAJOR CONTRIBUTIONS

Through his popular books, convinced Americans they had a personal stake in the preservation of ecological zones. Convinced them to support national parks and conservation movements. Emphasized that nature provides meaning for those who seek it.

BIBLIOGRAPHY

''Edwin Way Teale Is Dead at 81.'' *New York Times*, 21 October 1980 (obit.).

Miller, David Stuart. ''An Unfinished Pilgrimage: Edwin Way Teale and American Nature Writing.'' *Dissertations Abstracts International* 43. 8. (1983).

Zwinger, Ann, and E. W. Teale. *A Conscious Stillness*: *Two Naturalists on Thoreau's Rivers*. New York (1982).

UNPUBLISHED SOURCES

Several letters are in Cornell University Library, Collection of Regional History and University Archives; Library of Congress Manuscript Division; and Staten Island Institute of Arts and Sciences (N.Y.).

George A. Cevasco

TERRILL, LEWIS (MCIVER). Born Montreal, Quebec, 30 October 1878; died Ulverton, Quebec, 22 December 1968. Ornithologist, conservationist, lecturer, photographer.

FAMILY AND EDUCATION

One of five children of Frederick William Terrill, a lawyer whose family was among the original settlers of Sherbrooke, Quebec, and his wife, Alexandrina

McIver Terrill. Educated at Westmount Academy and Montreal High School. Mostly self-taught in natural history and ornithology. Married Elizabeth Edith Abbott in 1937; no children.

POSITIONS

Statistician at S. F. McIver Co., Bury, Quebec, 1899–1903. Montreal Sand and Gravel Company, 1904–1920. Joined Merchant's Bank as statistician in 1920. On that bank's merger with the Bank of Montreal in 1922 he became employed in the Foreign Department, Head Office, Montreal, where he remained until 1942. Although retirement age, he joined the British Metals Corporation, where he remained till 1950. One of the founders and first president of the Province of Quebec Society for Protection of Birds, 1917–1925.

CAREER

While student at Montreal High School, he was greatly encouraged in his natural history interest by John William Dawson, principal of McGill University. His position in Bury enabled him to familiarize himself with the avifauna in parts of the Eastern Township region of Quebec. His first papers, the result of fieldwork in the Bury region, "Remarks on Some Marsh-Dwellers" and "Summer Warblers in Compton County, Quebec," appeared in the *Ottawa Naturalist* (1903, 1904). He rarely traveled outside the province of Quebec. His observations of Quebec birdlife were published in the *Ottawa Naturalist* and its successor, the *Canadian Field Naturalist*, the *Wilson Bulletin*, and the *Auk*. He contributed considerably to life history studies of birds and to studies of the migration of the birds of Quebec. He became an associate of the American Ornithologists' Union in 1907 and was elected member in 1947. He joined both the Wilson Ornithological Club and Cooper Ornithological Club by invitation in 1911. He vigorously supported conservation efforts and the establishment of many federal, provincial, and private sanctuaries in Quebec and Ontario. He contributed data to the publications of W. W. Cooke, E. H. Forbush, and A. C. Bent and wrote the account of the fox sparrow for Bent's, "Life Histories of North American Birds." He prepared the bird report for the Province of Quebec Society for Protection of Birds' *Annual Report* from 1940 to 1947 and published sixty-seven popular weekly articles entitled "Outdoor Calendar" in the *Montreal Star* in 1925–1926, which impressed even professional ornithologists of the Smithsonian Institution. His ornithological publications appeared in the *Auk* (no. 15), *Wilson Bulletin* (no. 2), and *Ottawa* (later, *Canadian Field) Naturalist* (no. 25). A keen nature photographer before the advent of widespread color photography Terrill hand-colored more than 1,000 slides of birds and plants. As a lecturer he was in great demand and delivered more than 200 lectures to schools and various nature and conservation associations.

A patient, careful observer of birdlife, Terrill was a transitional figure between nineteenth-century naturalists and collectors and twentieth-century students of the "living bird." His early collection of bird skins and nearly half his egg collection were deposited at the Redpath Museum, McGill University. After his death his collection of more than 100 bird's nests and 650 sets of eggs was sent to the National Museum of Natural Sciences, Ottawa.

MAJOR CONTRIBUTIONS

Founder of Quebec's major conservation association. Popularizer of ornithology and of nature study in general in the Montreal area. He established high standards of field observation at a time when the study of the living bird in its environment was still a novelty. His meticulous notes on over sixty-five years of fieldwork are in the National Museum, Ottawa.

BIBLIOGRAPHY

Ainley, M. G. "Lewis McIver Terrill, Naturalist, Ornithologist, Conservationist." *Tchébec* (1981).

Godfrey, W. E. "In Memoriam: Lewis McIver Terrill." *Canadian Field Naturalist* (1972).

Montgomery, G. H. Jr. "Lewis McIver Terrill, Obituary." *Auk* (1972).

UNPUBLISHED SOURCES

See Archives, Bank of Montreal Head Office, Montreal Letter and Minutes, P.Q.S.P.B. Archives, Montreal Lewis McIver Terrill Correspondence, National Museum, Ottawa.

Marianne Gosztonyi Ainley

THOMPSON, DAVID. Born Westminster, London, England, 30 April 1770; died Longueuil, Quebec, 10 February 1857. Fur trader, mapmaker, explorer, surveyor.

FAMILY AND EDUCATION

Born in London of Welsh immigrants, Ann and David Thompson. Father died two years after his birth, causing financial hardships. Entered Grey Coat Hospital, charity school for the poor, at age seven. Learned mathematics and simple navigation. Seven-year apprentice to Hudson's Bay Company (20 May 1784). Married thirteen-year-old Charlotte Small (10 June 1799), in the custom of the country, fur trade marriage in the absence of clergy. Marriage regularized (Montreal, 30 October 1812). Charlotte was his lifelong companion through wilderness travels and later poverty. Seven sons and six daughters.

POSITIONS

Hudson's Bay Company (HBC) apprentice (1784–1791); HBC surveyor, 1794–1797; surveyor, North West Company (NWC), 1797; NWC clerk trader,

1799–1806; NWC partner, 1806–1812. Surveyor, private practice. Astronomer and surveyor, Boundary Commission, 1817–1827. Appointed justice of the peace, Glengarry County (1820). Various survey commissions, 1830s.

CAREER

Arrived at Fort Churchill in 1784 and wintered with Samuel Hearne. Fieldwork at hunting camps, gained experience traveling the land. Worked on the Saskatchewan, building a post near Batoche (1786). Spent next two years inland, wintering with Peigan Indians (1787–1788). A broken leg next winter changed his life as he and Peter Fidler studied applied mathematics, surveying, and astronomy with Philip Turnor, official HBC surveyor (1789–1790). Lost sight in one eye. Made surveying his vocation. Surveyed Nelson, Churchill Rivers, and part of Athabasca region, key to commercial struggle between HBC and NWC. Made a dangerous trip to Reindeer Lake. Friction with employers or ambition led to decision to leave HBC without notice (1797). Spent next fifteen years working for NWC. Asked to work on NWC region now subject to Jay's Treaty (1794), west of Lake Superior. Embarked on amazing two years: mapped NWC posts and the routes between them; first mapping of upper Red River Valley; mapped Mandan territories in Missouri; headwaters of the Mississippi; and gave first accurate information on the western regions. By 1798 mapping and trading on an arc up from the Saskatchewan Rivers to Lake Athabasca. Became trading clerk and married in the country. Spent first years of new century in the Peace River country, then with Duncan McGillivary seeking an economic route to the Pacific and opening trade with the Kootenay Indians. Promoted to partner (1804). Sent to the Rockies in 1806 as a response to Meriwether Lewis and William Clark's expedition to the Pacific. In 1807, with wife and family, crossed the heights of Howse Pass and descended to the upper Columbia. Spent next three years among the Kootenay and Flathead Indians, mapping Columbia's headwaters. Asked to return in 1810, either as part of an NWC deal with John Jacob Astor or as Astor's competitor, to complete the mapping and follow the Columbia. Delayed by hostile Peigan intermediaries, angry over NWC's direct trade with Kootenay and Flathead Indians. Took roundabout route and moved by canoe and horseback along the Columbia basin. Arrived mouth of Columbia (15 July 1811) to find Astor's Pacific Fur Company had established Fort Astoria. Finished mapping Columbia before his retirement in 1812. Brought family out with him to Montreal. Bought farm in 1815 at Williamstown, upper Canada (Ontario). Worked as a private surveyor and contract mapmaker. Finished maps of western Canada and the Columbia. Took contract to map controversial Canada–United States border from the St. Lawrence to Lake of the Woods. Victim, like many others, of NWC Agent McGillivary's 1825 bankruptcy. Late 1820s and 1830s brought financial reversals and decline into poverty. Minor survey work undertaken by him in the Ottawa Valley during 1830s. Living off his children, began writing his now famous narrative, left incomplete as he went blind. Died a pauper, in obscurity. Ultimately vindicated for his achievements by the historical work of Joseph Burr Tyrrell.

MAJOR CONTRIBUTIONS

The key cartographer of the western regions of Canada and intimately associated with the history of mountain passes, lakes, and the rivers of the plains and the Pacific, particularly the Columbia, Mississippi, and Athabasca. His work as an explorer and mapmaker reached from the St. Lawrence to the Pacific and from the Columbia to the Arctic. His narrative offers a rare alternative to the usual HBC interpretation of life, exploration, and fur trade competition. A refreshing change from the ''heroic'' explorer.

BIBLIOGRAPHY

David Thompson Sesquicentennial, 1809–1959 Symposium (1959).

Dempsey, Hugh. ''David Thompson on the Peace River.'' *Alberta Historical Review* 14 (1966).

Morton, A. S. *David Thompson* (1930).

Morton, A. S. ''The North West Company's Columbian Enterprise and David Thompson.'' *Canadian Historical Review* 17 (1936).

Nicks, John. ''David Thompson.'' *Dictionary of Canadian Biography.*

Parker, Elizabeth. ''Explorers of the West—David Thompson.'' *Canadian Alpine Journal* 29 (1946).

Smith, J. K. *David Thompson: Fur Trader, Explorer, Geographer* (1971).

Tyrrell, J. B. *David Thompson's Narrative* (1916; repr. 1962).

Tyrrell, J. B. ''The Rediscovery of David Thompson.'' *Transactions of the Royal Society of Canada* 22 (1928).

Vesiling, Priit J. ''David Thompson: The Man Who Measured Canada.'' *National Geographic* (May 1996).

White, M. C. *David Thompson's Journals Relating to Montana and Adjacent Regions, 1808–1812* (1950).

UNPUBLISHED SOURCES

Little has been done on the post-1812 years. Begin with John Nicks, in Bibliography. Heavily collected and published by J. R. Tyrrell. Collections at National Archives of Canada, Ottawa; Glenbow Museum, Calgary; University of British Columbia Library, University of Toronto Library, and Provincial Archives of British Columbia.

Lorne Hammond

THOREAU, HENRY DAVID. Born Concord, Massachusetts, 12 July 1817; died Concord, Massachusetts, 6 May 1862. Naturalist philosopher.

FAMILY AND EDUCATION

Son of John Thoreau, storekeeper and pencil maker, and Cynthia Dunbar. Studied at Concord Academy, entered Harvard in 1833, and graduated 1837. Never married.

POSITIONS

Taught school briefly at Canton, Massachusetts, in 1835; taught for a fortnight at Concord in 1837 but resigned because of opposition to corporal punishment in school; 1838–1841 formed a private school with his brother John and introduced practice of field trips for nature study; 1841–1843 lived in home of Ralph Waldo Emerson. Studied, wrote, and meditated at Walden Pond, 1845–1847. From 1849 to 1862 lived at family home, kept journals, and manufactured pencils.

CAREER

Encouraged to take an interest in the out-of-doors from an early age. Strongly influenced by R. W. Emerson from about age twenty, and became for some years a trancendentalist with Emerson and others. Emerson encouraged the younger man in his writing, and for some time Thoreau accepted Emerson's invitation to live with his family, later living with and tutoring a son of Emerson's brother William on Staten Island. Found teaching unproductive and was unsuccessful in breaking into the New York literary world. From 1843 until his death from tuberculosis in 1862, spent most of his time in New England, focusing on his personal writings, travels, the family pencil business, and occasional lecturing. Jailed for one night in 1846 for failure to pay a poll tax, which he justified by his opposition to the Mexican War and to slavery, he eventually became an outspoken abolitionist.

As a student of nature, Thoreau's twenty-six months at Walden Pond were formative and provided him with the leisure not only to write his first book, *A Week on the Concord and Merrimac Rivers* (1849), but also the opportunity to examine and to think about the complex relationships between man and nature. Thoreau initially prized wilderness because it helped to preserve something of the wilderness in man and because it offered an opportunity to get away from the demands of modern civilization. After his first trip to Maine (1846), however, he concluded that the wilderness was not an unmixed blessing; he found some of its lessons sobering. He determined that it was best for man to balance his wild and civilized natures. Wilderness was valuable in and of itself, but it could also help people address their deeper personal needs, including a "higher, or, as it is named, spiritual life" and a "primitive, rank and savage one." Civilization, on the other hand, fostered opportunities for "intellectual and moral growth." Ideally, Thoreau felt, man should draw from both extremes and hold the middle ground. Sensitive and perceptive about his surroundings, Thoreau gave much thought to a philosophy of the environment in his writings. His later travels took him to Cape Cod, and several more times to Maine, as well as Minnesota and Canada. The publication of *Walden, or Life in the Woods* (1854) was his last major study, although *Excursions* (1863), *The Maine Woods* (1864), and *A Yankee in Canada* (1866) were published posthumously from his numerous manuscripts and journals. His works now fill twenty volumes.

MAJOR CONTRIBUTIONS

As his journals indicate, Thoreau could be a serious, unsentimental student of nature. Advocated meticulous observations on field trips to see the annual progress of natural phenomena. His talents as a naturalist are demonstrated in his studies of tree succession. Thoreau's appreciation of man's impact upon nature and his ability to explain the intricate relationships between Americans and their natural environment anticipate the later development of ecological awareness.

BIBLIOGRAPHY

Angelo, Ray. *Botanical Index to the Journal of Henry David Thoreau* (1984).
Bennett, Jane. *Thoreau's Nature: Ethics, Politics, and the Wild* (1994).
Borst, Raymond R. *Henry David Thoreau: A Reference Guide, 1835–1899* (1987).
Borst, Raymond R., comp. *The Thoreau Legacy: A Documentary Life of Henry David Thoreau, 1817–1862* (1992).
Buell, Lawrence. *The Environmental Image: Thoreau, Nature Writing, and the Formation of American Culture* (1995).
Canby, Henry S. *Thoreau* (1939).
Dassow, Laura. *Seeing New Worlds: Henry David Thoreau and Nineteenth Century Natural Science* (1995).
Harding, Walter. *The Days of Henry Thoreau* (1965).
Harding, Walter, ed. *Walden: An Annotated Edition* (1995).
Harding, Walter, and Carl Bode, eds. *The Correspondence of Henry David Thoreau* (rev. ed., 1964).
Krutch, Joseph David. *Henry David Thoreau* (1948).
Nash, Roderick. *Wilderness and the American Mind* (3d ed., 1982).
Payne, Daniel G. *Voices in the Wilderness: American Nature Writing and Environmental Politics* (1996).
Richardson, Robert D. *Henry Thoreau: A Life of the Mind* (1986).
Salt, Henry S. *Life of Henry David Thoreau* (1896).
Van Doren Stern, Philip. *The Annotated Walden* (1970).

UNPUBLISHED SOURCES

Thoreau manuscripts are in the New York Public Library; Henry E. Huntington Library, San Marino, California; J. Pierpont Morgan Library, New York City; Harvard University Library; University of Texas Library, Austin; and Concord (Massachusetts) Library. *The Thoreau Society Bulletin* (1941–), *Thoreau Journal Quarterly* (1969–), and *The Thoreau Quarterly* (1982–) all offer useful current scholarship on Thoreau and his work. Since the mid-1970s, Princeton University Press has been publishing a definitive set of his writings in a series of scholarly volumes edited by Joseph Moldenhauer, which will supersede the older twenty-volume collection edited by Bradford Torrey and others (1906).

Phillip D. Thomas

TODD, WALTER EDMOND CLYDE. Born Smithfield, Ohio, 6 September 1874; died Beaver, Pennsylvania, 25 June 1969. Ornithologist, museum official.

FAMILY AND EDUCATION

Eldest son of William Todd, public school and seminary principal, and Isabella Hunter Todd. Graduate of Beaver High School, 1891. Attended Geneva College, Beaver Falls, Pennsylvania, for several months in the fall of 1891, left to accept a government position. Did not take a degree. Later declined offer of several honorary degrees from higher institutions in the Pittsburgh area on grounds that he was not an educated man. Married Leila E. Eason, 1907 (died 1927); no children.

POSITIONS

Messenger and later, assistant and clerk, U.S. Biological Survey, Washington, D.C., 1891–1899; assistant in ornithology, later, curator, Carnegie Museum, Pittsburgh, 1899–1944; curator emeritus, 1945–1969.

CAREER

A collector of birds as a teenager, he published his first paper, concerning the bird species of his home county, in the *Oologist* in 1887, at age thirteen. Began sending records of specimens (his first was a magnolia warbler), to C. Hart Merriam, then chief of the U.S. Biological Survey (USBS) in Washington, D.C., in 1889. Later sent Merriam bird migration reports and subsequently specimens of bird stomachs. Wrote Merriam and J. A. Allen, curator of birds at the American Museum of Natural History, seeking possible employment. Employed as messenger by USBS at fifty dollars per month, but also performed clerical tasks, labeling and cataloging specimens. Continued his own collecting and kept meticulously accurate and precise specimen collections and records for the remainder of his active career. All but two of his many published papers (on the mammal and bird fauna of his native Beaver County [1904], and the ''Life Zones of Western Pennsylvania'' [1924] dealt with ornithological subjects. His first major contribution, ''The Birds of Erie and Presque Isle, Erie County, Pennsylvania,'' appeared in 1904. Though he regretted never traveling there, Todd became an authority on tropical American species, and his *Birds of the Santa Marta Region of Colombia* (with M. A. Carriker, Jr.), published in 1922, won both authors the American Ornithologists' Union (AOU) Brewster Medal. Todd published *Birds of Western Pennsylvania* (1940), based in large part upon his own collections dating back to 1893. His last major work, *Birds of the Labrador Peninsula and Adjacent Areas*, published in 1963, won him a second Brewster Medal in 1967, the first time in AOU history that this had occurred. Again, his work was based in large measure on fieldwork done by Todd extending back to 1901. A very religious person, Todd would not travel on Sundays. He was also careful with his money. On one occasion, he declined a wedding invitation in Pittsburgh because his reduced-rate commutation ticket from his hometown of Beaver was not good on Saturdays. Todd's daily commuting, including the quarter century of his retirement, added up to more than 750,000 miles during his lifetime, so that his reluctance on this occasion was perhaps understandable.

MAJOR CONTRIBUTIONS

A meticulous, largely self-trained scientist who devoted himself to his craft for eighty years and became an authority on the birds of western Pennsylvania, Labrador, and the tropical Americas. His three published works on the avifauna of these regions are considered classics to this day.

BIBLIOGRAPHY

American Men of Science (1949).
Parkes, Kenneth C. *Auk* (October 1970) (obit.).
Who Was Who in America.

UNPUBLISHED SOURCES

Memoirs covering the years 1874–1907, are at Carnegie Museum, Pittsburgh.

Keir B. Sterling

TORREY, JOHN. Born New York City, 15 August 1796; died New York City, 10 March 1873. Botanist, chemist, educator.

FAMILY AND EDUCATION

Son of Capt. William Torrey (an original member of the Society of the Cincinnati) and Margaret Nichols Torrey. Received M.D. at College of Physicians and Surgeons in 1818; A.M. (honorary) from Yale in 1823; LL.D. (honorary) from Amherst College in 1845. Married Eliza Shaw in 1824; at least four children.

POSITIONS

Between 1824 and 1827 was professor of chemistry, mineralogy, and geology at U.S. Military Academy. From 1822 to 1855 was professor of chemistry at College of Physicians and Surgeons; and professor (during summers only) of chemistry and natural history at College of New Jersey (later, Princeton) from 1830 to 1854. Also taught for short periods at Williams College and New York University. Occasional public lecturer. In 1854 resigned from Princeton faculty and retired from active work at Columbia (and was named professor emeritus). Appointed trustee of Columbia University, 1856. Between 1853 and 1873 served as U.S. assayer. Appointed botanist of New York State in 1836.

CAREER

One of group of eleven students at College of Physicians and Surgeons who, under leadership of Samuel Latham Mitchill, established, in 1817, the Lyceum of Natural History, forerunner of New York Academy of Sciences. Did most of the work on *A Catalogue of Plants Growing Spontaneously within Thirty Miles of the City of New York* (1819); presented to Lyceum in 1817.

Though he practiced medicine for a time and was a medical school and college professor for decades, his real interest lay in other fields. Government exploring expeditions in West offered him opportunity to do work in botany. Members of expeditions turned over plants they collected to Torrey to study and report upon. He first reported on plant collections of Daniel Bates Douglas made in 1820. Other reports were prepared on collections of John Fremont made in 1842 and 1843–1844; W. H. Emory, 1846–1847; R. B. Macy, 1852; J. C. Ives 1857–1858. All of these and other reports—eighteen in all—appeared in the official reports of the various expeditions. Published *A Compendium of the Flora of the North and Middle States* in 1826. Edited U.S. edition of *Introduction to the Natural System of Botany* by John Lindley (1831). Asa Gray, who studied plants with Torrey, became his associate in preparation of *Flora of North America*. Seven parts, constituting all of one volume and most of another, appeared in 1838–1843. Appointment as state botanist resulted in preparation of *Flora of New York*, 2 vols. (1843). During career, Torrey built up a large and valuable botanical library of 600 volumes and herbarium of 40,000 species, which he transferred to Columbia College in 1860 and which in 1899 Columbia deposited with newly established New York Botanical Garden. Worked in New York with Smithsonian botanical collections in the 1860s. Asa Gray edited Torrey's report of plants collected by the Wilkes Expedition in the Pacific northwest during Torrey's final months of life. In 1839 elected foreign member of Linnaean Society of London; in 1841 chosen member of American Academy of Arts and Sciences; and in 1863 was one of incorporating members of National Academy of Sciences.

MAJOR CONTRIBUTIONS

Wrote reports on plants collected by numerous government exploring expeditions in West before Civil War. Studies by later scientists demonstrated thoroughness of these pioneering botanical studies. Also led American botanists in adopting natural system of classification developed by Antoine-Laurent de Jussieu and A. P. de Candolle. Had important influence on careers of Asa Gray and other young botanists. Central figure in a botanical society named, even in his lifetime, the Torrey Botanical Club, which in 1870 established first botanical monthly in the United States, *Bulletin of the Torrey Botanical Club*.

BIBLIOGRAPHY
Dictionary of American Biography. Vol. 9 (1927–1936).
Dictionary of Scientific Biography. Vol. 13 (1974).
Gray, Asa. ''John Torrey.'' *Biographical Memoirs, National Academy of Sciences*. Vol. I (1877).
Robbins, Christine C. ''John Torrey (1796–1873).'' *Bulletin of the Torrey Botanical Club* 95 (1968).
Rodgers, Andrew D. *John Torrey: A Story of North American Botany* (1942).
Who Was Who in America, 1607–1896 (1963).
UNPUBLISHED SOURCES
Principal collection of papers is at New York Botanical Garden, together with Torrey's herbarium. There is a small collection of Torrey Papers at Yale University Library. The William H. Emory, Daniel C. Eaton, and Silliman family collections at Yale also contain Torrey correspondence. Torrey's letters are scattered throughout various collections, including the Alexander Winchell Papers at the University of Michigan; William Eggleston Papers at Dartmouth College Library; Charles W. Short Papers in Filson Club, Louisville, Ky.; and Henry R. Schoolcraft Papers at the Library of Congress, Manuscript Division.

Richard Harmond

TRACY, CLARISSA. Born Jackson, Susquehanna County, Pennsylvania, 12 November 1818; died Ripon, Wisconsin, 13 November 1905. Botanist.

FAMILY AND EDUCATION

Daughter of Stephen Tucker, pioneer and probably a farmer, and Lucy Harris Tucker. Attended local schools at age three. Student and teacher, Franklin Academy, Harford, Pennsylvania, 1835–1840. Assistant, Ladies' Seminary, Honesdale, Pennsylvania, while continuing studies, 1840–1842. Student at Troy Seminary, New York, one term, 1844. Married Horace Hyde Tracy (died 1848); two children.

POSITIONS

Began teaching at age fourteen while at the same time she continued studies. Teacher and student, Franklin Academy, 1835–1840; assistant, 1840–1842, head, 1842–c.1846, Ladies' Seminary, Honesdale, Pennsylvania; directed a private school in Honesdale, c.1849–c.1851; connected with another academy in Honesdale, c.1851–1856. Directed a private school on Neenah, Wisconsin, 1856–1859; matron in charge of domestic operations, Head of Ladies' Department, and teacher, Ripon College, Wisconsin, 1859–1893.

CAREER

For perhaps a dozen years, from age fourteen to age twenty-six, pursued her education at several institutions in Pennsylvania while concurrently working as teacher. Lost husband at age thirty, following which she taught or operated schools and academies in Pennsylvania and Wisconsin for eleven years. Finally established in three capacities (two of them administrative) at Ripon College, 1859, where she remained until end of career. Taught a variety of subjects at Ripon over period of thirty-four years, including Latin, algebra, arithmetic, English literature, and composition. Best known for instruction in botany. Evidently retired from all formal institutional commitments in 1893, but bought a home in area and tutored students privately. Continued her affiliation with college until end of her life. Studied and collected plants in vicinity of Ripon for three decades. In 1889, published *Catalogue of Plants Growing without Cultivation in Ripon and the Near Vicinity*, a twenty-six-page booklet based on specimens that either she or her students had collected. Realizing that this compilation was not complete, projected supplements, which were not published. Also wrote poetry, some of which was published several years after her death.

MAJOR CONTRIBUTIONS

Collection of plants over a period of thirty years and publication of her *Catalogue*, which constituted most thorough listing of regional plants completed until that time.

BIBLIOGRAPHY

Brown, Victoria. *Uncommon Lives of Common Women* (1975).
Elliott, Clark A. *Biographical Dictionary of American Science* (1979).
Merrell, Ada C. *Life and Poems of Clarissa Tucker Tracy* (1908).
Siegel, Patricia Joan, and Kay Thomas Finley. *Women in the Scientific Search: An American Bio-Bibliography, 1724–1979* (1979).
Who's Who in American Education. 3d ed.

Barbara Lovitts and Keir B. Sterling

TRAILL, CATHERINE PARR (STRICKLAND). Born London, England, 9 January 1802; died Lakefield, Ontario, 29 August 1899. Pioneering naturalist, botanist, science writer, popularizer of science.

FAMILY AND EDUCATION

Fifth daughter of Thomas Strickland, a businessman, and Elizabeth Homer Strickland. Received an excellent private education. Became interested in nature at an early age. Authored several books for children as an adolescent. Sister to Susanna Moodie, a well-known fiction writer, who immigrated to Canada. Married Lt. Thomas Traill of the Scotch Fusiliers in 1832 and immigrated to Canada the same year; nine children.

POSITIONS

Homesteaded and pursued natural history, especially botany, in the Rice Lake district near present-day Peterborough, Ontario.

CAREER

Published about a dozen articles on botany and natural history in journals such as *Anglo-American Magazine* and the *Horticulturalist* during the 1850s. Her books include *The Backwoods of Canada; Letters from the Wife of an Emigrant Officer* (1836); *Female Emigrant's Guide* (1854), reprinted as *The Canadian Settler's Guide* (1855); and *The Canadian Crusoes: Tales of the Rice Lake Plains* (1852). Her two major scientific works, representative of her long-term field studies, were *The Canadian Wildflowers* (1868), illustrated by Agnes FitzGibbon, and *Studies of Plant Life in Canada, or Gleanings from Forest, Lake and Plain* (1885). The former was ''the only widely accessible scientific botany book on Canadian plants'' at the time (Ainley 1995). Traill's last natural history book was *Pearl and Pebbles, or Notes of an Old Naturalist* (1894). Received a grant of £100 during Lord Palmerston's administration in recognition of her work as a naturalist. Presented with an island in the Otonobee River by the dominion government. After her husband died in 1859 and before embarking on her most significant scientific work, she was able to support herself and her dependent children as a writer, although never earning more than a subsistence income. Living and writing under extreme hardships, she remained optimistic, cheerful, and motivated. Familiar with taxonomic classifications and at ease with scientific terminology and the use of a microscope. Aware of plant succession, animal behavior, and ecological systems. Interested in traditional native knowledge, medicines, and beliefs. ''Her [scientific] work was well received by the elite of the mid-to-late 19th century Canadian scientific community'' (Ainley 1995).

MAJOR CONTRIBUTIONS

Wrote for naturalists as well as the general public. Explored new areas and identified, named, and documented what she found and/or observed. Concerned about conservation and habitat destruction in upper Canada as early as the mid-1830s. Her works provide an outstanding environmental/historical record of the flora of the Rice Lake plains.

BIBLIOGRAPHY

Ainley, Marianne G. ''Science from the Backwoods: Catherine Parr Traill.'' In *Using Nature's Language: Women and the Popularization of Science*, ed. Ann B. Shteir and Barbara Gates (1995).

Ainley, Marianne G. ''Women and the Popularization of Science: 19th Century Women Science Writers in Canada.'' (1993).

Crossfield, E. Tina. ''Experience and Perception: A Comparison of the Environmental Outlooks of Catherine Parr Traill and Susanna Moodie in 19th Century Canada.'' [Institut Simone de Beauvoir Institute] *Bulletin* 15 (1995).

Fowler, Marian. ''Traill, Catherine Parr.'' *The Canadian Encyclopedia* (1988).

Morgan, Henry James. *Sketches of Celebrated Canadians* (1862).

Peterman, Michael A. "Strickland, Catherine Parr (Traill)." *Dictionary of Canadian Biography*. Vol. 12 (1990).

E. Tina Crossfield

TRANSEAU, EDGAR NELSON. Born Williamsport, Pennsylvania, 21 October 1875, died Columbus, Ohio, 25 January 1960. Botanist, ecologist, educator.

FAMILY AND EDUCATION

Son of Samuel and Martha Edith Zimmerman Transeau. Primary and secondary education in the public schools of Williamsport. B.A. in classical course, Franklin and Marshall College (1897); studied principally with Henry C. Cowles, pioneer plant ecologist, University of Chicago (1900–1901); Ph.D. in botany under Frederick C. Newcombe, University of Michigan (1904). Honorary D.Sc. from Franklin and Marshall College (1941) and Ohio State University (1949). Married Gertrude Hastings (1906); one daughter.

POSITIONS

Taught science, Williamsport High School (1897–1900) and Colorado Springs High School (1901–1902); professor of biology, Alma College (1904–1906); investigator, Biological Laboratory of Cold Spring Harbor (1906–1907); professor of botany, Eastern Illinois Teachers College (1907–1915); professor of plant physiology and ecology. The Ohio State University (1915–1946) served as chairman of the Department of Botany (1918–1946); professor emeritus (1946–1960). Summer teaching at College of Chatauqua, New York (1902–1903) and Biological Laboratory of Cold Spring Harbor (1906, 1907, 1923, 1924); researcher, U.S. Department of Agriculture, Bureau of Entomology, in Europe (1927); collaborator, Ohio Agricultural Experiment Station (1926–1928) and Central States Forest Experiment Station, Columbus (1928–1931).

CAREER

Studied Greek with his father and with mother's encouragement took an early interest in the natural world by making collections of minerals and fossils and later of birds, butterflies, and plants. During summers traveled widely to study ecological relationships of vegetation, Biological Laboratory of Cold Spring Harbor (1899), northern Michigan and Kansas (1900), Eagle (Winona) Lake Laboratory, Indiana, and glacial geology trip in Iowa (1901), Prairie and Great Plains states (1902), western United States and British Columbia (1904), Yarmouth, Nova Scotia (1907). Studied plant communities from geographic, historic, and climatic viewpoints which resulted in the mapping of four great forest centers in eastern North America and the Prairie Peninsula of the Midwest and showed that the major climatic determinant was the ratio of rainfall to evaporation, a concept that was extended to the forests, grasslands, and deserts throughout the country. With his colleague, Homer C. Sampson, and graduate students, reconstructed and mapped the original vegetation of the east-central states with particular reference to Ohio; the *Natural Vegetation Map of Ohio*, in eight colors at a scale of 1:500,000 and the first of its kind in scope and methodology for any state, was completed by his student, Robert B. Gordon (1966). With the discovery that the ponds, ditches, and small streams of the Prairie Peninsula had the richest algal flora then known, his early significant periodicity studies of algae in relation to underwater factors gave way to taxonomic studies culminating in his book, *The Zygnemataceae* (1951).

Instituted practical reforms in the teaching of general botany by developing a highly successful course based on plant processes and ecological thought using the demonstration-discussion method; the students making original observations and drawing inferences that incorporated the discoveries of the science and their various applications to human welfare. As chairman for twenty-eight years of the Department of Botany, The Ohio State University, Transeau exercised strong leadership abilities and transformed a small, unrecognized department into one of the largest and most prestigious in the country; supervised sixty graduate students in the disciplines of ecology, algology, and physiology.

Wrote three general botany textbooks and contributed over sixty articles to professional journals, including "On the Geographic Distribution and Ecological Relations of the Bog Plant Societies of Northern North America" (1903), "Climatic Centers and Centers of Plant Distribution" (1905), "Forest Centers of Eastern North America" (1905), "The Passing of the Teleological Explanation" (1913), "The Periodicity of Freshwater Algae" (1916), *Science of Plant Life* (1919, repr. 1921, 1922), *General Botany* (1923, repr. annually, 1924–37), "The Accumulation of Energy by Plants" (1926), "Vegetation Types and Insect Devastations: Distribution of the Mexican Bean Beetle and European Corn Borer in Ohio" (1927), "The Prairie Peninsula" (1935), *Textbook of Botany* (1940, rev. ed. with H. C. Sampson and L. H. Tiffany, 1953), "Prehistoric Factors in the Development of the Vegetation of Ohio" (1941), "The Golden Age of Botany" (1942), "Natural Vegetation: Its Characteristics and Distribution" (1957), chapter in *Global Geography* (by E. W. Miller, G. T. Renner & Assoc. 2d ed.). President of Ecological Society of America (1924), of Botanical Society of America (1940), and of Phycological Society of America (1951); vice-president for botany, American Association for the Advancement of Science (1941); Certificate of Merit from the Botanical Society of America (1956).

MAJOR CONTRIBUTIONS

Gave a lifetime of support and encouragement to botanical science in its broadest sense, making substantial contributions to plant ecology, algology, and botanical education at all levels, from high school to graduate school. B. S.

Meyer, his successor, wrote that his success as a leader was not of the extrovertive or aggressive type, but one of a reserved manner using great talent for generating loyalty and respect from students, coworkers, and friends; botany as a worthwhile field for human endeavor was instilled in all who came under his influence. Thanks to Transeau's rigorous, often Socratic method of teaching, his students enjoyed a remarkable discipline in clarity of thought and expression, a service whose benefits range far beyond the confines of botany; this attribute was rated by P. B. Sears as Transeau's greatest contribution.

BIBLIOGRAPHY

Baskin, J. M., R. L. Stuckey, and C. C. Baskin. "Variations in Transeau's Maps of the Prairie Peninsula." In D. C. Hartnett, ed. *Prairie Biodiversity* (1995).

McQuate, A. G. "Edgar Nelson Transeau," *Ohio Journal of Science* 60 (1960) (obit.).

Meyer, B. S. "Botany at The Ohio State University: The First 100 Years." *Ohio Biological Survey Bulletin* 6, 2 (1983).

Sears, P. B. "Resolution of Respect: Edgar Nelson Transeau, 1875–1960." *Bulletin of the Ecological Society of America* 41 (1960).

Stuckey, R. L. "Original Development of the Concept of the Prairie Peninsula." In R. L. Stuckey and K. J. Reese, eds. *The Prairie Peninsula—In the Shadow of Transeau, Ohio Biological Survey Biology Notes* 15 (1981).

Stuckey, R. L. "Edgar N. Transeau and His Natural Vegetation Maps of Ohio." *Ohio Journal of Science* 95, 2 (1995).

Taft, C. E. "Seventy Years of Phycology at The Ohio State University." *Ohio Journal of Science* 73 (1973).

Thomas, E. S. "Reflections on Transeau the man." *Ohio Biological Survey, Biology Notes* 15 (1981).

UNPUBLISHED SOURCES

Correspondence and unpublished materials were dispersed among his students, colleagues, and family members. Many of these archival items have been returned to The Ohio State University and are now retained in the archives of the university herbarium under the care of Ronald L. Stuckey. Selected items include original maps and notes accompanying the Ohio Natural Vegetation Survey Project, field notebooks, selected correspondence, family history, and an unpublished bibliography of Transeau's writings compiled by Ronald L. Stuckey.

Ronald L. Stuckey

TRUE, FREDERICK WILLIAM. Born Middletown, Connecticut, 8 July 1858; died Washington, D.C., 25 June 1914. Zoologist, marine mammalogist, museum administrator.

FAMILY AND EDUCATION

Son of Rev. Charles Kittredge and Elizabeth Bassett (Hyde) True. An older brother, Alfred Charles True, was a leader in American agricultural education. Received B.S., 1878; M.S., 1881; and LL.D., 1897, from New York University. Married Louise Elvina Prentiss, 1887; two children, including Webster Prentiss True, who for many years was editor at the Smithsonian Institution.

POSITIONS

Clerk, U.S. Fish Commission, 1878–1881; librarian, U.S. National Museum (USNM), 1881–1883; acting curator, Division of Mammals, USNM, 1881–1883; curator, Division of Mammals, USNM, 1883–1909; curator, Division of Comparative Anatomy, USNM, 1883–1890; executive curator, USNM, 1892–1901; head curator, Department of Biology, USNM, 1897–1911; assistant secretary in charge of library and exchanges, Smithsonian Institution, 1911–1914.

CAREER

Began his zoological studies with the lower groups of animals, but poor vision forced him to revise his plans and turn to the study of mammals, with particular emphasis on the cetaceans and allied groups. After joining the Fish Commission, he served as special agent investigating turtle and terrapin fisheries for the tenth census, 1880–1881. Organized exhibits for the Fish Commission at Berlin International Fisheries Exhibition, 1880. Studied Cetacea and other mammals in European museums, 1883–1884. Conducted investigations of the bottle-nosed dolphin at Cape Hatteras, North Carolina, 1886. First major publication *Contributions to the Natural History of the Cetaceans. A Review of the Family Delphinidae*, published in 1889. Traveled to Alaska and the Pribilof Islands, under the auspices of the Fish Commission, to study seal rookeries of the region, 1895. Directed Smithsonian and USNM exhibits at several international expositions between 1897 and 1905. Visited whaling stations in Newfoundland to study the sulphur-bottom whales, 1901. Published *The Whalebone Whales of the Western North Atlantic* (1904), in which he concluded that species found in the western North Atlantic were identical to those occurring in the northeastern section of the ocean. Served as U.S. delegate at the Seventh International Zoological Congress, 1907. In later years he focused his research on fossil cetaceans and collected extensively at Calvert Cliffs, Maryland, and other areas of the Chesapeake Bay. Published last major work, *An Account of the Beaked Whales of the Family Ziphiidae in the Collections of the United States National Museum*, in 1910.

MAJOR CONTRIBUTIONS

Research on the dolphins, whalebone whales, and beaked whales was significant contributions to the study of cetaceans. Was instrumental in building the U.S. National Museum's marine mammal collections into one of the finest in the world.

BIBLIOGRAPHY

Dictionary of American Biography. Vol. 10 (1964).
National Cyclopedia of American Biography. Vol. 19 (1926).
Washington Post, 29 June 1914 (obit.).
Who Was Who in America. Vol. 1 (1966).

UNPUBLISHED SOURCES

See Smithsonian Institution Archives, Record Unit 7181, Frederick William True Papers, 1887–1910 (includes correspondence, notebooks, research notes, drawings, and photographs); Smithsonian Institution Archives, Record Unit 208, Division of Mammals, 1882–1971 (includes True's correspondence as curator of mammals, USNM; Smithsonian Institution Archives, Record Unit 242, Department of Biology, 1897–1943 (includes True's correspondence as head curator, Department of Biology, USNM).

William Cox

TUFTS, ROBIE (WILFRED). Born Wolfville, Nova Scotia, 11 August 1884; died Wolfville, Nova Scotia, 7 November 1982. Ornithologist, conservationist.

FAMILY AND EDUCATION

Son of John Freeman Tufts, a professor of history and political economy at Acadia University, and Marie Woodworth Tufts, botanist and principal of Acadia Ladies' Seminar. One brother (Harold) and sister (Hilda). Attended Acadia University, 1901–1902, but did not graduate. Honorary degrees from Dalhousie (LL.D. 1966) and Acadia (D.C.L. 1973). Married Evelyn S. Tufts (divorced) and Lillian Thompson. One daughter.

POSITIONS

Bank of Montreal, 1902–c.1914. Brokerage firm in Boston, c.1914–1919. Chief federal migratory birds officer for the Maritime Provinces, 1919–1947.

CAREER

Upon completion of his freshman year at Acadia, accepted position with the Bank of Montreal, with which he worked in Wolfville, Halifax, Ottawa, and Montreal. Subsequently accepted job offer from Boston brokerage company. Inspired by his older brother Harold, studied local birds in his spare time and wrote papers for *Auk, The Oologist*, and the *Transactions of the Nova Scotian Institute of Science*. Interest in birds led him to apply for the position of chief federal migratory birds officer for the Maritime Provinces, an appointment newly created by the passing of the Migratory Birds Convention Act of 1919. Duties included educational activities and enforcement of the law with the aid of five full-time assistants.

While employed by the Canadian government, wrote *Some Common Birds of Nova Scotia* (1934), prepared approximately forty ornithological papers (mostly published in the *Canadian Field-Naturalist*), gave more than 500 lectures, and spoke many times on radio programs. After retirement wrote *The Birds of Nova Scotia* (1962; 2d ed. 1975; 3d ed. 1986, revised by members of the Nova Scotia Bird Society), *Birds and Their Ways* (1972), *20 Favourite Birds of Nova Scotia*

(1975), *Looking Back* (1975), *Nova Scotia Birds of Prey* (1978), and a dozen or more papers. A proficient taxidermist, he also developed a bird collection for the Nova Scotia Museum.

Although a keen hunter himself, he was criticized by some people for being too hard on hunters in the Maritimes. The 679 convictions he secured between 1919 and 1932 ruffled many feathers and resulted in the transfer of enforcement responsibilities to the Royal Canadian Mounted Police (who were less enthusiastic about migratory bird laws). However, he was both fair and generous. It was said that he "would have fined his own grandmother and paid the fine himself." At the same time many young offenders did not go to court but were given a lesson in conservation. Some of these people went on to become keen ornithologists, for example, W. Earl Godfrey (see Bibliography).

MAJOR CONTRIBUTIONS

His vigorous enforcement of the Migratory Birds Convention Act, together with his education of hunters by lecturing and writing, made a valuable contribution to the conservation of migratory birds and to wildlife conservation in general. For ornithologists, *The Birds of Nova Scotia* is still the standard reference on the subject.

BIBLIOGRAPHY

Elliot, Wendy. *Atlantic Advocate* 78. 9 (May 1988).

Godfrey, W. Earl. *Canadian Field-Naturalist* 98 (1984).

Tufts, R. *Looking Back* (1975).

UNPUBLISHED SOURCES

See Autobiographical note, Acadia University Archives; Radio scripts, Public Archives of Nova Scotia (MG1 vol 2439 #2–5); Radio talks on birds, Public Archives of Nova Scotia (O/S V/F v.14 #14).

Terence Day

TYRRELL, JOSEPH BURR. Born Weston, Ontario, 1 November 1858; died Toronto, Ontario, 26 August 1957. Geologist, explorer, mining engineer, historian.

FAMILY AND EDUCATION

Son of Elizabeth (Burr) and William Tyrrell. High School at Weston. Attended Upper Canada College, Toronto. University of Toronto, B.A. (1880), M.A. (1889), B.Sc., Victoria University (1889); LL.D., University of Toronto (1930); LL.D., Queen's University (1940). Married Mary Edith Carey, 1914 (died 1945), daughter of Baptist minister; three children.

POSITIONS

Explorer, Canadian Geological Survey (1881–1898). Mining consultant, Dawson, Yukon (1899–1905); Ottawa (1906), Toronto (1907–1920). Agent, Anglo-

French Exploration Company (1910–1924). Board of Directors, Lake Shore Mine. President, Kirkland Lake Gold Mining Company (1925–1955). Fellow, Royal Society of Canada (1910), president of Section IV (1915–1916). President, Royal Canadian Institute (1913). Founder and first president, Canadian Geographical Society. President, Champlain Society (1927–1932). Fellow in all British and American geological, geographical, mining, and metallurgical societies. Distinctions: Daly Gold Medal, Murchison and Wollaston Medals; Flavelle Gold Medal. Numerous honorary memberships.

CAREER

Explorer, Canadian Geological Survey (1881). Assisted G. M. Dawson in 1883 survey of Rocky Mountains. Explored north of Calgary between the Bow and Saskatchewan Rivers (1884–1887). Discovered dinosaur bones and coalfields, Drumheller, Alberta (1884). Report published (1888). Explored northwest Manitoba (1887–1889) and Lake Winnipeg region (1890–1891) and the unexplored area northeast of Lake Athabasca (1892). Following year with his brother James, traveled 3,200 miles, gathering information on glaciation and wildlife. Traveled the Athabasca, Chipman, and Dubawnt Rivers, across the Barren Lands to Chesterfield Inlet, canoed down Hudson's Bay to Churchill, ending with 600 miles on snowshoes inland to Winnipeg. Following year traveled the same amount, crossing the Barrens from Reindeer Lake to a point 200 miles southwest of Chesterfield Inlet, down to Churchill and overland to Lake Winnipeg. Located mineral outcrops, including copper, in 1895 explorations east of Lake Winnipeg. Short trip to the Yukon. Refused promotion on return. Resigned 1 January 1899.

Became Klondike mining consultant. Partner with Tom Green, a Brantford Mohawk and McGill civil engineer. Acted as intermediary in transition from small-scale panning to large-scale capital-intensive mining. Returned to Ottawa on demise of mining boom. Advised William Mackenzie on northern Ontario cobalt mining boom. Opened Toronto office when they fell out. In 1910 began sixteen-year relationship with the Anglo-French Exploration Company, veterans of South African mining. Financial freedom allowed latitude for academic pursuits and contacts. Led northern exploration party for Ontario following settlement of Manitoba border dispute, visiting old forts around York Factory. Late entry into Kirkland Lake gold boom, joining Board of Harry Oakes's Lake Shore Mine (1920). In 1924 Tyrrell, Osler and Company (Toronto) and the Anglo-French Company bought out the Kirkland Lake Gold Mine. Tyrrell directed operations personally, risking everything to deepen the mine. Success brought with it a battle for control, decided in Tyrrell's favor. Heart attack in 1928. Became millionaire and company president. Retired in 1954, died age ninety-nine.

Studied and promoted early Canadian explorers, gathering the papers of David Thompson, publishing *David Thompson, Explorer* (1910), Samuel Hearne's *A Journey from Prince of Wales' Fort in Hudson Bay to the Northern Ocean,*

1769, 1770, 1771 and 1772 (1911); *Thompson's Narrative of His Explorations in Western America* (1916); *Documents Relating to the Early History of Hudson's Bay* (1931); as well as over eighty articles on geology, mining, and history. Donor of the Royal Society of Canada's Tyrrell Gold Medal for historical writing on Canada (1927). Several geographic features named in his honor. His discovery of dinosaur bones commemorated with opening of Tyrrell Museum of Paleontology in 1985.

MAJOR CONTRIBUTIONS

Explored large areas of Canada. Provided new knowledge on geography, geology, botany, mammalogy, and paleontology. Preserved aspects of Canada's historical past. Yet his later career went hand in hand with a resource extraction ideology that saw the introduction of hydraulic mining onto the Yukon's river systems, the establishment of environmentally damaging mines without consideration of those costs, and a lack of interest in the impact of such developments on native peoples. That aspect of his life is the antithesis of an environmentalist. Yet the two halves, like the coalfields and dinosaur bones of Drumheller, are intertwined.

BIBLIOGRAPHY

Alcock, F. J. "Joseph Burr Tyrrell." *Proceedings of the Royal Society of Canada* 52 (1959).

Canadian Who's Who. Vol. 6 (1952–1954).

Eagan, W. E. "Joseph Burr Tyrrell, 1858–1957." M.A. thesis, University of Western Ontario (1971).

Inglis, A. *Northern Vagabond* (1978).

LeBourdais, D. M. "Tyrrell." *Beaver* (December 1952).

Loudon, W. J. *A Canadian Geologist* (1930).

Martyn, Katharine. *J. B. Tyrrell; Explorer and Adventurer: The Geological Survey Years, 1881–1898* (1993).

McNicholl, Martin K. "Joseph Burr Tyrrell." *Canadian Encyclopedia*.

"Obituary." *Canadian Historical Association Report* (1958).

Tyrrell, Edith. *I Was There* (1938).

Tyrrell, James William. *Across the Sub-Arctics of Canada* (1897, 1973).

Zaslow, Morris. *Reading the Rocks: The Story of the Geological Survey of Canada, 1842–1972* (1975).

UNPUBLISHED SOURCES

See University of Toronto Library, Toronto, Ontario. Left an eccentric, unpublished autobiography. National Archives of Canada, Ottawa, holds his Geological Survey reports and field books (RG 45).

Lorne Hammond

U

UHLER, PHILIP REESE (RHEES). Born Baltimore, 3 June 1835; died Baltimore, 21 October 1913. Geologist, entomologist, librarian.

FAMILY AND EDUCATION

Son of George Washington Uhler, merchant in Baltimore, and Ann Reese Uhler. Educated at Latin School of Daniel Jones and then Baltimore College. Student at Harvard College, 1863–1866, but did not take a degree. Later awarded B.S. by Harvard according to L. O. Howard. Honorary LL.D. from New York University (1900). Married Sophia Werdebaugh, 1869 (died 1883); one son. Married Pearl Daniels, 1886; one daughter.

POSITIONS

Was ''placed in business'' by his father for several years after completing his secondary schooling. Appointed assistant to John G. Morris, librarian of the Peabody Institute, 1862. From 1863 to 1866, Uhler was assistant to Louis Agassiz at the Museum of Comparative Zoology, Harvard University, while also studying and teaching. Studied the geology and entomology of Haiti for Agassiz. Returned to the Peabody, where he spent forty-five years developing the library. Provost, Peabody Institute, 1880–1911; associate in natural history, Johns Hopkins University, 1876–1913 (L. O. Howard suggests he was an associate professor).

CAREER

Interest in insects began when he started to collect specimens at a farm purchased by his family. Was further encouraged in this area by John Gottlieb Morris. Author of numerous works on geology, entomology, archeology, and libraries. Presented papers in the United States, England, and Canada. Among his works are "A List of the Fishes of Maryland" for the "Report of the Maryland Fisheries Commission" (with Otto Lugger, 1876); "Summary of the Hemiptera of Japan, Presented by Professor Milzukuri to the U.S. National Museum" (1889); and "Descriptions of a Few Species of Coleoptera Supposed to Be New" (*Proceedings of the Academy of Natural Sciences of Philadelphia*, vol. 7, 1856). He published in 1860 his writings on Hemiptera (a large order of insects that make up the true bugs such as bedbugs and squash bugs). In 1861 he translated from the Latin of Herman August Hagen the *Synopsis of the Neuroptera of North America*. His last work was the *Capsidae* (a large family of small leaf bugs that feed on the juices of plants), a major monograph that was published only in sections. Member of American Entomological Society, Boston Society of Natural History, Academy of Natural Sciences of Philadelphia, Geological Society and Entomological Society of Washington, Royal Society of Arts (London, England), and the University Club of Baltimore.

Fellow of the American Association for the Advancement of Science. President of the Maryland Academy of Science. Donated his large insect collection to the U.S. National Museum. Lost most of his eyesight, 1907; largely inactive the last six years of his life.

MAJOR CONTRIBUTIONS

Despite his relatively small output, made notable contributions to American entomology, particularly the study of Hemiptera (true bugs), some fossil forms, and to the geology of the Cretaceous period. Introduced new method (printed titles of books as well as author's name and used varied typesets) of cataloging of book and promoted an extended analysis of periodicals. Published five volume catalog of the Peabody collections (with Molison, 1883–1892).

BIBLIOGRAPHY

Baltimore Sun, 22 October 1913 (obit.).
Biographical Index to American Science (1920).
Dictionary of American Biography.
Essig, E. O. *A History of Entomology* (1931).
Howard, L. O. *Entomological News* (December 1913).
Index to Scientists of the World (1962).
Mallis, Arnold. *American Entomologists* (1971).
National Cyclopedia of American Biography.
New York Times, 22 October 1913 (obit.).
Proceedings of the Entomological Society of Washington 16 (1914).

Susan Ignaciuk

V

VAN HISE, CHARLES RICHARD. Born Fulton, Wisconsin, 29 May 1857; died Milwaukee, Wisconsin, 19 November 1918. Geologist, university professor and president, conservationist, and publicist.

FAMILY AND EDUCATION

Son of William Henry and Mary Goodrich Van Hise. Attended University of Wisconsin, where he received B.M.E. (bachelor of metallurgy) in 1879, B.S., 1880, M.S., 1882 and Ph.D., 1892. Married Alice Bushnell Ring, 1881; three children.

POSITIONS

Instructor, 1879–1883, assistant professor, 1883–1886, professor of mineralogy and geology, 1886–1903, University of Wisconsin. Nonresident professor of structural geology at University of Chicago, 1892–1903. Assistant geologist, 1883–1888, geologist in charge, Lake Superior Division, 1888–1900; geologist in charge of Division of Pre-Cambrian and Metamorphic Geology, 1900–1908, U.S. Geological Survey (USGS), and consulting geologist, USGS, 1909–1918. Consulting geologist, Wisconsin Geological and Natural History Survey, 1897–1903. President of University of Wisconsin, 1903–1918. Chairman, Wisconsin State Conservation Commission, 1908–1915 and State Board of Forestry, 1905–1918.

CAREER

Collaborated on first systematic geological studies of the Lake Superior iron and copper region published by the USGS. Pioneering in the use of the petrographic microscope and quantitative methods of analysis, he advanced geological knowledge explaining Precambrian rock in his widely accepted ''Principles of North American Pre-Cambrian Geology'' (1896) and *A Treatise on Metamorphism* (1904). His view was that vulcanism was not the sole means of ore deposition. Other publications by Van Hise included ''Crystalline Rocks of the Wisconsin Valley'' (with R. D. Irving, 1882); ''The Pre-Cambrian Rocks of the Black Hills'' (1890); ''The Pernokee Iron-Bearing Series of Michigan and Wisconsin'' (1892); ''The Marquette Iron-Bearing District of Michigan'' (with W. S. Bayley and H. L. Smyth, 1897); ''Metamorphism of Rocks and Rock Flowage'' (1898); ''The Iron-Ore Deposits of the Lake Superior Region'' (1901); ''Pre-Cambrian Geology of North America'' (with C. K. Leith, 1909); and ''The Influence of Applied Geology and the Mining Industry Upon the Economic Development of the World'' (1912).

After appointment as president (1903) of his alma mater he embraced the principles of Senator La Follette's ''Wisconsin idea,'' calling for participation of university faculty in public service. Thus his scientific career predisposed him to participation in the contemporary conservation movement, to which his most signal contribution was *The Conservation of Natural Resources*, published in 1910. This comprehensive work became the bible of this, the first (U.S.) conservation movement. It would run through six reprintings and a second edition in 1930. As a publicist Van Hise lectured widely on the scarcity of natural resources, insisted on the efficient use of these resources, and advocated public control where necessary to assure these aims. His book *Concentration and Control . . .* (1912), almost as popular as the conservation work, described the evils of industrial concentration and advocated a federal commission to control monopolistic and wasteful practices by American business. At the time of his death he was working on a manuscript titled ''Mineral Resources and the History of Civilization.''

MAJOR CONTRIBUTIONS

Prototype of the scientist who successfully bridged the ''two cultures'' of science and the humanities. As scientist, he was able to enlist the services of other scientists in collecting information on natural resource use, which he publicized through his popularly written textbook for the conservation movement. As a member of the National Conservation Commission he was able to condense the findings of its massive three-volume report within the pages of ''this most valuable book'' in the first conservation movement. As a respected scientist, the public accepted on his authority the basic theme of scarcity advocated by the movement. As humanist and publicist he was active in the conservation and antitrust movements. Regarded as an important progressive president of the University of Wisconsin.

BIBLIOGRAPHY

Chamberlin, T. C. ''Biographical Memoir of Charles Richard Van Hise, 1857–1918.'' *Memoirs of the National Academy of Sciences* (1924).

Clepper, Henry. *Professional Forestry in the United States* (1971).

Curti, M., and V. Carstensen. *The University of Wisconsin: A History, 1848–1925*. 2 vols. (1949).

Dictionary of American Biography.

Dictionary of Scientific Biography.

Hove, Arthur. *The University of Wisconsin: A Pictorial History* (1991).

Leith, Charles K. ''Memorial of Charles Richard Van Hise.'' *Bulletin of the Geological Society of America* 31 (1920).

Oleson, A., and J. Voss. *The Organization of Knowledge in Modern America, 1860–1920* (1979).

Vance, Maurice M. *Charles Richard Van Hise: Scientist Progressive* (1960).

UNPUBLISHED SOURCES

See letters in Presidents' Papers, University of Wisconsin.

Lawrence B. Lee

VENNOR, HENRY GEORGE. Born Montreal, lower Canada (Quebec), 30 December 1840; died Montreal, 8 June 1884. Geologist, ornithologist, weather prognosticator.

FAMILY AND EDUCATION

Son of a hardware merchant, Henry Vennor, and Marion Paterson. Graduated from the High School of Montreal. Graduated with honors from McGill College, Montreal, in 1860, where he studied geology, mineralogy, zoology, and civil engineering. Not actually known if he married or had any children, although the 1885 edition of his almanac claims that he was survived by a wife and three children.

POSITIONS

Employed by Montreal mercantile firm of John Frothingham and William Workman, 1860–1865. Member of survey group in the Geological Survey of Canada, worked in region of southeastern Ontario and Pontiac County, Quebec, 1865–1881. Opened mining consulting office in Montreal, 1881.

CAREER

Was closely associated with the first recorded identification of gold in Madoc Township, Ontario, 1866. Directed attention to phosphate deposits of Ottawa County, Quebec, 1872, which led to the development of profitable mines. Published three articles concerning the birds of the Montreal region between 1860 and 1864, continuing an interest begun while at McGill. Published *Our Birds of Prey* (1876); photographs by William Notman. Correctly predicted a green Christmas and a muddy New Year's Day in 1875, which aroused public interest and led to the publication in 1877 of the first *Vennor's Almanac*. The annual almanacs were widely read throughout eastern North America. In 1881 a separate American edition was issued, supplemented in 1882 and 1883 by monthly *Vennor's Weather Bulletin*.

Published some fourteen surveys and articles, 1860–1882, and one book, as well as his almanac and weather bulletin, 1877–1885. Was elected to the Geological Society of London (GSL) in 1870, following publication of a paper in the *Quarterly Journal* of the GSL in 1867 on the stratigraphy of the Precambrian Shield in the Hastings County area.

His departure from the Geological Survey of Canada in 1881 came about in part because his private mining interests, under new legislation then pending in the Canadian Parliament, would have conflicted with his public responsibilities. The Survey never printed his accounts of the work he had done in the late 1870s owing to disputes with his superiors over the circumstances under which he had been working and differences about the caliber of his reports. Though he had done fine field work in a geologically challenging region of the country, some younger associates within the Survey were making more rapid progress due to their more intensive scientific preparation, and this created friction within the organization.

MAJOR CONTRIBUTIONS

His seven published reports for the Geological Survey of Canada dealt with the distribution of various mineral deposits and their economic value. He achieved considerable public acclaim through publication of his popular weather almanacs, which he based on a long-term study of weather patterns, from which he claimed to have discovered a ''law of recurrences.''

BIBLIOGRAPHY

Appleton's Cyclopedia of American Biography (1889).
Dictionary of Canadian Biography (1982).
Gazette (Montreal), 9 June 1884 (obit.).
New York Times, 9 June 1884 (obit.).

UNPUBLISHED SOURCES

See Blacker Wood Library, McGill University, J. G. Vennor Papers.

Brooke Clibbon

VERRILL, ADDISON EMERY. Born Greenwood, Maine, 9 February 1839; died Santa Barbara, California, 10 December 1926. Naturalist, invertebrate zoologist, museum curator, university professor.

FAMILY AND EDUCATION

Son of George Washington and Lucy Hilborn Verrill. From old New England stock; family moved to Norway, Maine, when Addison was young. Student, Norway Liberal Institute. A boy naturalist who later studied with Louis Agassiz. B.A., Harvard University, 1862 (Lawrence Scientific School). Honorary A.M. (Yale) 1867. Married Flora Louisa Smith, 1865; six children.

POSITIONS

Assistant, Museum of Comparative Zoology, 1860–1864; curator, Essex Institute, 1864; professor of zoology, Sheffield Scientific School, Yale University, 1864–1907; also curator of zoology, 1867–1907, and instructor in geology, 1870–1894; professor of entomology and comparative anatomy, University of Wisconsin, 1867–1870; in charge of dredging for marine invertebrates, U.S. Fish Commission, 1871–1887; professor emeritus, Yale, 1907–1926.

CAREER

Field trip to coast of Maine with Alpheus Hyatt and N. S. Shaler to collect marine invertebrates, summer, 1860; expedition to Anticosti Island with Hyatt and Shaler to collect marine invertebrates and fossils, summer, 1861; reorganized "Natural History Cabinet" of Yale Natural History Society into Museum of Zoology, 1864; in charge of instruction, Sheffield Scientific School, 1864–1907; associate editor, *American Journal of Science*, 1869–1920; elected National Academy of Sciences, 1872; president, Connecticut Academy of Arts and Sciences, 1872.

Published more than 350 works on invertebrates, including *The Bermuda Islands* (1903); *Zoology of the Bermuda Islands*, vol. 2 (1903); *Geology and Paleontology of the Bermudas* (1906); *Coral Reefs of the Bermudas* (1907); *Monograph of the Shallow Water Starfishes of the North Pacific Coast* (1914); *Report on West Indian Starfishes* (1915); *Reports on Alcyonaria and Actinaria of Canadian Arctic Expedition* (1921); *Crustacea of Bermuda*, 3 pts. (1923); *Alcyonaria of the Blake Expedition* (1925). See also bibliography in Coe (1932). A tireless, constant worker, shy and retiring, but sociable to close friends. An empirical investigator not given to theorizing.

MAJOR CONTRIBUTIONS

Described about 1,000 new species of invertebrates, mostly marine. Best known for research on taxonomy and natural history of the radiates (coelenterates and echinoderms). A pioneer in marine ecology. Most famous for *Report of the Invertebrate Animals of Vineyard Sound and Adjacent Waters* (1873).

BIBLIOGRAPHY
Chittenden, R. H. *History of the Sheffield Scientific School, 1846–1922*, 2 vols. (1928).
Coe, W. R. "Addison Emery Verrill and His Contributions to Zoology." *American Journal of Science* 13 (1927).
Coe, W. R. "Biographical Memoir of Addison Emery Verrill, 1839–1926." *Biographical Memoirs, National Academy of Science* 14 (1932).
Dexter, R. W. "Three Young Naturalists Afield: The First Expedition of Hyatt, Shaler, and Verrill." *Scientific Monthly* 79 (1954).
Dictionary of Scientific Biography.
UNPUBLISHED SOURCES
Portions of diary and account of expedition to Anticosti Island (1861) are in Archives, Harvard University.

Ralph W. Dexter

VOGT, WILLIAM. Born Mineola, New York, 15 May 1902; died 11 July 1968. Organizational official, ecologist, ornithologist.

FAMILY AND EDUCATION

Son of William and Frances Belle (Doughty) Vogt and a descendant of Rev. Francis Doughty, first Episcopalian rector in New Amsterdam. Won poetry prize and edited college literary magazine at St. Stephens (now Bard) College, from which he received B.A. in 1925 and Honorary Sc.D. in 1952.

POSITIONS

Assistant editor for New York Academy of Sciences, 1930–1932; curator of Jones Beach State Bird Sanctuary, 1932–1935; field naturalist and lecturer with National Audubon Society and editor of *Bird Lore*, 1935–1939; consulting ornithologist to Peruvian Guano Administration, 1939–1942; associate director of Division of Science and Education, coordinator of Inter-American affairs, 1942–1943; chief of Conservation Section, Pan American Union, 1943–1950; national director of Planned Parenthood Federation of America, 1951–1961; secretary of the Conservation Foundation, 1964–1967; and until his death, served as representative of International Union for the Conservation of Nature and Natural Resources to the United Nations.

CAREER

Became interested in ornithology through books of Ernest Thompson Seton. In 1942, on a fellowship from Committee for Inter-American Artistic and Cultural Relations, studied climatology in Chile. His studies, revealing a decrease in natural resources in South America, roused his interest in human populations in relation to their environment, and he turned his attention from ornithology to ecology. In 1946, prepared a number of reports for Pan American Union on population and natural resources in several Latin American countries. In 1948, was secretary-general for first Inter-American Conference on Conservation of Renewable Natural Resources, which recommended that American countries establish agencies to protect and enrich their resources and warned that some 55 percent of U.S. land suffered from erosion. That same year his book *Road to Survival* stressed that by abuse of land and especially by ''excessive breeding,'' humankind had fallen into an ''ecological trap.'' Believing that population control was part of the solution to ''ecological trap,'' Vogt, in 1951, accepted post of national director of Planned Parenthood Federation of America. As secretary of Conservation Foundation, he maintained his interest in consequences of unchecked population.

Contributed articles to natural history and general-interest periodicals. Author of Audubon's *Birds of America* (1937); *El Hombre y la Tierra* (1944); *Road to Survival* (1948); *People!* (1960). Knew a great deal about people and environments they inhabited. His knowledge was enriched by frequent travel to many parts of world and by friends and continuous reading.

MAJOR CONTRIBUTIONS

Was a passionate student of world population and effects of human society and culture on the natural environment. Perhaps his most important legacy was *Road to Survival*. Translated into nine languages and read by several million people throughout the world, it was recognized by a number of contemporary critics as the best book in its field.

BIBLIOGRAPHY

''Vogt, William.'' *Current Biography* (1953).

''William Vogt.'' *Geographic Review* 59 (April 1969).

''William Vogt.'' *New York Times*, 12 July 1968 (obit.).

Richard Harmond

W

WAKSMAN, SELMAN ABRAHAM. Born Novaya Priluka, Russia, 22 July 1888; died Woods Hole, Massachusetts, 16 August 1973. Microbiologist, soil scientist, pharmacologist.

FAMILY AND EDUCATION

Son of Jacob and Fradia Waksman. Attended gymnasium in Zhitomir and Odessa, passed examinations in 1910. Left Russia, lived with cousin in Metuchen, New Jersey. Naturalized, 1916. B.S., 1915, M.S. Rutgers, 1916, Ph.D. University of California, Berkeley, 1919. Honorary D.Sc., Rutgers, 1943, Princeton, 1947, University of Madrid, 1950, Rhode Island State College, 1950, Pennsylvania Military College, 1953, Philadelphia College of Pharmacy and Science, 1953, Brandeis, 1954, Jacksonville University, 1959, University of Göttingen, 1962, University of Brazil, 1963; honorary M.D., University of Liege, 1946; Honorary LL.D., Yeshiva, 1949, Keio University, Japan, 1953; honorary D.H.L., Hebrew Union, 1951; *Dr. honoris causa,* University of Athens, 1952; Hebrew University, Israel, University of Strassbourg, 1958; and others. Married Bertha Deborah Mitnik, 1916; one son.

POSITIONS

Research assistant, soil microbiology, University of California, 1915; research biochemist, Cutter Laboratories, 1917–1918; bacteriologist, Takamine Laboratories, 1919–1920; lecturer in soil microbiology, 1918–1924; microbiologist, New Jersey Agricultural Experimental Station, 1918–1925; associate professor, 1925–1929, professor, 1929–1940, professor and chairman, 1940–1958, director,

Institute of Microbiology (now Waksman Institute of Microbiology), 1949–1958, Rutgers University; professor emeritus, Rutgers, 1958–1973. Marine bacteriologist, Woods Hole Oceanographic Institute, 1931–1942.

CAREER

As a young man, supported his academic work through scholarships, as a night watchman, tutor of English and science, with summer work on a California ranch, and as head of biochemistry section at Cutter Laboratories. Income from his early teaching career was supplemented by moonlighting at New Jersey State Experiment Station and another laboratory. Investigated nature, distribution, and properties of soil microorganisms, creation of humus, and enzymes for much of his life from the 1920s on. Gave increasing attention to activities of national and international professional organizations concerned with microbiology and soil science from the mid-1920s on. Attended International Conference on Soil Science in Soviet Union (1924), after which he produced ''Soil Science in 1924,'' a preliminary to his *Principles of Soil Microbiology* (1927, 2d ed. 1932). This volume was key text in field for several decades and sold well. Organized an international soil science congress in Washington, D.C. (1927); also participated in subsequent congresses in Russia (1930) and England (1935). Attended other international conferences and traveled in Europe, 1929, 1930, 1933, 1935, 1938, and went to Holy Land, 1938.

In 1939, began research on organisms that might have impact on infectious disease. In 1941, proposed that term ''antibiotic'' be confined to ''microbial products with antimicrobial properties.'' With very limited institutional resources but with assistance of graduate students and postdoctoral fellows at his Rutgers laboratory, Waksman worked out procedures for identifying and isolating some eighteen antibiotics developed under his superintendence, several of which proved to have medical applications. Streptomycin (which he discovered in 1944 with several colleagues) was the most important, and this won him the Nobel Prize in medicine and physiology (1952). Among the others were actinomycin (1940) and neomycin (1949). This research activity was triggered in part by work of others, including his former student Rene Dubos, also by fact that he was extremely knowledgeable about actinomycetes, group of microorganisms from which most antibiotics later came. Onset of World War II was yet another factor. As one who was not a medical scientist, Waksman was amazed by the growing value of his discoveries in treatment of tuberculosis and tubercular meningitis in children and in control of other diseases. Spent much time in post–World War II period traveling in the interests of enhancing methods of scientific exchange and encouraging widespread production of antibiotics. Majority of royalties from sale of streptomycin and neomycin were devoted to establishment of Institute of Microbiology at Rutgers (1951) and to creation of a Foundation for Microbiology, which supported courses, fellowships, research, prizes, and exchange programs in microbiology and related fields. Similar foundations were subsequently established in France and Japan, also underwritten by royalties from pharmaceutical industry.

Waksman's work in microbiology had made him notable figure in scientific circles even before he won world fame for his accomplishments with antibiotics. Published some 400 papers and twenty-eight books, some of them as coauthor, dealing with enzymes, humus, peats, soil microorganisms, and antibiotics, but ultimately focusing on streptomycin and neomycin. Also authored an autobiography, *My Life with the Microbes* (1954). Noted collector of books on a variety of subjects, which were given to Brandeis and Yale, together with Burndy Science Library following his death. Won many awards, including Order of the Rising Sun (Japan), Commander of the Legion of Honor (France), Commendatore of the Order of the Southern Cross (Brazil); Grand Cross of Public Health (Spain); corresponding membership (subsequently foreign associate), French Academy of Sciences; and many others. Waksman never lost his excitement for the microorganisms that were his consuming interest for six decades.

MAJOR CONTRIBUTIONS

Waksman's sixty years of devotion to microorganisms produced several antibiotics, which have continued to be crucial to the maintenance of human health. In addition, his research into the many beneficent and destructive properties of microorganisms has been invaluable.

BIBLIOGRAPHY

American Men of Science. 12th ed.

Bryson, V. "Selman A. Waksman (1888–1973)." *ASM News* 40 (1974).

Lechevalier, H. A. *The Development of Applied Microbiology at Rutgers Waksman Institute of Microbiology* (1982).

Sakula, A. "Selman Waksman (1888–1973), Discoverer of Streptomycin: A Centenary Review." *British Journal of Diseases of the Chest* 82 (1988).

Walsman, Byron, and H. A. Lechevalier. "Selman Abraham Waksman." *Dictionary of Scientific Biography* 18, Supplement 2 (1970–1990).

Who Was Who in America.

Keir B. Sterling

WALCOTT, CHARLES DOOLITTLE. Born New York Mills, New York, 31 March 1850; died Washington, D.C., 9 February 1927. Geologist, paleontologist, scientific administrator.

FAMILY AND EDUCATION

Walcott was the grandson of an exceptionally rich woolen mill owner. His father, Charles D. Walcott died when he was two, leaving a wife, Mary Lane Walcott, and four children, of whom Charles was the youngest. Educated in the Utica, New York, public schools and then at Utica Academy. Ended formal schooling at eighteen without graduating. Made arrangements to attend Harvard as a special student under Louis Agassiz, following the sale of his fossil collection to Harvard, but the latter's death prevented him from doing so. An early marriage (in 1871 or 1872) to his employer's youngest daughter, Lura Ann Rust, ended with her early death after sixteen months. There were no children. Married Helena B. Stevens, 1888 (died 1911); four children; married Mary Morris Vaux, 1914; no children. Awarded twelve honorary doctorates, including LL.D.'s from Hamilton, 1898; Chicago, 1901; John Hopkins, 1902; Pennsylvania, 1903; Yale, 1910; St. Andrews, 1911; and Pittsburgh, 1912; two D.Sc.'s from Cambridge University, 1909, and Harvard, 1913; and a Ph.D. from Konelige Frederiks Universitet, Oslo, 1911.

POSITIONS

Hardware store clerk, Utica, New York, 1868–1869; employed as farmworker by William Rust, Trenton Falls, New York, 1871–1876. Assistant to state geologist of New York, 1876–1879; assistant geologist, U.S. Geological Survey (USGS), 1879 (as employee number 20); in charge of Geological Survey's Division of Invertebrate Paleozoic Paleontology, 1882–1883. Paleontologist, USGS, 1883–1893; concurrent appointment as honorary assistant curator in charge of Paleozoic fossils, U.S. National Museum, 1892–1893, and de facto director, USGS, 1892–1894; geologist in charge of geology and paleontology, 1893–1894; director, USGS, 1894–1907. Concurrent appointment as assistant secretary of the Smithsonian Institution, 1896–1907; secretary of the Carnegie Institution of Washington, 1902–1905; director, U.S. Reclamation Service, 1905–1907; secretary of the Smithsonian Institution, 1907–1927, Founder, National Advisory Committee for Aeornautics, 1915.

CAREER

Carried out fieldwork in paleontology for half a century; as director, USGS, instituted a geological mapping project for the southeastern and northwestern United States, the latter in response to concerns of the mining industry; also began a program of engineering geology, in testing natural materials for roadbeds and in studies of slope stability. Interested in the efficient use of ores and of coal and established USGS branch to investigate these subjects. Indirectly responsible for establishment of U.S. national forests. Urged President Grover Cleveland to significantly increase the size of forest reserves, as had been recommended by National Academy of Sciences Forestry Commission. Forest reserves administered by USGS from 1897 until U.S. Forest Service established 1906. Urged establishment of national park in the Grand Tetons, 1897. Testified in favor of establishing Glacier National Park and, in general, was a strong supporter of the national park movement. Began separate Hydrographic Branch of USGS in 1902. Had charge of Reclamation Service until it became a separate agency in 1907. Walcott continued his fieldwork well into his seventies and continued publishing contributions to paleontology until his death. Responsible for establishment of National Research Council as a wartime measure during World War I.

Elected to the National Academy of Sciences, 1896, vice president, 1907–1917; president, 1917–1922. President, American Association for the Advancement of Science, 1923–1925. He has been described as having been, during the last twenty years of his life, "the most powerful scientific administrator in America."

MAJOR CONTRIBUTIONS

An indefatigable geological field-worker from his early days, Walcott was primarily a paleontologist. At the time of his death, it was estimated that he had written about 70 percent of the published data on the Precambrian and Cambrian of North America. His work on the Cambrian was of fundamental importance, despite major modifications by later workers. He produced a number of important papers concerning trilobites, arthropod appendages, and kindred subjects for more than half a century. His discovery of numerous soft-bodied fossils in the Burgess Shale, located in the Canadian Rockies of eastern British Columbia in 1909, was perhaps his most outstanding accomplishment. One modern authority, Stephen Gould, has termed them "the world's most important animal fossils." Very conservative in his thinking, Walcott posited that most of the fossils there fit into modern worm and arthropod groups. His conclusions were based in large part upon his long-held convictions concerning evolutionary progress, but he never found the time, due to heavy administrative burdens, to adequately study what he had uncovered. Modern authorities, notably Harry Whittington of Cambridge University and several of his students, have concluded that most of these organisms were unique to science, and, indeed, many remain undescribed today. The majority of them flourished and died out during the Middle Cambrian and have no modern descendants. As Gould has stated, they "probably exceed, in anatomical range, the entire spectrum of invertebrate life in today's oceans." Played a vital role in the administration of several federal and private scientific agencies for more than thirty-five years.

BIBLIOGRAPHY

Briggs, Derek E. G., D. H. Erwin, and F. J. Collier. The Fossils of the Burgess Shale (1994).

Conway Morris, S[imon], and H[arry] B. Whittington. "Fossils of the Burgess Shale. A National Treasure in Yoho National Park, British Columbia." Geological Survey of Canada, *Miscellaneous Reports* 43 (1985).

Gould, Stephen J. *Wonderful Life: The Burgess Shale and the Nature of History*. New York: Norton (1989).

Massa, W. R., Jr. *Guide to the Charles D. Walcott Collection* [Smithsonian Institution], *1851–1940*. Washington, D.C.: Archives and Special Collections, Smithsonian Institution, (1984).

Nelson, Clifford M., and Ellis Yochelson. "Organizing Federal Paleontology in the United States, 1858–1907." *Journal of the Bibliography of Natural History* 9, no. 4 (1980).

Whittington, H. B. *The Burgess Shale*. New Haven, Conn.: Yale University Press (1985).

Yochelson, Ellis L. "Charles Doolittle Walcott." *Biographical Memoirs, National Academy of Sciences*. Vol. 39. New York: Columbia University Press (1967).

Ellis L. Yochelson

WALCOTT, FREDERIC COLLIN. Born New York Mills, New York, 19 February 1869; died Stamford, Connecticut, 27 April 1949. Businessman, U.S. Senator, conservationist.

FAMILY AND EDUCATION

Son of William Stuart and Emma Alice Welch Walcott. Attended public schools, Utica, New York. Graduate of Lawrenceville (New Jersey) School, 1886; Phillips Academy, Andover, Massachusetts, 1887. B.A., Yale, 1891, honorary M.A., Yale, 1917. Honorary D.Sc., Trinity College, Hartford, Connecticut, 1928; honorary L.L.D., Hamilton College, 1940. Married Frances Dana Archbold, 1899 (died 1899); Mary Hussey Guthrie, 1907 (died 1931); two children by second marriage.

POSITIONS

Various positions in manufacturing and banking, 1907–1922, including president of New York Mills; treasurer, Aragon Mills, Aragon, Georgia; vice president, C. C. Kellogg and Sons, Utica, New York; vice president, Knickerbocker Trust Company, New York; vice president, Arizona Power Company. Retired from business, 1922. Assisted Herbert Hoover, Federal Food Administrator, in relief efforts, 1917–1919. President, Connecticut Board of Fisheries and Game, 1923–1928; chairman, Connecticut Water Commission, 1925–1928; Connecticut state senator, 1925–1929 (president pro tempore, 1927–1929); U.S. senator, Connecticut, 1929–1935; commissioner of welfare, Connecticut, 1935–1939; president, American Wildlife Institute, 1935; president, American Wildlife Foundation, 1945–1948; regent, Smithsonian Institution, 1941–1948.

CAREER

During his career in the U.S. Senate, Walcott served on Committees on Agriculture and Forestry, Banking and Currency, Education and Labor and as chair of the Committee on Indian Affairs. Also chaired Special Committee for Wildlife, U.S. Senate, 1930–1935. Sponsored Walcott-Kleberg Duck Stamp Act, establishing policy-forming programs and special funding for wildlife refuges. During his tenure, his committee was instrumental in passage of Forest Wildlife Refuge Act, Pittman-Robertson Federal Aid in Wildlife Restoration Act, Whaling Treaty, Coordination Act of 1934, making possible Cooperative Wildlife Research Unit Program, and Migratory Bird Treaty with Mexico. Instrumental in creation of Patuxent Wildlife Refuge Center in Maryland for U.S. Biological

Survey and in consolidation of Biological Survey and Bureau of Fisheries in new U.S. Fish and Wildlife Service, placed in Interior Department, 1939–1940. Through his efforts, President F. D. Roosevelt was led to convene first North American Wildlife Conference in 1936. The much-traveled Walcott made hunting and exploration expeditions to Singapore, Japan, North Africa, Chile, and Bolivia. Contributed chapter on private game preserves to William T. Hornaday's *Wildlife Conservation in Theory and Practice: Lectures Delivered before the Forestry School of Yale University* (1914). Collected flamingos and other rare birds in South America for American Museum of Natural History, 1925. His article on the expedition to the Laguna Colorado in southern Bolivia, published in *Geographical Review* later that same year, dealt primarily with the birdlife but was richly suggestive of the geography of a little-known part of Latin America. Subsequently made several hunting trips to Alaska. Helped found American Game Protection and Propagation Association, 1911, serving on its board until 1935. Honorary member of Wildlife Foundation, 1948.

MAJOR CONTRIBUTIONS

Played a principal role in framing much of the basic federal conservation legislation during the Hoover and first F. D. Roosevelt administrations. Longtime activist in American Game Protection and Propagation Association and other conservation organizations and activities.

BIBLIOGRAPHY

Biographical Directory of the United States Congress, 1774–1989 (1989).

Geographical Review (July 1949) (obit.).

Palmer, T. S. *Auk* (October 1949) (obit.).

Shoemaker, Carl, D., Ira N. Gabrielson, and W. L. McAtee. *Journal of Wildlife Management* (January 1950) (obit.).

Trefethen, James B. In R. H. Stroud, ed., *National Leaders of American Conservation* (1985).

Who Was Who in America.

Joan Ryan

WALCOTT, MARY VAUX. Born Philadelphia, 31 July 1860; died New Brunswick, Canada, 22 August 1940. Naturalist, artist.

FAMILY AND EDUCATION

Daughter of George and Sarah Morris Vaux of Philadelphia. Descendant (maternally) of Thomas Lloyd, a colonial governor and Thomas Wynne, physician and friend of William Penn. Graduated from Friends Select School, Philadelphia, 1879. Mother died in 1879. Ran her family's household and dairy farm. Married Charles Doolittle Walcott, secretary of the Smithsonian Institution, in 1914.

Walcott had three sons and a daughter by his first wife, Helena Burrows Stevens (deceased).

POSITIONS

Member of the Board of Indian Commissioners, appointed by President Coolidge, reappointed by President Hoover; clerk of the Twelfth Street Meeting of Quakers, Philadelphia; national president of the Society of Woman Geographers, 1933–1939.

CAREER

Began art lessons at age ten and eventually concentrated on rendering of wildflowers. With her family, visited the Canadian Rockies to study glaciers and rocks in summer 1887; she returned almost every summer. Became avid mountain climber and often rode horseback into the Rockies. Was first woman to scale Mt. Stephen, British Columbia, in 1900. Climbed Mt. Robson, 13,700 feet above sea level and the highest peak of British Columbia, in 1913. After marriage, led a public life and took her appointment to Board of Indian Commissioners seriously by visiting over 100 reservations. Helped to form Freer Gallery at the Smithsonian, which concentrated on Oriental art. Published the five-volume *North American Wild Flowers* in 1925 and *Illustrations of North American Pitcherplants* in 1935. In 1930, along with Mrs. Hoover, was instrumental in planning Florida Avenue Friends Meeting House in Washington, D.C., and took part in its dedication ceremony.

MAJOR CONTRIBUTIONS

Considered an accomplished watercolor artist, her *North American Wild Flowers* was seen as true culmination of artistry, as the paper-making and color-printing processes were revolutionary for their time. Her many public appearances as lecturer helped to promote scientific ideas to wide audience. An avid explorer and mountain climber, Mt. Mary Vaux, British Columbia, is testament to her achievements.

BIBLIOGRAPHY

American Men of Science. 6th ed.

Huntington, F. C. *Bulletin of the Society of Woman Geographers* (September 1940) (obit.).

New York Times, 25 August 1940 (obit.).

Shelter, S. G. *Notable American Women*.

Younger, H. C. *Science* (25 October 1940) (obit.).

UNPUBLISHED SOURCES

See biographical sketch, files of the Friends Meeting House, Florida Ave., Washington, D.C.; miscellaneous files of the Smithsonian Institution, Washington, D.C.

Geri E. Solomon

WALKER, EDMUND MURTON. Born Windsor, Ontario, 5 October 1877; died Toronto, Ontario, 14 February 1969. Entomologist, naturalist.

FAMILY AND EDUCATION

Eldest son of Byron Edmund Walker, president of Canadian Bank of Commerce, chairman of Board of Governors and chancellor of the University of Toronto. Grandfather and father both amateur naturalists, stimulated Walker's boyhood interest in entomology. Educated at the University of Toronto, received B.A. (natural science) in 1900, M.B. in 1903. Also studied invertebrate zoology at the University of Berlin, 1905. Honorary D.Sc. from Carleton University, 1963. Married Norma Ford, professor of human genetics, University of Toronto (died 1968); four children.

POSITIONS

Lecturer in invertebrate zoology, University of Toronto, 1906–1913. Assistant professor, 1913–1917. Associate professor, 1917–1926. Professor of invertebrate zoology and head of Department of Zoology, 1926–1948. Professor emeritus, 1948–1969. Editor, *Canadian Entomologist*, 1910–1920. Assistant director, Royal Ontario Museum of Zoology, 1918–1931. Honorary curator of the museum, 1931–1969.

CAREER

Supervised Georgian Bay Biological Station, 1907, 1908, 1912. Traveled widely in Canada to collect invertebrates, including to the Pacific Biological Station, Vancouver Island, 1913 and 1926, Godbout, Quebec, 1918, western Canada, 1921, St. Andrews, New Brunswick, 1925. Became world authority on Orthoptera and Odonata, with monographs and shorter papers dealing with their morphology, taxonomy, habits, life history, and geographical distribution. Noted for discovery in 1913, with T. B. Kurata, and subsequent study of ''ice-bug,'' *Grylloblatta campodeiformis*, which led to naming of a new insect order. Discovery stimulated him to study evolution of orthopteroids. Founded in 1914, and largely responsible for, the extensive invertebrate collection of the Royal Ontario Museum. Noted as successful teacher of both advanced and general students. Maintained interest in geographic distribution and comparative ecology.

Published over 130 articles and several monographs, including *The North American Dragon-Flies of the Genus Aeshna* (1912), described as ''an exceptionally penetrating analysis of a genus which had hitherto proved troublesome''); ''Insects and Their Allies,'' in *The Natural History of the Toronto Region* (J. H. Faull, ed.) (1913); *The North American Dragonflies of the Genus Somatochlora* (1925); ''Presidential Address—Grylloblatta, a Living Fossil,''

Transactions of the Royal Society of Canada (1937); "Changes in the Insect Fauna of Ontario," in *Changes in the Fauna of Ontario* (F. A. Urquhart, ed.) (1957); *The Odonata of Canada and Alaska*, 3 vols. (with P. S. Corbet, 1953, 1958, 1975), described as "one of the most comprehensive works ever published on a group of Canadian insects."

Noted for a fascination with all living things and an ability to communicate his excitement about them to others. Considered by contemporaries a naturalist in the tradition of Darwin. Stressed in systematic work description of the ecology, geographic distribution, and behavior of the species and the variability within populations.

MAJOR CONTRIBUTIONS

Assisted in development of Royal Ontario Museum as major center for study of systematics of invertebrates. Through rigorous systematic research and detailed descriptions of insect habitats and behavior, laid foundation for later research in insect ecology, behavior, and evolution in Canada.

BIBLIOGRAPHY

The Canadian Encyclopedia.

Craigie, E. H. *A History of the Department of Zoology of the University of Toronto* (1967).

Johnstone, K. *The Aquatic Explorers: A History of the Fisheries Research Board of Canada* (1977).

Wiggins, G. B., ed. *Centennial of Entomology in Canada, 1863–1963: A Tribute to Edmund C. Walker* (1966).

UNPUBLISHED SOURCES

Miscellaneous materials are at University of Toronto Archives and at Royal Ontario Museum Archives.

Stephen Bocking

WALTER, THOMAS. Born Hampshire, England, c.1740; died Berkeley County, South Carolina, 17 January 1789. Botanist, planter.

FAMILY AND EDUCATION

No information on parentage. Emigrated to eastern South Carolina as a young man from his native Hampshire, England. Acquired a plantation on the banks of the Santee River, where he remained for the rest of his life. Details of schooling unknown but seemingly received a strong educational foundation. Married Anne Lesesne, 1769 (died 1769), Ann Peyre, 1777 (died 1780), and later, Dorothy Cooper. Two daughters by second marriage and one by third; left numerous descendants. Buried in small botanical garden on his South Carolina estate.

POSITIONS

Botanized in Berkeley County (probably St. John's Parish), South Carolina, from young adulthood until 1789.

CAREER

Undertook a detailed survey of the vegetation within a fifty-mile radius of his South Carolina home. Prepared in Latin a descriptive manuscript dated 30 December 1787, summarizing all of the flowering plants of the region. It was taken to England by fellow botanist and friend John Fraser, where it was published under the title *Flora Caroliniana*. Work contained abbreviated descriptions of over 1,000 species constituting 435 genera. *Flora*, based on specimens from Walter's vast herbarium and Fraser's collections, included over 200 new species and 32 new genera; only 4 of the latter are given distinctive names. This work is of great significance; it is the first reasonably complete account of the flora of any definite locality in eastern North America employing the Linnaean binomial system of nomenclature. His herbarium (believed to have originally contained all the species addressed in *Flora*) was taken to England, as were Fraser's collections, along with the manuscript. The herbarium remained in the possession of the Fraser family until 1849, when it was obtained by the Linnaean Society of London. Later acquired by the British Museum, 1863, in a state of neglect, having suffered damage and loss. Maintained botanical garden on estate, where numerous species treated in *Flora* were cultivated. Made an unsuccessful attempt to introduce a Carolina grass, *Agrostis perennans*, into England. Failure was set forth in a rare volume by Fraser, entitled *A Short History of the Agrostis Cornucopiae: Or the New American Grass* (1789).

MAJOR CONTRIBUTIONS

A respected scholar in early American botany, managed to produce a sound, relevant work of enduring scientific merit despite the upheaval caused by local warfare during the Revolutionary period. Classical in text, Walter's *Flora* has been useful in aiding many American botanists in their interpretations of his brief descriptions.

BIBLIOGRAPHY

Coker, W. C. "A Visit to the Grave of Thomas Walter." *Journal of the Elisha Mitchell Science Society* (April 1910).

Dictionary of American Biography. Vol. 19 (1936).

Elliot, Clark A. *Biographical Dictionary of American Science: The Seventeenth through the Nineteenth Centuries* (1979).

Maxon, W. R. "Thomas Walter, Botanist." *Smithsonian Miscellaneous Collections* (April 1936).

Who's Who in American History. Historical vol. 1607–1896 (1989).

Lynn M. Haut

WARDER, JOHN ASTON. Born West Philadelphia, Pennsylvania, 19 January 1812; died Cincinnati, Ohio, 14 July 1883. Physician, forester, horticulturalist.

FAMILY AND EDUCATION

Son of Jeremiah and Ann Aston Warder. M.D., Jefferson Medical College, Philadelphia, 1836. Married Elizabeth Bowne Haines, 1836; at least four sons.

POSITIONS

In private medical practice, Cincinnati, 1837–1855. Professor of Chemistry and toxicology, Medical College of Ohio, 1854–1857. Elected to Ohio State Board of Agriculture, 1871, reelected, 1873 and 1875, served until 1877. Appointed one of the U.S. commissioners to the International Exposition in Vienna (in his capacity as a forester), 1873. Elected first (and only) president, American Forestry Association, 1876–1882; vice president, American Forestry Congress, 1882–1883. Appointed forestry agent by U.S. commissioner of agriculture, May 1883, but died before he could assume post.

CAREER

Knew many leading naturalists during his Philadelphia boyhood, including J. J. Audubon, William Bartram, William Darlington, François Michaux, Thomas Nuttall, and others. Helped his father develop large farm near Springfield, Ohio, 1830–1834. Published translation of a French work, *A Practical Treatise on Laryngheal Phthisis . . . and Diseases of the Voice*, by Trousseau and Belloc (1839). Director of both the Ohio and Miami Medical Colleges for some years. Cofounder of a number of scientific organizations, including the Western Academy of Natural Sciences and the Cincinnati Society of Natural History (of which he was president for five years). Began publication of *Western Horticultural Review*, 1850, which was later (1853) merged with another horticultural periodical. In its pages he first described the western catalpa (*C. speciosa*), not until that time recognized as a full species. Purchased part of former President William Henry Harrison's estate near North Bend, Ohio, 1855, and developed what was, for all practical purposes, an agricultural experiment station. Published *Hedges and Evergreens: A Complete Manual for the Cultivation, Pruning, and Management of All Plants Suitable for American Hedging* (1858). His valuable essay on strawberries appeared in the *Report of the U.S. Commissioner of Agriculture* for 1861. His series of articles on "Practical Entomology" published in *Farmer's Home Journal* (1864). Coauthor of report concerning work of the Flax and Hemp Commission for U.S. Department of Agriculture (USDA)

(1865). Published *American Pomology: Apples* (1867), covering some 1,500 varieties. Edited and annotated Alphonse De Breuil's *Vineyard Culture* (1867). President, Ohio Pomological Society, 1863, reelected annually until it was re-organized as Ohio Horticultural Society. Elected president of that organization, remaining in office until his death. Promoted forestry movement in the United States and was regarded as the leading American authority on the subject during the last decade of his life. Published report on forests and forestry growing out of the Vienna conference in *Report of the Commissioners*, vol. 1 (1876). Convened a citizen's conference in Chicago that resulted in creation of American Forestry Association (1875), of which he became first president. Petitioned Congress for creation of a commission to study forestry in Europe, 1879 and 1880, but no action taken. Was an organizer of the American Agricultural Association, 1881 (senior vice president, 1881–1883). As forestry agent for the USDA, would have examined forest resources from the Great Lakes to the Gulf and from the Alleghenies to the Mississippi River. Onset of his final illness precluded his doing so.

MAJOR CONTRIBUTIONS

Enjoyed international reputation for his work on apples. An early leader of the park and cemetery beautification movement. His actions in support of forest protection are regarded by some as the beginning of the conservation movement in the United States.

BIBLIOGRAPHY

Bailey, Liberty Hyde, ed. *Cyclopedia of American Horticulture* (1902).
Clepper, Henry, ed. *Origins of American Conservation* (1966).
Commercial Gazette (Cincinnati), 18 July 1883 (obit.).
Dictionary of American Biography. Vol. 19.
Dictionary of American Medical Biography (1928).
Hedrick, U. P. *A History of Horticulture in America* (1950).
National Cyclopaedia of American Biography. Vol. 4.

John W. Frederick and Keir B. Sterling

WETMORE, (FRANK) ALEXANDER. Born North Freedom, Wisconsin 18 June 1886; died High Point, Maryland, 7 December 1978. Ornithologist, avian paleontologist, administrator.

FAMILY AND EDUCATION

Son of Nelson Franklin and Emma Amelia (Woodworth) Wetmore. Received A.B. from University of Kansas in 1912 and M.S. in 1916, and Ph.D. in 1920 from George Washington University. Received honorary doctorates from University of Wisconsin, Centre College, George Washington University, and Ripon

College. Married Fay Holloway in 1912 (died 1953); one daughter, and Anne Beatrice Thielen in 1953.

POSITIONS

Assistant at University of Kansas Museum, 1905–1910, except for one year, when he filled similar position at Colorado Museum of Natural History. Agent for U.S. Biological Survey, U.S. Department of Agriculture, 1910–1912; assistant biologist, 1913–1924, and biologist in 1924. Superintendent of National Zoological Park, 1924–1925. Assistant secretary of Smithsonian Institution, 1925–1944, acting secretary, 1944–1945, secretary, 1945–1952; and research associate, 1952–1978.

CAREER

Early assignments with Biological Survey provided valuable experience as field biologist and collector. Became expert in skinning and preparing bird specimens. Investigated and published on waterfowl around Great Salt Lake, Utah. In 1920 went to South America to study North American birds that migrated south and in 1923, on behalf of Biological Survey and Bernice P. Bishop Museum of Honolulu, led *Tanager* exploring expedition to islands of mid-Pacific. Other investigations made to most of states of United States and Canada. As secretary of Smithsonian fostered laboratory and field researches in natural history, anthropology, and industrial arts; gave special encouragement to fundamental research by staff; and laid foundation for expanding exhibits, buildings, and programs that came to fruition in later administrations. During his tenure, National Air Museum and Canal Zone Biological Area were added to Smithsonian organization.

Bibliography contains over 700 entries, including 150 papers and monographs on fossil birds. Among larger works were *Observations on the Birds of Argentina, Paraguay, Uruguay, and Chile* (1926); *The Migration of Birds* (1927); *Birds of Porto Rico and the Dominican* Republic (1931); and *Fossil Birds of North America* (1931). Also proposed and published *Systematic Classification for the Birds of the World* (published in several editions between 1934 and 1960). Chief work was *The Birds of the Republic of Panama* (1965–1972), in three volumes (a fourth volume, completed by Smithsonian colleagues, was published posthumously in 1984). Also edited and in part wrote *Book of Birds* (2 vols., 1937) for National Geographic Society (N.G.S.) and later revision *Water, Prey, and Game Birds of North America* and *Song and Garden Birds* (1964–1965). Trustee of N.G.S. for nearly half a century, and active as vice-chair and later chair of N.G.S. Committee for Research and Exploration 1937–1974. Member of board, Gorgas Memorial Institute of Tropical and Preventative Medicine. Member of advisory committee on International Wildlife Protection, member of

International Commission for Bird Protection, member of commission of Institute of National Parks, Belgian Congo, U.S. representative to International-American Commission of Experts on National Protection of Wildlife Preserves. Home Secretary, National Academy of Sciences, 1951–1955. Recipient of many awards for his work in ornithology and conservation. Fifty-six genera, species and subspecies of animals (and one plant) named in his honor, together with a glacier in the Antarctic.

Member from 1908, president (1926–1929), and honorary president (1975–1978) of Ornithologists' Union. President of Washington Biologists' Field Club, 1928–1931. In 1940 named secretary-general of Eighth American Scientific Congress. In 1948 appointed by President Truman chairman of Interdepartmental Committee on Research and Development. Member of Board of Directors of Gorgas Memorial Institute of Tropical and Preventive Medicine. Home Secretary of National Academy of Sciences. President of Washington Academy of Sciences, Cosmos Club of Washington, Explorers Club of New York, Baird Ornithological Club, and the Tenth International Ornithological Congress at Uppsala, Sweden, in 1950.

MAJOR CONTRIBUTIONS

Has been described as the doyen of twentieth-century American ornithology. Work in ornithology has been described as "staggering." Described 189 species and subspecies of birds previously unknown. Enriched national collections with over 26,000 mammal and bird skins from North America, Puerto Rico, Argentina, Chile, Venezuela, Spain, Central America, and Panama. Contributed notably to study of fossil birds. On the occasion of his ninetieth birthday, S. Dillon Ripley stated, "Truly the incessant and intensive zeal which [Wetmore] has single-mindedly given to the study of birds over the years, often at very considerable personal expenditure in time and energy, will mark [his] career . . . as one of the most memorable in the entire history of American ornithology."

BIBLIOGRAPHY

American Men and Women of Science: Physical and Biological Sciences. 13th ed. (1976).

Current Biography. Vol. 40 (March 1979).

Locher, F. C., ed. *Contemporary Authors.* Vols 85–88 (1980).

Oehser, Paul H. "In Memoriam: Alexander Wetmore." *Auk* 97 (July 1980).

Olson, Storrs L., ed. *Collected Papers in Avian Paleontology Honoring the 90th Birthday of Alexander Wetmore* (1976).

Ripley, S. Dillon, and James A. Steed. "Memoir of Alexander Wetmore." In *Memoirs of the National Academy of Sciences* 56 (1987).

Washington Post, 9 December 1978 (obit.).

UNPUBLISHED SOURCES

The basic collection, consisting of 237 boxes of correspondence, manuscripts, field notes, photographs, and other material, is located at the Archives and Special Collections of the Smithsonian Institution, Washington, D.C. See also Victoria Cooper and William

E. Cox, *Guide to the Papers of Alexander Wetmore, c.1948–1979 and Undated*, Archives and Special Collections of the Smithsonian Institution, no. 11 (1990).

Richard Harmond

WHEELER, ARTHUR OLIVER. Born Kilkenny, Ireland, 1 May 1860; died Banff, Alberta, 20 March 1945. Surveyor, phototopographer, founder of the Alpine Club of Canada, conservationist.

FAMILY AND EDUCATION

Parents were Capt. Edward Oliver Wheeler and Josephine Helsham, landed Irish gentry. Attended private school in Dublin, Ballinasloe College in County Galway, and Dulwich College in London. Family immigrated to Collingwood, Ontario, in 1876. Father became harbormaster. Wheeler apprenticed as land surveyor with Hamilton and Ryley, 1877–1880. Qualified as land surveyor: Ontario, 1881, Dominion, 1882, Manitoba, 1882, British Columbia, 1891, Alberta, 1911. Married Clara Macoun, 1888 (died 1923) and Emmeline Savatard, 1925.

POSITIONS

Technical officer, Dominion Department of Interior Topographical Surveys Branch, Ottawa, 1885–1890, lieutenant in Dominion Land Surveyors' Intelligence Corps during 1885 Manitoba Metis Uprising. Surveyor in private practice, New Westminster, B.C., 1890–1893. Officer, Topographical Surveys Branch, Calgary, 1893–1910, topographer, 1903–1910. Private practice in B.C., 1910–1913. Commissioner, Alberta-British Columbia Boundary Commission, 1913–1925. Operator, Mt. Assiniboine Walking and Riding Tours, Banff, 1920–1927. Elected honorary member of Dominion Land Surveyors' Association, 1929. President, Alpine Club of Canada (ACC), 1906–1910, director, 1910–1926, honorary president, 1926–1945, editor of *Canadian Alpine Journal*, 1907–1927. Secretary, Canadian National Parks Association, 1923.

CAREER

Joined his first western survey in 1878, walking return trip from Winnipeg to Prince Albert, Northwest Territories. In 1885, learned methods of photogrammetry from Canada's surveyor general Edouard-Gaston Deville. Surveyed B.C. timber berths and mining sites and Alberta townships and subdivisions, 1890–1893. Worked on irrigation and phototopographic surveys of southern Alberta foothills and mountain passes, 1895–1900. Made notable survey of B.C. Selkirk Mountains, 1901–1902, began mountaineering. Sent to Alaska on secret reconnaissance related to Alaska-Yukon boundary dispute, 1903. Represented Canada at International Geographic Congress, Washington, D.C., in 1904. Published *The Selkirk Range*, 2 vols. (1905), a classic work in Canadian geography. Cofounded

ACC with Elizabeth Parker, 1906. Surveyed B.C. Northwest, 1910–1913. Led ACC expedition with the Smithsonian Institution in Washington to Mt. Robson and the Rainbow Mountains, 1911. As the B.C. commissioner on the interprovincial boundary commission, 1913–1925, coordinated the surveys and mapping of the watershed line between Alberta and B.C. from the 49th parallel 600 miles north to its intersection with the 120th meridian and produced three atlases. During 1920s, became outspoken advocate of protecting national parks.

MAJOR CONTRIBUTIONS

National leader in the field of phototopographic land surveys. Wheeler was a key figure in the advancement of Canadian mountaineering, and in the development of Canada's national parks. A pioneer surveyor and alpinist who promoted the Canadian mountain heritage and middle-class recreation in the national parks. Did much to promote an appreciation of Canada's mountain ranges and bring them to worldwide attention.

BIBLIOGRAPHY

"Arthur Oliver Wheeler." *Canadian Alpine Journal* (1940).

"Arthur Oliver Wheeler." *Canadian Alpine Journal* (1944–1945).

"The Arthur O. Wheeler Hut." *Canadian Alpine Journal* (1948).

Fraser, Esther. *Wheeler* (1978).

Johnston, Margaret, and John Marsh. "The Alpine Club of Canada, Conservation and Parks, 1906 to 1930." *Canadian Alpine Journal* (1986).

Karamitsanis, Aphrodite. *Place Names of Alberta: Mountains, Mountain Parks and Foothills* (1991).

UNPUBLISHED SOURCES

See Whyte Museum of the Canadian Rockies Archives, Banff, Alberta, Alpine Club of Canada Collection: correspondence, club records, photographs; National Archives of Canada, Ottawa, RG 84 National Parks Collection: correspondence; Canadian Permanent Committee on Geographical Names, Ottawa, Geographic Board of Canada correspondence.

PearlAnn Reichwein

WHEELER, GEORGE MONTAGUE. Born Hopkinton, Massachusetts, 9 October 1842; died New York City, 3 May 1905. Topographer, geographer, army officer, engineer.

FAMILY AND EDUCATION

Son of John and Miriam P. Daniels Wheeler. Grew up in Hopkinton, Massachusetts, but appointed to U.S. Military Academy from Colorado, where two of his brothers had moved. Graduated from West Point in 1866, standing sixth in class of thirty-nine. Married Lucy Blair of Francis Preston Blair family, 1874; she died in February, 1902. No children.

POSITIONS

With U.S. Army Corps of Engineers, 1866–1888. Promoted to 1st lieutenant, 1867, to captain, 1879, and retroactively to major in 1890 by special act of Congress. Military engineer in California, 1866–1871. Director, U.S. Geographical Surveys West of the 100th Meridian, 1871–1888; summer field work, 1871–1879, and office work in Washington throughout. Private engineer and consultant in Washington, D.C., and in New York City on medical retirement from the army in June, 1888.

CAREER

Sent by General E.O.C. Ord of the Department of California on an 1869 reconnaissance through Nevada. Proposed to General A. A. Humphries, Army Chief of Engineers, that he map United States west of the 100th Meridian, "the main object of this exploration to obtain correct topographical knowledge of the country traversed." Leader and chief administrator of field surveys that employed twenty-five to forty-eight scientists and topographers for nine field seasons. Personally led fourteen trips of from three to eight and a half months duration in areas from Death Valley to the Cascades and east to Colorado and New Mexico. Surveyed 359,065 square miles, one third of the mountainous West, and published maps of 326,891 square miles at a cost of $618,644.05. In addition to topographic mapping, the survey's civilian scientists did pioneering studies in geology, botany, paleontology, and archaeology. The geographer Grove Karl Gilbert named the Basin and Range region in a Wheeler publication. Timothy O'Sullivan photographed classic scenes for Wheeler. Among the other better-known civilians engaged by Wheeler were E. D. Cope, paleontologist; Henry Wetherbee Henshaw, ornithologist and later second chief of the U.S. Biological Survey; and Henry C. Yarrow, ornithologist and zoologist. In 1871, Wheeler was able to get further down the Colorado River (to Diamond Creek) than his rival Major John W. Powell managed to do in his several trips (1869, 1871–1872) down the river. Competition and duplication with the (Clarence) King, (F. V.) Hayden, and Powell surveys led to establishment of the U.S. Geological Survey in 1879, with King as first director. Was last army explorer of the West. Supplanted because civilian scientists chafed at military discipline and Wheeler's quick pace, and because geologists wanted more accurate base maps. Wheeler loyally served under King and Powell until ill health compelled his retirement in 1888 at age forty-six. Was War Department representative to Third International Geographic Congress in Venice, 1881.

Wheeler was author and editor for his survey's fifteen annual reports, eight final reports or monographs, sixteen miscellaneous publications, fifty topographic atlas sheets, thirty-three land classification atlas sheets, eleven geologic atlas sheets, and a total of 164 maps of all types and scales, published between 1872 and 1889. Of the eight final reports done by Wheeler and others, volume

2 dealt with paleontology, volume 3 with geology, volume 5 with birds, insects, other vertebrates, and mollusks, and volume 6 with botany. Wheeler's men collected 61,659 archaeological, geological, botanical, zoological, and paleontological specimens, including one species of bird new to science, together with eight reptiles, thirty-two fishes, one mollusk, and sixty-four insects. Prepared major report on Venice Geographical Congress, and on status and progress of mapping worldwide (1885). Argued for complete topographic coverage of the West with emphasis on ''map delineations of all natural objects, means of communication, artificial and economic features, the geologic and natural history branches being treated as adjunctive.'' Wheeler had promised complete coverage of the West by the late 1880s at scales of 1:253,440 or 1:506,880. Complete coverage of the United States at comparable scales (1:350,000) was not achieved until the 1950s, ironically by U.S. Army Map Service. Wheeler's ''Memoir Upon the Voyages, Discoveries, Explorations and Surveys to and at the West Coast of North America and Interior of the United States West of the Mississippi River Between 1500 and 1880,'' published as an appendix to volume I of his 1889 final report, was a valuable summary of accomplishments to that time.

MAJOR CONTRIBUTIONS

Wheeler insisted that topographic mapping should be of benefit for the people, should be readily available, and could be later refined as needed. He argued that military engineers rather than geologists should be entrusted with this vital government task. He conceived the contour map, a vital tool in American topographic work.

BIBLIOGRAPHY

Bartlett, Richard A. *Great Surveys of the American West* (1962).

Brown, F. M. ''Wheeler Expeditions.'' *Lepidoptera News* 9 (1955–1956); 20 (1966); and in *Journal of the New York Entomological Society* 65 (1957).

Cullum, G. W. *Biographical Register of Officers and Graduates of the United States Military Academy*. 3d ed. (1891).

Dawdy, D. O. *George Montague Wheeler: The Man and the Myth* (1993).

Dictionary of American Biography.

Dictionary of Scientific Biography.

Ewan, Joseph, and Nesta Dunn Ewan. *Biographical Dictionary of Rocky Mountain Naturalists* (1981).

Goetzmann, William H. *Exploration and Empire: The Explorer and the Scientist in the Winning of the American West* (1966).

New York Times, 5 May 1905 (obit.).

Rabbitt, Mary. *Minerals, Lands and Geology for the Common Defense and General Welfare*. Vol. 1: Before 1879 (1979).

Schmeckbier, Lawrence F. *Catalogue and Index of the Hayden, King, Powell, and Wheeler Surveys*, 58th Congress, 2d Session, House of Representatives Document No. 606 (1904).

Stegner, Wallace. *Beyond the Hundredth Meridian* (1953).

Wheeler, George M. "Memoir upon the Voyages, Discoveries, Explorations and Surveys to and at the West Coast of North America and Interior of the United States West of the Mississippi River between 1500 and 1800." *U.S. Geographical Survey West of the 100th Meridian.* Vol. 1 (1889) Appendix F.

UNPUBLISHED SOURCES

Official survey records, copies of general orders, letterpress books, and field notebooks, some 540 volumes in all, in Record Group 77, National Archives, Washington. Additional important survey records (31 volumes) in Yale University Libraries. Personal scrapbook in Bancroft Library, University of California, Berkeley.

Peter L. Guth

WHEELER, WILLIAM MORTON. Born Milwaukee, Wisconsin, 19 March 1865; died Cambridge, Massachusetts, 19 April 1937. Entomologist, zoologist.

FAMILY AND EDUCATION

Son of Julius Morton Wheeler, tanner and land speculator, and his second wife, Caroline Georgiana Anderson Wheeler. Attended German-English Academy, graduate of the German-American College (Normal School), 1884; Ph.D., Clark University, 1892. Honorary Sc.D., University of Chicago, 1916, Harvard, 1930, Columbia, 1933; LL.D., University of California, 1928. Married Dora Bay Emerson, 1898; two children.

POSITIONS

Employee, Ward's Natural Science Establishment, Rochester, New York, 1884–1885; taught in local high school, Milwaukee, 1885–1887; curator (custodian), Milwaukee Public Museum, 1887–1890; fellow and assistant in morphology, Clark University, 1890–1892; instructor in embryology, University of Chicago, 1893–1897 (in Europe studying at University of Wurzberg, occupying Smithsonian Table at Naples Zoological Station, and studying at Institut Zoologique, Liege during 1893–1894), assistant professor, University of Chicago, 1897–1899; professor of zoology, University of Texas, 1899–1903; curator of invertebrate zoology, American Museum of Natural History, 1903–1908, (continuing as honorary fellow and research associate until his death); professor of economic entomology, 1908–1926, and then professor of entomology, Harvard, 1926–1934; dean, Bussey Institute for Research in Applied Biology, Harvard, 1915–1929; exchange professor, University of Paris, 1925. Professor emeritus, Harvard, 1934–1937.

CAREER

Met Carl Akeley in Milwaukee when Henry Ward bought some of his specimens to the city, trying to get city fathers to buy collection, combine it with old academy collection, and start a public museum. Wheeler identified and listed birds and mammals, later arranged specimens and prepared catalogs. Studied spiders and, later, embryology under local specialists and also at Allis Lake Laboratory Biological Station, where he met C. O. Whitman. Was persuaded by Whitman to follow him to Clark and secure Ph.D., then followed Whitman to the new University of Chicago, where Wheeler was given an instructorship. At the University of Texas, Wheeler was sole professor, with an instructor and laboratory assistants to aid him. Began his study of ants while at Texas. While at the American Museum, continued study of ants while developing new exhibit Hall of Invertebrates. Published his book *Ants: Their Structure, Development, and Behavior* (1910). His twenty-six-year tenure at the Bussey Institution, a graduate research arm of Harvard (1908–1934), represented opportunity to train outstanding graduate students and continue his work on ants. Gave some thought to moving to California Institute of Technology with friend Thomas Hunt Morgan in 1928, but President Lawrence Lowell would not hear of his retirement. Moved to new laboratory in Cambridge, 1929, leading to his relinquishing the deanship at Bussey Institution. He took collecting trips to Mexico, Cuba, the Bahamas, several Central American nations, British Guiana, the Galapagos, Australia, Morocco, the Canary Islands, and various parts of the United States. Published some 500 titles, including *A Contribution to Insect Embryology* (1893); *The Fungus-Growing Ants of North America* (1907); *Social Life among the Insects* (1923); a translation of Rene de Reaumur's (1683–1757) *The Natural History of Ants* (1926); *Foibles of Insects and Men* (his most popular book) (1928); *The Social Insects, Their Origin and Evolution* (1928); (also published in England and France); *Emergent Evolution and the Development of Societies* (1928); *Demons of the Dust, a Study in Insect Behavior* (1930); *The Lamarck Manuscripts at Harvard* (with Thomas Barbour, 1933); *Essays in Philosophical Biology* (1939); and *Studies of Neotropical Ant-Plants and Their Ants* (1942). President of the East Branch, American Society of Zoologists, 1908; president, Entomological Society of America, 1908; fellow or member of many scientific societies in the United States and in Europe. Received numerous honors, including Cross of the French Legion of Honor.

MAJOR CONTRIBUTIONS

A major and enthusiastic authority on ants and other insects. Saw ants "as exemplars of . . . many biological principles." Effective scientific and popular writer on these and other subjects. Had great respect for amateurs in science and did much to highlight accomplishments of naturalists of the past and importance of the history of biology generally. Wheeler felt that his taxonomic work would have more lasting value than his essays, but though "he laid foundations of myrmecology in North America and did much to improve knowledge of ants in other parts of the world" (Evans and Evans), many of his insect names have been relegated to synonymy.

BIBLIOGRAPHY
Dictionary of American Biography.
Dictionary of Scientific Biography.
Essig, E. O. *A History of Entomology* (1931).
Evans, Mary Alice, and Howard Ensign Evans. *William Morton Wheeler, Biologist* (1970).
Mallis, Arnold. *American Entomologists* (1970).
Parker, George H. "William Morton Wheeler." *Biographical Memoirs, National Academy of Sciences.* Vol. 19 (1938).
World Who's Who in Science.
UNPUBLISHED SOURCES
Principal collection of Wheeler Family Papers is in possession of family.

Keir B. Sterling

WHERRY, EDGAR (THEODORE). Born Philadelphia, 10 September 1885; died 19 May 1982. Ecologist.

FAMILY AND EDUCATION

Son of Albert C. and Elizabeth S. (Doll) Wherry. Attended the Friends Central School, the Wagner Free Institute of Science, and University of Pennsylvania. From last-named institution received B.S. in chemistry in 1906 and Ph.D. in mineralogy in 1910. Married E. G. Smith in 1914.

POSITIONS

Instructor in mineralogy at Lehigh University, 1908–1910, and assistant professor, 1910–1913. Also instructor in chemistry, Wagner Free Institute, Philadelphia, 1908–1913. Assistant curator of mineralogy, U.S. National Museum, 1913–1917; crystallographer, U.S. Bureau of Chemistry, 1917–1923; chemist in charge of crop chemistry, U.S. Bureau of Chemistry, 1923–1930; associate professor of botany, 1930–1941, professor of botany, University of Pennsylvania, 1941–1955, professor emeritus from 1955.

CAREER

While engaged in fieldwork for Department of Agriculture, became seriously interested in study of plants and their relationship to soil, which led to invention of series of indicator dyes to determine measure of acidity and alkilinity of soils.

Following retirement in 1955, devoted himself to several projects, including books on ferns, wildflowers, and genus *Phlax* (on which he was a recognized authority). Also dedicated to development of rock gardens and was ardent conservationist. Wrote numerous technical articles and several books, including *The Fern Guide* (1960).

Fellow of Geological Society of America; member and, in 1923, president of

Mineral Society of America; member and, in 1927, president of Washington section of American Chemical Society; member and, in 1935, president of Pennsylvania Academy of Science. Editor of *American Mineralogist*, 1919–1921, and *Bulletin of American Rock Garden Society*, 1943–1947.

MAJOR CONTRIBUTIONS

Contributions in several fields. In geology, identified deposits of volcanic dust in bentonite in mid-Paleozoic and late Mesozoic ages; in plant ecology, application of a pH indicator to determine the reaction of soils; in mineralogy found mineral canotite in eastern Pennsylvania; and chemistry, developed simple method of determining boron in silicate. Also discovered several species of plants that received his name and identified a mineral, also named after him, wherrite.

BIBLIOGRAPHY

Bioscience 32 (November 1982).
Taxon (November 1982).
Who Was Who in America. Vol. 7, 1977–1981 (1981).

Richard Harmond

WHITEAVES, JOSEPH FREDERICK. Born Oxford, England, 26 December 1835; died 8 August 1909, Ottawa, Canada. Paleontologist.

FAMILY AND EDUCATION

Son of Joseph Whiteaves and his wife Sarah. Student in various private schools, Brighton, Oxford, and London, England. Did not pursue a regular college course, but attended some lectures at Oxford given to "advanced students." Emigrated to Canada, 1862. Married Julia Wolff, 1863 (divorced, 1868), married for a second time (name of wife and date of marriage not available); three children. Honorary LL.D., McGill University, 1900.

POSITIONS

Began his long tenure with the Geological Survey of Canada as parttime assistant, 1874–1876 (some sources give 1875–1876); subsequently held various overlapping responsibilities within the survey as its second paleontologist, 1876–1909, one of four assistant directors, 1877–1909; zoologist, 1883–1909, and museum director from the late 1870s until his death.

CAREER

Developed an interest in natural history in boyhood. Began independent studies of animals and plants in vicinity of his Oxford home, c.1855; later added strong and active interest in paleontology. First visited Canada, 1861, then emigrated there, 1862. First published paper, ''On the Land and Fresh Water Molluscs inhabiting the neighborhood of Oxford,'' appeared in 1858. Elected fellow of the Geological Society of London, 1859. Published several articles in 1861 based on study of fossils he had collected, one concerning the invertebrate fauna of the Lower Oölites, and a second on Coralline Oölite fossils. Following his arrival in Canada, continued with his studies of fossils in vicinity of Montreal. Was active member of the Natural History Society of Montreal from 1863 to 1875, serving as recording secretary and supervising the society's museum. Dredged waters of the Gulf of St. Lawrence, collecting living and fossil forms of marine invertebrates from 1867 to 1873, and published a number of reports on his findings. This dredging activity was underwritten by the Dominion government. Assisted Elkanah Billings, his predecessor as paleontologist for the Geological Survey, on a parttime basis from 1874 (some sources suggest 1875) to 1876, and succeeded to the post on Billings' death in 1876. Published reports on *Mesozoic Fossils* (1876–1884), and edited *Contributions to Canadian Paleontology*, (1885–1891), each in three volumes, containing descriptions of fossils typical of various Canadian rock formations. Also had responsibility for the survey's annual reports. Was instrumental in identifying fossils brought in by Geological Survey field men from various parts of the country and assisted in their proper placement in the stratigraphic succession. Had responsibility for direction of the survey's museum and oversaw transfer of the collections when the survey's offices were moved from Montreal to Ottawa in 1881. Was an authority on recent marine invertebrates, including mollusks, of both the Atlantic and Pacific coasts, and on the fossil fishes of Quebec and New Brunswick. Published nearly 150 papers, containing descriptions of more than 450 genera, species, and subspecies in various Canadian journals. Some of these focused on new species which he had found.

Whiteaves was a fellow of, and received the Lyell Medal from, the Geological Society of London (1907). He was a member of the Montreal Natural History Society and a founding member (1882) and an original fellow of the Royal Society of Canada. He contributed articles to, and was editor of, the *Canadian Naturalist and Geologist* (1868–1874). He was also a contributor to *Canadian Record of Science*, the *Transactions of the Royal Society of Canada*, *Ottawa Naturalist*, *American Journal of Science*, and to *Annals and Magazine of Natural History* in London. Certain of his colleagues were known to have felt that some of his work was hastily done because he held too many concurrent responsibilities within the Geological Survey.

MAJOR CONTRIBUTIONS

Whiteaves contributed many valuable titles to the literature concerning recent and fossil Canadian invertebrates during his thirty-four-year tenure with the Geological Survey of Canada. He was an able paleontologist and zoologist, and a major supporter of Canadian science during a very important period of its development in the late nineteenth century.

BIBLIOGRAPHY

Canadian Men and Women of the Time (1898).
Dictionary of Canadian Biography.
"Eminent Living Geologists: Joseph Frederick Whiteaves, LL.D., F.G.S., F.R.S. (Canada)." *Geological Magazine* (London), new series, 3 (1906).
Ottawa Evening Citizen, 9 August 1909 (obit.).
Ottawa Naturalist 26, 6 (1909) (obit.).
Proceedings and Transactions of the Royal Society of Canada, 3d series, 4 (1911) (obit.).
Zaslow, Morris. *Reading the Rocks: The Story of the Geological Survey of Canada, 1842–1972* (1975).

Keir B. Sterling

WILEY, FARIDA. Born Sydney, Ohio, 23 May 1887; died Melbourne, Florida, 15 November 1986. Naturalist, teacher.

FAMILY AND EDUCATION

Born on family horse farm. By age of twelve was regularly corresponding with Bureau of Biological Survey on nesting birds of her locality. Death of parents caused her to end formal schooling with high school. Never married, no children.

POSITIONS

Joined staff of American Museum of Natural History in 1919, teaching botany to blind children. In 1923 became assistant in Lantern Slide Division. Subsequently, was staff assistant, instructor, and assistant chairman in Department of Public Instruction (now Department of Education). Retired from museum in 1955 but remained honorary associate with museum and continued to teach. Also taught natural science courses to student teachers at Pennsylvania State College, Audubon Camp in Maine, and New York University branch on Long Island. Led final bird walks to Central Park at age 94.

CAREER

Best known for decades of leading "Morning Walks in Central Park" in spring and fall (during bird migrations) and all-day "Natural Science for the Layman" weekend trips. Also involved with various programs offered by Department of Education, including courses for New York City schoolteachers, a course for nature counselors and youth leaders, and behind-the-scenes tours and classes for New York City schools visiting the museum. Took part in planning of Felix Warburg Hall and Hall of North American Forests at Museum. Publications include *Ferns of Northeastern United States* (1936); ed., *John Bur-*

roughs' America (1951); coauthor, *The Story of Landscape* (1952); ed., *Ernest Thompson Seton's America* (1954); ed., *Theodore Roosevelt's America* (1955).

MAJOR CONTRIBUTIONS

Contributed greatly to increase and broadening of museum's role in science education and personally provided thousands of youth leaders and teachers with training in content and methodology of teaching science.

BIBLIOGRAPHY

New York Times, 18 November 1986 (obit.).
Washington Post, 26 December 1986 (obit.).

UNPUBLISHED SOURCES

The American Museum of Natural History has material relevant to Wiley's career, including correspondence and administrative files.

Note: I wish to thank Joel Sweimler of the American Museum of Natural History Special Collections Department for his help in acquiring information on Farida Wiley.

Richard Harmond

WILSON, ALEXANDER. Born Paisley, Scotland, 6 July 1766; died Philadelphia, 23 August 1813. Naturalist, poet, teacher.

FAMILY AND EDUCATION

Born of poor, pious Scotch peasants who wanted their only son, Alexander, to become a minister. After his mother died in 1776, his father remarried. Attended grammar school in Paisley, but not too successful a scholar, though voracious reader. In 1779 became an apprentice weaver; when apprenticeship expired, became a journeyman weaver and peddler, 1782–1789. Contributed some early poems to *Glasgow Advertiser*. In 1812 elected member of the American Philosophical Society of Philadelphia and the Society of Artists of the United States.

POSITIONS

After spending three years traveling in Scotland as a peddler with his brother-in-law, published in 1790 an account of his journeys, *Poems, Humorous, Satirical and Serious*. In the following year he published *Laurel Disputed, Watty and Meg* and a second edition of *Poems*. Having little success with his books, reverted to being a traveling weaver. In a sharp attack on Paisley's businessmen, placed responsibility for the plight of weavers on their capitalistic greed. Found guilty of libel and forced to burn copies of his pieces and to pay a fine. Disgraced, he left Belfast for the United States in 1794.

CAREER

In July 1794, landed at Newcastle, Delaware, then moved to Philadelphia, where he worked as printer with John Aiken. In a letter to his father, he commented on the birds and squirrels he saw on the road to Philadelphia. In 1794–1795, he worked as weaver in Pennsylvania and Virginia and as teacher at Frankfort, Pennsylvania. Between 1795 and 1801, taught school at Milestone, Pennsylvania, and did some surveying. Taught at Bloomfield, New Jersey, 1801–1802; then accepted teaching post at Union School, Gray's Ferry, near Philadelphia. While teaching at Kingessing, Pennsylvania, met celebrated botanist William Bartram, who gave him access to his library; at the same time he became acquainted with the engraver Alexander Lawson, who taught Wilson to draw, color, and sketch. In October 1804, Wilson took ornithological trip to Niagara Falls. Published essays in the *Literary Magazine* (Philadelphia). Determined to make a collection of bird drawings. In 1806, applied unsuccessfully to President Thomas Jefferson for place with Zebulon Pike expedition to the Southwest in hopes of expanding his study. Jefferson later contended that he had not received Wilson's application. Letter may have been withheld from Jefferson by staff, inasmuch as Pike mission was not supposed to be public knowledge. Wilson then became assistant editor of Rhee's *New Cyclopedia*. In 1807, began bird observation travels throughout Pennsylvania. In 1808, published first volume of *American Ornithology*. In 1808–1809, journeyed to the South doing fieldwork for his second volume of *American Ornithology* and selling subscriptions. Second volume published 1810. Interestingly, John J. Audubon declined to buy book on advice of his business partner in Hendersonville when Wilson solicited a subscription from him. Third and fourth volumes of *American Ornithology* published in 1811; fifth and sixth in 1812. Wilson's last ornithological trip took him as far north as Maine. Seventh volume appeared in 1813. An attack of dysentery, complicated by a cold, brought on his death. His friend George Ord, relying on Wilson's notes, completed an eighth volume of *American Ornithology*; that same year Ord brought out volume nine and included in it a life of Wilson.

MAJOR CONTRIBUTIONS

Wilson's life reflected his love of natural science and a thirst for fame. A pioneer in his work on American birds, he described some 278 species, of which 48 were new to science. On the basis of his *American Ornithology*, he has been ranked by some authorities as a rival to Audubon among the greatest American ornithologists. That his nine-volume work was undoubtedly the leading contribution on American birds in the century between publication of Catesby's *Natural History of Carolina, Florida, and the Bahama Islands* and Audubon's *Birds of America* is undisputed.

BIBLIOGRAPHY
Adams, Alexander B. *Eternal Quest* (1969).
Cantwell, Robert. *Alexander Wilson, Naturalist and Pioneer: A Biography* (1961).
Elman, Robert. *First in the Field* (1977).
Hunter, Clark, ed. *Life and Letters of Alexander Wilson* (1953).
Kastner, Joseph. *Species of Eternity* (1977).
Maurizi, D. "Alexander Wilson: Father of American Ornithology." *Conservationist* 31 (1977).
Plate, Robert. *Alexander Wilson: Wanderer in the Wilderness* (1966).
UNPUBLISHED SOURCES
See Alexander Wilson Papers, University of Oregon Library (Eugene).

Thomas J. Curran

WIRTH, CONRAD LOUIS. Born Hartford, Connecticut, 1 December 1899; died Williamstown, Massachusetts, 25 July 1993. Landscape architect, conservationist, U.S. government official.

FAMILY AND EDUCATION

Son of Theodore Wirth, horticulturist and park administrator, and Leonie Augusta Mense Wirth. Moved with family to Minneapolis, Minnesota, 1906. Student at St. John's Military Academy, Delafield, Wisconsin, 1916–1919. B.S., University of Massachusetts, 1923. Honorary doctor of landscape architecture, University of Massachusetts; honorary D.C.L., New England College, 1955; honorary Hum.D., University of North Carolina, 1958. Married Helen Augusta Olson, 1926 (died 1990); two sons.

POSITIONS

Landscape architect, MacRorie and McLaren, San Francisco, 1923–1925; partner, Neal and Wirth, landscape architects, 1925–1928; landscape architect, National Capital Park and Planning Commission, Washington, D.C., 1928–1931; assistant director for planning, National Park Service, 1931–1951; Advisory Board member, Civilian Conservation Corps (CCC), 1933–1942; associate director, National Park Service, 1951; director, National Park Service, 1951–1964.

CAREER

Influenced by his father's involvement in the park service, Wirth started his professional career as landscape architect. Most of his career was spent in federal service. During tenure with the National Capital Park and Planning Commission and later, wrote several parks reports, including *Estimates for Development and Maintenance of the National Capital Regional Park System: Preliminary Report* (1930) and *Boundary Lane Report: Big Bend National Park Project, Texas* (1935). Served as assistant director of the National Park Service for twenty years. Some of the great achievements of this period are related to his work with the CCC and to his representation of Interior Department on the CCC Advisory Council. In these capacities, Wirth was responsible for building hundreds of roads, bridges, and CCC camps in state and national parks, many of which keep serving the nation today. He initiated the Park, Parkway, and Recreational-Area Study Act of 1936, which facilitated a closer cooperation between the federal government and the several states and bolstered the state park planning program. Marginal lands were acquired and developed as parks as a result. This legislation also provided the foundation for the creation of the Bureau of Outdoor Recreation.

Between 1951 and 1964, served as director of the National Park Service. Focal point of his career was the ''Mission 66'' project, initiated in 1956 to stop deterioration of the parks. During post–World War II and Korean War periods, when there were no funds for maintenance and repair of parks, the number of park visitors increased dramatically. Lack of roads, insufficient tourist accommodations, and private ownership of land inside the parks complicated the situation. Between 1956 and 1966, Wirth secured funds for park preservation by lobbying successfully Congress and the administration. This resulted in increased protection for park visitors, the wilderness, and wildlife. During this period the idea of the ''visitor center'' was born. Exhibits and park museums were created as well. This landmark project resulted in increased recreational use of parks and provided protection for the environment.

His main work, *Parks, Politics, and the People* (1980), autobiographical in nature, provides a wealth of information on history of the National Park Service and other agencies. Wirth left the federal service only once, when he became a land policy adviser to the World War II U.S. Allied Council in Vienna, Austria, from 1945 to 1946.

Was director, National Conference on State Parks; director, American Shore and Beach Preservation Association; fellow, American Society of Landscape Architects; cofounder, White House Historical Association; trustee, National Geographic Society; president, American Institute of Park Executives; founder, National Recreation and Park Association, 1965. Presented Distinguished Service Award. Department of the Interior, 1956. The National Park Foundation created the Wirth Environmental Award in honor of Wirth and his father, and Wirth was the first Wirth Award honoree. Also received a number of other conservation awards. Only his untimely death prevented President John F. Kennedy from awarding Wirth the Citation of Merit. President Lyndon B. Johnson termed Wirth ''one of the greatest, finest, best public servants anywhere in the world.''

MAJOR CONTRIBUTIONS

Spent entire professional life serving the nation's parks. Changed the role the National Park Service played in conservation and preservation of the nation's resources. Helped to secure and develop parks, forests, recreation sites, wilderness and wildlife areas for future generations as these resources became more precious with the encroachment of industry and the population explosion.

BIBLIOGRAPHY

Current Biography (1952).

Frome, Michael. "A Kind of Special Breed." *American Forests* 70 (January 1954).

New York Times, 28 July 1993 (obit.).

Washington Post, 28 July 1993 (obit.).

Who's Who in America (1970–1971).

Wirth, Conrad L. "The Mission Called 66." *National Geographic* 130 (July 1966).

Wirth, Conrad L. *Parks, Politics, and the People* (1980).

UNPUBLISHED SOURCES

See U.S. National Park Service: National Park Service General Records; American Heritage Center, University of Wyoming, Laramie: Conrad Louis Wirth Papers, 1929–1982. Contains materials relating to Wirth's career with the National Park Service, including biographical and genealogical information; correspondence with government officials (1931–1982); subject files containing correspondence, photographs, reports, and speeches regarding legislation, National Park Service policies, budgets, annual reports, Mission 66, public relations, and statistics on park usage (1915–1969); photographs of Wirth and various national parks (1963–1972); a scrapbook of Wirth's World War II service (1945–1946); miscellaneous maps; speeches on the national parks and conservation issues (1930–1966); six photograph albums on Wirth, national parks, and the Civilian Conservation Corps (1934–1964); and a collection of commemorative stamps (1956–1972).

Andrew Sankowski

WOODHOUSE, SAMUEL WASHINGTON. Born Philadelphia, Pennsylvania, 27 June 1821; died Philadelphia, Pennsylvania, 23 October 1904. Ornithologist, army surgeon.

FAMILY AND EDUCATION

Son of Samuel Woodhouse, commodore, U.S. Navy, and H. Matilda Roberts Woodhouse. Received early education in private schools, Philadelphia and West Haven, Connecticut. M.D., University of Pennsylvania, 1847. Married Sara A. Peck, 1872; two children.

POSITIONS

Farmer in Chester County, Pennsylvania, c.1840–c.1844. Assistant resident physician, Philadelphia Hospital, 1847–1848. Acting assistant surgeon, U.S. Army, 1849–1856. Surgeon on several expeditions to Central America; for the Pennsylvania Militia; and at the Eastern Penitentiary. Surgeon on vessels of Cope's Line traveling between Philadelphia and Liverpool, and surgeon for Washington Grays Regiment, 1859–1860. Resident physician, Eastern Penitentiary, during Civil War. Reappointed acting assistant surgeon, U.S. Army, 1862, served during Civil War, then returned to private practice of medicine in Philadelphia.

CAREER

Began bird collecting, c.1840. Career as farmer terminated by illness. Soon after completion of medical school, participated in Creek and Cherokee Boundary Survey as surgeon-naturalist, 1849 and 1850. Was in Texas and in Zuni and Colorado River regions of present-day New Mexico, Arizona, and southeastern California, 1851–1852. Zoological results of his observations and specimens collected published in "Report on the Natural History of the Country Passed over by the Exploring Expedition Under the Command of Brevet Captain L[orenzo] Sitgreaves, U.S. Topographical Engineers, during the Year 1851." Published eight papers on birds and mammals in *Proceedings of the Academy of Natural Sciences of Philadelphia*, 1852–1853. Three of the birds and four mammals Woodhouse described as new are still valid species. Member of expedition led by Ephraim G. Squier to explore possible route for railway in Honduras, 1853. Made corresponding fellow, American Ornithologists' Union, 1903, and Life Member, Academy of Natural Sciences of Philadelphia. Several species of plants and animals named in Woodhouse's honor by Baird, Asa Gray, John LeConte, and others, who also published descriptions of a number of specimens Woodhouse collected in the southwest. Many of Woodhouse's plant specimens are in U.S. National Herbarium and in New York Botanical Garden Herbarium, and bird specimens are in National Museum of Natural History.

MAJOR CONTRIBUTIONS

Pioneering physician-naturalist in the decade preceding the Civil War and one of the first to report on the natural history of the southwest and Indian territories.

BIBLIOGRAPHY

Goetzman, W. H. *Army Exploration in the American West, 1803–1863* (1959).

Hume, E. E. *Ornithologists of the United States Army Medical Corps* (1942).

Stone, Witmer. "Samuel Washington Woodhouse." *Cassinia: Proceedings of the Delaware Valley Ornithological Club* (1904).

Tomer, John S., and Michael J. Brodhead, eds. *A Naturalist In Indian Territory: The Journals of S. W. Woodhouse, 1849–1850* (1992) (has excellent bibliography).

Woodhouse, James. *Samuel Washington Woodhouse* (1901).

UNPUBLISHED SOURCES

See Records of the Chief of Engineers, U.S. Army, National Archives, Washington; Woodhouse's Report on 1849–1850 Creek Indian Expedition, National Anthropological Archives and S. F. Baird Collection, Smithsonian Archives. Notebooks of 1849–1850 expedition in Academy of Natural Sciences of Philadelphia and Historical Society of Pennsylvania. See also Michael J. Brodhead, "Engineer-Naturalists of the United States Army: Contributions to Natural History by Officers of the Corps of Engineers and Corps of Topographical Engineers in the Nineteenth Century,'' unpublished manuscript, Historical Division, Office Chief of Engineers, Fort Belvoir, Virginia, 1985.

Keir B. Sterling

WRIGHT, ALBERT HAZEN. Born Monroe County, New York, 15 August 1879; died 4 July 1970. Zoologist, ecologist, local historian.

FAMILY AND EDUCATION

Son of Delos C. and Emily Hazen Wright. Received A.B. (1904), M.A. (1905), and Ph.D. (1908) from Cornell University. Married Anna Allen in 1910.

POSITIONS

Assistant in zoology, 1905–1908, Instructor, 1908–1915, assistant professor, 1915–1925, professor, 1925–1948, at Cornell.

CAREER

Associated for fifty years with a well-known course in vertebrate zoology at Cornell, using manual written by David Starr Jordan as introduction to world's fauna, naturalists, and natural history literature. Made collections of 150 varieties of peonies and 300 varieties of dwarf bearded iris. Explorations and publications on Great Okefenokee Swamp of Georgia contributed to its establishment as a national park. Reputation as zoologist rested on such works as *Field Note-Book of Fishes, Amphibians, Reptiles, and Mammals* (with Arthur A. Allen, 1913); *North American Anura: Life Histories of the Anura of Ithaca, New York* (1914); *Life Histories of the Frogs of Okefenokee Swamp, Georgia*, North America Salentia (Anura) no. 2 (1931); *Handbook of Frogs and Toads: The Frogs and Toads of the United States and Canada* (with Anna A. Wright, 1933; 2d ed., 1942; 3d ed., 1949); *Our Georgia-Florida Frontier: The Okefenokee Swamp, Its History and Cartography* (1945); *Scientific and Popular Writers on American Snakes, 1517–1944: A Check List and Short Biographies*, Herpetologica Series, no. 5, supplement 1 (1949); *Handbook of Snakes of the United States and Canada* (with Anna A. Wright, 1957, 2 vols.; 3d ed., 1979). The handbook on frogs and toads and on snakes, each several times revised, have remained standard references in the literature down to the present time. Was also author of many papers on Revolutionary and early Cornell history and local genealogy, including *Our Parents, Delos C. and Emily A. (Hazen) Wright: Their Ancestry* (1964).

Member of Conservation Committee of Division of Biology and Agriculture on National Research Council; president of Gamma Alpha Scientific Society. Served on editorial board of *Ecological Monographs*. Was honorary member of American Ornithologists' Union, Herpetologists' League, American Society of Ichthyologists and Herpetologists, and Academy of Zoology in India.

MAJOR CONTRIBUTIONS

One of great pioneers in science of ecology, several decades before term became household word. Was awarded title of Eminent Ecologist in 1955, by Ecological Society of America. Inspiring teacher, with great grasp of entire field of natural history, who was generous with his time, resulting in impressive roster of former students and scientists throughout world.

BIBLIOGRAPHY

"Alfred Hazen Wright." *Necrology of the Faculty*. Cornell University Library, Rare and Manuscript Collections, Carl A. Koch Library.

American Men of Science. 10th ed. (1961).

UNPUBLISHED SOURCES

See Albert Hazen Wright Papers, 1820–1960, consisting of correspondence, field notes, slides, photographs, transcripts of oral history interviews with Wright, and other material, located at Division of Rare and Manuscript Collections, Cornell University Library.

Richard Harmond

WRIGHT, MABEL OSGOOD. Born Charlestown, Massachusetts, 26 January 1859; died Fairfield, Connecticut, 16 July 1934. Nature writer, bird protector, novelist.

FAMILY AND EDUCATION

Third of three daughters of Rev. Samuel Osgood, descendant of one of the founders of Andover, Massachusetts, and a Unitarian minister, who served as pastor of New York Church of the Messiah (1850–1870) and later became Episcopal priest. Osgood was educated at home, then in private school on lower Fifth Avenue in New York City. She married James Osborne Wright, English-born rare book dealer, 1884 (died 1920). They settled in Fairfield, Connecticut. Record is uncertain, but it is thought they had no children.

POSITIONS

Associated with the magazine *Bird-Lore*, serving in Executive Department from 1899 to 1910; then became a contributing editor. Helped establish the Connecticut Audubon Society in 1898 and was elected president. Was a director, National Association of Audubon Societies, 1905–1928, and associate member of American Ornithologists' Union in 1895; full member in 1901.

CAREER

In 1893 published "A New England May Day" in *New York Evening Post*. Essay appeared subsequently with some others in *The Friendship of Nature* (1894). Following year published *Birdcraft: A Field Book of Two Hundred Song, Game and Water Birds*. All her books were published by Macmillan and Company. Along with noted naturalist Elliot Coues, she coauthored *Citizen Bird* (1897), which enjoyed great popularity. Her *Four-Footed Americans* (1898) was a series of stories for children conveying factual information about mammals. Ten years later wrote the *Gray Lady and the Birds*. Pioneered protection for birds, setting up the Birdcraft Sanctuary near Connecticut home. Was also a novelist who wrote some ten novels under the pseudonym "Barbara," as well as under own name, most notably *The Garden of a Commuter's Wife* (1901) and *The Stranger at the Gate* (1913). Today viewed as limited works, hardly worthy of notice, for they reveal upper-class bias against the changes she saw transforming New York and Connecticut. She lamented increasing number of suburbanites in Connecticut, as well as growing role of the "New Woman." In *The Woman Errant* (1904) she frontally attacked feminism. Semiautobiographical work *My New York* (1926) reflected what she considered the tragic changes that buffeted New York City and helped create urban tensions. Was an active member of exclusive Connecticut Society of Colonial Dames. Her work in ornithology displayed love of nature as well as accurate observations and descriptions of birdlife. As novelist and social commentator, however, her focus was narrow and quite provincial.

MAJOR CONTRIBUTIONS

Ornithological observations are first-rate. Love of birds led her to stress the need to provide protection for them. Her pioneering effort in Birdcraft Sanctuary may well be main claim to fame.

BIBLIOGRAPHY

American Women Writers: A Critical Reference Guide from Colonial Times to the Present. Vol. 4.

James, Edward T., ed. *Notable American Women, 1607–1950: A Biographical Dictionary* (1971).

"Mabel Osgood Wright." *Auk* (October 1934) (obit.).

"Mabel Osgood Wright." *Bird-Lore* (July–August 1934) (obit.).

"Mabel Osgood Wright." *New York Times*, 18 July 1934 (obit.).

National Cyclopaedia of American Biography. Vol. 12.

Strom, Deborah, ed. *Birdwatching with American Women* (1986).

Welker, Robert H. *Birds and Men: American Birds in Science, Art, Literature, and Conservation, 1800–1900* (1955).

Judith Moran Curran

WRIGHT, ROBERT RAMSAY. Born Aloa, Scotland, 23 September 1852; died Droitwich, Worcestershire, England, 5 September 1933. Zoologist, marine biologist.

FAMILY AND EDUCATION

Youngest son of Rev. John and Christian Wright. Early education at Aloa Academy and Edinburgh High School. Studied at University of Edinburgh, received M.A. in 1871, B.Sc. in 1873. Honorary LL.D. from University of Toronto, 1889. Married Katharine Octavia Smith, 1876 (died 1930); no children.

POSITIONS

Professor of natural history, later of biology, University of Toronto, 1874–1887. Curator, Museum of Natural History, 1876–1887. Professor of biology, 1887–1912. Dean of the Faculty of Arts, 1901–1906. Vice president, University of Toronto, 1902–1906. Member, Board of Management of the Biological Board of Canada, 1901–1912. Supervised Atlantic research station, Canso, Nova Scotia, and Malpeque, Prince Edward Island, 1901–1904, 1911. President, Royal Society of Canada, 1912.

CAREER

Came to Toronto at age twenty-two, from an assistant professorship in zoology at the University of Edinburgh and membership in the Challenger Deep-Sea Expedition, to become professor at the University of Toronto. Efforts to promote a scientific basis in medical education led to reestablishment of the Faculty of Medicine in 1887, when he became professor of biology. Studied parasitology and fish anatomy during the late 1870s, 1880s. Began development in 1889 of zoological museum, described as "perhaps the foremost instructional museum in the world for its time." In "Preliminary Report on the Fish and Fisheries of Ontario," *Commissioners' Report, Ontario Game and Fish Commission* (1892), summarized available information on the natural history of Ontario fish and urged a systematic survey of Ontario waters. Promoted, from 1897, marine biological research related to the fisheries. While supervisor of Atlantic station, 1901–1904, focused on plankton research. Served on Royal Society of Canada committee to consider increased biological research facilities in Canada, 1902, on committee of Biological Board to decide location for permanent Atlantic research station, 1906, and on committee to determine site for Pacific research station, 1907. Retired in 1912, returned to Oxford to study classical Greek.

Published widely in zoology, including natural history, comparative vertebrate anatomy, parasitology, and protozoology. Published an account in 1884 of the anatomy of *Amiurus*, the freshwater catfish, in *Proceedings of the Canadian Institute, Standard Natural History* (with A. B. Macallum, 1885), *High School Zoology* (1888), "On the Plankton of the Nova Scotian Coast," *Report of the Marine Biological Station of Canada* (1906). His last publication, "The Progress of Biology," *Proceedings of the Royal Society of Canada* (1912), summarizes his conception of biology.

Important research contributions, particularly in parasitology and fish anatomy, including introduction of term "neuromast" for sense organs in the lateral line system. Inspiring teacher, able to attract highly capable individuals to his department and to support their research efforts. His influence important factor in establishment of the Biological Board of Canada and its creation of research stations on Georgian Bay and the Atlantic and Pacific Coasts. Largely responsible for placing biological research at the University of Toronto upon a modern footing, particularly in marine biology and in the application of biology to medicine. Established Department of Zoology of the university as a major center for marine fisheries research. Its graduates were a main source of supply for the scientists of the Fisheries Research Board of Canada during its first forty years. Noted for wide-ranging abilities as a linguist, musician, artist, and popular lecturer.

MAJOR CONTRIBUTIONS

Development of University of Toronto as major center for zoological and marine research, promotion of the application of biology to medicine, and stimulus toward marine and fisheries research and marine biological stations in Canada.

BIBLIOGRAPHY

The Canadian Encyclopedia.

Craigie, E. H. *A History of the Department of Zoology of the University of Toronto* (1967).

Huntsman, A. G. *Proceedings of the Royal Society of Canada* (1912) (obit.).

Johnstone, K. *The Aquatic Explorers: A History of the Fisheries Research Board of Canada* (1977).

Macallum, A. B. "Publications by Members of Staff." In *The University of Toronto and Its Colleges, 1827–1906* (1906).

UNPUBLISHED SOURCES

Miscellaneous papers are in Archibald Macallum Papers, James Loudon Papers, and Robert Falconer Papers, University of Toronto Archives; papers relating to fishery and marine biology are in Public Archives of Canada, Ottawa; letters are in Royal Ontario Museum Archives.

Stephen Bocking

WYMAN, JEFFRIES. Born Chelmsford, Massachusetts, 11 August 1814; died Bethlehem, New Hampshire, 4 September 1874. Comparative anatomist, naturalist, anthropologist.

FAMILY AND EDUCATION

Son of Rufus Wyman, a prominent physician, and Ann Merrill Wyman. Family moved to Charlestown in 1818, when father became director of McLean Asylum. Graduated from Harvard, 1833, and from Harvard Medical School, 1837. Clinical apprenticeship under John Call Dalton, Sr. Studied comparative anatomy and natural history at the Museum National d'Histoire Naturelle in Paris, 1841–1842, and with Richard Owen in London, 1842. Married Adeline

Wheelwright, 1850 (died 1855); two daughters, and Annie Williams Whitney, 1861 (died 1864); a son.

POSITIONS

Demonstrator for John Collins Warren at Harvard Medical School, c.1837–1838. Curator of the Lowell Institute, 1839–1842. Professor of anatomy and physiology in the Medical Department, Hampden-Sidney College, Richmond, Virginia, 1843–1848. Hersey Professor of Anatomy, Harvard College, 1847–1874. Faculty of Lawrence Scientific School, 1847–1866. Faculty of Museum of Comparative Zoology, 1860–1874. Faculty of Harvard Medical School, 1866–1874. Curator of the Peabody Museum of Archaeology and Ethnology, Harvard University, 1866–1874.

CAREER

First introduced to the study of comparative anatomy by John Collins Warren, professor of anatomy at Harvard Medical School. Invited by John Amory Lowell to give series of Lowell lectures, 1840, on comparative anatomy and used proceeds to study abroad. Became member of the Boston Society of Natural History, 1837, and was an active member throughout his life. Served as corresponding secretary, as a curator, and on the Publications Committee. Served as president, 1854–1870, during the period when the society began a new series of memoirs and moved into a new museum building in Back Bay. At Harvard taught comparative anatomy and physiology, embryology, and zoology. Began an extensive Museum of Comparative Anatomy, consisting of his private collection. Museum was moved in 1858 to the newly built Boylston Hall at Harvard. Traveled frequently because of ill health and to increase his collections. Went to Labrador, 1849, to Paramaribo, Surinam, 1857, to La Plata, Argentina, 1858, and took eight trips to Florida, 1852–1874, mostly to the area around the St. Johns River. Visited Europe and toured museums, 1853, 1870.

Wrote some 200, mostly short, papers on comparative anatomy, embryology, parisitology, curious habits of animals, paleontology, bacteriology, anthropology, and archeology. Published mainly in *American Journal of Sciences*, the publications of the Boston Society of Natural History, and the *American Naturalist*. Took a morphological view of nature and, along with Agassiz, introduced "philosophical" anatomy to America. Known for his thoroughness in research and care in writing but also for his philosophical mind. Was sympathetic from the first to Darwin but tended to believe in a theistic, morphological form of evolution rather than natural selection. His most noted contribution is the first scientific description of the gorilla (1847), based on remains sent by the American missionary Thomas Savage. Gave the gorilla its name. Other important papers include descriptions of organ of vision in the blind fish of Mammoth Cave (1843, 1854); Lowell lectures of 1849; the anatomy of *Rana pipiens*

(1852); the embryology of the skate (1854); homologies of limbs in vertebrates (1868); experiments on spontaneous generation in response to Pasteur (1862, 1867); demonstration that the cells of the bee were not perfectly designed (1868); measurements of human craniums (1868); the discovery that the crocodile's range extended to Florida (1870); and the discovery of Indian shell mounds in Florida and evidence that they were made by man (1875).

Remembered as a modest but self-possessed man, a friend to all, always ready to help others. Avoided taking controversial stands and wrote in a terse, factual manner. Seen as not ambitious. In religion, a Unitarian and believer in the Deity. Biographers emphasized his character and considered him a near-ideal exemplar of a scientist.

MAJOR CONTRIBUTIONS

Important less for particular achievements than for his character, the breadth of his knowledge, and the thoroughness and competence of his descriptive and theoretical studies. Considered the foremost comparative anatomist in America. Important as a museum curator. Teacher of many leaders of next generation of naturalists, including F. W. Putnam, E. S. Morse, N. S. Shaler, A. Hyatt, B. G. Wilder, A. S. Packard, A. E. Verrill. Acted as a balance to Agassiz and probably helped students to make the transition to evolution. Also did much to encourage experimental physiology by promoting careers of H. P. Bowditch and S. W. Mitchell.

BIBLIOGRAPHY

Appel, T. A. ''Jeffries Wyman, Philosophical Anatomy and the Scientific Reception of Darwin in America.'' *Journal of the History of Biology* 21 (1988).

Appel, T. A. ''A Scientific Career in the Age of Character: Jeffries Wyman and Natural History at Harvard.'' In C. A. Elliott and M. W. Rossiter, eds., *Science at Harvard University: Historical Perspectives* (1992).

Doetsch, R. N. ''Early American Experiments on Spontaneous Generation, by Jeffries Wyman (1814–1874).'' *Journal of the History of Medicine* 17 (1962).

Dupree, A. H. *Dictionary of Scientific Biography.*

Dupree, A. H. ''Jeffries Wyman's Views on Evolution.'' *Isis* 44 (1953).

Dupree, A. H. ''Some Letters from Charles Darwin to Jeffries Wyman.'' *Isis* 42 (1951).

Gifford, G. E., Jr. ''An American in Paris, 1841–1842: Four Letters from Jeffries Wyman.'' *Journal of the History of Medicine* 22 (1967).

Gifford, G. E., Jr. ''Twelve Letters from Jeffries Wyman, M.D., Hampden-Sydney Medical College, Richmond, Virginia, 1843–1848.'' *Journal of the History of Medicine* 20 (1965).

Gray, Asa. *Proceedings of the Boston Society of Natural History* and *American Journal of Science* 109 (1875) (obit.).

Holmes, O. W. *North American Review* (1874) (obit.).

Packard, A. S. ''Jeffries Wyman.'' *Biographical Memoirs of the National Academy of Sciences.* Vol. 2 (1886) (obit.).

UNPUBLISHED SOURCES

See Wyman Papers in Countway Library, Harvard Medical School; see articles by Appel for other locations of manuscripts.

Toby A. Appel

Y

YARD, ROBERT STERLING. Born Haverstraw, New York, 1 February 1861; died Washington, D.C., 17 May 1945. Writer, editor, conservationist, association executive.

FAMILY AND EDUCATION

Son of Robert and Sarah (Purdue) Yard. His father was Methodist minister. Prepatory education at Freehold (New Jersey) Institute. Graduated from Princeton with B.A. in 1883. Married Mary Belle Moffat, 1895; one daughter.

POSITIONS

With W. R. Grace and Company, New York City, as public information officer, and was in charge of foreign cables and correspondence, 1883–1886. Reporter with *New York Sun*, 1887–1890; with *New York Herald* 1891–1900 as reporter and later Sunday editor. In charge of R. H. Russell Publisher, 1900–1901. Book advertising manager, Charles Scribner's Sons Publishers, 1901–1905, also editing Scribner publication *The Lamp*, 1903–1905. Co-founder (with William D. Moffat) of Moffat, Yard, and Company, serving as editor-in-chief and vice president, 1905–1911. Ended this partnership, 1911, appointed organizing secretary, National Citizens' League for Promotion of Sound Banking, served until 1913. Editor-in-chief, *Century Magazine*, New York, 1913–1914. Did promotional and publicity work for National Park Service within U.S. Department of the Interior, and served from 1916 to 1919 as chief educational secretary, National Park Service. General secretary, National Park Association, 1919–1934. Editor, *National Parks Bulletin*, 1919–1936; editor, *National Parks*

New Bulletin, 1926–1936. With Robert Marshall and half a dozen others founded Wilderness Society, 1935. Executive secretary, Wilderness Society, 1935–1937. President and permanent secretary, Wilderness Society, 1937–1945.

CAREER

Yard spent latter half of his life promoting the beauty and importance of national parks and other wilderness areas. From 1914, when he came to Washington, D.C., at the age of fifty-three, till his death in 1945 he fought for preserving the wilderness areas of the United States. He was a prolific writer of newspaper and magazine articles. Yard also authored a series of beautifully illustrated books, which helped generate national interest and enthusiasm for the national parks. The first of these was *National Parks Portfolio* (1916), which contained descriptions and photographs of nine largest parks, among them Yosemite, Yellowstone, and the Grand Canyon. This was followed by *Glimpses of Our National Parks* (1916) and by *Top of the Continent: The Story of a Cheerful Journey through Our National Parks* (1917), which Yard said was for ''children of all ages—young, old and in between.'' *The Book of the National Parks* (1919) was thorough examination of the national park system and the unique features that each park contained. Yard also wrote *Our Federal Lands: A Romance of American Development* (1928), which was survey of the country's public land. In addition to writing, Yard lobbied for wilderness preservation in his positions as head of the National Parks Association and the Wilderness Society. He was also involved with number of federal advisory groups. From 1925 to 1930 he served as secretary of the Joint Committee on Survey of Federal Lands; and from 1927 to 1932 as secretary of the Advisory Board of the Educational and Inspirational Use of National Parks. He also took personal interest in preserving the American elk, especially in the Yellowstone area. From 1941 to 1945 he was trustee and the manager of Robert Marshall Wilderness Fund. Yard was a member of American Geographical Society and the California Academy of Sciences.

MAJOR CONTRIBUTIONS

Yard was an enthusiastic advocate for the national park system and wilderness areas. His numerous writings helped stimulate national interest in these areas. He also played a vital role in finding and developing the National Parks Association and the Wilderness Society.

BIBLIOGRAPHY

Davis, Richard, ed. *Encyclopedia of American Forest and Conservation History* (1983).
National Cyclopedia of American Biography.
New York Times, 19 May 1945 (obit.).
Stroud, Richard H., ed. *National Leaders of American Conservation* (1985).
The Living Wilderness (July and December 1945).

Paul Cammarata

YOUNG, STANLEY PAUL. Born Astoria, Oregon, 30 October 1889; died Washington, D.C., 15 May 1969. Mammalogist, wildlife biologist, scientific administrator.

FAMILY AND EDUCATION

Son of Benjamin Youngquist, Swedish-born cabin boy on Confederate ship *Alabama*, later U.S. Merchant Mariner, sugar plant foreman, and owner of Pacific Coast fish canneries, and Christina Swanson Youngquist, also of Swedish birth. Father changed name to Young. Young was seventh of eight children. Mother died when he was twelve; father died when he was twenty-two. Raised from age twelve by married sister Carol, who survived him. B.S. (mining engineering), University of Oregon, 1911; M.S., biology, University of Michigan, 1915. Married Nydia Marie Acker, 1921; at least two children.

POSITIONS

Ranger, U.S. Forest Service, 1917; U.S. government hunter, Bureau of Biological Survey (BBS), 1917–1919; assistant inspector of control work, BBS, 1919; coyote control crew member, southeastern New Mexico, BBS, 1920; assistant Leader, Predatory Animal Control, BBS, 1920–1921; leader, Colorado-Kansas district, BBS, 1921–1927; assistant head, Division of Economic Investigations, BBS, 1927–1928; principal biologist in charge, Division of Predatory Animal and Rodent Control, BBS, 1928–1934; chief, Division of Game Management, BBS, 1934–1938; chief, Division of Predatory and Rodent Control, BBS, 1938–1939; senior biologist, Branch of Wildlife Research, Fish and Wildlife Service, 1939–1957; director, Bird and Mammal Laboratories, Fish and Wildlife Service, 1957–1959.

CAREER

Began his professional career as a ranger with the U.S. Forest Service in 1917 and later that year transferred to the Bureau of Biological Survey, where he spent the rest of his career. First position at BBS was as a government hunter, specializing in predator control in the Southwest, advancing to position as a leader in predator control activities in Colorado, Kansas, and New Mexico. During the 1930s served in various positions in Washington, D.C., including chief of the Division of Game Management, 1924–1938, and chief, Division of Predator and Rodent Control, 1938–1939. Served as senior biologist, Branch of Wildlife Research, Fish and Wildlife Service, 1939–1957, a position that freed him for research and writing.

Conducted field studies of wolves in northeastern Louisiana, where, with Tappen Gregory, he took some of the earliest flash photographs of wolves in their natural habitat. In September 1937 led a field expedition to Carmen Mountains,

Mexico, where mountain lions were photographed in their natural habitat. Other major fieldwork included a trip to Alaska, 1938, to study the behavior of wolves and coyotes.

Main interest was the large predators of North America, on which he published well over 100 articles and several major monographs, 1924–1970. Included are *The Last Stand of the Pack* (with Arthur W. Carhart, 1929), *The Wolves of North America* (with E. A. Goldman, 1944), *The Puma, Mysterious American Cat* (with E. A. Goldman, 1946), *Sketches of American Wildlife* (1946), *The Wolf in North American History* (1946), *The Clever Coyote* (with H. H. T. Jackson, 1951), and *The Bobcat of North America* (1958). *Leloo, the Last of the Loners*, published posthumously (1970).

Young was made honorary member of Wildlife Society, 1959, and of American Society of Mammalogists, 1964. Received Department of Interior Distinguished Service Award, 1957. Member of several scientific, writers', and other organizations.

MAJOR CONTRIBUTIONS

Pioneering work in the self-photography (using tripwires) of animals in the wild, particularly wolves and mountain lions. Work on the large predators of North America broke new ground in life history studies of mammals and remains in use today. Developed special techniques in predator control and wildlife management.

BIBLIOGRAPHY

Craig, James B. *American Forests* 76 (1970) (obit.).
Manville, Richard H. *Journal of Mammalogy* 51 (February 1970) (obit.).
Presnall, Clifford C. *Journal of Wildlife Management* 33 (1969) (obit.).

UNPUBLISHED SOURCES

See Smithsonian Institution Archives, Record Unit 7174, Stanley Paul Young Papers, 1921–1965; Smithsonian Institution Archives, Record Unit 7171, Bird and Mammal Laboratories, Fish and Wildlife Service, U.S. Department of the Interior Records, c.1885–1971; National Archives, Record Group 22, Records of the Fish and Wildlife Service.

William A. Deiss

Z

ZAHL, PAUL ARTHUR. Born Bensenville, Illinois, 20 March 1910; died Greenwich, Connecticut, 16 October 1985. Natural scientist, medical physiologist, explorer.

FAMILY AND EDUCATION

Son of Arthur Hermann and Barbara Diebre Zahl. Student, Marquette University, 1928–1929. A.B., North Central College of Illinois, 1932; A.M., 1934, Ph.D., Harvard, 1936. Honorary D.Sc., North Central College, 1972. Married Eda Seasongood Field, 1946; two children.

POSITIONS

Assistant in comparative anatomy and histology, Harvard, 1934–1936; research fellow in endocrinology, Parke Davis and Company, 1936–1937; research associate, Union College, 1937–1939; guest investigator, Memorial Hospital, New York City, 1939–1941; research associate, American Museum of Natural History, from 1951; staff physiologist, 1937–1946, associate director and secretary of the corporation, 1946–1958, research associate, Haskins Laboratories, from 1958; during World War II, with Office of Science Research and Development. Adjunct professor of biology, Fordham University, from 1964; senior scientist, 1958–1975, member, Committee for Research and Development, National Geographic Society, 1975–1985.

CAREER

With Harvard friend, established Haskins Laboratories in New York, which focused on research in biochemistry, biophysics, and medical physiology but had long-standing interests in natural history, travel, and exploration. Member of expeditions to Panama, British Guiana, Brazil, and Venezuela, 1937–1939; published first book, *To the Lost World* (1939), concerning search for giant ants in jungle areas of South America. On return trip for *National Geographic* (1957–1958), discovered some of longest beetles in world, with strong pincers capable of snapping pencils. Other expeditions to Mexico, 1945, Bahamas, 1946–1947, 1949. In *Flamingo Hunt* (1952), discussed search for Bahamas flamingo, and in *Coro-Coro, the World of the Scarlet Ibis* (1954), his search for that bird's breeding grounds. Additional expeditions to Sicily, 1953, where he researched deep-sea fauna in Straits of Messina; Australia, Ceylon, and Hawaii, 1955–1956; Canada, 1956; East Africa, 1958–1959; Trinidad and Columbia, 1963; Philippines, Singapore, and Malaysia, 1963–1965; West Africa, 1966–1967. Wrote of tallest California redwood, photographed albino gorilla in Central Africa and three-foot long, seven-pound frogs for various *National Geographic Magazine* projects; in Malaysia, located bathtub-sized Rafflesia flower with leathery petals more than two feet wide. When not traveling abroad, undertook various research projects for National Cancer Institute, National Science Foundation, and Atomic Energy Commission. Member of a number of scientific and research organizations. Held research grants from U.S. Public Health Service and Department of Health, Education, and Welfare.

MAJOR CONTRIBUTIONS

Indefatigable medical researcher, explorer, and naturalist. His articles and photographs enlivened a number of National Geographic publications.

BIBLIOGRAPHY

American Men and Women of Science. 12th, 13th ed.
Newsweek, 28 October 1985 (obit.).
Washington Post, 17 October 1985 (obit.).
Who's Who in America (1984).

Keir B. Sterling

ZAHNISER, HOWARD CLINTON. Born Franklin, Pennsylvania, 25 February 1906; died Hyattsville, Maryland, 5 May 1964. Conservationist, writer, editor.

FAMILY AND EDUCATION

Son of Rev. A. H. M. Zahniser, evangelical minister, and Bertha Belle Newton Zahniser, of Seneca Indian descent. Family lived in various parts of western

Pennsylvania when Zahniser was a child, then briefly in Greenville, Illinois. Attended public schools in Pennsylvania; graduated A.B. from Greenville (Illinois) College, 1927, where he edited the student newspaper. Honorary Litt.D. from Greenville, 1960. Married Alice Bernita Hayden, 1935; four children.

POSITIONS

After graduation taught high school English and worked as reporter on *Pittsburgh Press* (1928) before coming to Washington, D.C., in 1929 to work with the U.S. Department of Commerce and then the Bureau of Biological Survey in the U.S. Dept. of Agriculture. In 1945, became executive secretary and editor of the Wilderness Society; later named executive director, which position he held until his death.

CAREER

Served as an editor and speechwriter and prepared radio programs during his government service. Also served as book review editor for *Nature Magazine* before its merger with *Natural History*. Wrote annual entries on ''Conservation'' for the *Encyclopedia Britannica Yearbook* during the 1950s and early 1960s. During those years he delivered papers at the Sierra Club Wilderness Conferences, later collected in *Wilderness: America's Living Heritage* (1961) and other titles in that series. Also collected papers for *Voices for the Wilderness* (1969). His writings appeared regularly, both signed and unsigned, in *The Living Wilderness* from 1945 until his death. He served as its editor for most of the period from 1945 until 1960. He also published in the environmental periodicals of that period, largely on the topic of the need for national legislative protection of America's wilderness heritage. During his lifetime he received little public credit for his nearly single-handed role in drafting and pressing for passage of the Wilderness Act, because of the legislative nature of that work and the fact that the Wilderness Society did not want to jeopardize its tax-exempt status. Recent historical writing on that legislation and on the wilderness aspect of the national park system reflects the nature of his role.

MAJOR CONTRIBUTIONS

Most noted for his tireless advocacy of wilderness protection and conservation, culminating in the Wilderness Act of 1964.

BIBLIOGRAPHY

Dictionary of American Biography. Supplement 7.
''Howard Clinton Zahniser, 1906–1964.'' *Living Wilderness* (Winter 1964).
National Cyclopedia of American Biography.
National Parks Magazine (June 1964) (obit.).
New York Times, 29 June 1964 (obit.).

Scott, Douglas W. ''The Visionary Role of Howard Zahniser.'' *Sierra* 69 (May–June 1984).

''Zahnie.'' *Living Wilderness* (Summer 1964).

Edward D. Zahniser

Appendix

Entrants are categorized according to their field or fields of specialization. Up to three fields are given for each individual, although some were involved in four or five fields. An asterisk after an individual's name indicates that he or she had university training on an undergraduate, graduate, or professional level. The designation in parentheses after a name identifies the entrant's country of origin as follows: Br—Brazil; C—Canada; Cz—Czech Republic; De—Denmark; Es—Estonia; Fr—France; Ge—Germany; GB—Great Britain; Ho—Holland; In—India; Me—Mexico; No—Norway; Pa—Panama; Po—Poland; Pl—Portugal; Ru—Russia; SD—Santo Domingo; Sw—Sweden; Sz—Switzerland; Tu—Turkey. No designations appear after the names of individuals born in the United States.

ADMINISTRATORS

Baird, Spencer Fullerton [1823–1887]*

Baker, John Hopkinson [1894–1973]*

Beebe, (Charles) William [1877–1962]*

Birge, Edward Asahel [1851–1950]*

Buchheister, Carl William [1901–1986]*

Callison, Charles Hugh [1913–1993] (C)*

Clepper, Henry Edward [1901–1987]*

Grant, Madison [1865–1937]*

Hayden, Ferdinand Vandiveer [1829–1887]*

Jérémie, Nicolas [1669–1732] (C)

King, Clarence Rivers [1842–1901]*

MacKaye, Benton [1879–1975]*

Mather, Stephen Tyng [1867–1930]*

Miller, Alden Holmes [1906–1965]*
Nelson, Edward William [1855–1934]
Osborn, (Henry) Fairfield, (Jr.) [1887–1969]*
Prince, Edward Ernest [1858–1936] (GB)*
Ruthven, Alexander G(rant) [1882–1971]*
Vogt, William [1902–1968]*
Walcott, Charles Doolittle [1850–1927]
Wetmore, Alexander [1886–1978]*
Yard, Robert Sterling [1861–1945]*
Young, Stanley Paul [1889–1969]*

AGRICULTURALISTS
Bain, Francis [1842–1894] (C)*
Criddle, Stuart [1877–1971] (GB)
Kalm, Pehr [1716–1779] (Sw)*
Randall, Thomas Edmund [1886–1984] (GB)
Ravenel, Henry William [1814–1887]*
Street, Maurice George [1910–1966] (GB)

ANATOMISTS
Coues, Elliott [1842–1899]*
Godman, John Davidson [1794–1830]*
Harlan, Richard [1796–1843]*
Wyman, Jeffries [1814–1874]*

ANTHROPOLOGISTS
Eiseley, Loren C. [1907–1977]*
Henshaw, Henry Wetherbee [1850–1930]*
McGee, William John (W. J.) [1853–1912]*
Morse, Edward Sylvester [1838–1925]*
Newcombe, Charles Frederick [1851–1924] (GB)*
Powell, John Wesley [1834–1902]*
Wyman, Jeffries [1814–1874]*

ARCHEOLOGIST
Putnam, Frederic Ward [1839–1915]*

ARTISTS/PAINTERS/PHOTOGRAPHERS
Abbot, John [1751–1841] (GB)
Adams, Ansel Easton [1902–1984]

Audubon, John James [1785–1851] (SD)
Bartram, William [1739–1823]
Bodmer, (Johann) Karl [1809–1893] (Sz)
Brooks, Allan (Cyril) [1869–1946] (In)
Comstock, Anna Botsford [1854–1930]*
Criddle, Norman [1875–1933] (GB)
Darling, Jay Norwood "Ding" [1876–1962]*
Davidson, Florence Edenshaw (Jadał q' egənga) [1896–1993] (C)*
Fuertes, Louis Agassiz [1874–1927]*
Furbish, (Catherine) Kate [1834–1931]
Glover, Townend [1813–1883] (Br)
Gosse, Philip Henry [1810–1888] (GB)
Hochbaum, Hans Albert [1911–1988] (C)*
Hood, Robert [1797–1821] (GB)
Jackson, William Henry [1843–1942]
Jaques (Jacques), Francis Lee [1887–1969]*
Krieger, Louis Charles Christopher [1873–1940]*
LeConte, John Eatton, Jr. [1784–1860]
Lesueur, Charles-Alexandre [1778–1846] (Fr)
Peale, Charles Willson [1741–1827]
Peale, Titian Ramsay [1799–1885]
Peterson, Roger Tory [1908–1996]
Pittman, Harold Herbert [1889–1972] (GB)*
Raine, Walter [1861–1934] (GB)
Ridgway, Robert [1850–1929]
Teale, Edwin Way [1899–1980]
Terrill, Lewis (McIver) [1878–1968] (C)
Walcott, Mary Vaux [1860–1940]

ASTRONOMERS
de Bonnécamps, Joseph-Pierre [1707–1790] (Fr)*
Gilliss, James Melville [1811–1865]
Hariot (Harriot), Thomas [c.1560–1621] (GB)*

AUTHORS, see WRITERS

BIOLOGISTS
Bates, Marston [1906–1974]*

Carr, Archie Fairly [1909–1987]*
Carson, Rachel (Louise) [1907–1964]*
Cottam, Clarence [1899–1974]*
Day, Albert M. [1897–1979]*
Harvey, Ethel Browne [1855–1965]*
Hubbs, Carl Leavitt [1894–1977]*
Leopold, Aldo Starker [1913–1983]*
Morgan, Ann Haven [1882–1966]*
Murie, Olaus Johan [1889–1963]*
Nigrelli, Ross Franco [1903–1989]*
Olson, Sigurd Ferdinand [1899–1982]*
Pearson, T(homas) Gilbert [1873–1943]*
Preble, Edward Alexander [1871–1957]
Prince, Edward Ernest [1858–1936] (GB)*
Rousseau, Jacques [1905–1970] (C)*
Rowan, William Robert [1891–1957] (Sz)
Stefansson, Vilhjalmur [1879–1962] (C)*
Waksman, Selman Abraham [1888–1973] (Ru)*
Wright, Robert Ramsay [1852–1933] (GB)*
Young, Stanley Paul [1889–1969]*

BOTANICAL BREEDERS
Alexander, Annie Montague [1867–1950]
Coville, Frederick Vernon [1867–1937]*
Criddle, Stuart [1877–1971] (GB)

BOTANICAL COLLECTORS
Alexander, Annie Montague [1867–1950]
Bradbury, John [1768–1823] (GB)
Drummond, Thomas [c.1780–1835] (GB)
DuPratz, Antoine Simon Le Page [1689–1775] (Ho)
Hartweg, (Carl) Theodor [1812–1871] (Ge)

BOTANISTS (INCLUDING ETHNOBOTANISTS)
Bailey, Liberty Hyde [1858–1954]*
Bailey, William Whitman [1843–1914]*
Banister, John [1650–1692] (GB)*
Barton, Benjamin Smith [1766–1815]*
Bartram, John [1699–1777]

Bourgeau, Eugene [1813–1877] (Fr)
Brandegee, Mary Curran [1844–1920]*
Braun, (Emma) Lucy [1888–1971]*
Breitung, August Julius [1913–1987] (C)*
Brewer, William Henry [1828–1910]*
Britton, Elizabeth Gertrude (Knight) [1858–1934]*
Britton, Nathaniel Lord [1859–1934]*
Brodie, William [1831–1909] (GB)*
Cain, Stanley Adair [1902–1995]*
Carver, George Washington [1865–1943]*
Catesby, Mark [1679–1749] (GB)
Clayton, John [1694–1774] (GB)
Clements, Frederic Edward [1874–1945]*
Colden, Cadwallader [1867–1776] (GB)*
Colden, Jane [1724–1766]*
Coville, Frederick Vernon [1867–1937]*
Criddle, Norman [1875–1933] (GB)
Custis, Peter [1781–1842]*
Cutler, Manasseh [1742–1823]*
Darlington, William [1782–1863]*
Davidson, Florence Edenshaw (Jadał q' egənga) [1896–1993] (C)*
Dawson, John William [1820–1899] (C)*
Derick, Carrie (Matilda) [1862–1941] (C)*
Douglas, David [1799–1834] (GB)
Drummond, Thomas [c.1780–1835] (GB)
Eastwood, Alice [1859–1853] (C)
Eaton, Amos [1776–1842]*
Engelmann, George [1809–1884] (Ge)*
Fassett, Norman Carter [1900–1954]*
Fernald, Merritt Lyndon [1873–1950]*
Furbish, (Catherine) Kate [1834–1931]
Ganong, William Francis [1864–1941] (C)*
Garden, Alexander [1730–1791] (GB)*
Gray, Asa [1810–1888]*
Haenke, Thaddeus Peregrinus Xavierius [1761–1816] (Cz)*
Harshberger, John William [1869–1929]*
Hitchcock, Albert Speer [1865–1935]*
James, Edwin [1797–1861]*

Jérémie, Catherine-Gertrude [1662–1744] (C)
Logan, Martha Daniell [1704–1779]
Macoun, James Melville [1862–1920] (C)
Macoun, John [1831–1920] (GB)*
Marie-Victorin, (Conrad Kirouac) [1885–1944] (C)*
Masson, Francis [1741–1805] (GB)
Menzies, Archibald [1754–1842] (GB)*
Michaux, André [1746–1802] (Fr)*
Michaux, François André [1770–1855] (Fr)
Mociño, José Mariano [1757–1821] (Me)*
Mousley, (William) Henry [1865–1949] (GB)
Porsild, (Alf) Erling [1901–1977] (De)*
Porsild, Robert (Thorbjorn) [1898–1977] (De)*
Provancher, Léon [1820–1892] (C)*
Pursh, Frederick [1774–1820] (Ge)
Rafinesque (-Schmaltz), C(onstantine) S(amuel) [1783–1840] (Tu)
Ravenel, Henry William [1814–1887]*
Robinson, Harry [1900–1990] (C)
Rousseau, Jacques [1905–1970] (C)*
Sargent, Charles Sprague [1841–1927]*
Sarrazin, Michel [1659–1734] (Fr)*
Schweinitz, Lewis David von [1780–1834]*
Short, Charles Wilkins [1794–1863]*
Sullivant, William Starling [1803–1873]*
Traill, Catherine Parr (Strickland) [1802–1899] (GB)
Transeau, Edgar Nelson [1875–1960]*
Walter, Thomas [1740–1789] (GB)

BUSINESSMEN/ENTREPRENEURS
Albright, Horace Marden (Madden) [1885–1955]*
Baker, John Hopkinson [1894–1973]*
Bent, Arthur Cleveland [1866–1954]*
Binney, Amos [1803–1847]*
Broley, Charles (Lavelle) [1879–1959] (C)*
Callin, Eric Manley [1911–1985] (C)
Cartwright, Bertram William [1890–1967] (GB)
Downs, Andrew [1811–1892]
Dutcher, William [1846–1920]
Elliott, Stephen [1771–1830]*
Mather, Stephen Tyng [1867–1930]*

Robertson, A(bsalom) Willis [1887–1971]*
Sullivant, William Starling [1803–1873]*
Walcott, Frederic Collin [1869–1949]*

CARCINOLOGISTS
Rathbun, Mary Jane [1860–1943]
Schmitt, Waldo La Salle [1887–1977]*

CHEMISTS
Brewer, William Henry [1828–1910]*
Croft, Henry Holmes [1820–1883] (GB)*
Mitchill, Samuel Latham [1764–1831]*

CLERGYMEN
Bachman, John [1790–1874]*
Bethune, Charles James Stewart [1838–1932] (C)*
Cutler, Manasseh [1742–1823]*
Forster, Johann Reinhold [1729–1798] (Po)*
Provancher, Léon [1820–1892] (C)*
Schweinitz, Lewis David von [1780–1834]*

COLONIZERS/EXPLORERS/FRONTIERSMEN/MOUNTAINEERS/
TRAVELERS
Adams, Ansel Easton [1902–1984]
Alexander, Annie Montague [1867–1950]*
Andrews, Roy Chapman [1884–1960]*
Bartram, William [1739–1823]
Blakiston, Thomas Wright [1832–1891] (GB)
de Bonneville, Benjamin Louis Eulalie [1796–1878] (Fr)*
Boyd, Louise Arner [1887–1972]
Bradbury, John [1768–1823] (GB)
Cartier, Jacques [1491–1557] (Fr)
Champlain, Samuel de [1570–1635] (Fr)
Clark, William [1770–1838]
Custis, Peter [1781–1842]*
Cutler, Manasseh [1742–1823]*
Denys, Nicolas [1598–1688] (Fr)
DuPratz, Antoine Simon Le Page [1689–1775] (Ho)
Emory, William Hemsley [1811–1887]*
Featherstonhaugh, George William [1780–1866]*

Fidler, Peter [1769–1822] (GB)
Forster, Johann Reinhold [1729–1798] (Po)*
Fox (Foxe), Luke [1586–1635] (GB)
Franklin, John [1786–1847] (GB)
Fraser, Simon [1776–1862]
Frémont, John Charles [1813–1890]*
Fuertes, Louis Agassiz [1874–1927]*
Hantzsch, Bernhard Adolph [1875–1911] (Ge)
Hariot (Harriot), Thomas [c.1560–1621] (GB)*
Hayden, Ferdinand Vandiveer [1829–1887]*
Hearne, Samuel [1745–1792] (GB)
Herndon, William Lewis [1813–1857]
Hind, Henry Youle [1823–1908] (GB)*
Jackson, William Henry [1843–1942]
Josselyn, John [c.1608–1675] (GB)
Kalm, Pehr [1716–1779] (SW)*
Kane, Elisha Kent [1820–1857]*
Kennicott, Robert [1835–1866]
Lewis, Meriwether [1744–1809]
Long, Stephen Harriman [1784–1864]*
Lorquin, Pierre Joseph Michel [1797–1873] (Fr)*
Mackenzie, Alexander [1764–1820] (GB)
Macoun, John [1831–1920] (GB)
Marshall, Robert [1901–1939]*
Mexia, Ynes Enriquetta Julietta [1870–1938]
Michaux, Francois André [1770–1855] (Fr)
Page, Thomas Jefferson [1808–1899]
Palliser, John [1817–1887] (GB)
Peale, Titian Ramsay [1799–1885]
Pike, Zebulon Montgomery [1779–1813]
Powell, John Wesley [1834–1902]*
Rae, John [1813–1893]*
Rand, Austin Loomer [1905–1982] (C)*
Richardson, John [1787–1865] (GB)*
Smith, Jedediah Strong [1799–1831]
Smith, John [1580–1631] (GB)
Soper, J. Dewey [1893–1982] (C)*
Stansbury, Howard [1806–1863]

Stefansson, Vilhjalmur [1879–1962] (C)*
Stimpson, William [1832–1872]
Stöhler (Steller), Georg Wilhelm [1709–1746] (Ge)*
Thompson, David [1770–1857] (GB)
Tyrell, Joseph Burr [1858–1957] (C)
Walcott, Frederic Collin [1869–1949]*
Zahl, Paul Arthur [1910–1985]*

CONCHOLOGISTS
Anthony, John Gould [1804–1877]
Gould, Augustus Addison [1805–1866]*
Hyatt, Alpheus [1838–1902]*
Morse, Edward Sylvester [1838–1925]*

CONSERVATIONISTS
Akeley, Carl Ethan [1864–1926]
Albright, Horace Marden (Madden) [1880–1987]*
Anable, Gloria Elaine ''Glo'' Hollister [1901–1988]*
Atwood, Wallace Walter [1892–1949]*
Baker, John Hopkinson [1894–1973]*
Belaney, Archibald Stansfeld [1888–1938] (GB)
Bowers, Edward Augustus [1857–1924]*
Braun, (Emma) Lucy [1888–1971]
Buchheister, Carl William [1901–1986]*
Burgess, Thornton Waldo [1871–1965]*
Cahalane, Victor Harrison [1901–1993]*
Cassels, Elsie (Eliza McAlister) [1864–1938] (GB)
Coolidge, Harold Jefferson [1904–1985]*
Cottam, Clarence [1899–1974]*
Darling, Jay Norwood ''Ding'' [1876–1962]*
Dilg, Will H. [1867–1927]
Drury, Newton Bishop [1889–1978]*
Edge, Mabel Rosalie [1877–1962]
Evermann, Barton Warren [1853–1932]*
Gabrielson, Ira Noel [1889–1977]*
Gannett, Henry [1846–1914]*
Grant, Madison [1865–1937]*
Green, Charlotte Hilton [1889–1992]*
Grinnell, George Bird [1849–1938]*

Hall, Eugene Raymond [1902–1986]*
Harshberger, John William [1869–1929]*
Hickey, Joseph James [1907–1993]*
Hochbaum, Hans Albert [1911–1988]*
Hornaday, William Temple [1854–1937]*
Hubbs, Carl Leavitt [1894–1977]*
Ickes, Harold LeClair [1874–1952]*
Lawrence, (Alexander George) "Lawrie" [1888–1961] (GB)
Leopold, Aldo Starker [1913–1983]*
Lloyd, Hoyes [1888–1978] (C)*
Lutz, Frank Eugene [1898–1943]*
Manning, Ernest Callaway [1890–1941] (C)*
Mills, Enos Abijah [1870–1922]
Miner, (John Thomas) Jack [1865–1944]
Morgan, Ann Haven [1882–1966]*
Muir, John [1838–1914] (GB)*
Olson, Sigurd Ferdinand [1899–1982]*
Osborn, (Henry) Fairfield, (Jr.) [1887–1969]*
Parker, Elizabeth [1856–1944] (C)*
Pearson, T(homas) Gilbert [1873–1943]*
Pinchot, Gifford [1865–1946]*
Preble, Edward Alexander [1871–1957]
Roosevelt, Franklin Delano [1882–1945]*
Roosevelt, Robert Barnwell [1829–1906]
Rowan, William Robert [1981–1957] (Sz)
Saunders, William Edwin [1836–1941] (GB)*
Sears, Paul Bigelow [1891–1990]*
Shaler, Nathaniel (Southgate) [1841–1906]*
Sheldon, Charles [1867–1928]*
Stegner, Wallace Earle [1909–1993]*
Taverner, Percy (Algernon) [1875–1947] (C)
Terrill, Lewis (McIver) [1878–1968] (C)
Tufts, Robie (Wilfred) [1884–1982] (C)*
VanHise, Charles Richard [1857–1918]*
Wheeler, Arthur (Oliver) [1860–1945] (GB)*
Yard, Robert Sterling [1861–1945]*
Zahniser, Howard Clinton [1906–1964]*

ECOLOGISTS, see NATURALISTS

EDITORS, see WRITERS

EDUCATORS

Bailey, Loring Woart [1839–1925]*
Bailey, William Whitman [1843–1914]*
Barton, Benjamin Smith [1766–1815]*
Bates, Marston [1906–1974]*
Bickmore, Albert Smith [1839–1914]*
Bonnécamps, Joseph-Pierre de [1707–1790] (Fr)*
Britton, Elizabeth Gertrude (Knight) [1858–1934]*
Britton, Nathaniel Lord [1859–1934]*
Bumpus, Hermon Carey [1862–1946]*
Call, Richard Ellsworth [1856–1917]*
Comstock, Anna Botsford [1854–1930]*
Dana, James Dwight [1813–1895]*
Dawson, John William [1820–1899] (C)*
Derick, Carrie (Matilda) [1862–1941] (C)*
Dice, Lee Raymond [1887–1977]*
Dobie, J(ames) Frank [1888–1964]*
Dorney, Robert (Starbird) [1928–1987]*
Drake, Daniel [1785–1852]*
Dunn, Emmett Reid [1894–1956]*
Eaton, Amos [1776–1842]*
Eigenmann, Carl H. [1863–1927] (Ge)*
Eiseley, Loren C. [1907–1977]*
Elliott, Stephen [1771–1830]*
Errington, Paul L(ester) [1902–1962]*
Evermann, Barton Warren [1853–1932]*
Fernow, Bernhard Edward [1851–1923] (Ge)*
Gray, Asa [1810–1888]*
Grinnell, Joseph [1877–1939]*
Hamilton, William John, Jr. [1902–1990]*
Harshberger, John William [1869–1929]*
Hind, Henry Youle [1823–1908] (GB)*
Hubbs, Carl Leavitt [1894–1977]*
Jackson, Hartley Harrod Thompson [1881–1976]*
Jefferson, Thomas [1743–1826]*

Jordan, David Starr [1851–1931]*
Just, Ernest Everett [1883–1941]*
Laing, Hamilton Mack [1883–1982] (C)*
Leopold, A(ldo) Starker [1913–1983]*
Lowery, George Hines, Jr. [1913–1978]*
Marie-Victorin, (Conrad Kirouac) [1885–1944] (C)*
Miller, Alden Holmes [1906–1965]
Moseley, Edwin Lincoln [1865–1948]*
Newberry, John Strong [1822–1892]*
Nicholson, Henry Alleyne [1844–1899] (GB)*
Packard, Alpheus Spring, Jr. [1839–1905]*
Palmer, E(phraim) Laurence [1888–1970]*
Phelps, Almira Hart Lincoln [1793–1884]
Rudolph, Emanuel David [1927–1992]*
Short, Charles Wilkins [1794–1863]*
Simpson, George Gaylord [1902–1984]*
Stegner, Wallace Earle [1909–1993]*
Transeau, Edgar Nelson [1875–1960]*
VanHise, Charles Richard [1857–1918]*
Wiley, Farida [1887–1986]
Wilson, Alexander [1776–1813] (GB)

EMBRYOLOGIST

Harvey, Ethel Browne [1855–1965]*

ENGINEERS

Klauber, Laurence M. [1883–1968]*
LeConte, John Eatton, Jr. [1784–1860]
Long, Stephen Harriman [1784–1864]*
Strong, Harriet Williams Russell [1844–1926]

ENTOMOLOGISTS

Abbot, John [1751–1841] (GB)
Banister, John [1650–1692] (GB)*
Bethune, Charles James Stewart [1838–1932] (C)*
Brodie, William [1831–1909] (GB)*
Comstock, Anna Botsford [1854–1930]*
Comstock, John Henry [1849–1931]*
Couper, William [?–c.1890] (GB?)

Criddle, Norman [1875–1933] (GB)
Croft, Henry Holmes [1820–1883] (GB)*
Fitch, Asa [1809–1879]
Fletcher, James [1852–1908] (GB)
Glover, Townend [1813–1883] (Br)
Harris, Thaddeus William [1795–1856]*
Hentz, Nicholas Marcellus [1797–1856] (Fr)*
Hewitt, Charles Gordon [1885–1920] (GB)*
Howard, Leland Ossian [1857–1950]*
Lorquin, Pierre Joseph Michel [1797–1873] (Fr)*
Lutz, Frank Eugene [1898–1943]*
Mousley, (William) Henry [1865–1949] (GB)
Packard, Alpheus Spring, Jr. [1839–1905]*
Riley, Charles Valentine [1843–1895] (GB)
Saunders, William [1836–1941] (GB)
Scudder, Samuel Hubbard [1837–1911]*
Uhler, Philip Reese [1835–1913]*
Walker, Edmund Murton [1877–1969] (C)*
Wheeler, William Morton [1865–1937]*

ENTREPRENEURS, see BUSINESSMEN

ETHNOLOGISTS

Bodmer, (Johann) Karl [1809–1893] (Sz)
Dawson, George Mercer [1849–1901] (C)*
Grinnell, George Bird [1849–1938]*
Maximilian, Alexander Philip [1782–1867] (Ge)*
Rafinesque (-Schmaltz), C(onstantine) S(amuel) [1783–1840] (Tu)
Stefansson, Vilhjalmur [1879–1962] (C)*

EXPLORERS, see COLONIZERS

FORESTERS/SILVICULTURALISTS

Clepper, Henry Edward [1901–1987]*
Fernow, Bernhard Eduard [1851–1923] (Ge)*
Hough, Franklin Benjamin [1822–1885]*
Leopold, (Rand) Aldo [1887–1948]*
MacKaye, Benton [1879–1975]*
Manning, Ernest Callaway [1890–1941] (C)*
Marshall, Robert [1901–1939]*
McArdle, Richard Edwin [1899–1983]*

Michaux, André [1746–1802] (Fr)*
Michaux, François André [1770–1855] (Fr)
Pinchot, Gifford [1865–1946]*
Warder, John Aston [1812–1883]*

FRONTIERSMEN, see COLONIZERS

FUR TRADERS

Denys, Nicolas [1598–1688] (Fr)
Fidler, Peter [1769–1822] (GB)
Fraser, Simon [1776–1862]
Graham, Andrew [c.1733–1815] (GB)
Gunn, Donald [1797–1878] (GB)
Hearne, Samuel [1745–1792] (GB)
Hutchins, Thomas [c.1742–1790] (GB?)
Isham, James [1717–1761] (GB)
MacFarlane, Roderick Ross [1833–1920] (GB)
Mackenzie, Alexander [1764–1820] (GB)
Rae, John [1813–1893] (GB)*
Ross, Bernard Rogan [1827–1874] (GB)*
Smith, Jedediah Strong [1799–1831]
Thompson, David [1770–1857] (GB)

GENETICIST

Dice, Lee Raymond [1887–1977]*

GEOGRAPHERS

Boyd, Louise Arnes [1887–1972]
Champlain, Samuel de [1570–1635] (Fr)
Cooper, James Graham [1830–1902]*
Gannett, Henry [1846–1914]*
MacArthur, Robert Helmer [1930–1972]*
Sauer, Carl Ortwin [1889–1975]*
Schmidt, Karl Patterson [1890–1957]*
Shaler, Nathaniel (Southgate) [1841–1906]*
Thompson, David [1770–1857] (GB)

GEOLOGISTS

Abbot, Charles Conrad [1843–1919]
Agassiz, (Jean) Louis (Rodolphe) [1807–1873] (Sz)*

Atwood, Wallace Walter [1892–1949]*
Bailey, Loring Woart [1834–1925]*
Brewer, William Henry [1828–1910]*
Call, Richard Ellsworth [1856–1917]*
Dana, James Dwight [1813–1895]*
Dawson, George Mercer [1849–1901] (C)*
Dawson, John William [1820–1899] (C)*
Dutton, Clarence Edward [1841–1912]*
Eaton, Amos [1776–1842]*
Emmons, Ebenezer [1799–1863]*
Featherstonhaugh, George William [1780–1866] (GB)*
Hayden, Ferdinand Vandiveer [1829–1887]*
Hind, Henry Youle [1823–1908] (GB)
Hood, Robert [1797–1821] (GB)
James, Edwin [1797–1861]*
King, Clarence Rivers [1842–1901]*
Logan, William (Edmund) [1798–1875] (C)*
Maclure, William (James) [1763–1840] (GB)
McGee, William John [1853–1912]*
Mitchill, Samuel Latham [1764–1831]*
Moore, Raymond Cecil [1892–1974]*
Muir, John [1838–1914] (GB)*
Newberry, John Strong [1822–1892]*
Powell, John Wesley [1834–1902]*
Rathbun, Richard [1852–1918]
Scott, William Berryman [1858–1947]*
Shaler, Nathaniel (Southgate) [1841–1906]*
Tyrell, Joseph Burr [1858–1957] (C)
Uhler, Philip Reese [1835–1913]*
VanHise, Charles Richard [1857–1918]*
Vennor, Henry George [1840–1884] (C)*
Walcott, Charles Doolittle [1850–1927]

GOVERNMENT OFFICIALS/POLITICIANS
Baird, Spencer Fullerton [1823–1887]*
Bowers, Edward Augustus [1857–1924]*
Clark, William [1770–1838]
Colden, Cadwallader [1867–1776] (GB)*

Darling, Jay Norwood ''Ding'' [1876–1962]*
Day, Albert M. [1897–1979]*
Drake, Daniel [1785–1852]*
Fletcher, James [1852–1908] (GB)
Fortin, Pierre (Etienne) [1823–1888] (C)*
Fothergill, Charles [1782–1840]
Frémont, John Charles [1813–1890]
Glover, Townend [1813–1883] (Br)
Gunn, Donald [1797–1878] (GB)
Harkin, James Bernard [1875–1955] (C)
Henshaw, Henry Wetherbee [1850–1930]*
Hewitt, Charles Gordon [1885–1920] (GB)*
Ickes, Harold LeClair [1874–1952]*
Jefferson, Thomas [1743–1826]*
Lacey, John Fletcher [1841–1913]
Lewis, Meriwether [1774–1809]
Lloyd, Hoyes [1888–1978] (C)*
Merriam, Clinton Hart [1855–1942]*
Nelson, Edward William [1855–1934]*
Olmsted, Frederick Law [1822–1903]*
Robertson, A(bsalom) Willis [1887–1971]*
Roosevelt, Franklin Delano [1882–1945]*
Roosevelt, Robert Barnwell [1829–1906]
Roosevelt, Theodore [1858–1919]*
Walcott, Frederic Collin [1869–1949]*
Wirth, Conrad Louis [1899–1993]

HERPETOLOGISTS

Dunn, Emmett Reid [1894–1956]*
Klauber, Laurence M. [1883–1968]*
Pope, Clifford Hillhouse [1899–1974]*
Schmidt, Karl Patterson [1890–1957]*
Stejneger, Leonhard (Hess) [1851–1943] (No)*

HISTORIANS

Clepper, Henry Edward [1901–1987]*
Coues, Elliott [1842–1899]*
Dobie, J(ames) Frank [1888–1964]*

DuPratz, Antoine Simon LePage [1689–1775] (Ho)
Eckstorm, Fannie Pearson Hardy [1865–1946]*
Ganong, William Francis [1864–1941] (C)*
Hough, Franklin Benjamin [1822–1885]*
Mitchell, John [1711–1768]*
Rousseau, Jacques [1905–1970] (C)*
Rudolph, Emanuel David [1927–1992]*
Tyrell, Joseph Burr [1858–1957] (C)
Wright, Albert Hazen [1879–1970]*

HORTICULTURALISTS
Bailey, Liberty Hyde [1858–1954]*
Hartweg, (Carl) Theodor [1812–1871] (Ge)
Logan, Martha Daniell [1704–1779]
Sessions, Katherine (Kate) Olivia [1857–1940]*
Strong, Harriet Williams Russell [1844–1926]
Walter, Thomas [1740–1789] (GB)
Warder, John Aston [1812–1883]*

HYDROLOGIST
McGee, William John [1853–1912]*

ICHTHYOLOGISTS
Anable, Gloria Elaine ''Glo'' Hollister [1901–1988]*
Eigenmann, Carl H. [1863–1927]*
Evermann, Barton Warren [1853–1932]*
Fortin, Pierre (Etienne) [1823–1888] (C)*
Gilchrist, Frederick Charles [1859–1896] (C)
Gill, Theodore Nicholas [1837–1914]
Goode, George Brown [1851–1896]*
Prince, Edward Ernest [1858–1936] (GB)*
Putnam, Frederic Ward [1839–1915]*
Storer, David Humphreys [1804–1891]*

INVENTORS
Klauber, Laurence M. [1883–1968]*
Strong, Harriet Williams Russell [1844–1926]

JOURNALISTS, see WRITERS

LAWYERS

Albright, Horace Marden (Madden) [1880–1987]*
Bowers, Edward Augustus [1857–1924]*
Eisenmann, Eugene [1906–1981] (Pa)*
Lacey, John Fletcher [1841–1913]
Lorquin, Pierre Joseph Michel [1797–1873] (Fr)*
Pearse, Theed [1871–1971] (GB)
Perley, Moses Henry [1804–1862] (C)*
Robertson, A(bsalom) Willis [1887–1971]
Roosevelt, Robert Barnwell [1829–1906]

LICHENOLOGIST

Rudolph, Emanuel David [1927–1992]*

LIMNOLOGISTS

Birge, Edward Asahel [1851–1950]*
Juday, Chancey [1871–1944]*
Rawson, Donald Strathearn [1905–1961] (C)*

MALACOLOGISTS

Banister, John [1650–1692] (GB)*
Call, Richard Ellsworth [1856–1917]*
Cooper, James Graham [1830–1902]*
Dall, William Healey [1845–1927]*

MAMMALOGISTS

Anderson, Rudolph (Martin) [1876–1961]*
Anthony, Harold Elmer [1890–1970]*
Bailey, Vernon Orlando [1864–1942]*
Bigg, Michael Andrew [1939–1990] (GB)*
Burt, William Henry [1903–1987]*
Coolidge, Harold Jefferson [1904–1985]*
Criddle, Stuart [1877–1971] (GB)
Dixon, Joseph Scattergood [1884–1952]*
Goldman, Edward Alphonso [1873–1946]*
Hall, Eugene Raymond [1902–1986]*
Hamilton, William John, Jr. [1902–1990]*
Jackson, Hartley Harrad Thompson [1881–1976]*

Kalmbach, Edwin Richard [1884–1972]
Lowery, George Hines, Jr. [1913–1978]*
Mearns, Edgar Alexander [1856–1916]*
Merriam, Clinton Hart [1855–1942]*
Miller, Gerrit Smith, Jr. [1869–1956]*
Morton, Robert Allen [1953–1986] (C)
Peterson, Randolph Lee [1920–1989]*
Preble, Edward Alexander [1871–1957]*
Soper, J. Dewey [1893–1982] (C)*
True, Frederick William [1858–1914]*
Young, Stanley Paul [1889–1969]*

METEOROLOGISTS
Engelmann, George [1809–1884] (Ge)*
Hutchins, Thomas [c.1742–1790] (GB?)
Nelson, Edward William [1855–1934]*
Vennor, Henry George [1840–1884] (C)*

MILITARY FIGURES
de Bonneville, Benjamin Louis Eulalie [1796–1878] (Fr)*
Cartier, Jacques [1491–1557] (Fr)
Clark, William [1770–1838]
Dutton, Clarence Edward [1841–1912]*
Emory, William Hemsley [1811–1887]*
Franklin, John [1586–1635] (GB)
Frémont, John Charles [1813–1890]*
Gilliss, James Melville [1811–1865]
Herndon, William Lewis [1813–1857]
Jaques (Jacques), Francis Lee [1887–1969]
Kane, Elisha Kent [1820–1857]*
Lacey, John Fletcher [1841–1913]
La Galissoniere, Roland-Michel Barrin, Marquis de [1693–1756] (Fr)
Lewis, Meriwether [1774–1809]
Long, Stephen Harriman [1784–1864]*
Menzies, Archibald [1754–1842] (GB)*
Page, Thomas Jefferson [1808–1899]
Pike, Zebulon Montgomery [1779–1813]
Shufeldt, Robert Wilson [1850–1934]*
Stansbury, Howard [1806–1863]

Wheeler, George Montague [1842–1905]*
Woodhouse, Samuel Washington [1821–1904]*

MOUNTAINEERS, see COLONIZERS

MUSEUM CURATORS/DIRECTORS/FOUNDERS

Adams, Charles Christopher [1873–1955]*
Allen, Joel Asaph [1838–1921]*
Andrews, Roy Chapman [1884–1960]*
Anthony, Harold Elmer [1890–1970]*
Bailey, Alfred Marshall [1894–1978]*
Bartsh, Paul [1871–1960] (Ge)*
Bendire, Charles Emil (Karl Emil Bender) [1836–1897]
Bickmore, Albert Smith [1839–1914]*
Coolidge, Harold Jefferson [1904–1985]*
Dionne, Charles-Eusèbe [1846–1925] (C)
Dunn, Emmett Reid [1894–1956]*
Fleming, J. H. ''Harry'' (James Henry) [1872–1940] (C)*
Goode, George Brown [1851–1896]*
Grinnell, Joseph [1877–1939]*
Kennicott, Robert [1835–1866]
Lutz, Frank Eugene [1898–1943]*
Mociño, José Mariano [1757–1821] (Me)*
Osborn, Henry Fairfield [1857–1935]*
Peale, Charles William [1741–1827]
Putnam, Frederic Ward [1839–1915]*
Rathbun, Richard [1852–1918]
Ridgway, Robert [1850–1929]
Stejneger, Leonhard (Hess) [1851–1943] (No)*
Stimpson, William [1832–1872]
Taverner, Percy (Algernon) [1875–1947] (C)
Todd, W(alter) E(dmond) Clyde [1874–1969]
True, Frederick William [1858–1914]*
Verrill, Addison Emery [1839–1926]*

MYCOLOGISTS

Carver, George Washington [1865–1943]*
Krieger, Louis Charles Christopher [1873–1940]*
Schweinitz, Lewis David von [1780–1834]*

NATURALISTS/ECOLOGISTS

Abbott, Charles Conrad [1843–1919]*
Adams, Charles Christpher [1873–1955]*
Agassiz, (Jean) Louis (Rodolphe) [1807–1873] (Sz)*
Akeley, Carl Ethan [1864–1926]
Alexander, Annie Montague [1867–1950]
Allee, Warder Clyde [1885–1955]*
Audubon, John James [1785–1851] (SD)
Bachman, John [1790–1874]*
Bailey, Loring Woart [1839–1925]*
Bailey, Vernon Orlando [1864–1942]*
Bain, Francis [1842–1894] (C)*
Bartram, William [1739–1823]
Bartsch, Paul [1871–1960] (Ge)*
Baynes, Ernest Harold [1868–1925] (In)*
Beebe, (Charles) William [1877–1962]*
Bickmore, Albert Smith [1839–1914]*
Bigg, Michael Andrew [1939–1990] (GB)*
Binney, Amos [1803–1847]*
Blakiston, Thomas Wright [1832–1891] (GB)
Bonaparte, Charles Lucien Jules Laurent [1803–1857] (Fr)
Bradbury, John [1768–1823] (GB)
Braun, (Emma) Lucy [1888–1971]*
Brodie, William [1831–1909] (GB)*
Broley, Charles (Lavelle) [1879–1959] (C)
Brooks, Allan (Cyril) [1869–1946] (In)
Buchheister, Carl William [1901–1986]*
Burgess, Thornton Waldo [1871–1965]*
Burroughs, John [1837–1921]*
Cahalane, Victor Harrison [1901–1993]*
Cain, Stanley Adair [1902–1995]*
Carson, Rachel (Louise) [1907–1964]*
Cartwright, Bertram William [1890–1967] (GB)
Cassels, Elsie (Eliza McAlister) [1864–1938] (GB)
Catesby, Mark [1679–1749] (GB)
Clements, Frederic Edward [1874–1945]*
Coues, Elliott [1842–1899]*
Coville, Frederick Vernon [1867–1937]*

Custis, Peter [1871–1842]*
Dall, William Healey [1845–1927]*
DeKay, James Ellsworth [1792–1851] (Pl)*
Dionne, Charles-Eusèbe [1846–1925] (C)
Dixon, Joseph Scattergood [1884–1952]*
Dorney, Robert Starbird [1928–1987]*
Downs, Andrew [1811–1892]
Elliott, Stephen [1771–1830]*
Errington, Paul L(ester) [1902–1962]*
Eschscholtz, Johann Friedrich [1793–1831] (Ru, Es)*
Fleming, J. H. ''Harry'' (James Henry) [1872–1940]*
Forster, Johann Reinhold [1729–1798] (Po)*
Fortin, Pierre (Etienne) [1823–1888] (C)*
Fothergill, Charles [1782–1840]
Fuertes, Louis Agassiz [1874–1927]*
Ganong, William Francis [1864–1941] (C)*
Garden, Alexander [1730–1791] (GB)*
Gilchrist, Frederick Charles [1859–1896] (C)
Godman, John Davidson [1794–1830]*
Goldman, Edward Alphonso [1873–1946]*
Goode, George Brown [1851–1896]*
Gosse, Philip Henry [1810–1888] (GB)
Gould, Augustus Addison [1805–1866]*
Graham, Andrew [c.1733–1815] (GB)
Green, Charlotte Hilton [1889–1992]*
Grinnell, George Bird [1849–1938]*
Gunn, Donald [1797–1878] (GB)
Hamilton, William John, Jr. [1902–1990]*
Hantzsch, Bernhard Adolph [1875–1911] (Ge)
Hearne, Samuel [1745–1792] (GB)
Henshaw, Henry Wetherbee [1850–1930]*
Hentz, Nicholas Marcellus [1797–1856] (Fr)
Hochbaum, Hans Albert [1911–1988]*
Hood, Robert [1797–1821] (GB)
Hornaday, William Temple [1854–1937]*
Huntsman, Archibald Gowanlock [1883–1973] (C)*
Hyatt, Alpheus [1838–1902]*
Jackson, Hartley Harrad Thompson [1881–1976]*

Jordan, David Starr [1851–1931]*

Kalm, Pehr [1716–1779] (Sw)*

Kennicott, Robert [1835–1866]

Kieran, John Francis [1892–1981]*

de Kiriline Lawrence, Louise [1894–1992] (Sw)

Laing, Hamilton Mack [1883–1982] (C)

Lapham, Increase Allen [1811–1875]

Lawrence, (Alexander George) ''Lawrie'' [1888–1961] (GB)

LeConte, John Eatton, Jr. [1784–1860]

LeConte, Joseph [1823–1901]*

Leidy, Joseph [1823–1891]*

Leopold, Aldo Starker [1913–1983]*

Leopold, (Rand) Aldo [1887–1948]*

Lesueur, Charles-Alexandre [1778–1846] (Fr)

Lincecum, Gideon [1793–1873]

Lord, John Keast [1818–1872] (GB)*

MacArthur, Robert Helmer [1930–1972]*

MacFarlane, Roderick Ross [1833–1920] (GB)

Macoun, James Melville [1862–1920] (C)*

Macoun, John [1831–1920] (GB)*

Marie-Victorin, (Conrad Kirouac) [1885–1944] (C)*

Marsh, George Perkins [1801–1882]*

Maximilian, Alexander Philip [1782–1867] (Ge)*

Maxwell, Martha Ann Dartt [1831–1881]*

Miller, Harriet Mann [1831–1918]*

Mills, Enos Abijah [1870–1922]

Morgan, Ann Haven [1882–1966]*

Morse, Edward Sylvester [1838–1925]*

Morton, Robert Allen [1953–1986] (C)

Moseley, Edwin Lincoln [1865–1948]*

Nelson, Edward William [1855–1934]*

Newcombe, Charles Frederick [1851–1924] (GB)*

Nuttall, Thomas [1786–1859] (GB)

Olmsted, Frederick Law [1822–1903]*

Ord, George [1781–1866]

Osborn, (Henry) Fairfield, (Jr.) [1887–1969]*

Packard, Alpheus Spring, Jr. [1839–1905]*

Palmer, E(phraim) Laurence [1888–1970]*

Palmer, Theodore Sherman [1868–1955]*
Peale, Charles William [1741–1827]
Peale, Titian Ramsay [1799–1885]
Perley, Moses Henry [1804–1862] (C)*
Pickering, Charles [1805–1878]*
Pittman, Harold Herbert [1889–1972] (GB)*
Potter, Laurence Bedford [1883–1943] (GB)
Preble, Edward Alexander [1871–1957]
Priestly, Isabel M. [1893–1946] (GB)
Provancher, Léon [1820–1892] (C)*
Rafinesque (-Schmaltz), C(onstantine) S(amuel) [1783–1840] (Tu)
Randall, Thomas Edmund [1886–1984] (GB)
Richardson, John [1787–1865] (GB)*
Roosevelt, Theodore [1858–1919]*
Ross, Bernard Rogan [1827–1874] (GB)*
Sarrazin, Michel [1659–1734] (Fr)*
Say, Thomas [1787–1834]
Sears, Paul Bigelow [1891–1990]*
Seton, Ernest (Evan) Thompson [1860–1946] (GB)*
Sheldon, Charles [1867–1928]*
Shelford, (Ernest) Victor [1877–1968]*
Simpson, George Gaylord [1902–1984]*
Soper, J. Dewey [1893–1982] (C)*
Spreadborough, William [1856–1931] (GB)
Sternberg, Charles Hezelius [1850–1943]
Stimpson, William [1832–1872]
Stöhler (Steller) Georg Wilhelm [1709–1746] (Ge)*
Storer, David Humphreys [1804–1891]*
Street, Maurice George [1910–1966] (GB)
Swallow, Ellen Henrietta (Ellen Richards) [1842–1911]*
Traill, Catherine Parr (Strickland) [1802–1899] (GB)
Teale, Edwin Way [1899–1980]
Thoreau, Henry David [1817–1862]*
Transeau, Edgar Nelson [1875–1960]*
Verrill, Addison Emery [1839–1926]*
Vogt, William [1902–1968]*
Walcott, Mary Vaux [1860–1940]
Walker, Edmund Murton [1877–1969] (C)*

Wherry, Edgar (Theodore) [1885–1982]*
Wiley, Farida [1887–1986]
Wilson, Alexander [1766–1813] (GB)
Wright, Albert Hazen [1879–1970]*
Wyman, Jeffries [1814–1874]*

OCEANOGRAPHERS
Beebe, (Charles) William [1877–1962]*
Huntsman, Archibald Gowanlock [1883–1973] (C)*
Murphy, Robert Cushman [1887–1973]*

OOLOGISTS
Bendire, Charles Emil (Karl Emil Bender) [1836–1897] (Ge)
Bent, Arthur Cleveland [1866–1954]
Brewer, Thomas Mayo [1814–1880]*
Raine, Walter [1861–1934] (GB)
Randall, Thomas Edmund [1886–1984] (GB)

ORNITHOLOGISTS
Abbot, John [1751–1841] (GB)
Allen, Arthur Augustus [1885–1964]*
Anderson, Rudolph (Martin) [1876–1961]*
Audubon, John James [1785–1851] (SD)
Bailey, Alfred Marshall [1894–1978]*
Bailey, Florence Merriam [1863–1948]*
Bendire, Charles Emil (Karl Emil Bender) [1836–1897] (Ge)
Bent, Arthur Cleveland [1866–1954]*
Brewer, Thomas Mayo [1814–1880]*
Brewster, William [1851–1919]
Broley, Charles (Lavelle) [1879–1959] (C)*
Brooks, Allan (Cyril) [1869–1946] (In)
Callin, Eric Manley [1911–1985] (C)
Cassels, Elsie (Eliza McAlister) [1864–1938] (GB)
Cassin, John [1813–1869]
Catesby, Mark [1679–1749] (GB)
Chapman, Frank Michler [1864–1945]
Cooper, James Graham [1830–1902]*
Couper, William [?–c.1890] (GB?)
Dixon, Joseph Scattergood [1884–1952]*

Dutcher, William [1846–1920]
Eckstorm, Fannie Pearson Hardy [1865–1946]*
Eisenmann, Eugene [1906–1981] (Pa)*
Fleming, J. H. ''Harry'' (James Henry) [1872–1940] (C)*
Forbush, Edward Howe [1858–1929]
Hickey, Joseph James [1907–1993]*
Kalmbach, Edwin Richard [1884–1972]
de Kiriline Lawrence, Louise [1894–1992] (Sw)
Lawrence, (Alexander George) ''Lawrie'' [1888–1961] (GB)
Leopold, A(ldo) Starker [1913–1983]*
Lincoln, Frederick Charles [1892–1960]
Lowery, George Hines, Jr. [1913–1978]*
Mearns, Edgar Alexander [1856–1916]
Merriam, Clinton Hart [1855–1942]*
Miller, Alden Holmes [1906–1965]
Mitchell, Margaret Howell [1901–1988] (C)*
Mousley, (William) Henry [1865–1949] (GB)
Murphy, Robert Cushman [1887–1973]*
Nice, Margaret Morse [1883–1974]*
Palmer, Theodore Sherman [1868–1955]*
Pearse, Theed [1871–1971] (GB)
Pearson, T(homas) Gilbert [1873–1943]*
Peterson, Roger Tory [1908–1996]
Preble, Edward Alexander [1871–1957]
Ridgway, Robert [1850–1929]
Rowan, William Robert [1891–1957] (Sz)
Saunders, William Edwin [1861–1943] (C)*
Schreiber, Ralph W. [1942–1988]*
Shufeldt, Robert Wilson [1850–1934]*
Soper, J. Dewey [1893–1982] (C)*
Stejneger, Leonhard (Hess) [1851–1943] (No)*
Taverner, Percy (Algernon) [1875–1947] (C)
Terrill, Lewis (McIver) [1878–1968] (C)
Todd, W(alter) E(dmond) Clyde [1874–1969]
Tufts, Robie (Wilfred) [1884–1982] (C)*
Vennor, Henry George [1840–1884] (C)*
Vogt, William [1902–1968]*
Wetmore, Alexander [1886–1978]*

Bailey, Liberty Hyde [1858–1954]*
Bain, Francis [1842–1894] (C)*
Baynes, Ernest Harold [1868–1925] (In)*
Belaney, Archibald Stansfeld [1888–1938] (GB)
Brewer, Thomas Mayo [1814–1880]*
Britton, Elizabeth Gertrude (Knight) [1858–1934]*
Britton, Nathaniel Lord [1859–1934]*
Burgess, Thornton Waldo [1871–1965]*
Burroughs, John [1837–1921]*
Callison, Charles Hugh [1913–1993] (C)*
Carr, Archie Fairly [1909–1987]*
Carson, Rachel (Louise) [1907–1964]*
Cassin, John [1813–1869]
Darlington, William [1782–1863]*
DeKay, James Ellsworth [1792–1851] (P1)*
Denys, Nicolas [1598–1688] (Fr)
Dilg, Will H. [1867–1927]
Dobie, J(ames) Frank [1888–1964]*
Doubleday, Neltje DeGraff [1865–1918]
Edge, Mabel Rosalie [1877–1962]
Eiseley, Loren C. [1907–1977]*
Fernow, Bernhard Eduard [1851–1923] (Pr)*
Fothergill, Charles [1782–1840]
Godman, John Davidson [1794–1830]*
Gosse, Philip Henry [1810–1888] (GB)
Green, Charlotte Hilton [1889–1992]*
Harkin, James Bernard [1875–1955] (C)
Hentz, Nicholas Marcellus [1797–1856] (Fr)
Herbert, Henry William (Frank Forester) [1807–1858] (GB)*
Hubbs, Carl Leavitt [1894–1977]*
Ickes, Harold Le Clair [1874–1952]*
Jackson, Hartley Harrod Thompson [1881–1976]
Jordon, David Starr [1851–1931]*
Josselyn, John [c.1608–1675] (GB)
Kieran, John Francis [1892–1981]*
King, Clarence Rivers [1842–1901]*
de Kiriline Lawrence, Louise [1894–1992] (Sw)
Laing, Hamilton Mack [1883–1982]*

Woodhouse, Samuel Washington [1821–1904]*

PAINTERS, see ARTISTS

PALEONTOLOGISTS

Cope, Edward Drinker [1840–1897]*
Dall, William Healey [1845–1927]*
Harlan, Richard [1796–1843]*
Hyatt, Alpheus [1838–1902]*
Lambe, Lawrence Morris [1863–1919]*
Leidy, Joseph [1823–1891]*
Marsh, Othniel Charles [1831–1899]*
Moore, Raymond Cecil [1892–1974]*
Newberry, John Strong [1822–1892]*
Nicholson, Henry Alleyne [1844–1899] (GB)*
Osborn, Henry Fairfield [1857–1935]*
Scott, William Berryman [1858–1947]*
Simpson, George Gaylord [1902–1984]*
Sternberg, Charles Hezelius [1850–1943]
Walcott, Charles Doolittle [1850–1927]
Wetmore, Alexander [1886–1978]*
Whiteaves, Joseph Frederick [1835–1909] (GB)

PHARMACOLOGIST

Waksman, Selman Abraham [1888–1973] (Ru)*

PHARMACISTS

Saunders, William [1836–1941] (GB)
Saunders, William Edwin [1861–1943] (C)*

PHILOLOGIST

Ord, George [1781–1866]

PHOTOGRAPHERS, see ARTISTS

PHYSICIANS/SURGEONS

Darlington, William [1782–1863]*
Drake, Daniel [1785–1852]*
Emmons, Ebenezer [1799–1863]*
Engelmann, George [1809–1884] (Ge)*
Eschscholtz, Johann Friedrich [1793–1831] (Ru, Es)*

Garden, Alexander [1730–1791] (GB)*

Gould, Augustus Addison [1805–1866]*

Harlan, Richard [1796–1843]*

Hough, Franklin Benjamin [1822–1885]*

Hutchins, Thomas [c.1742–1790] (GB?)

Kane, Elisha Kent [1820–1857]*

LeConte, Joseph [1823–1901]*

Leidy, Joseph [1823–1891]*

Lincecum, Gideon [1793–1873]

Mearns Edgar Alexander [1856–1916]

Menzies, Archibald [1754–1842] (GB)*

Mitchell, John [1711–1768]*

Mitchill, Samuel Latham [1764–1831]*

Mociño, José Mariano [1757–1821] (Me)*

Newcombe, Charles Frederick [1851–1924] (GB)*

Pickering, Charles [1805–1878]*

Rae, John [1813–1893] (GB)*

Richardson, John [1787–1865] (GB)*

Robinson, Harry [1900–1990] (C)

Sarrazin, Michel [1659–1734] (Fr)*

Short, Charles Wilkins [1794–1863]*

Shufeldt, Robert Wilson [1850–1934]*

Stöhler (Steller) Georg Wilhelm [1709–1746] (Ge)*

Storer, David Humphreys [1804–1891]*

Warder, John Aston [1812–1883]*

Woodhouse, Samuel Washington [1821–1904]*

POLITICIANS, see GOVERNMENT OFFICIALS

PUBLISHERS, see WRITERS

SILVICULTURALISTS, see FORESTERS

SOCIAL WORKERS

Bailey, Florence Merriam [1863–1948]*

Mexia, Ynes Enriquetta Julietta [1870–1938]

SURGEONS, see PHYSICIANS

SURVEYORS

Dawson, George Mercer [1849–1901] (C)*

Featherstonhaugh, George William [1780–1866] (GB)*

Fidler, Peter [1769–1822] (GB)

Lambe, Lawrence Morris [1863–1919] (GB)*

Wheeler, Arthur (Oliver) [1860–1945] (GB)*

SYSTEMATISTS

Bonaparte, Charles Lucien Jules Laurent [1803–1857] (Fr)

Stejneger, Leonhard (Hess) [1851–1943] (No)*

Simpson, George Gaylord [1902–1984]*

TAXIDERMISTS

Akeley, Carl Ethan [1864–1926]

Couper, William [?–c.1890] (GB?)

Dionne, Charles-Eusèbe [1846–1925] (C)

Gilchrist, Frederick Charles [1859–1896] (C)

Hornaday, William Temple [1854–1937]*

Maxwell, Martha Ann Dartt [1831–1881]*

TAXONOMISTS

Carr, Archie Fairly [1909–1987]*

Gill, Theodore Nicholas [1837–1914]

Porsild, (Alf) Erling [1901–1977] (De)*

Schmidt, Karl Patterson [1890–1957]*

TOPOGRAPHERS

Gannett, Henry [1846–1914]*

Kalm, Pehr [1716–1779] (Sw)*

Wheeler, Arthur (Oliver) [1860–1945] (GB)*

Wheeler, George Montague [1842–1905]*

TRAVELERS, see COLONIZERS

VETERINARIAN

Lord, John Keast [1818–1872] (GB)*

WRITERS/EDITORS/JOURNALISTS/PUBLISHERS

Bailey, Alfred Marshall [1894–1978]*

Leopold, (Rand) Aldo [1887–1948]*
Lord, John Keast [1818–1872] (GB)*
MacKaye, Benton [1879–1975]*
Marsh, George Perkins [1801–1882]*
Miller, Harriet Mann [1831–1918]
Mills, Enos Abijah [1870–1922]
Miner, (John Thomas) Jack [1865–1944]
Muir, John [1838–1914] (GB)*
Murphy, Robert Cushman [1887–1973]*
Olmsted, Frederick Law [1822–1903]*
Olson, Sigurd Ferdinand [1899–1982]*
Parker, Elizabeth [1856–1944] (C)*
Perley, Moses Henry [1804–1862] (C)*
Pittman, Harold Herbert [1889–1972] (GB)*
Pope, Clifford Hillhouse [1899–1974]*
Rand, Austin Loomer [1905–1982] (C)*
Riley, Charles Valentine [1843–1895] (GB)
Roosevelt, Robert Barnwell [1829–1906]
Roosevelt, Theodore [1858–1919]*
Smith, John [1580–1631] (GB)
Stefansson, Vilhjalmur [1879–1962] (C)*
Stegner, Wallace Earle [1909–1993]*
Teale, Edwin Way [1899–1980]
Traill, Catherine Parr (Strickland) [1802–1899] (GB)
Wilson, Alexander [1766–1813] (GB)
Wright, Mabel Osgood [1859–1934]
Yard, Robert Sterling [1861–1945]*
Zahniser, Howard Clinton [1906–1964]*

ZOOLOGISTS

Agassiz, (Jean) Louis (Rodolphe) [1807–1873] (Sz)*
Allen, Joel Asaph [1885–1964]*
Anable, Gloria Elaine ''Glo'' Hollister [1901–1988]*
Anderson, Rudolph (Martin) [1876–1961]*
Andrews, Roy Chapman [1884–1960]*
Anthony, Harold Elmer [1890–1970]*
Baird, Spencer Fullerton [1823–1887]*
Barton, Benjamin Smith [1766–1815]*

Bartsch, Paul [1871–1960] (Ge)*
Bates, Marston [1906–1974]*
Binney, Amos [1803–1847]*
Bumpus, Hermon Carey [1862–1943]*
Burt, William Henry [1903–1987]*
Cope, Edward Drinker [1840–1897]*
Dana, James Dwight [1813–1895]*
Dice, Lee Raymond [1887–1977]*
Dixon, Joseph Scattergood [1884–1952]*
Eigenmann, Carl H. [1863–1927] (Ge)*
Gill, Theodore Nicholas [1837–1914]
Grant, Madison [1865–1937]*
Grinnell, Joseph [1877–1939]*
Hall, Eugene Raymond [1902–1986]*
Huntsman, Archibald Gowanlock [1883–1973] (C)*
Hyman, Libbie Henrietta [1888–1969]*
Just, Ernest Everett [1883–1941]*
Kellogg, (Arthur) Remington [1892–1969]*
LeConte, Joseph [1823–1901]*
Lincoln, Frederick Charles [1892–1960]
Morgan, Ann Haven [1882–1966]*
Nice, Margaret Morse [1883–1974]
Rand, Austin Loomer [1905–1982] (C)*
Rathbun, Richard [1852–1918]
Ruthven, Alexander G(rant) [1882–1971]*
Schmitt, Waldo La Salle [1887–1977]*
Soper, J. Dewey [1893–1982] (C)*
Stejneger, Leonhard (Hess) [1851–1943] (No)*
Simpson, George Gaylord [1902–1984]*
True, Frederick William [1858–1914]*
Verrill, Addison Emery [1839–1926]*
Wheeler, William Morton [1865–1937]*
Wright, Albert Hazen [1879–1970]*
Wright, Robert Ramsay [1852–1933] (GB)*

Selected Bibliography

BOOKS

Abbott, R. Tucker. *American Malacologists* (1974 [Supplement, 1975])
Adams, Alexander B. *Eternal Quest: The Story of the Great Naturalists* (1969)
Adler, Kraig ed. *Early Herpetological Studies and Surveys in the Eastern United States* (1978)
Adler, Kraig, ed. *Herpetological Explorations of the Great American West*. 2 vols. (1978)
Adler, Kraig, ed. *Contributions to the History of Herpetology* (1989)
Alcock, Frederick J. *A Century of the History of the Geological Survey of Canada* (1947)
Allee, Warder Clyde, et al. *Principles of Animal Ecology* (1950) [see especially "The History of Ecology," by W. C. Allee and T. Park, 13–72]
Allen, Elsa. *History of American Ornithology before Audubon* (1951)
Allen, Thomas B. *Guardian of the Wild: The Story of the National Wildlife Foundation, 1936–1986* (1987)
Allin, Craig W. *The Politics of Wilderness Preservation* (1982)
Altsheler, Brent. *Natural History Index Guide* (1940)
American Men of Science: A Biographical Directory [Now *American Men and Women of Science*], 1st–19th eds. (1906–)
Arthur, F. W. *The American Biologist through Four Centuries* (1982)
Bailey, James A., William Elder, and Ted D. McKinney, eds. *Readings in Wildlife Conservation* (1974)
Bailey, Martha J., ed. *American Women in Science: A Biographical Dictionary* (1994)
Barber, Lynn. *The Heyday of Natural History, 1820–1870* (1980)
Barr, E. S., ed. *An Index to Biographical Fragments in Unspecialized Scientific Journals* (1973)
Bartlett, Richard A. *Great Surveys of the American West* (1962)
Benson, Keith R., Jane Maienschein, and Ronald Rainger, eds. *The Expansion of American Biology* (1991)

Biographical Directory of the United States Congress, 1774–1989: Bicentennial Edition (1989)

Bocking, Stephen. *Ecologists and Environmental Politics: A History of Contemporary Ecology* (1997)

Bonta, Marcia M. *Women in the Field: America's Pioneering Naturalists* (1991)

Bramwell, Anna. *Ecology in the 20th Century: A History* (1989)

Bridges, William. *Gathering of Animals: An Unconventional History of the New York Zoological Society* (1974)

Brooks, Paul. *Speaking for Nature: How Literary Naturalists from Henry Thoreau to Rachel Carson Have Shaped America* (1980)

Brown, George W., Marcel Trudeau, et al. *Dictionary of Canadian Biography*. 12 vols. to date (1966–)

Buckman, Thomas R., ed. *Bibliography and Natural History* (1966)

The Canadian Encyclopedia 2d ed. (1988)

Canadian Men and Women of the Time

Canadian Who's Who (1910–)

Carroll, John A., ed. *Reflections of Western Historians* (1969)

Catalogue of the Transylvania University Medical Library [1799–1859] (1987)

Catlett, J. Stephen, ed. *A New Guide to the Collections in the Library of the American Philosophical Society* (1987)

Chartrand, Luc, Yves Gingras, and Raymond Duchesne, *Histoire des Sciences du Québec* (1987)

Cittadino, Eugene. "Ecology and the Professionalization of Botany in America, 1890–1905." In W. Coleman and C. Limoges, eds., *Studies in History of Biology*. Vol. 4 (1980), 171–98

Clepper, Henry. *Origins of American Conservation* (1966)

Coats, Alice M. *The Plant Hunters* (1969)

Cochrane, Rexmond C. *The National Academy of Sciences: The First Hundred Years, 1863–1963* (1978)

Cohen, Michael P. *The History of the Sierra Club, 1892–1970* (1988)

Colbert, Edwin H. *Men and Dinosaurs: The Search in Field and Laboratory* (1968)

Coleman, William, and Camille Limoges, eds. *Studies in History of Biology*. 7 vols. (1977–1983)

Contemporary Authors. 147 vols. (1962–); new revision of the preceding, 48 vols. and indexes (1981–)

Coues, Elliott, ed. *Birds of the Colorado Valley* (1878), bibliographical appendix

Coues, Elliott. *Second Installment of American Ornithological Bibliography* (1879)

Coues, Elliott. *Third Installment of American Ornithological Bibliography* (1880)

Cox, William E. *Guide to the Field Reports of the United States Fish and Wildlife Service* [including the U.S. Biological Survey] *Circa 1860–1961*, Smithsonian Archives (1986).

Craigie, Edward H. *A History of the Department of Zoology at the University of Toronto Up to 1962* (1966)

Cronon, William, ed. *Uncommon Ground: Rethinking the Human Place in Nature* (1995)

Cullum, George, et al., eds. *Biographical Register of Officers and Graduates of the U.S. Military Academy at West Point, New York*. 3 vols. and 8 supplements (1891–1950)

Dagg, Anne Innis. *Canadian Wildlife and Man* (1974)

Daniels, George H. *American Science in the Age of Jackson* (1968)

Daniels, George H. *Science in American Society* (1971)
Davis, Richard Beale. *Intellectual Life in the Colonial South, 1585–1763.* 3 vols. (1978)
Davis, Richard C., ed. *North American Forest History: A Guide to Archives and Manuscripts in the United States and Canada* (1977)
Davis, Richard C., ed. *Encyclopedia of American Forest and Conservation History.* 2 vols. (1983)
Debus, Allen G., ed. *World Who's Who in Science* (1968)
Dictionnaire Genealogique Des Familles Du Quebec
Drake, Ellen T., and William M. Jordan. *Geologists and Ideas: A History of North American Geology* (1985)
Drake, Francis S. *Dictionary of American Biography including Men of the Time* (1872)
Dunlap, Thomas R. *Saving America's Wildlife* (1988)
Dupree, A. Hunter. *Science in the Federal Government: A History of Policies and Activities to 1940* (1959)
Egerton, Frank N. "Ecological Studies and Observations before 1900." In B. J. Taylor and T. J. White, eds., *Issues and Ideas in America* (1976), 311–51
Egerton, Frank N., ed. *Early Marine Ecology* (1977)
Egerton, Frank N., ed. *History of American Ecology* (1977)
Egerton, Frank N. "The History of Ecology: Achievements and Opportunities" *Journal of the History of Biology* 16 (1983): 259–310; 18 (1985): 103–143
Ekirch, Arthur A., Jr. *Man and Nature in America* (1963)
Elliott, Clark, A., ed. *Biographical Dictionary of American Science: The Seventeenth through the Nineteenth Centuries* (1979)
Elliott, Clark A., comp. *Biographical Index to American Science, the Seventeenth Century to 1920* (1990)
Elman, Robert. *First in the Field* (1977)
Environmental Law Insitute. *The Evolution of National Wildlife Law.* Prepared for the Council on Environmental Quality (1977)
Essig, E. O. *History of Entomology* (1931)
Ewan, Joseph, ed. *A Short History of Botany in the United States* (1969)
Ewan, Joseph, and Nesta Dunn Ewan. *Biographical Dictionary of Rocky Mountain Naturalists: A Guide to the Writings and Collections of Botanists, Zoologists, Geologists, Artists, and Photographers, 1682–1932* (1981)
Fahl, Ronald J., ed. *North American Forest and Conservation History: A Bibliography* (1977)
Farber, Paul L. *The Emergence of Ornithology as a Scientific Discipline, 1760–1850* (1982)
Foster, Janet. *Working for Wildlife: The Beginning of Preservation in Canada* (1978, new edition, 1998)
Fox, Stephen R. *The American Conservation Movement: John Muir and His Legacy* (1981)
Gerstner, Patsy A. "Vertebrate Paleontology, an Early Nineteenth Century Transatlantic Science." *Journal of the History of Biology* 3 (1970): 137–48
Gillispie, Charles C., et al., eds. *Dictionary of Scientific Biography.* 18 vols. (1970–1980)
Goetzmann, William C. *Army Exploration in the American West, 1803–1863* (1959)
Goetzmann, William C. *Exploration and Empire: The Explorer and the Scientist in the Winning of the American West* (1966)
Goetzmann, William C., and Kay Sloan. *Looking Far North: The Harriman Expedition to Alaska, 1899* (1982)

Goetzmann, William C. *New Lands, New Men: America and the Second Great Age of Discovery* (1986)
Goode, George B., ed. *The Smithsonian Institution: 1846–1896: The History of the First Half-Century* (1897)
Goode, George B. ''The Beginnings of American Science: The Third Century.'' *Annual Report of the Board of Regents of the Smithsonian Institution etc., for the Year Ending June 30, 1897* (1901)
Goode, George B. ''Beginnings of Natural History in America.'' *Annual Report of the Board of Regents of the Smithsonian Institution, etc., for the Year Ending June 30, 1897* (1901)
Graham, Frank Jr., with Carl Buchheister. *The Audubon Ark: A History of the National Audubon Society* (1990)
Greene, John C. *American Science in the Age of Jefferson* (1984)
Gruson, Edward. *Words for Birds: A Lexicon of North American Birds with Biographical Notes* (1972)
Hanley, Wayne. *Natural History in America* (1977)
Harkányi, Katalin, comp. *The Natural Sciences and American Scientists in the Revolutionary Era: A Bibliography* (1990)
Harshberger, John W. *The Botanists of Philadelphia and Their Work* (1899)
Hay, O. P. *Bibliography and Catalogue of the Fossil Vertebrata of North America* (1901)
Hayes, Samuel P. *Conservation and the Gospel of Efficiency* (1959)
Hays, H. R. *Birds, Beasts, and Men: A Humanist History of Zoology* (1972)
Hazen, Robert M., and M. H. Hazen, eds. *American Geological Literature, 1669 to 1850* (1980)
Heitman, Francis B., ed. *Historical Register and Dictionary of the United States Army.* 2 vols. (1903)
Hellman, Geoffrey. *Bankers, Bones and Beetles: The First Century of the American Museum of Natural History* (1968)
Herbert, Miranda C., and Barbara McNeil, eds. *Biography and Genealogy Master Index: A Consolidated Index to More than 1,200,000 Biographical Sketches* 2d ed. (1980) (with later supplements)
Hewitt, Gordon C. *The Conservation of Wildlife in Canada* (1921)
Hindle, Brooke. *The Pursuit of Science in Revolutionary America, 1735–1789* (1986)
Howard, L. O. *A History of Applied Entomology* (1931)
Hume, E. E. *Ornithologists of the United States Army Medical Corps* (1942)
Index to Personal Names in the National Union Catalog of Manuscript Collections, 1959–1984. 2 vols. (1988)
Ireland, Norma O. *Index to Scientists of the World from Ancient to Modern Times: Biographies and Portraits* (1962)
James, E. T., Barbara Sicherman, et al. *Notable American Women*. 4 vols. (1971–1980)
James, Preston E., and Geoffrey J. Martin, with contributions by Harm J. de Blij and Clyde F. Kohn. *The Association of American Geographers: The First Seventy-Five Years, 1904–1979* (1978)
Johnson, Allen, John Garraty, et al. *Dictionary of American Biography*. 30 vols. (1927–)
Johnson, Rossiter, and John H. Brown, eds. *The Twentieth Century Biographical Dictionary of Notable Americans*. 10 vols. (1904)
Kastner, Joseph. *A Species of Eternity* (1977)
[Kessel, Edward L., ed.] *A Century of Progress in the Natural Sciences: 1853–1953* (1955)

Kline, Marcia B. *Beyond the Land Itself: Views of Nature in Canada and the United States* (1970)
Kohlstadt, Sally G. *The Formation of the American Scientific Community: The American Association for the Advancement of Science, 1848–1860* (1976)
Koppes, Clayton R. "Shifting Themes in American Conservation." In Donald Worster, ed., *The Ends of the Earth* (1988)
Lacey, Michael J., ed. *Government and Environmental Politics: Essays on Historical Developments since World War II* (1991)
Larkin, R. P., and G. L. Peters, eds. *Biographical Dictionary of Geographers* (1993)
Lee, R. Alton, et al., eds. *Encyclopedia USA*. 22 vols. to date (1983–)
Lindsay, Debra. *Science in the Subarctic: Trappers, Traders, and the Smithsonian Institution* (1993)
Linsley, E. Gorton, ed. *Beetles from the Early Russian Explorations of the West Coast of North America, 1815–1878*, with introduction by K. B. Sterling (1978)
Lutts, Ralph H. *The Nature Fakers: Wildlife, Science and Sentiment* (1990)
Mainero, Lisa, ed. *American Women Writers: A Critical Reference Guide from Colonial Times to the Present*. 4 vols. (1979–)
Mallis, Arnold. *American Entomologists* (1971)
Manning, Thomas G. *Government in Science: The U.S. Geological Survey, 1867–1894* (1967)
Mathiessen, Peter. *Wildlife in America* (1959) [rev. ed. 1987]
McIntosh, Robert P. "Ecology since 1900." In B. J. Taylor and T. J. White, *Issues and Ideas in America* (1976), 353–72
McKelvey, Susan Delano. *Botanical Exploration of the Trans-Mississippi West, 1790–1850* (1955, repr. 1991)
McNicholl, Martin K., and J. L. Cranmer-Byng, et al. *Ornithology in Ontario* (1994)
Meisel, Max. *A Bibliography of American Natural History: The Pioneer Century, 1769–1865*. 3 vols. (1924–1929)
Mengel, Robert M. *A Catalogue of the Ellis Collection of Ornithological Books in the University of Kansas Libraries*. 2 vols. (1972–)
Merchant, Carolyn. "Women of the Progressive Conservation Movement, 1900–1916." *Environmental Review* 8 (1984): 57–85
Merriam, C. Hart, ed. *Harriman Alaska Expedition*. 12 vols. (1902–1914)
Merrill, George P. *The First One Hundred Years of American Geology* (1924)
Miller, Charles A. *Jefferson and Nature* (1988)
Mirsky, Jeannette. *To the Arctic!: The Story of Northern Exploration from Earliest Times to the Present* (1948; repr. 1970)
Mitchell, Lee Clark. *Witnesses to a Vanishing America: The Nineteenth-Century Response* (1981)
Mitman, Gregg. *The State of Nature: Ecology, Community, and American Social Thought, 1900–1950* (1992)
Murray, E. J., ed. *Notable Twentieth-Century Scientists* (1995)
Nash, Roderick, ed. *The American Environment: Readings in the History of Conservation*. 2d ed. (1976)
Nash, Roderick. *Wilderness and the American Mind*. 3d ed. (1983)
Nash, Roderick F. *The Rights of Nature: A History of Environmental Ethics* (1989)
National Cyclopaedia of American Biography. 63 vols. and Index (1891–)
Neu, John, ed. *ISIS Cumulative Bibliography 1966–1975*. 2 vols. (1980) (same for 1976–1985, 2 vols., 1989; see also annual ISIS bibliographical compilations, also edited by John Neu)

Newman, Peter C. *Company of Adventurers* [1st vol. of his history of the Hudson's Bay Company] (1985)

Newman, Peter C. *Caesars of the Wilderness* [2d vol. of his history of the Hudson's Bay Company] (1987)

Newman, Peter C. *Merchant Princes* [3d vol. of his history of the Hudson's Bay Company] (1991)

Nixon, Edgar B., ed. *Franklin D. Roosevelt and Conservation, 1911–1945.* 2 vols. (1957)

Norwood, Vera. *Made from This Earth: American Women and Nature* (1993)

Oehser, Paul. "A Handlist of American Naturalists, Based on the Dictionary of American Biography." *The American Naturalist* 72 (1938): 534–46

Oehser, Paul. *Sons of Science: The Story of the Smithsonian Institution and Its Leaders* (1949)

Ogden, Gerald. *The United States Forest Service: A Historical Bibliography, 1876–1972* (1976)

Ogilvie, Marilyn B., ed. *Women in Science: Antiquity through the Nineteenth Century: A Biographical Dictionary with Annotated Bibliography* (1986)

Oleson, Alexandra, and Sanborn C. Brown, eds. *The Pursuit of Knowledge in the Early American Republic: American Scientific Societies from Colonial Times to the Civil War* (1976)

Overmier, Judith A. *The History of Biology: A Selected, Annotated Bibliography* (1989)

Paehlke, Robert, ed. *Conservation and Environmentalism: An Encyclopedia* (1995)

Palmer, Theodore S., ed. *Biographies of Members of the American Ornithologists' Union* (1954)

Payne, Daniel C. *Voices in the Wilderness: American Nature Writing and Environmental Politics* (1996).

Pelletier, Paul A. *Prominent Scientists: An Index to Collective Biographies* (1980) (and later revision)

Penick, James Jr. *Progressive Politics and Conservation* (1968)

Pennant, Thomas. *Arctic Zoology*. 2 vols. (1784, 1787)

Phillips, John C. *American Game Mammals and Birds: A Catalogue of Books, 1582 to 1925: Sport, Natural History, and Conservation* (1930)

Phillips, Venia T., and Maurice E. Phillips. *Guide to the Manuscript Collections in the Academy of Natural Sciences of Philadelphia* (1963)

Porter, Charlotte M. *The Eagle's Nest: Natural History and American Ideas, 1812–1842* (1986)

Preston, Douglas J. *Dinosaurs in the Attic: An Excursion into the American Museum of Natural History* (1986)

Rainger, Ronald, Keith R. Benson, and Jane Maienschein. *The American Development of Biology* (1988)

Reiger, John F. *American Sportsmen and the Origins of Conservation* (1975)

Rhees, W. J., ed. *The Smithsonian Institution: Documents Relative to Its Origin and History, 1835–1899.* 2 vols. (1901)

Richardson, Elmo. *The Politics of Conservation: Crusades and Controversies, 1897–1913* (1962)

Richardson, John, et al. *Fauna Boreali-Americana; Or the Zoology of the Northern Parts of British America, Containing Objects of Natural History Collected by the Late Northern Land Expeditions under Command of Captain Sir John Franklin, R.N.* Part I: *Quadrupeds*, 1829; Part II (with William Swainson): *The Birds*, 1831; Part III: *The Fish*, 1836; Part IV (with William Swainson and William Kirby): *Insecta*, 1837

Ripley, S. Dillon, and Lynette L. Scribner, comps. *Ornithological Books in the Yale University Library, Including the Library of William Robertson Coe* (1961)
Robbins, Roy M. *Our Landed Heritage: The Public Domain, 1776–1936* (1942)
Rodgers, Andrew Denny, III. *American Botany, 1873–1892: Decades of Transition* (1944)
Rosenberg, Kenneth A. *Wilderness Preservation: A Reference Handbook* (1994)
Rossiter, Margaret W. *Women Scientists in America: Struggles and Strategies to 1940* (1982)
Rothenberg, Marc. *The History of Science and Technology in the United States: A Critical, Selective Bibliography* (1982)
Schmitt, Peter J. *Back to Nature: The Arcadian Myth in Urban America* (1969)
Shor, Elizabeth Noble. *The Fossil Feud* (1974)
Short, C. Brant. *Ronald Reagan and the Pubic Lands: America's Conservation Debate, 1979–1984* (1989)
Siegel, Patricia J., and Kay Thomas Finley. *Women in the Scientific Search: An American Bio-Bibliography, 1724–1979* (1979)
Simonian, Lane. *Defending the Land of the Jaguar: A History of Conservation in Mexico* (1995)
Smallwood, W. M., and M.S.C. Smallwood. *Natural History and the American Mind* (1941)
Smith, Darrell H. *The Forest Service: Its History Activities and Organization* (1930)
Smith, Frank. *The Politics of Conservation* (1966)
Spears, Bordon, ed. *Wilderness Canada* (1970)
Stearns, Raymond P. *Science in the British Colonies of America* (1970)
Steen, Harold K. *The United States Forest Service: A History* (1976)
Stegner, Wallace. *Beyond the Hundredth Meridian: John Wesley Powell and the Second Opening of the West* (1954)
Stegner, Wallace. "It All Began with Conservation." *Smithsonian* 21 (1990): 35–43
Stephen, Leslie, Sidney Lee, et al., eds. *Dictionary of National Biography*. 30 vols. (1917–)
Sterling, Keir B., ed. *American Natural History Studies: The Bairdian Period* (1974)
Sterling, Keir B., ed. *Contributions to American Systematics* (1974)
Sterling, Keir B., ed. *Contributions to the Bibliographical Literature of American Mammalogy* (1974)
Sterling, Keir B., ed. *Contributions to the History of American Natural History* (1974)
Sterling, Keir B., ed. *Contributions to the History of American Ornithology* (1974)
Sterling, Keir B., ed. *Early Nineteenth Century Studies and Surveys* (1974)
Sterling, Keir B., ed. *Selected Works by Eighteenth-Century Naturalists and Travellers* (1974)
Sterling, Keir B., ed. *Selected Works in Nineteenth-Century North American Paleontology* (1974)
Sterling, Keir B., ed. *Selections from the Literature of American Biogeography* (1974)
Sterling, Keir B., ed. *Selected Works of Clinton Hart Merriam* (1974)
Sterling, Keir B., ed. *Selected Works of Joel Asaph Allen* (1974)
Sterling, Keir B., ed. *Rafinesque: Autobiography and Lives* (1978)
Sterling, Keir B., ed. *United States Exploring Expedition during the Years 1838, 1839, 1840, 1841, 1842 under the Command of Charles Wilkes, U.S.N.: Mammalogy and Ornithology*, by Titian R. Peale (1848) (1978)

Stetler, Susan L., ed. *Biography Almanac*. 3d ed., 3 vols. (1987)
Story, Norah, ed. *Oxford Companion to Canadian History and Literature* (1967)
Stresemann, Erwin. *Ornithology: From Aristotle to the Present*. Trans Hans J. and Cathleen Epstein; ed. G. W. Cottrell; with concluding chapter by Ernst Mayr, ''Materials for a History of American Ornithology'' (1975)
Strong, Douglas H. *Dreamers and Defenders: American Conservationists* (1988)
Stroud, Richard A., ed. *National Leaders of American Conservation* (1985)
Struik, Dirk. *Yankee Science in the Making* (1948)
Stuckey, Ronald L., ed. *Development of Botany in Selected Regions of North America before 1900* (1978)
Stuckey, Ronald L., ed. *Essays in North American Plant Geography from the Nineteenth Century* (1978)
Stuckey, Ronald L., ed. *Scientific Publications of Charles Wilkins Short* (1978)
Swain, Donald C., *Federal Conservation Policy, 1921–1933* (1963)
Thrapp, Dan L. *Encyclopedia of Frontier Biography*. 3 vols. (1993)
Uglow, Jennifer S., comp. and ed. *The Continuum Dictionary of Women's Biography, New Expanded Edition* (1989)
Union List of Manuscripts in Canadian Repositories
Van Hise, Charles R. *Conservation of Our Natural Resources* (1910)
Wallace, W. Stewart, and W. A. McKay. *Macmillan Dictionary of Canadian Biography* (1978)
Weiner, Douglas R. *Models of Nature: Ecology, Conservation, and Cultural Revolution in Soviet Russia* (1988)
Welker, Robert H. *Birds and Men: American Birds in Science, Art, Literature, and Conservation, 1800–1900* (1955)
Whitaker, John C. *Striking a Balance: Environment and Natural Resources Policy in the Nixon–Ford Years* (1976)
Whitrow, Magda, ed. *ISIS Cumulative Bibliography: A Bibliography of the History of Science Formed from ISIS Critical Bibliographies 1–90, 1913–1965*. 3 vols., 1971–1976 [see under Neu, John, for continuation]
Whittle, Tyler. *The Plant Hunters* (1988)
Who Was Who in America. 11 vols. (1943–)
Who's Who in America. (1899–)
Who's Who in Canada (various dates)
Wilson, James G., and John Fiske, eds. *Appleton's Cyclopaedia of American Biography*. 7 vols. (1888–1901) (see also Dearborn, L. E., ed., *A Supplement to Appleton's Cyclopaedia of American Biography*, 6 vols. [1918–1931])
Wood, Casey. *An Introduction to the Literature of Vertebrate Zoology* (1931)
Wood, Samuel. *Naturalists of the Frontier* (1948)
Worster, Donald. *Nature's Economy* (1977, 2d ed., 1994)
Wright, John Kirkland. *Geography in the Making: The American Geographical Society, 1851–1951* (1952)
Youmans, William Jay, ed. *Pioneers of Science in America*. Rev. ed. (1896)
Zaslow, Maurice. *Reading the Rocks: The Story of the Geological Survey of Canada* (1975)
Zeller, Suzanne. *Inventing Canada: Early Victorian Science and the Idea of a Transcontinental Nation* (1987)
Zimmer, J. T., ed. *Catalogue of the Ayer Library of Ornithology* (1928)

NEWSPAPERS AND PERIODICALS

Acadiensis
American Forests
American Heritage
American Journal of Botany
American Midland Naturalist
Annals of the Entomological Society of America
Annals of Science
The Archives of Natural History [formerly *Journal of the Society for the Bibliography of Natural History*]
Audubon [formerly *Audubon Magazine*]
Auk
Beaver
Biographical Memoirs of the National Academy of Sciences
Biography Index (1946–)
BioScience
Bulletin of the American Museum of Natural History
Bulletin of the Geological Society of America
Bulletin of the Nuttall Ornithological Club (1876–1883)
Canadian Alpine Journal
Canadian Entomologist
Canadian Field-Naturalist
Canadian Historical Review
Condor
Conservation Biology
Entomological News
Environmental Review (1976–1989), succeeded by *Environmental History Review* (1989–1995), merged with *Journal of Forest and Conservation History* to form *Environmental History* (1996–)
Forest History Newsletter (1957–1958), succeeded by *Forest History* (1959–1974), succeeded by *Journal of Forest History* (1975–1989), succeeded by *Journal of Forest and Conservation History* (1989–1995), merged with *Environmental History Review* (see previous item) to form *Environmental History* (1996–)
History of Geography Newsletter (1981–1986), succeeded by *History of Geography Journal* (1986–)
History of Science
ISIS
Journal of Conchology
Journal of Ecology
Journal of Herpetology
Journal of the History of Biology
Journal of Mammalogy
Journal of Mycology
Journal of Southern History
Journal of Wildlife Management
Mycologia
Natural History

Nature Magazine
New York Times
New York Times Obituary Index
Proceedings of the Academy of Natural Sciences of Philadelphia
Proceedings of the American Philosophical Society
Proceedings and Transactions of the Royal Society of Canada
Proceedings of the Royal Society of London
Quarterly Review of Biology
Science
Sierra Club Bulletin
Systematic Zoology
Transactions of the American Philosophical Society
Washington Post
Wilderness, formerly *The Living Wilderness*
Wildlife Abstracts [Bibliography and Index to *Wildlife Review*]
Wildlife Conservation
Wildlife Review
Wilson Bulletin

Index

Page numbers in **boldface** refer to main entries.

About the Editors and Contributors

MARIANNE GOSZTONYI AINLEY, Ph.D., is Professor of Women's Studies at the University of Northern British Columbia, Prince George, British Columbia.

DEAN C. ALLARD, Ph.D., is the retired Head of the Historical Archives Branch, United States Naval Historical Center, Washington, D.C.

TOBY A. APPEL, Ph.D., is Historical Librarian at the Cushing/Whitney Medical Library, Yale University, New Haven, Connecticut.

DOMENICA BARBUTO is the Reference Librarian at Hofstra University, Hempstead, New York.

ARTHUR A. BELONZI, Ph.D., is Professor of History and Chair of the Liberal Arts Department, College of Aeronautics, New York City, New York.

LYNN BERRY is a doctoral candidate in History at the University of Toronto, Toronto, Ontario.

STEPHEN BOCKING, Ph.D., is Assistant Professor of Environmental and Resource Studies at Trent University, Peterborough, Ontario.

MICHAEL J. BOERSMA is Prospect Research Coordinator at the Museum of Science and Industry, Chicago, Illinois.

CHARLES BOEWE, Ph.D., is Editor of the C. S. Rafinesque Papers and former Research Professor of History at Transylvania University, Louisville, Kentucky.

JOSEPH BONGIORNO, Ph.D., is an Assistant Professor of History at St. John's University, Jamaica, New York.

GUNNAR BROBERG, Ph.D., is Professor of Ideas and Sciences at the University of Lund, Lund, Sweden.

MICHAEL J. BRODHEAD, Ph.D., is Emeritus Professor of History, University of Nevada, Reno, and Archivist with the National Archives and Records Administration, Kansas City, Kansas.

WILLIAM R. BURK is Biological Sciences Librarian, University of North Carolina at Chapel Hill.

DAVID K. CAIRNS, Ph.D., is a Research Scientist with the Department of Fisheries and Oceans of Canada at Charlottestown, Prince Edward Island, and Adjunct Professor of Biology at the Université de Moncton in Moncton, New Brunswick.

DAVID CALVERLY is a graduate student in History at the University of Ottawa, Ottawa, Ontario.

PAUL CAMMARATA is Reference Librarian at the State University of New York at Stony Brook, New York.

GEORGE A. CEVASCO is Associate Professor of English at St. John's University, Jamaica, New York.

CYNTHIA D. CHAMBERS is Cataloguing Librarian, St. John's University, Jamaica, New York.

EUGENE CITTADINO, Ph.D., was a Professor of History at the State University of New York at Potsdam, New York.

BROOKE CLIBBON is an Information Specialist with NOVA Corporation of Alberta, Calgary, Alberta.

EUGENE COAN is Senior Advisor to the Executive Director of the Sierra Club, San Francisco, California.

ALISON MURPHY CONNER (deceased) was a resident of Seattle, Washington.

PETER L. COOK is a doctoral candidate in History at McGill University, Montréal, Québec.

STEVEN COOPER has most recently been a teacher of History at Hackensack High School, Hackensack, New Jersey.

WILLIAM COX is an Associate Archivist with the Smithsonian Institution, Washington, D.C.

E. TINA CROSSFIELD holds an interdisciplinary M.A. from Concordia University and is a resident of Okotoks, Alberta.

JUDITH MORAN CURRAN, Ph.D., is an Adjunct Professor of Education at the College of Staten Island, State University of New York.

THOMAS J. CURRAN, Ph.D., is an Associate Professor of History at St. John's University, Jamaica, New York.

JOAN D'ANDREA is an Instructional Librarian at St. John's University, Jamaica, New York.

NORMAND DAVID is Directeur général of the Association québécoise des groupes d'ornithologiques, Montréal, Québec.

ELLEN DAVIGNON is a resident of Whitehorse, Yukon.

RICHARD DAY is a Research Assistant in Paleobiology, Canadian Museum of Nature, Ottawa, Ontario.

TERRENCE DAY, Ph.D., is Director of the Atlantic Canada Centre for Environmental Science, Saint Mary's University, Halifax, Nova Scotia.

WILLIAM A. DEISS is a retired Associate Archivist and Research Collaborator, Smithsonian Institution Archives, Washington, D.C.

RALPH W. DEXTER, Ph.D. (deceased) was Professor of History at Kent State University, Kent, Ohio.

JOHN S. DOSKEY was a resident of Clayton, California.

JOHN A. DROBNICKI is Reference Librarian at York College of the City University of New York.

CLARK A. ELLIOTT, Ph.D., is Associate Archivist at Harvard University, Cambridge, Massachusetts.

DOROTHY TANCK DE ESTRADA, Ph.D., is a Professor of History at El Colegio de Mexico, Mexico City, Mexico.

CAROL FAUL (deceased) was Curator of Paleontology and Director of the Map Library at the Department of Geology, University of Pennsylvania.

JULIETTE M. FERNAN is a graduate student at St. John's University, Jamaica, New York.

DAN FLORES, Ph.D., is a Professor of History at the University of Montana, Missoula, Montana.

JOHN W. FREDERICK is on the staff of the Department of Botany, The Ohio State University, Columbus, Ohio.

FRITIOF M. FRYXELL, Ph.D. (deceased) was a Professor of History at Augustana College, Rock Island, Illinois.

JOHN ALLEN GABLE is the Executive Director of the Theodore Roosevelt Association, Oyster Bay, New York.

PETER GELDART is associated with Project Services International, Hull, Québec.

JAMES R. GLENN is a retired Archivist, Smithsonian Institution Archives, Washington, D.C.

SUSAN W. GLENN is a retired Assistant Archivist, Smithsonian Institution Archives, Washington, D.C.

JIM GLOVER, Ed.D., is a Professor of Education at Southern Illinois University, Carbondale, Illinois.

DAVID R. GRAY, Ph.D., is an Historian with the National Museums of Canada, Ottawa, Ontario.

ANITA GUERRINI, Ph.D., is Associate Professor of History, University of California at Santa Barbara, Santa Barbara, California.

PETER L. GUTH, Ph.D., is Associate Professor of Oceanography at the United States Naval Academy, Annapolis, Maryland.

CHRISTOPHER HAMLIN, Ph.D., is Associate Professor of History at the University of Notre Dame, South Bend, Indiana.

LORNE HAMMOND, Ph.D., formerly a Social Sciences and Humanities Research Council Postdoctoral Fellow in the Department of History, University of Victoria, now teaches there, and is an Historian with the Royal British Columbia Museum, Victoria, British Columbia.

RICHARD HARMOND, Ph.D., is Associate Professor of History at St. John's University, Jamaica, New York.

LYNN M. HAUT holds an M.A. in History from St. John's University, Jamaica, New York.

PAMELA M. HENSON is an Historian and Division Director, Institutional History Division, Smithsonian Institution Archives, Washington, D.C.

ERIC S. HIGGS, Ph.D., is a Professor of Anthropology at the University of Alberta, Edmonton, Alberta.

C. STUART HOUSTON, M.D., is a Professor of Diagnostic Radiology, College of Medicine, University of Saskatchewan, Saskatoon, Saskatchewan.

MARY I. HOUSTON is a resident of Saskatoon, Saskatchewan.

JENNIFER M. HUBBARD, Ph.D., teaches at the Institute of History and Philosophy of Science and Technology at the University of Toronto, Toronto, Ontario.

SUSAN IGNACIUK holds an M.A. in History from St. John's University, Jamaica, New York.

AARON J. IHDE, Ph.D., is Professor Emeritus of Chemistry at the University of Wisconsin-Madison, Madison, Wisconsin.

RICHARD A. JARRELL, Ph.D., is a Professor of Science Studies, Atkinson College, York University, North York, Ontario.

LIZ BARNABY KEENEY, Ph.D., is Dean of Academic Advising at Kenyon College, Gambier, Ohio.

ERIK KIVIAT, Ph.D., is Director of Hudsonia Limited, Bard College, Annandale-on-Hudson, New York.

SALLY GREGORY KOHLSTEDT, Ph.D., is Professor of the History of Science and Associate Dean of the Institute of Technology, University of Minnesota, Minneapolis, Minnesota.

DOROTHY I. LANSING, M.D., is a physician in West Chester, Pennsylvania.

WILLIAM E. LASS, Ph.D., is Professor of History at Mankato State University, Mankato, Minnesota.

LAWRENCE B. LEE, Ph.D., is Professor of History, Emeritus, at San Jose State University, San Jose, California.

DAVID L. LENDT, Ph.D., is Director of University Relations, University of Missouri System, Columbia, Missouri.

G. CARROLL LINDSAY was formerly associated with the Museum Services Department, New York State Museum, Albany, New York.

DAVID N. LIVINGSTONE, Ph.D., is a Professor of History at Queen's University of Belfast, Northern Ireland.

W. FERGUS LOTHIAN (deceased) was Historian Emeritus on the staff of Parks Canada, Ottawa, Canada.

BARBARA LOVITTS is completing a Ph.D. in Sociology at the University of Maryland, College Park, Maryland.

RICHARD MANGERI is Dean of the Latin School, Kollenberg Memorial High School, Uniondale, L.I., New York.

KATHY MARTIN, Ph.D., is a Biologist with the Canadian Wildlife Service, Delta, British Columbia.

RODNEY MARVE is Reference Librarian, Bay Shore-Brightwaters Public Library, Brightwaters, New York.

WILLIAM R. MASSA, JR. is Public Service Assistant, Yale University Archives, Sterling Memorial Library, New Haven, Connecticut.

PATRICK J. McNAMARA is a graduate student in Church History at the Catholic University of America, Washington, D.C.

MARTIN K. McNICHOLL, Ph.D., is a professional ornithologist and a resident of Burnaby, British Columbia.

CURT MEINE, Ph.D., is a biologist and historian on the staff of the International Crane Foundation, Baraboo, Wisconsin.

GREGG A. MITMAN, Ph.D., is Associate Professor of the History of Science at the University of Oklahoma, Norman, Oklahoma.

SASHA MULLALLY is a graduate in History from the University of Ottawa and a resident of Prince Edward Island.

MARGARET E. MURIE is an author and conservationist living in Moose, Wyoming.

JEFFREY S. MURRAY is an Archivist with the National Archives of Canada, Ottawa, Ontario.

CLIFFORD M. NELSON, Ph.D., is a Geologist with the United States Geological Survey, Reston, Virginia.

JANE DAVIS NELSON is a resident of Montréal, Québec.

MICHAEL A. OSBORNE, Ph.D., is Associate Professor of the History of Science, University of California at Santa Barbara, Santa Barbara, California.

CHERIE PARSONS is a graduate student in History at St. John's University, Jamaica, New York.

ROBERTA PESSAH is a Reference Librarian at St. John's University, Jamaica, New York.

ELLY PORSILD is a resident of Whitehorse, Yukon.

MICHAEL S. QUINN, Ph.D., is Associate Professor of History, Lakehead University, Thunder Bay, Ontario.

MARY C. RABBITT, Ph.D., is a retired Geologist, United States Geological Survey, Reston, Virginia.

THERESA REDMOND is a professional researcher and a resident of Ottawa, Ontario.

LINDA L. REESMAN is a graduate student in English at St. John's University, Jamaica, New York.

PEARLANN REICHWEIN, Ph.D., is a recent History graduate from Carleton University, Ottawa, Ontario.

JOHN F. REIGER, Ph.D., is Professor of History at the University of Ohio, Chillicothe, Ohio.

PAUL W. RIEGERT, Ph.D., is a Professor Emeritus of Biology, University of Regina, Regina, Saskatchewan.

GEORGE A. ROGERS, Ph.D., is Professor Emeritus of History at Georgia Southern University, Statesboro, Georgia.

VIVIAN ROGERS-PRICE, Ph.D., is Assistant Professor of History, Armstrong Atlantic State University, Savannah, Georgia.

NINA J. ROOT is the Librarian of the American Museum of Natural History, New York City, New York.

EMANUEL D. RUDOLPH, Ph.D. (deceased) was Professor of Botany, The Ohio State University, Columbus, Ohio.

JOAN RYAN is a Reference Librarian at St. Joseph's College, Patchogue, New York.

ANDREW SANKOWSKI is an Acquisitions Librarian at St. John's University, Jamaica, New York.

RALPH G. SCHWARZ was formerly the Director of Historic New Harmony, Inc., New Harmony, Indiana.

SUSAN SHEETS-PYENSON, Ph.D., is Professor of Geography, Concordia University, Montréal, Québec.

ARTHUR SHERMAN is an Acquisitions Librarian at St. John's University, Jamaica, New York.

JEHESKEL (HEZY) SHOSHANI, Ph.D., is affiliated with the Department of Zoology at Wayne State University, Detroit, Michigan.

GERI E. SOLOMON is an Archivist with the Hofstra University Archives, Hempstead, New York.

JAMES H. SOPER, Ph.D. (deceased) was on the staff of the National Museum of Natural Science, Ottawa, Canada.

JIM STAPLETON is Associate Director of Hudsonia Limited in Joyce, Washington.

LESTER D. STEPHENS, Ph.D., is Professor of History, University of Georgia, Athens, Georgia.

KEIR B. STERLING, Ph.D., is Ordnance Branch Historian, U.S. Army, Fort Lee, Virginia. He has been Professor of History, Pace University New York City and Pleasantville, New York, a National Science Foundation grantee, and visiting Professor of the History of Science, University of Wisconsin, Madison.

DAVID B. STEWART, M.D., LL.D., F.R.C.O.G., is Emeritus Professor of Obstetrics and Gynecology at the University of the West Indies, Kingston, Jamaica, and Emeritus Professor of Zoology at Brandon University, Brandon, Manitoba.

RONALD L. STUCKEY, Ph.D., is Professor of Botany Emeritus, The Ohio State University, Columbus, Ohio.

ROLF SWENSEN, Ph.D., is Reference Librarian, Queens College of the City University of New York, Queens, New York.

FELICITAS TANGERMANN is Archival Assistant, Yukon Provincial Archives, Whitehorse, Yukon.

PHILLIP D. THOMAS, Ph.D., is Professor of History at Wichita State University, Wichita, Kansas.

CONNIE THORSEN is an Instructional Material Center Librarian at St. John's University, Jamaica, New York.

CARMELA TINO is a Reference Librarian at St. John's University, Jamaica, New York.

ANTHONY TODMAN is a Reference Librarian at St. John's University, Jamaica, New York.

ARLEEN M. TUCHMAN, Ph.D., is Associate Professor of History at Vanderbilt University, Nashville, Tennessee.

RONALD S. VASILE is the Historian for the Canal Corridor Association, Chicago, Illinois. He was formerly Archivist-Historian at the Chicago Academy of Sciences, Chicago, Illinois.

KAREN M. VENTURELLA is a Periodicals Librarian at St. John's University, Jamaica, New York.

NOEL DORSEY VERNON, Ph.D., is Professor of Landscape Architecture at California State Polytechnic University, Pomona, California.

JOHN HENRY WADLAND, Ph.D., is Professor of Canadian Studies, Frost Centre for Canadian Heritage and Development, Trent University, Peterborough, Ontario.

WILLIAM A. WAISER, Ph.D., is Professor of History, University of Saskatchewan, Saskatoon, Saskatchewan.

JEAN WASSONG is an Archivist at the Diocese of Rockville Centre, Rockville Centre, New York.

RICHARD R. WESTCOTT, Ph.D., was Professor of History at Monmouth University, Monmouth, New Jersey.

FRANK C. WHITMORE, JR., Ph.D., is a retired Geologist with the United States Geological Survey, attached to the National Museum of Natural History, Washington, D.C.

WENDY WICKWIRE, Ph.D., is Associate Professor of History, University of Victoria, Victoria, British Columbia.

ELLIS L. YOCHELSON, Ph.D., is a retired Geologist with the United States Geological Survey, attached to the National Museum of Natural History, Washington, D.C.

EDWARD D. ZAHNISER was on the staff of Atlantis Rising, Shepherdstown, West Virginia.